Lecture Notes in Networks and Systems 752

The series "Lecture Notes in Networks and Systems" publishes the latest developments in Networks and Systems—quickly, informally and with high quality. Original research reported in proceedings and post-proceedings represents the core of LNNS.

Volumes published in LNNS embrace all aspects and subfields of, as well as new challenges in, Networks and Systems.

The series contains proceedings and edited volumes in systems and networks, spanning the areas of Cyber-Physical Systems, Autonomous Systems, Sensor Networks, Control Systems, Energy Systems, Automotive Systems, Biological Systems, Vehicular Networking and Connected Vehicles, Aerospace Systems, Automation, Manufacturing, Smart Grids, Nonlinear Systems, Power Systems, Robotics, Social Systems, Economic Systems and other. Of particular value to both the contributors and the readership are the short publication timeframe and the world-wide distribution and exposure which enable both a wide and rapid dissemination of research output.

The series covers the theory, applications, and perspectives on the state of the art and future developments relevant to systems and networks, decision making, control, complex processes and related areas, as embedded in the fields of interdisciplinary and applied sciences, engineering, computer science, physics, economics, social, and life sciences, as well as the paradigms and methodologies behind them.

Indexed by SCOPUS, INSPEC, WTI Frankfurt eG, zbMATH, SCImago.

All books published in the series are submitted for consideration in Web of Science.

For proposals from Asia please contact Aninda Bose (aninda.bose@springer.com).

Thi Dieu Linh Nguyen · Elena Verdú ·
Anh Ngoc Le · Maria Ganzha
Editors

Intelligent Systems and Networks

Selected Articles from ICISN 2023, Vietnam

Springer

Editors
Thi Dieu Linh Nguyen
Hanoi University of Industry
Bac Tu Liem, Hanoi, Vietnam

Anh Ngoc Le
Swinburne University of Technology
Hanoi, Vietnam

Elena Verdú
School of Engineering and Technology
Universidad Internacional De La Rioja
Logroño (La Rioja), Spain

Maria Ganzha
Warsaw University of Technology
Warsaw, Poland

ISSN 2367-3370 ISSN 2367-3389 (electronic)
Lecture Notes in Networks and Systems
ISBN 978-981-99-4724-9 ISBN 978-981-99-4725-6 (eBook)
https://doi.org/10.1007/978-981-99-4725-6

This Springer imprint is published by the registered company Springer Nature Singapore Pte Ltd.
The registered company address is: 152 Beach Road, #21-01/04 Gateway East, Singapore 189721, Singapore

Preface

The International Conference on Intelligent Systems & Network (2023), ICISN 2023, was held on March 18–19, 2023, at Swinburne Vietnam Alliance Program, Hanoi Location, Vietnam. The ICISN 2023 provides an international forum that brings together the researchers and the industry practitioners actively involved in the research in fields of intelligent computing, data science, or any other emerging trends related to the theme covered by this conference.

This conference has technical paper sessions, invited talks, and panels organized around the relevant theme. On ICISN 2023, audiences can meet some of the world's leading researchers, learn about some innovative research ideas and developments around the world, and become familiar with emerging trends in Science & Technology. The ICISN is also a perfect forum and platform for presenting ideas and achievements to researchers from multiple countries.

ICISN 2023 received a massive response in terms of the submission of papers around the world. We received papers from various countries outside Vietnam, such as India, China, South Korea, the USA, the UK, Iraq, Bangladesh, Pakistan, and Japan. The organizing committee of ICISN 2023 constituted an international solid program committee for reviewing papers. A peer-reviewed process has been adopted. The decision system adopted by EasyChair has been employed, and 77 papers have been selected after a thorough peer-reviewed process. The conference proceedings will be published in Lecture Note in Network Systems (LNNS) by Springer publisher and indexed by Scopus. We convey our sincere gratitude to the authority of Springer for providing the opportunity to publish the proceedings of ICISN 2023.

We appreciate Swinburne Vietnam for agreeing to host the conference and continuously supporting the organization team during the preparation. With their support, this conference was successful.

Our sincere gratitude to all keynote speakers, presenters, session chairs, and high officials in Vietnam for their gracious presence on the campus on the occasion.

We want to thank the plenary talks as Dr. Vijender Kumar Solanki, CMR Institute of Technology, Hyderabad, India, ICISN 2023 Chair, and Dr. Hoang Viet Ha, Director, Swinburne Vietnam Alliance Program, ICISN 2023 Co-chair.

We want to thank the keynote speakers as Prof. Alex Stojcevski, Dean, School of Science, Computing & Engineering Technologies, Swinburne University of Technology, Australia; Dr. Seok-Joo Koh, Director of Software Education Institute, Kyungpook National University, Korea; and Dr. Shiqi Yu, Associate Professor in the Department of Computer Science and Engineering, Southern University of Science and Technology, Shenzhen, China, for giving their excellent knowledge in the conference.

Furthermore, we extend our appreciation to the following industry keynote speakers for their valuable contributions: Mr. Anders Pettersson, Director of Sales Enablement, Silicon Labs; Mr. Alan Ho, Vice President of International Marketing of DataStax, Singapore; Dr. Hoang Hung Hai, Staff Manager in Product Marketing, Qualcomm Vietnam;

Dr. Le Nhan Tam, Chief Technology Officer, Microsoft Vietnam; Mr. Le Hong Viet, Chief Executive Officer, FPT Smart Cloud; Mr. Ngo Thanh Hien, CTO & Technical Sales Leader, IBM; Mr. Duc Dang, Head of Partnership and Business Development, Intel; Mr. Tran Huu Cong, Director of Embedded Software, GAM.DAP2 for giving their experience at the conference. We want to thank the reviewers for completing a big reviewing task in a short period.

ICISN 2023 was accompanied by leading technology companies both domestically and internationally, including Microsoft, IBM, Silicon Labs, Qualcomm, Intel, DataStax, FPT Software (GAM.DAP), FPT Smart Cloud, VNPT, VCCorp, Avada, and others. We want to express our gratitude for the participation and support of these technology corporations and companies in this important international industry discussion, which focuses on the role of businesses in promoting scientific research in technology at universities, particularly in core technologies such as AI, IoT, robotics, and other related technologies in artificial intelligence.

We want to thank the reviewers for their diligent and timely reviews and the program committee members for their valuable contributions to organizing the conference. Special thanks to: Dr. Hoang Viet Ha, Director of the Swinburne Vietnam Alliance Program; Dr. Truong Cong Duan, Head of the Academic Department, Swinburne Vietnam Alliance Program; Dr. Le Anh Ngoc, Director of Swinburne Innovation Space, Swinburne Vietnam Alliance Program-ICISN chair; Dr. Nguyen Thi Dieu Linh, Deputy of Head of Science and Technology Department, Hanoi University of Industry, Vietnam-ICISN chair; Dr. Tran Duc Tan, Vice Dean of Faculty of Electrical and Electronics Engineering, Phenikaa University-ICISN chair; Mr. Lai Hong Anh, Head of Admission & Marketing Department, Swinburne Vietnam Alliance Program; the event management support staffs from Swinburne Vietnam: Ms. Dang Phuong Thao, Academic Coordinator, Swinburne Vietnam Alliance Program; Ms. Tran Khanh Ly, Marketing Officer, Swinburne Vietnam Alliance Program for their efforts to make congress success.

We hope that the papers published in ICISN 2023 proceedings will be helpful for the researchers' pursuing studies in computer science, information technology, and related areas. Industrial engineers would also find this volume a good reference source.

We look forward to welcoming you all at the International Conference on Intelligent Systems & Networks held on March 23–24, 2024, at the Swinburne University of Technology Vietnam. We hope ICISN 2024 will be an exciting forum and the perfect platform for presenting your ideas and achievements.

Thi Dieu Linh Nguyen
Elena Verdú
Anh Ngoc Le
Maria Ganzha

Contents

MRI Brain Tumor Segmentation Using Bidimensional Empirical Mode Decomposition and Morphological Operations

Giang Hong Nguyen[1,2](✉), Yen Thi Hoang Hua[1], and Liet Van Dang[1]

[1] Vietnam National University, Ho Chi Minh City, Vietnam
nguyenhonggiang@caothang.edu.vn

[2] Department of General Education, Cao Thang Technical College, Ho Chi Minh City, Vietnam

Abstract. Image thresholding is a simple but effective image segmentation technique that is widely used in medical image analysis to detect tumors. The input of the method consists of a grayscale enhancement image and a threshold. In conventional approaches, the image enhancement is computed by the CLAHE method, and the threshold is computed by the Otsu method. The purpose of this article is to improve the quality of an MRI brain tumor image by using the Bidimensional Empirical Mode Detection (BEMD) algorithm and morphological operations to create an enhanced image for segmentation without blurring the edges of objects, as well as a threshold derived from the Neighborhood Valley-Emphasis (NVE) method that is appropriate for the brain tumor image because the object is smaller than the background. Analysis of the computational results of the proposed method on five MRI brain tumors from the Figshare database shows that: *(i)* convolution of the first trend function of BEMD and the combination of the top-hat and bottom-hat transforms adequately improve the image quality for segmentation; the metric used to measure image quality shows that MSE = 0.009, PSNR = 68.6731, EME = 2.8339, and EMEE = 0.1957; *(ii)* segmentation using the proposed algorithm's image enhancement and the NVE method's threshold for an accuracy of 99.61% and a precision of 96.1%.

Keywords: MRI · Segmentation · Morphological Operations · Empirical Mode Decomposition (BEMD) · Neighborhood Valley-Emphasis (NVE)

1 Introduction

The brain is a complex organ that controls bodily functions and thus plays an important role in life. Brain tumors are classified into two types based on the growth of abnormal cells in the brain. Benign brain tumors are not cancerous, which means they grow slowly and do not spread. Malignant tumors are types of cancer that grow more quickly and aggressively than benign tumors; They frequently spread and cause damage to other areas of the brain and spinal cord and must be treated as soon as possible in order to prolong life [11]. However, because the brain is shielded by the skull, it is difficult to diagnose a tumor; therefore, magnetic resonance imaging (MRI) of the head is the

T. D. L. Nguyen et al. (Eds.): ICISN 2023, LNNS 752, pp. 1–11, 2023.
https://doi.org/10.1007/978-981-99-4725-6_1

first choice in diagnosing brain tumors. Head MRI frequently has low contrast, making radiologist's analysis difficult and time-consuming. The use of computer-aided image processing algorithms in recent decades has made brain tumors analysis more convenient to aid in clinical diagnosis.

This article investigates thresholding of MRI brain image for tumor detection. Image thresholding is the simplest segmentation technique to convert a grayscale image into a binary image suitable for analysis. The Otsu threshold is widely used to detect brain tumors. A. Divya et al. (2022) used MRI images and three segmentation techniques, Otsu Threshold, Active Contour, and Fuzzy C-Media, to determine that Otsu Threshold is best suited for tumor extraction [2]. Anirban Patra et al. (2021) extracted a brain tumor using the Otsu method. The authors isolated the tumor using two different thresholds and then calculated the pixels of the tumor's area of influence to determine whether or not the tumor has progressed to an alarming stage [8]. E. Murali et al. (2020) used frequency domain filtering to remove high frequency components before extracting tumors using adaptive threshold and level sets. Finally, the images of the brain were classified as Feed Forwarded Artificial Neural Network [5].

The structure of the paper is as follows. Section 1 provides an overview of image thresholding. Section 2 presents the proposed algorithm for image enhancement. Section 3 shows the outcome of image enhancement for segmentation and tumor detection using the theshold of NVE. Section 4 contains a discussion and a conclusion.

2 Proposed Algorithm

The proposed algorithm consists of three steps: *(i)* Pre-processing; *(ii)* Image enhancemet and *(iii)* Segmentation. Figure 1 displays the suggested method.

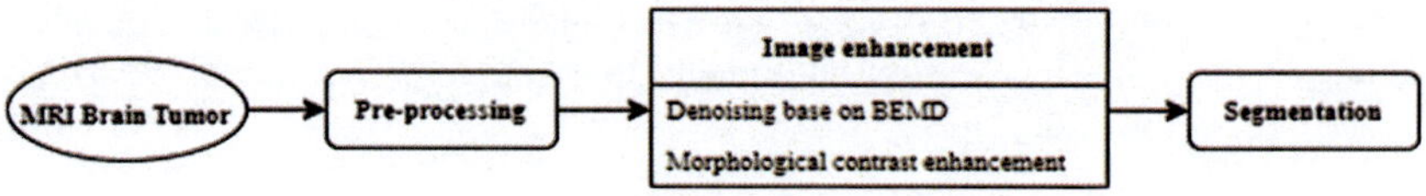

Fig. 1. The proposed algorithm.

2.1 Overview of BEMD

Huang et al. (1998) [3] developed Empirical Mode Decomposition (EMD), which uses a sifting process to decompose a stationary or non-stationary signal into a series of oscillating components at high to low frequencies, known as intrinsic functions (IMFs), without using basic functions. Each intrinsic function must satisfy two conditions: (i) the number of extrema and the number of zero crossing must be equal or differ by one; (ii) The average of the envelopes found by local maxima and minima must be zero. Nunes et al. (2003, 2005) extended the EMD algorithm for two-dimensional signals called the Bidimensional Empirical Mode Decomposition (BEMD) algorithm.

The BEMD of a signal $f(x, y)$ is implemented by a sifting process that decomposes the signal $f(x, y)$ into intrinsic mode functions $BIMFs(x, y)$ with frequencies ranging from high to low ($BIMF1(x, y)$ has the highest frequency) and residual $r(x, y)$. To reconstruct the original data, the intrinsic mode functions and the residual function are used as follows [6]:

$$f(x, y) = \sum_{j=1}^{n} b_j(x, y) + r_n(x, y) \tag{1}$$

The BEMD method was first used to analyze textures (Nunes 2003, 2005) [6, 7]. N. Vaijayanthi et al. (2018) used BEMD to decompose mammograms into BIMFs and applied adaptive global thresholding on BIMF4 and residual to obtain coarse segmented image, and then applied adaptive local thresholding to separate fine segmentation [9]. Vaijayanthi Nagarajan et al. (2019) applied MBEMD (Modify BEMD) to extract BIMFs from the ROI of the mammogram image and then used extracted features as input to the Support Vector Machine (SVM) and Linear Discriminant Analysis (LDA) classifiers to determine whether the mammogram was normal or abnormal [10].

2.2 Pre-processing

The pre-processing step improves the quality of the brain tumor ready for processing. This article uses the Gaussian function has a standard deviation of 10×10 and a kernel size of 3×3 pixels to remove noise and some built-in functions of Matlab are used to select the inner portion of the brain to preserve, the others correspond to the skull, and the label is removed.

The results are shown in Fig. 2, where (a)–(j) correspond to five input MRIs of the brain and (a1)–(j1) correspond to five MRI brain images after noise removal and skull removal.

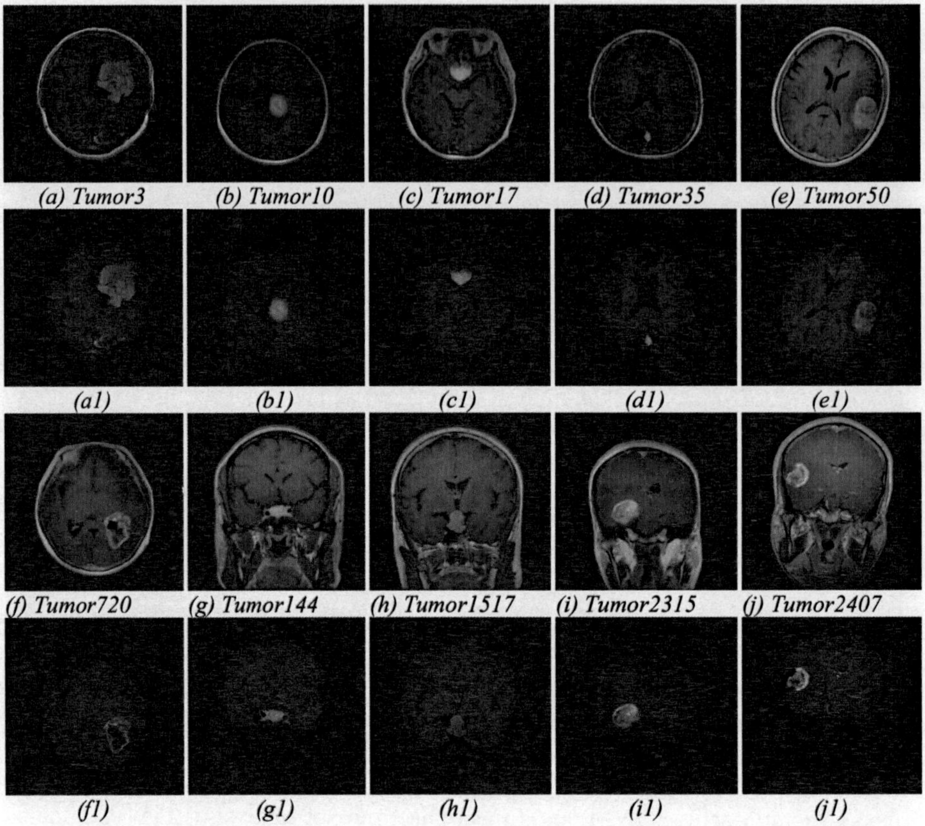

Fig. 2. Preprocessing, *(a)–(j)*: input MRI data, *(a1)–(j1)*: preprocessed images.

2.3 Image Enhancement

Pre-processed images frequently have low contrast, making segmentation difficult, so contrast needs to be enhanced. There are many contrast enhancement methods such as Histogram equalization (HE), Adaptive histogram equalization (AHE) and Contrast-Limited Adaptive Histogram Equalization (CLAHE). However, using these contrast enhancement techniques, the details of the object become blurred, resulting in inaccurate segmentation. This article addresses of this issue by combining the Bidimensional Mode Detection (BEMD) algorithm with morphological operations to improve brain MRI quality.

Morphological Contrast Enhancement

Morphological contrast enhancement employs two transforms: the top-hat transform to extract the bright areas and the bottom-hat transform to extract the dark areas. In order to obtain a contrast-enhanced image M_{EN} preserving the edges of objects, bright areas were added to the pre-processed image and dark areas were subtracted; the appropriate structuring element is used. This method can be defined [1, 4]:

$$M_{TH}(i,j) = OI(i,j) - (OI \circ SE)(i,j) \tag{2}$$

$$M_{BH}(i,j) = OI(i,j) - (OI \bullet SE)(i,j) \tag{3}$$

$$M_{EN} = (M + M_{TH}) - M_{BH} \tag{4}$$

where, OI denotes the original image, SE = 30 denotes disk structural element
° denotes the Morphological Opening Operation:

$$A(i,j) \circ B(i,j) = \left[A(i,j)\Theta B(i,j)\right] \oplus B(i,j) \tag{5}$$

and • denotes the Morphological Closing Operation:

$$A(i,j) \bullet B(i,j) = \left[A(i,j) \oplus B(i,j)\right]\Theta B(i,j) \tag{6}$$

Image Denoising Base on BEMD

BEMD is used to decompose a pre-processed image into different frequency components ranging from high to low frequencies, called Bi-Dimensional Intrinsic Mode Functions (BIMFi): BIMF1, BIMF2, BIMF3, BIMF4. Each intrinsic function has a corresponding trend function: Res1, Res2, Res3, Res4. The trend function, also called the residual function, is the remainder of the signal after extracting this intrinsic function and the intrinsic functions that preceded it. Therefore, any residual signal is a signal that has removed the high frequencies in the original signal, and is considered a smoothed signal. So, it is used to improve the contrast of the image.

Image Enhancement Using BEMD in Conjunction with Morphological Contrast Enhancement

Image enhancement based on BEMD and morphological operations is calculated by:

$$EI = Multiply(\text{Res}, M) \tag{7}$$

EI denotes image contrast enhancement using BEMD and morphological operations.

In this article, each residual function from Res1 to Res3 is used in conjunction with morphological enhancement to improve image contrast enhancement to find the optimal residual function in the proposed algorithm.

Figure 3 depicts the analysis of the proposed algorithms on Bain MRI number 720. *(a):* pre-processed image, *(b):* top-hat transform, *(c):* bottom-hat transform, *(d):* morphological image enhancement, *(e)~(g):* BIMF1 ~ BIMF3, *(h)~(j)*: Res1 ~ Res3, *(k)~(m):* Image enhancement using the proposed algorithm: $\mathrm{EI_i} = \mathrm{Multiply}(\mathrm{Res_i}, \mathrm{M})$ $EI_i =$ *Multiply(*Res_i*, M),* where $i = (1,2,3)$.

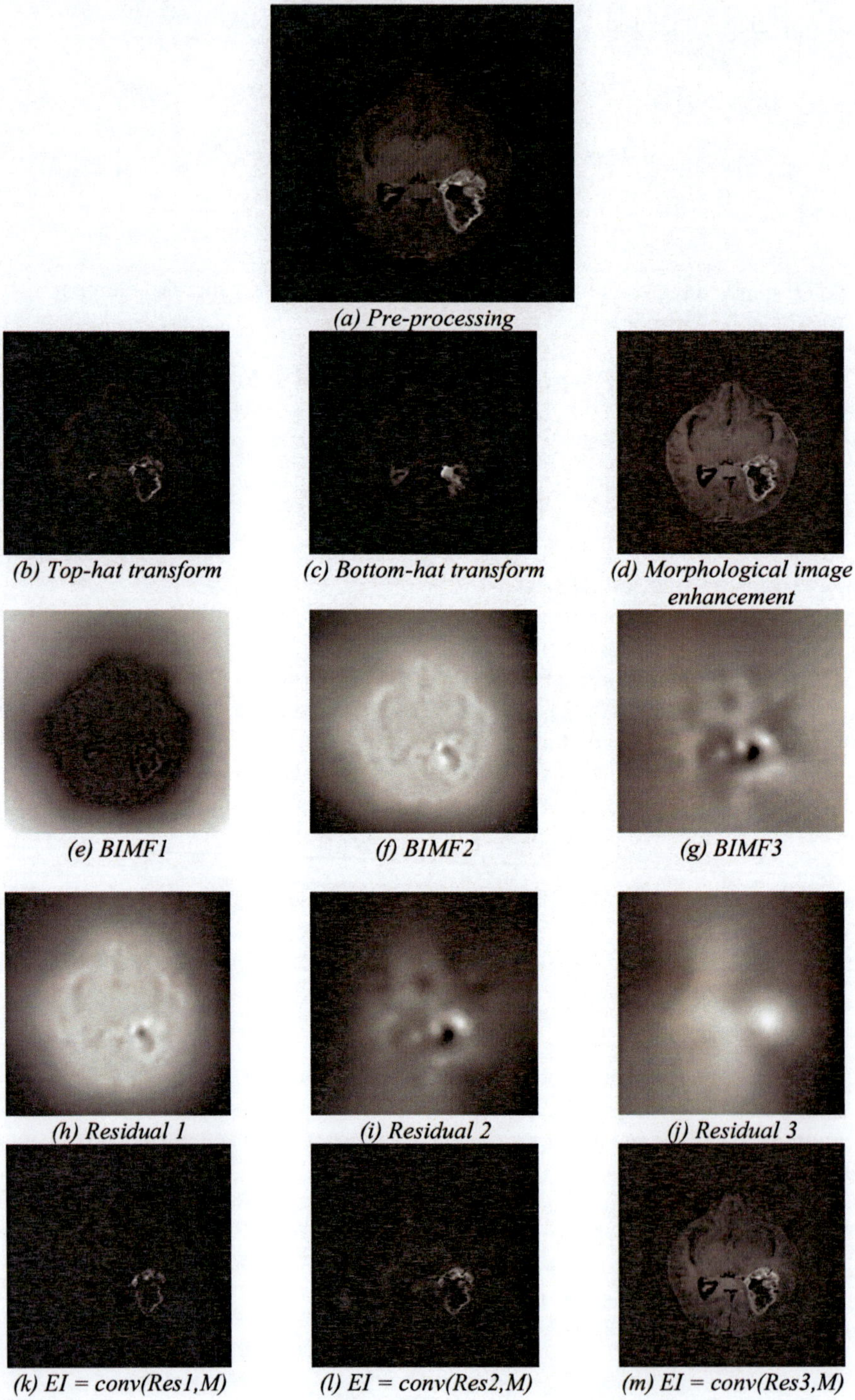

Fig. 3. The result of the analysis on Bain MRI Number 720 using the proposed algorithm's image enhancement: EI = multiply(Res*i*, M), *i* = *1, 2, 3*.

3 Results

3.1 Image Enhancement

The proposed image enhancement algorithm was tested on five pre-processed brain MRI images taken from the Figshare database published by Jun Cheng in 2015 [12]. Each image was used to generate three enhanced images corresponding to three residual 1, 2 and 3. In order to compare the results of the proposed algorithm with the results of other methods, the enhanced images of these five images were computed using three well-known techniques: HE, AHE and CLAHE.

Table 1 shows the average scores of the five enhanced images as calculated using the quality parameters such as Mean squared error (MSE), Peak Signal to Noise Ratio (PSNR), Enhancement Measurement Error (EME) and Measure of Enhancement by Entropy (EMEE).

Table 1. Comparison of Proposed method with other methods using Average values of all 5 images.

ON AVER	EI1 conv(Res1,M)	EI2 conv(Res2,M)	EI3 conv(Res3,M)	HE	AHE	CLAHE
MSE	0.00900	0.00940	0.00900	0.48870	0.00740	0.01380
PSNR	68.6131	68.5279	68.7392	51.2835	69.6817	66.7993
EME	2.83390	2.53280	2.45750	0.41710	3.51460	2.44300
EMEE	0.19570	0.17980	0.17590	0.22170	0.38410	0.32740

Table 1 shows that the AHE method gives the best image contrast enhancement; the second is the result of the proposed algorithm using residual image 1. However, due to a variety of factors, the quality parameter is unlikely to represent an image with the best segmentation.

Based on the measured quality parameters, an image with the best result in the three methods HE, AHE, and CLAHE is selected to compare it with the best result selected by the proposed algorithm from three residual images (Table 1).

3.2 Image Segmentation

Image Segmentation
Based on the results presented in Table 1, the two best contrast-enhanced images from the AHE method and the proposed residual image 1 method are selected for thresholding.

Because the tumor is small compared to the other brain, the histogram of two selected images is not bimodal, so the Neighborhood Valley-Emphasis (NVE) threshold is used to segment the tumor.

In Fig. 4, column (a) shows the input image; column (b) shows the ground truth image from the Figshare database; column (c) shows the results of tumor extraction by

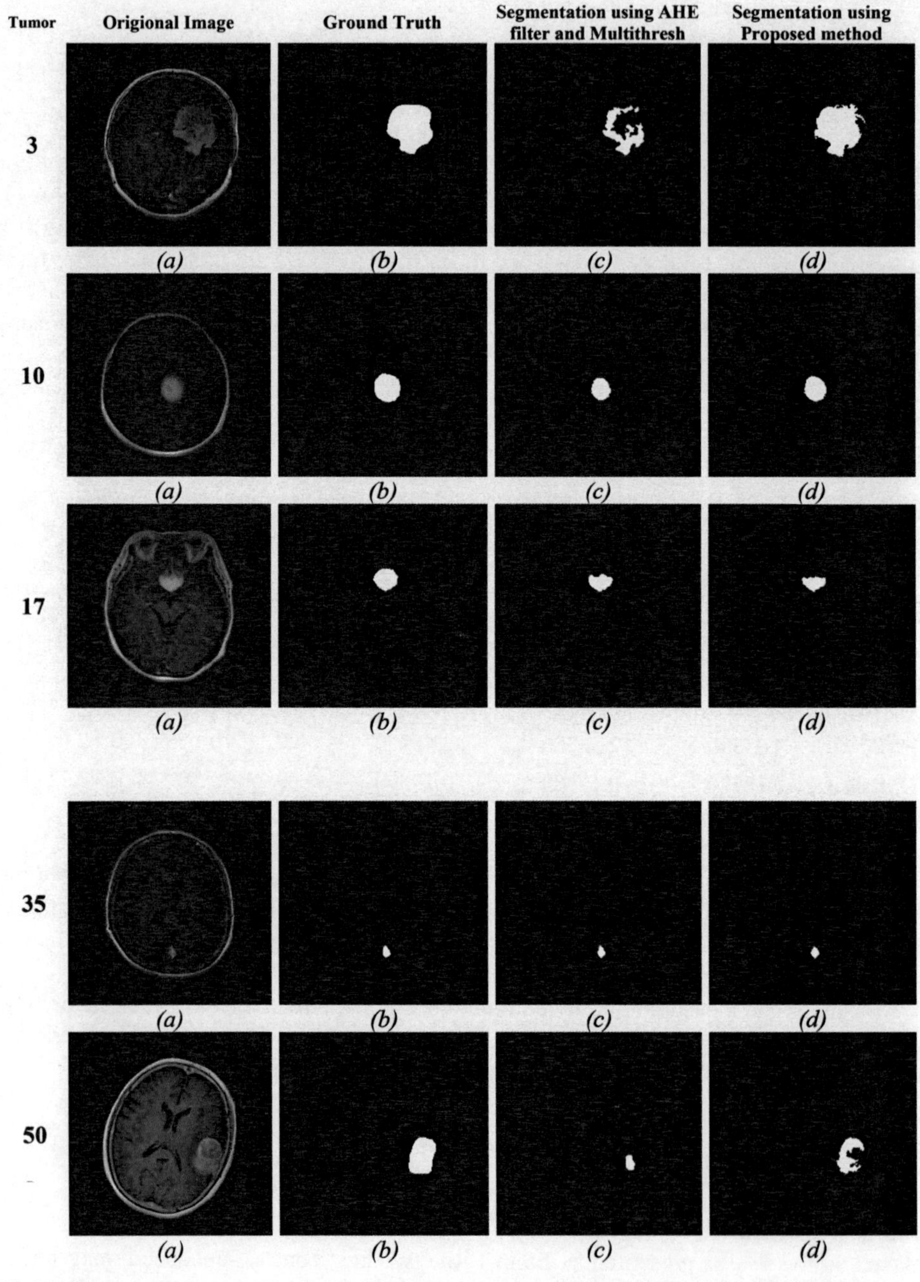

Fig. 4. Image segmentation, column (a): input images; column (b): ground truth image; column (c): tumor extraction using multi-thresholding for AHE enhancement and column (d): tumor extraction by NVE threshold for the proposed algorithm contrast enhancement image using residual 1

Table 2. Comparison of Proposed method with other methods using Average values of all 5 images

Tumor	TP (pixels)	TN (pixels)	FP (pixels)	FN (pixels)	Specificity (%)	Sensibility (%)	ACC (%)	Precision (%)
3	9833	250931	702	678	99,72	93,55	99,47	93,34
10	2735	258437	0	972	100	73,78	99,63	100
17	1555	259546	51	992	99,98	61,05	99,60	96,82
35	394	261648	37	65	99,99	85,84	99,96	91,42
50	3403	257111	44	1586	99,98	68,21	99,38	98,72
Average					**99,93**	**74,13**	**99,61**	**96,1**

multi-thresholding for contrast enhanced AHE image and column (d) shows the results of tumor extraction by NVE threshold for the proposed algorithm contrast-enhancement image using residual 1.

Performance Assessment

Based on the results of the proposed method of five brain tumors and the ground truth images from the Figshare database, some performance metrics such as specificity, sensitivity, accuracy and precision are calculated:

$$Specificity = TN/(TN + FP) \tag{7}$$

$$Sensitivity = TP/(TP + FN) \tag{8}$$

$$Accuracy(ACC) = (TP + TN)/(TP + FP + FN + TN) \tag{9}$$

$$Precision = TP/(TP + FP) \tag{10}$$

where: TP (True Positive): Both the detected pixels and the ground thruth pixels are true. FP (False Positive): The detected pixels are true, while the ground thruth pixels are false, TN (True Negative): The detected pixels are false, while the ground thruth pixels are true, FN (False Negative): Both the detected pixels and the ground thruth pixels are false.

The performance metrics are presented in Table 2. The tumor and non-tumor brains were extracted almost perfectly with accuracy = 99.61%, specificity = 99.93%, precision = 96.1% and sensitivity was quite good with sensitivity = 76.49%.

Two averages metrics of sensitivity (76.49) and accuracy (99.61) from Table 2 were compared to the analysis results of five brain tumor MRI images using Otsu's threshold of Divya et al., which have an average sensitivity of 85.69 and an accuracy of 94.22 [2]. The findings indicate that the proposed method is superior in accuracy but inferior in sensitivity. This result demonstrates the feasibility of the proposed algorithm.

4 Conclusion

In this paper, a detection of brain tumors by image processing is proposed. The proposed system is helpful in the automatic detection of brain tumors. Here: *(i)* in the processing step, the first contrast-enhanced image (M) was obtained by using the top-hat transform to extract the bright areas and the bottom-hat transform to extract the dark areas in the original MRI brain tumors. Then we applied the convolution of the first trend function (Res1) of the BEMD with M to adequately improve the image quality for segmentation. Since the highest frequencies are present in the first mode (imf1) during the BEMD sifting process, so the associated trend (residual) function 1 (Res1) is image denoising. Therefore, *Res1* was selected for image enhancement; *(ii)* in the segmentation step, the NVE threshold was used to segment the tumor, the results (Fig. 4) showed that the tumors are well detected. The obtained results were also compared to brain tumors segmented using multi-threshold enhanced images AHE, resulting in two very consistent results.

The limitation of this paper is that it only investigates image enhancement for segmentation (unsuitable for humans) and tumor detection using thresholding method because thresholding is a simple but very effective method for tumor detection when the image is appropriately enhanced and the threshold is chosen correctly. The classification of brain tumor and a method for selecting thresholds by empirical mode decomposition will be presented in a future article.

Acknowledgments. We would like to thank the Faculty of Physics and Physics Engineering - University of Science - VNU HCM for facilitating the completion of this paper.

Conflicts of Interest. The authors declare no conflicts of interest regarding the publication of this paper.

References

1. Anitha, J., Peter, J.D., Pandian, S.I.A.: A dual stage adaptive thresholding (DuSAT) for automatic mass detection in mammograms. Computer Methods and Programs In Biomedicine **138**, 93–104 (01 Oct 2016)
2. Divya, A., Dharshini, V.D., Manjula, G., Regin, R., Hussein, S.M., Al-Attar, W.M.A.M.: Brain Tumor Detection Using Image Processing Technique from MRI Images Based on OTSU Algorithm. Central Asian Journal Of Thẻoetical And Applied Sciences **3**(5), 45–65 (May 2022)
3. Huang, N.E., et al.: The empirical mode decomposition and the Hilbert spectrum for nonlinear and non-stationary time series analysis. Proceedings Of The Royal Society A Mathematical, Physical and Engineering Sciences, vol. 454(1971), pp. 903–995 (08-March-1998)
4. Li, H., Wang, Y., Liu, K.J.R., Lo, S.C.B., Freedman, M.T.: Computerized radiographic mass detection—Part I: Lesion site selection by morphological enhancement and contextual segmentation. IEEE Transactions On Medical Images **20**(4), 289–301 (2001). Apr
5. Murali, E., Meena, K.: Brain Tumor Detection from MRI using Adaptive Thresholding and Histogram based Techniques. Scalable Computing: Practice and Experience **21**(1), 3–10 (2020). March

6. Nunes, J.C., Bouaoune, Y., Delechelle, E., Niang, O., Bunel, Ph.: Image analysis by bidimensional empirical mode decomposition. Image and Vision Computing **21**(12), 1019–1026 (09 May 2003)
7. Nunes, J.C., Guyot, S., Del'echelle, E.: Texture a nalysis based on local analysis of the Bidimensional Empirical Mode Decomposition. Machine Vision and Applications **16**, 177–188 (May 2005)
8. Patra, A., et al.: Study on Brain Tumor Detection using Otsu Method. International Research Journal of Engineering and Technology (IRJET) **8**(4), 756–759 (2021). Apr
9. Vaijayanthi, N., Caroline, B.E., Murugan, V.S.: Automatic Detection of Masses in Mammograms Using Bi-Dimensional Empirical Mode Decomposition. Journal of Medical Imaging and Health Informatics **8**, 1326–1341 (2018). Sep
10. Vaijayanthi, N., Caroline, B.E., Murugan, V.S.: Feature extraction based on empirical mode decomposition for automatic mass classification of mammogram images. Medicine in Novel Technology and Devices **1** (Mar 2019)
11. https://www.abta.org/about-brain-tumors/brain-tumor-education/
12. https://figshare.com/articles/dataset/brain_tumor_dataset/1512427

A Study on Fuzzy Nonsingular Fast Terminal Sliding Mode Control for Pendubot with Uncertainties

Van-Truong Nguyen[1](✉), Hai-Binh Giap[1], Ngoc-Tien Tran[1], Ngo-Huu Manh[2], and Van-Anh Nguyen[3,4]

[1] Faculty of Mechanical Engineering, Ha Noi University of Industry, Ha Noi, Vietnam
nguyenvantruong@haui.edu.vn

[2] Science-Technology and International Cooperation Department, Sao Do University, Hai Duong, Vietnam

[3] Research and Development Department, Murata Welding Laboratories, Osaka, Japan

[4] Welding Engineering and Laser Processing Centre, Cranfield University, Cranfield, UK

Abstract. This paper proposes a new fuzzy nonsingular fast terminal sliding mode control (FNFTSMC) for pendubot with uncertainties. In the suggested method, the nonsingular fast terminal sliding mode controller (NFTSMC) is designed to get wonderful features namely speedy response, finite time converge, and singularity avoidance. Fuzzy logic control is utilized to optimize NFTSMC for rejecting chattering phenomenon and reducing errors. Moreover, the genetic algorithm is adopted to determine unknown constants of NFTSMC. The global stability of the system is guaranteed by using the Lyapunov synthesis method. Finally, some simulations are conducted to prove the efficiency of the FNFTSMC compared with NFTSMC.

Keywords: Fuzzy logic control · nonsingular fast terminal sliding mode control · pendubot · genetic algorithm

1 Introduction

The technique of sliding mode control (SMC) has been extensively used in many uncertain systems because of its advantages such as robustness, easy application, and simply construction [1]. In conventional SMC, a linear hyperplane is chosen as the sliding surface. Nevertheless, SMC can guarantee convergence from asymptotic error and neglect finite-time. Besides, the chattering phenomenon is a big problem for SMC systems [1].

To solve these drawbacks, the researchers proposed various approaches [2–5]. In [2], a nonlinear sliding surface is used to design terminal sliding mode controller (TSMC). But TSMC's problem is a singularity because forms of terms with negative fractional powers can exist. Therefore, nonsingular terminal sliding mode control (NTSMC) is designed in [3] to deal with the singularity in TSMC. However, the NTSMC controller suffers from a weakness which is the convergence time. The NTSMC approach leads

T. D. L. Nguyen et al. (Eds.): ICISN 2023, LNNS 752, pp. 12–18, 2023.
https://doi.org/10.1007/978-981-99-4725-6_2

to the push of the fast terminal sliding mode control (FTSMC) controller [4]. Unfortunately, NTSMC and FTSMC only handle an individual weakness of SMC. To solve the singularity problem and get fast convergence time simultaneously, the nonsingular fast terminal sliding mode (NFTSM) controller is developed [5]. The NFTSM method ensures convergence in finite time, rapid response when states' errors are huge, avoids singularities, and reduces noise. However, the chattering phenomenon in NFTSM still happens and responds against uncertain noises not well.

This paper proposes a new controller combining NFTSMC and fuzzy logic control for pendubot with uncertainties. The unknown parameters of FNFTSMC are determined by a genetic algorithm. This suggested scheme brings a finite time convergence without a singular problem, faster response, lower error, and chattering free. The pendubot system's stability is ensured by using the Lyapunov criteria. The simulation outcomes turn out that the proposed controller reduces the chattering phenomenon and has good performance against uncertain noises.

2 The Proposed Method

2.1 The Dynamic Model of the Pendubot

The state variables of pendubot system are defined as $y_1 = \varphi_1 - \frac{\pi}{2}$, $y_2 = \dot{\varphi}_1$, $y_3 = \varphi_2$, $y_4 = \dot{\varphi}_2$. The dynamic model of the pendubot is detemined as follows [6]:

$$\begin{cases} \dot{y}_1 = y_2 \\ \dot{y}_2 = M_1(Y) + N_1(Y)u \\ \dot{y}_3 = y_4 \\ \dot{y}_4 = M_2(Y) + N_2(Y)u \end{cases} \tag{1}$$

where:

$$u = \tau_1 \tag{2}$$

$$M_1 = \frac{1}{B_1 B_2 - B_3^2 \cos^2(\varphi_2)} \begin{bmatrix} B_2 B_3 (\dot{\varphi}_1 + \dot{\varphi}_2)^2 \sin\varphi_2 + B_3^2 \dot{\varphi}_1^2 \sin\varphi_2 \cos\varphi_2 \\ -B_2 B_4 g \cos\varphi_1 + B_3 B_5 g \cos\varphi_2 \cos(\varphi_1 + \varphi_2) \end{bmatrix} \tag{3}$$

$$N_1 = \frac{B_2}{B_1 B_2 - B_3^2 \cos^2(\varphi_2)} \tag{4}$$

$$M_2 = \frac{1}{B_1 B_2 - B_3^2 \cos^2(\varphi_2)} \begin{pmatrix} -B_3 (B_2 + B_3 \cos\varphi_2)(\dot{\varphi}_1 + \dot{\varphi}_2)^2 \sin\varphi_2 \\ +B_4 g (B_2 + B_3 \cos\varphi_2) \cos\varphi_1 \\ -B_5 g (B_1 + B_3 \cos\varphi_2) \cos(\varphi_1 + \varphi_2) \\ -B_3 \dot{\varphi}_1^2 \sin\varphi_2 (B_1 + B_3 \cos\varphi_2) \end{pmatrix} \tag{5}$$

$$N_2 = \frac{-B_2 - B_3 \cos\varphi_2}{B_1 B_2 - B_3^2 \cos^2(\varphi_2)} \tag{6}$$

where: $B_1 = m_1 d_{c1}^2 + m_2 d_1^2 + R_1, B_2 = m_2 d_{c2}^2 + R_2, B_3 = m_2 d_1 d_{c2}, B_4 = m_1 d_{c1} + m_2 d_1, B_5 = m_2 d_{c2} m_1$ and m_2 are the link's masses; d_1 and d_2 are the length of each link; d_{c1} and d_{c2} are the distances from the links' centroids to the joints; R_1 and R_2 are respectively the moments of inertia of two links at their centroids of gravity; g is the acceleration due to gravity.

2.2 Fuzzy Nonsingular Fast Terminal Sliding Mode Design

Sliding surfaces of link 1 and link 2 are chosen as follows:

$$\begin{cases} s_1 = y_1 + \theta_1 y_1^{\omega_1} + \theta_2 y_2^{\gamma_1} \\ s_2 = y_3 + \theta_3 y_3^{\omega_2} + \theta_4 x_4^{\gamma_2} \end{cases} \tag{7}$$

where $\theta_1 , \theta_2 , \theta_3 , \theta_4 > 0;\ 1 < \gamma_1, \gamma_2 < 2;\ \omega_1 > \gamma_1;\ \omega_2 > \gamma_2$.

The time derivative of sliding surfaces is:

$$\begin{cases} \dot{s}_1 = y_2 + \theta_1 \omega_1 |y_1|^{\omega_1 - 1} y_2 + \theta_2 \gamma_1 |y_2|^{\gamma_1 - 1} \dot{y}_2 \\ \dot{s}_2 = y_4 + \theta_3 \omega_2 |y_3|^{\omega_2 - 1} y_4 + \theta_4 \gamma_2 |y_4|^{\gamma_2 - 1} \dot{y}_4 \end{cases} \tag{8}$$

The system is stable when $\dot{s}_1 = 0$ và $\dot{s}_2 = 0$, the equivalent control law of subsystems is expressed as:

$$\begin{cases} \tau_{eq1} = -(\gamma_1 \theta_2 y_2^{\gamma_1 - 1} N_1)^{-1} (y_2 + \omega_1 \theta_1 y_1^{\omega_1 - 1} y_2 + \gamma_1 \theta_2 y_1^{\omega_1 - 1} M_1) \\ \tau_{eq2} = -(\gamma_2 \theta_4 y_4^{\gamma_2 - 1} N_2)^{-1} (y_4 + \omega_2 \theta_3 y_3^{\omega_2 - 1} y_4 + \gamma_2 \theta_4 y_3^{\omega_2 - 1} M_2) \end{cases} \tag{9}$$

For an underactuated system, the overall control law has to consist of all subsystems' equivalent control [6]. The overall control law is expressed as:

$$\tau = \tau_{eq1} + \tau_{eq2} + \tau_{sw} \tag{10}$$

where τ_{sw} is the switch control component.

The general surface of pendubot is as follows:

$$S = \xi s_1 + s_2 \tag{11}$$

where $\xi > 0$

The switch control is designed as:

$$\tau_{sw} = -(\alpha + \beta)^{-1} (\beta \tau_{eq1} + \alpha \tau_{eq2} + \eta.sign(S) + K.S) \tag{12}$$

The overall control law τ is determined:

$$\tau = (\alpha + \beta)^{-1} (\alpha \tau_{eq1} + \beta \tau_{eq2} - \eta.sign(S) - K.S) \tag{13}$$

where $\alpha = \xi \theta_2 \omega_1 |y_2|^{\omega_1 - 1} N_1,\ \beta = \theta_4 \omega_2 |y_4|^{\omega_2 - 1} N_2$

According to the above control law, the chattering phenomenon is most affected by the control parameter η. If this parameter is big, the steady-state time will be fast but

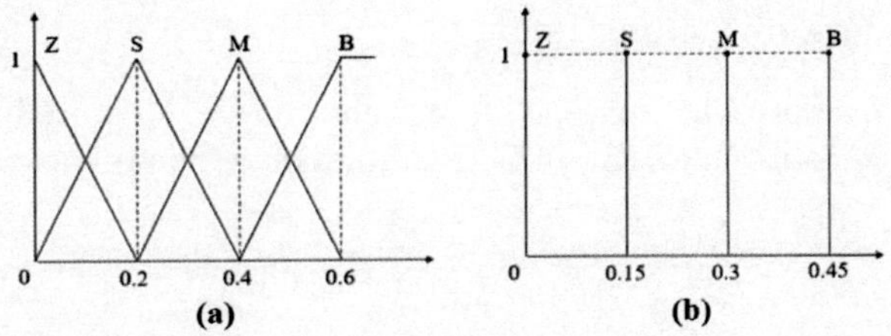

Fig. 1. A. Input membership functions of $|S|$, b. Output membership functions of η

Table 1. Rule base of the proposed method

Rule	1	2	3	4
$\|S\|$	Z	S	M	B
η	Z	S	M	B

the "chattering" phenomenon will also be large and vice versa. Therefore, the proposed method uses a fuzzy algorithm to generate the adaptive law of the η parameter according to the absolute value of the sliding surface S. The input and output membership functions are described in Fig. 1. a and Fig. 1. b. The rule base is shown in Table 1.

Combing the control law of NFTSMC (13) and the fuzzy controller, the FNFTSMC law is as:

$$\tau = (\alpha + \beta)^{-1}(\alpha\tau_{eq1} + \beta\tau_{eq2} - \hat{\eta}.sign(S) - K.S) \tag{14}$$

System's stability is demonstrated by the Lyapunov theory with Lyapunov function:

$$V = \frac{1}{2}S^2 \tag{15}$$

The time derivative of V is:

$$\begin{aligned}
\dot{V} &= S\dot{S} \\
&= \dot{V}(t) = S\dot{S} = S(\xi\dot{s}_1 + \dot{s}_2) \\
&= S[\xi y_2 + \xi\theta_1\omega_1|y_1|^{\omega_1-1}y_2 + \xi\theta_2\gamma_1|y_2|^{\gamma_1-1}M_1 + \xi\theta_2\gamma_1|y_2|^{\gamma_1-1}N_1(\tau_{eq1} + \tau_{eq2} + \tau_{sw}) \\
&\quad + y_4 + \theta_3\omega_2|y_3|^{\omega_2-1}y_4 + \theta_4\gamma_2|y_4|^{\gamma_2-1}M_2 + \theta_4\gamma_2|y_4|^{\gamma_2-1}N_2(\tau_{eq1} + \tau_{eq2} + \tau_{sw})] \\
&= S[\xi\theta_2\gamma_1|y_2|^{\gamma_1-1}N_1(\tau_{eq2} + \tau_{sw}) + \theta_4\gamma_2|y_4|^{\gamma_2-1}N_2(\tau_{eq1} + \tau_{sw})] \\
&= S[-\hat{\eta}.Sign(S) - K.S] = -\hat{\eta}|S| - KS^2 \leq 0
\end{aligned} \tag{16}$$

where K, $\hat{\eta} > 0$.

From (17), the Lyapunov theory is satisfied. Therefore, the stability of the overall system is ensured.

2.3 Genetic Algorithm Optimization

In the FNFTSMC approach, the genetic algorithm (GA) [6] is used to determine the unknown parameters in control law (13). The fitness function is chosen as follow:

$$J = \sum_{j=1}^{n} \left[E_1(j).E_1^T(j) + E_2(j).E_2^T(j) \right] \tag{17}$$

where E_1, E_2 are the error value matrixes between actual q_1 and desired and between actual q_2 and desired q_2.The GA generations is 200. The population size is 20. Whereas, the crossover and mutation co efficient is 0.7 and 0.3, respectively.

3 Results

In order to prove the efficiency of the proposed appoach, some simulations are conducted compared to the NFTSMC [5]. Two kinds of noise used to simulate are disturbance noise and fault noise. Disturbance noise is defined as the equation follows:

$$e(t) = 0.9 * 23 * \frac{\pi}{180} * exp\left(-\frac{(t-3)^2}{2 * 0.25^2} \right) \tag{18}$$

where t is time(s). Whereas, fault one is turning off the control signal for a period of time. The parameters of FNFTSMC and NFTSMC [5] under training GA are listed in Table 2.

Table 2. The controllers' parameters

θ_1	θ_2	θ_3	θ_4	ω_1	ω_2	ξ	η	K
2.5	1.2	2.1	1.3	1.1737	1.0659	1.84	0.747	9.305

The simulation outcomes are described in Fig. 2 and Fig. 3. The responses of link 1 and link 2 suffered disturbance fault are shown in Fig. 2. The RMSE, MAE comparisons in disturbance noise case are described in Table 3. Suffered fault noise time are 0.8 s and 1.2 s, respect to NFTSMC and FNFTSMC. Compared simulation results of two controller, it can be seen that the suggested scheme has better performance: lower error-type values and longer time subjected to fault noise. The control input is shown in Fig. 3. It can be shown that the chattering phenomenon had been almost eliminated by combining with the fuzzy logic controller.

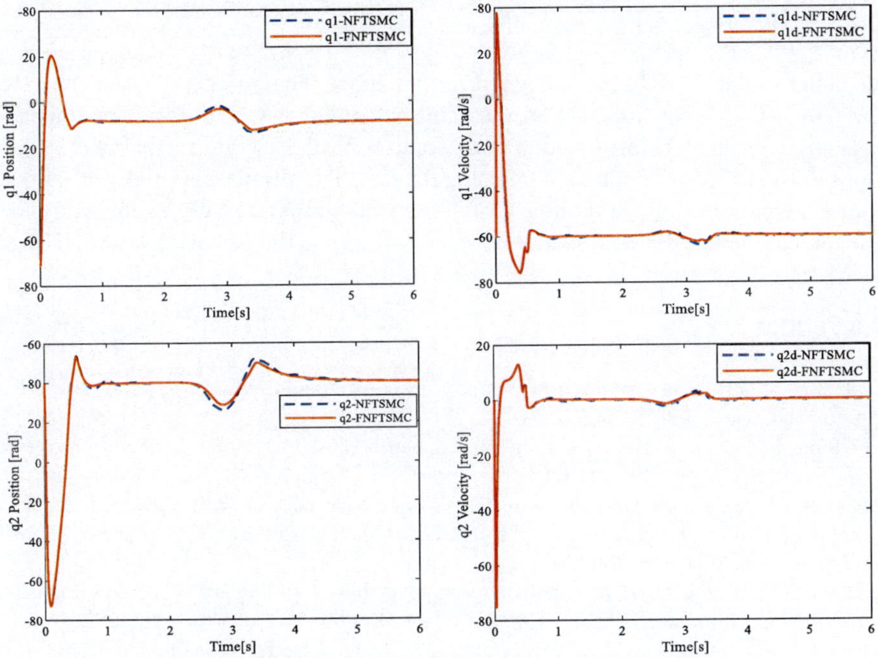

Fig. 2. Responses of two links suffered disturbance fault

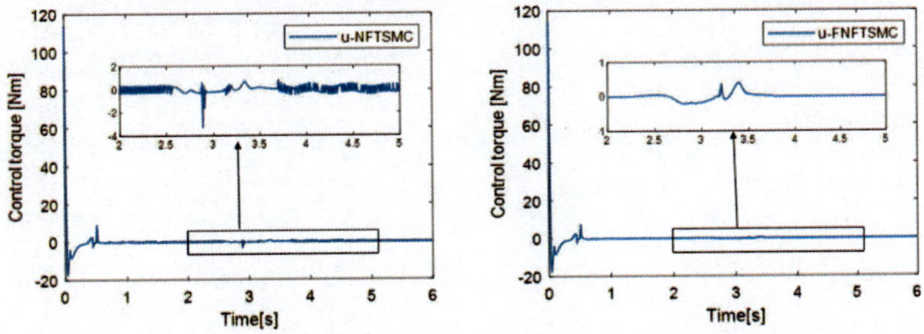

Fig. 3. NFTSMC and FNSTSMC's control torque in suffered disturbance fault case

Table 3. Error comparison between NFTSMC [6] and FNFTSMC

	RMSE q1	RMSE q2	MAE q1	MAE q2
NFTSMC	0.2761	0.4125	0.0828	0.1141
FNFTSMC	0.2634	0.4069	0.0812	0.1136

4 Conclusion

This paper studies FNFTSMC designed for pendubot. The NFTSMC law is utilized to make the system have fast convergence and nonsingular. Chattering phenomena and errors are reduced by using adaptive fuzzy logic control. Besides, the parameters of the proposed controller are optimized by using the GA. The simulation process results turn out that the proposed method outperforms conventional NFTSMC. In the future, other artificial intelligence methods can be adopted to improve the performance of NFTSMC.

References

1. Incremona, G.P., et al.: Sliding mode control of constrained nonlinear systems. IEEE Trans. Automat. Control. **62**(6), 2965–2972 (2017)
2. Pietrala, M., et al.: Terminal sliding mode control of second order systems with velocity constraint. ICCC, pp. 223–227 (2018)
3. Abrazeh, S., et al.: Nonsingular terminal sliding mode control with ultra-local model and single input interval type-2 fuzzy logic control for pitch control of wind turbines. IEEE/CAA J. Autom. Sin. **8**(3), 690–700 (2021)
4. Nguyen, V.T.: Non-negative adaptive mechanism based sliding mode control for parallel manipulators with uncertainties". Comput. Mater. Contin. **74**(2), 2771–2787 (2023)
5. Nguyen, V.T., et al.: Nonlinearities Output-Feedback Adaptive Nonsingular Fast Terminal Sliding Mode Control for Redundant Parallel Manipulators. ICSSE, pp. 1–5 (2020)
6. Nguyen, V.T., et al.: Adaptive neural network hierarchical sliding-mode control for pendubot based genetic algorithm optimization. ICISN, pp. 574–580 (2022)

Fluid Pipeline Leak Localization Relying on Acoustic Emission Signal Analysis

Thang Bui Quy[1](✉) and Jong-Myon Kim[2]

[1] Institute of System Integration, Le Quy Don Technical University, Hanoi, Vietnam
bqthangcndt@gmail.com

[2] Department of Electrical, Electronics and Computer Engineering, University of Ulsan, Ulsan 44610, Republic of Korea

Abstract. This paper introduces a leak–localization method for fluid pipeline through monitoring bursts in acoustic emission signals. First, the algorithm seeks bursts in individual signals using the Neyman–Pearson signal detection theorem, groups adjacent bursts from two signals captured by two acoustic emission sensors mounted at two pipeline ends and localizes those burst sources through the time difference of arrival technique. Then, a coordinate histogram is constructed from a set of resulting burst source coordinates in a period to indicate a suspicious leaky position on the testing pipeline. The experimental results reveal that the proposed technique can determine single leaks with a mean relative location error of 1.2% while conventional methods return relative errors of greater than 7.5%. Furthermore, the proposed method can properly localize two simultaneous leaks whereas others cannot.

Keywords: Acoustic emission · Impulse detection · Leak localization

1 Introduction

Pipeline plays a crucial role in a system transporting water, petrol, or industrial fluid. Despite designing and assembling pipelines using technical instructions [1, 2], a leak could still happen due to material aging [3, 4], leading to fire and explosion which would pollute environment and even cause human injuries and deaths. Thus, early pipeline leak localization is necessary to reduce such serious consequences.

Even though the cross-correlation function (CCF), which returns time difference of arrival (TDOA) between two signals to implement acoustic emission (AE) source localization [5], has been applied to leak localization for pipeline, it strongly depends on background noise [6]. Moreover, there may be many concurrently existing AE sources on a pipeline including multiple–leaks and leakage–unrelated sources such as flow–induced vibrations and external collisions. Consequently, the CCF–based method is greatly restricted to pipeline leak localization. Another approach using a generalized cross-correlation (GCC) can enhance the CCF shape by combining the CCF with a pre-filter such as Roth, Smoothed Coherence Transform, Phase Transform (PHAT), Eckart Filter [7], thus resulting in a more accurate TDOA estimator. However, implementing

T. D. L. Nguyen et al. (Eds.): ICISN 2023, LNNS 752, pp. 19–26, 2023.
https://doi.org/10.1007/978-981-99-4725-6_3

a prefilter requires a priori information about signal sources and noise background and the GCC–based method may be inappropriate for localizing several simultaneous signal sources as well. To address the issue, this paper proposes monitoring bursts in AE signals. First, bursts are detected from individual signals by an adaptive threshold depending on their noise levels and a given false alarm probability using the Neyman–Pearson signal detection theorem [8]. Second, based on arrival time of a pair of bursts from two signal channels, a source coordinate is computed. Finally, leakage can be speculated at high density positions on a source location histogram drawn along the testing pipeline. Experimental results show that the proposed method outperforms the CCF– and GCC–based localization techniques and it can accurately identify single leaks and multiple concurrent leaks as well.

2 Methodology

We assume that a leak is existing at a coordinate x on a pipeline with length d and two AE sensors S_1 and S_2 are mounted at its two ends (see Fig. 1 (a)). The general AE burst–based localization approach is illustrated in Fig. 1 (b).

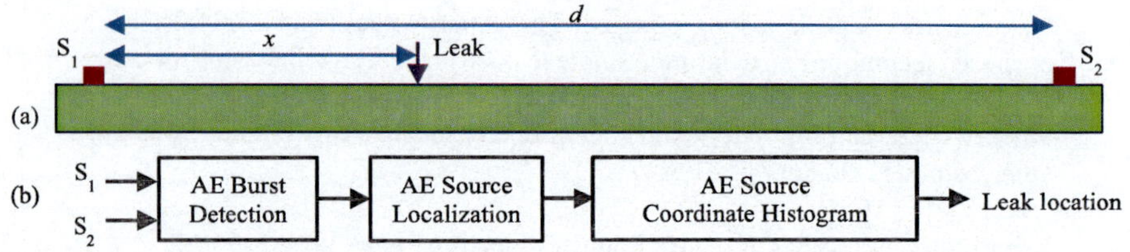

Fig. 1. Pipeline leak localization diagram based on AE burst monitoring: (a) AE sensor installation, (b) Localization algorithm.

2.1 AE Burst Detection

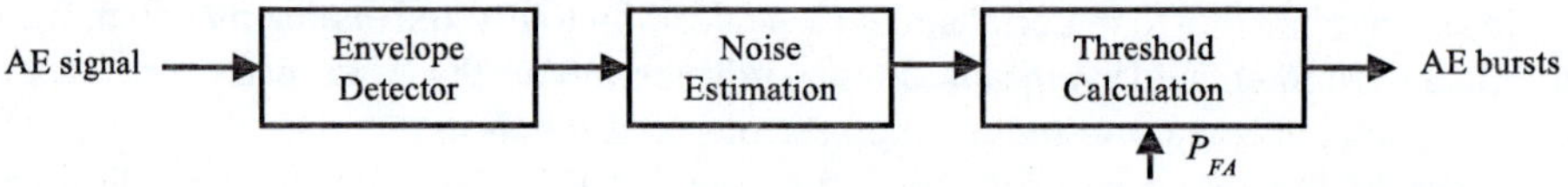

Fig. 2. AE burst detection algorithm.

Figure 2 illustrates the whole AE burst detection process in a raw AE signal. First, we conduct an AE signal to an envelope detector and then seek bursts from the output signal envelope using an adaptive threshold γ with a given false alarm probability P_{FA}. The detection theorem [8] states that to obtain a maximum detection probability P_D, it is essential to select if H_1 satisfies

$$L(z) = \frac{p(z|H_1)}{p(z|H_0)} > \gamma \tag{1}$$

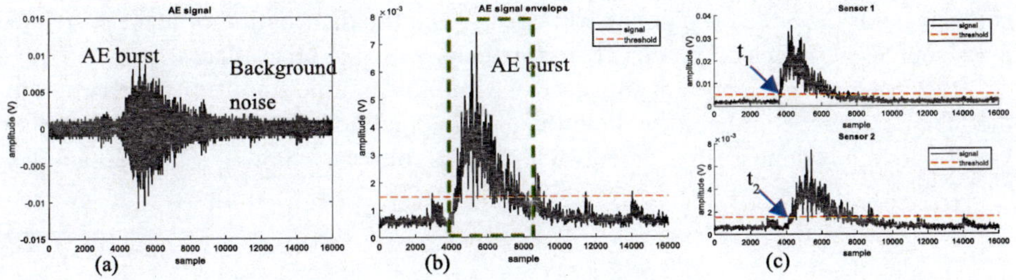

Fig. 3. (a) Raw AE signal, (b) Burst detected in AE signal envelope using an adaptive threshold, (c) Two bursts from signals acquired by two sensors S_1 and S_2.

where the detection threshold γ is given by

$$P_{FA} = \int_{\{z:L(z)>\gamma\}} p(z|H_0)dz \tag{2}$$

where $L(z)$ is the likelihood ratio function, H_0 is the null hypothesis (no burst or background noise), H_1 is the alternative hypothesis (a burst), z is an observation signal envelope, and p(z) is the probability density function. The H_1 hypothesis is decided if the sample value surpasses the threshold γ whereas the H_0 hypothesis is decided. Because the threshold is calculated on the background noise of input signal, it is frequently updated, thus can adapt to any change of working pipeline conditions (pressure and flow rate). Subfigures 3 (a, b) show an AE burst detected in a real AE signal acquired from a leaky pipeline.

2.2 AE Source Localization

Based on the wave propagation theorem [9, 10], AE source coordinate is given by:

$$x = \frac{d - C\Delta t}{2}, \Delta t = t_2 - t_1 \tag{3}$$

where C is the AE wave speed and $t_{1,2}$ (see Fig. 3 (c)) are the arrival times of AE bursts from the two signal channels, respectively, Δt is the TDOA value.

In other words, the source coordinate must comply with the condition $0 < x < d$ because we only consider regularities belonging to the testing pipeline and ignore what is outside it. Therefore, we obtain a constraint on the TDOA value as follows:

$$|\Delta t| < \frac{d}{C} \tag{4}$$

2.3 AE Source Coordinate Histogram

Due to the existence of leakage–unrelated events, we should construct a source location histogram in time along pipeline length and then leak can be speculated around high–density points on this histogram. This speculation is reasonable because a leak itself is always non-movable on pipeline if it occurs, and the distribution of leakage–related burst sources at the leaky region is therefore denser than at other places.

Because leak can happen at any place, we initially assume a uniform distribution for the leak occurrence probability along the testing pipeline and then divide the pipeline length into equivalent small segments with size as follows:

$$b = \frac{d}{n} \tag{5}$$

where b is the segment length, n is the number of divided segments. Call Δx_m as the expected source location error (this value is usually predefined in practice to evaluate the result). Hence, we have a constraint on the length of segments to guarantee the coordinate histogram resolution, given by (6).

$$b \leq \frac{\Delta x_m}{2} \tag{6}$$

As a result, we can choose the value n as follows:

$$n \geq \frac{2d}{\Delta x_m} \tag{7}$$

The AE burst source density per segment is referred to as an accumulated number of detected burst sources falling in the segment. We further utilize a weight when constructing the histogram as follows:

$$h_i = h_i + M_s \times N_s \tag{8}$$

where h_i is the contemporary source density in the ith element, and M_s and N_s are magnitude and duration of a burst, respectively. The term $M_s \times N_s$ in Eq. (8) improves the simple counting, which is $h_i = h_i + 1$. With the assistance of the AE burst source coordinate histogram, we can suspect a leaking point if it has a high AE burst source density (see Fig. 4).

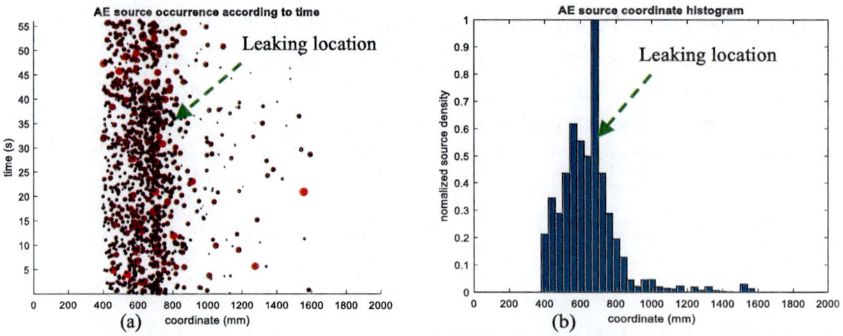

Fig. 4. AE burst source location histogram: (a) time–variant AE burst source occurrence, (b) AE burst source coordinate histogram.

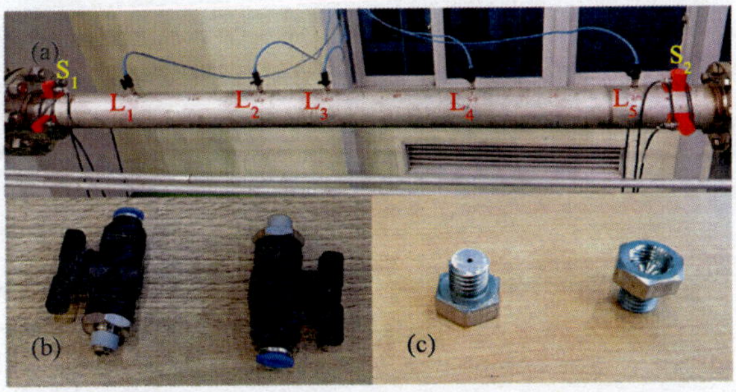

Fig. 5. The experimental configuration: (a) pipeline testbed, (b) valve, (c) orifice.

3 Experimental Setup

To verify the proposed approach, we established a water pipeline model with two R15I sensors installed at the testing pipe ends (see Fig. 5). These sensors can detect signals from 50 kilohertz (kHz) to 400 kHz [11], which can cover the frequencies of AE waves propagating in metal objects from 100 kHz to 300 kHz as specified in the BSI standard BS EN 15856 [12]. We digitized AE signals at 1 megahertz (MHz) by a NI-9223 module. After completing the hardware setup, we applied the pencil lead break technique [13] to test the data acquisition, thus ensuring the proper installation. To simulate pipeline leakage, we created a leak tool composed of a valve and an orifice (see Fig. 5 (b, c)). Thus, the normal/leaky pipeline states are linked to closed/open valve. In the experiment, the five leakage points were arranged along the pipeline (see Fig. 5 (a)). The testbed design parameters are detailed in Table 1.

Table 1. Experimental parameters

No	Parameter	Detail
1	Thickness of pipelines	6.02 [mm]
2	Outer diameter of pipelines	114.3 [mm]
3	Material of pipelines	Stainless steel 304
4	Location of leak L_1 (x_1)	200 [mm]
5	Location of leak L_2 (x_2)	800 [mm]
6	Location of leak L_3 (x_3)	1000 [mm]
7	Location of leak L_4 (x_4)	1400 [mm]
8	Location of leak L_5 (x_5)	1800 [mm]
9	Sensor installation distance (d)	2000 [mm]
10	Orifice diameter	0.5 [mm]
11	Inner flow pressure	700 kPa
12	Inner water temperature	27 °C

4 Experimental Results

Apart from the proposed method, we implemented the CCF– and GCC–PHAT–based ones on the same signals acquired in Sect. 3 for comparison. Additionally, we define relative leak coordinate error to facilitate comparison as follows:

$$e_r = 100 \times \frac{\Delta x}{d} \tag{9}$$

where e_r refers to as the relative error and Δx is the absolute coordinate error between the values estimated by the localization methods and its true leak coordinate. Furthermore, the AE wave velocity C is examined and approximated to be 3250 *m/s* by using signals generated by the pencil lead break technique.

Figure 6 illustrates AE burst source coordinate histograms resulting from CCF–, GCC–PHAT–based, and proposed leak localization methods, respectively, applied in various experiments. We can see that the proposed method is superior to the others because most CCF– and GCC–PHAT–based histograms show the highest AE burst source density locations far from the true leak coordinates while the proposed–method histograms precisely indicate these locations. Moreover, the implemented GCC–PHAT–based method does not improve the CCF–based one to locate a leak as expected. In every experiment, the GCC–PHAT–based method virtually turns out the dense AE burst source distribution at the center region however a leak is not always present there. Table 2 shows the relative leak coordinate errors in leak experiments if the leaky location is determined to be the position with the topmost density. The mean errors for the single leak cases (L_1 – L_5) demonstrate that the suggested leak localization mechanism outperforms the others. In addition, the multi–leaks case can be addressed by the proposed method as

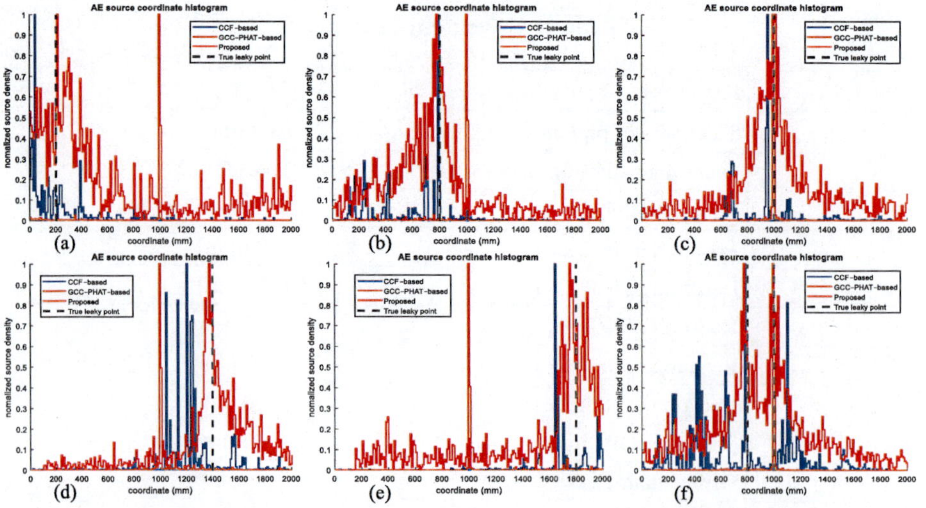

Fig. 6. Results of different leak localization methods: (a) L_1, (b) L_2, (c) L_3, (d) L_4, (e) L_5, (f) two concurrent leaks (L_2 and L_3).

presented in Fig. 6 (f) and the relative coordinate error in this case is small, 1.25% (see Table 2).

Table 2. Relative leak coordinate error (%)

Leakage experiment	CCF	GCC	Proposed method
L_1	8	39.5	0.5
L_2	1	9.5	1.5
L_3	2.5	0.5	0.5
L_4	8	25	1
L_5	10	40.5	2.5
Average (single leaks)	7.5	23	1.2
Multi–leaks ($L_{2,3}$)	x	x	1.25

5 Conclusions

The paper proposed a mechanism of AE burst monitoring–based pipeline leak localization outperforming conventional methods using cross–correlation and generalized cross–correlation functions. In this paper, the proposed and traditional localization methods were implemented and compared in the same condition of various leak experiments involving single leaks and even multi–leaks in a water pipeline testbed. The results show that the proposed algorithm brings about the mean relative coordinate error much smaller than the others do. Moreover, the proposed leak localization can be applied for the multi–leaks situation while this can be challenging if utilizing cross–correlation and generalized cross–correlation functions.

Acknowledgments. This work was supported by the Korea Institute of Energy Technology Evaluation and Planning (KETEP) and the Ministry of Trade, Industry & Energy (MOTIE) of the Republic of Korea (No. 20192510102510). It was also funded by the Science Research Program of Le Quy Don Technical University, Hanoi, Vietnam (No. 22.1.78).

References

1. Antaki, G.A.: Piping and Pipeline Engineering Design, Construction, Maintenance, Integrity, and Repair. Marcel Dekker, Inc., USA (2003)
2. McAllister, E.W., (ed.): Pipeline Rules of Thumb Handbook: Quick and accurate solutions to your everyday pipeline problems. Elsevier Inc. All rights reserved (2009)
3. Murvay, P.S., Silea, I.: A survey on gas leak detection and localization techniques. J. Loss Prev. Process Ind. **25**(6), 966–973 (2012)
4. Group EGPID. EGIG GAS PIPELINE INCIDENTS (2020)

5. Muravin, B.: Acoustic emission science and technology. The J. Build. Infrastru. Eng. Israeli Asso. Eng. Archit. **1**, 4–5 (2009)
6. Gao, Y., Brennan, M.J., Joseph, P.F., Muggleton, J.M., Hunaidi, O.: A model of the correlation function of leak noise in buried plastic pipes. J. Sound Vib. **277**(1–2), 133–148 (2004)
7. Knapp CaC, G.: The generalized correlation method for estimation of time delay. IEEE Trans. Acoust. Speech Signal Process. **24**(4), 320–327 (1976)
8. Kay, M.S.: Detection Theory. Fundamentals of Statistical Signal Processing. Upper Saddle River, New Jersey 07458: Prentice Hall PTR (1998)
9. Miller, R.K., Hill, E.v.K., Moore, P.O.: Acoustic Emission Testing. Columbus, OH: American Society for Nondestructive Testing (2005)
10. Rienstra, S.W., Hirschberg, A.: An Introduction to Acoustics. Eindhoven University of Technology (2018)
11. Group, M.: R15I-AST Sensor. Princeton Junction, NJ, USA (2015)
12. 15856 BE. Non-Destructive Testing – Acoustic Emission – General Principles of AE Testing for the 416 Detection of Corrosion with Metallic Surrounding Filled with Fluid. European Standard (2010)
13. Ranganayakulu, S.V., Goud, B.S., Sastry, P.V., Kumar, B.R.: Calibration of acoustic emission system for materials characterization. Univers J Mater Sci. **3**, 62–69 (2015)

Design of Measuring and Warning System Based on Sound Intensity in High-Traffic Areas

Phat Nguyen Huu[1(✉)], Quyen Nguyen Thi[1], Dinh Dang Dang[1], Thanh Le Thi Hai[1], and Quang Tran Minh[2,3]

[1] Hanoi University of Science and Technology, Hanoi, Vietnam
{phat.nguyenhuu,thanh.lethihai}@hust.edu.vn,
{quyen.nt182748,dinh.dd200152}@sis.hust.edu.vn

[2] Department of Information Systems, Faculty of Computer Science and Engineering, Ho Chi Minh City University of Technology (HCMUT), 268 Ly Thuong Kiet, District 10, Ho Chi Minh City, Vietnam
quangtran@hcmut.edu.vn

[3] Vietnam National University Ho Chi Minh City (VNU-HCM), Linh Trung Ward, Thu Duc District, Ho Chi Minh City, Vietnam

Abstract. At the intersection of green and red lights, people often honk their horns to urge people. Besides, noise pollution in urban areas due to the abuse of car horns leads to high sound intensity that affects the health of people. Although many solutions have been given, this problem has not been completely solved. Therefore, we present a solution to measure the sound intensity at the intersection of red and green lights in this paper. When the sound intensity is too great (the threshold), the waiting time for the red light will increase. It will directly affect the perception of people and thus reduce the noise. The results show that the system has recovered noise with an accuracy of up to 96%.

Keywords: Sound intensity · STM32 · warning system · high-traffic · cut-off frequency

1 Introduction

Sound intensity is the amount of energy transmitted per unit of time through an area placed perpendicular to the direction [1]. Humans can only bear up to one hour of sound intensities greater than 90 dB per day. It must start to be aware of how dangerous from 80 to 90 dB. Street noise falls around 70 dB [2]. Therefore, selecting a sound intensity threshold of 80 dB as the limit level exceeded will increase the time to wait for the red light.

The goal is to design a sound intensity meter circuit to control the led display time. If the intensity exceeds 80 dB, the red light display time will increase to 10 s. The system can be divided into two large blocks. The first block captures the sound intensity and filters the signal, amplifying the signal to compensate for the lost energy. The second part controls the display block based on the received

T. D. L. Nguyen et al. (Eds.): ICISN 2023, LNNS 752, pp. 27–33, 2023.
https://doi.org/10.1007/978-981-99-4725-6_4

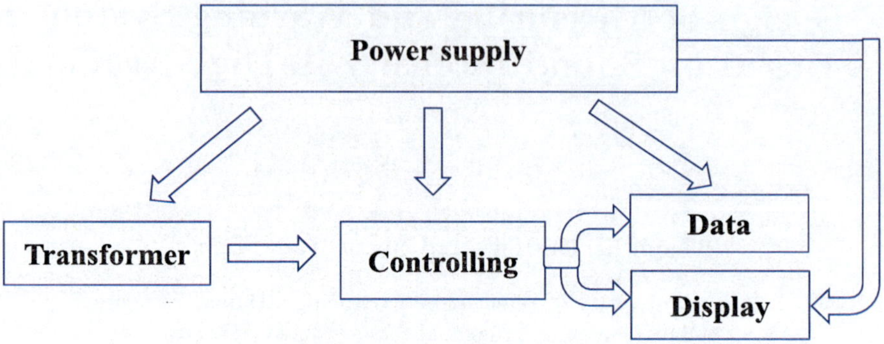

Fig. 1. Overview of the proposed system.

sound intensity. Currently, the system is being tested at a low sound intensity and achieving relatively high accuracy.

The rest of the paper includes four parts and is organized as follows. Section 1 will discuss the related work. In Sect. 2, we present the proposal system. Section 3 will evaluate the proposed system and analyze the results. In the final section, we give conclusions and future research directions.

2 Related Work

Currently, noise pollution in urban areas is increasing and there are many studies to reduce noise [3–8]. Several studies on urban noise measurement and solutions such as urban noise discomfort measurement based on RNN [3], research on equipment with noise reflective and absorbing materials combined with filters to process input noise and generate an anti-interference set [4], estimate road traffic noise in urban areas [5] or the most recent study is the study of using plants with noise absorption capacity [6]. One of the main causes of urban sound pollution is caused by vehicles. Therefore, there have been many studies to reduce vehicle noise, such as controlling the noise of electric vehicles [7] or identifying the source of noise in cars [8]. However, these studies focus on reducing and preventing noise caused by vehicles and equipment that ignore the problems of humans. This research is a solution that directly affects the consciousness of people to reduce urban sound pollution. As analyzed above, we present a solution to measure the sound intensity at the intersection of green and red lights in the paper.

3 Proposal System

In this paper, the display times of the green, yellow, and red lights are 30, 3, and 45 s, respectively. During the standby time, when the noise is between 75dB and 80 dB, the red warning light will flash. If the sound intensity is greater than 80 dB for 3 consecutive seconds, they will increase the red light waiting time by 10 s. The measured intensity value is saved on an SD memory card for 30 s and this data can evaluate the traffic volume at this location. The block diagram of the proposed system is shown in Fig. 1.

Table 1. Attenuation depends on distance.

Distance (meters)	1	2	4	5	8	16	32	64
Attenuation (dB)	0	−6	−12	−14	−18	−24	−30	−36

In Fig. 1, the control block uses two stm32f103c8t6 microcontrollers [9] to control the operation of the circuit. The control block will receive the audio signal from the optical environment. Data is saved to the memory card in storage for 30 s. The display block counts the display time of each lead, announcing this time through the 7-segment 2-digit LED.

3.1 Display Block

The signal used to signal the moving state at the intersection is a traffic light with three colors, green, yellow, and red. The display times corresponding to the three lights (green, yellow, and red) are 30, 3, and 45 s, respectively. The three lights will work one after the other and repeat. We use another red light to signal an upcoming sound intensity greater than 80 dB. When the sound intensity is between 75 and 80 dB, this red light will continuously blink. The LED is super bright with a voltage between 2 and 3.5 V and current consumption of 10 to 25 mA.

3.2 Data

Besides, it is necessary to use the LED to count the time for operating and displaying the devices. The led uses two numbers with a voltage drop of 2.2 V and the maximum current flowing through each led is 25 mA. We use the Micro SD/SDHC memory card module to store data. The power is between 4.5 and 5.5 V. SD Card supports FAT16 and FAT32 formats and communicates via SPI standards.

3.3 Transformer

We need to measure the sound emitted by vehicles. However, the measuring circuit and indicator lights are usually placed on traffic light poles over 6 m. Therefore, the sound has been attenuated while traveling through the air with an attenuation distance of approximately 5 m. The attenuation of sound while traveling in air depends on the distance as shown in Table 1 [10]. It is approximately 14 dB at a distance of 5 m. Therefore, the gain of the switch block is 5. Since the horn frequency of cars and motorcycles is in the range of 400 to 1000 Hz, we need to use the converter block. The diagram of the converter block is shown in Fig. 2.

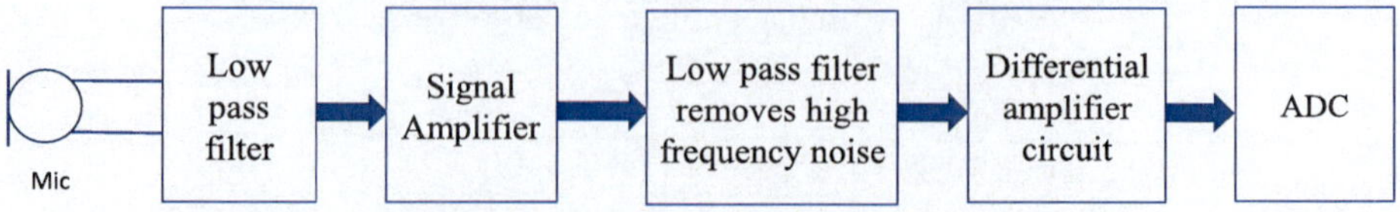

Fig. 2. Detailed diagram of the switch block.

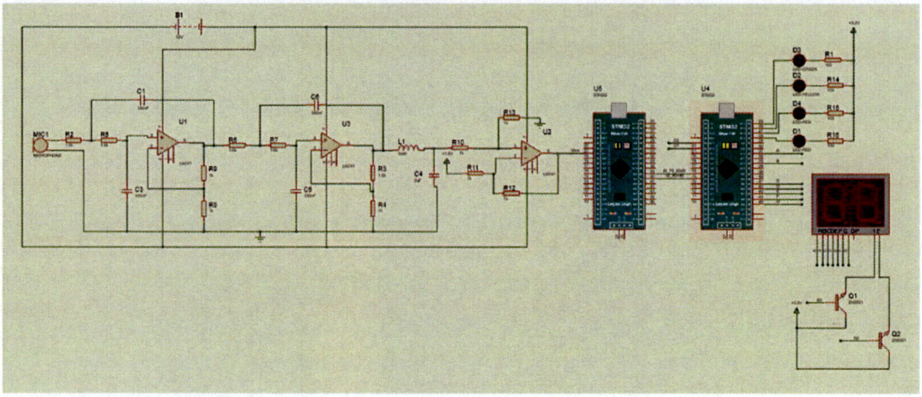

Fig. 3. Schematic diagram of the proposed circuit.

3.4 Data Filter Block

We use an active low-pass Butterworth filter circuit that has the advantages of gain control, good efficiency, and high accuracy with low noise capability [11,12]. We use a *4th* order Butterworth low-pass circuit to filter frequencies in the range from 0 to 1000 Hz and amplify the received signal by 5 times.

In an LC filter, the cut-off frequency of the filter is determined by the formula:

$$f = \frac{1}{2\pi\sqrt{LC}}. \tag{1}$$

Therefore, we have cut-off frequency $f = 3560\,\text{Hz}$ with $L = 1\,\text{mH}$ and $C = 2\,\mu\text{F}$.

The analog-to-digital converter (ADC) of stm32f103c8t6 has a default resolution of 12 bits and there are 4096 posts on the output. The output has a value from $0 \div 4095$ corresponding to a voltage level from $0 \div 3.3$ V. We use 8 Mhz internal quartz to configure the chip at 36 Mhz. Therefore, the number of samples obtained is 27×10^6 (sample).

4 Experiment

The circuit diagram is shown in Fig. 3. The sound of the environment will be recorded by the microphone. This signal will be amplified and filtered in the frequency range from 0 to 1000 Hz by a data filter block. The device will generate noise at high frequencies while operating. Therefore, we use the LC filter to remove these noises. The signal after the subtraction circuit will go into one pin of the stm32f103c8t6. The microcontroller will read the received signal to the ADC and determine the sound intensity that exceeds the threshold through the two output pins. The signal reading from these two pins will control the LEDs and their display time.

4.1 Result

Due to the limitation of the components and equipment, the measurement results should be evaluated at a low voltage amplitude as shown in Figs. 4 and 5.

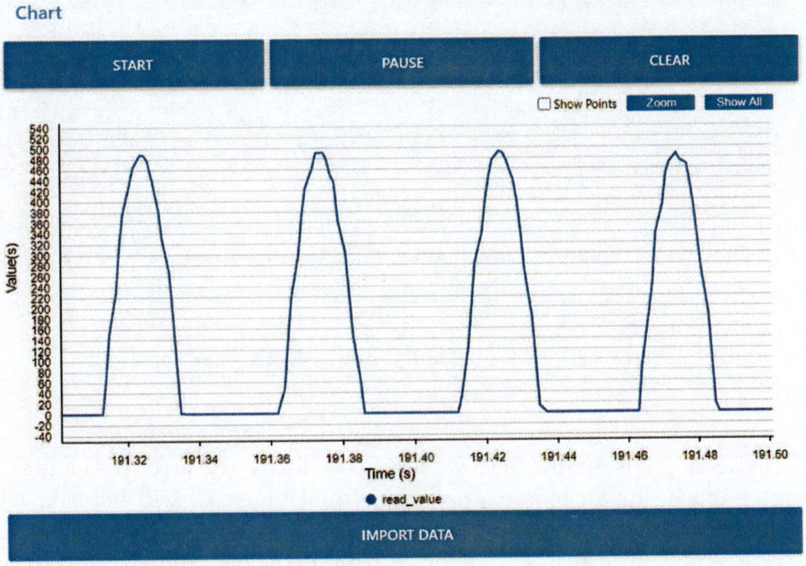

Fig. 4. Input signal diagram.

Instead of reading the value after conversion in digital form, STM32Cube Monitor software will be used to display these values in graph form. Therefore, we can observe and evaluate them. We have an input signal with a voltage amplitude of approximately 0.403 V (corresponding to 500) 20 Hz. Since the input voltage is small, a subtraction circuit is not used.

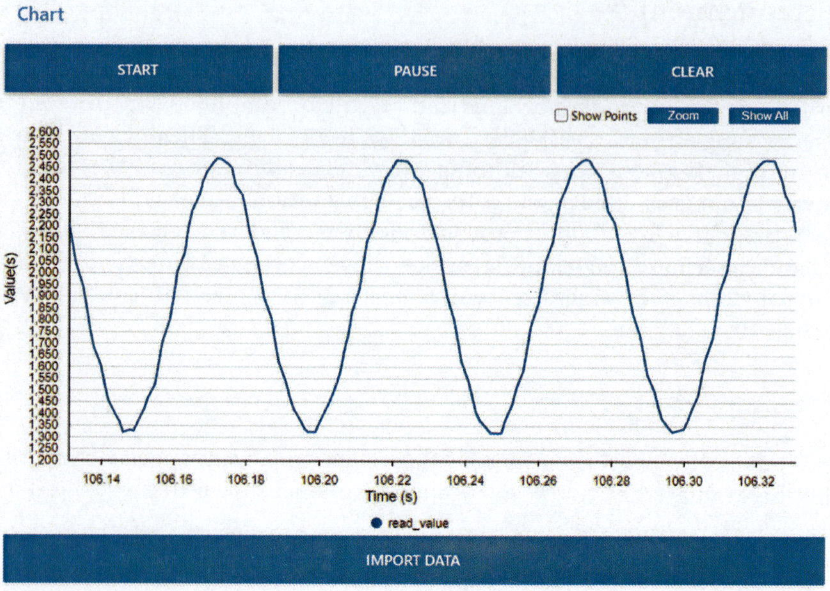

Fig. 5. Output signal diagram.

4.2 Evaluation

Table 2. Attenuation depends on distance.

Frequency (Hz)	20	100	440	500	1000
Error (%)	2.8589	2.4827	4.614	4.888	5.4741

The received signals are discrete values. To evaluate the error of the input source converter block, a sinusoidal signal with equal phase 0^0 will be used. Since the circuit is mainly concerned with magnitude values, the measured error will be the average at the peaks. The error at a point is the comparison of the actual measured value. The output signal has a maximum value of approximately 250 corresponding to a gain of 5 times 20 Hz. The errors are shown in Table 2. The results show that the mean error of the system is about 4.0635%.

5 Conclusion

In the paper, we design the circuit and focus on measuring sound intensity to minimize noise. The results show that the mean error of the device is $\approx$ 4.0635% that is suitable for real applications.

However, there are still many limitations in terms of input sound intensity, noise handling, and data storage on the memory card to evaluate traffic. Therefore, we will apply the AI model to increase accuracy as well as improve quality that it can be applied in the future.

Acknowledgment. This research is funded by Vietnam National University Ho Chi Minh City (VNU-HCM) under grant number NCM2021-20-02.

References

1. Madejewski, B.: Sound intensity measurement on electrical machines. In: 1991 Fifth International Conference on Electrical Machines and Drives (Conf. Publ. No. 341), pp. 241–244 (1991)
2. Bçkowski, A., Radziszewski, L., Dekýš, V., Świetlik, P.: Frequency analysis of urban traffic noise. In: 2019 20th International Carpathian Control Conference (ICCC), pp. 1–6 (2019)
3. Shu, H., Song, Y., Zhou, H.: RNN based noise annoyance measurement for urban noise evaluation. In: TENCON 2017 - 2017 IEEE Region 10 Conference, pp. 2353–2356 (2017)
4. Agarwal, A., et al.: A hybrid approach to noise-reduced pods in urban areas. In: 2020 IEEE MIT Undergraduate Research Technology Conference (URTC), pp. 1–5 (2020)
5. Radadi, A., Rehimi, F.: Implementation of a numerical tool for estimating the sound noise caused by road traffic in urban area. In: 2013 International Conference on Advanced Logistics and Transport, pp. 211–214 (2013)
6. Jamaludin, S.M., Hashim, W., Mohamad, Z., NL, M.K., Lim, A.Y.K.: Plants and their noise absorption: identifying most suitable plants to reduce noise via impedance tube data analysis. In: 2021 2nd International Conference on Artificial Intelligence and Data Sciences (AiDAS), pp. 1–6 (2021)
7. Wu, S.: External acceleration noise control of electric vehicles based on sound intensity measurement method. In: 2021 13th International Conference on Measuring Technology and Mechatronics Automation (ICMTMA), pp. 414–418 (2021)
8. Wan, S., Kong, L., Wang, H., Ma, F., Liu, Q., Zhao, F.: Application of measurement technology of sound intensity in noise source identification of automotive interior noise. In: 2011 Second International Conference on Mechanic Automation and Control Engineering, pp. 7208–7212 (2011)
9. Abdrakhmanov, V.K., Salikhov, R.B., Vazhdacv, K.V.: Development of a sound recognition system using STM32 microcontrollers for monitoring the state of biological objects. In: 2018 XIV International Scientific-Technical Conference on Actual Problems of Electronics Instrument Engineering (APEIE), pp. 170–173 (2018)
10. Collman, R.: Distance attenuation (2015). https://www.acoustical.co.uk/distance-attenuation/distance-attenuation/
11. Maji, K.B., Sree, U.S., Kar, R., Mandal, D., Ghoshal, S.P.: Butterworth filter design using seeker optimization algorithm. In: 2015 International Conference on Science and Technology (TICST), pp. 298–302 (2015)
12. Md Idros, M.F.B., Abu Hassan, S.F.B.: A design of butterworth low pass filter's layout basideal_filter approximation on the ideal filter approximation. In: 2009 IEEE Symposium on Industrial Electronics Applications, vol. 2, pp. 754–757 (2009)

Cooking Recipe Generation Based on Ingredients Using ViT5

Khang Nhut Lam[1(✉)], Y-Nhi Thi Pham[1], and Jugal Kalita[2]

[1] Can Tho University, Can Tho, Vietnam
lnkhang@ctu.edu.vn
[2] University of Colorado, Colorado Springs, USA
jkalita@uccs.edu

Abstract. This study proposes an ingredient-based recipe recommendation to make cooking easier and more convenient. Given the ingredients, the system can recommend classic recipes from the database or generate new recipes using the ViT5 model. We fine-tune ViT5 and train the model on our CookyVN-recipe dataset, consisting of 26,752 recipes in Vietnamese. The ROUGE-1, ROUGE-2, and ROUGE-L scores of our model are 64.45, 35.92, and 38.21, respectively.

Keywords: recipe generation · ViT5 · ingredients · CookyVN-recipe dataset · ingredient-based recipe generator · Vietnamese

1 Introduction

Information technology is currently used in every aspect of daily life, including business, education, entertainment, and cooking. Artificial intelligence could assist people in recommending foods, calculating calories, and providing recipes. Smart products with artificial intelligence are acquiring market traction. For example, the Samsung Refrigerators with Family Hub[1] autonomously manage foods, provide multimedia entertainment, and automatically recommends recipes; SmartyPans[2] help users track the nutrition of foods in real-time and support fitness applications; and Hello Eggs[3] provide users with shopping lists, weekly meal plans, and the ability to order groceries.

Platforms for ordering and delivering meals, such as Foody[4] and ShopeeFood[5], have substantially benefited users. However, there are many benefits to cooking your own food, including cost savings and personal satisfaction for the cook. People who cook their own meals, especially young people, frequently ask questions like "What shall we have for dinner today?", "What dish should we cook with giant gourami and pineapple?", and "Are there any more recipes

[1] https://www.samsung.com/us/explore/family-hub-refrigerator/overview/.
[2] https://smartypans.io/.
[3] http://helloegg.net/.
[4] https://www.foody.vn/.
[5] https://shopeefood.vn/.

T. D. L. Nguyen et al. (Eds.): ICISN 2023, LNNS 752, pp. 34–39, 2023.
https://doi.org/10.1007/978-981-99-4725-6_5

than the fried giant gourami?". This study aims to simplify recipe suggestions for users. The system recommends users with common cuisines or generates new recipes based on the input ingredients. We make the following contributions:

- We build a recipe recommendation system that suggests common recipes from our database or creative recipes based on ingredients.
- We fine-tune the ViT5 model to perform recipe recommendation. To the best of our knowledge, our work is the first study of a recipe recommendation model for Vietnamese foods.
- We construct a collection of Vietnamese recipes, named CookyVN-recipe dataset, and are working to make it accessible to the research community.

The remaining sections of our paper are structured as follows. Section 2 presents related work. Our approach is proposed in Sect. 3. Experimental results are discussed in Sect. 4. Section 5 concludes the paper.

2 Related Work

Some studies have been conducted to create recipes from ingredients. Cookpad[6], which is available on both website and mobile application, offers delicious and simple recipes from the community. The users of Cookpad can send messages to the dish's maker directly to get further details. A smart cooking program, called Yummly[7], supports daily meal dishes, each of which includes a video with detailed directions. In addition, Yummly provides diet and allergy information, as well as recipe scheduling.

Homemade RecipeBowl[8] is a recipe generator using neural networks. In particular, CNN is used to create recipes from food images, and LSTM [3] is used to generate recipes from ingredients. RecipeNLG[1] produces recipes by using an NER model for identifying and extracting food entities. The authors fine-tune GPT-2 [7] for recipe generation. Chef Transformer[9] is a recipe generator using T5 [8]. This model is trained on the RecipeNLG dataset[10]. Chef Transformer is better than RecipeNLG, which only achieves a ROUGE-2 score of 0.135, with a ROUGE-2 score of 0.247.

Lee et al. [4] propose a cooking recipe generation system, called RecipeGPT, by fine-tuning the GPT-2 model [7] on the Recipe 1M dataset [5]. RecipeGPT produces a cooking instruction from a recipe title and ingredients, as well as creates a recipe from a recipe title and cooking instructions. The best ROUGE-L score of RecipeGPT is 0.37. RecipeGM [9] is a recipe generation model using a hierarchical CNN with self-attention. According to the authors, RecipeGM occasionally performs better than RecipeGPT. A cooking recipe generation model

[6] https://cookpad.com/vn?via=jp.
[7] https://www.yummly.com/.
[8] https://github.com/ankitgoyal0301/Homemade-RecipeBowl-Your-Ingredients-Our-Recipes.
[9] https://github.com/chef-transformer/chef-transformer.
[10] https://recipenlg.cs.put.poznan.pl/.

using an encoder-decoder framework with reinforcement learning is presented by Fujita et al. [2]. Their model improves the ingredient matching by 21%. A model for creating recipes from food images is presented by Wang et al. [13]. Their model encodes food images to tree embeddings using graph attention networks, and generates recipe trees using RNN. A food cross-modal retrieval module combines the recipe trees with the original recipe features to generate new recipes. Their model achieves the best recall scores at top-1, top-5, and top-10 are 23.4, 48.8, and 60.1, respectively. Rokon et al. [10] identify ingredients from food images using CNNs, and then recommend recipes based on ingredients extracted using a linear search in their database. The authors report that ResNet [11] helps the model achieve the best accuracy value of 94%.

In this work, when a user inputs ingredients, our system provides two functions: recommending food recipes, that use these ingredients, from our database, and generating new recipes using a generative model ViT5.

3 ViT5 Model for Recipe Generation

The ViT5 model [6] follows the Transformer [12] and T5 [8] architectures. ViT5 is trained on a Vietnamese monolingual dataset, named CC100[11]. The ViT5 model for generating recipes consists of 12 stacked encoders and 12 stack decoders. Figure 1 presents ViT5 for food recipe generation based on ingredients.

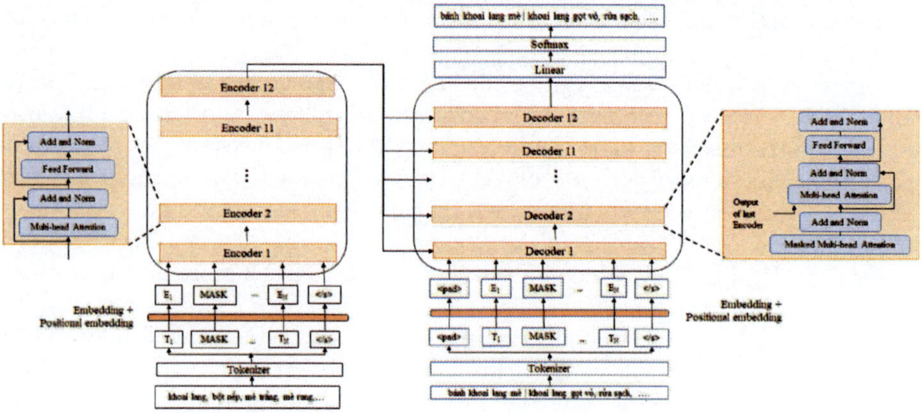

Fig. 1. ViT5 for food recipe generation.

The ingredients are tokenized, embedded into vectors, added to positional embeddings, and fed to the encoder of the ViT5 model. We use pre-trained VietAI/vit5-base provided by Hugging Face[12]. Similarly, the recipes are tokenized and embedded into vectors, added to positional embeddings are fed to

[11] https://autonlp.ai/datasets/cc100-vietnamese.

[12] https://huggingface.co/VietAI/vit5-base.

the decoder of ViT5. In addition, each decoder receives the output of the last encoder as inputs. We apply a linear layer and a softmax layer on the output of the last decoder to generate the output.

4 Experimental Results

We perform experiments in the Google Colaboratory environments with GPU Tesla T4, 16GB RAM. To the best of our knowledge, there has been no study on Vietnamese recipe generation. Consequently, there is no Vietnamese recipe dataset available to train the recipe generator. First, we construct a dataset, named CookyVN-recipe dataset, for training Vietnamese food recipe generation model. Selenium[13] is used to extract data from several cooking websites on the Internet. Then, we pre-process the extracted data by removing special characters and converting uppercase to lowercase. The CookyVN-recipe dataset consists of 26,752 recipes with 6 elements {title, serving, ingredients, quantity, recipes, links}. We use titles, ingredients, and recipes to train the cooking recipe generation model. In particular, the ViT5 recipe generation model encodes ingredients, and decodes titles and recipes. Statistics on the number of words in ingredients, and titles and recipes are presented in Table 1.

Table 1. Statistics on the number of words in ingredients, titles and recipes.

Number of words	Ingredients	Titles and recipes
Max	89	888
Min	4	22
Average	17	183

The CookyVN-recipe dataset is divided into 3 parts including 86% for training, 6% for validation, and 8% for testing. The ViT5 model for recipe generation is trained using a batch size of 2, epoch count of 5, learning rate of 10^{-5}, and AdaFactor optimizer. During the training process, the max length of the ingredients, and the titles and recipes are 128 words and 1,024 words, respectively.

The ROUGE-1, ROUGE-2, and ROUGE-L scores of the ViT5 model for food recipe generation from ingredients are 64.45, 35.92, and 38.21, respectively. In addition, we request volunteers to cook dishes using the recipes generated by ViT5, and evaluate dishes using a 5-point scale: 1-bad, 2-not good, 3-average, 4-good, and 5-tasty. The average score is 3.6. An example of a recipe generated from ingredients, and an image of a dish cooked by a volunteer is presented in Fig. 2. However, our recipe generation model occasionally generates a recipe without showing a clear cooking instruction. For example, from the ingredients including "cá điêu hồng" (one type of fish), the ViT5 recipe generation model

[13] https://www.selenium.dev/documentation/webdriver/.

Ingredients	Title	Recipe	Image	Evaluation
cá tra, cà chua, đậu bắp, giá đỗ	Canh chua cá tra	cá tra làm sạch, rửa sạch, cắt khúc vừa ăn. đậu bắp cắt xéo. cà chua cắt múi cau. me ngâm với nước ấm cho mềm, lọc lấy nước cốt me, bỏ bã. đun sôi nồi nước, cho cà chua, đậu bắp, giá đỗ vào, nấu sôi nhỏ lửa khoảng 10 phút. cuối cùng, cho cá tra vào, nêm gia vị muối, hạt nêm, đường trắng cho vừa ăn, nấu thêm 5 phút. tắt bếp, múc canh ra tô, trang trí thêm ngò rí nếu thích và thưởng thức cùng với cơm nhé.		3- Bình thường (3- average)

Fig. 2. An example of a title and a recipe generated from the ingredients using ViT5.

Ingredients	Title	Recipe
cá điêu hồng, cà chua, hành lá	Canh chua cá điêu hồng	cà chua rửa sạch, cắt múi cau. hành lá bỏ rễ rửa sạch rồi cắt khúc khoảng 3-4cm, phần đầu hành trắng cắt khúc, phần lá xanh cắt nhỏ. cho nồi lên bếp, cho dầu ăn vào, dầu nóng cho đầu hành vào phi thơm. cho cà chua và xào sơ, cho nước vào nấu sôi. khi cà chua đã mề, nêm nếm gia vị cho vừa ăn, cho cá đã làm sạch vào. đợi canh sôi lại, cho hành lá cắt khúc vào tắt bếp. múc canh ra tô, rắc thêm ít tiêu lên trên và dùng với cơm nóng

Fig. 3. An example of a recipe generated without showing a clear cooking instruction

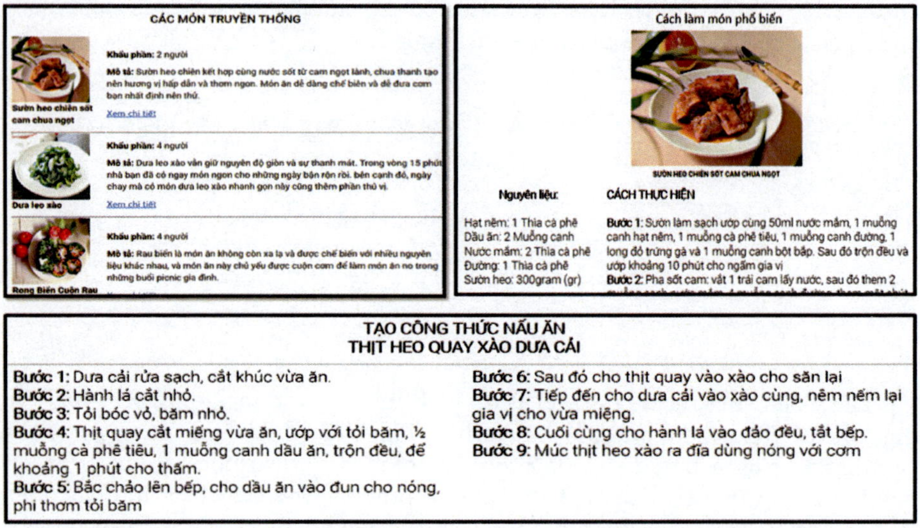

Fig. 4. Example of the recipe recommendation website: listing common recipes from the database (top left), a recipe based on ingredients queried from the database (top right), a recipe created using ViT5 (bottom).

creates a recipe that simply instructs users to put the cleaned fish into a pot as presented in Fig. 3. For future work, we will study an approach to let the model pay attention on the main ingredients, and provide clearly cooking instructions to users. Next, we build a website to recommend users with common recipes queried from our database, or new recipes created using ViT5. Figure 4 shows an example of our recipe recommendation website.

5 Conclusion

We construct a ViT5-based recipe recommendation system from ingredients. The system provides users with common recipes selected from the database and newly creative recipes generated using ViT5. The results show that the model performs fairly well. However, the cooking instructions are occasionally not specific. Besides, we have built the CookyVN-recipe-dataset including 26,752 different recipes in Vietnamese. We are discovering a reinforce learning method to improve the quality of recipes, and make the model pay attention on the main ingredients.

References

1. Bień, M., Gilski, M., Maciejewska, M., Taisner, W., Wisniewski, D., Lawrynowicz, A.: RecipeNLG: a cooking recipes dataset for semi-structured text generation. In: Proceedings of the 13th International Conference on Natural Language Generation, pp. 22–28 (2020)
2. Fujita, J., Sato, M., Nobuhara, H.: Model for cooking recipe generation using reinforcement learning. In: 2021 IEEE 37th International Conference on Data Engineering Workshops (ICDEW), pp. 1–4 (2021)
3. Hochreiter, S., Schmidhuber, J.: Long Short-Term Memory. Neural Comput. **9**, 1735–1780 (1997)
4. Lee, H.H., et al.: RecipeGPT: generative pre-training based cooking recipe generation and evaluation system. In: Companion Proceedings of the Web Conference, pp. 181–184 (2020)
5. Marín, J., et al.: Recipe1M+: a dataset for learning cross-modal embeddings for cooking recipes and food images. IEEE Trans. Pattern Anal. Mach. Intell. **43**, 187–203 (2021)
6. Phan, L., Tran, H., Nguyen, H.C., Trinh, T.H.: ViT5: pretrained text-to-text transformer for vietnamese language generation. In: NAACL (2022)
7. Radford, A., Wu, J., Child, R., Luan, D., Amodei, D., Sutskever, I., et al.: Language models are unsupervised multitask learners. OpenAI blog **1**(8), 9 (2019)
8. Raffel, C., et al.: Exploring the limits of transfer learning with a unified text-to-text Transformer. J. Mach. Learn. Res. **21**(140), 1–67 (2020)
9. Reusch, A., Weber, A., Thiele, M., Lehner, W.: RecipeGM: a hierarchical recipe generation model. In: 2021 IEEE 37th International Conference on Data Engineering Workshops (ICDEW), pp. 24–29 (2021)
10. Rokon, M.S.J., Morol, M.K., Hasan, I.B., Saif, A.F.M., Khan, R.H.: Food recipe recommendation based on ingredients detection using deep learning. In: Proceedings of the 2nd International Conference on Computing Advancements (2022)
11. Targ, S., Almeida, D., Lyman, K.: Resnet in Resnet: generalizing residual architectures. arXiv:abs/1603.08029 (2016)
12. Vaswani, A., et al.: Attention is all you need. In: Advances in Neural Information Processing Systems, vol. 30 (2017)
13. Wang, H., Lin, G., Hoi, S.C.H., Miao, C.: Learning structural representations for recipe generation and food retrieval. IEEE Trans. Pattern Anal. Mach. Intell. **PP** (2022)

Proposing Lung Abnormality Detection Model Using AI

Phat Nguyen Huu[1(✉)], Bach Le Gia[2], Bang Nguyen Anh[1], Dinh Dang Dang[1], Thanh Le Thi Hai[1], and Quang Tran Minh[3,4]

[1] Hanoi University of Science and Technology, Hanoi, Vietnam
{phat.nguyenhuu,thanh.lethihai}@hust.edu.vn,
{bang.na182377,dinh.dd200152}@sis.hust.edu.vn
[2] Hanoi Amsterdam High School for Gifted, Hanoi, Vietnam
[3] Department of Information Systems, Faculty of Computer Science and Engineering, Ho Chi Minh City University of Technology (HCMUT), 268 Ly Thuong Kiet, District 10, Ho Chi Minh City, Vietnam
quangtran@hcmut.edu.vn
[4] Vietnam National University Ho Chi Minh City (VNU-HCM), Linh Trung Ward, Thu Duc District, Ho Chi Minh City, Vietnam

Abstract. The diagnostic imaging industry in many countries is currently severely under-resourced. The reason is that people know about it is not too much. Besides, the demand for radiologists has increased because of the impact of the Covid-19 epidemic. There are situations that the general practitioner or the clinician must diagnose x-rays. Therefore, the article proposes a solution to save time by assisting doctors in diagnosing problems by giving suggestions on abnormal areas and creating an AI application to recognize images that can be detected abnormal areas or not. The results show that although the accuracy of the proposed solution only achieves about 30% due to the small number of samples, it promises to be highly feasible when applied in practice.

Keywords: Lung abnormality · AI · Object detection · Deep learning · CNN

1 Introduction

Despite going through a boom and peak period, more than 1 million people are infected with Covid-19 every day and over 2,000 people are dying from this virus. This virus directly attacks the lung organs causing both arteries, alveolar capillaries, and even veins to become blocked. It can be recognized and treated promptly if we find it as soon as possible. However, the doctors could not respond in time due to the time limit and many patients need to be treated.

The diagnostic imaging industry in many countries is currently severely under-resourced. The reason is that people know about it is not too much. Therefore, the number of students in this field is limited. In addition, the demand for radiologists has increased due to the impact of the Covid-19 epidemic. There are

T. D. L. Nguyen et al. (Eds.): ICISN 2023, LNNS 752, pp. 40–46, 2023.
https://doi.org/10.1007/978-981-99-4725-6_6

many situations in which a general practitioner or clinician must diagnose radiographs even though it is not their specialty. Many doctors have just graduated from school and their reading experience is not high which will be very difficult to diagnose the disease. This is time-consuming and causes inevitable errors.

Therefore, the goal of the article is to build an AI algorithm to save time and assist doctors in diagnosing by giving suggestions about abnormal areas based on You only look once (YOLO). The algorithm implements based on the principle that when we put the chest X-ray image file into it, it returns pneumonia results if the image has an abnormal area.

Our paper includes four parts. In Sect. 1, we present the related work. Section 2 will propose the system. In Sect. 3, we develop the proposed system and evaluate the results. Finally, we give works for the future.

2 Related Work

Currently, there are many applications of AI for healthcare that are able to save a lot of time for patients [1–7]. Virtual nurse assistants are able to answer and monitor patients quickly. The applications allow patients and healthcare professionals to communicate more frequently among visiting hospitals.

Today, doctors diagnose and treat organ-related problems using imaging techniques such as CT scans, PET scans, MRIs, and X-rays. However, when working with such a wide variety of medical tools, there are many challenges, especially with chest X-ray images. Diagnostic work with X-rays can be fraught with errors, even for experienced practitioners. Existing methods of interpreting chest X-ray images categorize them into a list of findings. The data used is taken from [3]. The data includes 15,000 X-ray Dicom images of adults including 14 types of abnormalities appearing in the lungs. The data were collected at two hospitals (108 and Medical University Hospital Hanoi) from 2018 to 2020 [3].

The data provided in the contest include 205.96 GB of Dicom images including 15,000 images of training and 3,000 images of test files and one file train.csv is the label of data. There are 14 types of abnormalities as shown in Fig. 1. We applied the VinDr algorithm to convert from Dicom to PNG image format and transformed the given data labels from the contest from tabular format into training data for the YOLO model. After collecting the chest X-ray image dataset, we analyze the data fields about the abnormal areas using the python library as shown in Fig. 1.

As a result, we can conclude that conventional machine learning methods cannot solve the problem of recommending outlier regions. Therefore, we propose to use a deep learning model (YOLO v5 and YOLO v7) [4–7] to detect abnormal areas in the image.

3 Algorithm Proposal

We propose to apply the object detection method to localize and label different anomalies in this paper. The models used in this paper are YOLOv5 and YOLOv7 [4–7]. The proposed algorithm is shown in Fig. 2.

YOLO is a convolutional neural network (CNN) model that is designed based on the combination of convolutional layers [8–10]. The convolutional layers will

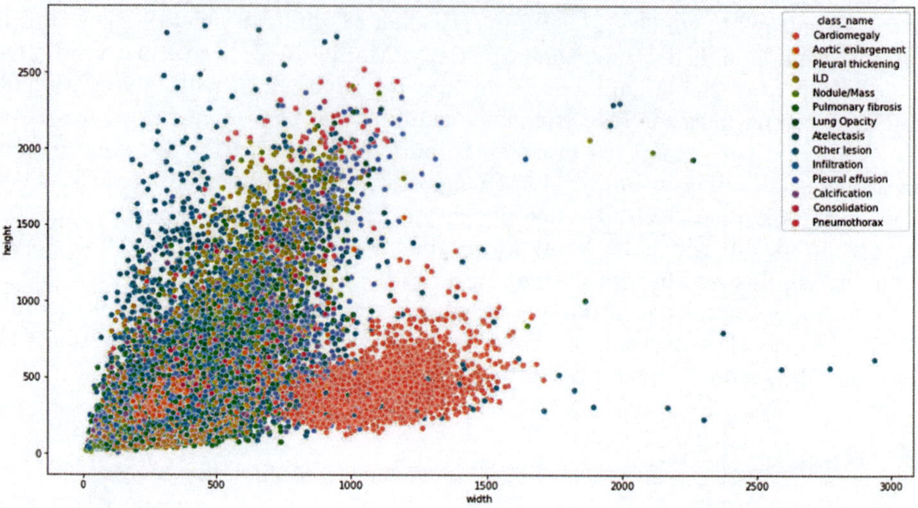

Fig. 1. Histograms of the **length and width distributions** of different types of abnormal regions.

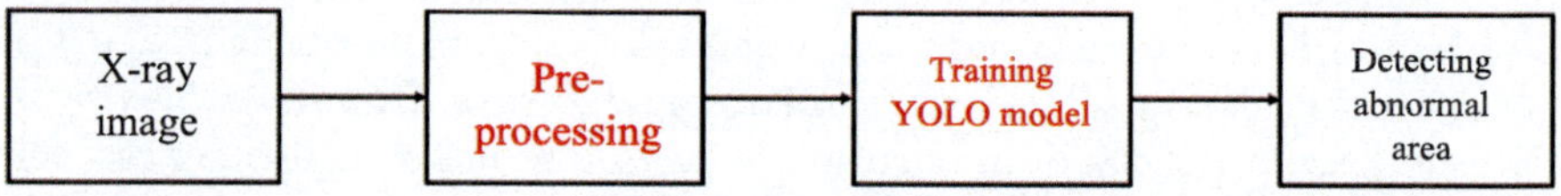

Fig. 2. The block diagram of the proposed method.

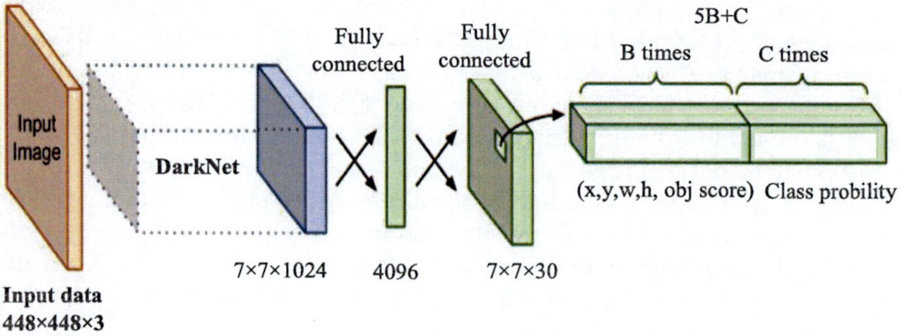

Fig. 3. Architectural diagram of YOLO [8,9,11,12].

detect the features and the full-connected layers will predict the probability as shown in Fig. 3.

The darknet architecture component called the base network has a feature extraction effect. Its output is a $7 \times 7 \times 1024$ feature map. YOLO is able to predict many feature maps that help detect large objects. To predict small objects, we use the anchor box that is kept fixed .

YOLO version 3 is used darknet-53 for network feature extractor. It includes 53 layers, namely batch normalization and Relu. To reduce the output, we down samples 2 times that minimizes the size of the model. It is calculated as follows:

$$y^T = [p_o, (t_x, t_y, t_w, t_h), (p_1, p_2, ..., p_c)], \quad (1)$$

where p_o and (t_x, t_y, t_w, t_h) are the predicted probability and bounding box, respectively. t_x, t_y and t_w, t_h are the center and width and length of the bounding box, respectively. The predictive probability distribution is $(p_1, p_2, ..., p_c)$.

Understanding the output is quite important to configure the correct parameters while training the model through open sources such as the darknet. Therefore, the output will be determined by the number of classes according to the formula $(n_{class} + 5)$.

We select three anchor boxes (Box 1, Box 2, Box 3) that concatenated with the output of YOLO. More details of attributes of a bounding box can be seen in [8,9,11,12].

4 Result

4.1 Preparing and Training Data

From the collected dataset in Sect. 2, we divided them into two parts, training (12257 images) and testing (2743 images) parts. Its ratio is 80:20.

The training process requires the use of a computer with high hardware configuration to meet the parallel computation of the parameters. In this paper, we use a server computer with the configuration as Table 1.

Table 1. Configure the model training hardware.

CPU	GPU	RAM
Intel Xeon Silver 4216	Nvidia Quadro RTX	
@2.1 GHz 16 Cores	5000 VRAM	125 GB
32 Threads	16 GB	

4.2 Result

Both YOLOv5 and YOLOv7 models were trained on the server with 200 epochs. The results are shown in Fig. 4.

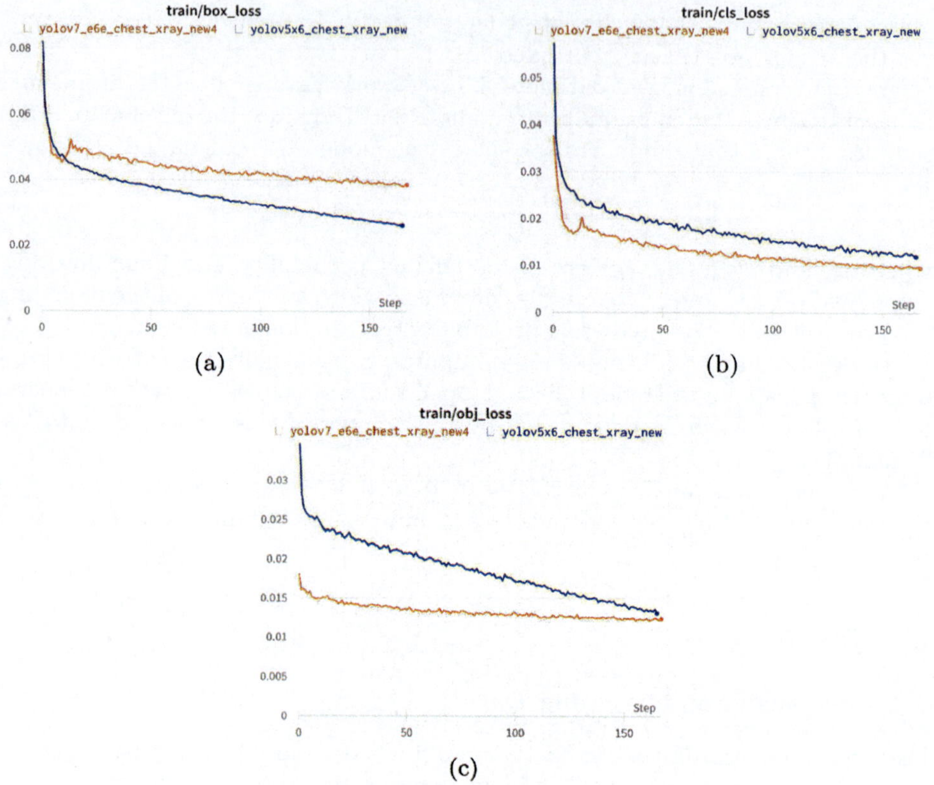

Fig. 4. The loss value of (a) the bounding box, (b) class, and (c) the object over each epoch.

We use the mean average precision (mAP) parameter to evaluate the accuracy of the model. The mAP is calculated as

$$mAP = \frac{1}{N}\sum_{i=1}^{N} AP_i. \tag{2}$$

The results show that precision, recall, and AP are always between 0 and 1. The results are shown in Fig. 5.

We perform on the same image after training the YOLOv5 and YOLOv7 models. The results are shown in Fig. 6.

The results of the training and testing dataset show that the learning model in the trainer is stable and good. However, the accuracy of training and testing is only $mAP \approx 30\%$. It shows that diagnosis and detection with radiographs are a big challenge.

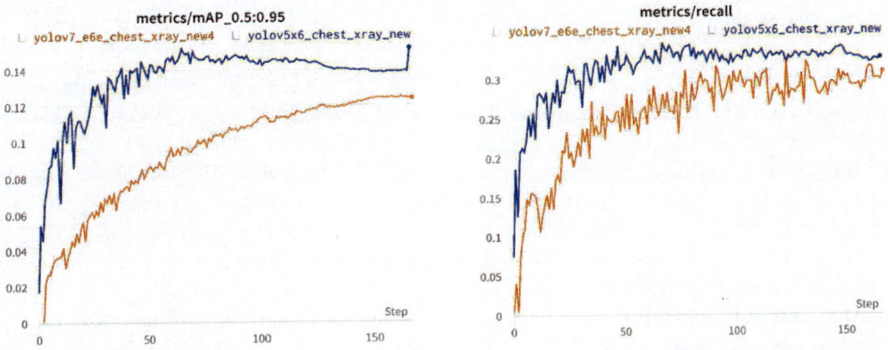

(a) (b)

Fig. 5. Results of (a) accuracy of the model and (b) detecting anomalies across each epoch.

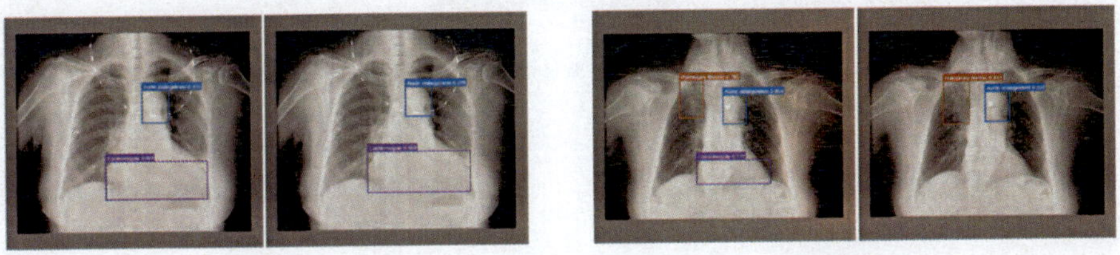

(a) (b)

Fig. 6. Results of YOLOv5 (left) and YOLOv7 (right) for (a) aortic enlargement and (b) pulmonary fibrosis.

5 Conclusion

In this paper, we have proposed a method to detect regions on chest radiographs. This method can assist doctors in diagnosing abnormalities in the lungs, saving time and manpower for the medical industry. Through experiments with an accuracy of about 30%, it can be concluded that this problem still faces many difficulties. However, the results can also be considered a foundation for future development studies.

In the future, we will find the way to improve the model that can be applied in practice to solve many other problems in the future.

Acknowledgment. This research is funded by the Hanoi University of Science and Technology (HUST) under project number T2022-PC-013. The authors would like to thank the HUST for the financial support.

Work by Quang Tran Minh (HCMUT) acknowledges the support of time and facilities from Ho Chi Minh City University of Technology (HCMUT), VNU-HCM.

References

1. Kumar, A., Joshi, S.: Applications of AI in healthcare sector for enhancement of medical decision making and quality of service. In: 2022 International Conference on Decision Aid Sciences and Applications (DASA), pp. 37–41 (2022)
2. Raj, S.D., Karthiban: Applications of artificial intelligence in healthcare. In: 2022 International Conference on Computer Communication and Informatics (ICCCI), pp. 1–2 (2022)
3. Nguyen, H.Q., et al.: VinDr-CXR: an open dataset of chest x-rays with radiologist's annotations. Sci. Data **9**, 1–8 (2022)
4. Katsamenis, I., et al.: TraCon: a novel dataset for real-time traffic cones detection using deep learning, May 2022
5. Nepal, U., Eslamiat, H.: Comparing YOLOv3, YOLOv4 and YOLOv5 for autonomous landing spot detection in faulty UAVs. Sensors **22**(2) (2022). https://www.mdpi.com/1424-8220/22/2/464
6. Wang, C.Y., Bochkovskiy, A., Liao, H.Y.: YOLOv7: trainable bag-of-freebies sets new state-of-the-art for real-time object detectors, July 2022
7. Xu, R., Lin, H., Lu, K., Cao, L., Liu, Y.: A forest fire detection system based on ensemble learning. Forests **12**, 217 (2021)
8. Ammar, A., Koubaa, A., Ahmed, M., Saad, A.: Aerial images processing for car detection using convolutional neural networks: comparison between faster R-CNN and YOLOv3, October 2019
9. Artamonov, N., Yakimov, P.: Towards real-time traffic sign recognition via yolo on a mobile GPU. J. Phys: Conf. Ser. **1096**, 012086 (2018)
10. Redmon, J., Farhadi, A.: YOLO9000: better, faster, stronger. In: 2017 IEEE Conference on Computer Vision and Pattern Recognition (CVPR), pp. 6517–6525 (2017)
11. Pham, M.T., Courtrai, L., Friguet, C., Lefèvre, S., Baussard, A.: YOLO-fine: one-stage detector of small objects under various backgrounds in remote sensing images. Remote Sens. **12**, 2501 (2020)
12. Redmon, J., Divvala, S., Girshick, R., Farhadi, A.: You only look once: Unified, real-time object detection. In: 2016 IEEE Conference on Computer Vision and Pattern Recognition (CVPR), pp. 779–788 (2016)

Ring-Based Hybrid GPON Network with Inter-oNU Transmission Capability

Shalini Khare[1], Amit Kumar Garg[2], Aditi Phophaliya[1], Vijay Janyani[1] (✉), and Ghanshyam Singh[1]

[1] Department of ECE, MNIT, Jaipur, India
{vjanyani.ece,gsingh.ece}@mnit.ac.in

[2] Department of ECE, IIIT Kota (MNIT Campus), Jaipur, India

Abstract. This paper proposes the Hybrid Optical Fiber (OF)-Free Space Optics (FSO) based Inter-Optical Network Unit (ONU) transmission capable Wavelength Division Multiplexing (WDM)-Passive Optical Network (PON) architecture. This architecture can transmit up to 40 Gbps for downstream, upstream, and inter-ONU transmission. The downstream transmission is flowing from the center office (CO) to respective ONUs while Inter-ONU transmission is performed from one ONU to another ONU which are connected in a ring manner. The upstream transmission is performed from a particular ONU and travels towards the CO via a circular ring. The upstream signal can be transmitted on the same wavelengths which are used for the downstream transmission and therefore it utilizes the available bandwidth efficiently. In the proposed architecture, four 10 Gbps optical line terminals (OLT) transmit the signal through 20 km of single-mode fiber (SMF). Inter-ONU feature increases the reliability and survivability of the optical network. The bit error rate (BER) performance of the proposed architecture for the main and Inter-ONU transmission is reasonably good for the combination of 20 km feeder fiber (FF) and 200 m distributed fiber (DF)/FSO link. The architecture also provides better receiver sensitivity for a hybrid OF-FSO optical network with Inter-ONU transmission.

Keywords: Free space optics · WDM-PON · Inter-ONU · Hybrid OF-FSO

1 Introduction

In the last decade, the demand for high data rate, and security have increased exponentially. Explosive demand for the internet in daily life increases access to the Internet of Things (IoT). PON is a low-cost point to multipoint broadband technology. For broadband access services, the WDM-PONs are gaining interest with the growing internet usage. The WDM-PON is a promising broadband technique that can be used to fulfill the end-user bandwidth requirements. In the WDM-PON network, each ONU utilize the dedicated wavelengths in order to access the secured and high-speed transmission [1, 2]. Today, OF technology is one of the most suitable technologies to transmit data at a higher speed. The ring based PON architecture can provide direct inter-ODN transmission along

T. D. L. Nguyen et al. (Eds.): ICISN 2023, LNNS 752, pp. 47–53, 2023.
https://doi.org/10.1007/978-981-99-4725-6_7

with broadcasting and intra-ODN transmission. The direct inter-ODN, intra-ODN transmission seems to be better than the conventional transmission in terms of energy saving, transmission latency, and the reduction of harmful gases [3]. Furthermore, all the optical distribution networks (ODN) can communicate with other ODN using NxN array waveguide grating (AWG) and reduces transmission latency, energy consumption, and operational expenditure at low bandwidth requirements [3, 4]. In WDM-PON, downstream data is sent from the OLT of the CO to the ONUs which are located at the user end. Similarly, for the upstream transmission, the data is transmitted from the ONUs to the CO by incorporating the power splitters or power combiners [5]. These combiners incur insertion losses which may be compensated by making the improvement in the link power budget [6, 7]. Transmission through OF is difficult in geographically restricted areas like mountains, rivers, etc. Earthquakes and other disaster situations may also create severe challenges for getting reliable transmission. FSO is the best solution to overcome such a situation due to its large bandwidth, license-free, cost effective and quick installation [8, 9]. Therefore, for reliable transmission, the integration of OF and FSO seems to be a good solution.

In this paper, a ring-based hybrid architecture with inter-ONU capability is investigated. Integration of OF and FSO provides fault protection capability. The proposed architecture not only enhances the network reliability but also increases survivability. The major advantage of the proposed network architecture is to provide direct inter-ONU transmission so each ONU can communicate with other ONUs directly without transmitting data to the OLT. Direct inter-ONU transmission avoids the need for complex processing of data at the CO, and therefore provide the utilization the available resources efficiently.

2 Working and Operation

The proposed 40 Gbps GPON inter-ONU access network architecture works on unidirectional transmission, hence the same wavelengths can be used for both downstream and upstream transmission with 10 Gbps. The network has a CO which holds the responsibility of distributing 40 Gbps of data transmission via four 10 Gbps wavelengths λ_1, λ_2, λ_3 and λ_4 to four different ONUs via a 20 km long FF1. The spectrum of transmitted signal operating with frequencies 193.1 THz, 193.2 THz, 193.3 THz, and 193.4 THz corresponding to the wavelengths λ_1, λ_2, λ_3 and λ_4 respectively. The main highlighting part of this architecture is the backup protection at the CO side and the distribution side along with the inter-ONU transmission. At the CO side, if any fault occurs in FF1, optical switch (OS) reroutes the transmission via FF2. At ONU side, the hybrid OF/FSO transmission link provides the link protection, if any fault is occurring in 200 m OF link, the OS reroutes the transmission via 200 m FSO link. FF1 is used for primary transmission and FF2 is used as backup protection. For the downstream transmission, wavelengths λ_1, λ_2, λ_3 & λ_4 are used in primary transmission and λ_{11}, λ_{12}, λ_{13} & λ_{14} are used in the backup transmission for ONU1, ONU2, ONU3, and ONU4 respectively. The distance between any two consecutive ONUs are kept 1km. For e.g., ONU1 and ONU2 are separated by a distance of 1km through DF1. For the upstream transmission, wavelengths λ_1, λ_2, λ_3 & λ_4 are used for both primary and backup transmission.

For inter-ONU transmission, wavelengths λ_5 to λ_{10} are used for transmission. Each ONU uses a different wavelength to communicate with other ONUs. To avoid interference between the signals, different wavelengths are used for downstream and inter-ONU transmission. Therefore, at the ONU1 side, the receiver is tuned to receive the four wavelengths λ_1, λ_5, λ_8, λ_{10}. For this purpose, we use either a tunable receiver or two dedicated receivers: one for the main downstream transmission, and another for inter-ONU transmission. Tunable FBG (T.FBG) are used for inter-ONU transmission. Therefore, each ONU can communicate with only one ONU at a time (Fig. 1).

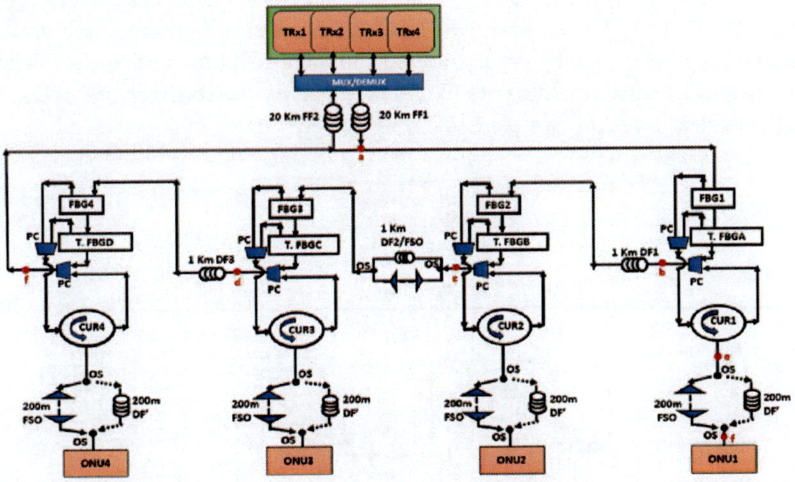

Fig. 1. Proposed ring-based inter-ONU hybrid WDM-PON architecture

The Bragg wavelengths assigned to FBG1, FBG2, FBG3, and FBG4 are λ_1, λ_2, λ_3, and λ_4 respectively. For example, for transmission towards ONU1, FBG1 is tuned at wavelength λ_1. The designated wavelength for communication between ONU1 and ONU2 is λ_5, ONU1 and ONU3 is λ_8, ONU1 and ONU4 is λ_{10}, ONU2 and ONU3 is λ_6, ONU2 and ONU4 is λ_7, ONU3 and ONU4 is λ_9. However, the same wavelength is used for two-way communication between the ONUs.

For inter-ONU transmission, the transmitter and the receiver of ONU1 is tuned to wavelengths λ_5, λ_8, and λ_{10} for transmitting and receiving the signal to communicate with ONU2, ONU3, and ONU4 respectively. For this purpose, the transmitted wavelength which is generated from ONU1 is passed to the power combiner (PC) via CUR1, OF/FSO link. At PC, the transmitted wavelength from FBG1 and FBGA are combined with the wavelength which is transmitted from the CO and then the combined signal is passed to the next FBGs and ONUs via DF1 of length 1km, as shown in Fig. 2. The working of this architecture is similar for all the ONUs and inter-ONU communication. Inter-ONU transmission along with downstream transmission is shown in Fig. 2. The frequencies 193.5 THz, 193.6 THz, 193.7 THz, 193.8 THz, 193.9 THz and 194 THz corresponding to the wavelengths λ_5, λ_6, λ_7, λ_8, λ_9, and λ_{10} are used for Inter-ONU transmission. Initially, the wavelengths λ_1, λ_2, λ_3 & λ_4 are coming from the CO, flow towards the FF1 and reaches FBG1. FBG1 then reflects the wavelength λ_1 and passes other wavelengths λ_2, λ_3 & λ_4 towards DF1. After reflection, the wavelength λ_1 reaches to the 200 m FSO/DF link via PC, and CUR1. If no fault occurs in 200 m OF, the signal travels via OF link and the wavelength λ_1 is received by the ONU1 receiver. The remaining wavelengths λ_2, λ_3 & λ_4 passes from T.FBGA and move towards DF1. The wavelengths λ_2 and λ_3 are received by the receivers at ONU2 and ONU3 respectively. At the same time, if ONU3 wants to communicate with ONU1, it transmits wavelength λ_8. Therefore, at the point 'd', we get two wavelengths λ_8 and λ_4. Wavelength λ_8 for inter-ONU and λ_4 for downstream transmission. After DF3, wavelength λ_4 is received by ONU4 receiver and wavelength λ_8 reach to the FBG1. FBG1 passes the wavelength λ_8 to T.FBGA. T.FBGA is tuned at λ_8, therefore it reflects this wavelength towards PC. At PC wavelengths λ_1 and λ_8 are combined and then travel towards CUR1. The wavelengths λ_1 and λ_8 flows via 200 m OF and received by the receiver of ONU1. Similar mechanism follows for other inter-ONU transmission.

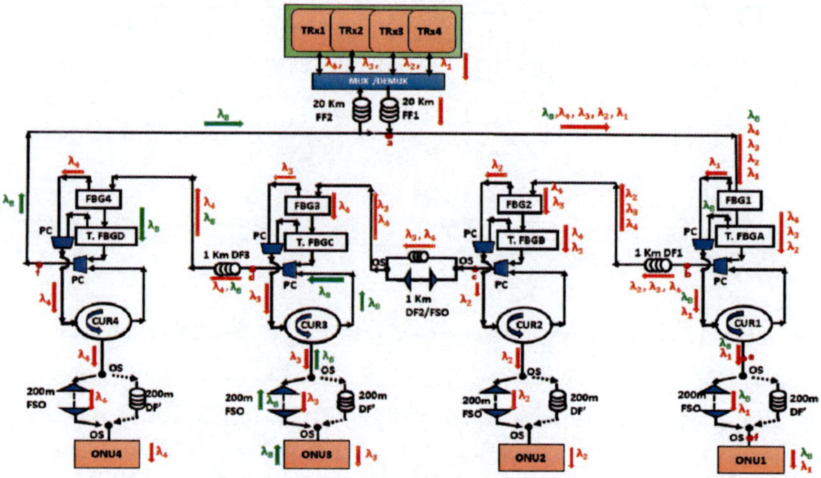

Fig. 2. Inter-ONU and downstream signal flow of proposed ring-based inter-ONU hybrid WDM-PON architecture.

3 Discussion of Results

The simulation analysis is performed on OPTISYSTEM 16.1 simulation tool at 0 dBm laser power. For FSO and OF transmission, the considered value of attenuation is 0.233 and 0.2 dB/km respectively, the transmitter & receiver aperture diameter is 5 cm & 20 cm respectively, the beam divergence is 1.5 mrad, insertion loss in circulator is 0.5 dB, insertion loss in 2 × 1 power combiner is 3.5 dB, gain of OF is 20 dB, and channel model is gamma-gamma.

Figure 3, (a) & (b) shows the received optical power sensitivity of downstream transmission, where OF and FSO links are working as an active link at ONU side, and values of received optical power sensitivities are given in Table 1.

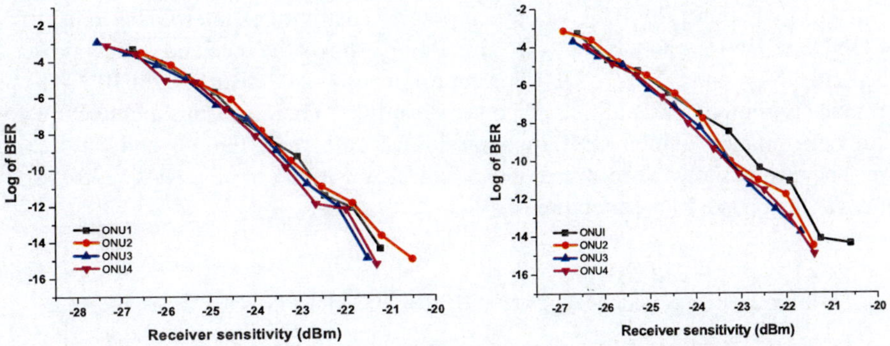

(a) (b)

Fig. 3. Received optical power sensitivity for downstream signal transmission in proposed ring architecture for (a) OF and (b) FSO link at ONU side

Table 1. Received optical power at BER $= 10^{-9}$ for downstream at ONU.

Received optical power (dBm)	FSO (dBm)	OF (dBm)
ONU1	-23.06	-23.22
ONU2	-23.49	-23.41
ONU3	-23.61	-23.51
ONU4	-23.79	-23.68

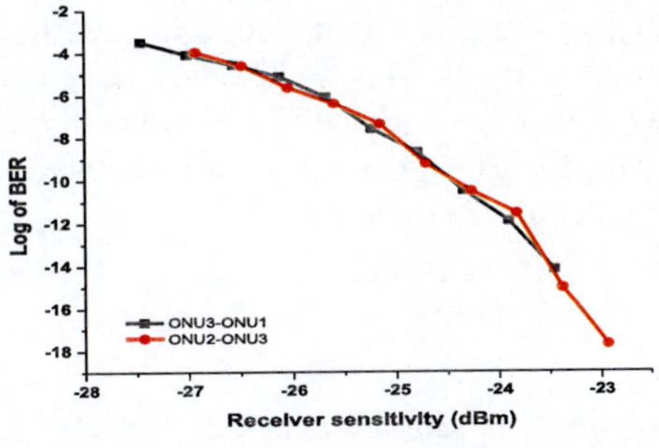

Fig. 4. BER Vs received optical power sensitivity of inter-ONU transmission.

The Fig. 4 shows the BER vs optical power sensitivity of inter-ONU transmission for ONU3 to ONU1 and ONU2 to ONU3. Table 2 shows the received optical power for ONU3 to ONU1 and ONU2 to ONU3 is approximately -24.7 dBm at BER 10^{-9}. The performance is good in terms of optical power sensitivity. Therefore, this architecture gives better performance for inter-ONU transmission as compared to the conventional architecture. The results show the received optical power sensitivity of all ONUs approximately same for the proposed architecture.

Table 2. Received optical power at BER = 10^{-9} for Inter-ONU transmission.

Received optical power (dBm) at BER 10^{-9}	
ONU3 to ONU1	-24.7
ONU2 to ONU3	-24.7

4 Conclusion

In this paper, the simulation study of ring-based Inter-ONU transmission in DWDM PON network has been performed. Inter-ONU transmission capability of the proposed architecture is extremely attractive to transmit low latency signals among ONUs and offers reliable transmission performance. Simulation results are performed for 20 km FF1, FF2, and 200 m OF/FSO distributed links. A hybrid OF-FSO link provides reliability against the faults occurring in the OF. The simulation results show that the received optical power sensitivity for the given architecture is good for downstream, upstream as well as Inter-ONU transmission. This architecture is beneficial for geographically restricted areas such as mountains and rivers. Moreover, the proposed architecture is beneficial in the areas where, local signal transmission between the users with high data rate, high security, and reliability is required.

References

1. Wey, J.S.: The Outlook for PON Standardization: A Tutorial. J. Lightwave Technol. **38**(1), 31–42 (2020). https://doi.org/10.1109/jlt.2019.2950889
2. Zhao, Q., Chan, C.K: A wavelength-division-multiplexed passive optical network with flexible optical network unit internetworking capability. J. Lightw. Technol. **25**(8), 1870–1977 (2007). https://doi.org/10.1109/jlt.2007.901537
3. Garg, A.K., Janyani, V., Batagelj, B.: Ring based latency-aware and energy-efficient Hybrid WDM TDM-PON with ODN interconnection capability for smart cities. Optical Fiber Technology **58**, 102242 (2020). https://doi.org/10.1016/j.yofte.2020.102242
4. Garg, A.K., Janyani, V., Singh, G., Ismail, T., Selmy, H.: Dedicated and broadcasting downstream transmission with energy-efficient and latency-aware ONU interconnection in WDM-PON for smart cities. Opt. Fiber Technol. **52**, 101949 (2019). https://doi.org/10.1016/j.yofte.2019.101949

5. Khare, S., Garg, A.K., Janyani, V.: Ring based hybrid FSO- fiber optic architecture utilizing same wavelengths for downstream and upstream transmission with FN/DN protection capability. Opt. Fiber Technol. **73**, 103050 (2022). https://doi.org/10.1016/j.yofte.2022.103050
6. Garg, A.K., Janyani, V.: Power budget improvement in energy efficient long reach hybrid passive optical network. 2015 International Conference on Microwave and Photonics (ICMAP) (2015). https://doi.org/10.1109/icmap.2015.7408789
7. Vaughn, M., Kozischek, D., Meis, D., Boskovic, A., Wagner, R.: Value of reach-and-split ratio increase in FTTH Access Networks. J. Lightwave Technol. **22**(11), 2617–2622 (2004). https://doi.org/10.1109/jlt.2004.836741
8. Ghassemlooy, Z., Popoola, W., Rajbhandari, S.: Optical Wireless Communications: System and Channel Modelling with MATLAB®, 1st ed., CRC Press (2012)
9. Singh, P., Bohara, V.A., Srivastava, A.: Reliable and cost effective all-optical wireless architecture for a broadband access network. Journal of Optical Communications and Networking **15**(2), 98–110 (2023). https://doi.org/10.1364/JOCN.463574

Improving Availability of Enterprise Blockchain Using Real-Time Supervisor

Hung Ho-Dac[1], Len Van Vo[1], Bao The Nguyen[1], Cuong Hai Vinh Nguyen[1], Phuong Cao Hoai Nguyen[1], Chien Khac Nguyen[2], Huy Bui Quang Tran[1], and Huu Van Tran[1](✉)

[1] Thu Dau Mot University, Binh Duong, Vietnam
{hunghd,lenvv,baont,cuongnhv,phuongnch,huutv}@tdmu.edu.vn
[2] People's Police University, Ho Chi Minh City, Vietnam

Abstract. Blockchain is one of the key technologies today and is increasingly being applied in many different fields. Blockchain in the enterprise is emphasized due to its ability to overcome the inherent limitations of centralized applications widely used in enterprises. The modern workplace trend after the COVID-19 pandemic has paved the way for the deployment of enterprise blockchain networks on end devices (nodes) provided by enterprises to employees. However, according to the above approach, ensuring the availability of nodes is very difficult because it heavily depends on employees' behaviors. To increase enterprise blockchain availability, we propose a novel approach that integrates a real-time supervisor with enterprise blockchain to monitor and ensure established policies.

Keywords: Enterprise Blockchain · Real-Time Supervisor · Availability Improvement

1 Introduction

The idea of activity-based working is likely to be included into the design of modern workplace. This design combines areas designated for particular activity with more open spaces. The traditional workplace model is turned on its head by the activity-based workplace model, which turns open workstations into quiet spaces and enclosed rooms into areas for seclusion and interaction.

Work has taken on new characteristics. Employees place a high value on work that enlightens and fulfills them and expect to work safely from anywhere, on any device. They are happier, more valuable, and more likely to remain in their current position when their productivity tools improve the caliber and efficacy of their work experience. While giving employees this empowerment, businesses must also safeguard important IT resources. Many key service providers have introduced solutions for the modern workplace such as Microsoft [1], and Google [2].

Following the trend of modern workplace, IT resources will be distributed among employees. This is one of the important resources that should be exploited. On that basis, we propose an approach using blockchain to exploit data and processing capacity of IT

T. D. L. Nguyen et al. (Eds.): ICISN 2023, LNNS 752, pp. 54–59, 2023.
https://doi.org/10.1007/978-981-99-4725-6_8

resources. However, IT resources are highly dependent on employees and how they are used, so we added a real-time monitoring component to ensure policies and increase efficiency of IT resources usage. The rest of the paper is organized as follows: Sect. 2 surveys the related work. Sect. 3 introduces the enterprise blockchain and real-time supervisor. Sect. 4 presents the proof-of-concept experiment. We conclude our work and propose future work in Sect. 5.

2 Related Work

There are many studies on enterprise blockchain and especially on blockchain availability. Regarding the agricultural domain, Agri-SCM-BIoT [3] (Agriculture Supply Chain Management Using Blockchain and Internet of Things) architecture is presented by Showkat Ahmad Bhat et al. to address storage and scalability optimization, interoperability, security and privacy concerns, as well as storage issues with existing single-chain agriculture supply chain systems. We also talked about the many categories of security risks associated with IoT infrastructure and potential blockchain-based defense mechanisms.

An end-to-end inter-bank payment systems prototype [4] built on the Hyperledger Fabric corporate blockchain platform is presented by Xin Wang et al. For inter-bank payment activity, the prototype enables gross settlement, impasse resolution, and reconciliation. This prototype has been demonstrated to offer a higher level of payment settlement service as part of the Ubin Project's preliminary study.

Camilo Denis González et al. [5] propose a system that ensures high reliability by applying enterprise blockchain technology to electronic voting, securing the secret ballot. They also discusses how the solution handles some of the security and dependability problems that electronic voting system solutions typically encounter. A flexible network setup is also shown.

A performance assessment of a private infrastructure supporting a blockchain-based application [6] is provided by Carlos Melo et al. While keeping an eye on the system, they noted a little rise in resource usage, which may be related to software aging problems on the Hyperledger Fabric platform or its fundamental parts. Additionally, the effect of this resource increase on the likelihood that the system will function has been assessed. When consumption problems were taken into account, one of the transaction types increased RAM usage by about 80% in less than 3 h, lowering system availability to 98.17%. The availability achieved 99.35% in circumstances where there were no infrastructure resource increase problems, with an annual downtime of 56.43 h.

3 Proposed Approach

3.1 Enterprise Blockchain

Blockchain uses a highly complex encryption technology to provide secure data storage. In a blockchain, information that already existed in earlier blocks is not changed; rather, information is only added to the current block. Each block is connected to the one before it and contains details about when it was created as well as transaction data [7]. Data cannot be changed after it has been approved by the network. Utilizing timestamps, open audit, and consensus procedures, data is kept in an immutable manner [8]. Blockchain can therefore safeguard data against fraud and modification.

Enterprise blockchain is a type of permissioned blockchain network that can be integrated and utilized for enterprise usages. This kind of blockchain is drastically altering how organizations conduct themselves by giving them control, accountability, immutability, scalability, security, and ease of use. Enterprise blockchain can be classified as private or consortium and are typically permissioned. A blockchain that is private operates on a closed network and has stringent permissions. Such a blockchain is typically utilized inside of a company where only specific employees can participate in a blockchain network. Businesses and corporations that desire to use blockchain solely internally will benefit from it the most. In situations where a group of two or more parties collectively set the network rules and each party can contribute, consortium blockchain is ideally suited for organizational collaboration. The main advantages are that it is highly secure and scalable.

3.2 Real-Time Supervisors

IT resources not only play the role of storage and processing, but also the entrance to the enterprise's systems. Therefore, enterprises need to monitor IT resources to closely monitor employee behavior as well as ensure security policies are ensured. 1E [9] is a real-time end-user management platform, with core modules that provide services from diagnostics and troubleshooting to monitoring and feedback on the user experience. From a supervisor's end point perspective, 1E provides improved insight into key problematic areas. In addition, 1E constantly collects data from endpoints and automatically analyzes and statistics the received data and then visualizes it in the form of tables and diagrams. From an employee perspective, 1E provides a more detailed view of job performance and enhances user interaction. Another popular platform for endpoint managers is NinjaOne [10]. NinjaOne provides IT departments with solutions to automate, manage, and remediate all their endpoint management tasks. NinjaOne provides transparency access and management of devices in the same system including Windows, Mac, Linux servers, workstations, laptops as well as virtual machines.

3.3 Integration

We propose using a private blockchain architecture including in-firm nodes in the enterprise location and auxiliary nodes in employee devices. Combining the two types of nodes will help increase the availability of the blockchain network as well as the ability to actively balance resources at the nodes. Besides, we also propose using a monitoring system to collect the IT resources status and employees' behaviors (Fig. 1). Based on that, the control system will apply appropriate policies on IT resources to ensure the availability of the blockchain network.

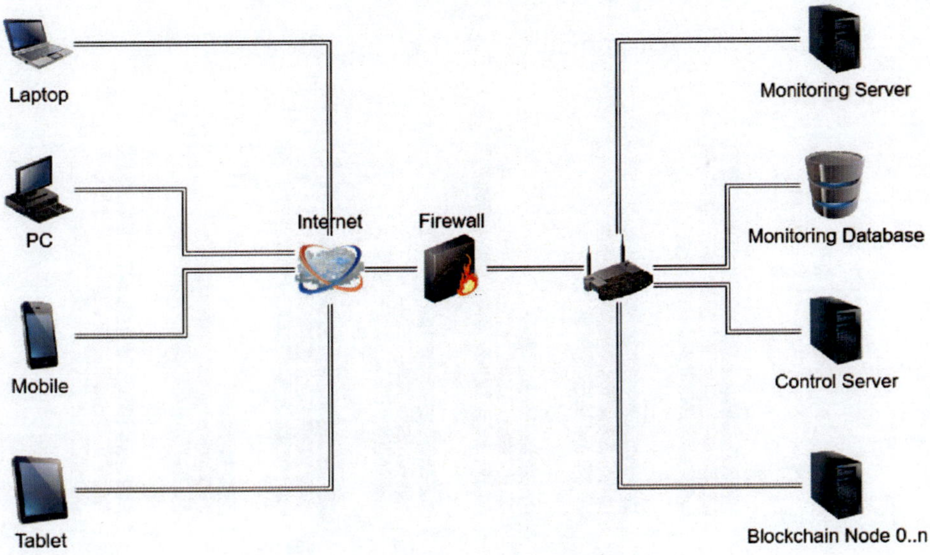

Fig. 1. Proposed overall architecture

4 Experiment

We conducted a proof-of-concept experiment on the I1.501 lab system at Thu Dau Mot University. The lab has 49 PCs with I5 gen 7th CPU, 8 GB RAM, and 200 GB SDD. In experiment 1, we respectively tested the availability of the blockchain network with size 20, 30, and 40 nodes with a duration of 4 h per day, continuously for 7 days. The up time results show that the higher the number of nodes, the higher the availability will be (Fig. 2).

However, in reality, employees' devices are always used to conduct business tasks, so it is necessary to balance resources to ensure the quality of employee experience and the availability of the blockchain network.. Therefore, we conduct experiment 2 on 2 separate 20-node blockchain networks and set a resource policy that exceeds the 90% threshold that will temporarily shut down that blockchain node. Each node host a website and receive random requests. Figure 3 demonstrates the result of the experiment. The monitoring and control mechanism show a more efficiency uptime result.

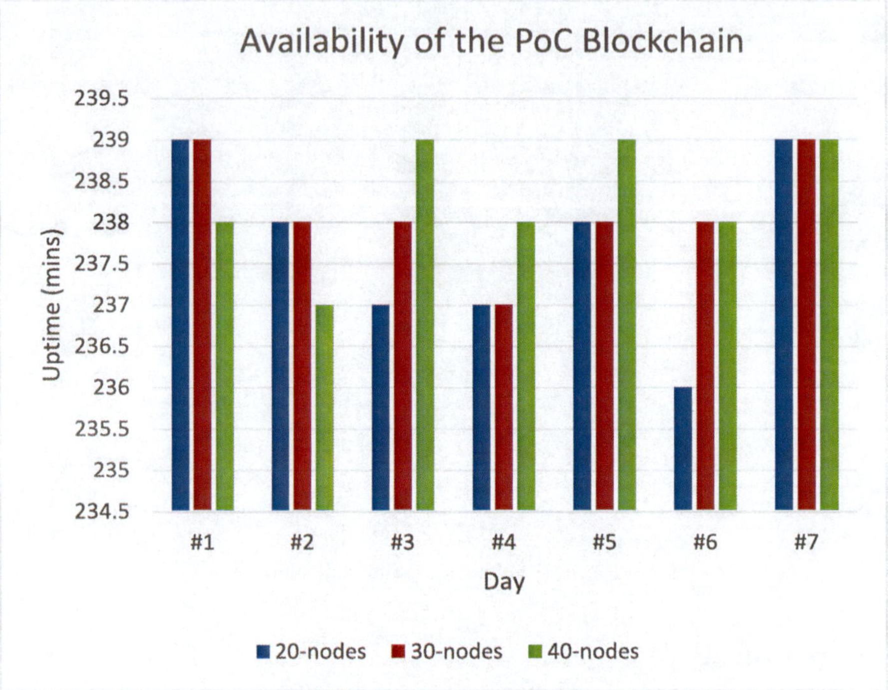

Fig. 2. Availability of the POC blockchain with different blockchain network size

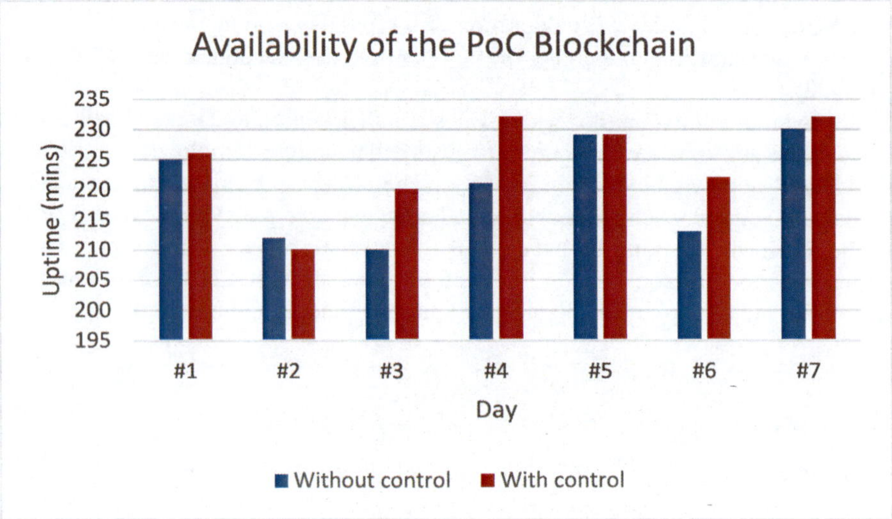

Fig. 3. Availability of the POC blockchain with control and without control

5 Conclusion

Applying blockchain to enterprise applications will limit the disadvantages caused by the factor of centralization. It is possible to deploy enterprise blockchain on employee devices. Combined with monitoring and control mechanisms, enterprise blockchain network availability is enhanced while ensuring user experience.

Acknowledgments. This research is funded by Thu Dau Mot University under grant number DT.22.1-009.

References

1. "Modern Workplace – Microsoft Cloud Partner Program." Partner.microsoft.com, https://partner.microsoft.com/en-us/solutions/modern-workplace
2. Google Workspace | Business Apps & Collaboration Tools, Google, https://workspace.google.com/
3. Bhat, S.A., et al.: Agriculture-food supply chain management based on blockchain and IoT: a narrative on enterprise blockchain interoperability. Agriculture **12**(1), 40 (2021)
4. Wang, X., et al.: Inter-bank payment system on enterprise blockchain platform. 2018 IEEE 11th international conference on cloud computing (CLOUD). IEEE (2018)
5. Denis González, C., et al.: Electronic voting system using an enterprise blockchain. Applied Sciences **12**(2), 531 (2022)
6. Altarawneh, A., et al.: Availability analysis of a permissioned blockchain with a lightweight consensus protocol. Computers & Security **102**, 102098 (2021)
7. Melo, C., et al.: Performance and availability evaluation of the blockchain platform hyperledger fabric. The Journal of Supercomputing, 1–23 (2022)
8. Bodkhe, U., et al.: Blockchain for industry 4.0: a comprehensive review. IEEE Access **8**, 99 (2020)
9. "1E Digital Employee Experience (DEX) Platform." 1E (23 Dec. 2022). https://www.1e.com/platform/
10. Director, B.K., et al.: Endpoint Management Software: Catch Attacks Early. NinjaOne, https://www.ninjaone.com/endpoint-management/

Optimized PID Controller for Two-Wheeled Self-balancing Robot Based on Genetic Algorithm

Van-Truong Nguyen[1](✉), Quoc- Cuong Nguyen[1], Dinh-Hieu Phan[1], Thanh-Lam Bui[1], and Xiem HoangVan[2]

[1] Department of Mechatronics Engineering, Hanoi University of Industry, Hanoi, Vietnam
nguyenvantruong@haui.edu.vn

[2] Faculty of Electronics and Telecommunications, VNU-University of Engineering and Technology, Vietnam National University, Hanoi 10000, Vietnam

Abstract. In this paper, a proportional integral derivative (PID) controller is developed to apply to the dynamics model of the two-wheeled self-balancing (TWSB) robot. The proposed control strategy is created using a PID controller in conjunction with a genetic algorithm, PID-GA for short. The parameters are selected based on the genetic algorithm method to help the system achieve the fastest stability. Therefore, the controller has a simple structure and is easy to implement in practice. Based on the simulation result, it can be seen clearly that the TWSB robot control system's output values closely match the target values and that the error is roughly zero, which demonstrates the proposed controller's effectiveness and performance.

Keywords: Two-wheeled self-balancing robot · PID control · Genetic algorithm

1 Introduction

The two-wheeled self-balancing (TWSB) robot is a unique mobile robot. The TWSB robot may find application in a variety of areas, including transportation and exploration [1]. Based on the inverted pendulum technique, anytime the angle of inclination away from the pivot point changes, the TWSB robot's wheels will either speed up or slow down. The TWSB robot is a mechanical system that is underactuated and is a high-order, multivariable, nonlinear, tightly coupled system. Hence, the TWSB robot has become a research subject attractive to many scientists for research and development [1, 2].

The PD/PID controller has been around for a long time and is currently the most used control approach. The PID controller has garnered favorable reviews from engineering and researcher personnel because of its simple design, the apparent physical significance of each controller parameter, and its conducive adjustment [3, 4]. In [3], Mudeng employed a proportional integral derivative (PID) controller as the control system for a self-balancing robot that operated on the principle of an inverted pendulum. Nguyen

T. D. L. Nguyen et al. (Eds.): ICISN 2023, LNNS 752, pp. 60–65, 2023.
https://doi.org/10.1007/978-981-99-4725-6_9

et al. [4] proposed an adaptive proportional integral derivative controller to combat external disturbances and complicated noises occurring in the parallel manipulator system. However, they encounter issues while standardizing these controllers.

In addition, many authors design controllers for TWSB robots using artificial intelligence. To achieve real-time balancing and velocity-tracking control goals for a TWSB vehicle, Chen et al. [5] proposed a hierarchical sliding mode control (SMC) controller with perturbation estimations. Nguyen [6] presented an adaptive SMC for parallel manipulators with uncertainties. In [7], Nguyen et al. developed a controller technique by combining the sliding mode control with the chattering-free neural network. This control technique compensates for the system's nonlinearity to eliminate tracking errors through online updating of the sliding mode condition parameter. In the aforementioned control methods, sliding mode control, fuzzy logic systems, and neural networks are effective tools for handling unknown nonlinearities. These systems run pretty steadily and meet the specified goals. However, it should be emphasized that they presupposed the system actuators were in good operating order. In the actual industrial system, solving the limits will need a lot of resources. Therefore, balancing control performance with resource-saving is a subject that calls for more research.

Within this paper, a PID-GA controller is proposed to control the nonlinear of a TWSB robot system. The advantages of a PID-GA controller has a straightforward construction, clearly defined physical meanings for each control parameter, and excels in transient response and disturbance rejection. The initial settings of the proposed controller are selected by a genetic algorithm [8]. The performance of the PID-GA controller and PID controllers developed by Mudeng [3] are compared to demonstrate efficacy. The experiments provide a full description of how to tune the control settings. The simulation results indicate that compared with the conventional PID controller, the proposed PID-GA controller can provide better trajectory tracking accuracy of the end-effector.

2 System Dynamic of the TWSB Robot

We denote the mechanical quantities applied to the system as shown in Fig. 1. The kinetic energy of the left, right, and body wheels are calculated as follows:

$$T_1 = \frac{1}{2}wv_l^2 + \frac{1}{2}J_l\left(\frac{v_l}{R}\right)^2 \tag{1}$$

$$T_2 = \frac{1}{2}wv_r^2 + \frac{1}{2}J_r\left(\frac{v_r}{R}\right)^2 \tag{2}$$

$$T_3 = \frac{1}{2}w_c v_c^2 + \frac{1}{2}J_c\dot{a}^2 \tag{3}$$

From the above calculations, the TWSB robot's total kinetic energy is determined as follows.

$$\begin{aligned} T &= T_1 + T_2 + T_3 \\ &= wR^2\left(\dot{c}_l^2 + \dot{c}_r^2\right) + \frac{w_c}{8}R^2(\dot{c}_l + \dot{c}_r)^2 + \frac{1}{2}w_c RL(\dot{c}_l + \dot{c}_r)\dot{a}\cos a + \frac{7}{6}w_c h^2 a^2 \end{aligned} \tag{4}$$

Consider the ground as the surface for comparison. The robotic system's total potential energy is determined as follows.

$$V = 2wgR + w_c g(L\cos a + R) \tag{5}$$

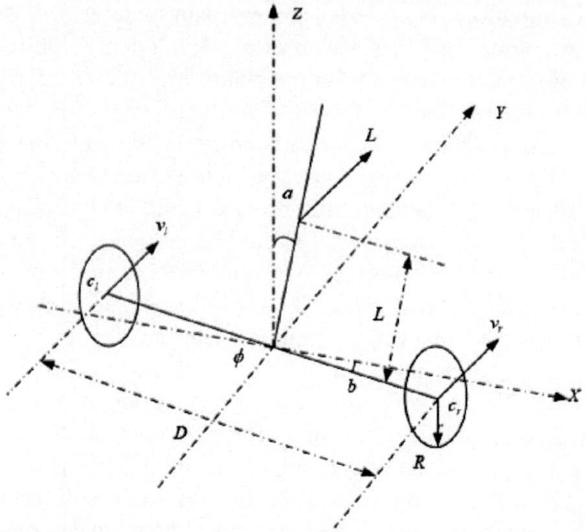

Fig. 1. The TWSB robot structure and mechanical quantities affect itself

The second Lagrange equation is

$$Q_1 = \frac{d}{dt}\left(\frac{\partial L}{\partial \dot{q}_1}\right) - \frac{\partial L}{\partial q_1}(i = 1, 2, 3..., k) \tag{6}$$

where $L = T - V$, place the (4) and (5) operators in the equation and then arrange the variables.

$$J(q)\ddot{q} + f(q, \dot{q})\dot{q} + G(q) = \tau \tag{7}$$

These are some of the equation's unknowable components.

$$J(q) = \begin{bmatrix} 2wR^2 + \frac{w_c}{4}R^2 & \frac{w_c}{4}R^2 & \frac{w_c}{2}RL\cos a \\ \frac{w_c}{4}R^2 & 2wR^2 + \frac{w_c}{4}R^2 & \frac{M_c}{2}RL\cos a \\ \frac{w_c}{2}Rh\cos a & \frac{w_c}{2}RL\cos a & \frac{7}{3}w_c L \end{bmatrix} \tag{8}$$

$$f(q, \dot{q}) = \begin{bmatrix} 0\,0 & -\frac{1}{2}w_c RL\dot{a}\sin a \\ 0\,0 & -\frac{1}{2}w_c RL\dot{a}\sin a \\ 0\,0 & 0 \end{bmatrix} \tag{9}$$

$$G(q) = \begin{bmatrix} 0 \\ 0 \\ -w_c gL\sin a \end{bmatrix} \tag{10}$$

$$\text{The generalized forces are } \tau^T = [M_l, M_r, 0] \tag{11}$$

$$\text{The generalized coordinates are } q^T = [c_1, c_2, a] \tag{12}$$

3 The PID-GA controller

3.1 The PID controller

There are three parameters for the PID controller, each of which affects it differently. The proportionate gain K_P can swiftly, promptly, and proportionately modify the deviation to increase the control sensitivity, but there is a steady-state error and the control accuracy is poor. The steady-state error can be removed using the integral gain K_I, but the system's stability will be impacted. The integral gain K_D has the ability to accelerate the system's responsiveness, decrease overshoot and oscillation, and "anticipate" the dynamic process. Therefore, choosing the optimal K_P, K_I, and K_D parameters is an important task.

The traditional PID controller has an input-output equation in the time domain that is:

$$e(t) = y(t) - r(t) \tag{13}$$

$$u(t) = K_P e(t) + K_I \int_0^t e(t)dt + K_D \frac{de(t)}{dt} \tag{14}$$

where $e(t)$, $y(t)$, $r(t)$, and $u(t)$ represent the error signal, controller output, system input, and system output, respectively.

3.2 Paramater Optimization

Genetic Algorithm (GA) is a particular class of algorithms that employ methods including inheritance, mutation, selection, and crossover that are inspired by evolutionary biology. In a number of disciplines including social science, computer science, biology, and engineering, GA has been used to identify the best answers to challenging issues. Unlike local search techniques, GA is built on a collection of independent computations that are managed by a probabilistic strategy. GA is started with a set of solutions called population. One chromosomal string per individual encodes the data genetic of that individual. In succeeding generations, how the best individuals among succeeding generations would be chosen through this simulation. The fitness of each chromosome is evaluated after each generation, and the chromosomes for the next generation are probabilistically selected based on their fitness scores. Random mating and procreation occur between some of the chosen chromosomes. Random of crossover and mutation occur during creating offspring. Because the fitness values of chromosomes higher are more likely to be picked, chromosomes in the current generation may have better midpoint-fitness values than those in the previous generation. Up until the final criterion is met, the evolutionary process is repeated.

In order for the robot TWSB to function properly, the appropriate KI, KP, and KD parameters are extremely important. In theory, GA can solve this problem. However, choosing the right requirements for GA to find the right parameter is equally important. We choose the following search conditions: The objective function is $J = \min[E_1(i) + E_2(i)]$; The maximum number of generations is 200; The number of individuals is 20; The hybridization coefficient is 0.2; The mutation coefficient is 0.8. The PID-GA controller structure for the TWSB robot is shown Fig. 2.

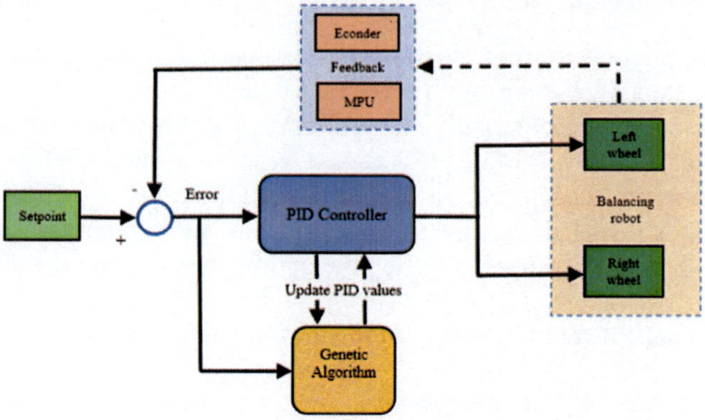

Fig. 2. The PID-GA controller structure for the TWSB robot

4 Demonstrative Example

In this section, in order to further confirm the efficiency of the proposed controller, we give certain simulations. The kinematic parameters of the TWSB robot system are $M = 0.03$(kg), $R = 0.007$(m), $M_c = 0.01$(kg), $R_c = 0.001$(m), $l = 0.009$(m), $h = 0.0012$(m), $g = 9.81$. In order to show how performance has improved, the proposed controller is contrasted with the PID controller [4]. The parameters are set according to page 16 of the paper [3] as follows: $K_P = 4.0602$, $K_I = 4.41$, $K_D = 1.1025$. The simulations are run in MATLAB/Simulink, the simulation time is set as 10 s, and the sampling time is set as 0.01 s.

The simulation outcomes are shown in Fig. 3. It is obvious to observe that both PID and PID-GA controllers are capable of meeting the balance requirements initially. However, Around the 15th second, the PID controller oscillates unstably and tends to collapse. Meanwhile, the PID-GA keeps the system stable and constant. Based on the simulation results, it has been shown that the PID-GA controller is optimal and more efficient than the PID controller.

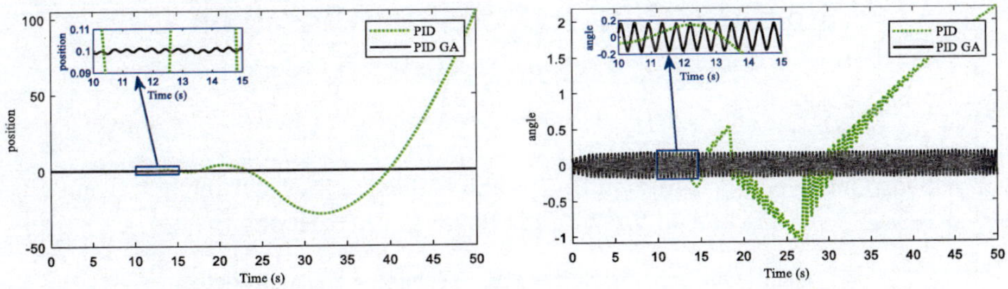

Fig. 3. The tracking response of PID-GA controller for the TWSB robot in simulation

5 Conclusion

This paper presents our study on the PID-GA control method for the dynamics model of the TWSB robot. The PID-GA controller is built based similar structure to the PID controller. The improvement of the PID-GA controller is that the parameters of the PID controller are selected based on the genetic algorithm. This combination has completely solved the problem of determining the most suitable K_P, K_I, and K_D parameters. The simulation results on the software Matlab show that PID-GA adheres to the planned trajectory well, with a quick and consistent reaction time.

References

1. Guo, L., et al.: Optimal control of a two-wheeled self-balancing robot by reinforcement learning. Int. J. Robust Nonlinear Control. **31**(6), 1885–1904 (2021)
2. Ünker, F.: Proportional control moment gyroscope for two-wheeled self-balancing robot. J. Vib. Control **28**(17–18), 2310–2318 (2022)
3. Mudeng, V., et al.: Design and simulation of two-wheeled balancing mobile robot with pid controller. Int. J. Sustain. Transport. Technol. **3**(1), 12–19 (2020)
4. Nguyen, V.T., et al.: Adaptive PID tracking control based radial basic function networks for a 2-DOF parallel manipulator. In: ICSSE, pp. 309–312 (2017)
5. Chen, L., et al.: Robust hierarchical sliding mode control of a two-wheeled self-balancing vehicle using perturbation estimation. Mech. Syst. Signal Process. **139**, 106584 (2020)
6. Nguyen, V.T.: Non-negative adaptive mechanism based sliding mode control for parallel manipulators with uncertainties. Comput. Mater. Contin. **74**(2), 2771–2787 (2023)
7. Nguyen, V.T., et al.: Adaptive chattering free neural network-based sliding mode control for trajectory tracking of redundant parallel manipulators. Asian J. Control **21**(2), 908–923 (2019)
8. Katoch, S., Chauhan, S.S., Kumar, V.: A review on genetic algorithm: past, present, and future. Multimed. Tools Appl. **80**(5), 8091–8126 (2020)

Pancreatic Cancer Detection Based on CT Images Using Deep Learning

Hoang Quang Huy[1], Ngo Tien Dat[1], Dinh Nghia Hiep[1], Nguyen Ngoc Tram[1], Tran Anh Vu[1(✉)], and Pham Thi Viet Huong[2]

[1] Hanoi University of Science and Technology, Hanoi, Vietnam
vu.trananh@hust.edu.vn
[2] Nternational School, Vietnam National University, Hanoi, Vietnam

Abstract. In this paper, we propose an automated deep learning Convolutional Neural Network (CNN) based detection method to distinguish cancerous tumors from benign tumors to identify pancreatic cancer in Computed Tomography (CT) images. We train and test the model using two datasets provided by The Cancer Imaging Archive (TCIA) and the Medical Segmentation Decathlon (MSD). The first dataset contains 18942 CT images of 82 patients with normal pancreas while the second dataset contains 15000 images of 280 patients with confirmed pancreatic cancer. Our proposed model is a modified extension of the current CNN network with the addition of the Densely Connected Convolutional layers. Our method has shown superior results compared to other research and deep learning method with an accuracy of 97.4%, sensitivity of 98.3%, and specificity of 96.6%.

Keywords: Deep learning · Pancreatic cancer · Convolutional neural networks · Computed tomography

1 Introduction

Pancreatic cancer is the 4th most common cause of death in the US [1] and the 7th common in the world [2], with a survival rate of 4% over 5 years. It is often not diagnosed in time for the tumor to be detectable. The American Cancer Society estimated that 31,270 out of 31,860 patients would die from the disease in 2006 [3] and the number surged to an estimation of 37,390 patients in 2012 [4]. Despite the tremendous effort of recent research to achieve early detection of pancreatic cancer through advanced imaging methods, diagnosis at a curable phase remains very difficult. Early stage of the tumor's metastasis is undetectable by imaging techniques such as Computed Tomography (CT) (approximately 40% of tumors smaller than 2 cm evade CT scan) [5]. This leads to distant metastasis present in more than one-half of the cases [6], resulting in an elimination of local therapies such as surgery and radiation. Furthermore, pancreatic cancers also carry a few mutated genes currently undruggable such as KRAS, CDKN2A (p16), TP53, and SMAD4 [7].

Pancreatic cancer has traditionally been identified manually by medical professionals using CT scans. However, because manual diagnosis is time consuming and involves inevitable human error, it is prone to inaccuracy.

T. D. L. Nguyen et al. (Eds.): ICISN 2023, LNNS 752, pp. 66–72, 2023.
https://doi.org/10.1007/978-981-99-4725-6_10

Current research using computer-aided diagnostic (CAD) systems and artificial intelligence (AI) techniques to anticipate the development of pancreatic cancer has shown promising results. Convolutional Neural Network (CNN) [8, 9] has been shown to be effective and powerful in addressing various image classification problems. This method relies on neurons made up of of activation functions and parameters to extract information from images by identifying the local spatial correlations they contain. CNN has shown promising in the imaging diagnosis of a number of illness and is effective in achieving pathologically certified pancreatic ductal adenocarcinoma classification using clinical CT images [9].

In [5], 370 patients with pancreatic cancer and 320 healthy controls from a Taiwanese center's contrast-enhanced CT scans were segmented using CNN. The results were promising, with patch-based accuracy of 0.856 and sensitivity of 0.0913 in the local dataset. Subsequently, our paper focuses on the construction and further investigation of the CNN model to detect pancreatic cancer on clinical CT scans.

In this paper, we propose a Convolutional Neural Network (CNN) based model to separate cancerous images from healthy ones using DICOM CT scans in NII format. First, the CT images are converted from DICOM to JPG format, then they will be preprocessed for further clarification. Preprocessing steps include Dilation, Contour Cropping, Denoise, Contrast, and Brightness adjustment, and resizing. Next, the CT images are labeled and divided into training, testing, and validation sets before the training stage. The performance of our proposed model will be assessed based on accuracy and confusion matrix. The overall results have shown to be superior compared to alternative state-of-the-art research in terms of Accuracy and Sensitivity.

Our paper is organized as follows: part I introduces the motivation of our research. Part II provides the proposed methodology. Part III presents our results and compares them with other articles and other models. The final part concludes the paper with experimental results, discussion, and evaluation.

2 Methodology

2.1 Datasets

In this research, we use The Cancer Image Archive (TCIA) dataset [10] and the Medical Segmentation Decathlon (MSD) dataset [11]. The TCIA dataset contains 3D CT images of 82 patients, provided by the National Institutes of Health Clinical Center in 2016. The patients were selected on the basis of having completely healthy pancreatic. 17 of the participants are healthy kidney donors while the remaining 65 were chosen by a radiologist from patients who neither had major abdominal pathologies nor pancreatic cancer lesions (18942 images). The CT images of those patients were then acquired on Philips and Siemens MDCT scanners and made available on TCIA for the public. The images are provided in DICOM format and have resolutions of approximately 512×512 pixels with slice thickness varying between $1.5 - 2.5$ mm. All ground-truth was manually extracted and obtained from the clinic that used them for treatment planning. The second dataset was provided by the Memorial Sloan Kettering Cancer Center, New York, US for MSD. The dataset contains 7983 portal-venous phase CT scans of 280 patients pathologically diagnosed with pancreatic cancer and undergoing resection of pancreatic masses (15000 images).

2.2 Data Preprocessing

Preprocessing is an important step in the data mining process. In this stage, we prepare the data by filtering the raw input images to optimize the training results of the deep learning network. To improve the overall classification result, we must remove the noise and focus the network to train only on the abdominal area. The steps for the proposed image conversion are as follows:

1. **Image extraction**: We extract patients' CT scans from the DICOM file dataset and convert them into 2D images under JPG format.
2. **Dilation operations**: Next, we use a 3x3 matrix to dilate the image, clarifying the pancreas and tumor area.
3. **Image contour:** We remove redundant parts around the abdomen such as only the abdominal Region of Interest (ROI) by utilizing the "find contour" algorithm according to measured upper, lower, left, and right extremes parameters.
4. **Result**: After contouring, there would still be unwanted noises which are white-colored regions surrounding the ROI. As such, we would denoise, apply contrast and brightness adjustment, then resize for optimal image conditions.

After pre-processing, the CT images are filtered through a mask leaving only images representing the ROI. The same procedure is applied for both datasets.

Figure 1 shows the preprocessing process.

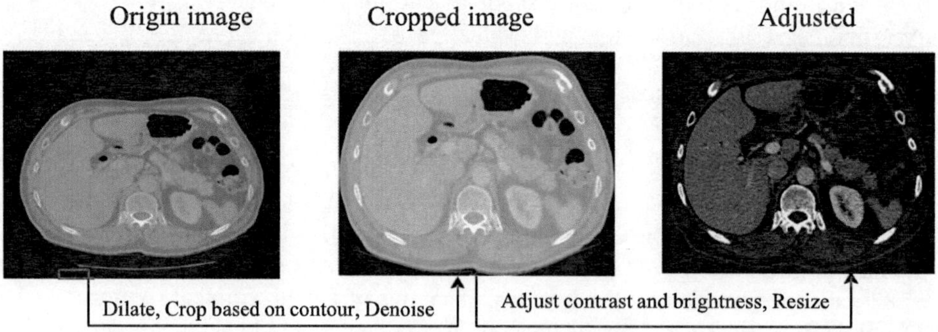

Fig. 1. Preprocessing process

From the input image dataset, we can see a lot of images that do not contain pancreas. We therefore removed these images and kept the images containing pancreatic images. Therefore, the input image set consists of only images containing the pancreas (9676 images). Then we split our dataset, both of non-cancerous patients and non-cancerous images of cancer patients with pancreatic cancer, randomly into the training, validation, and testing set (6400 images in the training set, 1700 images in the validation test, and 1576 images in the testing set). The Fig. 2 shows the datasets classification. After being pre-processed, the CT images will be compressed into two.npy files corresponding to its categorized cancer dataset.

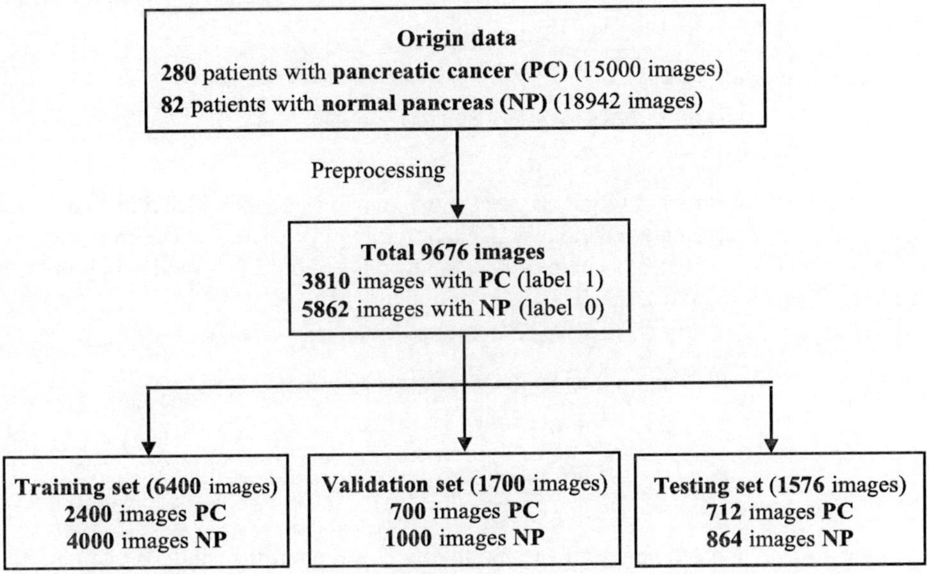

Fig. 2. Datasets classification

2.3 Training Model

Densely Connected Covolutional Network (DenseNet)

Inspired by the CNN framework, we apply the dense convolutions network (DenseNet) [12] to classify pancreatic CT images, an essential step in diagnosing pancreatic cancer. DenseNet works by cross-connecting multiple layers in the network. The blocks sequence of the last convolutional layer in the encoding path is shown in Fig. 3. Based on the concept of "collective knowledge", the feature maps iterate through the network, helping it to attain a diverse but not redundant set of feature maps, and improving its performance by encouraging information flow and recycling features. All of the previous layers' feature maps are utilized as inputs for each layer, and those layers' own feature maps are used as inputs for each next layer. Each layer's feature maps are used as inputs, and each layer's own feature maps are used as inputs for the subsequent layers. For L layers, there are $L(L + 1)/2$ direct connections.

Training and Testing

In the final step, we label normal and cancer patients' images "0" and "1" respectively. The label is 1 for cancer patients and 0 for non-cancerous patients. For each label, 60% of patients are assigned randomly to the training set, 20% to the validation set, and 20% to the test set.

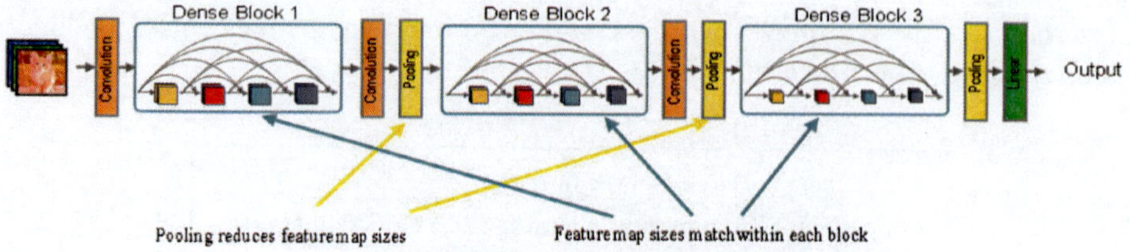

Fig. 3. Block sequence of the DenseNet model [10].

Based on the number of target classes (n), we train our model with a mini-batch size of 32. In order to calculate for the loss in each training cycle, we use the cross-entropy loss function (Eq. 1) [13]. This helps adjust the weights of our CNN model and improve the overall accuracy. We use Adam as the optimizer [14].

$$
\begin{aligned}
L(\theta) &= \frac{1}{N}\sum_{n=1}^{N} -\log p(\tilde{y} = \tilde{y}_{\mathrm{n}} | x_{\mathrm{n}}, \theta) \\
&= \frac{1}{N}\sum_{n=1}^{N} -log(\sum_{i}^{c} p(\tilde{y} = \tilde{y}_{\mathrm{n}} | y = i) p(y = i | x_{\mathrm{n}}, \theta))
\end{aligned} \quad (1)
$$

In order to evaluate efficiency of the method, we evaluated the model's accuracy on the validation set. 15 training epochs are performed on our model. We used 10-fold cross-validation procedure. In each step, we randomly divided ech set of images into 10 folds, 6 of them were used for training, 2 folds were used for validation, and the last fold for testing the model. We ran the whole process 10 times and recorded the average performance. We evaluate our model in term of accuracy and confusing matrix.

3 Results

We tested on 1576 images of the testing set and trained our proposed neural network on 6400 images of the training set and 1700 images of the validation with the corresponding given ground-truth as described in the above section. The network was trained from scratch for all datasets. We evaluated our returns based on Accuracy, Sensitivity, and Specificity. The results show that our proposed method outperforms state-of-the-art alternative models.

Figure 4 shows the confusion matrix evaluated on the testing set that we described above. The quantitative results of the proposed DenseNet method compared to other methods are reported in Table 1. Our proposed method outperforms all the other methods. We achieve an excellent accuracy of 97.4%, sensitivity of 98.3%, and specificity of 96.6%.

It can be seen that our experimental accuracy, sensitivity, specificity, and balanced accuracy are higher than that in other articles. Since different articles use different dataset, this comparison is for reference only.

From Table 2, it can be seen that the performance of our modified DenseNet model on the testing set surpassed that of other models. For the testing set, results gained from

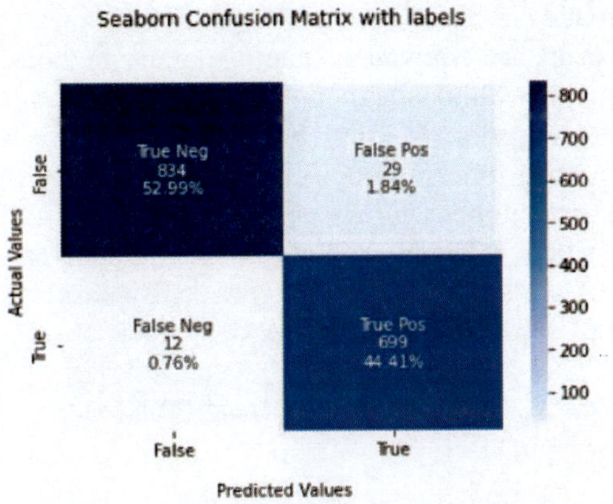

Fig. 4. Proposed method confusion matrix

Table 1. Compare the results with some articles

Author/Year	Method	Accuracy	Sensitivity	Specificity	Balanced accuracy
Han Ma et al. [15], 2020	CNN	95.47%	91.57%	98.27%	94.92%
Kao-Lang Liu et al. [5], 2020	CNN (VGG)	86.7%	91.2%	85.8%	88.5%
Our result	CNN (DenseNet)	**97.4%**	**98.3%**	96.6%	**97.5%**

Table 2. Compare the performance of the DenseNet model with another model

Model	MobileNet	DenseNet	EfficientNet
Training accuracy	0.951 (0.770 to 0.988)	0.978 (0.803 to 0.991)	0.969 (0.869 to 0.987)
Testing accuracy	0.948	**0.974**	0.945

our proposed model is predominant with the highest testing accuracy of 97.4%, which is the most significant factor. Despite not obtaining the leading training accuracy among three model, we attain an extremely high training accuracy of 92.3%.

4 Conclusions

In conclusion, with the introduction of the densely connected convolutional blocks, the CNN-based method has successfully distinguished cancerous tissues from benign tissues on portal venous CT with a vast improvement in classification accuracy. Our proposed

DenseNet for pancreatic cancer image classification has shown superior performance in relation to recently proposed equivalent state-of-the-art methods. Specifically, we were able to solve the issue of redundancy that other methods have stumbled upon by adding a modified pre-processing step. We implemented the mechanism of image dilation and image contour to remove unwanted noises and capture only the ROI of the images. This resulted in a much higher average accuracy of 97.4% as well as sensitivity and specificity. This is critical in diagnosing and assessing pancreatic tumors, which other extensions of CNN models were not able to do. It also displays great promise for the future development of computer-aided cancer detection.

Acknowledgment. This research is funded by Hanoi University of Science and Technology (HUST) under project number T2021-PC-018.

References

1. Costache, M., et al.: Which is the best imaging method in pancreatic adenocarcinoma diagnosis and staging – CT, MRI or EUS? Curr. Health Sci. J. **43**(2), 132–236 (2017)
2. Rawla, P., Sunkara, T., Gaduputi, V.: Epidemiology of pancreatic cancer: global trends, etiology and risk factors. World J. Oncol. **10**(1), 10–27 (2019)
3. Jemal, A., et al.: Cancer statistics, 2006. CA: A Cancer J. Clin. **56**(2), 106–130 (2006)
4. Siegel, R., Naishadham, D., Jemal, A.: Cancer statistics, 2012. CA: A Cancer J. Clin. **62**(1), 10–29 (2012)
5. Liu, K.-L., et al.: Deep learning to distinguish pancreatic cancer tissue from non-cancerous pancreatic tissue: a retrospective study with cross-racial external validation. Lancet Digit Health **2**(6), e303–e313 (2020)
6. Freelove, R., Walling, A.D.: Pancreatic cancer: diagnosis and management. Am. Fam. Phys. **73**(3), 485–492 (2006)
7. Vincent, A., Herman, J., Schulick, R., Hruban, R.H., Goggins, M.: Pancreatic cancer. Lancet Dig. Health **378**(9791), 607–620 (2011)
8. Kenner, B., et al.: Artificial intelligence and early detection of pancreatic cancer: 2020 summative review. Pancreas **50**(3), 251–279 (2021)
9. He, K., Zhang, X., Ren, S., Sun, J.: Deep residual learning for image recognition. In: Proceedings of the IEEE Conference on Computer Vision and Pattern Recognition, pp. 770–778, 27–30 June 2016
10. Roth, H.R., Farag, A., Turkbey, E.B., Lu, L., Liu, J., Summers, R.M.: Data From Pancreas-CT. The Cancer Imaging Archive (2016)
11. Antonellia, M., et al.: The Medical Segmentation Decathlon. 10 June 2021
12. Baldha, S.: Introduction to DenseNets (Dense CNN). Data Science Blogathon 14 Mar 2022
13. Zhang, Z., Sabuncu, M.: Generalized cross entropy loss for training deep neural networks with noisy labels. Advances in Neural Information Processing Systems 31 (NeurIPS 2018) (2018)
14. Kingma, D.P., Ba, J.: Adam: a method for stochastic optimization. In: The 3rd International Conference for Learning Representations (2015)
15. Ma, H., et al.: Construction of a convolutional neural network classifier developed by computed tomography images for pancreatic cancer diagnosis. World J. Gastroenterol. **26**(34), 5156–5168 (2020)

Robust Adaptive Control for Industrial Robots Using Sliding Mode Control and RBF Neural Network

Le Van Chuong[1](✉), Mai The Anh[1], and Ngo Tri Nam Cuong[2]

[1] Vinh University, Vinh City, Nghean, Vietnam
chuonglv@vinhuni.edu.vn

[2] Institute of Electrical Engineering and Automation, Hanoi, Vietnam

Abstract. This paper presents a method of synthesizing a robust adaptive control system for the industrial robot with six degrees of freedom. We describe the mathematical model of the robot in the form of nonlinear state equations that consider variable parameters and unmeasurable external disturbance. Uncertainty components in the dynamic robot model are identified based on adaptive control and the RBF neural network. We create a signal vector from the identification results to compensate for these components. The control law is built based on sliding mode control in which the chattering phenomenon is reduced to a minimum because the influence of the uncertainty components has been eliminated. The article's proposed control system is adaptable, robust, anti-interference, and ensures good control quality.

Keywords: Industrial robot · Adaptive control · Sliding mode control

1 Introduction

Nowadays, robots are widely used in industrial fields to perform simple to complex and repetitive operations to replace manual processes. In order to meet the increasing requirements of actual production, improving the control quality of industrial robots is an important issue. Along with the development of automatic control theory, specific methods of synthesizing control systems for industrial robots have been proposed. In recent years, methods of adaptive control, sliding mode control, and neural networks (NNs) have been studied with many outstanding advantages. The adaptive control methods for industrial robots: model reference adaptive control applied to simple robot systems with a small number of degrees of freedom [1, 2]; adaptive inverse dynamics control based on the model-based nonlinear control method [3–5]; adaptive Li-Slotine control applied to robots with a model with an uncertain parameter in the form of a constant but with poor noise resistance [6–8]. Adaptive control methods based on NNs for robots in [9–12]. However, when designing adaptive laws for NNs, it is necessary to pay attention to the training method and the convergence speed of the algorithm. The sliding mode control method is applied to industrial robots without knowing the exact dynamic parameters

T. D. L. Nguyen et al. (Eds.): ICISN 2023, LNNS 752, pp. 73–84, 2023.
https://doi.org/10.1007/978-981-99-4725-6_11

[13–16]. This control method is only effective when the upper bound of the uncertain component is known; in many cases, it cannot determine this limit directly. Besides, the main disadvantage of the sliding mode control method is the chattering phenomenon because the control signal generated from the sliding mode controller is a discontinuous function. Next, the paper proposed a method of synthesizing the sliding mode controller combined with the RBF NN for the industrial robot with six degrees of freedom (6-DOF). The proposed controller has high control quality, adaptability, robustness, good anti-interference, and minimal chattering phenomenon.

2 Mathematical Model of Industrial Robot

The dynamic model of a 6-DOF industrial robot is described by the equation [17]:

$$\mathbf{M}(\mathbf{q})\ddot{\mathbf{q}} + \mathbf{C}(\mathbf{q})\left[\dot{\mathbf{q}}\dot{\mathbf{q}}\right] + \mathbf{N}(\mathbf{q})\left[\dot{\mathbf{q}}^2\right] + \mathbf{g}(\mathbf{q}) = \tau, \tag{1}$$

where $\mathbf{M}(\mathbf{q}) \in \mathbb{R}^{6\times 6}$ is the positive definite inertia matrix; $\mathbf{C}(\mathbf{q}) \in \mathbb{R}^{6\times 15}$ is the Coriolis torques; $\mathbf{N}(\mathbf{q}) \in \mathbb{R}^{6\times 6}$ is centrifugal torques; $\mathbf{g}(\mathbf{q}) \in \mathbb{R}^6$ are the gravitational forces; $\mathbf{q} \in \mathbb{R}^6$, $\dot{\mathbf{q}} \in \mathbb{R}^6$, and $\ddot{\mathbf{q}} \in \mathbb{R}^6$ are vectors of the robot end effector's position, velocity, and acceleration; $\boldsymbol{\tau} \in \mathbb{R}^6$ is the generalized joint force vector.

$$\mathbf{M}(\mathbf{q}) = \begin{bmatrix} m_{11} & m_{12} & \dots & m_{16} \\ \vdots & \vdots & \ddots & \vdots \\ m_{61} & m_{62} & \dots & m_{66} \end{bmatrix}; \tag{2}$$

$$\mathbf{C}(\mathbf{q}) = \begin{bmatrix} c_{112} & \dots & c_{116} & c_{123} & \dots & c_{156} \\ \vdots & \ddots & \dots & \vdots & \ddots & \vdots \\ c_{616} & \dots & c_{616} & c_{623} & \dots & c_{656} \end{bmatrix}; \tag{3}$$

$$\mathbf{N}(\mathbf{q}) = \begin{bmatrix} n_{11} & n_{12} & \dots & n_{16} \\ \vdots & \vdots & \ddots & \vdots \\ n_{61} & n_{62} & \dots & n_{66} \end{bmatrix}; \tag{4}$$

$$\mathbf{g}(\mathbf{q}) = \left[g_1, g_2, \dots, g_6\right]^T. \tag{5}$$

The symbols $\left[\dot{\mathbf{q}}^2\right]$ and $\left[\dot{\mathbf{q}}\dot{\mathbf{q}}\right]$ are given by:

$$\left[\dot{\mathbf{q}}^2\right] = \left[\dot{q}_1^2, \dot{q}_2^2, \dots, \dot{q}_6^2\right]^T; \tag{6}$$

$$\left[\dot{\mathbf{q}}\dot{\mathbf{q}}\right] = \left[\dot{q}_1\dot{q}_2, \dot{q}_1\dot{q}_3, \dots, \dot{q}_1\dot{q}_6, \dot{q}_2\dot{q}_3, \dots, \dot{q}_5\dot{q}_6\right]^T. \tag{7}$$

Due to the invertibility of matrix $\mathbf{M}(\mathbf{q})$, we rewrite Eq. (1) as follows:

$$\ddot{\mathbf{q}} = \mathbf{M}^{-1}\left(-\mathbf{C}\left[\dot{\mathbf{q}}\dot{\mathbf{q}}\right] - \mathbf{N}\left[\dot{\mathbf{q}}^2\right] - \mathbf{g}\right) + \mathbf{M}^{-1}\boldsymbol{\tau}. \tag{8}$$

To facilitate the design process later, we set: $\mathbf{x} = [x_1, x_2, \ldots, x_{12}]^T$, $x_1 = q_1$, $x_2 = \dot{q}_1$, ..., $x_{11} = q_6$, $x_{12} = \dot{q}_6$; $\mathbf{u} = [u_1, u_2, \ldots, u_6]^T = [\tau_1, \tau_2, \ldots, \tau_6]^T$. Equation (8) is rewritten as:

$$\dot{\mathbf{x}} = \Psi(\mathbf{x}, \mathbf{u}). \tag{9}$$

We perform Taylor expansion (9) at the origin equilibrium point $(\mathbf{x}_0, \mathbf{u}_0)$, we have:

$$\dot{\mathbf{x}} = \mathbf{A}\mathbf{x} + \mathbf{B}\mathbf{u} + \mathbf{f}(\mathbf{x}), \tag{10}$$

where $\mathbf{A} \in \mathbb{R}^{12\times12}$, $\mathbf{B} \in \mathbb{R}^{12\times6}$ are Jacobian matrices:

$$\mathbf{A} = \left.\frac{\partial\Psi}{\partial\mathbf{x}}\right|_{(\mathbf{x}_0,\mathbf{u}_0)}; \quad \mathbf{B} = \left.\frac{\partial\Psi}{\partial\mathbf{u}}\right|_{(\mathbf{x}_0,\mathbf{u}_0)}; \tag{11}$$

and $\mathbf{f}(\mathbf{x}) \in \mathbb{R}^{12}$ is a higher order terms of the Taylor series expansion and unknown nonlinear smooth function vectors. In fact, it is difficult to accurately determine the parameters of the robot due to the complexity of determining the values of mass, torque, geometrical dimensions, and operating mode. Therefore, the robot's parameters can change unpredictably over a wide range. In many cases, the robot is often affected by unmeasurable external disturbances. So the Eq. (11) can be rewritten as:

$$\dot{\mathbf{x}} = [\mathbf{A} + \Delta\mathbf{A}]\mathbf{x} + [\mathbf{B} + \Delta\mathbf{B}]\mathbf{u} + \mathbf{f}(\mathbf{x}) + \mathbf{d}(t), \tag{12}$$

where $\mathbf{A}$, $\mathbf{B}$ are known matrices; $\Delta\mathbf{A}$, $\Delta\mathbf{B}$ are time-varying matrices; $\mathbf{d}(t) \in \mathbb{R}^{12}$ is the vector of unmeasured external disturbances and bounded.

3 Synthesis of the Robust Adaptive Control Law

To synthesize the controller for industrial robots with Eq. (12) tracking the desired trajectory, we propose the control signal as follows:

$$\mathbf{u} = \mathbf{u}_{AC} + \mathbf{u}_{SMC}, \tag{13}$$

where $\mathbf{u}_{AC}$ is a control law for the adaptive compensation of uncertain components $\mathbf{f}_{\sum}(\mathbf{x}, \mathbf{u}, t) = \Delta\mathbf{A}\mathbf{x} + \Delta\mathbf{B}\mathbf{u} + \mathbf{f}(\mathbf{x}) + \mathbf{d}(t)$; $\mathbf{u}_{SMC}$ is the sliding mode control law to the robot tracking the desired trajectory $\mathbf{x}_d$. The block diagram of the control system for the 6-DOF industrial robot is shown in Fig. 1.

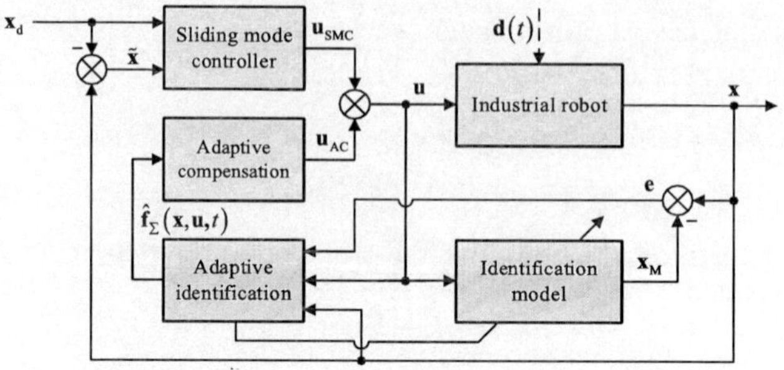

Fig. 1. The block diagram of the control system for the 6-DOF industrial robot.

3.1 Algorithm for Identification of Uncertain Parameters

Without loss of generality, we set: $\mathbf{f}^*(\mathbf{x}) = \Delta\mathbf{A}\mathbf{x} + \mathbf{f}(\mathbf{x})$ and substitute into (12), we have:

$$\dot{\mathbf{x}} = \mathbf{A}\mathbf{x} + [\mathbf{B} + \Delta\mathbf{B}]\mathbf{u} + \mathbf{f}^*(\mathbf{x}) + \mathbf{d}(t). \tag{14}$$

The identification model for uncertain parameters in (14) is as follows:

$$\dot{\mathbf{x}}_M = \mathbf{A}\mathbf{x}_M + \left[\mathbf{B} + \Delta\hat{\mathbf{B}}\right]\mathbf{u} + \hat{\mathbf{f}}^*(\mathbf{x}) + \hat{\mathbf{d}}(t), \tag{15}$$

where $\mathbf{x}_M$ is state vector of the model; $\Delta\hat{\mathbf{B}}$ is the estimated matrix of $\Delta\mathbf{B}$; $\hat{\mathbf{f}}^*(\mathbf{x})$ is the estimated vector of $\mathbf{f}^*(\mathbf{x})$; $\hat{\mathbf{d}}(t)$ is the estimated vector of $\mathbf{d}(t)$.

From (14) and (15), we have:

$$\dot{\mathbf{e}} = \mathbf{A}\mathbf{e} + \left[\Delta\tilde{\mathbf{B}}\right]\mathbf{u} + \hat{\mathbf{f}}^*(\mathbf{x}) + \hat{\mathbf{d}}(t), \tag{16}$$

where:

$$\mathbf{e} = \mathbf{x} - \mathbf{x}_M; \tag{17}$$

$$\Delta\tilde{\mathbf{B}} = \Delta\mathbf{B} - \Delta\hat{\mathbf{B}}; \tag{18}$$

$$\tilde{\mathbf{f}}^*(\mathbf{x}) = \mathbf{f}^*(\mathbf{x}) - \hat{\mathbf{f}}^*(\mathbf{x}) \tag{19}$$

$$\tilde{\mathbf{d}}(t) = \mathbf{d}(t) - \hat{\mathbf{d}}(t). \tag{20}$$

With $\mathbf{f}^*(\mathbf{x})$ is a smooth function vector, by using a RBF NN for the approximation [18]. The elements of $\mathbf{f}^*(\mathbf{x})$ can be written:

$$f_i^*(\mathbf{x}) = \sum_{j=1}^{L} w_{ij}^*\phi_{ij}(\mathbf{x}) + \varepsilon_i; \tag{21}$$

$i = \overline{1, 12}; j = \overline{1, L}$, where L is number of basis function with a large enough number to guarantee the error $|\varepsilon_i| < \varepsilon_i^{\mathrm{M}}$, $w_{ij}^* = \text{const}$ is the ideal weights. The basis functions:

$$\phi_{ij}(\mathbf{x}) = \exp\left(\left\|\mathbf{x} - \mathbf{c}_{ij}\right\|^2 / 2\sigma_{ij}^2\right), \tag{22}$$

where $\mathbf{c}_{ij}$ are the position of the center of the basis functions $\phi_{ij}(\mathbf{x})$, and σ_{ij} are the standard deviation of the basis functions. The evaluation vector $\hat{\mathbf{f}}^*(\mathbf{x})$:

$$\hat{f}_i^*(X) = \sum_{j=1}^{L} \hat{w}_{ij}\phi_{ij}(\mathbf{x}). \tag{23}$$

The approximation of the RBF NN is made by adjusting $\hat{w}_{ij}$ against w_{ij}^*:

$$\tilde{w}_{ij} = w_{ij}^* - \hat{w}_{ij}. \tag{24}$$

The identification process will converge if $\mathbf{e} \to 0$ which means that (16) is stable. For Eqs. (16), the Lyapunov function is selected as follows:

$$V = \mathbf{e}^T\mathbf{P}\mathbf{e} + \sum_{i=1}^{12}\sum_{j=1}^{6} \Delta\tilde{b}_{ij}^2 + \sum_{i=1}^{12}\sum_{j=1}^{L} \tilde{w}_{ij}^2 + \sum_{i=1}^{12} \tilde{d}_i^2, \tag{25}$$

where $\mathbf{P}$ is a positive definite symmetric matrix. Take the derivative of both sides (25):

$$\dot{V} = \dot{\mathbf{e}}^T\mathbf{P}\mathbf{e} + \mathbf{e}^T P\dot{\mathbf{e}} + 2\sum_{i=1}^{12}\sum_{j=1}^{6} \Delta\dot{\tilde{b}}_{ij}\Delta\tilde{b}_{ij} + 2\sum_{i=1}^{12}\sum_{j=1}^{L} \dot{\tilde{w}}_{ij}\tilde{w}_{ij} + 2\sum_{i=1}^{12} \dot{\tilde{d}}_i\tilde{d}_i. \tag{26}$$

Substitute (16) into (26):

$$\begin{aligned} \dot{V} = \mathbf{e}^T\left(\mathbf{A}^T\mathbf{P} + \mathbf{P}\mathbf{A}\right)\mathbf{e} + 2\mathbf{u}^T\Delta\tilde{\mathbf{B}}^T\mathbf{P}\mathbf{e} + +2\mathbf{e}^T P\tilde{\mathbf{f}}^*(\mathbf{x}) + 2\mathbf{e}^T P\tilde{\mathbf{d}}(t) + \\ +2\sum_{i=1}^{12}\sum_{j=1}^{6} \Delta\dot{\tilde{b}}_{ij}\Delta\tilde{b}_{ij} + +2\sum_{i=1}^{12}\sum_{j=1}^{L} \dot{\tilde{w}}_{ij}\tilde{w}_{ij} + 2\sum_{i=1}^{12} \dot{\tilde{d}}_i\tilde{d}_i. \end{aligned} \tag{27}$$

For (16) to be stable, from (27) to with draw the condition to $\dot{V} < 0$:

$$\mathbf{e}^T\left(\mathbf{A}^T\mathbf{P} + \mathbf{P}\mathbf{A}\right)\mathbf{e} + 2\mathbf{e}^T\mathbf{P}\varepsilon < 0; \tag{28}$$

$$2\mathbf{u}^T\Delta\tilde{\mathbf{B}}^T\mathbf{P}\mathbf{e} + 2\sum_{i=1}^{12}\sum_{j=1}^{6} \Delta\dot{\tilde{b}}_{ij}\Delta\tilde{b}_{ij} = 0; \tag{29}$$

$$2\mathbf{e}^T P\tilde{\mathbf{f}}^*(\mathbf{x}) + 2\sum_{i=1}^{12}\sum_{j=1}^{L} \dot{\tilde{w}}_{ij}\tilde{w}_{ij} = 0; \tag{30}$$

$$2\mathbf{e}^T P\tilde{\mathbf{d}}(t) + 2\sum_{i=1}^{12} \dot{\tilde{d}}_i \tilde{d}_i = 0. \tag{31}$$

Transform the inequality (28):

$$\mathbf{e}^T\left(\mathbf{A}^T\mathbf{P} + \mathbf{PA}\right)\mathbf{e} + 2\mathbf{e}^T\mathbf{P}\varepsilon = -\mathbf{e}^T\mathbf{Q}\mathbf{e} + 2\sum_{i=1}^{12} \varepsilon_i \overline{\mathbf{P}}_i\mathbf{e} < 0, \tag{32}$$

where $\mathbf{Q} = -\left(\mathbf{A}^T\mathbf{P} + \mathbf{PA}\right)$, suppose that $\mathbf{A}$ is Hurwitz matrix; $\overline{\mathbf{P}}_i$ is the i-th row of the matrix $\mathbf{P}$. Using inequality transformations [19] for (32), we have:

$$\|\mathbf{e}\| > 2\sum_{i=1}^{12} \varepsilon_i \left\|\overline{\mathbf{P}}_i\right\| / r_{\min}(\mathbf{Q}); \quad r_{\min}(\mathbf{Q}) \text{ is the smallest eigenvalue of the matrix } \mathbf{Q}. \tag{33}$$

Thus, if simultaneously satisfied (29), (30), (31), (33) then $\dot{V} < 0$ and (16) is stable. From (18), (29), notice that the parameter of the robot changes slowly, i.e.$\Delta\dot{b}_{ij} \approx 0$:

$$\Delta\dot{\hat{b}}_{ij} = u_j\overline{\mathbf{P}}_i\mathbf{e};\; i = \overline{1, 12};\, j = \overline{1, 6}. \tag{34}$$

From (19), (21), (23), (24) and (30), because of $w^*_{ij} = const$ so $\dot{w}^*_{ij} = 0$:

$$\hat{f}_i^*(\mathbf{x}) = \sum_{j=1}^{L} \hat{w}_{ij}\phi_{ij}(\mathbf{x}); \quad \dot{\hat{w}}_{ij} = -\overline{\mathbf{P}}_i\mathbf{e}\phi_{ij}(\mathbf{x}); \quad i = \overline{1, 12}; \quad j = \overline{1, 6}. \tag{35}$$

From (20), (31), notice that external disturbances changes slowly so $\dot{d}_i(t) \approx 0$:

$$\dot{\hat{d}}_i = -\overline{\mathbf{P}}_i\mathbf{e}; \quad i = \overline{1, 12}. \tag{36}$$

The identification results from (34), (35), and (36) are used to synthesis the compensation control law $\mathbf{u}_{\mathrm{AC}}$.

3.2 Algorithm for Compensation of Uncertain Parameters

We set: $\mathbf{f}_{\sum}(\mathbf{x}, \mathbf{u}, t) = \Delta\mathbf{Bu} + \mathbf{f}^*(\mathbf{x}) + \mathbf{d}(t) = \left[f_{\sum 1}, f_{\sum 2}, \ldots, f_{\sum 12}\right]^T$, and substitute into (14):

$$\dot{\mathbf{x}} = \mathbf{Ax} + \mathbf{Bu} + \mathbf{I}\mathbf{f}_{\sum}(\mathbf{x}, \mathbf{u}, t), \tag{37}$$

where $\mathbf{I} \in \mathbb{R}^{12\times 12}$ matrix has elements $I_{ij} = 1$ if $i = j$ and $f_{\sum i} \neq 0$; $I_{ij} = 0$ if $i \neq j$ and $f_{\sum i} = 0$; $i, j = \overline{1, 12}$.

Substitute (13) into (37):

$$\dot{\mathbf{x}} = \mathbf{Ax} + \mathbf{B}\mathbf{u}_{\mathrm{SMC}} + \mathbf{B}\mathbf{u}_{\mathrm{AC}} + \mathbf{I}\mathbf{f}_{\sum}(\mathbf{x}, \mathbf{u}, t). \tag{38}$$

From (38), $\mathbf{f}_{\sum}(\mathbf{x}, \mathbf{u}, t)$ will be compensated when:

$$\mathbf{B}\mathbf{u}_{\mathrm{AC}} + \mathbf{I}\mathbf{f}_{\Sigma}(\mathbf{x}, \mathbf{u}, t) = 0. \tag{39}$$

To create a signal vector $\mathbf{u}_{\mathrm{AC}}$ satisfying (39), we choose:

$$\mathbf{u}_{\mathrm{AC}} = -\mathbf{H}\mathbf{f}_{\Sigma}(\mathbf{x}, \mathbf{u}, t), \tag{40}$$

where $\mathbf{H}$ is the gain matrix. From (34), (35), (36), replace $\mathbf{f}_{\sum}(\mathbf{x}, \mathbf{u}, t)$ with $\hat{\mathbf{f}}_{\sum}(\mathbf{x}, \mathbf{u}, t)$:

$$\hat{\mathbf{f}}_{\sum}(\mathbf{x}, \mathbf{u}, t) = \Delta\hat{\mathbf{B}\mathbf{u}} + \hat{\mathbf{f}}(\mathbf{x}) + \hat{\mathbf{d}}(t). \tag{41}$$

From (40) and (41), we have:

$$\mathbf{u}_{\mathrm{AC}} = -\mathbf{H}\hat{\mathbf{f}}_{\Sigma}(\mathbf{x}, \mathbf{u}, t). \tag{42}$$

Substitute (42) into (39):

$$-\mathbf{B}\mathbf{H}\hat{\mathbf{f}}_{\Sigma}(\mathbf{x}, \mathbf{u}, t) + \mathbf{I}\mathbf{f}_{\Sigma}(\mathbf{x}, \mathbf{u}, t) = 0. \tag{43}$$

To satisfy (43), we must have:

$$\mathbf{B}\mathbf{H} = \mathbf{I}. \tag{44}$$

From (44), we choose $\mathbf{H} = \mathbf{B}^{+}$, where $\mathbf{B}^{+}$ is the pseudo-inverse matrix of $\mathbf{B}$[19]. Thus, with (42), the uncertainty elements are compensated then (38) becomes:

$$\dot{\mathbf{x}} = \mathbf{A}\mathbf{x} + \mathbf{B}\mathbf{u}_{\mathrm{SMC}}. \tag{45}$$

Next, the system (45) of the control law is built based on sliding mode control.

3.3 Synthesis of the Sliding Mode Control Law

The error vector between the state vector and the desired state vector $\mathbf{x}_{\mathrm{d}}$:

$$\tilde{\mathbf{x}} = \mathbf{x} - \mathbf{x}_{\mathrm{d}} \rightarrow \mathbf{x} = \tilde{\mathbf{x}} + \mathbf{x}_{\mathrm{d}}. \tag{46}$$

Substitute (46) into (45):

$$\dot{\tilde{\mathbf{x}}} = A\tilde{\mathbf{x}} + \mathbf{B}\mathbf{u}_{\mathrm{SMC}} + \mathbf{A}\mathbf{x}_{\mathrm{d}} - \dot{\mathbf{x}}_{\mathrm{d}}. \tag{47}$$

For (47), the hyper sliding surface is chosen as follows [20]:

$$\mathbf{s} = \mathbf{\Omega}\tilde{\mathbf{x}}, \tag{48}$$

where $\mathbf{s} = [s_1, s_2, ..., s_6]^T$; $\Omega \in \mathbb{R}^{6\times12}$ is the parameter matrix of hyper sliding surface and choose Ω is Hurwitz matrix such that $\det(\Omega\mathbf{B}) \neq 0$.

The control signal $\mathbf{u}_{\mathrm{SMC}}$ can be written by:

$$\mathbf{u}_{\mathrm{SMC}} = \begin{cases} \mathbf{u}_{\mathrm{eq}} \text{ if } \mathbf{s} = 0 \\ \mathbf{u}_{\mathrm{N}} \text{ if } \mathbf{s} \neq 0 \end{cases}, \tag{49}$$

$\mathbf{u}_{\text{eq}}$ is the equivalent control signal that keeps the system (47) on the hyper sliding surface (48); $\mathbf{u}_{\text{N}}$ is the control signal that moves the system (47) towards the hyper sliding surface (48). From (49), we can rewrite:

$$\mathbf{u}_{\text{SMC}} = \mathbf{u}_{\text{eq}} + \mathbf{u}_{\text{N}}. \tag{50}$$

$\mathbf{u}_{\text{eq}}$ is defined in [20]:

$$\mathbf{s} = \boldsymbol{\Omega}\tilde{\mathbf{x}} = 0. \tag{51}$$

From (47) and (51), the equivalent control signal can be defined as follows:

$$\mathbf{u}_{\text{eq}} = -[\boldsymbol{\Omega}\mathbf{B}]^{-1}\left[\boldsymbol{\Omega}\mathbf{A}\tilde{\mathbf{x}} + \boldsymbol{\Omega}\mathbf{A}\mathbf{x}_{\text{d}} - \boldsymbol{\Omega}\dot{\mathbf{x}}_{\text{d}}\right]. \tag{52}$$

Next, we define the control signal $\mathbf{u}_{\text{N}}$. For (48), the Lyapunov function selected by:

$$V = {}^{1}\!/_{2}\mathbf{s}^{T}\mathbf{s}. \tag{53}$$

Condition for the existence of slip mode can be written:

$$\dot{V} = \mathbf{s}^{T}\dot{\mathbf{s}} < 0. \tag{54}$$

From (47), (48), (50), (54), and attention to (52), we have:

$$\dot{V} = \mathbf{s}^{T}[\boldsymbol{\Omega}\mathbf{B}\mathbf{u}_{\text{N}}] < 0. \tag{55}$$

So to satisfy (54), from (55), we have:

$$\mathbf{u}_{\text{N}} = -[\boldsymbol{\Omega}\mathbf{B}]^{-1}\delta\text{sgn}(\mathbf{s}), \tag{56}$$

where δ is a small positive coefficient. Substituting (52) and (56) into (50), we have:

$$\mathbf{u}_{\text{SMC}} = -[\boldsymbol{\Omega}\mathbf{B}]^{-1}\left[\boldsymbol{\Omega}\mathbf{A}\tilde{\mathbf{x}} + \boldsymbol{\Omega}\mathbf{A}\mathbf{x}_{\text{d}} - \boldsymbol{\Omega}\dot{\mathbf{x}}_{\text{d}} + \delta\text{sgn}(\mathbf{s})\right]. \tag{57}$$

Thus, the article has synthesized the sliding mode control law (57) for the industrial robot to follow the desired trajectory. Besides, when the adaptive recognition algorithm converges, the components of uncertainty change are compensated for, making the $\mathbf{u}_{\text{SMC}}$ (57) independent of the uncertain components of the robot. Then we can choose the positive coefficient δ with a small value, which means that chattering phenomenon in the sliding mode control law is reduced to a minimum.

Finally, the control signals (42) and (57) are used for (13), and the control laws of the industrial robot (1) have been synthesized successfully.

4 Results and Discussion

By choosing the values for the parameters of the 6-DOF PUMA 560 robot manipulator proposed in [17] which the dynamic model is described by the Eq. (1).

Perform Taylor series expansion of Eq. (9) at the origin equilibrium point $(\mathbf{x}_0, \mathbf{u}_0) = (\mathbf{0}, \mathbf{0})$. We have the matrices $\mathbf{A} \in \mathbb{R}^{12\times 12}$ and $\mathbf{B} \in \mathbb{R}^{12\times 6}$ (11) as follows:

$$
\mathbf{A} = \begin{bmatrix}
0 & 1 & 0 & 0 & 0 & 0 & 0 & 0 & 0 & 0 & 0 & 0 \\
0 & 0 & -0.74 & 0 & 0.17 & 0 & 0 & 0 & 0 & 0 & 0 & 0 \\
0 & 0 & 0 & 1 & 0 & 0 & 0 & 0 & 0 & 0 & 0 & 0 \\
0 & 0 & 0.71 & 0 & 0.46 & 0 & 0 & 0 & 0 & 0 & 0 & 0 \\
0 & 0 & 0 & 0 & 0 & 1 & 0 & 0 & 0 & 0 & 0 & 0 \\
0 & 0 & 6.93 & 0 & 5.32 & 0 & 0 & 0 & 0 & 0 & 0 & 0 \\
0 & 0 & 0 & 0 & 0 & 0 & 0 & 1 & 0 & 0 & 0 & 0 \\
0 & 0 & 0 & 0 & 0 & 0 & 0 & 0 & 0 & 0 & 0 & 0 \\
0 & 0 & 0 & 0 & 0 & 0 & 0 & 0 & 0 & 1 & 0 & 0 \\
0 & 0 & 0.11 & 0 & 0.12 & 0 & 0 & 0 & 0.16 & 0 & 0 & 0 \\
0 & 0 & 0 & 0 & 0 & 0 & 0 & 0 & 0 & 0 & 0 & 1 \\
0 & 0 & 0 & 0 & 0 & 0 & 0 & 0 & 0 & 0 & 0 & 0
\end{bmatrix};
$$

$$
\mathbf{B} = \begin{bmatrix}
0 & 0 & 0 & 0 & 0 & 0 \\
0.254 & 0.003 & 0.029 & 0 & 0 & 0 \\
0 & 0 & 0 & 0 & 0 & 0 \\
0.003 & 0.149 & -0.044 & 0 & 0 & 0 \\
0 & 0 & 0 & 0 & 0 & 0 \\
0.029 & -0.044 & 0.876 & 0 & -0.006 & 0 \\
0 & 0 & 0 & 0 & 0 & 0 \\
0 & 0 & 0 & 4.98 & 0 & 0 \\
0 & 0 & 0 & 0 & 0 & 0 \\
0 & 0 & -0.006 & 0 & 5.55 & 0 \\
0 & 0 & 0 & 0 & 0 & 0 \\
0 & 0 & 0 & 0 & 0 & 5.26
\end{bmatrix}.
$$

Assume the nonlinear function vector $\mathbf{f}(\mathbf{x})$ and the external disturbance vector $\mathbf{d}(t)$ in Eq. (12) have the form:

$$
\mathbf{f}(\mathbf{x}) = \begin{bmatrix}
0 \\
\sin(x_2 + x_5) + 0.2\sin(x_7 + x_6) + 0.5\sin(x_9)\sin(x_{11}) \\
0 \\
\sin(x_3) + \sin(0.5x_8) + \sin(x_8)\sin(x_3) + \sin(0.5x_8) + \sin(x_8) \\
0 \\
\sin(0.2x_3) + \sin(x_5) + \sin(0.8x_7) + \sin(x_9) + \sin(0.5x_{10}) \\
0 \\
\sin(x_3 + x_5) + \sin(x_7) + \sin(0.8x_9) + \sin(x_{12}) \\
0 \\
\sin(0.8x_3) + \sin(x_5 + x_7) + \sin(x_9) + \sin(0.6x_{11}) \\
0 \\
\sin(x_3 + 0.5x_5) + \sin(x_7) + \sin(0.6x_9) + \sin(0.9x_{12})
\end{bmatrix};
$$

$$\mathbf{d}(t) = \begin{bmatrix} 0 \\ \sin(0.5t + \pi/2) \\ 0 \\ 0.8\sin(0.6t) \\ 0 \\ 0.6\sin(0.5t + \pi/3) \\ 0 \\ 0.7\sin(0.8t + \pi/4) \\ 0 \\ 0.9\sin(0.6t - \pi/3) \\ 0 \\ 0.5\sin(0.7t) + 0.3 \end{bmatrix}.$$

It is assumed that the parameters of the robot have a change of $\Delta\mathbf{A} = 25\%\mathbf{A}$, $\Delta\mathbf{B} = 25\%\mathbf{B}$, and the desired trajectory for each articulation has the following form:

$$q_{1d} = \sin(t) + 0.2; \quad q_{2d} = 0.5\sin(0.8t - \pi/2); \quad q_{3d} = 0.8\sin(0.9t + \pi/4);$$
$$q_{4d} = 1.5\sin(t + 3\pi/2); \quad q_{5d} = \sin(0.6t + \pi/3); \quad q_{6d} = 1.5\sin(0.9t + \pi/4).$$

Simulations were performed using Matlab software. The simulation results are shown in Fig. 2 and Fig. 3.

The results in Fig. 2 show that the robot's uncertainty components recognition algorithms (34), (35), and (36) have entirely converged. Next, with controller $\mathbf{u}$ (13) where $\mathbf{u}_{AC}$ (42) and $\mathbf{u}_{SMC}$ (57), the robot's trajectory has tracking to the desired trajectory shown in Fig. 3. These simulation results once again prove the correctness and effectiveness of the proposed control law.

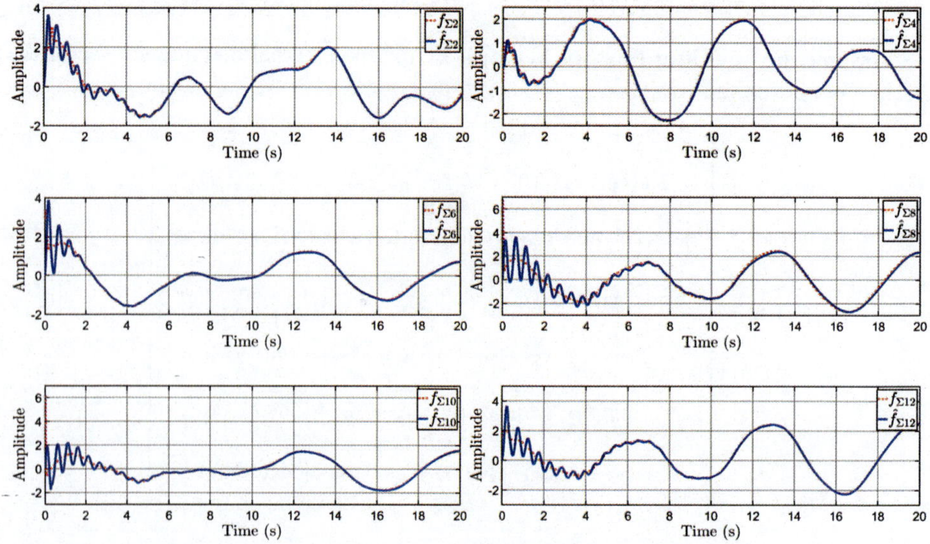

Fig. 2. The identification result of uncertainty vector $\hat{\mathbf{f}}_{\sum}(\mathbf{x}, \mathbf{u}, t)$.

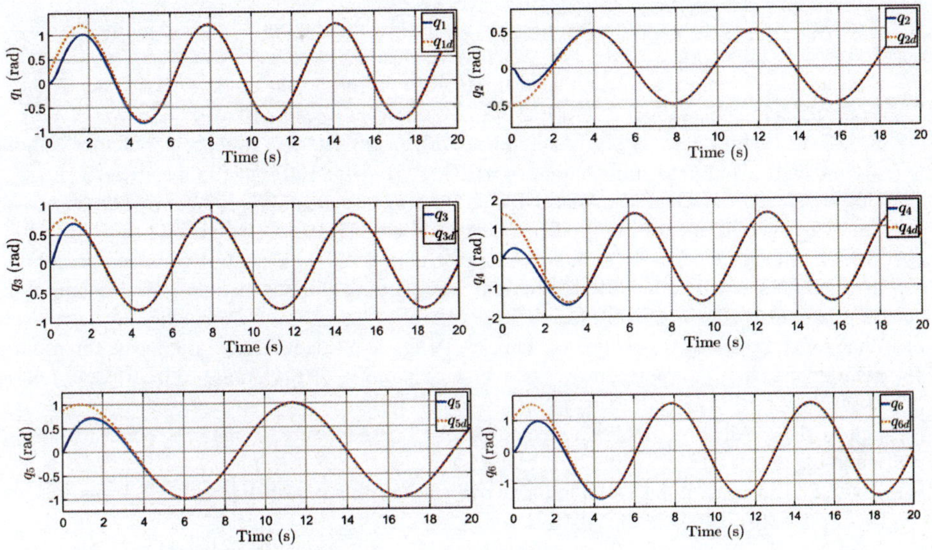

Fig. 3. Trajectory tracking of the 6-DOF industrial robot.

5 Conclusion

This paper has synthesized the 6-DOF industrial robot control systems based on sliding control and RBF neural network. Robot's mathematical model is described as nonlinear state equations considering variable parameters and unmeasured external disturbance using the Taylor expansion method. We synthesized the identification rule of uncertain components in the robot model based on adaptive control theory and RBF neural network; building a mechanism to compensate for uncertain elements from the identification results ensures the system is invariant with these components. The sliding mode control law is synthesized with a chattering phenomenon reduced to a minimum and overcomes the limitations of the basic sliding mode control method [13–16]. Our control system has good controllability, adaptability, and robustness. Simulation results confirm the correctness and effectiveness of the proposed method.

References

1. Zhang, D., Wei, B.: A review on model reference adaptive control of robotic manipulators. Annu. Rev. Control. **43**, 188–198 (2017)
2. Tung, P.C., Wang, S.R., Hong, F.Y.: Application of MRAC theory for adaptive control of a constrained robot manipulator. Int. J. Mach. Tools Manuf **40**(14), 2083–2097 (2000)
3. Wang, H., Xie, Y.: Adaptive inverse dynamics control of robots with uncertain kinematics and dynamics. Automatica **45**(9), 2114–2119 (2009)
4. Chen, Y., Mei, G., Ma, G., Lin, S., Gao, J.: Robust adaptive inverse dynamics control for uncertain robot manipulator. Int. J. Innovative Comput. Inform. Control **10**(2), 575–587 (2014)
5. Ghavifekr, A.A., Velázquez, R., Safari, A.: Multirate adaptive inverse dynamics control of 5 DOF industrial gryphon robot. In: 2021 9th RSI International Conference on Robotics and Mechatronics (ICRoM), pp. 255–260 (2021)

6. Slotine, J.J.E., Li, W.: Composite adaptive control of robot manipulators. Automatica **25**(4), 509–519 (1989)
7. Slotine, J.J., Weiping, L.: Adaptive manipulator control: a case study. IEEE Trans. Autom. Control **33**(11), 995–1003 (1988)
8. Kai, C.Y., Huang, A.C.: A regressor-free adaptive controller for robot manipulators without Slotine and Li's modification. Robotica **31**(7), 1051–1058 (2013)
9. He, W., Huang, H., Ge, S.S.: Adaptive neural network control of a robotic manipulator with time-varying output constraints. IEEE Trans. Cybern. **47**(10), 3136–3147 (2017)
10. Wu, Y., Huang, R., Li, X., Liu, S.: Adaptive neural network control of uncertain robotic manipulators with external disturbance and time-varying output constraints. Neurocomputing **323**, 108–116 (2019)
11. Zhang, S., Dong, Y., Ouyang, Y., Yin, Z., Peng, K.: Adaptive neural control for robotic manipulators with output constraints and uncertainties. IEEE Trans. Neural Netw. Learn. Syst. **29**(11), 5554–5564 (2018)
12. Fateh, M.M., Ahmadi, S.M., Khorashadizadeh, S.: Adaptive RBF network control for robot manipulators. J. AI Data Min. **2**(2), 159–166 (2014)
13. Islam, S., Liu, X.P.: Robust sliding mode control for robot manipulators. IEEE Trans. Industr. Electron. **58**(6), 2444–2453 (2010)
14. Adhikary, N., Mahanta, C.: Sliding mode control of position commanded robot manipulators. Control. Eng. Pract. **81**, 183–198 (2018)
15. Piltan, F., Sulaiman, N.B.: Review of sliding mode control of robotic manipulator. World Appl. Sci. J. **18**(12), 1855–1869 (2012)
16. Yen, V.T., Nan, W.Y., Van Cuong, P.: Recurrent fuzzy wavelet neural networks based on robust adaptive sliding mode control for industrial robot manipulators. Neural Comput. Appl. **31**(11), 6945–6958 (2018). https://doi.org/10.1007/s00521-018-3520-3
17. Lavín-Delgado, J.E., Solís-Pérez, J.E., Gómez-Aguilar, J.F., Escobar-Jiménez, R.F.: Trajectory tracking control based on non-singular fractional derivatives for the PUMA 560 robot arm. Multibody Sys. Dyn. **50**(3), 259–303 (2020). https://doi.org/10.1007/s11044-020-09752-y
18. Cotter, N.E.: The Stone-Weierstrass theorem and its application to neural networks. IEEE Trans. Neural Netw. **1**(4), 290–295 (1990)
19. Ortega, J.M.: Matrix Theory. Springer US, Boston, MA (1987)
20. Utkin, V.I.: Sliding Modes in Control and Optimization. Springer Berlin Heidelberg, Berlin, Heidelberg (1992)

Detecting Imbalance of Patients with Vestibular Diagnosis Using Support Vector Machine

Hang Dang Thuy[1], Hue Tran Thi[1], and Dinh Do Van[2](✉)

[1] Le Quy Don Technical University, Hanoi, Vietnam
[2] Sao Do University, Hải Dương, Vietnam
dodinh75@gmail.com

Abstract. Nowadays, vestibular is a common disease in Vietnam. The diagnosis of vestibular disorders is made using a variety of methods. In the content of the article, we research and detect vestibular disorders by testing balance. We use a data set to quantitatively measure the patient's body angle by using a camera and computer. Then, data analysis and logistic regression model were built to classify patients with vestibular disorders. This makes sense in the process of testing and diagnosing diseases more accurately and efficiently.

Keywords: Vestibular disorder · Romberg test · Support Vector Machine

1 Introduction

Vestibular disorder is a condition in which the process of transmitting and receiving information of the vestibule is disturbed or obstructed, causing an imbalance when changing positions, making the patient dizzy and lightheaded, blurred vision, tinnitus, nausea, unsteady walking, easy to fall [1]. There are many methods to identify vestibular disorders, but they are mainly divided into four main groups: group of eye movement test methods, group of methods of hearing test, group of imaging methods and group of methods. Balance test method [2].

In imaging studies such as magnetic resonance imaging (MRI), CT scans can detect tumors, complications, and other abnormalities that can cause vestibular symptoms [2, 3]. The balance test is often considered the common solution for screening people with suspected vestibular insufficiency. Balance testing methods include: Past pointing test, Time up and go test and Romberg test [4].

Some studies using the Romberg test had good results [6–8]. Studies [6] and [8] currently only focus on people over 40 years old. In the study [7], the volunteers had to carry out the survey for a long time and at a high cost, which is not really consistent with the current situation of our country. In Vietnam, there are some researches to assess the need for rapid balance for young people aged 20 to 30 [9–12]. For add more research on this topic, in our research, we propose the method to analysis Romberg test data by using support vector machine technique, which is a supervised data mining technique for classification, to support the quick diagnosis of vestibular disease.

T. D. L. Nguyen et al. (Eds.): ICISN 2023, LNNS 752, pp. 85–90, 2023.
https://doi.org/10.1007/978-981-99-4725-6_12

2 Methodology

2.1 Data Preparation and Processing

Sets of data from patients with and without balance disorder are collected with Romberg's test, which is shown in Fig. 1 [13]. The system consists of one camera connected to the computer. The camera was placed one to three meters away from the volunteer and captured two stickers on the volunteer's back. The two stickers are glued at a certain distance from each other and the color must not coincide with the shirt color for the best ability to distinguish. In the test model, the resolution is 1 Megapixel and the frame rate is 25 frames or more. The two stickers are the same size. Label color selection should be based on the patient's clothing color to maximize contrast.

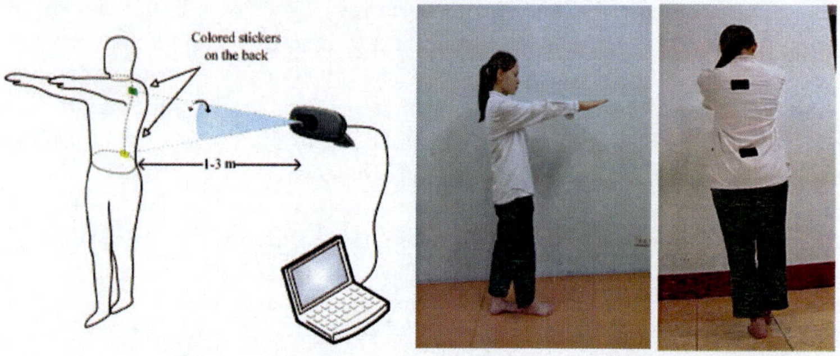

(a) (b)

Fig. 1. System connection model (a) and Standing position of the patient viewed from both sides (b)

The dataset has 458 samples, of which 213 patients and 245 normal [14] with attributes: Patient code, sample number, tilt angle and disease status. Here we pay attention to the angle of inclination and the condition of the disease (Fig. 2).

	PatientCode	Sample number	Tilt angle	Disease status
0	1001	1	8.770763	Normal
1	1001	2	8.770763	Normal
2	1001	3	9.090277	Normal
...	...	...	...	...
95192	40075	199	3.494639	Patient
95193	40075	200	3.494639	Patient

Fig. 2. Information of dataset

In the dataset, we consider the following features:

Value of mean deviation angle (mean):

$$mean = \frac{x_1 + x_2 + \ldots + x_{200}}{200} \tag{1}$$

Standard Deviation (std):

$$std = \sqrt{\frac{\sum (x - \overline{x})^2}{n - 1}} \tag{2}$$

Maximum deviation:

$$x_{\max} - x_{\min} \tag{3}$$

The standard deviation reflects the degree of sway of the person being tested, representing the degree of inclination relative to the person's standing position (mean value).

For patient, due to the strong swaying level, the maximum deviation will be very large compared to those who are normal. Number of times the tested person has a deviation angle greater than 3 degrees.

People without disease, the deviation angle is usually very small, the number of times the deviation angle is greater than 3 degrees is usually less than 3 times. Number of times the tested person oscillates back and forth to the vertical position (0 degrees).

With patient, they sway to the left and right many times continuously, so the number of times they oscillate through the vertical position is also very large.

The information about the person being tested will be displayed: Total variation, total deviation angle greater than 3 degrees, maximum deviation, standard deviation, mean value, measure of deviation angle and graph of the oscillation angle of the person being tested (Fig. 3).

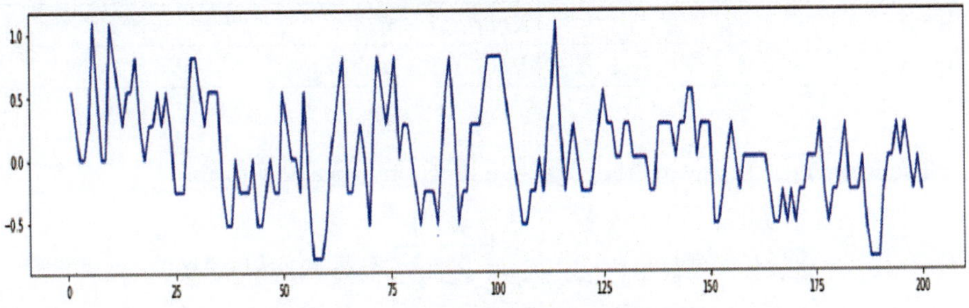

Fig. 3. Oscillation graph of the 320th patient

Patient number: 320
Total Oscillation: 47.0
Total deviation angle greater than 3: 0.0
Maximum deviation: 1.8854699729999997
Standard Deviation: 0.4060539765272586
Average value: 0.049983841939999996
Doctor's diagnosis: normal

2.2 Support Vector Ma chine

SVM is a logic-based machine learning algorithm statistical theory [13]. The basic problem of SVM is the classification problem two classes: Given r points in space n dimension (each point belongs to a class denoted by +1 or −1), the purpose of the SVM algorithm is to find an optimal partition hyperplane lets divide these points into two parts for points of the same class lying on one side with this super flat. The optimal hyperplane that divides the data into two classes is a hyperplane with data can be separated into two separate classes with maximum margin.

In this research, we need to find the hyperplane H0: y = w.x − b = 0 and two hyperplanes H+ H− support parallel to H0 and have the same distance to H0. Provided that there are no elements of the set the sample is between H+ and H−, then:

$$\begin{aligned} &\mathrm{H+: w.x - b >= +1 \quad with\ y = +1} \\ &\mathrm{H-:\ w.x - b >= -1 \quad with\ y = -1} \end{aligned}$$

The above optimal hyperplane solution can be extended in case the data is not linearly separable by mapping the data into a space with a larger number of dimensions using a kernel function K. Some commonly used multiplication functions are described in the following table will use in the paper.

Table 1. Four Kernel Functions.

Type of kernel function	Definition
Poly	$K(x_a, x_b) = (x_a \times x_b + 1)^p$
RBF	$K(x_a, x_b) = exp\left(-\frac{\lVert x_a - x_b \rVert^2}{2\sigma^2}\right)$
Linear	$K(x_a, x_b) = x_a \times x_b$
Sigmoid	$K(x_a, x_b) = \tanh(a \times x_a \times x_b - b)$

The Kernel trick converts the objective function into a new form:

$$\begin{aligned} &\lambda = \arg \underset{\lambda}{max} \sum_{n=1}^{N} \lambda_n - \frac{1}{2} \sum_{n=1}^{N} \sum_{m=1}^{N} \lambda_n \lambda_m y_n y_m k(x_n, x_m) \\ &\text{subject to} \sum_{n=1}^{N} \lambda_n y_n = 0;\ 0 \le \lambda_n \le C,\ \forall n = 1, 2, \ldots, N \end{aligned} \tag{4}$$

where N is the number of data point in the training set; x_n is the nth vector in the training set; y_n is the label of nth data point (y_n can be 1 or −1); λ_n is the Lagrange factor of nth data point; C is a constant described above. After this function is solved, support vectors will be found, and labelling can be performed next.

3 Results and Discussions

To perform classification, the initial data set is divided into training data (train dataset) and test data (test dataset). The data is divided at a ratio of 70/30 (70% for training dataset and 30% for testing dataset. 30% for testing dataset is taken randomly from the data set). These metrics will be used to evaluate the performance of the models.

Precision:

$$precision = \frac{TP}{TP + FP}$$

Recall

$$recall = \frac{TP}{TP + FN}$$

ACC:

$$ACC = \frac{2 \times precision \times recall}{precision + recall}$$

With the Scikit-learn library using the SVC function to initialize the model.

With the SVM model using a kernel function of the form 'kernel': ['rbf', 'poly', 'sigmoid', 'linear']. Apply algorithm in grid search engine GridSearchCV to find optimal parameters C and gamma. Where 'C': [0.1,1, 10, 100, 1000], 'gamma': [1,0.1.0.01,0.001,00001]. Then use the grid command. Best_estimator_ to find the best set of hyperparameters for the model:

{'C': 100, 'gamma': 0.01, 'kernel': 'rbf'}.
{'C': 0.1, 'gamma': 0.01, 'kernel': 'poly'}.
{'C': 0.1, 'gamma': 0.001, 'kernel': 'sigmoid'}.
{'C': 1000, 'gamma': 0.001, 'kernel': 'linear'}.

The results of running the SVM algorithm with kernel functions are shown in Table 1.

Table 2. Results of running SVM algorithm.

Kernel	Accuracy	Recall	Precision
RBF	**89%**	**78%**	**95%**
Poly	86%	76%	90%
Linear	87%	75%	92%
Sigmoid	71%	60%	73%

Table 2 show that the results are quite different when running the model with different kernel functions. Specifically, the RBF function gives the highest result (89%), followed by linear (87%), poly (86%) and the lowest is sigmoid (71%). The SVM algorithm was developed from the above algorithm by adding support vectors to create the separation domain between domains, so the results are higher than the other model [15, 16].

4 Conclusion

This article has presented the data classification method of SVM. Analysis and classification program for patients with vestibular disorders. The results obtained in the models are very promising, especially for the RBF model with high Precision (95%). The results achieved will be built to larger training data set, collect standard and combine more features with many methods. Other machine learning makes it possible to achieve even higher accuracy.

References

1. https://hahoangkiem.com/benh-than-kinh-tam-than/hoi-chung-tien-dinh-chan-doan-va-dieu-tri-1533.html. Last accessed 21 Nov 2022
2. Vestibular Disorders Association with Kelsey Hatton. Diagnostic Tests for Vestibular Disorders (2015)
3. Agrawal, Y., John, P.: Carey Disorders of Balance and Vestibular Function in US Adults, ©2009 American Medical Association. Vol. 169, no. 10, 25 May 2009
4. https://en.wikipedia.org/wiki/Vestibular_system. Last accessed 25 Jun 2022
5. https://en.wikipedia.org/wiki/Balance_disorder. Last accessed 25 Jun 2022
6. Agrawal, Y., Carey, J.P: Disorders of Balance and Vestibular Function in US Adults. American Medical Association, vol. 169, no. 10, 25 May 2009
7. Cohen, H.S., Kimball, K.T.: Usefulness of some current balance tests for identifying individuals with disequilibrium due to vestibular impairments. J. Vestib. Res. **18**(5–6), 295–303 (2008)
8. Agrawal, Y., Carey, J.P., Hoffman, H.J., Sklare, D.A., Schubert, M.C.: The modified Romberg Balance Test: normative data in U.S. adults. Otol. Neurotol. **32**, 1309–1311 (2011)
9. Huy, H.Q., et al.: A design of a Vestibular Disorder evaluation system. In: The 4th International Conference on Research in Intelligent and Computing in Engineering, pp. 1105–1117 (2019)
10. Vu, T.A., et al.: The models of Relationship Between Center of Gravity of Human and Weight, Height and 3 Body's Indicators (Chest, Waist and Hip). J. Sci. Technol. **139**, 57–61 (2019)
11. Vu, T.A., et al.: Predicting human's balance disorder based on center of gravity using support vector machine. In: ICCASA 2021. Lecture Notes of the Institute for Computer Sciences, Social Informatics and Telecommunications Engineering, vol. 409, pp. 38–47 (2021)
12. Vu, T.A., et al.: A novel fast-qualitative balance test method of screening for vestibular disorder patients. Indonesian J. Electr. Eng. Comput. Sci. **25**(2), 910–919 (2022)
13. Thao, N.T., Huyen, N.T., Ha, D.T.T., Huyen, T.T.T., Thuy, N.T.: Methods of using vectors and applications in bioinformatics. Sci. Technol. (2011)
14. https://figshare.com/articles/dataset/Data_Romberg/21218405
15. Rizzo, A.: Machine learning-based assessment tool for imbalance and vestibular dysfunction with virtual reality rehabilitation system. Comput. Methods Programs Biomed. **116**, 311–318 (2014)
16. Kabade, V., Hooda, R.: Machine learning techniques for differential diagnosis of vertigo and dizziness: a review. Adv. Mach. Learn. Tech. Biomed. Imaging Sens. Healthc. Appl. **21**(22), 7565 (2021)

Outage Constrained Robust Secure Transmission for a MISO SWIPT System

Phuong Anh Nguyen[1,2] and Anh Ngoc Le[1,2(✉)]

[1] Swinburne University of Technology, Melbourne, Australia
{phuonganhnguyen,nle}@swin.edu.au
[2] Swinburne Vietnam, FPT University, Hanoi, Vietnam
{anhnp75,ngocla2}@fe.edu.vn

Abstract. This paper investigates simultaneous wireless information and power transfer (SWIPT) in a multiple-input single-output (MISO) system for the secure beamforming design. Under knowledge of statistical channel state information to eavesdroppers and energy receivers, the considered problem aims to minimize the transmit power under the probabilistic constraints of outage secrecy rate and harvested energy. The problem is intractable to solve due to its nonconvexity nature and the statistical channel errors. Exploiting the special structure of these problems, we first transform the probabilistic ones into tractable form by the Bernstein inequalities and the non-convex constraints are transformed as DC (Difference of Convex functions) ones. The general DCA (DC Algorithm) based algorithms are developed to solve the proposed DC programs. Numerical results illustrate the efficiency and robustness of our proposed algorithms.

Keywords: DC programming · beamforming · physical layer secrecy · outage probability · SWIPT

1 Introduction

Simultaneous wireless information and power transfer (SWIPT)[7] system has shown pledge in breaking the wireless devices' battery capacity limitations. Potential eavesdroppers could be energy receivers, which are frequently placed closer to the transmitter than information receivers [10]. Thus, illegal receivers can get the transmitter's information. As a result, SWIPT systems have further strict security than traditional wireless networks. [8] proved that physical layer security can guarantee secure information transfer in wireless networks. Secrecy capacity, which quantifies the information transmission rate without information leaking to eavesdroppers, is one of the most crucial ideas in physical layer security [11,12]. With the aid of beamforming, a transmitter can steer its sig-

T. D. L. Nguyen et al. (Eds.): ICISN 2023, LNNS 752, pp. 91–97, 2023.
https://doi.org/10.1007/978-981-99-4725-6_13

nal in a certain direction. This will considerably increase the secrecy capacity based on beamforming. Channel models must be carefully studied while using beamforming.

In this article, we consider a practical script where eavesdroppers are not system users. We specifically investigate the MISO SWIPT system, where the transmitter has some imperfect knowledge of the energy receivers' channel and the eavesdroppers' channel is unknown. Specifically, the transmitter only knows the statistical channels of eavesdroppers and statistical channel errors of energy receivers. For this system, a secure beamforming scheme is developed with the PM problem.

We study the power minimization (PM) problem under probabilistic constraints of the secrecy rate and the harvested energy. To address this extremely difficult problem, we propose algorithms based on general DCA (DC Algorithm) [2–6] for solving it. The problem first is transformed into a general DC (Difference-of-Convex functions) program. However, this one contains exponential DC constraints. Thus, we study carefully its special structure, then, replace the exponential DC constraints by quadratic ones for which results in the second DC program. The efficient algorithms based on general DCA (called DCA_1 and DCA_2) are developed to deal with the corresponding DC problems. Numerical experiments validate the efficiency of our proposed algorithms.

2 System Model

We develop a secure beamforming scheme in a MISO system in which the transmitter has N antennas. $\mathbf{g}_s$, $\mathbf{g}_l$, $l \in M$, and $\mathbf{h}_i$, $i \in K$ respectively denote the channel vectors from the transmitter to legal information receiver, l-th power receiver, and i-th eavesdropper. $\mathbf{q}$ is the beamforming vector.

The channel of energy receivers and eavesdroppers are modeled as

$$\begin{aligned} \mathbf{g}_l = \overline{\mathbf{g}}_l + \Delta\mathbf{g}_l,\ \Delta\mathbf{g}_l \sim \mathcal{CN}(0, \mathbf{E}_l),\ l \in M, \\ \mathbf{h}_i \sim \mathcal{CN}(0, \mathbf{X}_i),\ i \in K. \end{aligned}$$

The secrecy rate, transmitted power at the transmitter, and received energy at the l-th energy receiver are written as

$$\begin{aligned} &R_r = \log\left(1 + \frac{|\mathbf{g}_s^H \mathbf{q}|^2}{\sigma_s^2}\right) - \max_i \log\left(1 + \frac{|\mathbf{h}_i^H \mathbf{q}|^2}{\sigma_{e,i}^2}\right), \\ &P_t = \|\mathbf{q}\|^2, \\ &E_{EH,l} = |\mathbf{g}_l^H \mathbf{q}|^2,\ l \in M. \end{aligned}$$

The PM problem with outage probability constraints of the secrecy rate and the harvested energy are investigated as follows.

$$\min_{\mathbf{q}} \|\mathbf{q}\|^2, \tag{1}$$

$$\text{s.t. } Pr\{|\mathbf{g}_l^H\mathbf{q}|^2 \geq \overline{E}\} \geq 1-\rho_l,\ l \in M, \tag{2}$$

$$Pr\Big\{\log\Big(1+\frac{|\mathbf{g}_s^H\mathbf{q}|^2}{\sigma_s^2}\Big) - \log\Big(1+\frac{|\mathbf{h}_i^H\mathbf{q}|^2}{\sigma_{e,i}^2}\Big) \geq R\Big\} \geq 1-p_{e,i},\ \forall i \in K. \tag{3}$$

3 Secure SWIPT of Beamforming Design

Reformulation of the Considered Problem. By applying two Bernstein [1,9] inequalities, the problem is reformulated as

$$\begin{aligned}
&\min_{\{a_l\},\{b_l\},\mathbf{q}} \|\mathbf{q}\|^2, \\
&\text{s.t. } b_l \geq 0,\ l \in M, \\
&\quad \mathrm{Tr}\{\mathbf{E}_l^{\frac{1}{2}}\mathbf{q}\mathbf{q}^H\mathbf{E}_l^{\frac{1}{2}}\} - \sqrt{2\ln(1/\rho_l)}a_l + \ln(\rho_l)b_l + \overline{\mathbf{g}}_l^H\mathbf{q}\mathbf{q}^H\overline{\mathbf{g}}_l - \overline{E} \geq 0, \\
&\quad \sqrt{\mathbf{q}^H\mathbf{E}_l\mathbf{q}\mathbf{q}^H(\mathbf{E}_l + 2\overline{\mathbf{g}}_l\overline{\mathbf{g}}_l^H)\mathbf{q}} \leq a_l,\ l \in M, \\
&\quad \sigma_{e,i}^2\Big(1-\frac{1}{2^R}\Big) - \mathbf{q}^H\Big(\mathbf{X}_i\ln p_{e,i} + \frac{\sigma_{e,i}^2}{\sigma_s^2 2^R}\mathbf{g}_s\mathbf{g}_s^H\Big)\mathbf{q} \leq 0,\ i \in K.
\end{aligned} \tag{4}$$

The first DCA scheme. The problem (4) is rewritten as

$$\min_{\{a_l\},\{b_l\},\{n_l\},\{m_l\},\mathbf{q}} \|\mathbf{q}\|^2, \tag{5}$$

$$\text{s.t. } b_l \geq 0,\ l \in M, \tag{6}$$

$$2^{n_l+m_l} \leq a_l,\ l \in M, \tag{7}$$

$$\mathbf{q}^H\mathbf{E}_l\mathbf{q} - 2^{2n_l} \leq 0,\ l \in M, \tag{8}$$

$$\mathbf{q}^H(\mathbf{E}_l + 2\overline{\mathbf{g}}_l\overline{\mathbf{g}}_l^H)\mathbf{q} - 2^{2m_l} \leq 0,\ l \in M, \tag{9}$$

$$\sqrt{2\ln(1/\rho_l)}a_l - \ln(\rho_l)b_l + \overline{E} - \mathbf{q}^H(\mathbf{E}_l + \overline{\mathbf{g}}_l\overline{\mathbf{g}}_l^H)\mathbf{q} \leq 0,\ l \in M, \tag{10}$$

$$\sigma_{e,i}^2\Big(1-\frac{1}{2^R}\Big) - \mathbf{q}^H\Big(\mathbf{X}_i\ln p_{e,i} + \frac{\sigma_{e,i}^2}{\sigma_s^2 2^R}\mathbf{g}_s\mathbf{g}_s^H\Big)\mathbf{q} \leq 0,\ i \in K. \tag{11}$$

Define $\mathbf{Y}_{s,i} = \mathbf{X}_i \ln p_{e,i} + \frac{\sigma_{e,i}^2}{\sigma_s^2 2^R}\mathbf{g}_s\mathbf{g}_s^H$, $\mathbf{x} = (\{a_l\}, \{b_l\}, \{n_l\}, \{m_l\}, \mathbf{q})$. We propose a DC program for the reformulated problem as follows.

$$\min_{\mathbf{x}} \|\mathbf{q}\|^2, \tag{12}$$
$$\begin{aligned}
\text{s.t. } & (6)\text{–}(7),\\
& S_{1l}(\mathbf{x}) - T_{1l}(\mathbf{x}) \le 0, S_{1l}(\mathbf{x}) = \mathbf{q}^H\mathbf{E}_l\mathbf{q},\ T_{1l}(\mathbf{x}) = 2^{2n_l},\\
& S_{2l}(\mathbf{x}) - T_{2l}(\mathbf{x}) \le 0, S_{2l}(\mathbf{x}) = \mathbf{q}^H(\mathbf{E}_l + 2\overline{\mathbf{g}}_l\overline{\mathbf{g}}_l^H)\mathbf{q},\ T_{2l}(\mathbf{x}) = 2^{2m_l},\\
& S_{3l}(\mathbf{x}) - T_{3l}(\mathbf{x}) \le 0,\ l \in M,\\
& S_{3l}(\mathbf{x}) = \sqrt{2\ln(1/\rho_l)}a_l - \ln(\rho_l)b_l + \overline{E},\ T_{3l}(\mathbf{x}) = \mathbf{q}^H(\mathbf{E}_l + \overline{\mathbf{g}}_l\overline{\mathbf{g}}_l^H)\mathbf{q},\\
& S_{4i}(\mathbf{x}) - T_{4i}(\mathbf{x}) \le 0, S_{4i}(\mathbf{x}) = \sigma_{e,i}^2\Big(1 - \frac{1}{2^R}\Big),\ T_{4i}(\mathbf{x}) = \mathbf{q}^H\mathbf{Y}_{s,i}\mathbf{q}.
\end{aligned}$$

We develop the DCA$_1$ scheme for solving the problem (12) as follows.

Algorithm 1.

Initialization: Let $(\mathbf{x}^0)$ be a first point, $v \leftarrow 0$.
Repeat
1. Solve the following convex subproblem with $\mathbf{x}^{v+1}$

$$\begin{aligned}
& \min_{\mathbf{x}} \|\mathbf{q}\|^2,\\
\text{s.t. } & (6)\text{–}(7),\\
& S_u(\mathbf{x}) - \overline{T}_u(\mathbf{x}^v) \le 0, u = \{1l, 2l, 3l, 4i\},\ l \in M,\ i \in K.
\end{aligned}$$

2. $v \leftarrow v + 1$.

Until Convergence.

This first DC formulation contains exponential DC constraints. Thus, we develop another DC program reformulate to avoid exponential constraints as follows.

The Second DCA Scheme. The problem (4) is rewritten as

$$\min_{\mathbf{x}} \|\mathbf{q}\|^2, \tag{13}$$
$$\text{s.t. } n_l \ge 0,\ m_l \ge 0,\ b_l \ge 0,\ a_l \ge 0,\ l \in M, \tag{14}$$
$$\mathbf{q}^H\mathbf{E}_l\mathbf{q} \le n_l,\ l \in M, \tag{15}$$
$$\mathbf{q}^H(\mathbf{E}_l + 2\overline{\mathbf{g}}_l\overline{\mathbf{g}}_l^H)\mathbf{q} \le m_l,\ l \in M, \tag{16}$$
$$\sqrt{2\ln(1/\rho_l)}a_l - \ln(\rho_l)b_l + \overline{E} - \mathbf{q}^H(\mathbf{E}_l + \overline{\mathbf{g}}_l\overline{\mathbf{g}}_l^H)\mathbf{q} \le 0,\ l \in M, \tag{17}$$
$$\sigma_{e,i}^2\Big(1 - \frac{1}{2^R}\Big) - \mathbf{q}^H\mathbf{Y}_{s,i}\mathbf{q} \le 0,\ i \in K,\ s \in \mathcal{S}, \tag{18}$$
$$n_l.m_l - a_l^2 \le 0,\ l \in M. \tag{19}$$

The DC constraints for (17)–(19) are expressed as follows.

$$
\begin{aligned}
&\min_{\mathbf{x}} \|\mathbf{q}\|^2, \\
\text{s.t. } &(14)-(16), \\
&S_{3l}(\mathbf{x}) - T_{3l}(\mathbf{x}) \le 0,\ l \in M, \\
&S_{4i}(\mathbf{x}) - T_{4i}(\mathbf{x}) \le 0,\ i \in K, \\
&S_{6l}(\mathbf{x}) - T_{6l}(\mathbf{x}) \le 0, S_{6l}(\mathbf{X}) = \frac{1}{2}(n_l + m_l)^2,\ T_{6l}(\mathbf{x}) = \frac{1}{2}(n_l^2 + m_l^2) + a_l^2.
\end{aligned} \tag{20}
$$

DCA_2 scheme is proposed to deal with the second DC progam as follows.

Algorithm 2.

Initialization: initialize $(\mathbf{x}^0)$ as an initial point, $v \leftarrow 0$.
Repeat

1. Solve the following convex subproblem with $\mathbf{x}^{v+1}$

$$
\begin{aligned}
&\min_{\mathbf{x},s} \|\mathbf{q}\|^2 + t_0 s_0 + t_1 s_1, \\
\text{s.t. } &(14)-(16), \\
&S_u(\mathbf{x}) - \overline{T}_u(\mathbf{x}^v) \le 0, u = \{3l, 4i, 6l\},\ l \in M,\ i \in K.
\end{aligned}
$$

2. $v \leftarrow v + 1$.

Until Convergence.

---

4 Numerical Experiments

The effect of the secrecy rate, received power, and number of transmit antenna are investigated. We implement the algorithms on 10 independent channel realizations with the large scale fading and small scale fading model.

Figure 1 shows that the received power increases as the transmitted power increases. Figure 2 presents that the increase of secrecy rate causes the transmit power increasing. Figure 3 shows that the transmit power declines with the increasing the transmit-antenna number.

The effect of DC programs is evaluated by the comparison between DCA_1 and DCA_2. DCA_2 is better than DCA_1 in terms of the transmit power and CPU time. Specifically, the gain of DCA_2 versus DCA_1 is from 29.1% to 32.3% (Fig. 1) in terms of the transmit power. DCA_2 runs faster than DCA_1. The ratio of gain is from 5.1 to 5.9 times for CPU time (Fig. 2). The reason is because the convex subproblems of DCA_1 have exponentially constraints while that of DCA_2 have quadratic ones.

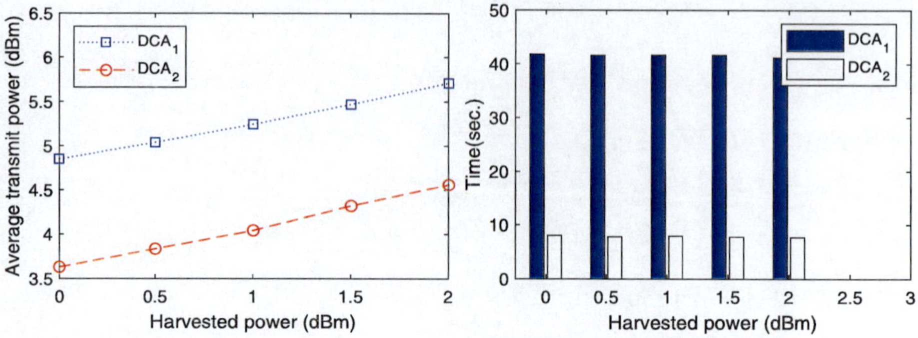

Fig. 1. Received power versus transmitted power and CPU(s)

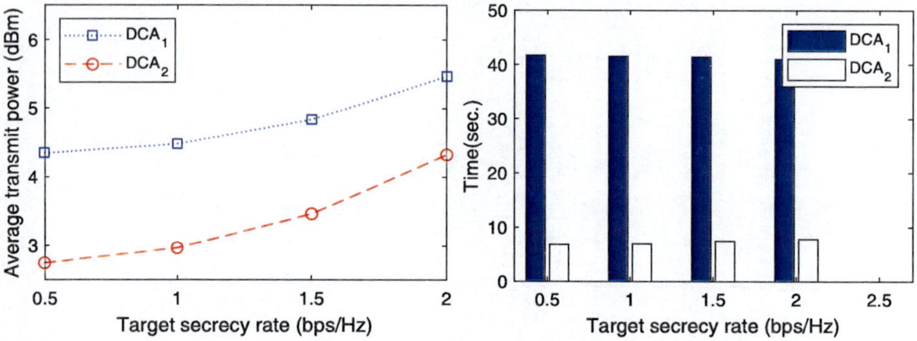

Fig. 2. Secrecy rate versus transmitted power and CPU(s)

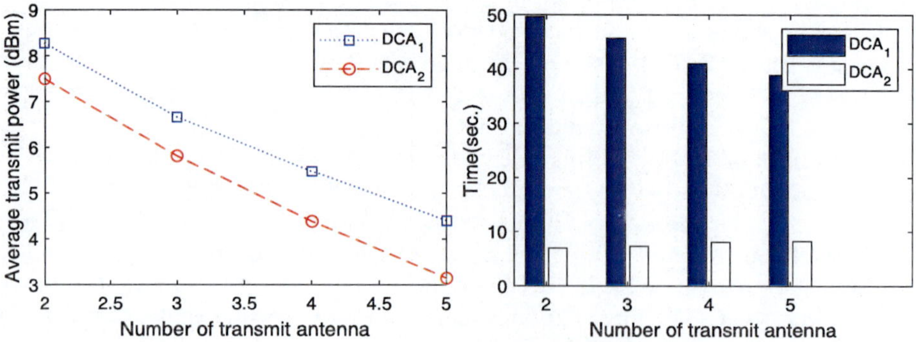

Fig. 3. Number of antenna versus transmitted power and CPU(s)

5 Conclusion

This paper studied a beamforming strategy in MISO SWIPT system, where the eavesdropper's statistical CSI and the statistical channel errors of energy receivers are known. We investigated the PM problem with the probability constraints of the required secrecy rate and energy harvesting. The considered problem is extremely difficult due to its nonconvexity nature. The considered problem was first transformed into a DC problem, which results in DCA_1 approach. However, the first DC problem leads to convex subproblems with exponential constraints. Then, the new DC program was studied to replace the exponential DC constraints by quadratic ones. DCA_2 algorithm was proposed for solving this second DC program. Numerical results verified the effectiveness of the proposed schemes on both quality and CPU time.

References

1. Huang, Y., Tan, C.W., Rao, B.D.: Outage balancing in multiuser MISO networks: network duality and algorithms. In: 2012 IEEE Global Communications Conference (GLOBECOM), pp. 3918–3923 (2012)
2. Le Thi, H.A., Pham Dinh, T.: The DC (Difference of convex functions) programming and DCA revisited with DC models of real world nonconvex optimization problems. Ann. Oper. Res. **133**(1), 23–46 (2005)
3. Le Thi, H.A., Pham Dinh, T.: DC programming and DCA: thirty years of developments. Math. Program. **169**(1), 5–68 (2018). https://doi.org/10.1007/s10107-018-1235-y
4. Pham Dinh, T., Le Thi, H.A.: Convex analysis approach to D.C. programming: theory, algorithm and applications. Acta Mathematica Vietnamica **22**(1), 289–355 (1997)
5. Pham Dinh, T., Le Thi, H.A.: D.C. optimization algorithms for solving the trust region subproblem. SIAM J. Optim. **8**(2), 476–505 (1998)
6. Pham Dinh, T., Le Thi, H.A.: Recent advances in DC programming and DCA **8342**, 1–37 (2014)
7. Varshney, L.R.: Transporting information and energy simultaneously. In: 2008 IEEE International Symposium on Information Theory, pp. 1612–1616 (2008)
8. Wyner, A.D.: The wire-tap channel. Bell Syst. Tech. J. **54**(8), 1355–1387 (1975)
9. Yuan, Y., Ding, Z.: Secrecy outage design in MIMO-SWIPT systems based on a non-linear EH model. In: 2017 IEEE Globecom Workshops (GC Wkshps), pp. 1–6 (2017)
10. Yuan, Y., Ding, Z.: Outage constrained secrecy rate maximization design with SWIPT in MIMO-CR systems. IEEE Trans. Veh. Technol. **67**(6), 5475–5480 (2018)
11. Zou, Y., Wang, X., Shen, W.: Eavesdropping attack in collaborative wireless networks: security protocols and intercept behavior. In: Proceedings of the 2013 IEEE 17th International Conference on Computer Supported Cooperative Work in Design (CSCWD), pp. 704–709 (2013)
12. Zou, Y., Wang, X., Shen, W.: Intercept probability analysis of cooperative wireless networks with best relay selection in the presence of eavesdropping attack. In: 2013 IEEE International Conference on Communications (ICC), pp. 2183–2187 (2013)

Check for updates

Framework for Digital Academic Records Management Using Blockchain Technology

Hung Ho-Dac[1], Len Van Vo[1], Bao The Nguyen[1], Cuong Hai Vinh Nguyen[1], Phuong Cao Hoai Nguyen[1], Chien Khac Nguyen[2], Son Le Pham[3], and Huu Van Tran[1(✉)]

[1] Thu Dau Mot University, Thu Dau Mot, Binh Duong, Vietnam
{hunghd,lenvv,baont,cuongnhv,phuongnch,huutv}@tdmu.edu.vn
[2] People's Police University, Ho Chi Minh City, Vietnam
[3] Binh Duong Center of Natural Resources and Environment Technical – Monitoring, Ho Chi Minh City, Vietnam

Abstract. The application of blockchain to centralized systems is one of the trends of improving the privacy, security, and transparency of the centralized approach. Blockchain is especially meaningful in trustless systems, where stakeholders are susceptible to tampering and consensus is required. In this work, we propose a practical framework for academic records management so that stakeholders can exploit records effectively. Based on the framework, we also conduct feasibility testing. The experiment is conducted at the computer lab I1.501 in the Thu Dau Mot University (TDMU). The results show that it is possible to apply blockchain to the academic records management system. The source code is available in Github with an open source license to make sure that institute can freely use this framework or modify to make it more suitable (https://github.com/vanhuudhsp/AcademicRecordsBlockChain).

Keywords: Blockchain · Academic Records · Technology Framework

1 Introduction

Most current systems have a centralized client-server design. In these systems, a client (user) has the ability to modify data that is stored on the centralized computer. The centralized authority is in charge of the server database as a whole and has control over the different access control policies that have been established for the information kept in the database. Users have the authority to validate their credentials before gaining entry to the database. A possible solution to the issues with traditional centralized systems is blockchain technology.

A blockchain is a network of connected blocks that can be used to distribute, openly exchange, and securely store data. Each block holds data and connects to other blocks using pointers. The integrity of blockchain and resistance to tampering are ensured by these links. Each time new data is added, the blockchain is expanded by one block or unit by establishing a connection to the free end. The blockchain grows longer and

T. D. L. Nguyen et al. (Eds.): ICISN 2023, LNNS 752, pp. 98–102, 2023.
https://doi.org/10.1007/978-981-99-4725-6_14

wider as new information is added to it. If one of the blocks in the chain is altered, the cryptographic connections are destroyed, disrupting the complete blockchain. The stored data's accuracy can also be verified by the individual.

This work proposes a framework for a blockchain-based academic records management system that will preserve a students' academic history. Along with the primary purpose, it also implements the ability to track each participant's transactional accomplishments. This work also discusses the design of a blockchain-based system architecture. Additionally, certain fundamental smart contracts that are created using the academic flow respecting business logic are also proposed in this work.

The organization of this work structured as follows: Sect. 2 reviews the relevant works. Section 3 presents the suggested approach. The proof-of-concept experiment is presented in Sect. 4. In Sect. 5, we wrap up our work and discuss about future work.

2 Related Work

There are numerous studies demonstrating the use of blockchain in different fields: education [1], human resource management [2], Internet of Things [3], healthcare [4], entrepreneurship and innovation [5], cyber-security [6].

Regarding the implementation in education, according to Rodelio Arenas et al. [7], the solution is built on the open source multi-chain, a permissioned blockchain technology. For the secrecy of the academic documents, the suggested Credence Ledger provides streaming over many chains. The suggested approach can grant easily verifiable digital credentials for students without requiring them to be exchanged on a public blockchain using a coin. It also gives third parties, such employers a chance to confidentially and independently confirm proofs of shared academic achievement. Arif Rachmat et al.'s [8] innovative blockchain-based technique for creating a setting in which individuals can be the supervisors of their official educational records and can easily share those data with others is presented in another work. By incorporating blockchain's decentralized data storage, the solution might make it straightforward for the involved individual or organization to determine who has permission to access the relevant parties' data.

3 Proposed Approach

3.1 Architecture

Based on multi-tier architecture, we propose a design with basic layers including but not limited to application layer, business layer and data layer (Fig. 1). This design will serve the user groups related to the academic records management system. The components are described in detail as follows:

- User groups related to the system include but not limited to students, lecturers, managers, enterprises, service providers, and public authorities. Each user can take the role as a node in the blockchain network.
- Applications pool contains application related education system include but not limited to student portal, lecture portal, university information system, education management system, learning management system, e-portfolio system. Applications pool may contains some queuing application to proactively receive data (Apache Kafka).

- Academic flow services basically are smart-contracts (Hyperledger Fabric) which implement the actual academic flows of the institution.
- Data services include internal nodes (devices of students, lecturers, supporting staffs, managers; and dedicated servers) and external nodes (devices of enterprises, service providers, and public authorities).

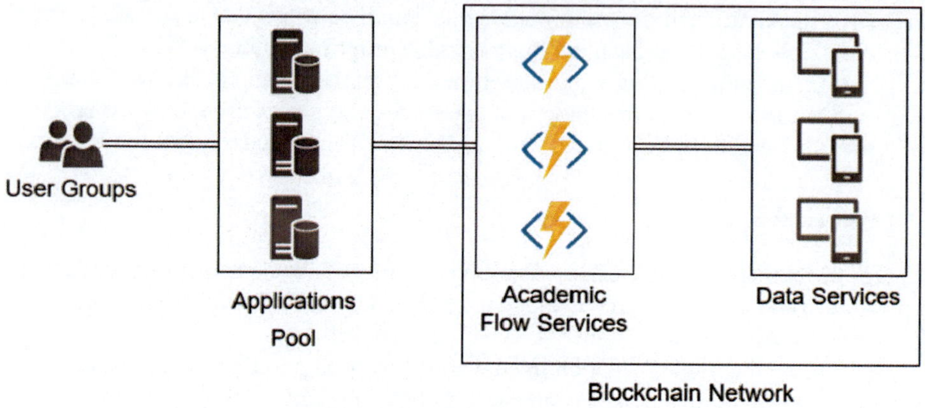

Fig. 1. Proposed overall design

In design, blockchain acts as both a place to store data in the form of NoSQL and as a place to execute operations in the form of smart-contracts. This is because modern blockchains are all enterprise blockchains and have evolved, not merely transaction storage like the first version. Node spread and consensus mechanism between nodes is the key to ensure privacy, security, and transparency [9]. Smart-contracts can include multiple parties to confirm the constraints, which will clearly improve fraud resistance.

3.2 Academic Flow

Academic flows heavily rely on the smart-contracts. Each institution will have different academic flows but they will share the basic ones. In this work, based on the actual situation at TDMU, we propose the basic lines as shown in Table 1.

Table 1. TDMU academic flows

#	Flow	Type
1	InitAcademicRecord	invoke
2	UpdateGrade	invoke

(*continued*)

Table 1. (*continued*)

#	Flow	Type
3	VerifyAcademicRecord	invoke
4	RetrieveAcademicRecord	query
5	InitCertificate	invoke
6	SignCertificate	invoke
7	VerifyCertificate	invoke
8	RetrieveCertificateInfo	query
9	InitDegree	invoke
10	SignDegree	invoke
11	VerifyDegree	invoke
12	RetrieveDegreeInfo	query
13	InitNode	invoke
14	VerifyNode	invoke
15	RetrieveNodeInfo	query
16	InitFlow	invoke
17	ApproveFlow	invoke
18	VerifyFlow	invoke
19	RetrieveFlow	query

4 Experiment

We conducted a proof-of-concept experiment on the I1.501 lab system at Thu Dau Mot University. The lab has 49 PCs with I5 gen 7th CPU, 8 GB RAM, and 200 GB SDD. In experiment, we set up based on the design with a Hyperledger Fabric blockchain including 20 nodes to run the following flows InitAcademicRecord, UpdateGrade, VerifyAcademicRecord, and RetrieveAcademicRecord. The latency demonstrated in the Table 2.

Table 2. Latency of academic flows

#	Flow	Type	Min. Lat. (second)	Max. Lat. (second)	Avg. Lat. (second)
1	InitAcademicRecord	invoke	1.1	1.9	1.7
2	UpdateGrade	invoke	2.1	2.4	2.2
3	VerifyAcademicRecord	invoke	3.4	3.7	3.5
4	RetrieveAcademicRecord	query	0.5	1.1	0.8

The flow RetrieveAcademicRecord basically read the data from hyperledger at nodes without performing any update operation so it takes smallest average latency. The flow InitAcademicRecord and UpdateGrade perfume update operation so it takes higher average latency. And the flow VerifyAcademicRecord perform multi-stakeholder verification including external systems so it takes the highest average latency. The average latency is less than 5 s so this result is acceptable in practice.

5 Conclusion

Applying blockchain to the academic records management system is an approach that improve the privacy, security, and transparency. Deployment does not require too many devices or technology. Experimental results show the feasibility of the approach with the acceptable latency.

Acknowledgments. This research is funded by Thu Dau Mot University under grant number DT.22.1-009.

References

1. Sun, H., Wang, X., Wang, X.: Application of blockchain technology in online education. Int. J. Emerg. Technol. Learn. **13**(10), 252 (2018)
2. Onik, M.M.H., Miraz, M.H., Kim, C.-S.: A recruitment and human resource management technique using blockchain technology for industry 4.0. In: Smart Cities Symposium 2018. IET (2018)
3. Kumar, N.M., Mallick, P.K.: Blockchain technology for security issues and challenges in IoT. Proc. Comput. Sci. **132**, 1815–1823 (2018)
4. Farouk, A., et al.: Blockchain platform for industrial healthcare: vision and future opportunities. Comput. Commun. **154**, 223–235 (2020)
5. Chen, Y.: Blockchain tokens and the potential democratization of entrepreneurship and innovation. Bus. Horiz. **61**(4), 567–575 (2018)
6. Fernandez-Carames, T.M., FragaLamas, P.: A review on the application of blockchain to the next generation of cybersecure industry 4.0 smart factories. IEEE Access **7**, 45201–45218 (2019). https://doi.org/10.1109/ACCESS.2019.2908780
7. Arenas, R., Fernandez, P.: CredenceLedger: a permissioned blockchain for verifiable academic credentials. In: 2018 IEEE International Conference on Engineering, Technology and Innovation (ICE/ITMC). IEEE (2018)
8. Rachmat, A.: Design of distributed academic-record system based on blockchain. In: 2019 International Conference on ICT for Smart Society (ICISS), vol. 7. IEEE (2019)
9. Ho-Dac, H., Tran, V.H., Nguyen, T.B., Nguyen, H.V.C., Vo, V.L.: Blockchain in enterprise applications: an introduction. In: Le Anh, N., Koh, S.-J., Nguyen, T.D.L., Lloret, J., Nguyen, T.T. (eds.) Intelligent Systems and Networks: Selected Articles from ICISN 2022, Vietnam, pp. 121–129. Springer Nature, Singapore (2022). https://doi.org/10.1007/978-981-19-3394-3_15

Novel Forest Height Extraction Method Based on Neuman Volume Scattering Model from PolInSAR Images

HuuCuong Thieu[1], MinhNghia Pham[2(✉)], NgocTan Nguyen[3], VanDung Nguyen[2], and DucHoc Tran[1]

[1] Telecommunications University, Khanh Hoa, Vietnam
[2] Le Quy Don Technical University, Ha Noi, Vietnam
nghiapm2018@mta.edu.vn
[3] CMC University, Ha Noi, Vietnam

Abstract. This article nominates a new manner based upon the Neuman volume scattering model to get better the correctness of forest elevation calculation using polarized interference UAV-SAR images. The forest parameters are extracted by the suggested solution which are carried out through 3 steps. First, the contribution of the scattering components is determined based on the Neuman volume scattering model. Next, constraint conditions are added to determine the optimal complex polarimetry interferometry coherence (CPI) coefficient for ground and volume scattering components. Then the sum of least squares algorithm is used to calculate the ground phase. Finally, the forest parameters is directly restored based on the optimal iterative method. The efficiency of the newly mentioned manner is assessed with UAV-SAR data received from the AfiSAR project of NASA/JPL.

Keywords: Forest height · complex polarimetry interferometry coherence (CPI) · RVoG model · surface phase

1 Introduction

Forest elevation is one of the essential specifications for assessing forest growth and especially it is an important parameter for supervising and conserving forest resources. PolInSAR technique has been showing outstanding advantages in determining forest height. Based on the advantages of PolInSAR, a large number of manners have been presented for forest parameter estimation. In general, these approaches can be classified into two main groups: (1) group based on RVoG model; (2) grouping based on target decomposition technical. For the forest parameter estimation group based on the target decomposition technical [1,3,7], although there have been certain successes, the forest parameter extraction is still based on two assumptions: (1) the scattered entities in the tree canopy are modeled shaped as bipolar cylindrical scatterers; (2) these scattered

T. D. L. Nguyen et al. (Eds.): ICISN 2023, LNNS 752, pp. 103–111, 2023.
https://doi.org/10.1007/978-981-99-4725-6_15

entities have a random distribution. In practice, the interaction process between radar waves and tree canopy is extremely complex, so the above two assumptions become inappropriate when modeling wave scattering with different types of tree canopy.

Another direction that is being prioritized for research and development at present is to develop methods for converting forest elevations derived from the 2-layer RVoG model [2,4,6]. This solution illustrates the scattering of ultrasonic waves in the natural environment as a mass of randomly oriented scattering objects placed on a relatively flat land surface. In 2018, Tayebe and colleagues proposed a volume optimization manner to enhance the effectiveness of estimating forest height of the 3-SI algorithm (VoM) [4]. However, this manner still uses inappropriate assumptions leading to the forest elevation extraction get error.

To overcome the above limitations, this article nominates a new manner for forest parameter calculation using the PolInSAR images. This manner is based upon combining the advantages of the adaptive 2-component decomposition technique with a general volume scattering model (GVM) and the parametric inversion technique based on the RVoG model. First, a GVM based upon the Neuman scattering model [6] is applied. In which, the parameters of the canopy are determined directly through the non-linear total least squares method and the interaction process between the radar wave and the canopy is restored through the power analysis of the components scattering. Next, the CPI coefficients for the mainland and volume scattering ingredients are determined through the search for an optimal polarization coefficient. Finally, the forest elevation is determined based upon the iterative optimization method.

2 Methodology

2.1 Determine Parameter of Volume and Ground Scattering Components

a. General volume scattering model

The data of PolInSAR system can be presented by a 6-dimension complex matrix for each pixel in the scene [5], given description in (1).

$$[T] = \langle \mathbf{k}.\mathbf{k}^{*T} \rangle = \begin{bmatrix} T_1 & \Omega \\ \Omega^{*T} & T_2 \end{bmatrix} \quad ; \quad \mathbf{k} = \begin{bmatrix} \mathbf{k}_1 \\ \mathbf{k}_2 \end{bmatrix} \tag{1}$$

The parameters of the matrix T are shown in the reference [5]. In practice, the scattering of polarized radar waves with forest canopy is extremely complex. Therefore, a common volume scattering model (GVM) that can accurately reflect the architecture and spatial distribution of the canopy of forest is essential. According in [6], a common volume scattering model is developed in which the parameters of the model can be adapted to each structure and spatial distribution of different forest canopies.

$$T_v = f_v \begin{bmatrix} 1 & X_{22}\delta & 0 \\ X_{22}\delta^* & X_{22}|\delta|^2 & 0 \\ 0 & 0 & (1 - X_{22})\,|\delta|^2 \end{bmatrix} \tag{2}$$

where $X_{22} \in (0,1)$ related to the random distribution of the target. When $X_{22} \to 1$, the scattering objects in the canopy have a uniform distribution and vice versa, they have a completely random distribution. δ is a complex number representing the anisotropy of the scatterer. In which, $|\delta| \to 0$ represent the target with disc shape, $|\delta| \to 1$ represent the target with dipole form and $|\delta| > 1$ represent the target with the form of dihedral structure. f_v represents the energy of the volume scattering ingredient.

b. Parameter estimation based on nonlinear total least squares algorithm

Based on the Freman decomposition technique [3], the coherence matrix $T = ((T_1 + T_2)/2)$ of the PolInSAR data is able to analyze into the sum of two submatrix corresponding to the volume and mainland scattering elements (3).

$$T = T_v + T_g \tag{3}$$

In which, T_v, T_g is the coherence matrix of the volume and mainland scattering elements, respectively.

In fact, the surface scattering component in the forest environment includes the direct scattering from the mainland and the scattering between the tree trunk and the mainland. Therefore, in order to best reflect this scattering process, a surface scattering model with constraints based on scattering power is developed. In the case of surface scattering plays a dominant role, where $T_{11} - T_{22} > 0$ with T_{ii} is the row i column i component in the matrix T_1. When, the coherence matrix of the surface scattering component has the following form:

$$T_g = \begin{bmatrix} f_g & f_g\beta^*\cos 2\theta & -f_g\beta^*\sin 2\theta \\ f_g\beta\cos 2\theta & f_d + f_g|\beta|^2\cos^2 2\theta & -\frac{1}{2}f_g|\beta|^2\sin 4\theta \\ -f_g\beta\sin 2\theta & -\frac{1}{2}f_g|\beta|^2\sin 4\theta & f_g|\beta|^2\sin^2 2\theta \end{bmatrix} \tag{4}$$

where f_g, f_d are the coefficients related to the scattering power of the scattering ground and the dihedral scattering components, respectively. θ is the orientation angle of the surface scattering and β is the parameter representing the properties of the surface. In contrast, when $T_{11} - T_{22} < 0$, the dihedral scattering component plays a dominant role, the surface scattering matrix is determined by formula (5).

$$T_g = \begin{bmatrix} f_g + f_d|\alpha|^2 & f_d\alpha\cos 2\theta & -f_d\alpha\sin 2\theta \\ f_d\alpha^*\cos 2\theta & f_d\cos^2 2\theta & -\frac{1}{2}f_d\sin 4\theta \\ -f_d\alpha^*\sin 2\theta & -\frac{1}{2}f_d\sin 4\theta & f_d\sin^2 2\theta \end{bmatrix} \tag{5}$$

where α is a parameter related to the reflectance of radar waves at the surface of the soil and tree trunks. Then expression (3) can be re-expressed as (6).

$$T = T_v + T_g + T_{residual} \tag{6}$$

Equation (6) is a system of non-linear equations with 9 real unknowns and 8 equations. To solve this problem, the proposed algorithm uses the optimal non-linear total least squares manner. The optimal parameters are determined based on the following condition:

$$\min = \|T_{residual}\|_2 = \|T - T_v - T_g\|_2 \tag{7}$$

From the optimal values determined in (7) it is possible to determine the matrices T_g and T_v the corresponding ones.

2.2 Determination of Surface Phase

The Ω is the non-Hermitain complex matrix [2] that encompasses information about the change of the interference phase and the polarization states of the object. When, cross-covariance matrix Ω can be analyzed as (8).

$$\Omega = \Omega_v + \Omega_g = e^{j\phi_v} T_v + e^{j\phi_g} T_g \tag{8}$$

where, ϕ_g is the center phase of the surface scattering element and ϕ_v is the center phase of the direct one from the tree canopy.

Most of the previous surface phase estimation algorithms [2,4] assume that the CPI coefficient of the ground scattering is always equal to 1. However, the scattering of microwaves in the nature forest environment is relatively complex and so this assumption often causes errors. To overcome this limitation as well as improve the accuracy in surface phase estimation, this paper proposes a method to eliminate the scattering mechanism to determine the CPI coefficient of the optimal surface scattering component. In the PolInSAR system [5], the polarization CPI is expressed as (9).

$$\tilde{\gamma}(\omega_1, \omega_2) = \frac{\omega_1^H \Omega \omega_2}{\sqrt{\omega_1^H T_1 \omega_1 \omega_2^H T_2 \omega_2}} = \tilde{\gamma}_v(\omega_1, \omega_2) + \tilde{\gamma}_g(\omega_1, \omega_2) \tag{9}$$

Expression (9) shows that if a pair of properly polarized vectors (ω_1, ω_2) is selected, the volume scattering component can be completely eliminated. When, $\tilde{\gamma}(\omega_1, \omega_2) \approx \tilde{\gamma}_g(\omega_1, \omega_2)$ that means:

$$\omega_1^H \Omega_v \omega_2 = e^{j\phi_v} \omega_1^H T_v \omega_2 = 0 \tag{10}$$

The optimal pair of polarization vectors (ω_1, ω_2) is selected based on the analysis of the eigenvalues of the matrix T_v. Then, this optimal pair of vectors is specifically defined as (11).

$$\omega_{1,2} = \begin{bmatrix} \frac{t_{V22} - t_{V11} \pm \sqrt{(t_{V22} - t_{V11})^2 + 4|t_{V12}|^2}}{2t_{V12}} & 1 & 0 \end{bmatrix}^T \tag{11}$$

In which $t_{V\mathrm{ij}}$, $(i, j = 1, 2)$ corresponds to column i, row j of the matrix T_v. When, the CPI coefficient representing the surface scattering component is determined according to equation (12).

$$|\tilde{\gamma}_g(\omega_1, \omega_2)| = \left| \frac{\omega_1^H \Omega \omega_2}{\sqrt{\omega_1^H T_1 \omega_1}\sqrt{\omega_2^H T_2 \omega_2}} \right| \tag{12}$$

After determine $|\tilde{\gamma}_g|$ the total least squares was used to determine the surface phase. ϕ_g is defined as the intersection point between the radius circle $|\tilde{\gamma}_g|$ and the line-fit where that point has the greatest distance to reach $\tilde{\gamma}_{HV}$.

3 Forest Height Determination Based on Optimal Iterative Method

The HV polarization channel of PolInSAR is the one that has a great contribution in the CPI factor of the direct scattering element from the canopy [2,4,6]. Therefore, the HV polarization channel is often used for forest elevation calculation and this value is determined as (13).

$$\tilde{\gamma}_{HV} = e^{j\phi_g} \frac{\tilde{\gamma}_v + |\tilde{\gamma}_g| m(\omega)}{1 + m(\omega)} \tag{13}$$

where $m(\omega)$ is the ratio between volume and ground scattering, $\tilde{\gamma}_v$ is the CPI factor for the direct scattering element from the canopy [2]. When, (13) can be restated as follows:

$$m(\omega) = \frac{\tilde{\gamma}_v - e^{-j\phi_g}\tilde{\gamma}_{HV}}{e^{-j\phi_g}\tilde{\gamma}_{HV} - |\tilde{\gamma}_g|} \tag{14}$$

According to radar theory, the interference phase of the component scattered directly from the canopy will be on the top of the tree top. However, in practice this component is usually located somewhere in the canopy. Moreover, the optimal CPI factor of the direct scattering component from the canopy will have the tiniest contribution of the other scattering components. Therefore, a proposed optimal iterative approach is applied to get better the accuracy of forest elevation calculation. By letting the parameter h_v , σ vary within its respective range $h_v \in \left(0 \div \frac{2\pi}{k_z}\right)$ and $\sigma \in (0 \div 1)$. Then, the forest height h_v and average mean extinction σ will be determined according to the condition (15).

$$\|m(\omega)\|_{\min} \tag{15}$$

4 Experimental Results

In this part, the new approach and VoM [4] will be applied to evaluate forest parameters for the PolInSAR dataset collected by the unmanned aerial vehicle (UAV-SAR) of NASA/JPL belonging to the project the AfriSAR project. The radar operates at L-band, baseline (0–160) (m) and angle of incidence

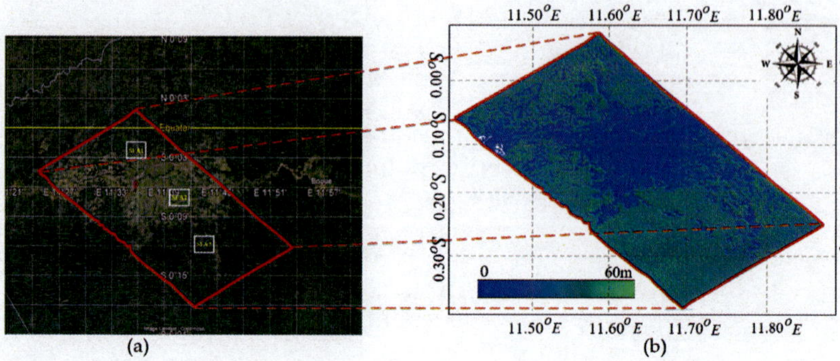

Fig. 1. Study forest area: (a) optical image, (b) Li-DAR image.

$(21^o - 65^o)$. This dataset supplies PolInSAR images for tree elevation calculation in Lope National Park, Gabon, Africa. Both data sets were measured on February 25, 2016 by NASA's coordination with ESA and the Gabon Space Agency.

The survey area is low mountainous terrain interspersed with rivers, lakes and roads. At the same time, this area also contains rich forest resources with many types of trees at different altitudes. The PolInSAR system placed on the UAV-SAR unmanned aerial vehicle is used to observe and measure forest parameters of the study area and extract Li-DAR measurements corresponding to each stand. Li-DAR data are received from the Distributed Active Archive Center for Biogeochemical Dynamics (ORNL DACC) that are shown in Fig. 1(b).

Figure 1(a) reveals optical image of Lope park that is extracted from Google Earth software. The data of the observed forest area provided by the UAV-SAR system has a very large size (8618×4922) pixels. In this paper, 3 small areas have been selected for analysis and evaluation of forest parameters. The study forest areas have been marked in Fig. 1(a), each area has size (600×600) pixels and the signal is SFA1, SFA2 and SFA3.

Figures 2(a), (b), (c) are 2D histograms showing the calculated forest elevation of the nominated method for the studied forest areas SFA1, SFA2 and SFA3, respectively. This result shows that the selected forest areas for analysis and assessment have clear differences in topography, density and tree height. In which, the results in Fig. 3(b) corresponding to SFA2 show that the forest height is highly concentrated at an altitude of approximately 20 m. SFA2 is an area with river topography and open land with sparse tree density, so the mean tree height of SFA2 is not as high as that of the rest of the study areas. While, SFA1 and SFA3 have similar topography and tree density.

Figures 2(a), (b) display the results of forest heights estimated by the proposed method in these two areas, which are usually concentrated at elevations of more than 20 m. In particular, at SFA3 appeared many pixels reaching a height of approximately 35 m. In addition, the results obtained in Figs. 2(a), (b) and (c)

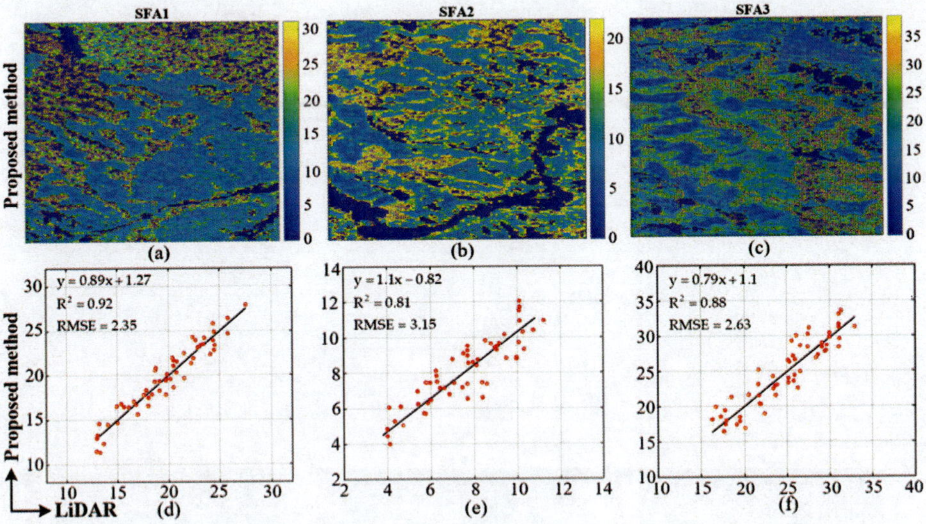

Fig. 2. (a), (b), (c) results of height forest estimated by the proposed manner; (d), (e), (f) comparison of height forest estimated by the proposed manner and LiDAR.

also show the similarity in topography compared with the selected study areas of optical images extracted from Google software Earth (Fig. 1(a)).

Moreover, Figs. 2(d), (e) and (f) exhibit the results of comparing the accuracy of the forest elevation calculated by the proposed manner and the Li-DAR dataset in the 3 survey areas. To perform this measurement, each survey forest area was divided into 60 sub-areas and averaged tree heights, corresponding to 60 red dots in Figs. 2(d), (e) and (f). In which, the coefficient of determination obtained R^2 are 0.92, 0.81 and 0.88, respectively, corresponding to SFA1, SFA2 and SFA3. In prediction theory, the efficiency of proposed manner get more better when these values closer to 1. In addition, the root mean square error (RMSE) values obtained for the survey areas are 2.35 m, 3.15 m and 2.63 m, respectively. These values reflect the accuracy of calculated tree elevation of the nominated manner.

The efficiency of the methods compared with Li-DAR data on the three surveyed forest areas is shown in Table 1. In which, the average tree height calculated by the nominated method at SFA1 is 22.7 m, SFA2 is 9.2 m and SFA3 is 27.6 m. Meanwhile, the estimated height results by VOM in these 3 areas are 21.5 m, 8.7 m and 26.9 m. In general, the results of the average forest height calculated by these two solutions are often lower than those extracted from the Li-DAR data for each respective study forest area. However, the forest height extracted by the new way reveals a higher accuracy when compared with this value of the Li-DAR.

From the results in Figs. 2 and Table 1, it can be concluded that the proposed method has the ability to compute diversely on different terrains with high efficiency and has a higher reliability than the VoM [4].

Table 1. Efficacy of methods compared with Li-DAR data.

Study Forest Areas	Forest parameters	LiDAR Data (m)	VoM [4] (m)	Proposed , method (m)
SFA1	$\bar{h}_v$ (m)	24.3	21.5	22.7
	ϕ_g (rad)	–	–0.22	–0.15
	σ (dB/m)	–	0.19	0.16
SFA2	$\bar{h}_v$ (m)	9.8	8.7	9.2
	ϕ_g (rad)	–	–0.21	–0.19
	σ (dB/m)	–	0.27	0.23
SFA3	$\bar{h}_v$ (m)	28.4	26.9	27.6
	ϕ_g (rad)	–	–0.16	–0.09
	σ (dB/m)	–	0.24	0.21

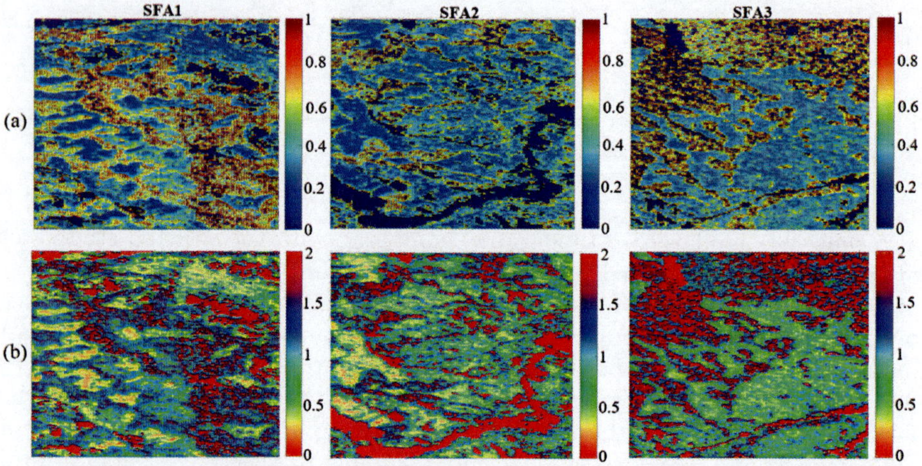

Fig. 3. Parameters estimated by the proposed manner: (a) Random distribution of the target $|X_{22}|$, (b) Varying shape factor $|\delta|$.

The results of Fig. 3(a) show the change of the random distribution $|X_{22}|$ across the three forest areas studied. The coefficient of $|X_{22}|$ varies in the range $(0 \div 1)$ and in forested areas this value is usually greater than 0.6. Moreover, Fig. 3(b) provides about the varying shape factor $|\delta|$. These values change randomly in the range $(0 \div 2)$ and it also shows the anisotropy of the scatterer. The results in Fig. 3 show that the newly introduced solution accurately describes the radar scattering process in the natural forest environment and thereby improves the accuracy of the relatively calculated forest height.

5 Conclusion

The article mention a new approach for forest parameter calculation using L-band polarimetric interference UAV-SAR images. The suggested method not only overcomes the limits of the previous ones, but also significantly increase the accuracy of calculated forest height. Through the evaluating process with UAV-SAR and Li-DAR data, it can be seen that the proposed solution has significantly improved the accuracy compared to the volume optimization method (VoM) [4]. At the same time, when analyzing and evaluating on UAV-SAR data with complex terrain, the proposed method also shows high efficiency and reliability.

References

1. Ballester-Berman, J.D., Lopez-Sanchez, J.M.: Applying the Freeman-Durden decomposition concept to polarimetric SAR interferometry. IEEE Trans. Geosci. Remote Sens. **48**(1), 466–479 (2009)
2. Cloude, S., Papathanassiou, K.: Three-stage inversion process for polarimetric SAr interferometry. IEE Proc. Radar, Sonar Navig. **150**(3), 125–134 (2003)
3. Li, H., Li, Q., Wu, G., Chen, J., Liang, S.: Adaptive two-component model-based decomposition for polarimetric SAR data without assumption of reflection symmetry. IEEE Trans. Geosci. Remote Sens. **55**(1), 197–211 (2016)
4. Managhebi, T., Maghsoudi, Y., Zoej, M.J.V.: A volume optimization method to improve the three-stage inversion algorithm for forest height estimation using polinsar data. IEEE Geosci. Remote Sens. Lett. **15**(8), 1214–1218 (2018)
5. Mette, T., Papathanassiou, K., Hajnsek, I., Pretzsch, H., Biber, P.: Applying a common allometric equation to convert forest height from Pol-InSAR data to forest biomass. In: IGARSS 2004. 2004 IEEE International Geoscience and Remote Sensing Symposium, vol. 1. IEEE (2004)
6. Neumann, M., Ferro-Famil, L., Reigber, A.: Estimation of forest structure, ground, and canopy layer characteristics from multibaseline polarimetric interferometric SAR data. IEEE Trans. Geosci. Remote Sens. **48**(3), 1086–1104 (2009)
7. Xie, Q., Zhu, J., Lopez-Sanchez, J.M., Wang, C., Fu, H.: A modified general polarimetric model-based decomposition method with the simplified Neumann volume scattering model. IEEE Geosci. Remote Sens. Lett. **15**(8), 1229–1233 (2018)

Q-Learning Based Multiple Agent Reinforcement Learning Model for Air Target Threat Assessment

Nguyen Xuan Truong[1](✉), Phung Kim Phuong[1], Hoang Van Phuc[1], and Vu Hoa Tien[2]

[1] Institute of System Integration, Le Quy Don Technical University, Hanoi, Vietnam
truongnx.isi@lqdtu.edu.vn

[2] Faculty of Control Engineering, Le Quy Don Technical University, Hanoi, Vietnam

Abstract. Air target threat assessment is an important issue in air defense operations, which is an uncertainty process to protect the valuable assets against potential attacks of the various hostile airborne objects such as aircraft, missiles, helicopters and drones/UAV. This paper proposes a method to solve the problem of threat assessment of air targets by presenting the process of air defense scenarios in the form of Markov decision process and using reinforcement learning with Deep Q-Learning to predict most dangerous enemy actions to provide more accurate threat assessment of air attacks. On the basis of information about typical air defense combat environment, parameters of binding target trajectory (speed limit, overload limit…) and capabilities of defensive units (number of target channels, fire zone limitation, burning time,…) a simulation environment is built to train and evaluate the optimal (most dangerous) trajectory model of the target based on the given environment. This optimal trajectory can provide input information that is closer to reality, such as real time of arrival; probability of aircraft being shot down by SAM; angle of attack… for threat assessment methods (fuzzy logic, Bayes network, neural network…). The proposed model has been tested on the OpenAI Gym tool using Python programming language. It was shown that the model is suitable to calculate the level of danger of the target with the object to be protected in the context of the general air attacking environment with dynamic and complex constrains.

Keywords: Reinforcement learning · Q-Learning · Air target · Threat assessment · Command - control

1 Introduction

Threat assessment of aerial objects is an essential input information for defensive as well as offensive against potentially dangerous enemy air targets such as bomber, fighter, helicopters, transporter and drones/UAV. Along with the development in modern weaponry, the theater of military operations becomes more and more complicated and diversified. Reasonable assessment of threat value is an important addition to target information to assist commanders in making decisions. This makes threat assessment one of the

T. D. L. Nguyen et al. (Eds.): ICISN 2023, LNNS 752, pp. 112–121, 2023.
https://doi.org/10.1007/978-981-99-4725-6_16

main topics of military researches [1, 2]. Since aerial operations have a vital role in the modern battlefield, the efficient threat assessment of incoming targets can be used as a reference for sorting the priority and distribution of countermeasures, including defensive firepower [3]. Air target threat assessment aims to analyze the track information and operational capabilities of incoming enemy air targets and obtain a quantitative description of the degree of the enemy target threat [4]. Researchers have proposed many difference methods for threat assessment problem.

Threat assessment problem considers a typical tactical situation where we have a set of defended assets $\mathbf{A} = \{A_1, A_2, ..., A_n\}$ that defensive side is responsible to protect (e.g. air bases, nuclear power plants, command post, harbors, radars, monuments, parliament's buildings, and etcetera). There is also a set of air targets $\mathbf{T} = \{T_1, T_2, ..., T_m\}$, which have been detected in the surveillance area and can be the potential threat to each of the protected assets with hostile movements and actions. The first step for each air target-defended asset pair (T_i, A_j), where $T_i \in \mathbf{T}$ and $A_j \in \mathbf{A}$, is to assign a threat value representing the degree of threat T_i poses to A_j, i.e., to define a function $f : \mathbf{T} \times \mathbf{A} \rightarrow [0.1]$, , assuming numbers between 0 (lowest possible threat value) and 1 (highest possible threat value). Secondly, based on the calculated threat values, we will create a prioritized threat list, ranging from the most severe threat to the least.

Most current threat assessment methods use distance, speed, directions, ETA, CPA, and TCPA... as quantitative input data [4–6]. These values were based on the assumption that aerial objects move with constant speed and toward the defended object. In reality, straight movement toward the defended object may not be the best choice, especially with the present of possible countermeasures and defensive firepower. In complicated and intensive air attack situation, the prediction of the most dangerous track of aerial objects will make the input data for threat assessment more accurate. The intent and future actions of aerial objects are unknown to defender, but for threat assessment, we need to predict potential actions that aerial objects can perform and assign the threat value based on the most dangerous possible scenario.

In this paper, we evaluate the applicability of the Reinforcement Learning (RL) algorithm with Deep Q-Learning in the problem of air target threat assessment, using current track parameters of the target (position, speed limit, overload limit...), and capabilities of defensive units (number of target channels, fire zone limitation, burning time,...), to predict the optimal track of the aerial object (the most dangerous attack on the protected asset), thereby assessing the danger of the target and their priority for countermeasures.

2 Related Works

In the conventional methods for solving threat assessment problem, a threat of air target $T_i \in \mathbf{T}$ is evaluated by a combination of attack capability, intent, and access level of the target [1, 4–7]. The attack capability of a target is the ability to inflict damage on protected assets, and the intention of a target means a will, or determination, to inflict such damage, and a plan of actions to perform the attack. The attack capability of a target is can be estimated using variables such as a type and speed of a target, while the intention of an air object is represented using motor mechanics, conduct of operation status, speed and further tactical information.

Intention of air objects is difficult to be estimated forward, but for threat assessment, most researches are based on the assumption that the intention is to approach the protected object. The approach parameter of a target is represented using mostly the variables such as closest point of approach (CPA) d_{CPA}^{ij}, Time to CPA (TCPA) T_{CPA}^{ij}, and Time before Hit (TBH), etc. The correlation between the target T_i and the protected object A_j is illustrated in Fig. 1.

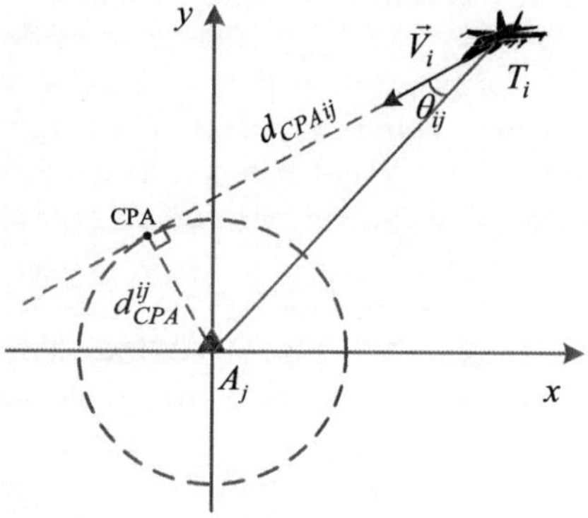

Fig. 1. Correlation diagram between the air target and the protected object

From the target track information, it is possible to determine the approach parameters of the target to the protected object, in which, the CPA distance of the target d_{CPA}^{ij} is the shortest distance to the protected object. Time to CPA is calculated as follows:

$$T_{CPA}^{ij} = \frac{d_{CPA\,ij}}{v_i}, \quad \forall i = 1, \ldots, n; \quad \forall j = 1, \ldots, m \tag{1}$$

where, $d_{CPA\,ij}$ is the distance from the target i to the CPA of the protected object j and v_i is the target velocity i.

The representative techniques related to a threat level evaluation includes artificial neural network-based technique [6, 8, 9], Bayesian inference-based technique [4, 10–12], and Fuzzy logic-based technique [1, 7, 10], etc. Bayesian inference based threat evaluation technique calculates the final threat level by combining conditional probability using the occurrence probability of each threat evaluation element based on conditional probability. This technique has a merit of comparatively higher accuracy, but it has the critical point that a thorough verification process is required when defining conditional probability between elements. Fussy logic-based threat evaluation technique is advantageous to the expression of a change in a value consequent on weighted value and associative relation of each variable, and it calculates a threat level by considering the influencing degree on a threat level consequent on the condition of required variables for calculation. This technique has a merit of low complexity and easy implementation, but it has the critical point that its accuracy is lower than Bayesian inference-based threat evaluation technique.

It can be seen that the above methods are considering the problem of air target threat assessment based on the analysis of information about the target: trajectory parameters (distance, speed, flight direction, type…) obtained from the sensor system (radar), there are not many studies that consider the problem as an interactive problem in which the actions of the target do not continuously change to complete the task due to the impact of the changing combat environment.

3 Deep Q-Learning for Air Target Threat Assessment

3.1 Deep Q-Learning Based Reinforcement Learning

Reinforcement learning is often defined as a Markov decision process, with the basic scheme proposed on Fig. 2. In the context of the Markov decision process, 'S' denotes the set of all possible states, 'A' signifies the set of all possible actions, and 'R' represents the distribution of potential rewards, which can be obtained from any sin-gle state and action (s, a) pair. 'P' is a transition probability distribution of the next state, and finally, the last term in the Markov decision process is the discount factor (γ).

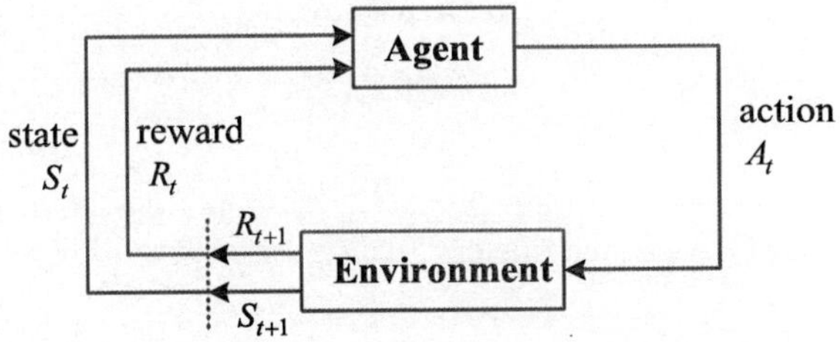

Fig. 2. Reinforcement learning scheme

In the field of reinforcement learning, the agent makes decisions based on the cur-rent state of the environment and a data structure known as the Q _table. The Q _table stores the values associated with actions taken in each state, enabling the agent to predict the values of different state-action pairs. This iterative update learning method is a model-free algorithm that approximates the objective function 1 using the value function, without relying on any prior knowledge of the environment. After each iteration, the update Q _table takes the form:

$$Q(s, a) = Q(s, a) + \alpha\big(r(s, a) + \gamma \max_{a' \in A} Q(s', a') - Q(s, a)\big), \quad (2)$$

In the given scenario, r represents the immediate payoff associated with taking a certain action s, s' represents the next state in the environment, a denotes the learning rate, which determines the extent of incorporating new knowledge and lies between 0 and 1, and γ signifies the discount rate, which reflects the weightage given to future payoffs and also ranges between 0 and 1 (Fig. 3).

The computation of all values in the Q _table becomes impractical when dealing with multi-dimensional environments. It is infeasible to calculate values for a large state space, such as a complete game state with pixels. To address this issue, a solution is to utilize a function estimator for $Q(s, a)$ value estimation. Neural networks are well-suited

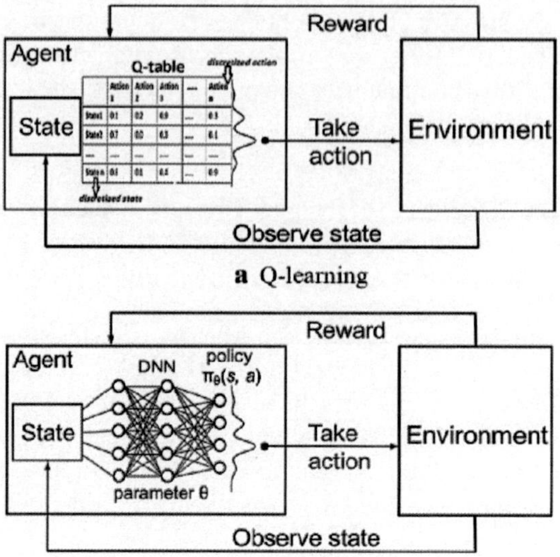

a Q-learning

b Deep Q-learning

Fig. 3. Differences between Q-learning using Q _table (**a**) and Deep Neural Network (**b**)

as estimators for this purpose. Consequently, an additional term denoted by (θ) should be incorporated into our Q _table parameters,, which represents all the weights associated with the neural network.

3.2 Definition of Learning Environment

Reinforcement learning models are defined by agents and an environment. The environment describes a task or simulation, and the Agent is an AI algorithm that interacts with the environment and tries to solve it. The goal of the agent is to solve the task by taking optimal actions. Every taken action will have a resulting value called reward. We will have to build a simulation environment to solve specific tasks. Q-Learning environment for target assignment problem is a very specific environment, and we will have to build them from basic elements. In general, the environment was built as a 2D space with the following elements: Bomber - moving object represent the military airplane, bomber mission is attacking the Object; SAM - surface-to-air missile system that intend for protection of the Object; Object - protected objects.

The environment model of the bomber should take into account the position of SAM and protected objects, and in some case the position of other bomber for coordinated attack. The challenge of convergence arises in reinforcement learning algorithms due to the continuous and high-dimensional nature of the state space, where the number of states and actions is exceedingly large. To tackle this issue, it becomes necessary to discretize the state space of the bomber environment, i.e., to partition it into discrete states or intervals, in order to facilitate the convergence of the algorithm. Bomber action space is a movement speed vector, discretized and defined with upper and lower boundaries. Agents send action to the environment, and the environment sends the observation and reward to the agent after executing each action received from the agent. Observation is nothing but an internal state of the environment. And reward signifies how good the action was (Fig. 4).

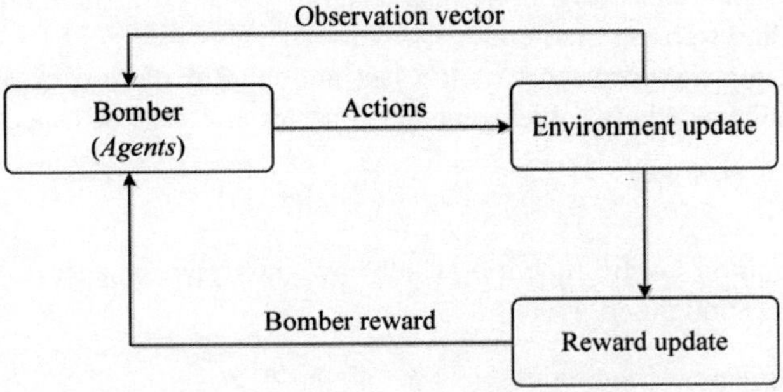

Fig. 4. Environment interaction scheme

The environment interacts with the agent by sending its state (as an observation) and a reward value to the agent after each step. Thus, the complete environment must have the following elements to be defined: A state vector which represents the internal state of the Simulation; A reward calculation mechanism to train the bombers' agent; A reward calculation mechanism to train the bombers' agent; A reward calculation mechanism to train the missile agent; Create step function definition that rules the environment behaviors and define how the state of the environment will be updated after each simulation step.

In order to evaluate the quality of each decision, the agent receives a value from the environment called reward. In common methods, agent only get reward when it succeeds or fail in the final results. However, our state space is too large and using only final result will lead to low learning efficiency. Besides, for faster convergence, the reward function also needs to provide gradient for every single action of the agent to lead the agent to the more reasonable actions. Therefore, we design a reward function to calculate reward that incorporate into a single value the following four factors:

$$R = R_D \times \omega_D + R_{fire} \times \omega_{fire} + R_M \times \omega_M + A \times \omega_A \tag{3}$$

(1) R represent the total reward. R_D is the reward based on current distance from bomber to PA. The purpose of this reward is to provide an artificial gravity that pulls the agent closer to PA.

$$R_D = log_{10}(D + \delta) \tag{4}$$

where D stand for the Euclidean distance between bomber and PA, and δ is a positive constant to avoid logarithm of zero.

(2) R_{fire} is calculated based on the time since the bomber enter fire range of SAM.

$$R_{fire} = (1 - P_{kill})^{(T_{fire}/T_{mc})} \tag{5}$$

P_{kill} is the probability of bomber being killed by a single firing circle of SAM (firing circle is a sequence of detection, acquisition, tracking, launch, and guidance). The value of $\left(T_{fire}/T_{mc}\right)$ can be floored to an integer value to add a realistic constraint that a bomber cannot be shot down if it stays in firing range of SAM for a period of time less than the firing circle of SAM, but we left it as is to maintain function derivation and training gradient.

(3) R_M is the maneuvering cost of the last action. Calculation of R_M is based on Euclidean distance between current speed vector and previous speed vector:

$$R_M = \left[d(V_T, V_{T-1})\right]^2 \tag{6}$$

(4) A is the common reward function, which have three states: success $A = R_A$; failure $A = --R_A$; undefined $A = 0$.

Success happens if the bomber gets inside the damage range of PA before the end of the training episode; failure event is when the episode ends without success; and undefined event success has not happened, but the training episode is still not finished. We add a weighting coefficient to each reward element: ω_d, ω_{fire}, ω_M, ω_A to balance between them and control each element effect on the model decision. Movement in real world always cost energy, so ω_M should be negatives. In our environment, ω_d and ω_{fire} should also be negatives because D – the distance to PA should be maximally reduced to achieve success and the longer T_{fire} – time the bomber stay in shooting range of missile, the more chances it can be shot down and the lesser reward it has.

3.3 Implement Method of Optimal Path Planning in Air-Attack Scenario

The RL environment was implemented using OpenAI gym extension library in Python language. After preprocessing tasks, we defined a Markov decision process for the environment. Figure 5 show a sample of an attack scenario, the big circle shows fire range of the SAM, small circle shows vulnerable range of protected objects, first bomber gets in the fire range to takes down first protected asset, but it has also taken down by SAM, when the second bomber gets in, it's also engaged by SAM. In the process of training RL model, bombers agents need to find a way to minimize the loss. If there is a constraint that a SAM can engage on only one bomber at a time, and bomber were "smarter", bombers can get in the fire range of SAM at the same time to maximize the step reward value and episode reward value.

Algorithm 1. Training iteration for Deep Q-Learning

Initialize memory, vectors, and spaces
For episode=1,M **do**
  Initialize positions of **bombers, SAMs** and **protected objects**
  For t=1,T **do**
    With probability e (exploration rate) select a random action a_t
    Otherwise $a_t = \max_a Q^*(s,t)$
    Execute action a_t and update status of the environment s_{t+1}
    Calculate reward r_t
    Store transition (s_t, a_t, s_{t+1}, r_t) in D
      Set $y_j = r_j + \gamma \max_a Q(s_{t+1}, a, \theta)$
      Perform Gradient descent step on $(y_j - Q(s_{t+1}, a, \theta))^2$
End for
End for

3.4 Training Results

The training process for RL models is divided into episodes. Each episode start with an initial status generated randomly, and the episode ends if a timeout event happens or all bomber has accomplished their mission. Figure 5 demonstrates the results of four final episodes, the dots represent protected objects, the circle represents the effective range of defensive firepower, and the movement of bombers during the training episodes will be logged as a track. These track represent the sequence actions performed by deep RL model.

For simplicity, we have several constrains to the environments: (1) effective firepower range is fixed; (2) number of protected objects is always 3; (3) number of bomber is always 2 and their initial positions are outside effective firepower range. The training and simulation results show that the trained bomber agent has the ability to accomplish the defined mission (attack all the defended objects) in different randomly generated situation. It proves that the method proposed in this paper can guide the bomber to follow a path to attack a protected object while reducing the time that it is inside the range of defensive fircpower.

We have also logged the reward value during training process to evaluate the effect of reward function on. As the training process goes on, reward value converge to a maximum value, while the time duration of each episode reduced, which mean it takes less time for the RL agent to complete the mission. The training process ends after 500000 episodes, reward value and duration of episodes still changing due to random initial state of each episode. As demonstrated in Fig. 6, the track generated by RL agent is still not the most optimized and smooth as real tracks because of the random exploration nature of a RL agent.

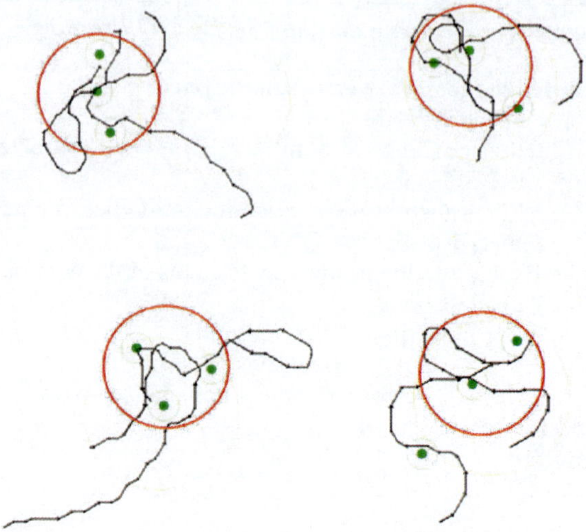

Fig. 5. Results of final episodes

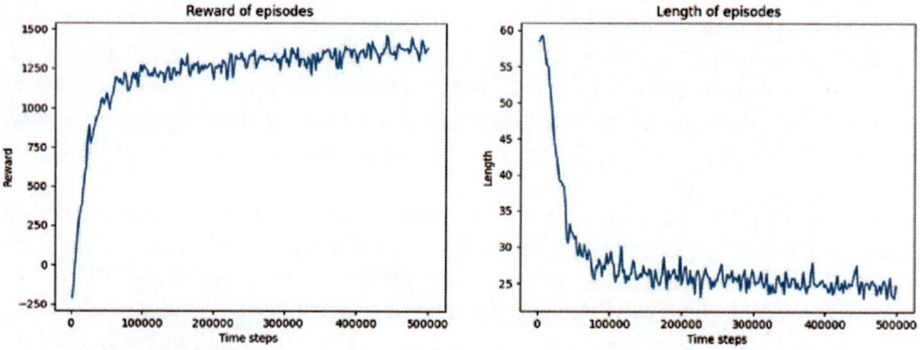

Fig. 6. Reward and length of episodes during the training process

4 Conclusion

Quick and automated prediction of most dangerous track of aerial attacker is a useful input information for threat assessment problems in a rapidly changing warfare. In this paper, a method to predict optimal attacking track is proposed, which uses reinforcement learning to train an action strategy to make aerial attacker to complete tasks by approaching protected objects. RL training environment was built based on objects provides detailed reward.

The convergence and generalization of the proposed reinforcement learning algorithm are evaluated using the OpenAI Gym platform. The convergence analysis demonstrates that as the number of reinforcement learning iterations increases, the algorithm enables the inspection robot to successfully accomplish the task goal of navigating through the detection point without collisions.

References

1. Johansson, F.: Evaluating the performance of TEWA systems, Swedish Defence Research Agency, January 2010
2. Roux, J.N., van Vuuren, J.H.: Real-time threat evaluation in a ground based air defence environment. ORiON **24**(1), 75–101. ISSN 0529-191-X
3. Johansson, F., Falkman, G.: SWARD: system for weapon allocation research & development. In: Information Fusion (FUSION). https://doi.org/10.1109/ICIF.2010.5712067
4. Kumar, S., Tripathi, B.K.: Modelling of threat evaluation for dynamic targets using Bayesian network approach. In: International Conference on Emerging Trends in Engineering, Science and Technology (ICETEST) (2015)
5. Ünver, S., Gürbüz, T.: Threat evaluation in air defense systems using analytic network process. J. Mil. Strategic Stud. **19**(4), 10–39 (2019). ISSN: 1488-559X
6. Lee, H., Choi, B.J., Kim, C.O., Kim, J.S.: Threat evaluation of enemy air fighters via neural network-based Markov chain modeling. Knowl. Based Syst. **116**, 49–57 (2017). https://doi.org/10.1016/j.knosys.2016.10.032
7. Chen, D., Feng, Y., Liu, Y.: Threat assessment for air defense operations based on intuitionistic fuzzy logic. In: 2012 International Workshop on Information and Electronics Engineering (IWIEE) (2012)
8. Yue, L., Yang, R., Zuo, J.: Air target threat assessment based on improved moth flame optimization-gray neural network model. Math. Probl. Eng., 14 (2019). Article no. 4203538. https://doi.org/10.1155/2019/4203538
9. Chen, X., Wang, T.: Threat assessment of aerial target based on modified PSO optimized BP neural network. Int. J. Control Autom. **10**(2), 103–104 (2017). https://doi.org/10.14257/ijca.2017.10.2.09
10. Johansson, F., Falkman, G.: A comparison between two approaches to threat evaluation in an air defense scenario. In: Torra, V., Narukawa, Y. (eds.) MDAI 2008. LNCS (LNAI), vol. 5285, pp. 110–121. Springer, Heidelberg (2008). https://doi.org/10.1007/978-3-540-88269-5_11
11. Yang, H., Han, C., Tu, C.: Air targets threat assessment based on BP-BN. J. Commun. **13**(1), 21–26 (2018)
12. Wang, X., Zuo, J., Yang, R.: Target threat assessment based on dynamic Bayesian network. J. Phys. Conf. Seri. **1302** (2019). https://doi.org/10.1088/1742-6596/1302/4/042023

Real-Time Multi-vessel Classification and Tracking Based on StrongSORT-YOLOv5

Quang-Hung Pham[1], Van-Sang Doan[2], Minh-Nghia Pham[1](✉), and Quoc-Dung Duong[3]

[1] Faculty of Radio-Electronics, Le Quy Don Technical University, Hanoi, Vietnam
nghiapm2018@mta.edu.vn
[2] Faculty of Communication and Radar, Vietnam Naval Academy, Nha Trang, Vietnam
[3] Faculty of Control Engineering, Le Quy Don Technical University, Hanoi, Vietnam

Abstract. Vessel detection, classification, and tracking are very important problems in maritime surveillance systems. In recent years, the field of computer vision has significantly developed, which allows its application to these systems. Accordingly, in this paper, a method based on a YOLOv5-based deep neural network combined with the Strong Simple Online Real-time Object Tracking (StrongSORT) algorithm is proposed for vessel detection, classification, and tracking. Specifically, the YOLOv5 model is trained by using a dataset of diverse images, which are collected from various public sources. The dataset contains several popular vessel types for the purpose of classification. Experimental results show that the proposed model gives high accuracy of vessel classification and high-speed tracking of approximately 16 frames per second, which is near real-time. The model has been embedded into a real small demonstrator to verify the potential implementation in maritime surveillance systems.

Keywords: Classification and tracking · deep learning · maritime surveillance system

1 Introduction

Due to increasingly dense maritime traffic, surveillance systems based on optical cameras have become more critical in operation on the sea for collision avoidance, navigation, and water area management. Accordingly, many research works propose using deep neural networks (DNN) for detection and classification through on-ship cameras [1] or satellite images [2]. Specifically, k-nearest neighbor (KNN), a traditional machine learning (ML) algorithm, is used in [3] to classify vessels based on the similarity of their shape features. The classification accuracy based on stratified ten-fold cross-validation is about 91%. However, the method in [3] needs an expert feature extraction that may consume a lot of time. Instead of manual feature extraction, a real-time vessel detector based on fast U-Net and remapping attention (FRSD) via a common camera is proposed in [4].

The fast U-Net provides compressed features to decrease the number of training parameters, which can boost the performance in various rain–fog weather conditions

T. D. L. Nguyen et al. (Eds.): ICISN 2023, LNNS 752, pp. 122–129, 2023.
https://doi.org/10.1007/978-981-99-4725-6_17

while maintaining the real-time speed. In [5], a deep residual network and cross-layer jump connections are proposed for vessel recognition and tracking. Experimentally, the proposed method has outperformed some other state-of-the-art algorithms on a large number of vessel video datasets. In another research, Huang et al. [6] propose to use a regressive deep convolutional network for vessel image/video detection and classification based on their selfbuilt dataset. The proposed network is inspired by four aspects: the feature extraction of YOLOv2 [7], the feature pyramid network layer of YOLOv3 [8], a clustering algorithm based on proper frame and scale, and an activation function. As a result, the proposed regressive deep convolutional network obtains an mAP of 0.92, a recall of 0.98, and an IoU of 0.80.

In spite of having successful deployment for vessel detection and classification, the above-mentioned works did not consider tracking the ships in real-time. Therefore, In this paper, a combination of vessel classification and tracking is proposed using YOLOv5 [9] and StrongSORT [10]. The YOLOv5 model is applied for the vessel classification; meanwhile, the StrongSORT algorithm is used for vessel tracking. In addition, the proposed method is embedded into a small real demonstrator, which can detect, classify and track vessels in real-time.

The rest of this paper is arranged as follows: Sect. 2 describes the proposed method for vessel classification and tracking. Our self-collected dataset is introduced in Sect. 3. Section 4 evaluates and discusses the performance results of the proposed models. The demonstration is presented in Sect. 5. Finally, Sect. 6 concludes our research work.

2 Proposed Method for Vessel Classification and Tracking

To date, there are many systems used for tracking vessels, but most of them can classify and track only one object. Therefore, in this paper, we propose a DNN model for real-time multi-vessel classification and tracking. In addition, we also build a small demonstration model to verify our approach. In this section, we describe the YOLOv5-based method for vessel classification and the StrongSORT-based method for vessel tracking.

2.1 YOLOv5-Based Vessel Classification

Many research works reported that YOLOv5 provides a fast speed and high accuracy in multi-object detection problems [9]. Accordingly, we propose to use YOLOv5 for the multi-vessel classification. There are 5 versions of YOLOv5, including YOLOv5n, YOLOv5s, YOLOv5m, YOLOv5l, and YOLOv5x. The smallest version is for YOLOv5n, and the largest is for YOLOv5x. The larger the YOLOv5 version, the better the accuracy, but the slower the speed.

YOLOv5 has three important parts: Backbone, Neck, and Head. The first part is the Backbone block, which is used to extract important features from the given input image. In YOLOv5, two CSPnets (Cross Stage Partial networks) [11] are used as a backbone to extract rich informative features from images. They have shown significant improvement in processing time with deeper networks. The second part is the Neck block, which helps to identify the same object with different sizes and scales. Indeed, the Neck block is applied to produce feature pyramids [12], which help models to generalize well on object scaling. They are very useful and help models to perform well on unseen data. There are different types of feature pyramid techniques like FPN (feature pyramid network) [12], BiFPN (bi-directional feature pyramid network) [13], PANet (path aggregation network) [14], etc.; nevertheless, only PANet is used in this paper. The last is the Head block, which is used to perform the final detection part. The Head block is applied anchor boxes on features and generates final output vectors with class probabilities, object scores, and bounding boxes. The same as the previous YOLOv3 and YOLOv4 versions, YOLOv5 Head uses three outputs.

2.2 StrongSort-Based Vessel Tracking Method

There are some techniques that can be used for real-time multi-object tracking. The first technique that can be mentioned here is SORT (Simple Online Realtime Object Tracking) [15]. Because the SORT technique focuses on the tracking problem; therefore, any detector can be used, such as YOLO [16], SSD [17] or RCNN [18]. In the SORT technique, the Hungary algorithm is used to measure the correlation coefficients between objects from consecutive image frames, and then the simple Kalman filter is used for processing those correlated objects. SORT can achieve a good performance at a high frame rate; however, it is suffered from the ID switch problem, which occurs when the object is obscured.

To mitigate the mentioned disadvantage, DeepSORT is proposed in [19] to resolve the ID switches. DeepSORT borrows the ReID model to extract features of the object in order to improve the accuracy of the data process. In addition, DeepSORT also turns the matching mechanism based on IoU (Intersection over Union) matrix cost into the mechanism of matching Cascade and IoU matching, which helps to match the objects after a long time disappearance. Therefore, DeepSort can reduce the number of ID switches up to 45% and decrease the error due to obscured objects or disappearance in time duration. Of course, the processing speed is a little slow down while ensuring near real-time processing if GPU (Graphic Processing Unit) is applied. Next, StrongSORT is developed from DeepSORT by applying new techniques to enhance tracking efficiency [10]. For the appearance branch, a stronger appearance feature extractor, named BoT, is applied to replace the original simple CNN and EMA (exponential moving average) replaces the bank feature. For the motion branch, the NSA Kalman filter increases the algorithm's accuracy on HOTA (higher order tracking accuracy) points.

The numerical comparison results in [10] indicated that StrongSORT has a lower FPS but gives higher accuracy than DeepSort and SORT. However, the processing speed is not the first priority in this case because vessels are targets that move relatively slowly.

3 Vessel Dataset Description

There are many criteria to classify vessels; however, in this work, we categorize the vessels into eight classes in accordance with their usage purpose, including Carrier, Coast guard, Container, Navy, passenger; Tanker, Cargo, and Boat. Each class has different types, colors, sizes, and shapes under different observation ranges and angles. Conventionally, to recognize a vessel we often observe it from a side view rather than a direct front or rear view. Because the front and rear views give less representative information for classification than other sides, and that information is very similar for many vessel types. Therefore, we collected diverse vessel images, which were observed from the right and left sides. As a result, the collected vessel dataset has different parameter distributions as shown in Fig. 1.

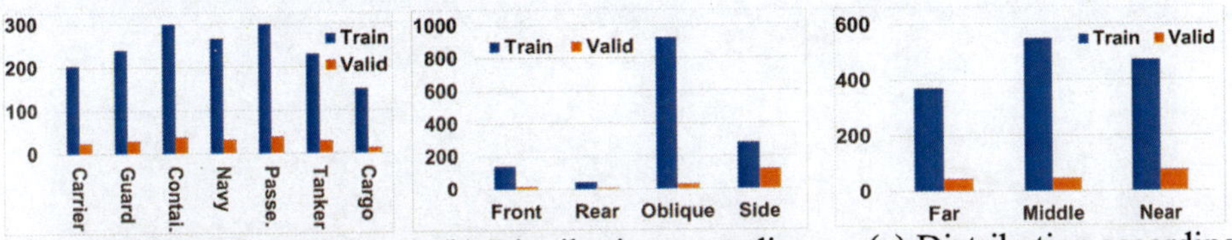

(a) Distribution according to classes (b) Distribution according to observation angles (c) Distribution according to distances

Fig. 1. Dataset with different statistical distributions

The annotation of vessels in images is conducted by box bounding with the fittest edges of those vessels and no vessel is missing annotation. Figure 2 illustrates a result of bounding box annotation for vessels in an image of the dataset. In order to make diversity for the dataset, augmentation using noise addition, blurring, and zooming in/out are used. Consequently, the dataset has 1389 images for training and 166 images for validation.

4 Experimental Results and Discussion

In this section, we evaluate the performance of three YOLOv5 models, including YOLOv5m, YOLOv5l and YOLOv5x on the labeled dataset. After training, the trained models are tested on an unseen dataset of 300 images. The train loss and accuracy are shown in Fig. 3, where we can see that all three models obtain good performance after 300 epochs.

Logically, the different sizes of objects in an image give us different confidence levels. It is also the same with the YOLOv5 model. Therefore, we perform a statistic to demonstrate the dependence of classification accuracy on the area (number of pixels that cover the vessel) of the vessel. The numerical results are shown in Table 1, where we can see that the bigger the vessel is, the higher the accuracy that the YOLOv5 models achieve. In addition, the YOLOv5x yields the highest performance in terms of classification accuracy due to the largest structure; meanwhile, the YOLOv5m provides the worst performance compared with the two others.

4.1 Performance Evaluation on Processing Speed

In this evaluation, the three YOLOv5 models are compared to each other in terms of processing speeds when they run with a video of 6 vessels in a sea area. The tracking result is illustrated in Fig. 4, and the processing speeds are reported in Table 2. We can observe that the YOLOv5m model gains the highest speed of 16 FPS due to the smallest structure compared to the two others. Despite having the highest classification accuracy, the YOLOv5X model executes the slowest.

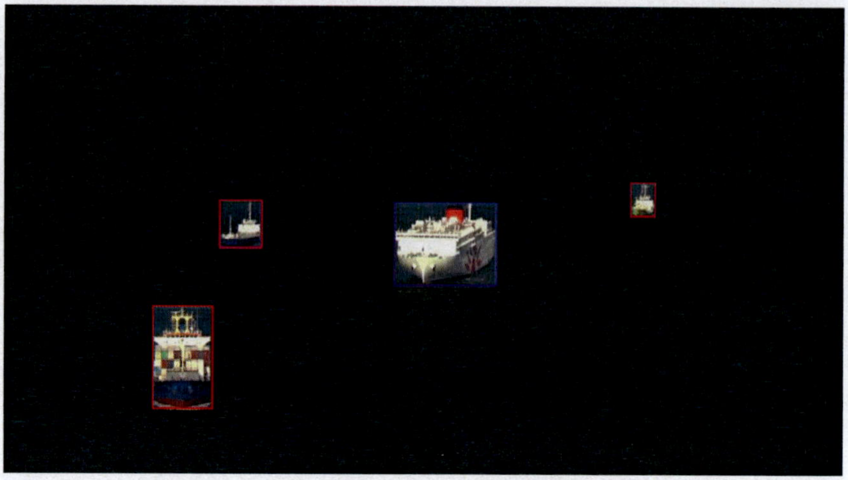

Fig. 2. Bounding box annotation for vessels in images

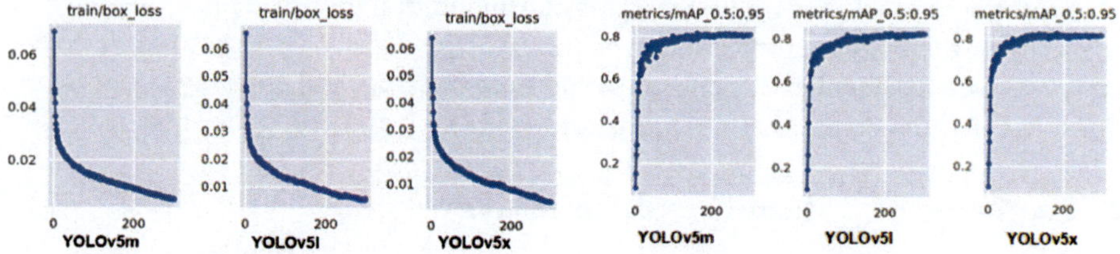

(a) Box Loss (b) Mean Average Precision

Fig. 3. Result of train

Table 1. Relationship between classification accuracy and object size in image

Area in pixel ($w \times h$)	YOLOv5m	YOLOv5l	YOLOv5x
0 ÷ 100000	0.8004	0.8837	0.9207
100000 ÷ 200000	0.8964	0.9484	0.9761
200000 ÷ 300000	0.9328	0.9561	0.9801
300000 ÷ 400000	0.9274	0.9551	0.9829
400000 ÷ 500000	0.9204	0.9610	0.9863
500000 ÷ 600000	0.9042	0.9693	0.9909
600000 ÷ 700000	0.8890	0.9689	0.9924
≥700000	0.9510	0.9697	0.9922
Average	**0.9069**	**0.9481**	**0.9752**

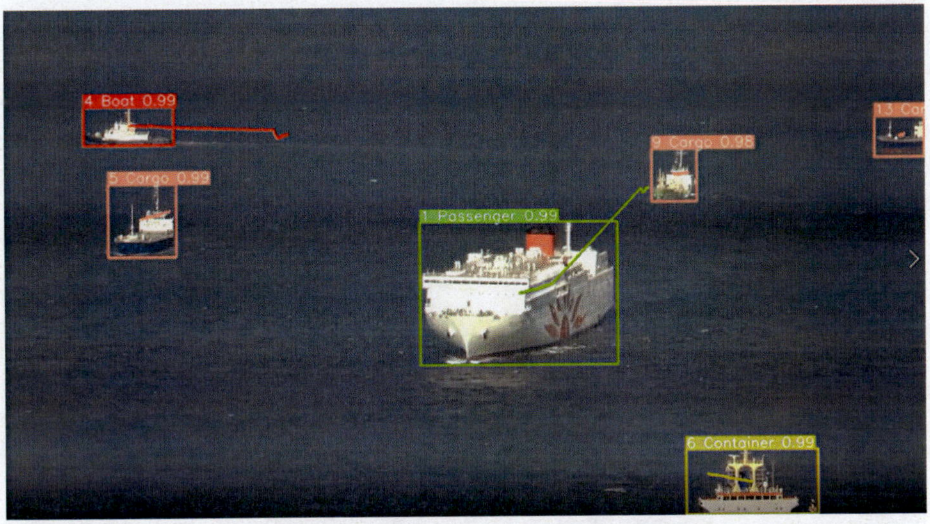

Fig. 4. Tracking trajectory of 6 vessels

Table 2. Processing speeds of three YOLOv5 models

Parameter	YOLOv5m	YOLOv5l	YOLOv5x
Time detect (ms)	32.8	60.2	128.4
Time track (ms)	28.1	27.6	28.6
FPS	16	11	6

5 Demonstration of Real-Time Implementation

In this section, we implement our proposed model to a real demonstrator, which consists of an optical camera (Imilab Xiaomi W88), two servos (MG-90S), an Arduino Mega, and a laptop (PC: Intel Core i7-8750h, RAM: 16 GB, GPU: Nvidia GeForce 1050Ti). The processing scheme of the demonstration is shown in Fig. 5(a), and the experiment workspace photography is shown in Fig. 5(b).

Images captured by a camera are transferred to the YOLOv5 model to detect and classify objects in each frame. After that, StrongSORT will assign an ID for each object and track all the identified objects. The output data of StrongSORT include bounding box, ID and center coordinates are sent to two blocks: display and control. The display block retrieves the object trajectory to display it on a monitor. Meanwhile, the user can choose an object for tracking in the control block. The system will calculate a deviation value between the image frame center and the chosen object's center to control two step-servos (vertical and horizontal). These servos will rotate the camera to reduce the deviation value. In this way, the camera continually rotates and tracks the chosen object. Figure 6 illustrates the tracking process in the real-time experiment.

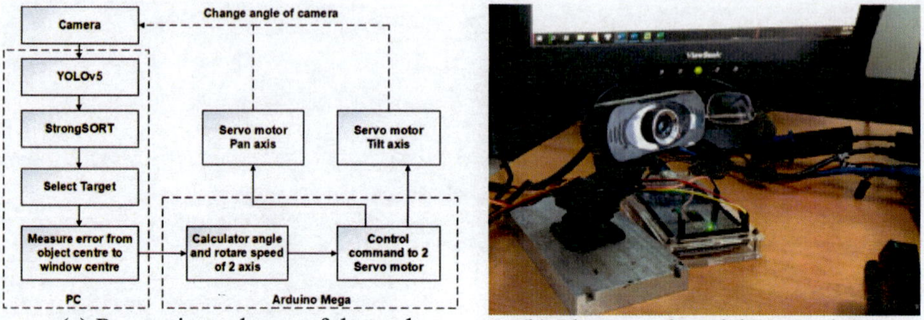

(a) Processing scheme of the real demonstration

(b) Photography of the experiment workspace

Fig. 5. Real-time implementation

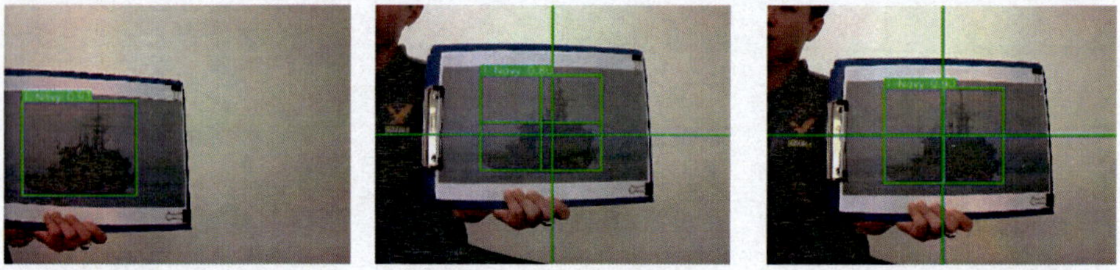

(a) The target is detected with an ID process

(b) The target is chosen

(c) Target tracking

Fig. 6. Target tracking process

6 Conclusion

In the paper, we have applied the YOLOv5 model for multi-vessel detection and classification, applied StrongSORT for multi-target tracking, and proved their operation in a real demonstrator. As a result, YOLOv5m can provide a faster processing speed than YOLOv5l and YOLOv5x due to its smaller structure, but it obtains lower classification accuracy than the two others. The demonstrator can track the chosen vessel in real time and provide good classification accuracy. In the future, we will build a system for practical application, as well as optimize the algorithm for faster execution.

References

1. Teixeira, E., Araujo, B., Costa, V., Mafra, S., Figueiredo, F.: Literature review on ship localization, classification, and detection methods based on optical sensors and neural networks. Sensors **22**(18), 6879 (2022)
2. Kanjir, U., Greidanus, H., Oštir, K.: Vessel detection and classification from spaceborne optical images: a literature survey. Remote Sens. Environ. **207**, 1–26 (2018)
3. Luo, Q., Khoshgoftaar, T., Folleco, A.: Classification of ships in surveillance video. In: 2006 IEEE International Conference on Information Reuse & Integration. IEEE, September 2006

4. Zhao, P., Xiaoyuan, Y., Chen, Z., Liang, Y.: A real-time ship detector via a common camera. J. Mar. Sci. Eng. **10**(8), 1043 (2022)
5. Liu, B., Wang, S.Z., Xie, Z.X., Zhao, J., Li, M.: Ship recognition and tracking system for intelligent ship based on deep learning framework. TransNav Int. J. Mar. Navig. Saf. Sea Transp. **13**(4), 699–705 (2019)
6. Huang, Z., Sui, B., Wen, J., Jiang, G.: An intelligent ship image/video detection and classification method with improved regressive deep convolutional neural network. Complexity **2020**, 1–11 (2020)
7. Redmon, J., Farhadi, A.: YOLO9000: better, faster, stronger. In: 2017 IEEE Conference on Computer Vision and Pattern Recognition (CVPR), pp. 6517–6525 (2017)
8. Redmon, J., Farhadi, A.: YOLOv3: an incremental improvement. arXiv, abs/1804.02767 (2018)
9. Zhang, X., Yan, M., Zhu, D., Guan, Y.: Marine ship detection and classification based on YOLOv5 model. J. Phys. Conf. Ser. **2181**(1), 012025 (2022)
10. Du, Y., Song, Y., Yang, B., Zhao, Y.: StrongSORT: make DeepSORT great again (2022)
11. Wang, C.-Y., et al.: CSPNet: a new backbone that can enhance learning capability of CNN. In: 2020 IEEE/CVF Conference on Computer Vision and Pattern Recognition Workshops (CVPRW). IEEE, June 2020
12. Lin, T.-Y., Dollár, P., Girshick, R., He, K., Hariharan, B., Belongie, S.: Feature pyramid networks for object detection (2016)
13. Xu, W., Gan, Y., Su, J.: Bidirectional matrix feature pyramid network for object detection. In: 2020 25th International Conference on Pattern Recognition (ICPR). IEEE, January 2021
14. Liu, S., Qi, L., Qin, H., Shi, J., Jia, J.: Path aggregation network for instance segmentation (2018)
15. Bewley, A., Ge, Z., Ott, L., Ramos, F., Upcroft, B.: Simple online and realtime tracking. In: 2016 IEEE International Conference on Image Processing (ICIP). IEEE, September 2016
16. Redmon, J., Divvala, S.K., Girshick, R.B., Farhadi, A.: You only look once: unified, real-time object detection. In: 2016 IEEE Conference on Computer Vision and Pattern Recognition (CVPR), pp. 779–788 (2016)
17. Liu, W., et al.: SSD: single shot multibox detector. In: Leibe, B., Matas, J., Sebe, N., Welling, M. (eds.) ECCV 2016. LNCS, vol. 9905, pp. 21–37. Springer, Cham (2016). https://doi.org/10.1007/978-3-319-46448-0_2
18. Qian, Y., Zheng, H., He, D., Zhang, Z., Zhang, Z.: R-CNN object detection inference with deep learning accelerator. In: 2018 IEEE/CIC International Conference on Communications in China (ICCC Workshops). IEEE, August 2018
19. Wojke, N., Bewley, A., Paulus, D.: Simple online and realtime tracking with a deep association metric (2017)

Power Management for Distributed DC Microgrid Under Minimum System Price

Tuan Nguyen Ngoc[1](✉), Luat Dao Sy[2], and Tinh Tran Xuan[3]

[1] Military Technical Academy of Vietnam, Hanoi, Vietnam
ngoctuanhvhn@gmail.com
[2] Dong Nai University, Bien Hoa, Vietnam
[3] Air Defence - Air Force Academy of Vietnam, Hanoi, Vietnam

Abstract. To reduce the DC microgrid (DCMG) system significantly, this paper proposes a power management for distributed DC microgrid system under minimum communication links. Only seven communication links are enough to achieve the power balance and voltage regulation in case of both the grid-connected and islanded modes for distributed DCMG system that consists of 5 agents. Several simulations have been implemented to verify the effectiveness of the proposed scheme under various conditions.

Keywords: Distributed DC microgrid · power management · voltage regulation · minimum communication link

1 Introduction

The DC microgrid (DCMG) system provides a more effective solution as compared with the AC microgrid due to neglecting the unnecessary power conversion stage and control issues such as the harmonics, frequency, and reactive power [1, 2]. Therefore, the DC microgrid which consists of a utility grid, an energy storage system (ESS), electric vehicle (EV), renewable energy sources (RESs), and loads, has been grasping more and more attention.

On the other hand, depending on the communication link, the microgrid control method is divided into centralized control, distributed control, and decentralized control. The disadvantages of centralized control are the single point of failure, high system cost, and low flexibility [3]. Contrary, the decentralized method provides high flexibility and optimal system cost due to eliminating the central controller and communication links among power agents. However, global system stability is one of the most negative issues of the decentralized approach since lacks information [4]. Therefore, the distributed method is a replacement solution by integrating the advantages as well as reducing the disadvantages of the above methods [5, 6].

In distributed DCMG system, the DC-link voltage control is used to regulate the DC-link voltage (DCV) at the desired value based on the data exchanged through the communication network among neighbor agents and local measurements [7]. However,

T. D. L. Nguyen et al. (Eds.): ICISN 2023, LNNS 752, pp. 130–137, 2023.
https://doi.org/10.1007/978-981-99-4725-6_18

the communication link (DCL) topology of the distributed DCMG system should be carefully designed to ensure the minimum system price [8].

Motivated by these concerns, this paper proposes a power management for distributed DC microgrid system under minimum communication links. Only seven communication links with binary data are enough to achieve the power balance and voltage regulation in case of both the grid-connected and islanded modes. Several simulations have been implemented to verify the effectiveness of the proposed scheme under various conditions.

This paper is organized as follows: Sect. 2 presents the system configuration and proposed communication topology. Section 3 includes the proposed control strategy of the distributed DCMG system. Section 4 and Sect. 5 are the simulation results and the conclusion, respectively.

2 System Configuration of a Distributed DCMG

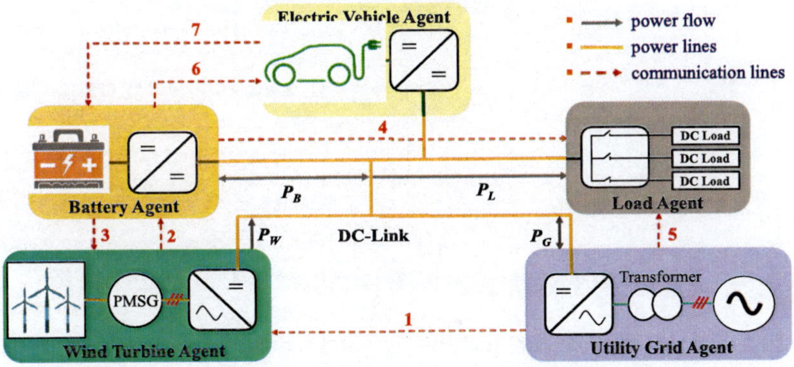

Fig. 1. Configuration of the proposed distributed DC microgrid

Figure 1 shows the configuration of the proposed distributed DCMG. There are five agents in this system: electric vehicle agent, battery agent, load agent, wind power agent, and grid agent. The communication links are utilized to exchange information among neighbor agents. In this proposed scheme, voltage regulation and power balance are achieved with only seven DCLs with binary data that reduces significantly the system price when compared to the study [7]. The DCV is regulated by only one agent at all times at the desired value. The other agents make the control decision locally based on the collecting data from the neighbor agents and local measurements.

Table 1 lists the exchange data, data type, and data information of each communication link used in this study.

Table 1. System communication topology

Communication Link	Exchange Data	Data Type	Data Information
1	D_{GW}	Binary	0: Grid agent does not regulate DCV 1: Grid agent regulates DCV
2	D_{WB}	Binary	0: Wind turbine and Grid agents do not regulate DCV 1: Wind turbine/Grid agents regulate DCV
3	D_{BW}	Binary	0: Battery and EV agents do not regulate DCV 1: Battery/EV agents regulate DCV
4	D_{BL}	Binary	0: Load agent is NOR/REC 1: Load agent is NOR/SHED
5	D_{GL}	Binary	0: Grid agent does not regulate DCV 1: Grid agent regulates DCV
6	D_{BE}	Binary	0: Wind, Grid and Battery agents do not regulate DCV 1: Wind/Grid/Battery agents regulate DCV
7	D_{EB}	Binary	0: EV agent does not regulate DCV 1: EV agent regulates DCV

3 Proposed Control Strategy of Distributed DC Microgrid

Figure 2 shows the proposed control strategy of the distributed DCMG system. In the grid-connected mode, the grid agent regulates the DCV at the desired value by DCV control by inverter mode ($GVCM_{inv}$) or converter mode ($GVCM_{con}$). In this situation, the wind turbine agent and load agent receive the information from the grid agent and then decide the corresponding operating modes: the wind turbine agent runs in the maximum power point tracking (MPPT) mode; the load agent runs in the normal mode. At the same time, the battery agent collects the data from the wind turbine agent and then operates in the charging at the maximum current control mode ($BCCM_{char}$) until the full state of charge (SOC_B). Finally, the EV agent receives the data from the battery agent and operates at the charging at the maximum current control mode ($ECCM_{char}$) until the full SOC_E.

In the islanded mode, the voltage regulation and power balance are achieved by the battery agent, EV agent, or wind turbine agent based on the communication network and local measurements as shown in Fig. 2. On the other hand, the local tracking control of all local agents in the distributed DCMG is implemented by using the cascaded loops of outer-loop voltage control and inner-loop current control with the PI controllers to ensure zero steady-state tracking errors.

Table 2 shows the detailed operating modes of the control strategy. Based on the relation of the generated power from the wind power agent, the demanded load, battery

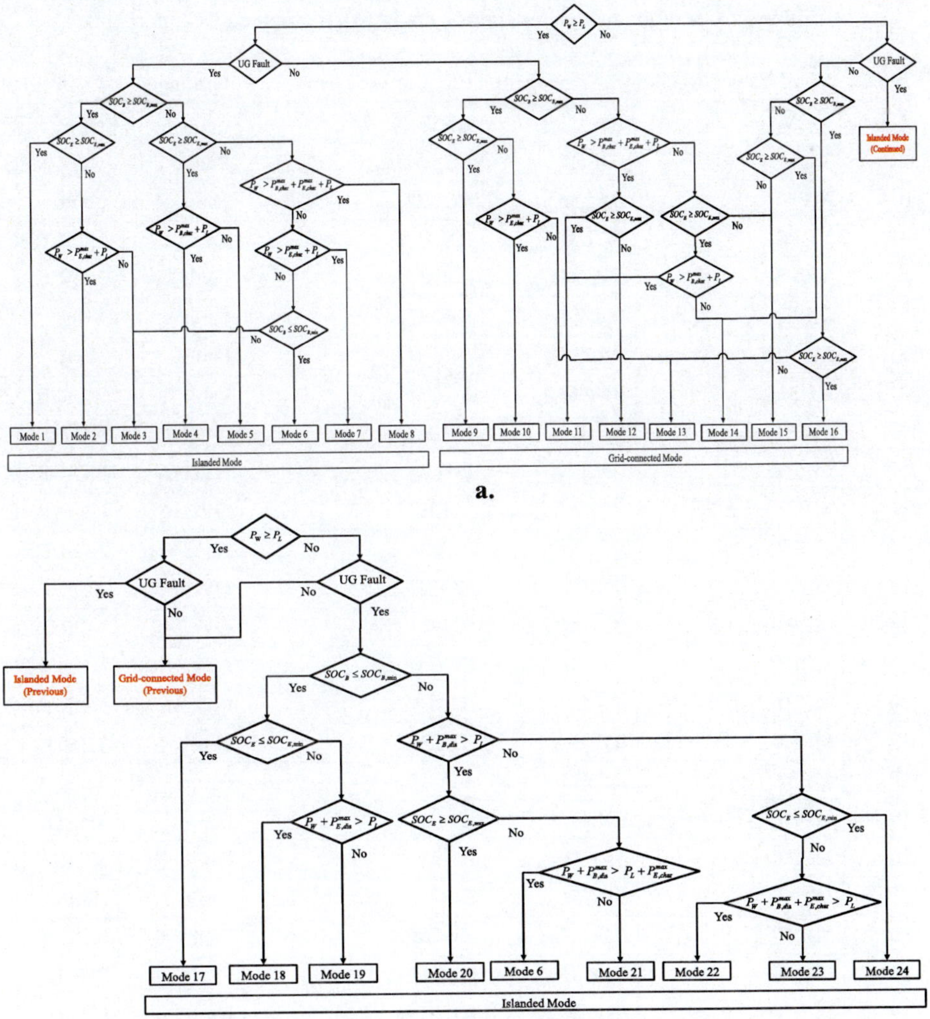

Fig. 2. Proposed power flow control strategy of distributed DC microgrid

state-of-charge (SOC_B) level, EV state-of-charge (SOC_{EV}) level, and availability of the grid, twenty-four operating modes are developed in this paper to maintain power management and voltage stabilization under minimum DCLs.

Table 2. Operation modes for control strategy of distributed DC microgrid

Mode	Wind Power Agent	Battery agent	Electric Vehicle Agent	Grid agent	Load agent
1	VCM	IDLE	IDLE	Fault	NOR/REC
2	VCM	IDLE	$ECCM_{char}$	Fault	NOR/REC
3	MPPT	$BVCM_{dis}$	$ECCM_{char}$	Fault	NOR
4	VCM	$BCCM_{char}$	IDLE	Fault	NOR/REC
5	MPPT	$BVCM_{char}$	IDLE	Fault	NOR
6	MPPT	IDLE	$EVCM_{char}$	Fault	NOR
7	MPPT	$BVCM_{char}$	$ECCM_{char}$	Fault	NOR
8	VCM	$BCCM_{char}$	$ECCM_{char}$	Fault	NOR/REC
9	MPPT	IDLE	IDLE	$GVCM_{inv}$	NOR/REC
10	MPPT	IDLE	$ECCM_{char}$	$GVCM_{inv}$	NOR/REC
11	MPPT	$BCCM_{char}$	IDLE	$GVCM_{inv}$	NOR/REC
12	MPPT	$BCCM_{char}$	$ECCM_{char}$	$GVCM_{inv}$	NOR/REC
13	MPPT	IDLE	$ECCM_{char}$	$GVCM_{con}$	NOR/REC
14	MPPT	$BCCM_{char}$	IDLE	$GVCM_{con}$	NOR/REC
15	MPPT	$BCCM_{char}$	$ECCM_{char}$	$GVCM_{con}$	NOR/REC
16	MPPT	IDLE	IDLE	$GVCM_{con}$	NOR/REC
17	MPPT	IDLE	IDLE	Fault	SHED
18	MPPT	IDLE	$EVCM_{dis}$	Fault	NOR
19	MPPT	IDLE	$EVCM_{dis}$	Fault	SHED
20	MPPT	$BVCM_{dis}$	IDLE	Fault	NOR
21	MPPT	$BCCM_{dis}$	$EVCM_{char}$	Fault	NOR
22	MPPT	$BCCM_{dis}$	$EVCM_{char}$	Fault	NOR
23	MPPT	$BCCM_{dis}$	$EVCM_{dis}$	Fault	SHED
24	MPPT	$BCCM_{dis}$	IDLE	Fault	SHED

4 Simulation Results

In this section, the simulations are implemented for a distributed DCMG to demonstrate the feasibility and reliability of the proposed control scheme under various conditions based on the PSIM software.

Figure 3 shows the simulation results for the transition from the islanded mode to the grid-connected. Initially, because the wind turbine agent supplies power higher than the sum of the demand load and the maximum EV charging, the wind turbine agent regulates DCV at VCM mode. In this situation, the battery agent operates at the IDLE mode due to the full SOC_B while the EV agent runs in $BCCM_{char}$ mode.

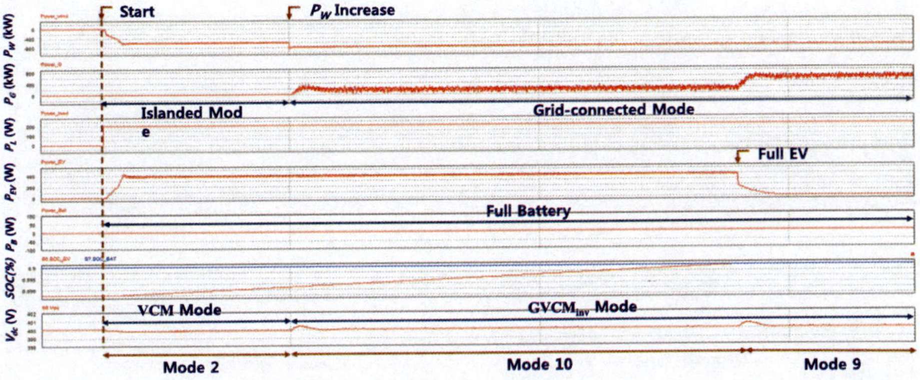

Fig. 3. Results of simulation result in case of the islanded mode to the grid-connected mode when SOC_B is full

Once the utility grid agent is connected to the distributed DCMG system, the DCV is regulated at the desired value by $GVCM_{inv}$. In this case, the wind turbine agent changes the operating mode from VCM mode to MPPT mode while the EV agent keeps running in $ECCM_{char}$ mode until the full SOC_E.

Similarly, Fig. 4 shows the simulation results for the transition from the islanded mode to the grid-connected when SOC_E is full.

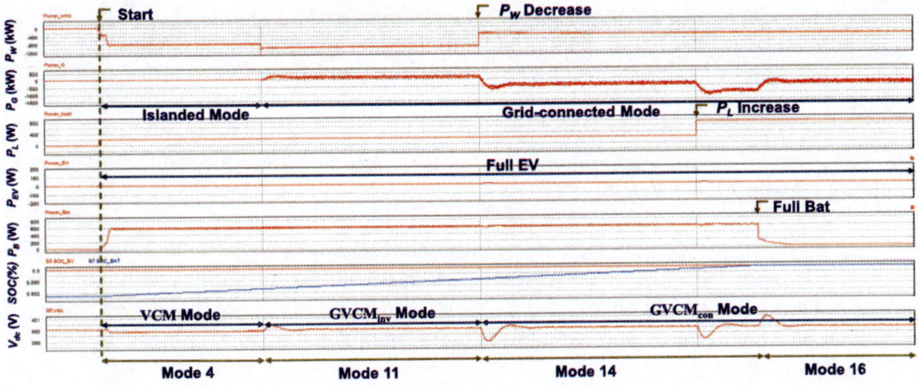

Fig. 4. Results of simulation result in case of the islanded mode to the grid-connected mode when SOC_E is full

Figure 5 shows the simulation results in case of the load shedding when SOC_B is low. Initially, because the wind turbine agent supplies power lower than the sum of the demand load and the operating mode of the battery agent is IDLE mode because of the low SOC_B, the EV agent regulates DCV at the desired value. In this situation, the wind turbine agent operates at the MPPT mode.

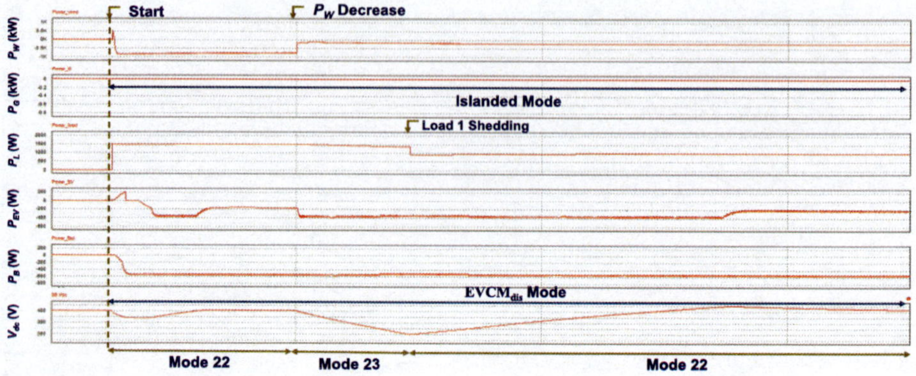

Fig. 5. Results of simulation result in case of load shedding when the wind turbine power decreased

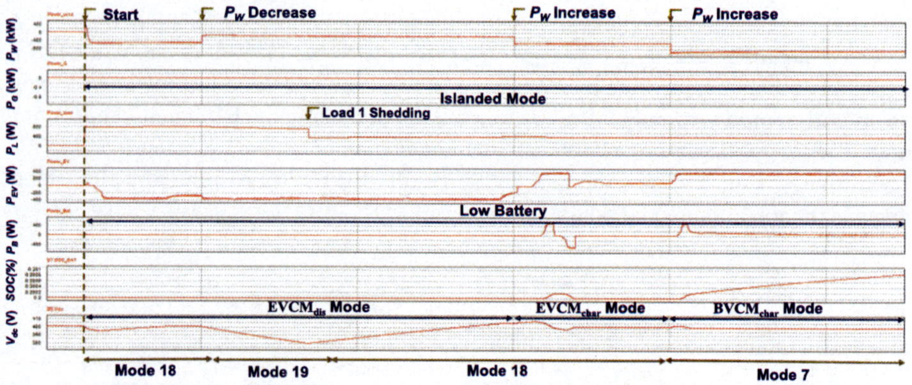

Fig. 6. Results of simulation result in case of load shedding when the wind turbine power decreased under the low SOC_B

Once the wind turbine power decreases due to uncertain conditions, which causes the power in DCMG to be lower than the demand load. As a result, the load shedding mode is activated to avoid the power imbalance. After that, the EV agent continues regulating DCV by $EVCM_{dis}$.

Similarly, Fig. 6 shows the simulation results in case of load shedding when the wind turbine power decreased under the low SOC_B.

5 Conclusions

In this paper, a topology of the distributed DCMG with minimum DCLs is presented to reduce the system price. By using this structure, the distributed DCMG system not only maintains the DC-link voltage reliably but also the power balance in both the grid-connected and islanded modes. Simulation results verified the performance of the proposed scheme.

References

1. Dragicevic, T., Lu, X., Vasquez, J.C., Guerrero, J.M.: DC microgrids -Part I: a review of control strategies and stabilization techniques. IEEE Trans. Power Electron. **31**(7), 4876–4891 (2016)
2. Weiss, M., Dekker, P., Moro, A., Scholz, H., Patel, M.K.: On the electrification of road transportation—a review of the environmental, economic, and social performance of electric two-wheelers. Transp. Res. D Transp. Environ. **41**, 348–366 (2015)
3. Mehdi, M., Kim, C.-H., Saad, M.: Robust centralized control for DC islanded microgrid considering communication network delay. IEEE Access **8**, 77765–77778 (2020)
4. Tucci, M., Riverso, S., Vasquez, J.C., Guerrero, J.M., Ferrari-Trecate, G.: A decentralized scalable approach to voltage control of DC islanded microgrids. IEEE Trans. Control Syst. Technol. **24**(6), 1965–1979 (2016)
5. Guo, F., Xu, Q., Wen, C., Wang, L., Wang, P.: Distributed secondary control for power allocation and voltage restoration in islanded DC microgrids. IEEE Trans. Sustain. Energy **9**(4), 1857–1869 (2018)
6. Guo, F., Wang, L., Wen, C., Zhang, D., Xu, Q.: Distributed voltage restoration and current sharing control in islanded DC microgrid systems without continuous communication. IEEE Trans. Ind. Electron. **67**(4), 3043–3053 (2020)
7. Nguyen, T.V., Kim, K.-H.: An improved power management strategy for MAS-based distributed control of DC microgrid under communication network problems. Sustainability **12**, 122 (2020)
8. Tran, D.T., Habibullah, A.F., Kim, K.-H.: Seamless power management for a distributed DC microgrid with minimum communication links under transmission time delays. Sustainability **14**, 14739 (2022)

Detection and Diagnosis of Atopic Dermatitis Using Deep Learning Network

Anh-Minh Nguyen[1], Van-Hieu Vu[2](✉), and Thanh-Binh Trinh[3]

[1] High School for Foreign Languages, Hanoi National University, Hanoi, Vietnam
anhminhnguyen4444@gmail.com

[2] Institute of Information Technology, Vietnam Academy of Science and Technology, Hanoi, Vietnam
vvhieu@ioit.ac.vn

[3] Faculty of Computer Science, Phenikaa University, Hanoi, Vietnam
binh.trinhthanh@phenikaa-uni.edu.vn

Abstract. This paper proposes a method to diagnose atopic dermatitis based on deep learning network by analyzing image data of infected skin areas. We use deep learning network to analyze the layers and get the suitable layer for diseased and non-diseased images. These images are further feature extracted through HOG feature extraction and SVM classifier to classify disease and non-disease. The proposed method proves to be effective in precision and recall, that can be considered as an adjunct to traditional diagnostic methods, and the results obtained are equivalent to that of a diagnostician while limiting the heterogeneity between the predictors.

Keywords: Atopic Dermatitis-AD · Deep Learning · Convolutional Neural Networks-CNNs

1 Introduction

Atopic Dermatitis-AD is a severe symptom of eczema (Eczema-Ecz) [1]; cause swelling, redness, stretch marks and itching on the skin. Most of the cases occur in children aged 70%, of which up to 15–20% of children aged 13 and under are affected by the disease [2,3]. In order to achieve high efficiency in treatment, there is a need for a method to estimate the severity of the disease to assess the progress of the patient. Currently in the field of information technology, deep learning technologies is developing very quickly in the digital era and is widely used in medical fields such as pathology and radiology [4,5]. Besides, the application of this technology in dermatology is still quite modest and error prone because of the complexity of the [6] variants.

AD is a severe form of chronic itchy eczema that affects 15–20% of children but can be infected at any age. It is estimated that up to 16.5 million adults in the US (7.3% respectively) have had AD at the age of 2 or older, with 40% of them having moderate or severe disease [7]. In addition to the skin effects,

T. D. L. Nguyen et al. (Eds.): ICISN 2023, LNNS 752, pp. 138–147, 2023.
https://doi.org/10.1007/978-981-99-4725-6_19

people with AD are at increased risk for respiratory allergies such as asthma or allergic rhinitis, or neurological disorders such as depression or anxiety disorders [8,9]. In addition, AD is often misdiagnosed because the symptoms are often distributed and variable, and are similar to other inflammatory skin diseases. Therefore, there is a need for a method to assist doctors and specialists in making a diagnosis, especially useful for young and inexperienced dermatologists.

Deep Learning is becoming a buzzword these days, especially for medical imaging tasks, such as diagnostics, due to the high performance achieved in image classification. Among the types of deep learning, the Convolutional Neural Network (CNN) is the most common type of DL model used for diagnostics and medical image analysis. Deep learning models, specifically CNNs, RNNs (Recurrent Neural Networks) or GANs (Generative Adversarial Networks) have shown to be effective in analyzing clinical image data. For example, CNN has been applied to aid in the diagnosis and early detection of signs of Alzheimer's disease, osteoarthritis or diabetic retinopathy through a database of imaging scans such as MRI, EGG or ultrasound [10]. The development of deep learning technology in skin diagnosis was initiated by the diagnosis of melanoma. Since then, there have been a number of studies using artificial intelligence to identify skin cancers based on the input of skin biopsies and dermatoscopes. The use of similar technology in the diagnosis of dermatological diseases, specifically AD, is absolutely practical and necessary.

Because CNN incurs high computational cost to implement and may require adjustment of large number of parameters. So to solve this problem some pre-trained models are set up with predefined network architecture. In this study, we use a transfer learning model based on Visual Geometry Group with a 16-layer deep model architecture (VGG16) [11]. We exploit feature maps on some deep layers, then use transform technique HOG [12] to extract features from feature maps. Finally, a binary classifier is selected to classify diseased and non-diseased images.

In the next sections, Sect. 2 will introduce some related studies. The Sect. 3 is the proposed method and the Sect. 4 is some conclusions.

2 Related Work

2.1 Automatic Assessment, Classification and Diagnosis of Dermatological Diseases

The method of using deep learning based on image data to automatically diagnose inflammatory skin diseases is introduced in [13], the authors have built an artificial intelligence assistant to diagnose skin diseases (AIDDA).) based on clinical skin imaging for the diagnosis of psoriasis (psoriasis-Pso), Ecz and AD used the CNN EfficientNet-b4 deep learning model from Google.

The team used Pytorch to develop the deep learning software platform, the ROC (Receiver Operating Characteristic curve) performance index linking the sensitivity index and AUC (Area Under the Curve) shows that AIDDA has an accuracy of 95.8% with error 0.09%; ROC and AUC are 97.6% with 0.12%

error and 95.89% with 0.06% error, respectively. Although the AUC with other ConvNets models compared in the study (including SE-ResNet101-32 × 4d, SE-ResNet101 [14] and Inception-v3 [15]) are quite similar, the team's model The study is still better at the ROC measure. The model is applied to create a mobile application and more than 7.000 doctors at hospitals have signed up to use this application. User data shows that nearly 100.000 images taken from doctors have entered the application to improve the diagnostic process, proving the effectiveness and necessity for such a tool to support using such deep learning networks, especially in the diagnosis of dermatological diseases in general and atopic dermatitis in particular. Similar to [13], in many works and trials have conducted using and combining CNN in diagnosing atopic dermatitis, the results show that the accuracy is increased, the calculation speed is high compute is also improved.

In [7], the authors experimented with CNN deep learning technology combined with a measure of severity of atopic dermatitis (SCORAD) to form an automatic SCORAD calculation algorithm (ASCORAD). They also used Pytorch to build models and three datasets from three different dermatologist sources, based on two criteria: segmentation of the injured surface and severity rating of visual cues. The obtained results showed that the overall RMAE achieved a decent threshold of 13.0% and achieved an excellent AUC of 0.93 and an IoU of 0.75 when estimating the lesion surface on light skin. This demonstrates that deep learning networks can be a favorable and objective alternative to automatic assessment of AD with much potential, comparable to human expert evaluation results, while reducing heterogeneity of multi-expert reviews and reduced time consuming.

In [16] used a similar method through two deep learning models CNN and LSTM (long short-term memory) to diagnose psoriasis (Psoriasis-Pso), a disease with many features similar to AD-the same group of inflammatory skin diseases, has a similar relationship with Ecz. The results show that the accuracy obtained in CNN and LSTM is 84.2% and 72.3% respectively, showing the potential of applying deep learning model in diagnosing dermatological diseases and is evidence for the significant improvement in model accuracy in [13].

Using deep learning like R-CNN using the Tensorflow library to diagnose diseases like AD, acne and scabies has been exploited by the [17] research team. The authors have built a web-based system for faster and more efficient diagnosis of skin diseases. The results showed that the model correctly predicted 88% for AD, 85% for acne and 84.7% for scabies. Through comparison, it can be seen that the results are similar to many other studies such as [13,16]. The study [18] introduces an alternative assessment that uses CNN and raw 3D RSOM images to classify AD based on traditional machine learning approaches such as SVM (support vector machine) and RF (Random Forest). The authors evaluated on the auto-generated index EVSI (Eczema Vascular Severity Index) including (i) healthy skin with AD and (ii) mild symptoms with moderate-severe AD. The results show that the CNN model has an accuracy of 97% and RF is 65%. This again proves the performance of deep learning, especially CNN.

2.2 Convolutional Neural Networks -CNN

CNN is a class of artificial neural networks commonly used in classification problems, most commonly applied to visual image analysis. CNN has many other popular applications in computer vision [20], recommendation systems, natural language processing or image classification, medical image data computation [21]. The CNN system usually consists of 3 parts described in Fig. 1 including: input, including data source prepared from a third party training model; hidden layer, including feature selection and classification classes; and output, including diagnostic data and data-related metrics.

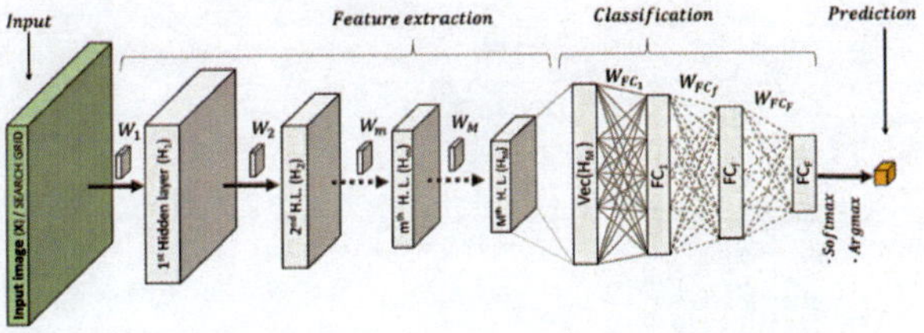

Fig. 1. Simple structure model of CNN network.

In particular, the classification and feature extraction classes often play a decisive role in giving the final result, which is divided into many sub-branches, including many classes and corresponding classifiers. In the tests, each CNN includes a significant number of classifiers, which help to see the index change at each layer, thereby giving the most accurate conclusions and application models. In this study, we propose to use VGG16 network in image feature extraction. We do not use output layer labels but use images filtered through several layers of the network layer at conv3, conv5 as shown in Fig. 2.

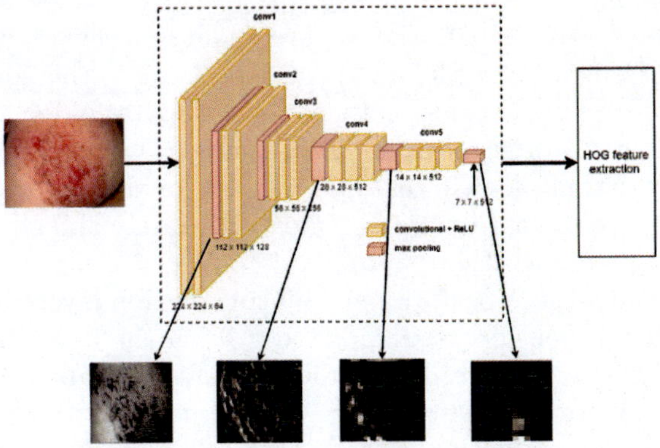

Fig. 2. VGG16 extract feature map

3 Proposed Model

The model consists of input from the filtered dataset from the CNN layer VGG16 (Fig. 3), a layer consisting of HOG (Histogram of Oriented Gradients) image classification and through SVM classification. Performance metrics used include accuracy index (Accuracy), measurement accuracy (Precision) and F1 score.

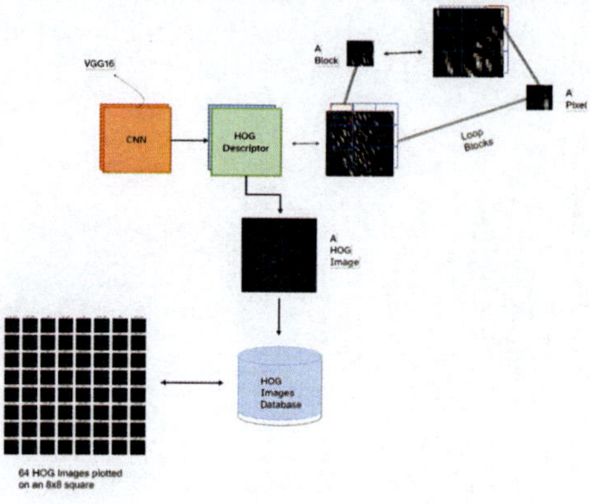

Fig. 3. Classification model based on HOG feature

Through the feature extraction model in [22], each image fed into the system (Fig. 5), to comprehensively evaluate the performance at each convolution layer, there will be 64 feature image layers represented in each last layer of each integration layer in the VGG16 model. The results for a dataset consisting of 320 feature images each divided into 5 folders are named after that integration class. There is also five folder containing images with size 8×8. The containing all the feature images of each layer for a comprehensive assessment of the quality of each layer (see Fig. 4).

To comprehensively evaluate the feature quality on several convolution layers (Conv1, Conv2, Conv3, Conv4, Conv5) corresponding to the integration layers of VGG16 as shown in Fig. 2 and structured according to folder. In each directory, there are two sub-directories labeled AD and not_AD representing the specificity of AD and other dermatological diseases. This is for the purpose of selecting the layer or directory whose features can be most accurately and efficiently diagnosed as input. The steps follow Algorithm 1.

To evaluate the feature performance by convolution layers for 30% of images on both diseased and non-diseased image sets. After the HOG feature extraction, we use the SVM classifier to get the results as shown in Table 1. The results show that the image feature at the 3rd layer (Conv3) achieves the best results.

After the feature evaluation and feature selection at the Conv3 layer, we proceed to extract the feature map at the output of Conv3 and extract the HOG feature, and finally use the SVM classifier to classify the images disease and no disease (AD and not_AD) as shown in Fig. 5.

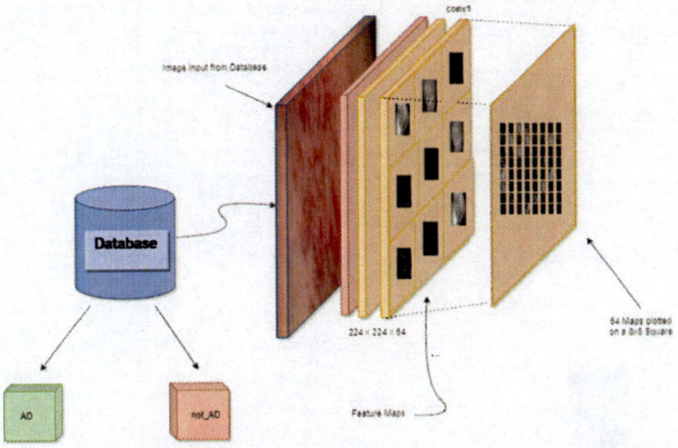

Fig. 4. Extract images of the convolution layers

Algorithm 1. Deep learning feature extraction and classification algorithm SVM

Input: Data set of diseased and non-atopic dermatitis
Output: model and evaluation results

```
Initialization: Three Models Level 1 and Composite Model
for each an image do
    for output per convolutional layer do
        feature extraction of image according to VGG16
        extract the HOG feature of each image
        save HOG feature in database
    end for
end for
load HOG database
Divide the district training, test set, evaluation set
initialize the SVM classifier
use GridSearchCV to find parameter for classifier
training and testing according to the SVM classifier
performance evaluation according to the evaluation set
Return: save evaluation results and models
```

Table 1. Feature performance evaluation by convolution layer

	Precision	Recall	F1-score
Conv1	0.60	1.00	0.75
Conv2	0.80	0.80	0.80
Conv3	1.00	0.75	0.86
Conv4	0.80	0.80	0.80
Conv5	0.83	0.83	0.83

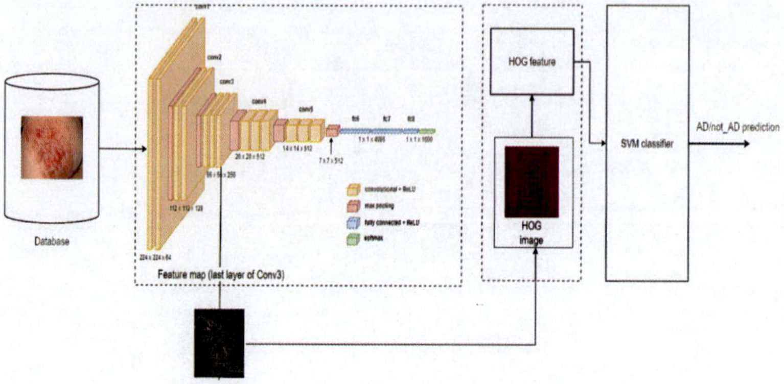

Fig. 5. System architecture model

4 The Experiment

4.1 Description of Data

The data set we used for the experiment consists of two main directories: AD (photographs with atopic dermatitis) and not_AD (photographs without atopic dermatitis). The original dataset has a total of 207 image samples divided into two folders AD and not_AD. In which, there are 90 images in AD and 117 images in not_AD. The AD folder includes photos of areas infected with atopic dermatitis in different parts of the body. Files in not_AD include photos of areas affected by other inflammatory or dermatological conditions such as psoriasis, eczema, and atopic dermatitis.

4.2 Performance Metrics

The ROC curve is a graph commonly used in testing, based on the sensitivity & specificity index (Sensitivity-TPR & Specificity-FPR) of a deep learning model. Most related studies use ROC to assess the objectivity of deep learning models according to some threshold. TPR and FPR are also two parameters on two axes of the graph calculated as follows:

- TPR (True Positive Rate), means that the Recall metrics is determined: $TPR = TP/(TP + FN)$
- FPR (False Positive Rate) is determined: $FPR = FP/(FP + TN)$

AUC is an index of the area under an ROC curve. The index measures the entire two-dimensional area under the curve, giving a comprehensive view of the performance of all classification thresholds. Precision, Recall and F1-score are all TPR and FPR based metrics identified above. Whereby,

- Precision is defined as: $Precision = TP/(TP + FP)$
- Recall same as TPR, is defined as: $Recall = TP/(TP + FN)$

In this model the higher the two indicators, the better. However, in reality, if we adjust the model to increase Recall too much, it can reduce Precision and vice versa. Hence we need these two quantities to be in balance. To solve this problem when choosing how to adjust the model, we have the F1-score is defined as: $F1 = 2\times(Precision \times Recall)/(Precision + Recall)$ In ideal conditions, F1-score = 1, from which we can base the model selection through F1-score.

4.3 Experimental and Evaluation

Compared with the results of related studies such as [10,19], F1-score metrics as well as our two indexes, precision and recall, especially in the first 3 layers, show an effective significantly improved. Particularly at the two floors Conv1 and Conv2, respectively, with accuracy, they all give F1, Precision and Recall results of 1.00.

However, we recommend using Conv3 for further research, because the layer has deeper features of the image for comparison, instead of just limiting the surface evaluation like the first two layers Conv1 & Conv2 as well as avoid a significant drop in accuracy in the last two layers Conv4 & Conv5. To see the efficiency, the proposed method is compared with the baseline methods such as classification method using VGG16 and SVM classifier method using HOG images shown in Table 2.

Table 2. Compare performance metrics proposed method with baseline methods (HOG+SVM and VGG16)

Method	Types	Precision	Recall	F1-score
HOG+SVM	AD	82.2%	86.4%	84.2%
	not_AD	79.5%	82.6%	81.0%
VGG16	AD	91.9%	87.6%	89.7%
	not_AD	93.6%	88.0%	90.7%
Proposed Model	AD	98.0%	95.0%	96.5%
	not_AD	96.0%	98.0%	97.0%

Experimental accuracy is better than other studies, such as [13] when the model also achieves threshold accuracy of 0.86. Some other experiments also apply VGG16 as the classifier integration layer before adding the image to the HOG feature classifier. One of them can be mentioned is [21], experimenting instead of classifying and evaluating images, the authors analyze brand logo images. Usually, the studies on the separation of symbols or labels of images such as logos will incorporate the use of SVM after having obtained the results from HOG image feature calculations such as [22]. This also proves the feasibility of applying SVM with DNN layer as CNN and classifying HOG similarly in the assessment and diagnosis of AD and dermatitis in particular, even including skin cancer and other diseases other skin in general.

5 Conclusion

Our research and experiments have shown certain conclusions in improving the efficiency of detecting and diagnosing atopic dermatitis using deep learning. The main deep learning network used in our research and experiments is CNN, which gives high-precision diagnostic results and highly accurate data images. This shows the potential of this technology not only in the detection of atopic dermatitis in particular but also dermatological diseases in general, and can even be applied in the diagnosis of skin cancer with a large enough data set produce indicators comparable to expert diagnoses.

Similarly, it is possible to combine CNN with many other classes of classifications and DNNs, not only in medical imaging diagnostics, but also in facial recognition or representative logos brand representation. Despite the above-mentioned positive points, there are certain limitations when using deep learning networks to diagnose AD. One of them is the limitation on the number and depth of datasets and the inconsistency in those image datasets. However, the technology has achieved the purpose of comprehensive efficiency improvement and has had practical applications in central hospitals as well as laid the foundation for similar studies.

References

1. Aoki, T., Fukuzumi, T., Adachi, J., Endo, K., Kojima, M.: Re-evaluation of skin lesion distribution in atopic dermatitis. Analysis of cases 0 to 9 years of age. Acta Derm. Venereol. Suppl. **176**, 19–23 (1992)
2. Larsen, F.S., Hanifin, J.M.: Epidemiology of atopic dermatitis. Immunol. Allergy Clin. **22**(1), 1–24 (2002)
3. Hong, S., et al.: The prevalence of atopic dermatitis, asthma, and allergic rhinitis and the comorbidity of allergic diseases in children. Environ. Health Toxicol. **27** (2012)
4. Fuxench, Z.C.C., et al.: Atopic dermatitis in America study: a cross-sectional study examining the prevalence and disease burden of atopic dermatitis in the us adult population. J. Investig. Dermatol. **139**(3), 583–590 (2019)
5. Luoma, R., Koivikko, A., Viander, M.: Development of asthma, allergic rhinitis and atopic dermatitis by the age of five years: a prospective study of 543 newborns. Allergy **38**(5), 339–346 (1983)
6. Silverberg, J., et al.: Symptoms and diagnosis of anxiety and depression in atopic dermatitis in us adults. Br. J. Dermatol. **181**(3), 554–565 (2019)
7. Medela, A., Mac Carthy, T., Robles, S.A.A., Chiesa-Estomba, C.M., Grimalt, R.: Automatic scoring of atopic dermatitis using deep learning: a pilot study. JID Innov. **2**(3), 100107 (2022)
8. Wang, S., Yang, D.M., Rong, R., Zhan, X., Xiao, G.: Pathology image analysis using segmentation deep learning algorithms. Am. J. Pathol. **189**(9), 1686–1698 (2019)
9. Mazurowski, M.A., Buda, M., Saha, A., Bashir, M.R.: Deep learning in radiology: an overview of the concepts and a survey of the state of the art with focus on MRI. J. Magn. Reson. Imaging **49**(4), 939–954 (2019)

10. Li, J.: Developing machine learning and statistical methods for the analysis of genetics and genomics. UCLA Electronic Theses and Dissertations, pp. 119–146 (2021)
11. Simonyan, K., Zisserman, A.: Very deep convolutional networks for large-scale image recognition. CoRR,abs/1409.1556 (2015)
12. Dalal, N., Triggs, B.: Histograms of oriented gradients for human detection. In: 2005 IEEE Computer Society Conference on Computer Vision and Pattern Recognition (CVPR 2005), vol. 1, pp. 886–893, vol. 1 (2005)
13. Wu, H., et al.: A deep learning, image based approach for automated diagnosis for inflammatory skin diseases. Ann. Transl. Med. **8**(9), 581 (2020)
14. Hu, J., Shen, L., Albanie, S., et al.: Squeeze-and-excitation networks. In: Proceedings of the IEEE Conference on Computer Vision and Pattern Recogniton, pp. 7132–7141 (2018)
15. Laitinen, E., Lohan, E.S., Talvitie, J., Shrestha, S.: Access point significance measures in WLAN-based location, pp. 24–29 (2012)
16. Aijaz, S.F., Khan, S.J., Azim, F., Shakeel, C.S., Hassan, U.: Deep learning application for effective classification of different types of psoriasis. J. Healthcare Eng. 2022 (2022)
17. Dwivedi, P., Khan, A.A., Gawade, A., Deolekar, S.: A deep learning based approach for automated skin disease detection using fast R-CNN. In: 2021 Sixth International Conference on Image Information Processing (ICIIP), vol. 6, pp. 116–120. IEEE (2021)
18. Park, S., et al.: Model learning analysis of 3d optoacoustic Mesos copy images for the classification of atopic dermatitis. Biomed. Opt. Express **12**(6), 3671–3683 (2021)
19. Jiang, Z., et al.: Accurate diagnosis of atopic dermatitis by combining transcriptome and microbiota data with supervised machine learning. Sci. Rep. **12**(1), 1–13 (2022)
20. Khan, S., Rahmani, H., Shah, S.A.A., Bennamoun, M.: A guide to convolutional neural networks for computer vision. Synth. Lect. Comput. Vision **8**(1), 1–207 (2018)
21. Kayalibay, B., Jensen, G., van der Smagt, P.: CNN-based segmentation of medical imaging data. arXiv preprint arXiv:1701.03056 (2017)
22. Brownlee, J.: How to visualize filters and feature maps in convolutional neural networks. Machine Learning Mastery (2019)

Prediction of the Welding Process Parameters and the Weld Bead Geometry for Robotic Welding Applications with Adaptive Neuro-Fuzzy Models

Minh Duc Vu[1], Chu Anh My[2(✉)], The Nguyen Nguyen[2], Xuan Bien Duong[3], Chi Hieu Le[4], James Gao[4], Nikolay Zlatov[5], Georgi Hristov[6], Van Anh Nguyen[7], Jamaluddin Mahmud[8], and Michael S. Packianather[9]

[1] Department of Aerospace Engineering, Le Quy Don Technical University, Hanoi, Vietnam
[2] Institute of Simulation Technology, Le Quy Don Technical University, Hanoi, Vietnam
myca@lqdtu.edu.vn
[3] Center of Advanced Technology, Le Quy Don Technical University, Hanoi, Vietnam
[4] Faculty of Science and Engineering, University of Greenwich, Gillingham, Kent ME4 4TB, UK
[5] Institute of Mechanics, Bulgarian Academy of Sciences, 1113 Sofia, Bulgaria
[6] University of Ruse "Angel Kanchev", 8 Studentska Street, 7004 Ruse, Bulgaria
[7] Welding Engineering Laser Processing Centre, Cranfield University, Cranfield, UK
[8] College of Engineering, Universiti Teknologi MARA, Shah Alam, Selangor, Malaysia
[9] School of Engineering, Cardiff University, Cardiff CF24 3AA, UK

Abstract. The weld bead geometry is the important information for determining the quality and mechanical properties of the weldment. The welding process parameters or variables that affect the weld bead geometry in the conventional arc welding process include the following: the welding voltage U, the welding current I, the wire feed speed WFS, the contact tip to work distance D, and the welding speed S. Modeling and predicting the weld bead geometry play an important role in welding process planning, to determine the optimal welding process parameters for achieving the improved weld quality. There have been lots of efforts and studies to develop modeling solutions and simulations to determine the weld bead geometry (Height H and Width W) from the welding process parameters (U, I, WFS, D, S) as the inputs. The welding process parameters can be determined based on the experiences, and the conventional analysis of variance (ANOVA); however, the high welding quality and accuracy are not always obtained. With the advancement of computer vision technologies, digital images from cameras and videos can be used for training the deep learning models, to accurately identify and classify objects. The digital images for evaluating the welding quality and the characteristics of welding objects can be captured via the use of the high-speed camera, and there are emerging data acquisition systems that can handle a huge dataset. In this paper, an adaptive neuro-fuzzy inference system (ANFIS) model is proposed to determine weld bead geometry from the main welding process parameters U, I and S. The proposed ANFIS model was successfully developed for the first basic investigations, as the foundation for further developments of innovative

T. D. L. Nguyen et al. (Eds.): ICISN 2023, LNNS 752, pp. 148–158, 2023.
https://doi.org/10.1007/978-981-99-4725-6_20

robotic welding systems which can be used for higher educations or research in Smart Manufacturing, with potentials for industrial applications.

Keywords: Welding robots · ANOVA · ANFIS · weld bead geometry · GMA welding · gas metal arc welding · GMAW · metal inert gas welding · MIG

1 Introduction

The welding robots have been playing an important role in manufacturing industries. The combination of the advanced welding and robotic technologies leads to the emerging wire-arc additive manufacturing (AM) [1], which is one of the key-enabling technologies for Rapid Manufacturing and Cloud Manufacturing, through the capability of making functional parts and added-value products with the short Time-to-Market, as well as personalized and mass-customized design features. Under the impacts of Smart Manufacturing and Industry 4.0 [2–4], there have been an emerging need of investigating optimizations of process parameters for manufacturing processes, including determination of optimal parameters in the wire arc additive manufacturing in particular and the welding and machining processes in general [1, 5–12, 16], to predict and enhance the quality of the welding, AM and machining processes, as well as developing the effective computer aided process planning (CAPP), based on the big data analytics, Artificial Intelligence (AI), machine learning and simulations.

The weld bead geometry is the important information for determining the quality and mechanical properties of the weldment. Figure 1 presents the typical weld bead geometry, including bead height, bead width, and the depth of penetration. The welding process parameters or variables that affect the weld bead geometry in the conventional arc welding process include the following: the welding voltage U, the welding current I, the wire feed speed WFS, the contact tip to work distance D, and the welding speed S. Modeling and predicting the weld bead geometry play an important role in welding process planning, to determine the optimal welding process parameters or welding parameters for achieving the improved weld quality.

There have been lots of efforts of innovating the welding technologies, and developing simulations and optimizations of the welding process. With the advancement of computer vision technologies, digital images from cameras and videos can be used for training the deep learning models, to accurately identify and classify objects. Therefore, the digital images for evaluating the welding quality and the characteristics of welding objects can be captured via the use of the high-speed camera, and there are emerging data acquisition systems that can handle a huge dataset. Zhuohua *et al.* (2018) proposed a vision-based method for three-dimensional (3D) control in the robotic welding with the steel sheets, based on the digital images from passive vision sensors, in which the process parameters can be monitored, and the role of process parameters and their relationships are investigated [7]. There are different sensing technologies that are used for data collection and evaluation of the weld pool state in the primary arc welding process for precision joining of metals [8]. 3D vision sensing technologies can be used to develop imaging and measurement systems for evaluation of welding quality and determination

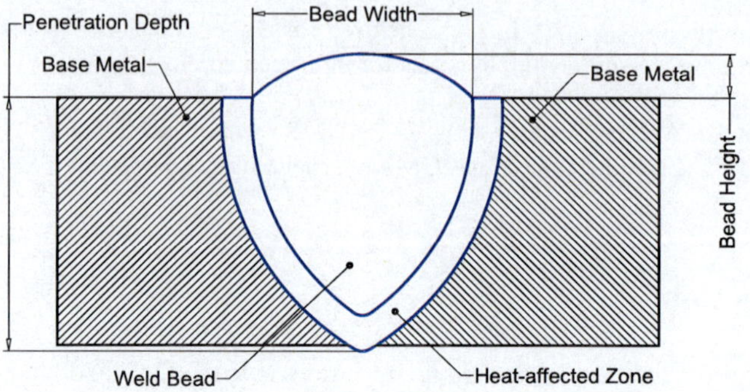

Fig. 1. A typical weld bead geometry which is affected by the welding process parameters, including the welding voltage, the welding current, the wire feed speed, the contact tip to work distance, and the welding speed.

of welding parameters, especially for optimal dynamic modeling and penetration control strategies in gas tungsten arc welding (GTAW), to predict the depth of penetration with the *adaptive neuro-fuzzy inference system* (ANFIS) models [9, 10]. Huabin *et al.* (2009) developed a neural network (NN) model to calculate the full penetration state for the closed-loop control of robotic arc welding systems [11]; in which the welding current, welding voltage, weld pool geometry and changing ratio of bead width act as input for the back side bead width output related to the depth penetration; and the neuron network (NN) works on 1,000 training datasets and 200 testing datasets. D. T. Thao *et al.* (2014) reported the use of experimental data and analysis of variance (ANOVA) to describe the formular of the bead width from different tool tip distance T, gas flow rate G, welding speed G, arc current I, welding voltage V, to predict process parameters on top-bead width in robotic Gas Metal Arc (GMA) welding process [12], in which the genetic algorithm (GA) was utilized to estimate the coefficients of an intelligent model. Jamie *et al.* (2016) proposed an architecture based on an artificial neural network (ANN), to learn welding skills automatically in industrial robots, in which an optic camera and a laser-based sensor were used to measure the bead geometry (width and height), and a real-time computer vision algorithm was developed to extract training patterns in order to acquire knowledge to later predict specific geometries of the weld bead [13]. A deep learning approach can be used to estimation of the bead parameters in welding tasks; Soheil *et al.* (2015) presented the ANN algorithm which was applied to capture the response of 2 outputs (depth of penetration, bead width) with 4 hidden layers, 4 input parameter (voltage, current, traveling speed, wire speed); and it was reported that, the deep learning approach could reduce the error of RMS (Root-Mean-Square) bead parameters remarkably [14].

This study aims to investigate solutions for predicting the welding process parameters and the weld bead geometry for robotic welding applications based on the adaptive neuro-fuzzy models. Firstly, the structure of a digital twin model mentioned in [15] is used for conducting the experiments and obtaining the datasets. Then, the ANFIS model is developed with the use of the obtained datasets, to estimate the weld bead width from the key welding process parameters as the inputs: the welding speed S, the welding voltage

U, and the welding current I. In this way, the relationships between the welding process parameters and the weld bead geometry can be determined, and the welding process parameters can be optimally computed for robotic welding process planning and related applications in Smart Manufacturing.

The rest of the paper is organized as follows. Section 2 presents the materials and methods, with the focus on obtaining the datasets for the experiments, and then predicting the weld bead geometry based on the ANFIS model. Section 3 presents the results and discussions. Finally, Sect. 4 presents conclusions and further work.

2 Materials and Methods

2.1 Datasets for the Bead Width Obtained from the Experiments

The welding experiments were done by using the welding robot system RV-12SD integrated with the welding equipment Jassic Mig 250 A (Fig. 2).

Fig. 2. The integrated welding robot system

For the experiments, the straight welding butt joints are taken into account. The two welded parts are put together in the same plane. The welding seam is 10 cm long. The main welding parameters for the experiments are as follows:

- Welding speed S: 18–30 cm/min
- Welding voltage U: 20–25 V
- Welding current I: 150–250 A
- Welding wire diameter: 1.6 mm
- The welding nozzle's distance: 10 mm

The output geometry of the weld bead includes the bead width W and the bead height H which are extracted by the vision-based edge detecting method with the use of the laser projection beam. In the Labview system, the output geometry can be conveniently calculated with the use of the Vision Assistant tools. The bead width is calculated as the average value computed along to the whole welding seam. Tables 1 and 2 presents the data obtained from the experiments, Table 1 is data from wide range of inputs and Table 2 is narrow range.

Table 1. The datasets obtained from the experiments with the large varied speed. S, I, U, W are the welding voltage, the welding current, the welding speed and welding bead width respectively.

S (Cm/min)	I (Ampe)	U (Volt)	W (mm)	S (Cm/min)	I (Ampe)	U (Volt)	W (mm)
18	150	20	6.48	22	220	25	8.50
18	180	20	7.07	22	250	25	9.04
18	200	25	9.02	19	150	20	6.30
18	220	25	9.44	19	180	20	6.88
18	250	25	10.04	19	200	25	8.77
30	150	20	4.97	19	220	25	9.18
30	180	20	5.42	19	250	25	9.76
30	200	25	6.91	24	150	20	5.58
30	220	25	7.24	24	180	20	6.09
30	250	25	7.69	24	200	25	7.76
20	150	20	6.14	24	220	25	8.13
20	180	20	6.70	24	250	25	8.64
20	200	25	8.54	26	150	20	5.35
20	220	25	8.93	26	180	20	5.84
20	250	25	9.50	26	200	25	7.45
22	150	20	5.84	26	220	25	7.80
22	180	20	6.37	26	250	25	8.29
22	200	25	8.12	22	220	25	8.50

Table 2. The datasets obtained from the experiments with the smaller varied speed. S, I, U, W are the welding voltage, the welding current, the welding speed and welding bead width respectively.

S (Cm/min)	I (Ampe)	U (Volt)	W (mm)
28	150	20	5.15
28	180	20	5.62
28	200	25	7.17
28	220	25	7.50
28	250	25	7.98
18	140	20	6.27
18	170	20	6.88
18	190	25	8.80
18	210	25	9.23
18	240	25	9.84
20	140	20	5.94
20	170	20	6.52
20	190	25	8.33
20	210	25	8.74

(continued)

Table 2. *(continued)*

S (Cm/min)	I (Ampe)	U (Volt)	W (mm)
20	240	25	9.32
28	150	20	5.15
28	180	20	5.62
28	200	25	7.17
28	220	25	7.50
28	250	25	7.98

2.2 Development of the ANFIS Model for Predicting the Weld Bead Geometry

The ANFIS model has the input membership functions, the output membership functions and the rules (the Sugeno rule). The datasets are divided for the training, in which 70% of the dataset are for training and 30% of the dataset are for testing.

Figure 3 presents the ANFIS model structure. The inputs of the ANFIS model include the following: the welding voltage U, the welding current I, and the welding speed S. The output of the ANFIS model is the welding bread width W.

Figures 4, 5 present the membership functions for the welding voltage U, the welding current I and the welding speed S respectively.

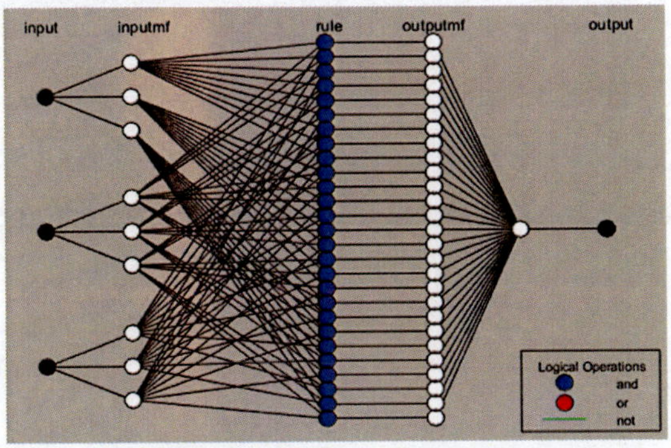

Fig. 3. The ANFIS model in which the output is the welding bread width W, and the inputs include the welding voltage U, the welding current I, and the welding speed S.

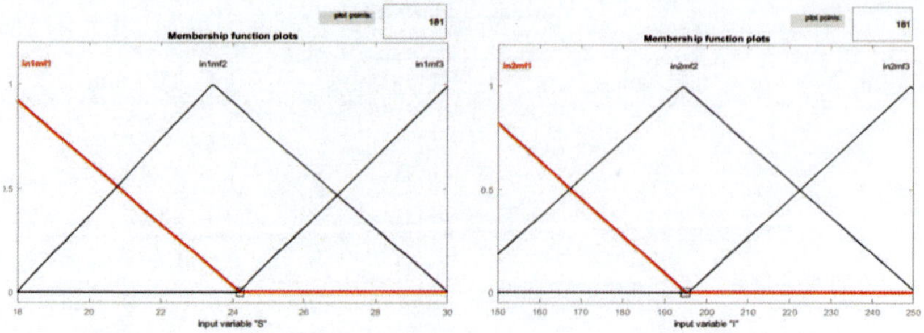

Fig. 4. The membership function for the welding speed S and current I

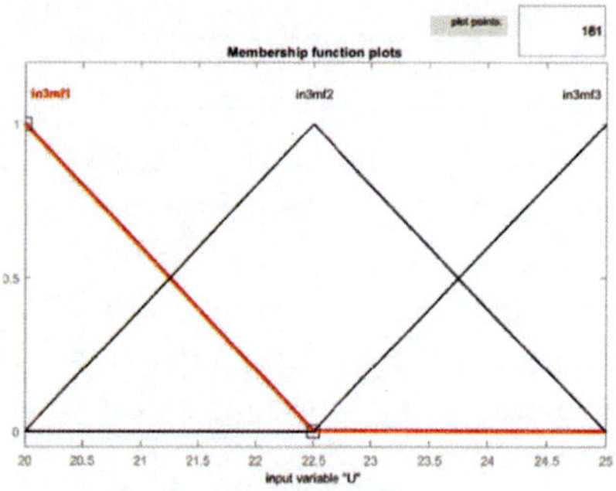

Fig. 5. The membership function for the welding voltage U

3 Results and Discussions

The designed ANFIS model has been trained with Epoch = 100. After training process, the training error value was reduced and less than 2.82 * 10^{-3}, which is acceptable and less than the allowable error of a welding seam geometry (Fig. 6).

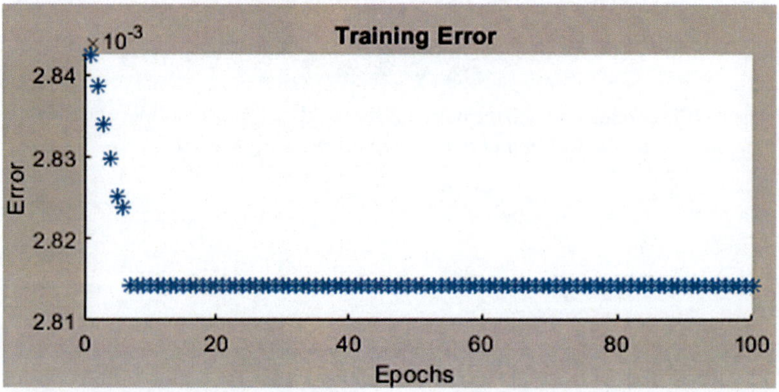

Fig. 6. Training error after 100 epochs

It has shown that, for the training data, the accuracy $R^2 = 0.9991$; for test data, $R^2 = 0.9757$. See Fig. 7.

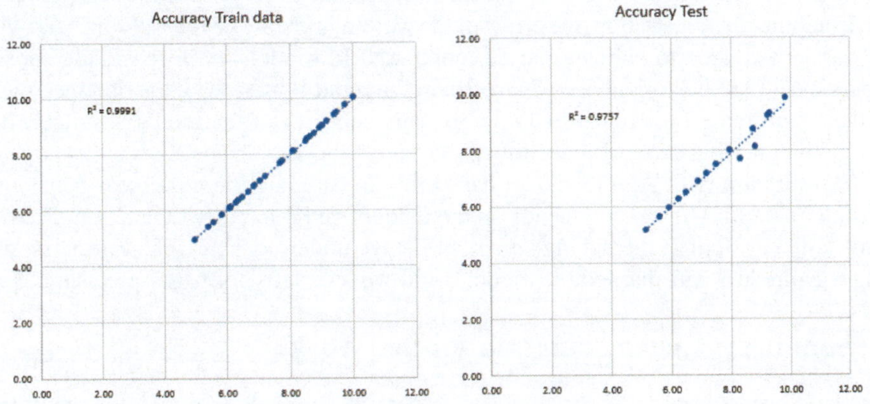

Fig. 7. Accuracy estimation

To validate the designed ANFIS model, the flowing demonstration was carried out and the performance of the trained model was evaluated with respect to background and knowledge of the welding process. In this manner, the inputs S = 25 cm/min and U = 24 V were kept as constants, and the welding current I was increased steps by steps as

Table 3. Effect of the welding current I and speed S on the welding bead width W

I (Ampe)	W (mm)
180	3.29
185	3.7
190	4.11
195	4.51
200	4.56
205	4.61
210	4.66
220	4.77
225	4.82
230	4.87

a

S (cm/min)	W (mm)
20	5.09
21	4.97
22	4.85
23	4.74
24	4.63
25	4.56
26	4.45
28	4.29
29	4.21
30	4.15

b

shown in Table 3a. As shown in Table 3a, when the welding current I increased from 180 A to 230 A, the bead width W of the welding seam also increased from 3.29 mm to 4.87 mm. It is clearly seen that when the input I is increasing the output W is increasing, accordingly, which is matching with the physical characteristics of a welding process.

In another demonstration, the effect of the welding speed S on the welding bead width W was investigated to validate the designed ANFIS model. In this case, the constant inputs were I = 200 A, and U = 24 V. When the input S increased, the obtained output values W decreased as shown in Table 3b. This behavior of the model is suitable with the welding performance of a welding process.

The solution with ANFIS model has ability to estimate the non-linear behavior of welding process. The result depends on the amount and reliability of data set. There are some small errors but the average error still stays under considerable. Furthermore, if data is calibrated and sharped the model could work exactly for better precision.

4 Summary, Conclusions and Future Work

The weld bead geometry is the important information for determining the quality and mechanical properties of the weldment. The welding process parameters or variables that affect the weld bead geometry in the conventional arc welding process include the following: the welding voltage U, the welding current I, the wire feed speed WFS, the contact tip to work distance D, and the welding speed S. Modeling and predicting the weld bead geometry play an important role in welding process planning, to determine the optimal welding process parameters for achieving the improved weld quality. In the robotic GMAW or MIG welding, a robotic welding system basically consists of two functional units: the robot arm and a GMAW or MIC welding station. A robot arm automatically moves the robotic welding torch, following the programmed tool-path only; meanwhile, the welding process is controlled independently by a GMAW welding station, with the relevant welding process parameters. There have been lots of efforts and studies to develop modeling solutions and simulations to determine the weld bead geometry (Height H and Width W) from the welding process parameters (U, I, WFS, D, S) as the inputs. The welding process parameters can be determined based on the experiences, and the conventional analysis of variance (ANOVA); however, the high welding quality and accuracy are not always obtained, especially when there are additional constraints and technical requirements. With the advancement of computer vision technologies, digital images from cameras and videos can be used for training the deep learning models, to accurately identify and classify objects. The digital images for evaluating the welding quality and the characteristics of welding objects can be captured via the use of the high-speed camera, and there are emerging data acquisition systems that can handle a huge dataset. This study presents an ANFIS model-based approach to determine weld bead geometry from the main welding process parameters U, I and S. The ANFIS model was successfully developed for the first basic investigations, as the foundation for further developments of innovative robotic welding systems which can be used for higher educations or research in Smart Manufacturing, with potentials for industrial applications.

Further studies will focus on enhancements of the ANFIS model-based solutions and predictive models for effectively determining the welding process parameters, with the real-time monitoring of the weld bead geometry and data collections for Digital Twins and Smart Manufacturing applications, as well as to extend the robotic welding applications to the emerging wire-arc AM, which is one of the key-enabling technologies for Rapid Manufacturing and Cloud Manufacturing.

Acknowledgement. This research is funded by Vietnam National Foundation for Science and Technology Development (NAFOSTED) under grant number 107.01-2020.15.

References

1. Petrik, J., et al.: Beyond parabolic weld bead models: AI-based 3D reconstruction of weld beads under transient conditions in wire-arc additive manufacturing. J. Mater. Process. Technol. **302**(2022), 117457 (2022)
2. Le, C.H., et al.: Challenges and conceptual framework to develop heavy-load manipulators for smart factories. Int. J. Mechatron. Appl. Mech. **2020**(8), 209–216 (2020)
3. Arey, D., et al.: Lean industry 4.0: a digital value stream approach to process improvement. Procedia Manuf. **54**, 19–24 (2021). ISSN 2351-9789
4. Daniel, A., et al.: An investigation into the adoption of automation in the aerospace manufacturing industry. In: Advances in Manufacturing Technology XXXIII, pp. 87–92. IOS Press (2019). ISBN: 1643680099. https://doi.org/10.1007/s00170-018-1897-x
5. Nguyen, T.-T., Le, C.-H.: Optimization of compressed air assisted-turning-burnishing process for improving machining quality, energy reduction and cost-effectiveness. J. Eng. Manuf. Proc. Inst. Mech. Eng. Part B **235**(6–7) (2020). https://doi.org/10.1177/0954405420976661
6. Singh, C.J., et al.: Optimization of FFF process parameters by naked mole-rat algorithms with enhanced exploration and exploitation capabilities. Polymers **13**(11), 1702, 2073–4360 (2021). https://doi.org/10.3390/polym13111702
7. Yu, Z., He, Y., Xu, Y., Chen, H.: Vision-based deviation extraction for three-dimensional control in robotic welding with steel sheet. Int. J. Adv. Manuf. Technol. **95**(9–12), 4449–4458 (2018). https://doi.org/10.1007/s00170-017-1546-9
8. Wang, X.: Three-dimensional vision-based sensing of GTAW: a review. Int. J. Adv. Manuf. Technol. **72**(1–4), 333–345 (2014). https://doi.org/10.1007/s00170-014-5659-0
9. Wang, Z.: An imaging and measurement system for robust reconstruction of weld pool during arc welding. IEEE Trans. Ind. Electron. **62**(8), 5109–5118 (2015). https://doi.org/10.1109/TIE.2015.2405494
10. Wang, X.: Three-dimensional vision applications in GTAW process modeling and control. Int. J. Adv. Manuf. Technol. **80**(9–12), 1601–1611 (2015). https://doi.org/10.1007/s00170-015-7063-9
11. Chen, H., et al.: Closed-loop control of robotic arc welding system with full-penetration monitoring. J. Intell. Robot. Syst. **56**, Article no. 565 (2009)
12. Thao, D.T., Kim, I.S., Na, H.H., Jung, S.M., Shim, J.Y.: Development of mathematical model with a genetic algorithm for automatic GMA welding process. Int. J. Adv. Manuf. Technol. **73**(5–8), 837–847 (2014). https://doi.org/10.1007/s00170-014-5842-3
13. Aviles-Viñas, J.F., Rios-Cabrera, R., Lopez-Juarez, I.: On-line learning of welding bead geometry in industrial robots. Int. J. Adv. Manuf. Technol. **83**(1–4), 217–231 (2015). https://doi.org/10.1007/s00170-015-7422-6
14. Keshmiri, S., et al.: Application of deep neural network in estimation of the weld bead parameters. In: 2015 IEEE/RSJ International Conference on Intelligent Robots and Systems (IROS) (2015). https://doi.org/10.1109/IROS.2015.7353868

15. Vu, M.D., et al.: A conceptual digital twin for cost-effective development of a welding robotic system for smart manufacturing. In: Long, B.T., Kim, Y.-H., Ishizaki, K., Toan, N.D., Parinov, I.A., Vu, N.P. (eds.) MMMS 2020. LNME, pp. 1018–1025. Springer, Cham (2021). https://doi.org/10.1007/978-3-030-69610-8_134
16. My, C.A., et al.: Inverse kinematic control algorithm for a welding robot-positioner system to trace a 3D complex curve. In: 2019 International Conference on Advanced Technologies for Communications (ATC). IEEE (2019). https://doi.org/10.1109/ATC.2019.8924540

Simplified Model Predictive Current Control to Balance Neutral-Point Voltage for Three-Level Sparse Four-Leg VSI

Dang Khoa Nguyen[1] and Huu-Cong Vu[2(✉)]

[1] Faculty of Applied Sciences, International School, Vietnam National University, Hanoi, Vietnam

[2] Department of Electrical Engineering, Hanoi University of Civil Engineering, Hanoi, Vietnam

congvh@huce.edu.vn

Abstract. This paper proposes a novel approach for balancing the neutral-point voltage (NPV) of the three-level sparse four-leg voltage source inverter (VSI) using a simplified model predictive current control (MPCC) method. By using a new cost function, the proposed method finds the best voltage vector without the need for current prediction calculations. The best switching state from the best voltage vector will be selected based on the deviation of two capacitor voltages (DTCV) to balance the NPV, without the use of weighting factor and capacitor voltage calculations. All results from simulation show that the proposed method significantly reduces computation time and eliminates the time-consuming weighting factor adjustment required by the conventional MPCC method.

Keywords: Current Control · Balance Capacitor Voltages · Four-Leg Voltage Source Inverter (VSI)

1 Introduction

Nowadays, three-level voltage source inverters (VSIs) have gained significant attention in various power conversion applications such as photovoltaic grid-connected, renewable energy generation, and motor drive systems [1–3]. When compared to 2-level VSIs, they have obvious advantages, such as better total harmonic distortion, better reliability, and less electromagnetic interference issues [4, 5].

On other hand, unbalanced load condition can destabilize the system and deteriorate power quality [6]. The three-level three-leg (TLTL) VSIs cannot address this issue since these topologies have lack available paths to circulate the zero-sequence currents. To address the drawback of the TLTL-VSIs under unbalanced load condition, the three-level neutral point clamped (TLNPC) four-leg VSI and the three-level T-type four-leg VSI were introduced [7, 8]. However, these topologies exhibit some drawbacks, including decreased reliability, increased system volume, and higher cost due to a larger number of power switches. To address these problems, a three-level sparse four-leg VSI is developed in [9] and is illustrated in Fig. 1. This topology offers a reduction of four active switches which is compared to the TLNPC four-leg VSI.

T. D. L. Nguyen et al. (Eds.): ICISN 2023, LNNS 752, pp. 159–168, 2023.
https://doi.org/10.1007/978-981-99-4725-6_21

Some methods have been proposed to regulate three-level four-leg VSIs (TLFL-VSIs), like carrier-based PWM [8], space vector pulse-width modulation (PWM) [7], and model-predictive current control (MPCC) [10]. MPCC has gained recognition as a compelling and effective current control technique for TLFL-VSIs owing to its simple and intuitive nature and quick dynamic response [10, 11]. In [9], an MPCC method was established to manage the three-level sparse four-leg VSI, but it does not account for balancing capacitor voltages, which can deteriorate the output voltage performance and raise the voltage across on power switches. Additionally, this technique necessitates a large number of computations, resulting in high computation time.

This paper proposes a simplified MPCC to control the three-level sparse four-leg TLSFL inverter to address these issues. Initially, the reference voltage vector is obtained using the prediction model of the output current. A new cost function is then is established to choose the best voltage vector. Finally, the optimal switching state is determined, taking into account the impact of redundant switching states on capacitor voltages, to balance the capacitor voltages. The proposed method eliminates the need for current and capacitor voltage prediction calculations, as well as the weighting factor adjustment, simplifying the fulfillment process and significantly reducing the computational burden.

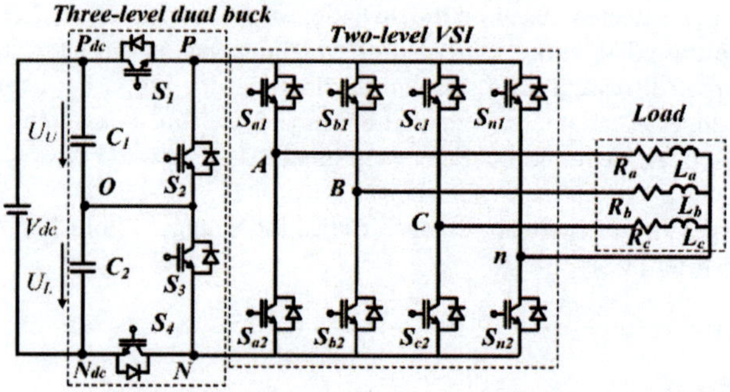

Fig. 1. Three-level sparse four-leg VSI.

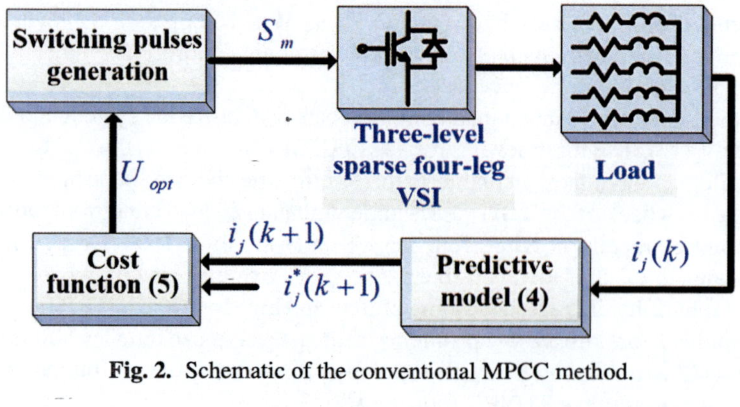

Fig. 2. Schematic of the conventional MPCC method.

2 Conventional MPCC Method for the TLSFL Inverter

Figure 1 depicts the circuit topology of a TLSFL inverter, which is composed of 12 active switches. In this configuration, the DC-bus voltage is provided through two DC-link capacitors connected in series.

Table 1. Selection of Best Switching Sate From The Best Voltage Vector in Group I

The best voltage vector	The DTCV	The best switching state	The best voltage vector	The DTCV	The best switching state
U_0	$\Delta U_{NP} > 0$ or $\Delta U_{NP} < 0$	PPPP or OOOO or NNNN	U_8	$\Delta U_{NP} > 0$ or $\Delta U_{NP} < 0$	PPNP
U_1	$\Delta U_{NP} > 0$ or $\Delta U_{NP} < 0$	PNNN	U_9	$\Delta U_{NP} > 0$ or $\Delta U_{NP} < 0$	NPPP
U_2	$\Delta U_{NP} > 0$ or $\Delta U_{NP} < 0$	NPNN	U_{10}	$\Delta U_{NP} > 0$ or $\Delta U_{NP} < 0$	PNPP
U_3	$\Delta U_{NP} > 0$ or $\Delta U_{NP} < 0$	NNPN	U_{11}	$\Delta U_{NP} > 0$ or $\Delta U_{NP} < 0$	PNNP
U_4	$\Delta U_{NP} > 0$ or $\Delta U_{NP} < 0$	PPNN	U_{12}	$\Delta U_{NP} > 0$ or $\Delta U_{NP} < 0$	NPNP
U_5	$\Delta U_{NP} > 0$ or $\Delta U_{NP} < 0$	NPPN	U_{13}	$\Delta U_{NP} > 0$ or $\Delta U_{NP} < 0$	NNPP
U_6	$\Delta U_{NP} > 0$ or $\Delta U_{NP} < 0$	PNPN	U_{14}	$\Delta U_{NP} > 0$ or $\Delta U_{NP} < 0$	NNNP
U_7	$\Delta U_{NP} > 0$ or $\Delta U_{NP} < 0$	PPPN			

The voltage between the output terminal X and the neutral-point O can be expressed as [5]:

$$U_{XO} = \frac{H_1(H_t + 1) + H_2(H_t - 1)}{2}\frac{V_{dc}}{2}; X = A, B, C, n \tag{1}$$

where

$$H_t = \begin{cases} 1 \text{ if } S_{t1} = 1 \textit{ and } S_{t2} = 0 \\ -1 \text{ if } S_{t1} = 0 \textit{ and } S_{t2} = 1 \end{cases}; (t = a, b, c, n)$$

$$H_1 = \begin{cases} 1, \ S_1 \text{ is } ON \text{ and } S_2 \text{ is } OFF \\ 0, \ S_1 \text{ is } OFF \text{ and } S_2 \text{ is } ON \end{cases}$$

$$H_2 = \begin{cases} 0, \ S_3 \text{ is } ON \text{ and } S_4 \text{ is } OFF \\ 1, \ S_3 \text{ is } OFF \text{ and } S_4 \text{ is } ON \end{cases}$$

According to Eq. (1), the output voltage exhibits a tripartite division ($V_{dc}/2$, 0, $-V_{dc}/2$), which is further symbolized by three distinct switching states, denoted as P, O, and N. The switching states of the three-level sparse four-leg VSI are represented by ordered sets $S_A S_B S_C S_n$. The complete catalog of all 45 possible switching states is presented in Table 1 and Table 2. Furthermore, these 45 switching states can be expressed as 29 space voltage vectors.

Table 2. Selection of Best Switching Sate From The Best Voltage Vector in Group II

The best voltage vector	The DTCV	The best switching state	The best voltage vector	The DTCV	The best switching state
U_{15}	$\Delta U_{NP} > 0$	POOO	U_{22}	$\Delta U_{NP} > 0$	POPO
	$\Delta U_{NP} < 0$	ONNN		$\Delta U_{NP} < 0$	ONON
U_{16}	$\Delta U_{NP} > 0$	OPOO	U_{23}	$\Delta U_{NP} > 0$	OPPO
	$\Delta U_{NP} < 0$	NONN		$\Delta U_{NP} < 0$	NOON
U_{17}	$\Delta U_{NP} > 0$	OOPO	U_{24}	$\Delta U_{NP} > 0$	POOP
	$\Delta U_{NP} < 0$	NNON		$\Delta U_{NP} < 0$	ONNO
U_{18}	$\Delta U_{NP} > 0$	PPOP	U_{25}	$\Delta U_{NP} > 0$	OPOP
	$\Delta U_{NP} < 0$	OONO		$\Delta U_{NP} < 0$	NONO
U_{19}	$\Delta U_{NP} > 0$	POPP	U_{26}	$\Delta U_{NP} > 0$	OOPP
	$\Delta U_{NP} < 0$	ONOO		$\Delta U_{NP} < 0$	NNOO
U_{20}	$\Delta U_{NP} > 0$	OPPP	U_{27}	$\Delta U_{NP} > 0$	PPPO
	$\Delta U_{NP} < 0$	NOOO		$\Delta U_{NP} < 0$	OOON
U_{21}	$\Delta U_{NP} > 0$	PPOO	U_{28}	$\Delta U_{NP} > 0$	OOOP
	$\Delta U_{NP} < 0$	OONN		$\Delta U_{NP} < 0$	NNNO

The output phase voltages U_{pn} ($p =$ A, B, C) according to the pole voltages are:

$$U_{pn} = U_{pO} - U_{nO} \tag{2}$$

The continuous-time dynamic equation of each phase can be obtained by applying Kirchhoff's voltage law as:

$$U_{pn} = L_z \frac{di_j}{dt} + R_z i_j \tag{3}$$

where i_j ($j =$ A, B, C) are the output current of phase j, while R_z ($z =$ a, b, c) and L_z are the inductances and resistances of the load, respectively.

A discrete-time prediction model for output current can be obtained by utilizing forward Euler approximation with the sampling period of T_s

$$i_j(k+1) = i_j(k) + \frac{T_s}{L_z}\big(U_{pn}(k) - R_z i_j(k)\big) \tag{4}$$

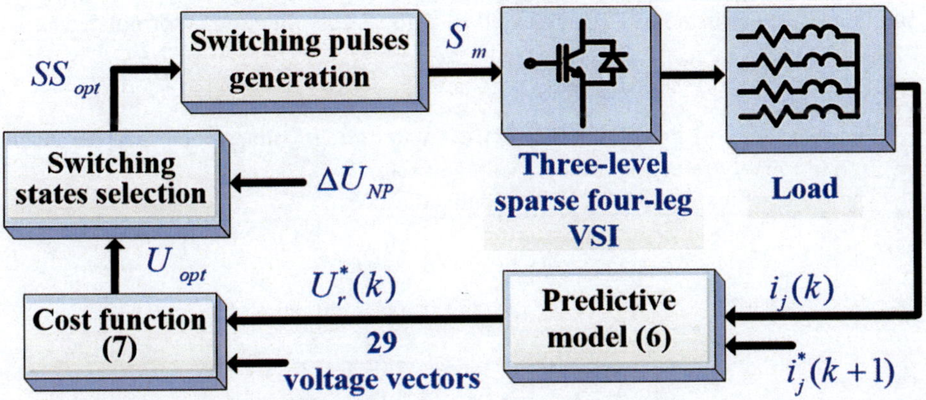

Fig. 3. Schematic of the proposed MPCC method.

In order to track the current reference, a predefined cost function is designed as [5]:

$$g = \sum_{j=A,B,C} \left| i_j^*(k+1) - i_j(k+1) \right| \tag{5}$$

where i_j^* are the output reference current.

The schematic of the conventional MPCC method for the TLSFL inverter is illustrated in Fig. 2.

3 Proposed MPCC Method for TLSFL Inverter

The conventional method for the TLSFL inverter requires a large number of calculations including 87 (29 * 3) current predictions and 29 cost function evaluations. Moreover, the balanced capacitor voltages are not considered in the conventional MPCC method. To address these problems, a simplified MPCC method is proposed in this section. The proposed method involves obtaining a reference voltage vector using an output current prediction model, selecting an optimal voltage vector based on a new cost function, and determining the best switching state that balances the capacitor voltages. The proposed method eliminates the need for current prediction and capacitor voltage prediction calculations, and avoids weighting factor adjustment, simplifying its fulfillment and significantly reducing execution time.

To avoid the need for current prediction calculations, the proposed MPCC method introduces the reference voltage vector $\mathbf{U}_{ABC}^* = \left[U_{An}^* \; U_{Bn}^* \; U_{Cn}^* \right]^T$ for the optimal voltage vector determination. Each component of the voltage vector U_r^* ($r = An, Bn, Cn$) is derived from the prediction model of output current in (4):

$$U_r^*(k) = \frac{L_Z}{T_s} i_j^*(k+1) + \frac{R_Z * T_s - L_Z}{T_s} i_j(k) \tag{6}$$

The selection of the best voltage vector in the proposed method is contingent upon its proximity to the voltage vector U_{ABC}^*, which is known to generate the most favorable output current performance. To achieve this, a novel cost function is formulated as

$$g_{new} = \left|U_{An}^{*}(k) - U_{An}\right| + \left|U_{Bn}^{*}(k) - U_{Bn}\right| + \left|U_{Cn}^{*}(k) - U_{Cn}\right| \quad (7)$$

The cost function, as stated in Eq. (7), evaluates all 29 voltage vectors of the TLSFL inverter, ultimately selecting the best voltage vector.

The NPV, which is the DTCV, is calculated as:

$$\Delta U_{NP} = U_U - U_L \quad (8)$$

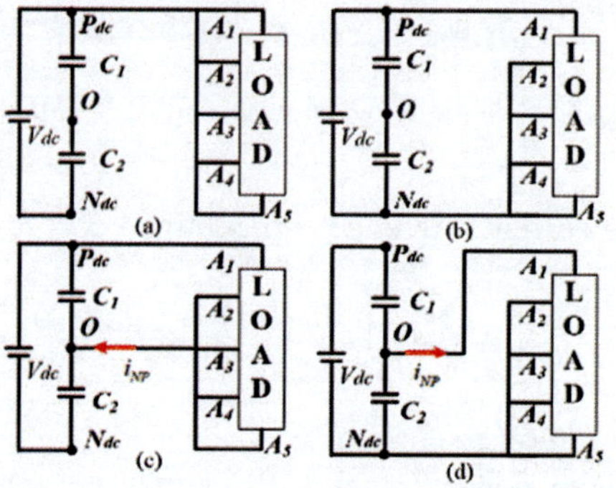

Fig. 4. The influences of switching states on the neutral-point voltage.

The TLSFL inverter can be characterized by a set of 29 voltage vectors. These vectors can be categorized into two distinct groups based on their impact on the NPV. The first group, referred to as Group I, consists of 15 voltage vectors ranging from $\boldsymbol{U}_0$ to $\boldsymbol{U}_{14}$, while the second group, labeled as Group II, encompasses 14 voltage vectors from $\boldsymbol{U}_{15}$ to $\boldsymbol{U}_{28}$.

Figure 4 provides an analysis of the impact of switching states on the capacitor voltages. The switching states from the voltage vectors in group I do not affect the capacitor voltages because the neutral point O is left unconnected, as in Figs. 4 (a), (b). On the other hand, each voltage vector in Group II comprises two switching states, namely the P-type and the N-type, that exhibit opposite effects on the NPV. For example, the P-type switching state [POOO] of the voltage vector $\boldsymbol{U}_{15}$ reduce the NPV, as illustrated in Fig. 4(c). Conversely, the negative switching state [ONNN] increases the NPV, as in Fig. 4(d). Thus, the switching states from group II are utilized to maintain balancing the capacitor voltages in the proposed MPCC method. If $U_U > U_L$, the P-type switching states are utilize maintain balancing the capacitor voltages, while the N-type switching states are chosen if $U_U < U_L$. The best switching state to maintain balancing the capacitor voltages are illustrated in Table 1 and Table 2 which uses the best voltage vector determined by the cost function in Eq. (7). The block diagram of the proposed MPCC method is presented in Fig. 3.

Table 3 shows an evaluation of computational complexity for two MPCC methods. As demonstrated in the table, the proposed method utilizes a mare three calculations of the

Table 3. Computational burden comparison

Method	Number of voltage vectors	Calculations of current	Calculations of voltage
Conventional	29	29 * 3	0
Proposed	29	0	1 * 3

reference voltage vector to eliminate the 87 calculations required for current prediction in the conventional method. This reduction in computational burden is substantial and noteworthy.

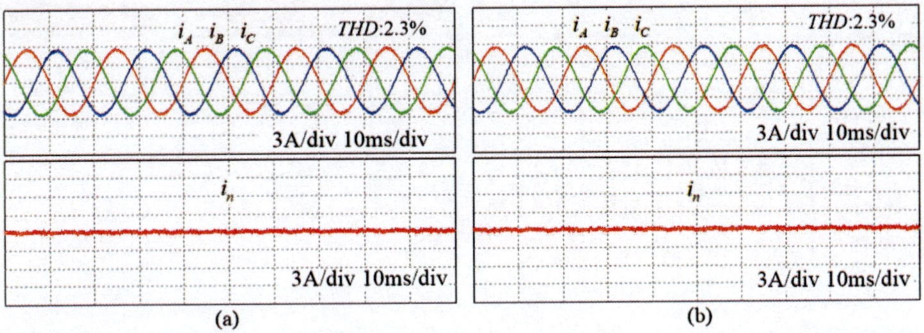

Fig. 5. Simulation outcomes of both methods under balanced load. (a) the conventional method. (b) the proposed method.

4 Simulation Results

To assess the credibility of the proposed MPCC method, a comparison with the conventional MPCC method [9] was conducted to regulate the three-level sparse four-leg VSI. Since the conventional method did not account for balanced NPV, two DC sources were used to maintain a safe and accurate operation of the inverter. The simulation parameters used in this comparison were a DC voltage of 150 V, sampling time ($T_s = 50\ \mu s$), and reference frequency of f = 50 Hz.

Figure 5 displays the simulation results of two methods: the conventional MPCC method [9] and the proposed MPCC method in a steady state with balanced references ($i_j^* = 3.5$ A) and balanced load ($R_a = R_a = R_c = 13\ \Omega$, $L_a = L_a = L_c = 15$ mH). As demonstrated in the figure, both methods exhibit accurate tracking of their references, with a total harmonic distortion (THD) value of 2.3%, and a neutral current i_n that is zero. Additionally, Fig. 6 illustrates the output voltage U_{AO} and the DTCV with the proposed method.

Notably, the output voltage exhibits three voltage levels, and the NPV is precisely regulated within a specific range of 1 V. To demonstrate the efficacy of the proposed method, Fig. 7 depicts its performance when the balanced capacitor voltages algorithm

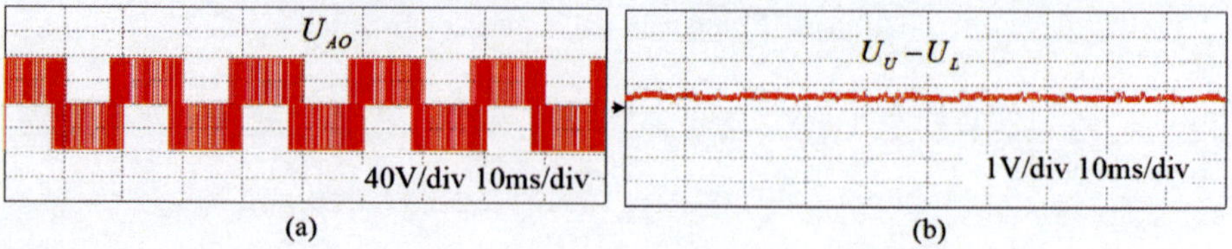

Fig. 6. Steady-state performances under the proposed method. (a) The output voltage U_{AO}. (b) DTVC.

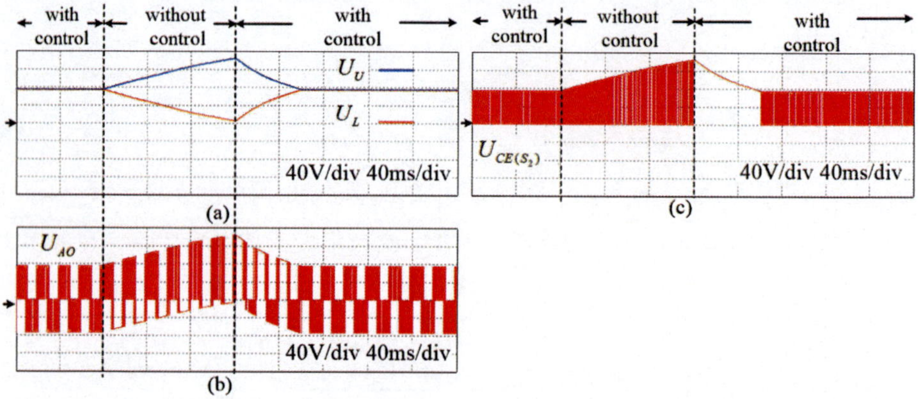

Fig. 7. Simulation outcomes of the proposed method when the balanced the neutral-point capacitor voltages is enabled and disabled. (a) upper and lower capacitor voltages, (b) the output voltage U_{AO}, and (c) the voltage across on the switch S_2.

is enabled and disabled. As depicted, deactivating the balanced capacitor voltages algorithm causes the capacitor voltages to diverge, leading to deteriorate the output voltage performance and raise the voltage across on power switches. Conversely, activating the balanced capacitor voltages algorithm make the capacitor voltages to converge to their desired value of 75 V.

Figure 8 presents a comparative analysis of the dynamic performance of the conventional method and the proposed method. The evaluation is conducted under the condition of a sudden change in the reference current from ($i_j^* = 3.5$ A) to ($i_j^* = 5.5$ A). The results demonstrate that the proposed method exhibits a desirable dynamic response similar to that of the conventional method.

Furthermore, to reinforce the effectiveness of the proposed method, Fig. 9 illustrates the simulation outcomes of the proposed method in steady state under both balanced references ($i_j^* = 3.5$ A) and unbalanced load ($R_a = R_b = 9\ \Omega$, $R_c = 13\ \Omega$, $L_a = L_b = L_c = 15$ mH). As evident from the figure, the output currents are able to track the references effectively, while the capacitor voltages remain balanced around their normal values.

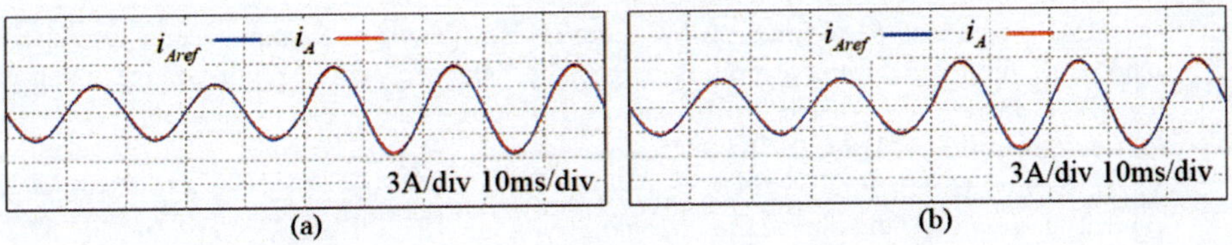

Fig. 8. Dynamic performance. (a) The conventional method. (b) The proposed method.

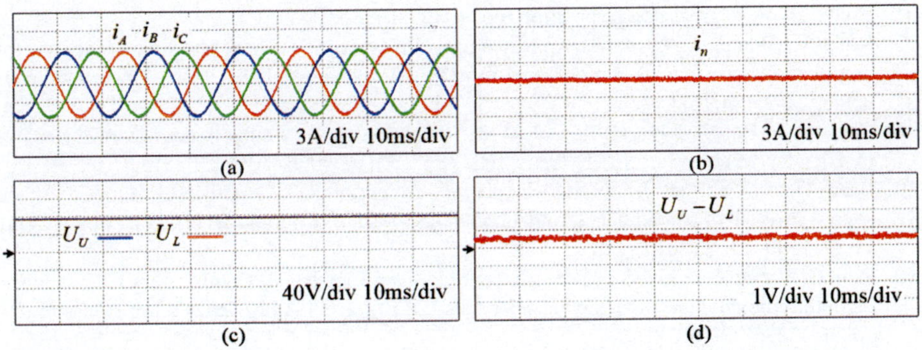

Fig. 9. Simulation outcomes of the proposed method under both balanced references and unbalanced load.

5 Conclusion

This study presented a novel method for controlling the voltages of a TLSFL inverter using simplified model predictive current control (MPCC). The proposed MPCC method is designed to regulate the output currents by introducing a reference voltage vector for determining the best voltage vector without the need for current prediction calculations. In addition, to avoid the complexity of capacitor voltage calculations and the time-consuming process of adjusting weighting factors, the best switching state is selected based on the deviation of the capacitor voltages to maintaining balanced capacitor voltages. This greatly simplifies the implementation process and significantly reduces the computational burden compared to conventional method.

References

1. Choi, U.-M., Blaabjerg, F., Lee, K.-B.: Control strategy of two capacitor voltages for separate MPPTs in photovoltaic systems using neutral-point-clamped inverters. IEEE Trans. Ind. Appl. **51**(4), 3295–3303 (2015)
2. Kim, S.-M., Won, I.J., Kim, J., Lee, K.-B.: DC-link ripple current reduction method for three-level inverters with optimal switching pattern. IEEE Trans. Ind. Electron. **65**(12), 9204–9214 (2018)
3. Schweizer, M., Friedli, T., Kolar, J.W.: Comparative evaluation of advanced three-phase three-level inverter/converter topologies against two-level systems. IEEE Trans. Ind. Electron. **60**(12), 5515–5527 (2012)

4. Vu, H.-C.,. Nguyen, T.D., Chun, T.-W., Lee, H.-H.: New virtual space vector modulation scheme to eliminate common-mode voltage with balanced neutral-point voltage for three-level NPC Inverters. In: IEEE 3rd International Future Energy Electronics Conference and ECCE Asia, pp. 313–318 (2017)
5. Vu, H.-C., Lee, H.-H.: Simple MPCC strategy for three-level sparse five-phase VSI to suppress current harmonics with balanced neutral-point voltage. IEEE Trans. Power Electron. **37**(1), 771–781 (2021)
6. Uddin, M., Mirzaeva, G., Goodwin, G., Rivera, M.: Computationally efficient model predictive control of a four-leg inverter for common mode voltage elimination. In: IEEE Energy Conversion Congress and Exposition, pp. 4032–4038 (2018)
7. Rojas, F., Kennel, R., Cardenas, R., Repenning, R., Clare, J.C., Diaz, M.: A new space-vector-modulation algorithm for a three-level four-leg NPC inverter. IEEE Trans. Energy Convers. **32**(1), 23–35 (2016)
8. Chee, S.-J., Ko, S., Kim, H.-S., Sul, S.-K.: Common-mode voltage reduction of three-level four-leg PWM converter. IEEE Trans. Ind. Appl. **51**(5), 4006–4016 (2015)
9. Basri, H.M., Mekhilef, S.: Digital predictive current control of multi-level four-leg voltage-source inverter under balanced and unbalanced load conditions. IET Electr. Power Appl. **11**(8), 1499–1508 (2017)
10. Roh, C., Kwak, S., Choi, S.: Three-phase three-level four-leg NPC converters with advanced model predictive control. J. Power Electron. **21**(10), 1574–1584 (2021). https://doi.org/10.1007/s43236-021-00283-z
11. Vu, H.-C., Lee, H.-H.: Simplified model predictive current control strategy for dual five-phase VSI-fed open end load to eliminate common-mode voltage and reduce current harmonics. J. Power Electron. **21**(8), 1155–1165 (2021). https://doi.org/10.1007/s43236-021-00266-0

An All-Digital Implementation of Resonate-and-Fire Neuron on FPGA

Trung-Khanh Le(✉), Trong-Tu Bui, and Duc-Hung Le

Faculty of Electronics and Telecommunications, The University of Science, Vietnam National University, Ho Chi Minh City, Vietnam
{ltkhanh,bttu,ldhung}@hcmus.edu.vn
http://www.fetel.hcmus.edu.vn/

Abstract. Spiking neurons and spiking neuron networks (SNN) have recently been considered the third generation of neuron models, replacing the perceptron neuron, which has been the most popular model. Spiking neurons can be emulated by both analog and digital implementations. Because of the complexity of mathematical models, integrate-and-fire (IAF) and leaky integrate-and-fire (LIF) neurons, which are simpler, are the common models for digital and VLSI implementations. However, these models lack the dynamic properties of the Hodgkin-Huxley model. In order to overcome these issues, the resonate-and-fire (RAF) model, which has more dynamic properties, has been proposed by Izhikevich. This paper presents the first all-digital design of the RAF model. The design does not require floating-point operations and multipliers. Its RTL structure can be realized on a VLSI system. The design uses 54 ALM units on a DE10-nano board based on a Cyclone V SoC FPGA. It has been verified at 20 kHz of subthreshold oscillation.

Keywords: spiking neuron · IAF · LIF · resonate-and-fire · dynamic properties

1 Introduction

The Hodgkin-Huxley model [1] is the most famous biophysically plausible neuron model. This kind of model uses four variables to calculate and simulate the electrical properties of the neural membrane. There are alternative models that reduce the complexity of the Hodgkin-Huxley model, such as Hindmarsh-Rose (three variables), FitzHugh-Nagumo (two variables), Morris-Lecar and Izhikevich (two variables). These reductions help to simulate neuron models on computers efficiently. However, it is still difficult to implement these models on real hardware. The integrate-and-fire (IAF) model and leaky integrate-and-fire (LIF) model [3] are simpler models that have become popular for creating spiking neuron hardware. IAF and LIF models, which are simple, do not have as many dynamic properties as real neuron cells. Hence, their networks do not have the same computational performance as biophysically inspired spiking neuron networks. The resonate-and-fire (RAF) model was introduced by Izhikevich to solve the dynamic limitations of the IAF model [4].

T. D. L. Nguyen et al. (Eds.): ICISN 2023, LNNS 752, pp. 169–175, 2023.
https://doi.org/10.1007/978-981-99-4725-6_22

In the past few years, most papers with hardware implementation have focused on synapses within networks of spiking neurons rather than on models of spiking neurons. Therefore, while research on this type of neuron model is still ongoing more than two decades later [8] (in 2018), [9] (in 2019), and [10] (in 2022), there have been few hardware realizations of this model, all of them are analog designs [5,6]. Although analog designs for spiking neuron cells can optimize power and chip area, they will encounter device mismatches [5], scaling technology [7], noise, and Process-Voltage-Temperature (PVT) variations. Moreover, storing weight values in analog voltage/current is not as easy as in digital forms. Some researchers used alternative materials, such as RRAM, memristor, biristor, and special FET devices, to implement neurons and synapses in SNNs; however, most of these designs still need analog circuits. For these reasons, this paper presents the first all-digital design of a resonate-and-fire neuron cell. The proposed design is made in RTL and verified on a DE10-nano board based on a Cyclone V SoC FPGA.

2 Resonator

According to papers [11,12], some kinds of neurons, for example, thalamic and neocortical neurons, have resonance features in their frequency responses. This feature is caused by subthreshold oscillations of membrane potential. Consequently, the subthreshold oscillation changes the distance between membrane potential and threshold level. When current pulses are applied to the membrane, action potentials are generated or not generated depending on the phase match between these impulses and the subthreshold oscillation. This mechanism is depicted in Fig. 1 by the appearance of excitatory impulses.

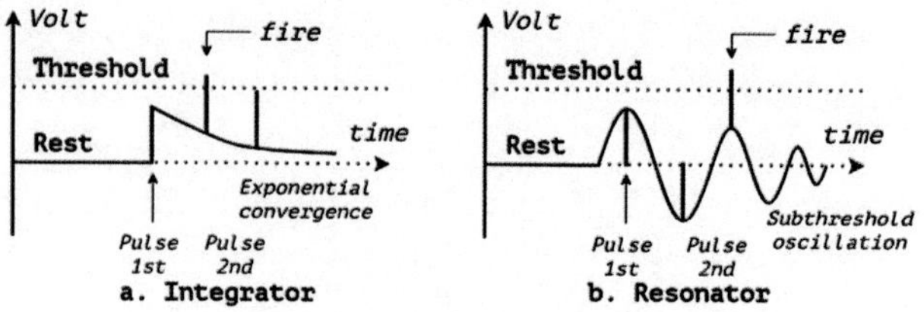

Fig. 1. Neuron membrane potentials in an integrator and a resonator [4].

In Fig. 1a, after the first impulse, the membrane potential of integrator neurons (IAF and LIF models) gradually returns to the resting level via the discharge of ion K^+ into the extracellular medium. If the second pulse comes as soon as the membrane potential is still high enough, the integrator will fire an action potential. The shorter the interval between these pulses, the higher the frequency of action potential that will be generated by the integrator. This operation is not the same as that in neurons, which have subthreshold oscillation.

In Fig. 1b, the neuron fires only when the phases of excitatory impulses match with the phases of subthreshold oscillation (resonance). This type of neuron is referred a resonator. This mechanism shows that the resonator has a function of frequency preference.

3 All-Digital Resonate-And-Fire Neuron

3.1 Idea for Realization

Based on the mechanism of resonators, Izhikevich introduced a resonate-and-fire neuron model, which can be written in two types of equations, the first equation was described in [4]:

$$\dot{z}_k = (b_k + i\omega_k)z_k + \sum_{j=1}^{n} c_{kj}\delta(t - t_j^*) \tag{1}$$

where $z_k \in \mathbb{C}$ is the state of the k-th neuron, $b_k < 0$ is the rate of attraction to the rest, ω is subthreshold oscillation frequency, $c_{kj} \in \mathbb{C}$ is synaptic coefficient, δ is Direc delta function, t_j^* is the nearest moment of firing of the k-th neuron. The RAF neuron will be an integrator if $\omega = 0$.
The second equation was mentioned in [2]:

$$\begin{aligned} v' &= 0.04v^2 + 5v + 140 - u + I \\ u' &= a(bv - u) \\ \text{if } v \geq 30 \text{ mV}, &\text{ then } \begin{cases} v \leftarrow c \\ u \leftarrow u + d \end{cases} \end{aligned} \tag{2}$$

where v is membrane potential, u is membrane recovery variable, $a = 0.1$, $b = 0.26$, $c = -60$ and $d = -1$. To convert to an integrator, the values of these parameters should be $a = 0.02$, $b = 0.2$, $c = -65$ and $d = 8$.

The last equation (2) is suitable for software simulation. In order to implement it by digital circuits, it is required to modify the equation to reduce the number of multipliers and adders [13,14]. Figure 1 suggests ideas for using a phase-frequency detector (PFD) or a sample-and-hold circuit to work as a neuron membrane. To keep the design simple and all-digital, a D flip-flop sample-and-hold circuit is proposed as presented in Fig. 2.

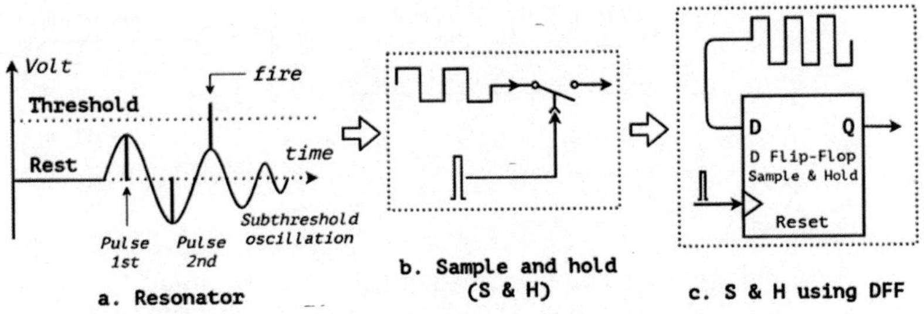

Fig. 2. A resonator based on a sample-and-hold circuit.

3.2 Circuit Designing

From equation (1), subthreshold oscillation is sampled by current impulses in a sample-and-hold circuit. After that, the sampled phase of the subthreshold oscillation should be integrated into a buffer. An action potential is only fired by the neuron when the phase of the current impulse matches with the phase of the neuron's oscillation. In the case of excitatory impulses, the matching phase should be the two continuous high states of subthreshold oscillation, as shown in Fig. 1. To keep the design of this buffer simple, we use a 2-bit shifter as a phase-accumulating buffer. An action potential is fired when both bits of the buffer are set. If there is no subthreshold oscillation, the buffer can be considered an integrate-and-fire neuron.

In Fig. 1, the membrane potential tends to come back to the resting state after receiving an impulse. In the LIF model, it is the repolarization phase of the action potential, also known as the leaky phase. The recorded phase must be stored at this moment, so the falling edge of the subthreshold oscillation should be used as the master clock for the shifter. Because of the repolarization progress, the oldest sampled phase should be cleared. Therefore, the sample-and-hold flip-flop should be cleared at the low state of the subthreshold oscillation, and the oldest bit of the 2-bit buffer should be cleared at the new cycle of the neuron's oscillation.

Finally, where does the subthreshold oscillation come from? The oscillation can be generated by a low-frequency clock located outside the neuron cell and used to synchronize all neurons in a network. However, to let each neuron have a unique oscillation, which will be useful for SNN training, this research uses a digital-controlled oscillator (DCO) inside a neuron cell to generate subthreshold oscillation. This DCO can be a controlled-ring oscillator or a numerically-controlled oscillator (NCO). The generic design based on the analysis above is shown in Fig. 3.

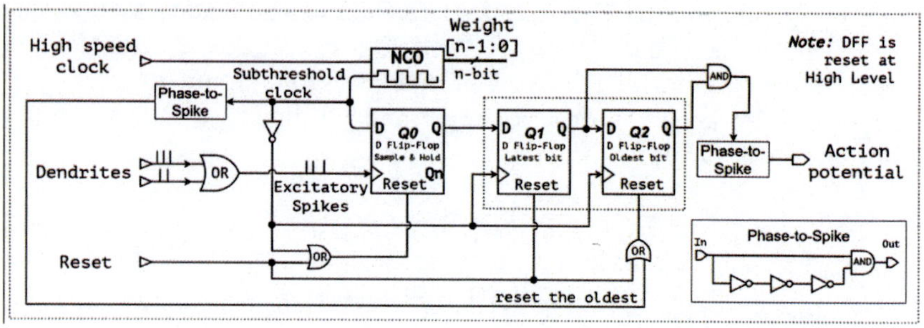

Fig. 3. Generic design of a simple all-digital RAF neuron.

In Fig. 3, input spikes coming from dendrites are summed via an OR gate. The subthreshold oscillation comes from an n-bit NCO. The "reset" signal is used to reset all flip-flops for simulation; it can also be used as the "enable" or "disable" signal. All the flip-flops in the design are assumed to use high-state resetting. The signal "reset the oldest" is used to clear the oldest sampled phase at the new cycle of the subthreshold oscillation. A phase-to-spike circuit is used to clip that signal to a reset spike. This type of circuit is also used to convert the action potential into a spike.

4 Experiment Result

A result of the Verilog simulation of the design in Sect. 3.2 is shown in Fig. 4.

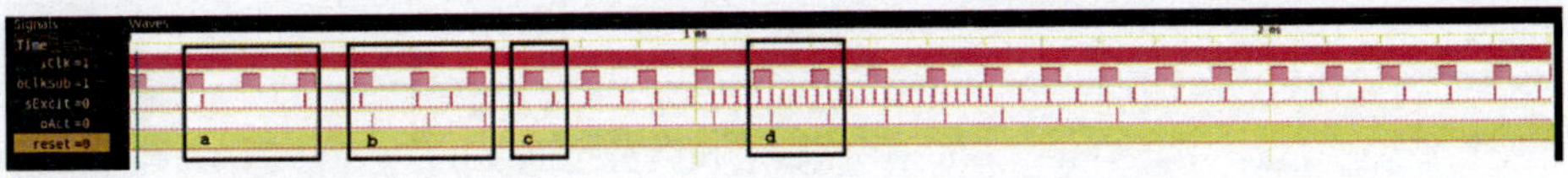

Fig. 4. Simulation result for a simple all-digital RAF neuron.

In Fig. 4, the high frequency clock is "iClk" which is 1 MHz for generating subthreshold oscillation. This neuron's oscillation is the "oClkSub" signal, its frequency is 10 kHz. The excitatory impulse is the multiple-frequency "sExcit" signal. The signal "oAct" is the output action potential. The "a" region shows no action spike when the impulse interval is too long; the "b" region shows spikes when phases are matched; the "c" region shows no spike when the impulse interval is half the eigenperiod of the neuron's oscillation; and the "d" region shows spikes for very high frequency impulses. These regions show that the design has resonance properties and can fire action potentials like an RAF neuron.

The Verilog project was synthesised using Quartus Prime Lite and tested on a DE10-nano Cyclone V development kit. The project only uses logic gates (73 basic logic gates and 22 D flip-flops) to implement the circuit as shown in Fig. 3 including an 8-bit NCO. The RTL structure is shown in Fig. 5. Synthesis results are shown in Table 1.

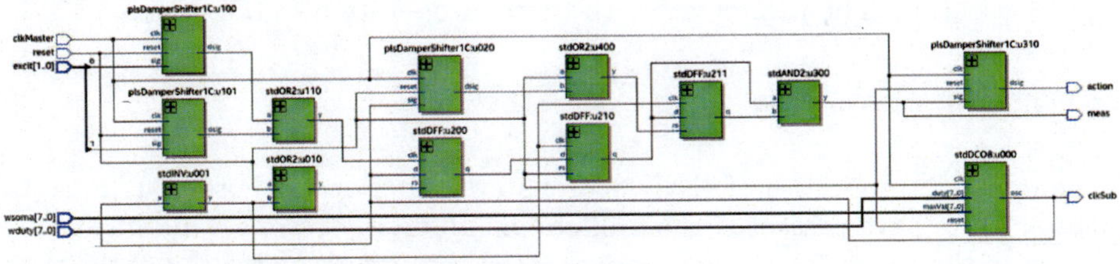

Fig. 5. RTL structure of an 8-bit simple all-digital RAF neuron on a DE10-nano board.

Table 1. Summary of the resource utilization on the DE10-nano development kit.

Resource	Available	Used	Utilization
Logic (in ALMs)	41,910	54	0.13%
Registers	166,036	0	0%
Block Memory bits	5,662,720	0	0%
DSP blocks	112	0	0%
18×18 Multiplier	224	0	0%
PLL	6	0	0%
DLL	4	0	0%

Target device is Cyclone V 5CSEBA6U23I7 (110K logic elements).

Table 1 shows that the implementation on the Cyclone V Soc FPGA requires 54 ALM units to emulate a simple all-digital RAF neuron. As a result, about 776 neurons can be implemented on the kit. An experiment has been set up as shown in Fig. 6a. The results are shown in Figs. 6b and 6c.

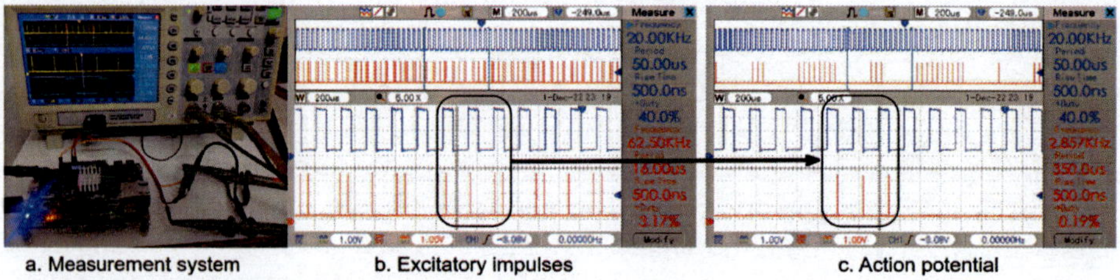

Fig. 6. Measurement system and results on the DE10-nano kit.

Figure 6 shows measurements on a neuron, which has a programmable subthreshold oscillation at a frequency of 20 kHz; the excitatory impulse is mixed between 10 kHz and 15 kHz spikes. At matched-phase moments, action potential spikes occurred. The results verify the correct operation of the design.

5 Conclusion

The dynamic properties of RAF neurons are superior to those of IAF or LIF neurons. We have presented the first simple all-digital resonate-and-fire neuron design in the paper. The design implemented the complex equation of the model with just a few flip-flops and logic gates. The design does not require any multiplier, adder, floating-point module, or even fixed-point operation. Using an n-bit NCO for subthreshold oscillation, this design allows creating SNNs that have programmable neurons' eigenperiods. Without behavior modules, the design's

RTL structure, which is only made up of basic logic gates (AND, OR, NOT, and XOR gates), can be turned into a VLSI chip using the standard digital IC design flow.

References

1. Hodgkin Alan, L., Huxley Andrew, F.: A quantitative description of membrane current and its application to conduction and excitation in nerve. J. Physiol. **117**(4), 500-544 (1952). https://doi.org/10.1113/jphysiol.1952.sp004764
2. Izhikevich, E.M.: Simple model of spiking neurons. IEEE Trans. Neural Netw. **14**(6), 1569–1572 (2003). https://doi.org/10.1109/TNN.2003.820440
3. Abbott, L.F.: Lapicque's introduction of the integrate-and-fire model neuron (1907). Brain Res. Bull. **50**(5), 303–304 (1999). https://doi.org/10.1016/S0361-9230(99)00161-6
4. Izhikevich, E.M.: Resonate-and-fire neurons. Neural Netw. **14**(6-7), 883–894 (2001). https://doi.org/10.1016/s0893-6080(01)00078-8
5. Nakada, K., Asai, T., Hayashi, H.: Analog VLSI implementation of resonate-and-fire neuron. Int. J. Neural Syst. **16**(6), 445–456 (2006). https://doi.org/10.1142/S0129065706000846
6. Hsieh, H.-Y., Tang, K.-T.: VLSI implementation of a bio-inspired olfactory spiking neural network. IEEE Trans. Neural Netw. Learn. Syst. **23**(7), 1065–1073 (2012). https://doi.org/10.1109/TNNLS.2012.2195329
7. Joubert, A., Belhadj, B., Temam, O., Héliot, R.: Hardware spiking neurons design: analog or digital? The 2012 International Joint Conference on Neural Networks (IJCNN), pp. 1–5 (2012). https://doi.org/10.1109/IJCNN.2012.6252600
8. Mankin, R., Paekivi, S.: Memory-induced resonance like suppression of spike generation in a resonate-and-fire neuron model. Phys. Rev. E. **97**(1-1), 012125 (2018). https://doi.org/10.1103/PhysRevE.97.012125
9. Sehgal, S., Patel, N.D., Malik, A., Roop, P.S., Trew, M.L.: Resonant model-a new paradigm for modeling an action potential of biological cells. PLoS One. **14**(5), e0216999 (2019). https://doi.org/10.1371/journal.pone.0216999
10. Mario Antoine Aoun: Resonant neuronal groups. Phys. Open. **13**, 100104 (2022). https://doi.org/10.1016/j.physo.2022.100104
11. Puil, E., Meiri, H., Yarom, Y.: Resonant behavior and frequency preferences of thalamic neurons. J. Neurophysiol. **71**(2), 575–582 (1994). https://doi.org/10.1152/jn.1994.71.2.575
12. Hutcheon, B., Miura, R.M., Puil, E.: Models of subthreshold membrane resonance in neocortical neurons. J. Neurophysiol. **76**(2), 698–714 (1996). https://doi.org/10.1152/jn.1996.76.2.698
13. Heidarpur, M., Ahmadi, A., Ahmadi, M., Azghadi, M.R.: CORDIC-SNN On-FPGA STDP learning with Izhikevich neurons. IEEE Trans. Circ. Syst. I Regul. Papers. **66**(7), 2651–2661 (2019). https://doi.org/10.1109/TCSI.2019.2899356
14. Wang, J., et al.: A high-accuracy and energy-efficient CORDIC based Izhikevich neuron with error suppression and compensation. IEEE Trans. Biomed. Circ. Syst. **6**(5), 807–821 (2022). https://doi.org/10.1109/TBCAS.2022.3191004

Extended State Observer-Based Backstepping Sliding Mode Control for Wheel Slip Tracking

Duc Thinh Le, The Anh Nguyen, Xuan Duc Pham, Quoc Manh Le, Nhu Toan Nguyen, Danh Huy Nguyen, Duc Chinh Hoang(✉), and Tung Lam Nguyen(✉)

Hanoi University of Science and Technology, Hanoi 100000, Vietnam
{chinh.hoangduc,lam.nguyentung}@hust.edu.vn

Abstract. The wheel slip controller (WSC) serves as the cornerstone of the anti-lock braking system (ABS). It is necessary to investigate and test a nonlinear robust WSC since the friction between the road and tire is a nonlinear function of wheel slip. In this research, a backstepping sliding mode controller (BSMC) is designed to control the wheel slip of a quarter-vehicle model. Extended state observer (ESO) is used in combination with the design of the BSMC in order to estimate the total uncertainty in the system. The viability of the proposed controllers is then verified in numerical simulation, and the performance of the proposed controllers is evaluated using three different key performance indicators (KPIs).

Keywords: Wheel slip control · Sliding mode control · Backstepping control · Extended state observer

1 Introduction

The ABS is one of the most essential components in modern vehicles for enhancing the safety of the driver and passengers [1]. To satisfy the demand, numerous onboard ABS is developed. For instance, the electronic stability controller, which uses the active braking system to increase lateral stability, has been used in [2]; the automatic emergency braking system can exert control over either the brake torque or the targeted wheel slip to improve the overall performance [3].

The implementations for wheel slip control fall into two categories: a rule-based strategy built around the thresholds of wheel slip and direct torque control techniques based on the vehicle or wheel model [4]. Challa et al. suggest a combined 3-phase rule-based Slip and Wheel Acceleration Threshold Algorithm for Anti-lock Braking in HCRVs and gives a method for determining the specific threshold values that make up the rule-based ABS algorithm [5]. Vignati et al. suggest a method based on the force derivative and wheel peripheral acceleration. Changes to the thresholds and gains of the control strategy are made to improve performance using the estimation of the normal load and the friction coefficient [6]. The wheel slip control strategy based on a model has fewer tuning parameters than the rule-based approach described in the preceding paragraph and

T. D. L. Nguyen et al. (Eds.): ICISN 2023, LNNS 752, pp. 176–185, 2023.
https://doi.org/10.1007/978-981-99-4725-6_23

may be able to achieve continuous tracking control for wheel slip. In [7], Solyom et al. introduce a gain-scheduled controller that controls tire slip is proposed along with a design model, and the proposed controller outperformed several other examined methods in terms of deceleration. An adaptive continuous wheel slip control system designed for high-dynamic decoupled electro-hydraulic brake systems in SUVs is introduced by Savitski et al. [8]. However, the impact of external disturbances is not discussed in any of the research studies mentioned above; only the impact of model uncertainty on system performance is examined. Designing a model-based WSC that includes model uncertainties and external disturbances are needed.

The backstepping method is an efficient and adaptable approach for controlling nonlinear systems due to its straightforward design process. It has been studied in [9–11]. However, the backstepping strategy is vulnerable to aggregated uncertainty. To deal with aggregated uncertainty, the backstepping method is expanded to include the sliding mode control algorithm [12–14].

In Zhang and Li's research [15], the authors propose an adaptive BSMC method that utilizes a RBFNN. The problem with Zhang and Li's work is that the proposed controller uses a derivative form of the wheel slip, even though using a derivative form in a controller is strongly discouraged. In order to overcome this weakness, we employ the ESO in place of the RBFNN to estimate the lumped uncertainty and the state of the model due to its simplicity and efficiency. In this research, the BSMC combined with ESO is introduced to design a WSC based on a quarter-car model. The novelty of this study is the use of ESO to estimate the state variables and total uncertainty of the model, without requiring the derivative form of the wheel slip. Finally, simulations are used to compare the performance of the controller.

The remaining sections are arranged as follows: the dynamic model quarter-vehicle is proposed in Sect. 2; the controller design is presented with two small parts: ESO design and BSMC in Sect. 3; Sect. 4 shows the results of simulations using Matlab/Simulink and KPIs comparison; and finally, we conclude the paper in Sect. 5.

2 Quarter-Vehicle Dynamic Model

The nominal model of this study is the quarter-vehicle model, due to its simplicity (it only has 2 degrees of freedom: longitudinal velocity and angular velocity) and effectiveness. Figure 1 illustrates the WSC design using a basic and efficient model of quarter-vehicle. The mathematical expressions of the quarter-car model are as follows [16]:

$$\begin{cases} J\dot{\omega} = rF_x - T_b \\ m\dot{v} = -F_x \end{cases} \tag{1}$$

where J is the wheel inertia, ω is the angular velocity of the wheel, r is the wheel radius, F_x is the longitudinal tire-road contact force, T_b is the brake torque, m

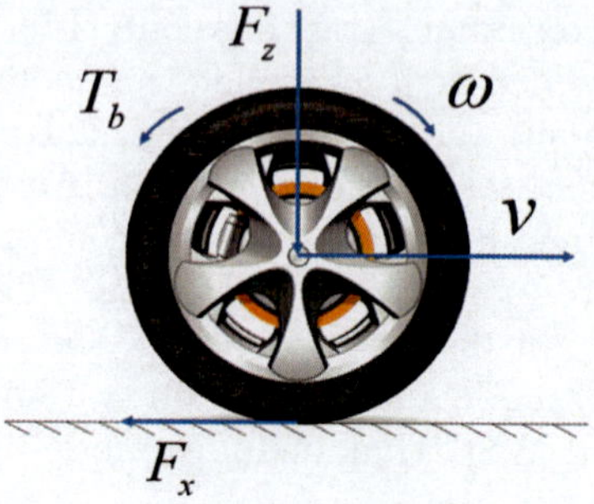

Fig. 1. Model of the quarter-vehicle

is mass of the quarter-car and v is the longitudinal speed. The wheel slip is defined as:

$$\lambda = \frac{v - \omega r}{v} \tag{2}$$

Burckhardt's tire model provides the mathematical equation between wheel slip and the coefficient of tire-road friction [16]:

$$\mu(\lambda) = \vartheta_1(1 - e^{-\lambda\vartheta_2}) - \lambda\vartheta_3 \tag{3}$$

where ϑ_1, ϑ_2, ϑ_3 are the coefficients of a specific road's condition. F_x can be rewritten as $F_x = F_z\mu(\lambda)$ with F_z is the vertical tire-road contact force. Assume that both model uncertainties and external disturbances are present in the model uncertainty, and their total disturbance is a torque represented as ΔD, added to the angular speed dynamic equations.

$$\begin{cases} J\dot{\omega} = rF_x - T_b + \Delta D \\ m\dot{v} = -F_x \end{cases} \tag{4}$$

After some transformations, (2) is rewritten as follows:

$$\dot{x}_2 = f(x_1, x_2) + Gu + D_x \tag{5}$$

where $x_1 = \lambda$, $x_2 = \dot{x}_1$, $u = \dot{T}_b$, $f(x_1, x_2) = \frac{-1}{v}[\frac{-2F_z\mu(x_1)}{m}x_2 + (\frac{1-x_1}{m} + \frac{r^2}{J})F_z\dot{\mu}(x_1)]$, $G = \frac{r}{Jv}$, $D_x = \frac{2\Delta Dr}{Jv^2} - \frac{\Delta\dot{D}r}{Jv}$ is the total uncertainty.

Assumption 1: Total disturbance D_x is bounded. Therefore, there is an upper positive bound L such that $|D_x| \leq L$.

From equation (5) , we can get the state equation of wheel slip to develop the controller:

$$\begin{cases} \dot{x}_1 = x_2 \\ \dot{x}_2 = f(x_1, x_2) + Gu + D_x \end{cases} \tag{6}$$

3 Design of the Controller

This part proposed the design of the ESO and BSMC. The control objective is to achieve the desired wheel slip and the observer assists in estimating the state variables and lumped uncertainty. The ESO and BSMC are designed as follows.

3.1 Extended State Observer

Based on the standard ESO design, from (6), set $x_3 = D_x$, then $\dot{x}_3 = \dot{D}_x$. The original system can be turned to:

$$\begin{cases} \dot{x}_1 = x_2 \\ \dot{x}_2 = x_3 + Gu + f(x_1, x_2) \\ \dot{x}_3 = \dot{D}_x \end{cases} \tag{7}$$

The ESO is designed such that:

$$\begin{cases} \dot{\hat{x}}_1 = \hat{x}_2 - \frac{k_1}{\varepsilon}(\hat{x}_1 - x_1) \\ \dot{\hat{x}}_2 = \hat{x}_3 - \frac{k_2}{\varepsilon^2}(\hat{x}_1 - x_1) + Gu + f(\hat{x}_1, \hat{x}_2) \\ \dot{\hat{x}}_3 = -\frac{k_3}{\varepsilon^3}(\hat{x}_1 - x_1) \end{cases} \tag{8}$$

The goal of the observer is to properly design parameters in such a way that

$$\hat{x}_1 \to x_1, \hat{x}_2 \to x_2, \hat{x}_3 \to D_x$$

as $t \to \infty$ where $\hat{x}_1, \hat{x}_2$ and $\hat{x}_3$ are the states of the ESO, $\varepsilon > 0, k_1, k_2, k_3$ are the positive constants, polynomial $s^3 + k_1 s^2 + k_2 s + k_3$ is Hurwitz. Define

$$\boldsymbol{\eta} = \left[\eta_1 \; \eta_2 \; \eta_3\right]^{\mathrm{T}} \tag{9}$$

where $\eta_1 = \frac{x - \hat{x}_1}{\varepsilon^2}$, $\eta_2 = \frac{x_2 - \hat{x}_2}{\varepsilon}$, $\eta_3 = D_x - \hat{x}_3$. The derivatives of η_1, η_2, η_3 is obtained as below:

$$\begin{aligned} \varepsilon\dot{\eta}_1 &= \frac{\dot{x}_1 - \dot{\hat{x}}_1}{\varepsilon} \\ &= -k_1\eta_1 + \eta_2 \\ \varepsilon\dot{\eta}_2 &= \dot{x}_2 - \dot{\hat{x}}_2 \\ &= -k_2\eta_1 + \eta_3 + [f(x_1, x_2) - f(\hat{x}_1, \hat{x}_2)] \\ \varepsilon\dot{\eta}_3 &= \varepsilon(\dot{D}_x - \dot{\hat{x}}_3) \\ &= -k_3\eta_1 + \varepsilon\dot{D}_x \end{aligned} \tag{10}$$

The observation error system can be written as:

$$\varepsilon\dot{\eta} = \mathbf{A}\boldsymbol{\eta} + \mathbf{B}\tilde{f} + \varepsilon\mathbf{C}\dot{D}_x \tag{11}$$

where $A = \begin{pmatrix} -k_1 & 1 & 0 \\ -k_2 & 0 & 1 \\ -k_3 & 0 & 0 \end{pmatrix}, B = \begin{pmatrix} 0 \\ 1 \\ 0 \end{pmatrix}, C = \begin{pmatrix} 0 \\ 0 \\ 1 \end{pmatrix}, \tilde{f} = f(x_1, x_2) - f(\hat{x}_1, \hat{x}_2)$. It can be seen that the characteristic equation of $\mathbf{A}$ is

$$|\lambda I - \mathbf{A}| = \begin{vmatrix} \lambda + k_1 & -1 & 0 \\ k_2 & \lambda & -1 \\ k_3 & 0 & \lambda \end{vmatrix} = 0 \tag{12}$$

From equation (12), we can get:

$$\lambda^3 + k_1\lambda^2 + k_2\lambda + k_3 = 0 \tag{13}$$

If we choose k_1, k_2, and k_3 such that $\mathbf{A}$ is Hurwitz, then there exists a unique symmetric positive definite matrix $\mathbf{P}$ satisfying the Lyapunov function for any given symmetric positive definite matrix $\mathbf{Q}$ as follows:

$$\mathbf{A}^{\mathrm{T}}\mathbf{P} + \mathbf{P}\mathbf{A} + \mathbf{Q} = 0 \tag{14}$$

Choose the Lyapunov function as:

$$V_0 = \varepsilon \boldsymbol{\eta}^{\mathrm{T}}\mathbf{P}\boldsymbol{\eta} \tag{15}$$

The derivation of V_0 is obtained as follows:

$$\begin{aligned} \dot{V}_0 &= \varepsilon\dot{\boldsymbol{\eta}}^{\mathrm{T}}\mathbf{P}\boldsymbol{\eta} + \varepsilon\boldsymbol{\eta}^{\mathrm{T}}\mathbf{P}\dot{\boldsymbol{\eta}} \\ &= [\mathbf{A}\boldsymbol{\eta} + \mathbf{B}\tilde{f} + \varepsilon\mathbf{C}\dot{D}_x]^{\mathrm{T}}\mathbf{P}\boldsymbol{\eta} + \boldsymbol{\eta}^{\mathrm{T}}\mathbf{P}[\mathbf{A}\boldsymbol{\eta} + \mathbf{B}\tilde{f} + \varepsilon\mathbf{C}\dot{D}_x] \\ &= \boldsymbol{\eta}^{\mathrm{T}}\mathbf{A}^{\mathrm{T}}\mathbf{P}\boldsymbol{\eta} + (\mathbf{B}\tilde{f})^{\mathrm{T}}\mathbf{P}\boldsymbol{\eta} + \varepsilon(\mathbf{C}\dot{D}_x)^{\mathrm{T}}\mathbf{P}\boldsymbol{\eta} + \boldsymbol{\eta}^{\mathrm{T}}\mathbf{A}\mathbf{P}\boldsymbol{\eta} + (\mathbf{B}\tilde{f})\mathbf{P}\boldsymbol{\eta}^{\mathrm{T}} + \varepsilon(\mathbf{C}\dot{D}_x)\mathbf{P}\boldsymbol{\eta}^{\mathrm{T}} \\ &= \boldsymbol{\eta}^{\mathrm{T}}(\mathbf{A}^{\mathrm{T}}\mathbf{P} + \mathbf{P}\mathbf{A})\boldsymbol{\eta} + 2\boldsymbol{\eta}^{\mathrm{T}}\mathbf{P}\mathbf{B}\tilde{f} + 2\boldsymbol{\eta}^{\mathrm{T}}\mathbf{P}\varepsilon\mathbf{C}\dot{D}_x \\ &\leq -\boldsymbol{\eta}^{\mathrm{T}}\mathbf{Q}\boldsymbol{\eta} + 2||\boldsymbol{\eta}||.||\mathbf{P}\mathbf{B}||.||\tilde{f}|| + 2\varepsilon||\boldsymbol{\eta}||.||\mathbf{P}\mathbf{C}||.|\dot{D}_x| \end{aligned} \tag{16}$$

and

$$\dot{V}_0 \leq -\lambda_{min}(\mathbf{Q})||\boldsymbol{\eta}||^2 + 2||\boldsymbol{\eta}||||\mathbf{P}\mathbf{B}|||\tilde{f}| + 2\varepsilon L||\mathbf{P}\mathbf{C}||||\boldsymbol{\eta}|| \tag{17}$$

in which $-\lambda_{min}(\mathbf{Q})$ the smallest eigenvalue of $\mathbf{Q}$. To get $\dot{V}_0 \leq 0$, the coefficient ε is designed to satisfy the following condition:

$$\lambda_{min}(\mathbf{Q})||\boldsymbol{\eta}||^2 - 2||\boldsymbol{\eta}||||\mathbf{P}\mathbf{B}|||\tilde{f}| - 2\varepsilon L||\mathbf{P}\mathbf{C}|||\boldsymbol{\eta}|| > 0 \tag{18}$$

then the observer error $\boldsymbol{\eta}$ is asymptotic convergence. In addition, to alleviate the peaking phenomenon of ESO, the gain ε is designed as follows:

$$\frac{1}{\varepsilon} = R = \sigma\frac{1 - e^{-\lambda_1 t}}{1 + e^{-\lambda_2 t}}, 0 \leq t \leq t_{max} \tag{19}$$

where $\sigma, \lambda_1, \lambda_2$ are positive constants. Finally, we can get

$$\lim_{\varepsilon\to 0}\hat{x}_1 = x_1, \lim_{\varepsilon\to 0}\hat{x}_2 = x_2, \lim_{\varepsilon\to 0}\hat{x}_3 = D_x$$

3.2 Backstepping Sliding Mode Controller

Define $e_1 = x_1 - x_d$, where x_d is the reference wheel slip. Consequently, we get:

$$\begin{aligned} \dot{e_1} &= \dot{x}_1 - \dot{x}_d \\ &= x_2 - \dot{x}_d \end{aligned} \tag{20}$$

Choose the Lyapunov candidate function $V_1 = \frac{1}{2}e_1^2$. The time derivative of V_1 is expressed as follows:

$$\begin{aligned} \dot{V}_1 &= e_1\dot{e}_1 \\ &= e_1(x_2 - \dot{x}_d) \end{aligned} \tag{21}$$

To get $\dot{V}_1 < 0$, choose sliding variable

$$\begin{aligned} s &= c_1 e_1 + \dot{e}_1 \\ &= x_2 + c_1 e_1 - \dot{x}_d \end{aligned} \tag{22}$$

where $c_1 > 0$ is positive constant. Then $x_2 = s - c_1 e_1 + \dot{x}_d$, from which (21) is rewritten as follows:

$$\dot{V}_1 = e_1 s - c_1 e_1^2 \tag{23}$$

If $s = 0$ then $\dot{V}_1 < 0$. Define another Lyapunov candidate function

$$V_2 = V_1 + \frac{1}{2}s^2 \tag{24}$$

The derivative of $\dot{s} = \dot{x}_2 + c_1\dot{e}_1 - \ddot{x}_d = f(x_1, x_2) + Gu + D_x + c_1\dot{e}_1 - \ddot{x}_d$, then

$$\dot{V}_2 = \dot{V}_1 + s\dot{s} = e_1 s - c_1 e_1^2 + s(f(x_1, x_2) + Gu + D_x + c_1\dot{e}_1 - \ddot{x}_d) \tag{25}$$

Define the estimated sliding mode variable be $\hat{s} = \hat{e}_2 + c\hat{e}_1$, with $\hat{e}_1 = \hat{x}_1 - x_d$ and $\hat{e}_2 = \hat{x}_2 - \dot{x}_d$. To get $\dot{V}_2 < 0$, the control signal u is selected as:

$$u = \frac{1}{G}(-f(x_1, x_2) - c_2\hat{s} - \hat{e}_1 - c_1\dot{\hat{e}}_1 + \ddot{x}_d - \gamma \mathrm{sgn}(\hat{s}) - \hat{x}_3) \tag{26}$$

where $c_2 > 0, \gamma > 0$ are positive constants. Therefore,

$$\dot{V}_2 = s(e_1 - \hat{e}_1) + s(D_x - \hat{x}_3) + sc_1(\dot{e}_1 - \dot{\hat{e}}_1) - c_1 e_1^2 - c_2 s\hat{s} - \gamma s \mathrm{sgn}(\hat{s}) < 0 \tag{27}$$

As a result, when $t \to \infty, e_1 \to 0$ and $e_2 \to 0$. The saturation function sat(s) is used instead of the sign function sgn(s) to reduce chattering effect [17].

4 Simulation Results

The designed controller is simulated using MATLAB/SIMULINK environment. Two scenarios - dry asphalt and wet asphalt road conditions - are used to test the developed controller. They both use a step function for the desired wheel slip, with the value of 0.1. The quarter-car model's parameters are $m = 35\,\mathrm{m}$, $J =$

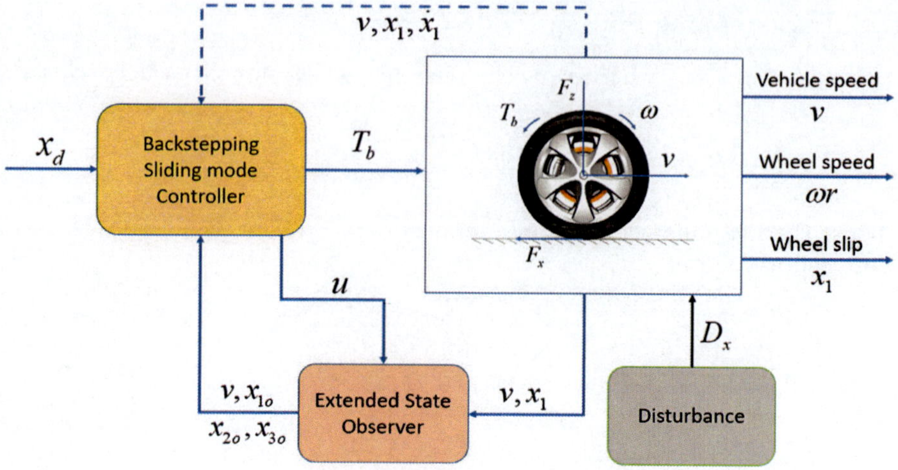

Fig. 2. Block diagram of the BSMC, which is tested with and without ESO

0.9 kgm^2 and $r = 0.310$ m. $F_z = 3450$ N. The car's initial speed is $v = 27.78$ m/s, and the simulation ends when it slows to 1 m/s. Assuming the disturbance is a torque represented as a sine function, added to the angular speed dynamic equations at a frequency of 25 rad/s and an attitude of 100 Nm (Fig. 2) .

Following are the parameters that were chosen for the designed ESO to obtain accurate tracking of the uncertainty: $\sigma = 5000, \lambda_1 = \lambda_2 = 3, k_1 = 6, k_2 = 11, k_3 = 6$. The designed BSMC's parameters are chosen as $c_1 = 300, c_2 = 200, \gamma = 10$.

Figure 3 displays the results of the simulation in the dry asphalt scenario, comparing the performance of the suggested controller with and without the ESO. Figure 3(a) demonstrates that the vehicle and wheel speed decrease over a reasonable time period. The suggested controller with ESO has superior wheel slip tracking precision than the controller without ESO, as shown in Fig. 3(b). The ESO can assist in precisely tracking wheel slip when high frequency, high attitude disturbances are present. Figure 3(c) shows the brake torque value in both methods. The effectiveness of the ESO in tracing the dynamic model's overall level of uncertainty is demonstrated in Fig. 3(d).

The simulation results for the wet asphalt scenario are shown in Fig. 4, which contrasts the performance of the suggested controller. Figure 3(a) depicts the vehicle and wheel speed decreasing over a reasonable time period. Figure 4(b) shows that the suggested controller with ESO has better tracking of the wheel slip, whereas without ESO, the fluctuation increases when the vehicle tends to slow down. Figure 4(c) shows the brake torque value in both methods. Figure 4(d) illustrates that ESO can precisely track the disturbances of the quarter car.

To verify the viability of the suggested controller in both scenarios, 3 different KPIs sets have been chosen. Mean Fully Developed Deceleration (MFDD) is specially made for measuring how well the car decelerates over the whole ABS

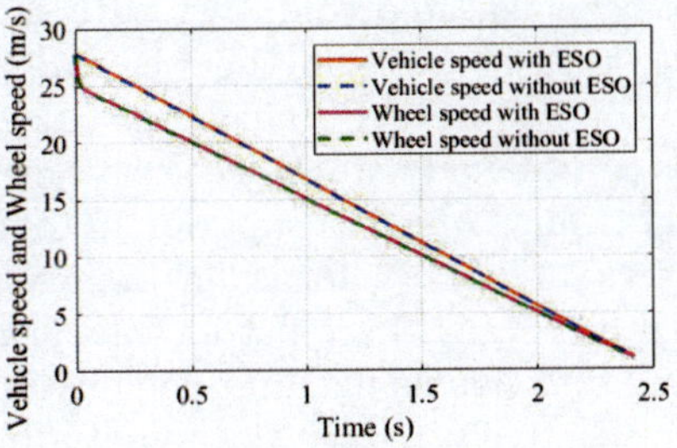

(a) Vehicle and wheel speed

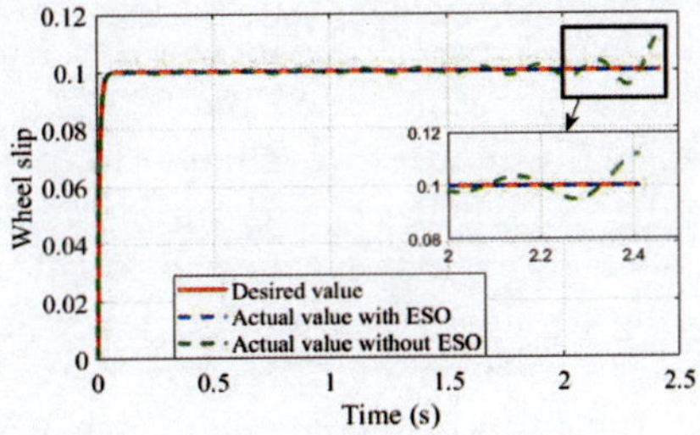

(b) Desired slip tracking

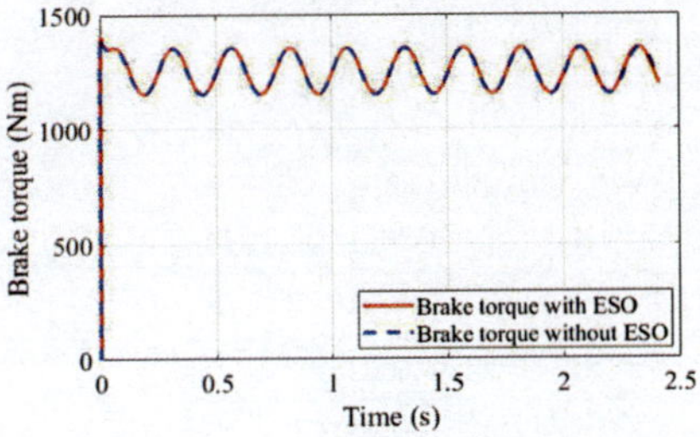

(c) Brake torque

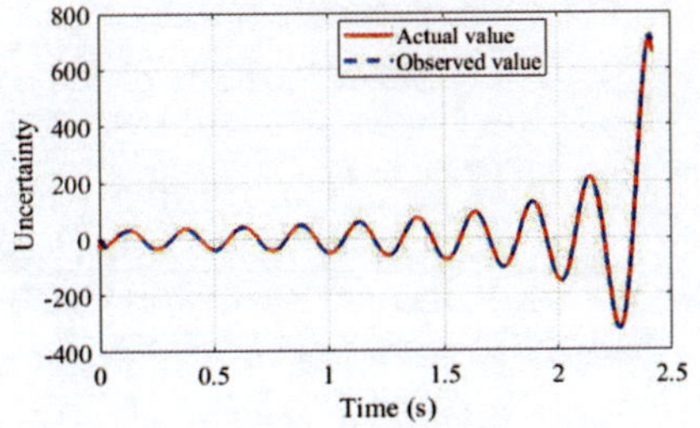

(d) Uncertainty and observed value

Fig. 3. Dry asphalt scenario

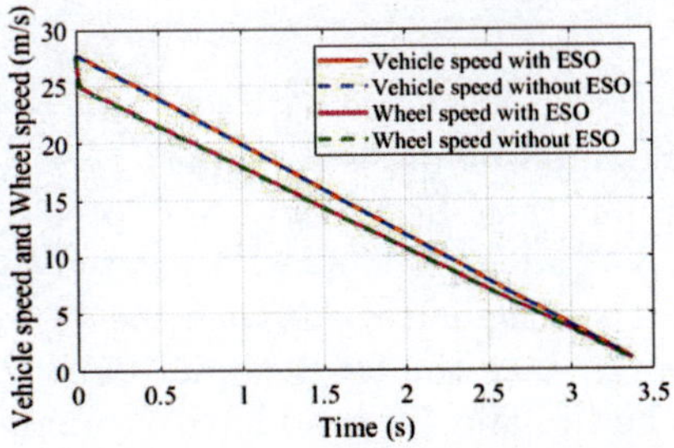

(a) Vehicle and wheel speed

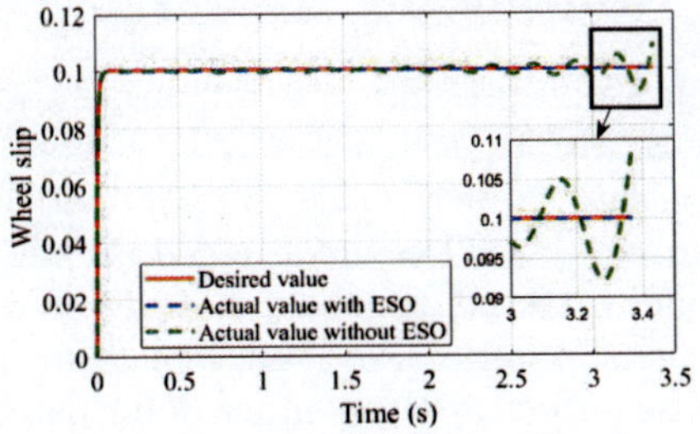

(b) Desired slip tracking

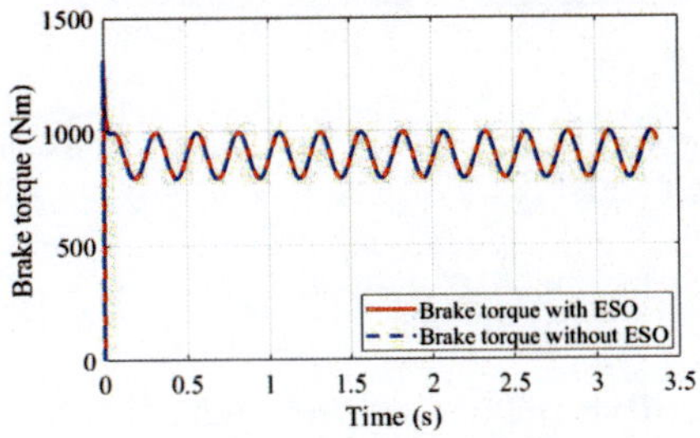

(c) Brake torque

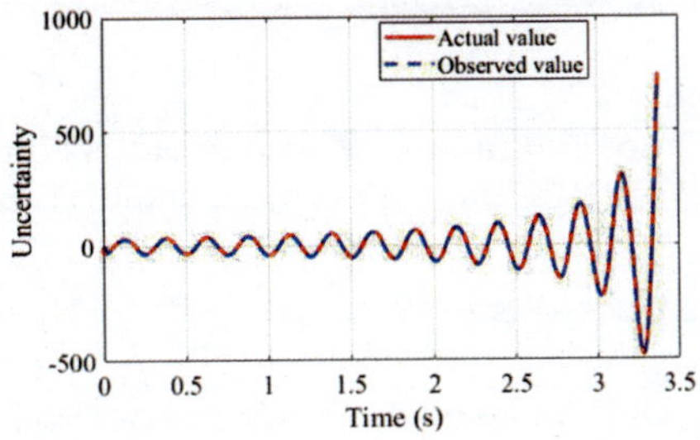

(d) Uncertainty and observed value

Fig. 4. Wet asphalt scenario

activation period. The mean acceleration of the suggested controller with ESO is increased by 0.62% in the dry asphalt test case, while in the wet asphalt test case, it is increased by 0.02%, as shown in Table 1. ABS efficiency (η_{ABS}) is specially made to evaluate the performance of steady-state deceleration. In the dry asphalt test case, η_{ABS} increases the value, which is superior. Integral time-weighted average of the longitudinal jerk ($ITAE_{J_x}$) focus on improving driving comfort. The comfort features offered by the ABS are better with fewer car jerks. The huge differences between using and not using ESO demonstrate that the proposed controller with ESO can effectively reduce the longitudinal jerk and provide more driving comfort in the presence of lumped uncertainty.

Table 1. KPIs of the control systems in 2 scenarios.

KPIs	Dry asphalt		Wet asphalt	
	With ESO	Without ESO	With ESO	Without ESO
MFDD	11.186	11.1171	7.9578	7.9566
η_{ABS}	1.0213	1.0209	1.0203	1.0203
$ITAE_{J_x}$	0.09275	2.785	0.04874	1.692

5 Conclusion

This study presents the comparison between using ESO and without using ESO for BSMC to design the WSC, where the nominal model is the quarter-vehicle with uncertainty. The total model uncertainty can be quickly and precisely tracked by ESO using designed parameters. The advantage of the suggested controller is tested, using simulations on dry and wet asphalt roads with the desired wheel slip is the step function. In future work, since the longitudinal force and vertical force at the contact point between the tire and the road have an unidentified scaling factor, an adaptive control technique is suggested to improve the controllers.

Acknowledgment. This research is funded by Hanoi University of Science and Technology (HUST) under project number T2022-PC-003.

References

1. Qiu, Y., Dai, Z.: Adaptive constrained antilock braking control under unknown time-varying slip-friction characteristics. Nonlinear Dyn. 1–18 (2022)
2. Youn, I., et al.: Combined effect of electronic stability control and active tilting control based on full-car nonlinear model. In: Proceedings of the Dynamics of Vehicles on Roads and Tracks: Proceedings of the 24th Symposium of the International Association for Vehicle System Dynamics, Graz, Austria (2015)

3. Pretagostini, F., Ferranti, L., Berardo, G., Ivanov, V., Shyrokau, B.: Survey on wheel slip control design strategies, evaluation and application to antilock braking systems. IEEE Access **8**, 10951–10970 (2020)
4. Yin, D., et al.: A multiple data fusion approach to wheel slip control for decentralized electric vehicles. Energies. **10**(4), 461 (2017). https://doi.org/10.3390/en10040461
5. Challa, A., et al.: A 3-phase combined wheel slip and acceleration threshold algorithm for anti-lock braking in heavy commercial road vehicles. Veh. Syst. Dyn. **60**(7), 2312–2333 (2022). https://doi.org/10.1080/00423114.2021.1903048
6. Vignati, M., Sabbioni, E.: Force-based braking control algorithm for vehicles with electric motors. Veh. Syst. Dyn. **58**(9), 1348–1366 (2020). https://doi.org/10.1080/00423114.2019.1621354
7. Solyom, S., Rantzer, A., Lüdemann, J.: Synthesis of a model-based tire slip controller. Veh. Syst. Dyn. **41**(6), 475–499 (2004). https://doi.org/10.1080/004231105123313868
8. Savitski, D., et al.: Robust continuous wheel slip control with reference adaptation: application to the brake system with decoupled architecture. IEEE Trans. Indus. Inform. **14**(9), 4212–4223 (2018). https://doi.org/10.1080/00423114.2019.1621354
9. Le, D., Dang, V., Dinh, B., Vu, H., Pham, V., Nguyen, T.: Disturbance observer-based speed control of interior permanent magnet synchronous motors for electric vehicles. Reg. Conf. Mech. Manuf. Eng. 244–259 (2022)
10. Manh, T., Le, D., Huy, P., Quang, D., Quang, D., Nguyen, T.: Nonlinear control of axial gap magnetic bearing motors: a disturbance observer-based method. Int. Conf. Eng. Res. Appl. 675–684 (2021)
11. Thi, H., Dang, V., Nguyen, N., Le, D., Nguyen, T.: A neural network-based fast terminal sliding mode controller for dual-arm robots. Int. Conf. Eng. Res. Appl. 42–52 (2023)
12. Dang, V., Nguyen, D., Tran, T., Le, D., Nguyen, T.: Model-free hierarchical control with fractional-order sliding surface for multi-section web machines. Int. J. Adapt. Control Signal Process. (2018)
13. Pham, V., et al.: Backstepping sliding mode control design for active suspension systems in half-car model. In: International Conference on Engineering Research and Applications, pp. 281–289 (2023)
14. Nguyen, D., Truong, V., Lam, N.: Others Nonlinear control of a 3-DOF robotic arm driven by electro-pneumatic servo systems. Measure. Control Autom. **3**, 51–59 (2022)
15. Zhang, J., Li, J.: Adaptive backstepping sliding mode control for wheel slip tracking of vehicle with uncertainty observer. Measur. Control. **51**(9-10), 396–405 (2018). https://doi.org/10.1177/0020294018795321
16. Le, D., Nguyen, D., Le, N., Nguyen, T.: Traction control based on wheel slip tracking of a quarter-vehicle model with high-gain observers. Int. J. Dyn. Control **10**, 1130–1137 (2022)
17. Dang, V., et al.: Adaptive control for multi-shaft with web materials linkage systems. Inventions **6**, 76 (2021)

Evaluation of Valued Tolerance Rough Set and Decision Rules Method for WiFi-Based Indoor Localization in Different Environments

Ninh Duong-Bao[1,2], Jing He[1], Luong Nguyen Thi[3], Seon-Woo Lee[4], and Khanh Nguyen-Huu[5](✉)

[1] College of Computer Science and Electronic Engineering, Hunan University, Changsha 410082, China
{duongbaoninh,Jhe}@hnu.edu.cn

[2] Faculty of Mathematics and Informatics, Dalat University, Dalat 66100, Vietnam

[3] Faculty of Information Technology, Dalat University, Dalat 66100, Vietnam
luongnt@dlu.edu.vn

[4] Division of Software, Hallym University, Chuncheon 24252, Korea
senu@hallym.ac.kr

[5] Department of Electronics and Telecommunications, Dalat University, Dalat 66100, Vietnam
khanhnh@dlu.edu.vn

Abstract. Among various technologies being applied for indoor localization, WiFi has become a common source of information to determine the pedestrian's position due to the widespread of WiFi access points in indoor environments. However, the fluctuation of the WiFi signals makes it difficult to achieve a good localization result. In this paper, to handle this problem, the valued tolerance rough set and decision rules method (VTRS-DR), which is firstly registered to WiFi fingerprinting-based localization, will be implemented and evaluated in big and complicated environments using two public datasets. The first one was conducted by a subject using a smartphone at a multi-floor library for several months. Furthermore, to evaluate the localization accuracy when WiFi data was collected from different pedestrians as well as different smartphones, a crowdsourced WiFi fingerprinting dataset was utilized. From the deep analyses of localization results, the VTRS-DR method shows high accuracy and high robustness when testing in different environments with a mean error of 3.84 m, which is 27.87% lower than other compared methods.

Keywords: Indoor localization · WiFi fingerprinting · VTRS-DR

1 Introduction

Over the last 20 years, the increase of mobile devices as well as the wide deployment of WiFi access points (APs) in modern buildings has opened the chance for WiFi-based indoor localization. WiFi fingerprinting is one of the most promising methods which conceptually includes two basic phases. The first one is known as the offline phase in

T. D. L. Nguyen et al. (Eds.): ICISN 2023, LNNS 752, pp. 186–194, 2023.
https://doi.org/10.1007/978-981-99-4725-6_24

which the received signal strength (RSS) values from available APs are collected at predefined positions (i.e. reference points - RPs) to build the fingerprinting database where each fingerprint is a set of pair {position, RSS values}. The position of the user is determined in the online phase. The RSS values collected at an unknown position will be compared to the fingerprints from the offline database to estimate the user's position. Even though this technique is bright to solve indoor localization, it still faces some challenges that need to be researched. Firstly, it requires lots of effort (i.e. time-consuming and labor-intensive) to build and maintain the database. Moreover, the fluctuation of the RSS values due to the multipath effects, the smartphone heterogeneity, the changes in environments, etc. is also a big problem.

The valued tolerance rough set and decision rule (VTRS-DR) method, proposed by Stefanowski and Tsoukias [1], was originally applied to WiFi-based indoor localization in the research of Duong-Bao et al. [2] to handle the variation of RSS values. The method was utilized not only in the offline phase but also in the online phase of the WiFi fingerprinting. Although they achieved great localization results, the authors did test their proposed method in only a small office room using one subject and one smartphone. In this paper, to extend and evaluate their methods in different environments such as bigger areas with different numbers of smart devices and users for data collection, we investigate the classification results of the VTRS-DR method in two different public datasets. Our work aims to find out how different setup environments affect the classification results.

The rest of this paper is presented as follows. Section 2 is about some related works while the Sect. 3 describes briefly the VTRS-DR method. In Sect. 4, the two datasets are reported. Section 5 shows the localization results of the VTRS-DR when using the two datasets. Finally, the conclusion, as well as future developments, are figured out in Sect. 6.

2 Related Works

Until now, many extensive surveys summed up various systems related to WiFi-based localization [3, 4]. They deeply analyzed and showed the advantages and disadvantages of the existing solutions, then offered some directions for future systems. There are two main approaches for WiFi fingerprinting-based localization. The first one is the deterministic matching algorithm which includes some well-known algorithms such as the nearest neighbor (NN)-based algorithms. In [5], Duong-Bao et al. compared the localization results of using weighted KNN (WKNN) with different distance measures. They showed that the Chi-squared measure achieved distinguished results, even when changing the grid size or the number of available APs. The second matching algorithm is probabilistic-based. This algorithm estimates the user's position by calculating the highest probability of the new observation to the most similar position. Li et al. [6] proposed a probabilistic fingerprint database to adapt to the crowdsourced data based on the multivariate Gaussian Mixture Model, then they applied the Hidden Markov Model in the online phase to predict the user's position. When testing with a smartphone that appeared for the first time in the database, the proposed method still achieved a good result with the accuracy of 94%. Furthermore, in recent years, deep learning-based approach is being focused. Some were summarized in the comprehensive review of Feng et al. [7]. Based on the combination between the convolutional neural network (CNN) and the auto-encoder, Qin et al. [8] proposed their localization systems which got a mean error of 12.4 m when testing in different datasets.

The previous research work [2], which attempted to handle the changes of RSS values, presented the outstanding performance of the VTRS-DR method in a small area. Encouraging by that work, in this paper, we evaluate the classification results of the method when the WiFi signals were collected in bigger areas such as multi-floor buildings, and using the crowdsourcing approach.

3 VTRS-DR Method

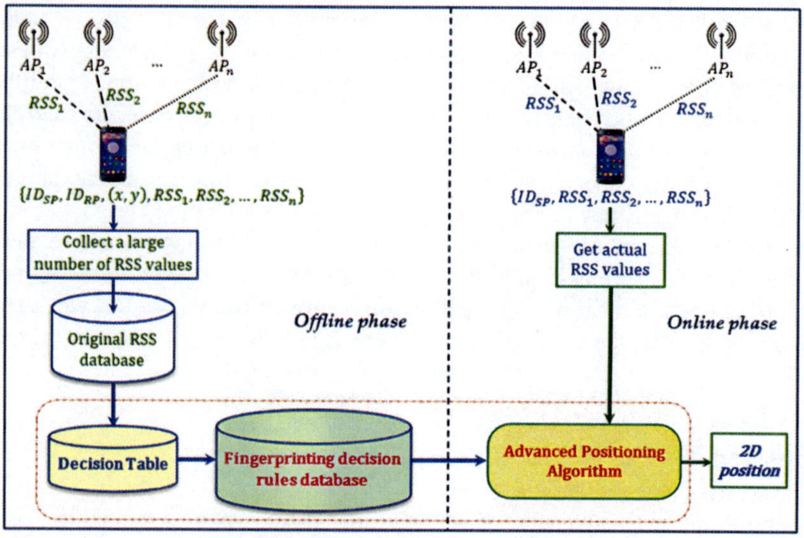

Fig. 1. Overview of the proposed system.

From the original perspective, the rough set theory [9] is designed to work with vague information by finding the equivalence relation between objects in a provided object set. Rather than choosing the relation values in only two levels (i.e. 0 and 1), the VTRS-DR method smoothens the relation values running from 0 to 1. This method, when carrying out WiFi fingerprinting-based indoor localization, is involved in both offline and online phases as shown in Fig. 1. In the former one, a fingerprinting decision rules database is constructed based on the decision table, which is adapted from the traditional fingerprinting database. The data structure of the new database is formatted as {Rule, Condition of Rule, Decision Attribute, Support Components}. Each decision rule represents an object which is a fingerprint of one RP. The condition of rule includes the vector of RSS values which are collected from the available APs. Meanwhile, each decision attribute is defined as one RP (including the 2D coordinate) in a group of RPs. At last, the support components with the credibility degree and the support object set value will be used as the criteria to classify one new fingerprint to one RP. In the latter phase, a multi-level comparison, which includes the valued tolerance relation, the support components, and the Euclidean distance, is considered between the RSS values collected at an unknown position and the fingerprinting decision rules database, then the best decision class (i.e. one RP) will be classified from the whole predefined RPs. More details of the VTRS-DR method can be found in [2].

4 Dataset Description

4.1 Dataset 1

The first WiFi fingerprinting dataset was created on the 3rd and 5th floors at the library building of Universitat Jaume I (UJI), Spain. There were 448 APs (with some APs not detected every month) and 96 RPs set up in the area of 308.4 m^2 for two floors. The distance between two adjacent RPs was about 2 m. The WiFi RSS values were collected over a long period (i.e. 15 consecutive months) by one subject using one Samsung Galaxy S3 smartphone. To create the fingerprinting database, the subject stood at each RP and made a scan in a certain direction. For each RP, the RSS values were collected six times, which means 576 scanning times in total for two floors. In addition, in the library, many things such as bookshelves, furniture as well as the different number of working people caused the variation of WiFi signals, thus, these directly affect the localization results. Figure 2 shows the changes in RSS values from some APs over the whole RPs on two floors. More details of this dataset can be found in [10].

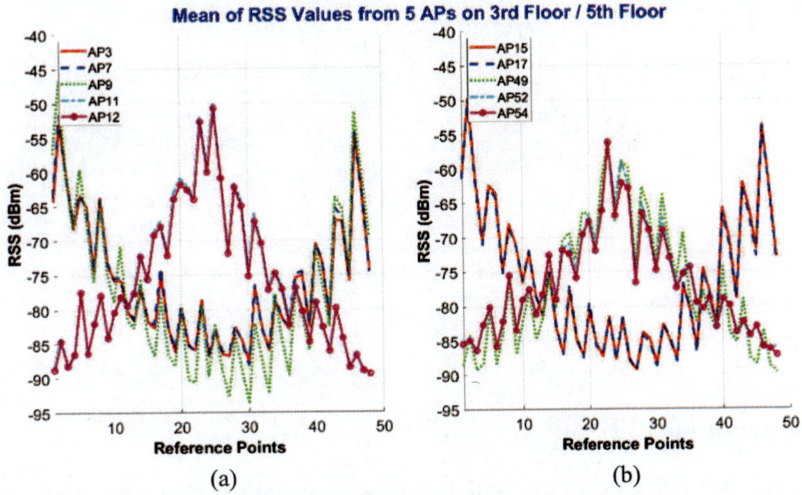

Fig. 2. The fluctuation of RSS values from some APs on (a) 3rd floor and (b) 5th floor.

4.2 Dataset 2

The crowdsourcing approach, which comprises many people as well as many smartphones, can be used to reduce the required efforts to make the fingerprint database. To validate the robustness of the VTRS-DR method when using this approach, we chose the dataset which was conducted on the 4th floor of the SYL building, China. The covered area was 2600 m^2 with many rooms and corridors. In this area, there were 23 APs (dual band) and 296 RPs. The grid size was 1.2 m. Three users (ID1 to ID3) using two smartphones (i.e. Xiaomi 8 and Xiaomi 11) were asked for data collection. For each RP, the RSS values were collected 30 times with the scanning frequency of 1 Hz, which means a total of 8880 scanning times. This dataset covered different factors when collecting the WiFi signals such as smartphone heterogeneity and human body loss (i.e. different devices and different people). Figure 3 displays the probability density function of RSS values collected from three users as well as two smartphones. Since the data were collected in a big area, there is a finite number of APs that can be scanned at one RP, thus, the APs that cannot be seen were assigned with the RSS value of -105 dBm which results in a high probability of this value. More details of this dataset can be found in [11].

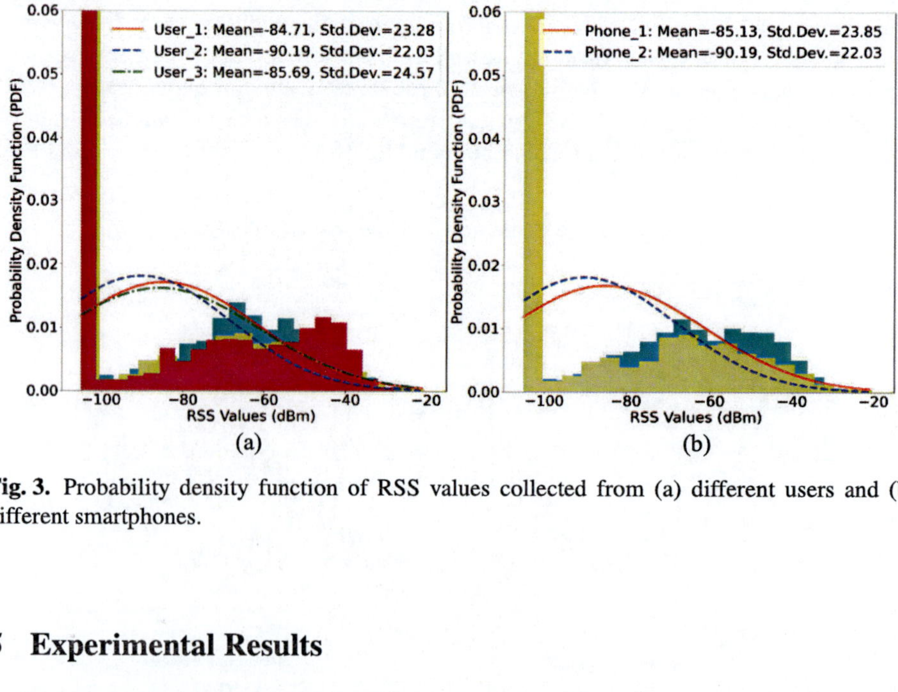

Fig. 3. Probability density function of RSS values collected from (a) different users and (b) different smartphones.

5 Experimental Results

In this section, the VTRS-DR method will be compared to three NN-based methods including the 1-NN, the WKNN with K = 3, and the WKNN with K = 7, respectively. The fingerprinting decision rules database will be built based on the conventional fingerprinting database. To do that, the valued tolerance relation matrix that contains the relations of all pairs of two fingerprints (i.e. two rows in the database) needs to be created. This is time-consuming since the number of relations is proportional to the number of fingerprints in the database. The VTRS-DR method is implemented and evaluated using MATLAB R2019a.

5.1 Dataset 1

The fingerprinting decision rules database is created by combining the whole WiFi data collected in the offline phase over 15 months. For each floor, the testing data includes five different tests: two tests at the same RP positions as in the training phase (i.e. Test-01 and Test-05) and three tests at random positions (i.e. Test-02, Test-03, and Test-04). The first four tests included 48 test positions while the last one had 68 test positions. For each position, the RSS values were collected six times. It is worth noting that the user collected the WiFi signals in different directions which can also be one factor that affects the changes of RSS values. The localization error is defined as the mean error over several scanning times (e.g. six times) at one testing position. Meanwhile, the mean localization error is described as the average of the localization errors over the whole testing positions.

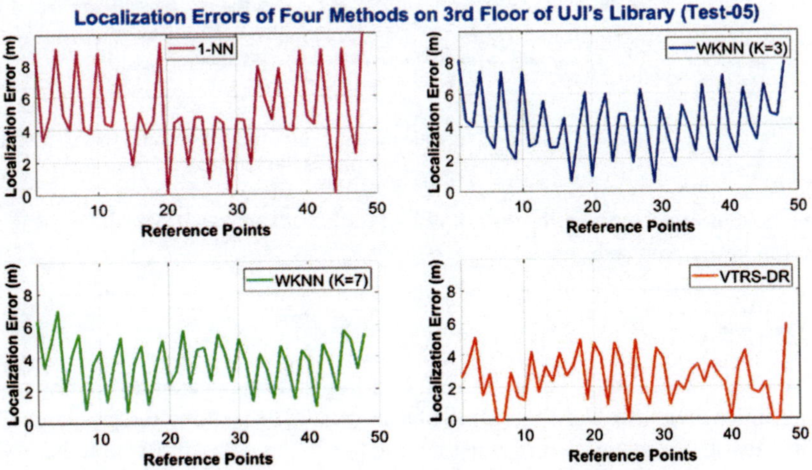

Fig. 4. Localization errors of four methods on the 3rd floor for Test-05.

Figure 4 shows the localization errors of four methods on the 3rd floor in Test-05. From this figure, the VTRS-DR achieves the best result since the errors are almost under 4 m. Furthermore, it has six positions that have the error of 0 m, which means the testing fingerprints are classified to the right position. Figure 5 displays the mean errors of four methods for five tests on both floors. Overall, these results of the VTRS-DR method are better than others, which reduces the localization errors by 20.52% up to 37.01%. The mean errors of the proposed method, when applied to five testing datasets, are 2.53 m, 3.58 m, 3.53 m, 2.71 m, and 2.57 m for the tests on the 3rd floor and 2.82 m, 2.92 m, 3.30 m, 3.12 m, and 2.42 m for the tests on the 5th floor, respectively. The mean errors of the first and the last tests are smaller than others since the RSS values in these were collected at the same RPs while others are collected in random positions, which

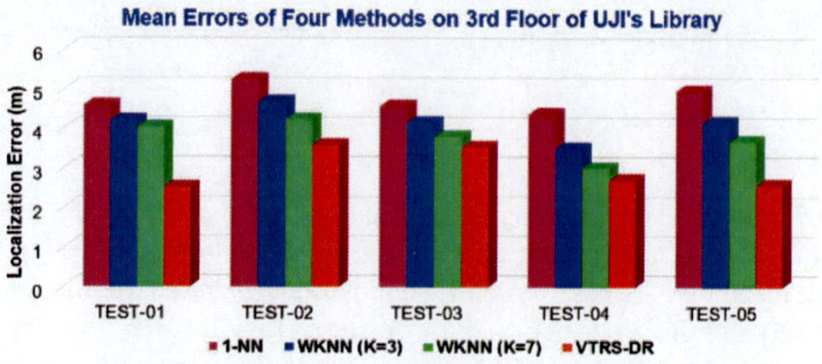

(a)

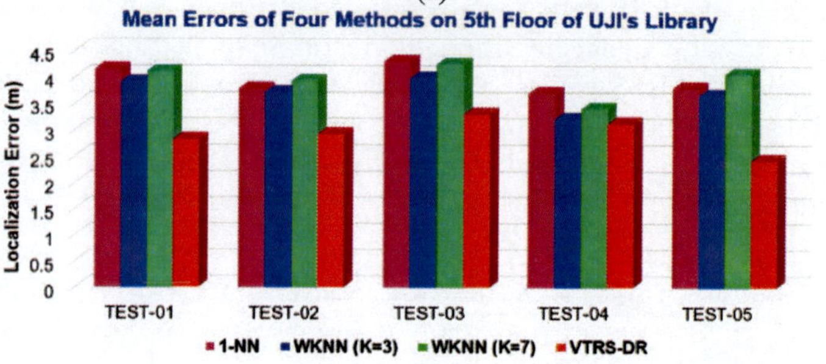

(b)

Fig. 5. Mean localization errors of four methods on (a) 3rd floor and (b) 5th floor.

means the classification results cannot return to their exact positions since these are not registered in the database.

5.2 Dataset 2

In this dataset, the training file of the SYL building is used to generate the fingerprinting decision rules database. For the online phase, there will be three testing data. The first one (i.e. Test-01) is from the provided testing file in the dataset in which the WiFi data was collected from 102 random positions, with 10 scanning times for each position. The next testing data (i.e. Test-02) is created by splitting every six random fingerprints from 30 fingerprints of each RP, which means this data has 1776 testing fingerprints over 296 RPs. The training database, after data extraction, has 7104 fingerprints, which means 80% for training and 20% for testing. This aims to test the localization results when the user stood at the same position of each predefined RP to scan the WiFi signals. The last testing data (i.e. Test-03) includes four sub-datasets which differ by the users as well as the smartphones used for data collection. Table 1 shows the involved RPs as well as the IDs of users and phones in each sub-dataset. The RSS values were collected 30 times for each RP, which result in 2640, 2610, 2010, and 1620 testing fingerprints for each sub-dataset, respectively. Each sub-dataset is split from the training database to evaluate how a new user or a new smartphone affects the localization performance of the VTRS-DR method.

Table 1. The four sub-datasets in the third testing data.

Sub-dataset	RP Range	UserID	PhoneID
1	1–88	1	1
2	89–175	2	2
3	176–242	3	1
4	243–296	2	2

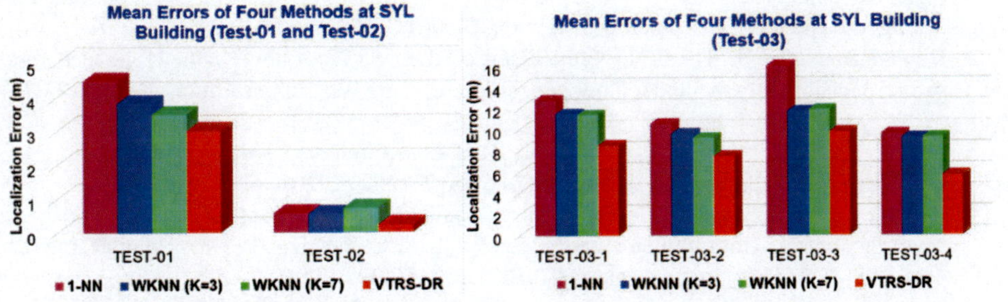

Fig. 6. Mean localization errors of four methods in three testing data.

Figure 6 (a) shows the mean localization errors of four methods in the first two testing data. As shown in this figure, the best results belong to the VTRS-DR even when testing at random positions (i.e. Test-01) or the same RPs' positions (i.e. Test-02). The mean errors of the proposed method for each test are only 2.99 m and 0.22 m. These results reduced the errors of other methods from 23.37% to 35.80%. Moreover, as aforementioned, the smartphone heterogeneity and the body of the user can affect the localization results. Figure 6 (b) presents the results when a new user or a new smartphone scan the RSS values to find the position. As can be seen in the figure, the mean errors dramatically increase for the whole four methods, which are about four times bigger than the previous tests. This points out that the new users/smartphones, that are not registered in the offline database, will degrade the localization accuracy. However, the performance of VTRS-DR is still the best with a mean error of 7.84 m, which is approximately 28.61% better than others.

6 Conclusion

In this paper, to evaluate the performance of the VTRS-DR method when applied to WiFi fingerprinting-based indoor localization, especially in big and complicated areas, we implemented and compared its localization results with popular deterministic methods in two different public datasets. These datasets were conducted in multi-floor buildings and used a crowdsourcing approach. The analyzed results prove the superiority of the VTRS-DR method when it achieves the mean error of 3.84 m, which reduces the localization error by 24.86% to 35.65% compared to other methods. The VTRS-DR, however, still cannot cover well the case when training and testing data are collected from different users or smartphones, which is more practical in real life. In the future, we will investigate to solve this issue. Furthermore, we will try to improve the method to reduce the computational cost which is essential in real-time applications.

References

1. Stefanowski, J., Tsoukiàs, A.: Valued tolerance and decision rules. In: Ziarko, W., Yao, Y. (eds.) Rough Sets and Current Trends in Computing. LNCS (LNAI), vol. 2005, pp. 212–219. Springer, Heidelberg (2001). https://doi.org/10.1007/3-540-45554-X_25
2. DuongBao, N., He, J., Thi, L.N., NguyenHuu, K., Lee, S.-W.: A novel valued tolerance rough set and decision rules method for indoor positioning using WiFi fingerprinting. Sensors **22**(15), 5709–5734 (2022)
3. Subedi, S., Pyun, J.-Y.: A survey of smartphone-based indoor positioning system using RF-based wireless technologies. Sensors **20**(24), 7230–7262 (2020)
4. Roy, P., Chowdhury, C.: A survey on ubiquitous WiFi-based indoor localization system for smartphone users from implementation perspectives. CCF Trans. Pervas. Comput. Interact. **4**(3), 298–318 (2022). https://doi.org/10.1007/s42486-022-00089-3
5. Duong-Bao, N., He, J., Thi, L.N., Nguyen-Huu, K.: Analysis of distance measures for wifi-based indoor positioning in different settings. In: 2022 2nd International Conference on Innovative Research in Applied Science, Engineering and Technology (IRASET), pp. 1–7. IEEE (2022)
6. Li, Y., Williams, S., Moran, B., Kealy, A.: A probabilistic indoor localization system for heterogeneous devices. IEEE Sens. J. **19**(16), 6822–6832 (2019)
7. Feng, X., Nguyen, K.A., Luo, Z.: A survey of deep learning approaches for WiFi-based indoor positioning. J. Inf. Telecommun. **6**(2), 163–216 (2022)
8. Qin, F., Zuo, T., Wang, X.: CCpos: WiFi fingerprint indoor positioning system based on CDAE-CNN. Sensors **21**(4), 1114–1131 (2021)
9. Pawlak, Z.: Rough sets. Int. J. Comput. Inform. Sci. **11**, 341–356 (1982)
10. MendozaSilva, G.M., Richter, P., TorresSospedra, J., Lohan, E.S., Huerta, J.: Long-Term WiFi fingerprinting dataset for research on robust indoor positioning. Data **3**(1), 3–20 (2018)
11. Bi, J., Wang, Y., Yu, B., Cao, H., Shi, T., Huang, L.: Supplementary open dataset for WiFi indoor localization based on received signal strength. Sat. Navig. **3**(1), 1–15 (2022)

Lyapunov-Based Model Predictive Control for 3D-Overhead Cranes: Tracking and Payload Vibration Reduction Problems

Chung Nguyen Van[1], Duong Dinh Binh[1], Hien Nguyen Thi[1], Hieu Le Xuan[1], Mai Hoang Thi[1], Thu Nguyen Thanh[1], Hue Luu Thi[2], Hoa Bui Thi Khanh[1,3], and Tung Lam Nguyen[1](✉)

[1] Hanoi University of Science and Technology, Hanoi, Vietnam
lam.nguyentung@hust.edu.vn
[2] Electric Power University, Hanoi, Vietnam
[3] Hanoi University of Industry, Hanoi, Vietnam

Abstract. In this work, we focus on controlling the 3-D overhead crane (3-DOC), the main problem when controlling the 3-DOC model is the problem of tracking the trajectory and reducing the load vibration. Therefore, the limitation of the system's state variables must be made explicit. To solve this problem, firstly, the problem of orbital tracking is solved using the Sliding Mode Control (SMC), but the system's states are not strictly controlled, so we propose a control method Lyapunov-based model predictive control (LMPC), which allows setting limits for state variables besides that, an auxiliary component is added based on the stability of the SMC to make the system globally stable. Finally, simulations are added to show the feasibility of the method for tracking, and anti-vibration for the payload.

Keywords: Model predictive control · Lyapunov-based model predictive control · Sliding mode control · 3-D overhead crane

1 Introduction

The invention of overhead cranes created a significant step forward in transporting and distributing goods in production warehouses, railway stations, ports, etc. Many studies worldwide have been published to improve and bring more outstanding features of the crane. The two main problems of controlling the crane that scientists are most interested in controlling the crane to bring the heavy object to the intended point quickly and limit the payload vibration during the transferring operations.

Overhead cranes have been evaluated as a complex nonlinear model where the swing angles of the load are closely related to the movement of the trolleys. There are two primary categories for controlling overhead cranes: open and closed-loop

T. D. L. Nguyen et al. (Eds.): ICISN 2023, LNNS 752, pp. 195–203, 2023.
https://doi.org/10.1007/978-981-99-4725-6_25

control. Closed-loop controls are preferred thanks to the advantages of being less affected by noise and changes in the system. Some typical closed-loop control methods such as linear feedback [1], methods based on Lyapunov's theorem [2], SMC [3–5], and model predictive control (MPC) [6–8] have been widely applied in crane control designs.

The SMC method has been used frequently because it possesses some good features such as noise rejection, and strongness to uncertainty in this system's parameters. Tuan (2013) et al. [1] proposed the second-order SMC for 3-DOC in a highly complicated operation though many system parameters were complex. The model is used to create a model-free control method, called the proportionality function with SMC (PD-SMC), designed for 3-D axis systems where disturbances are not modelled, to achieve the proposed trolley and the target load simultaneously by Menghua Zhang (2019) et al. [5]. Recent studies try to solve more practical problems to ensure system constraints such as thrust, transit time constraints [9], obstacle avoidance [10], and limit the swing angles with varying rope length during transportation [11], etc. Therefore, we expect these constraints to be incorporated into the controller. MPC with respect to time method optimization has a specific advantage owing to its distinct capability of system solving system constraints. However, with MPC, analysis of closed-loop stability is extremely challenging, particularly for complex nonlinear systems like overhead cranes. Scientists have combined the model predictive control method with other traditional control methods to make it easier to prove the stability. Marco Giacomelli (2018) et al. [8] proposed a speed controller for overhead cranes based on MPC combined with the PID control method. Some studies use LMPC scheme to control typical models [12–15] such as UAVs and mobile robots, etc.

In the present study, we proposed a novel LMPC scheme for 3-DOC systems. The LMPC algorithm is an improvement of the second-order sliding mode control motivated by Tuan et al. [1]. To ensure the closed-loop system's robust stability and recursive feasibility, a collection of stability constraints is provided. The major contributions of this paper can be summed up as follows:

- Solve the 3-DOC trajectory tracking control problem subject to input constraints and output constraints.
- Enhance the conventional SMC scheme in terms of stability and robustness by integrating the controller with the LMPC algorithm, thus equipping the capability of prediction for the system.
- A comprehensive numerical study between LMPC and SMC is provided.

The rest of this paper is organized as follows: The overall dynamic modelling of a 3-DOC is described in Sect. 2. Section 3 proposes a control scheme along with its stability and robustness properties. Simulation results and validation of the given control are demonstrated in Sect. 4. Finally, Sect. 5 concludes the paper.

2 Modeling of 3-D Overhead Cranes

The modelling of the 3-DOC is illustrated in Fig. 1, in which m_c, m_b, m_t and m_l are the equivalent masses of cargo, trolley, bridge, and hoist, respectively. The trolley is considered moving along the X-axis while the bridge is moving along the Y-axis. The position of cargo is determined by $[l\ \phi\ \theta]^T$. By using Lagrange equations, the system dynamics of 3-DOC can be written as follows:

$$\mathbf{M}(\mathbf{q})\ddot{\mathbf{q}} + \mathbf{D}\dot{\mathbf{q}} + \mathbf{C}(\mathbf{q}, \dot{\mathbf{q}})\dot{\mathbf{q}} + \mathbf{G}(\mathbf{q}) = \mathbf{F} \tag{1}$$

where $\mathbf{q} = \begin{bmatrix} x & y & l & \phi & \theta \end{bmatrix}^T$ and $\mathbf{F} = \begin{bmatrix} f_t & f_b & f_l & 0 & 0 \end{bmatrix}^T$ are respectively the state vector and driving force acting on the system. $\mathbf{M}(\mathbf{q})$ is the symmetric mass matrix, $\mathbf{C}(\mathbf{q}, \dot{\mathbf{q}})$ presents the Coriolis and centrifugal matrix, $\mathbf{D} = \text{diag}(D_t, D_b, D_l, 0, 0)$ and $\mathbf{G}(\mathbf{q})$ stand for the viscous damping matrix and gravitational force vector respectively.

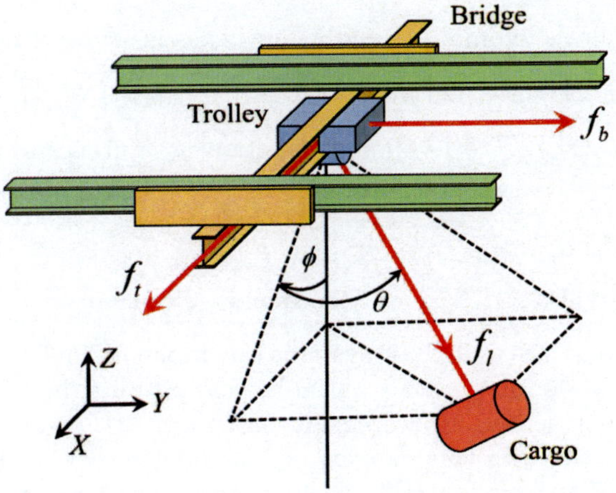

Fig. 1. Dynamic model of a 3-D overhead crane.

Since 3-DOC is an un-actuated system, we separate the system state vector into two parts: actuated states $\mathbf{q}_a = [x\ y\ l]^T$ and un-actuated states $\mathbf{q}_u = [\phi\ \theta]^T$ and establish the following equivalent dynamic model:

$$\overline{\mathbf{M}}\ddot{\mathbf{q}}_a + \overline{\mathbf{C}}_1\dot{\mathbf{q}}_a + \overline{\mathbf{C}}_2\dot{\mathbf{q}}_u + \overline{\mathbf{G}} = \mathbf{F}_a \tag{2}$$

In which, $\mathbf{F}_a = \begin{bmatrix} f_t & f_b & f_l \end{bmatrix}^T$ denotes the actuator input; $\overline{\mathbf{M}}, \overline{\mathbf{C}}_1, \overline{\mathbf{C}}_2$ and $\overline{\mathbf{G}}$ are the system's matrices defined according to [1]. We use this equation to design the control signals for the system.

3 Proposed Control Algorithm

3.1 Sliding Mode Control

Let $\mathbf{s} = \dot{\mathbf{q}}_a + \boldsymbol{\lambda}(\mathbf{q}_a - \mathbf{q}_{ar}) + \boldsymbol{\alpha}\mathbf{q}_u$ be the sliding surface, where $\mathbf{q}_{ar} = [x_r \;\; y_r \;\; l_r]^T$ is the reference trajectory. According to [1], Tuan (2013) et al. proposed the following second-order sliding mode control scheme:

$$\mathbf{F}_a = \overline{\mathbf{C}}_1\dot{\mathbf{q}}_a + \overline{\mathbf{C}}_2\dot{\mathbf{q}}_u + \overline{\mathbf{G}} + \mathbf{K}\text{sign}(\mathbf{s}) - \overline{\mathbf{M}}\left[2\lambda\dot{\mathbf{q}}_a + \boldsymbol{\lambda}^T\boldsymbol{\lambda}(\mathbf{q}_a - \mathbf{q}_{ad}) + \boldsymbol{\alpha}\dot{\mathbf{q}}_u + \boldsymbol{\lambda}\boldsymbol{\alpha}\mathbf{q}_u\right] \tag{3}$$

The Lyapunov candidate function can be chosen as:

$$V = \frac{1}{2}\mathbf{s}^T\mathbf{s} \tag{4}$$

With some calculation, its derivative can be illustrated by:

$$\dot{V} = \dot{\mathbf{s}}^T\mathbf{s} = -\mathbf{s}^T\boldsymbol{\lambda}\mathbf{s} - \mathbf{s}^T\overline{\mathbf{M}}^{-1}\mathbf{K}\text{sign}(\mathbf{s}) \tag{5}$$

Based on Barbalat's lemma, the sliding surface $\mathbf{s}$ is asymptotically stable, along with the chosen weighted parameters $\boldsymbol{\lambda} = \text{diag}(\lambda_1, \lambda_1, \lambda_3)$, $\boldsymbol{\alpha} = \begin{bmatrix}\alpha_1 & 0 & 0\\ 0 & \alpha_2 & 0\end{bmatrix}^T$ and $\mathbf{K} = \text{diag}(k_1, k_2, k_3)$. The Lyapunov function as well as its derivative is given in (6–7) and is used as input information to establish the LMPC for 3-DOC systems.

3.2 Lyapunov-Based Model Predictive Control

In reality, the structure of 3-DOC systems is complex, a control method is needed that sets tight limits for the state variables and outputs, besides ensuring global stability of the system. The system needs to pay attention to the position of the trolley and the vibrating angles of the cargo. So that it can ensure the small velocity of the trolley and of swing angles, in accordance with minimal requirement of driving forces.

A control method has been proposed in [6], which solves the problem by optimizing the cost function over each cycle as well as ensuring that the states of the system reach the desired values and the control signal converges after a period of operation. However, the control signal is still not subject to such constraints, and the stability of the system is not guaranteed through each cycle. Therefore, a new control method LMPC is proposed based on MPC and SMC methods, thereby not only ensuring the desired trajectory tracking, and the convergence of the control signal, but also ensuring system stability through each cycle due to Lyapunov auxiliary function. The proposed LMPC scheme is obtained by solving the following optimal problem:

$$\text{minimize} \quad J(k) = \sum_{j=1}^{N_p} \left\| \widehat{\mathbf{e}}(k+j) \right\|_P^2 + \sum_{j=1}^{N_p-1} \left\| \Delta\widehat{\mathbf{F}}(k+j) \right\|_Q^2$$

$$\text{subject to:} \quad \left|\widehat{\mathbf{x}}(k+j)\right| \leqslant \mathbf{x}_{\max} \qquad \forall j = 1, 2, ...N_p - 1$$

$$\left|\widehat{\mathbf{F}}(k+j)\right| \leqslant \mathbf{F}_{\max} \qquad \forall j = 1, 2, ...N_p$$

$$\dot{V}\left(\widehat{\mathbf{x}}(k), \widehat{\mathbf{F}}_a(k)\right) \leqslant \dot{V}\left(\widehat{\mathbf{x}}(k), h\left(\widehat{\mathbf{x}}(k)\right)\right)$$

where $\widehat{\mathbf{x}}(k+j)$, $\widehat{\mathbf{q}}(k+j)$ and $\widehat{\mathbf{F}}(k+j)$ are predicted trajectory vector, predicted output vector and predict control vector at sampling time $k+j$, respectively and given by current state $\mathbf{x}(k) = \left[\mathbf{q}(k) \;\; \dot{\mathbf{q}}(k)\right]^T$; $\widehat{\mathbf{e}}(k+j) = \widehat{\mathbf{q}}(k+j) - \widehat{\mathbf{q}}_r(k+j)$ and $\Delta\widehat{\mathbf{F}}(k+j) = \widehat{\mathbf{F}}(k+j) - \widehat{\mathbf{F}}(k+j-1)$ are respectively predicted error output and predicted change of input made at time k; N_p denotes the number of steps of prediction horizon; P and Q are positive definite weighted matrices; $h(\cdot)$ and $V(\cdot)$ are respectively the auxiliary Lyapunov-based SMC law and the corresponding Lyapunov function presented in (3–5).

These above-given constraints guarantee tight operation of the system, where the presence of the last constraint ensures the global stability of the system while solving the optimal problem at each prediction horizon. Note that this optimal problem is implemented in the domain of discrete time. In the simulation, the discrete-time model of 3DOC, as well as the optimal algorithm can be conducted by the usage of *Nonlinear MPC Toolbox* integrated with *MATLAB-Simulink*, or manually built-in *Python.*

Consider the Lyapunov candidate function:

$$V = \frac{1}{2}\mathbf{s}^T\mathbf{s} \tag{6}$$

where $\mathbf{s} = [s_1 \;\; s_2 \;\; s_3]^T = \dot{\mathbf{q}}_a + \boldsymbol{\lambda}(\mathbf{q}_a - \mathbf{q}_{ar}) + \boldsymbol{\alpha}\mathbf{q}_u$. Taking account of the second-order sliding mode control $\mathbf{F}_a = h(\mathbf{x})$ established in (3), the derivative of the Lyapunov function is:

$$\dot{V}\left(\widehat{\mathbf{x}}, h(\widehat{\mathbf{x}})\right) = \mathbf{s}^T\boldsymbol{\lambda}\mathbf{s} - \mathbf{s}^T\overline{\mathbf{M}}^{-1}\mathbf{K}\text{sign}(\mathbf{s}) \leqslant 0 \tag{7}$$

Taking utilization from the last constraint of LMPC optimal problem:

$$\dot{V}\left(\widehat{\mathbf{x}}, \widehat{\mathbf{F}}_a\right) \leqslant \dot{V}\left(\widehat{\mathbf{x}}, h(\widehat{\mathbf{x}})\right) \tag{8}$$

turns it out $\dot{V}\left(\widehat{\mathbf{x}}, \widehat{\mathbf{F}}_a\right) \leqslant 0$. This inequality implies the stabilization of the system under the LMPC method (Fig. 2).

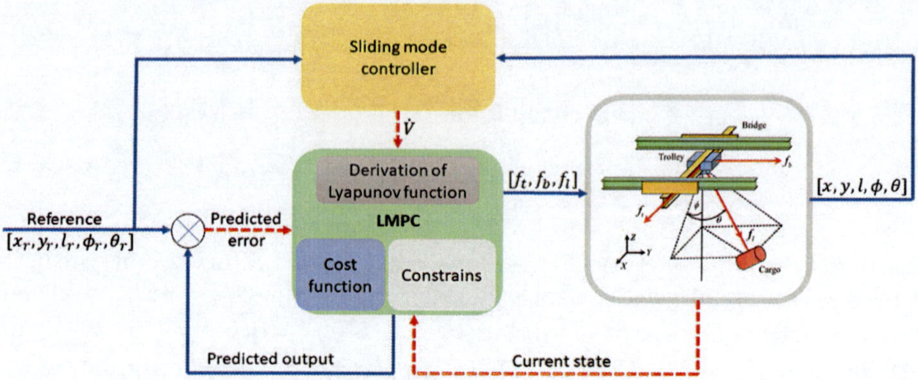

Fig. 2. Block diagram of LMPC controller

4 Simulation Results

In this section, we provide a simulation to verify the ability of the trajectory tracking problem respected by the LMPC method. In order to show the effectiveness of LMPC, comparison studies between LMPC and SMC are provided. We also investigated trajectory tracking over the prediction horizon of LMPC. Figure 3 displays the tracking trajectory of the systems, where the desired paths $\mathbf{q}_{ar} = [x_r \ y_r \ l_r]^T$ are designed as S-shape curves. It shows a better tracking trajectory of LMPC compared to SMC. When the system's set values change, there is a transitory period of the SMC so that the system can stick to the set values, while the transitory period of the LMPC is almost nonexistent. The system's trajectory under the control of the LMPC algorithm closely follows the set orbit, which is seen in the x and y orbitals. Trajectory tracking of l is almost no different but at the initial time, LMPC makes the system follow the desired value of l faster.

The vibrating angles are shown in Fig. 4, while the value of the reference vibrating angles ϕ, θ are set to 0 rad. Both control methods have small vibrating angles, which are almost not present in the implementation, the shaking of the SMC occurs because the values of x, and y do not follow the set values well, while that of the LMPC is the opposite.

Figure 5 shows the control input signals for 3-DOC according to two control methods SMC and LMPC. The difference between SMC and LMPC is, SMC has an oscillating control signal while reference changes but LMPC can solve this problem. The control input of the LMPC is a sample time signal. The great problems of SMC have been solved by LMPC such as the vibration of the load, the existence of a transitory period when the control signal is a trajectory, and the limitation of the control signal.

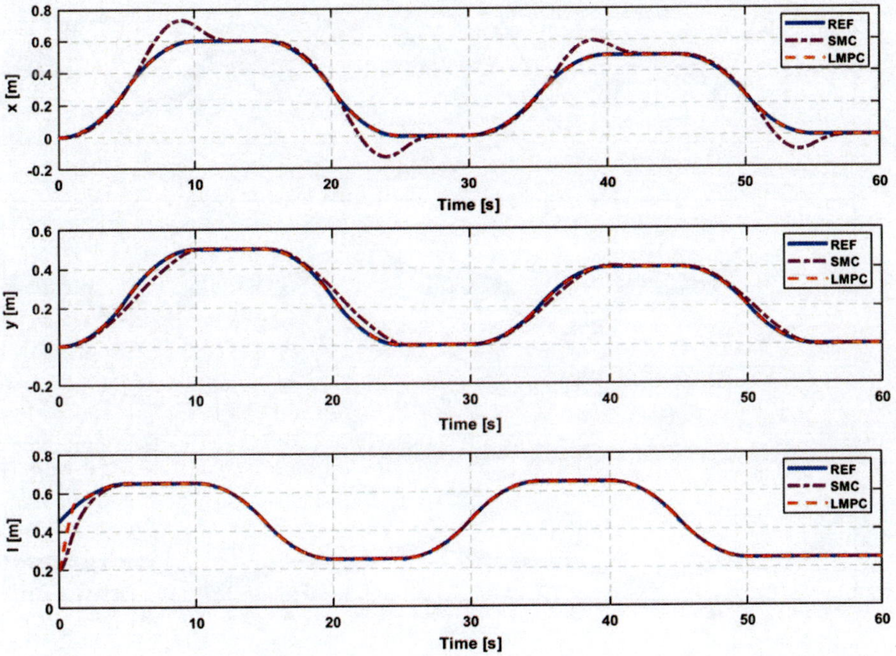

Fig. 3. The output x, y, l tracking performance

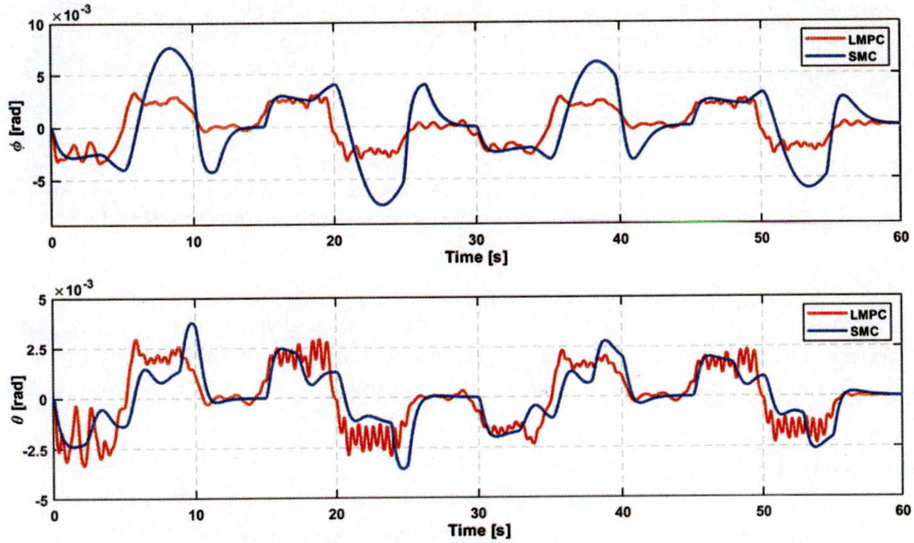

Fig. 4. The vibrating angles

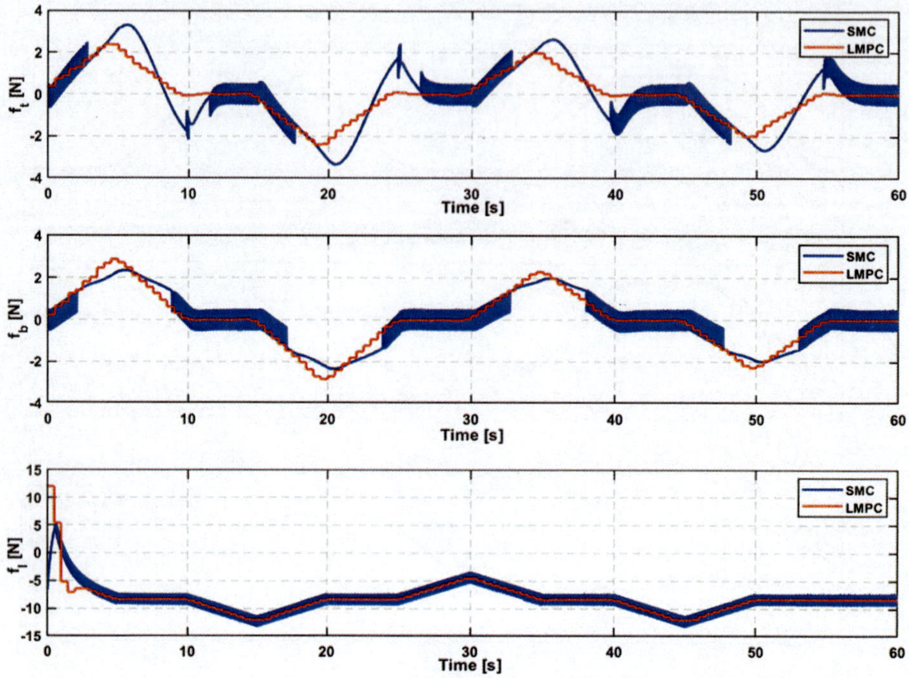

Fig. 5. The control input signals

Table 1. Simulation parameters

System parameters	Control parameters
$m_b = 7\text{kg}, m_t = 5\text{kg}, m_l = 2\text{kg}$, $m_c = 0.85\text{kg}, l_{\max} = 1\text{m}$, $D_t = 30\text{Nm/s}, D_b = 20\text{Nm/s}$, $D_l = 50\text{Nm/s}, g = 9.81\text{m/s}^2$	N_p=10, $T_s = 0.2$, $\mathbf{P} = \text{diag}(500, 500, 500, 200, 200)$, $\mathbf{Q} = \text{diag}(1, 1, 1)$, $\mathbf{F}_{\max} = [12\ 12\ 12]^T$, $\mathbf{q}_{\max} = [1\ 1\ 1\ 0.15\ 0.15]^T$, $\boldsymbol{\lambda} = \text{diag}(0.75, 0.75, 1)$, $\alpha_1 = -4$, $\alpha_2 = -3.5$, $\mathbf{K} = \text{diag}(0.5, 0.5, 1)$

Thanks to the satisfaction of LMPC in control design, the tracking effectiveness and reduction vibrating are obtained. In Figs. 3 and 4, the states are bounded and global stability is acquired (Table 1).

5 Conclusion

In this paper, a LMPC controller was purposed to control a 3-DOC, consisting of trajectory tracking and anti-vibration. The LMPC technique is applied to construct a contraction and guarantee closed-loop stability. These advantages are shown in the simulation results of different trajectories. In the future, experiments on real models will be performed to verify the designed controller. And advanced control methods will be further studied to integrate into LMPC to replace MPC and SMC to improve closed loop quality.

Acknowledgements. This research is funded by Hanoi University of Industry (HaUI) under project number 10-2023-RD.

References

1. Tuan, L.A., Lee, S.-G., Dang, V.-H., Moon, S., Kim, B.: Partial feedback linearization control of a three-dimensional overhead crane. Int. J. Control Autom. Syst. **11**(4), 718–727 (2013)
2. Makkar, C., Hu, G., Sawyer, W.G., Dixon, W.E.: Lyapunov-based tracking control in the presence of uncertain nonlinear parameterizable friction. IEEE Trans. Autom. Control **52**(10), 1988–1994 (2007)
3. Almutairi, N.B., Zribi, M.: Sliding mode control of a three-dimensional overhead crane. J. Vib. Control **15**(11), 1679–1730 (2009)
4. Cuong, H.M., Lee, S.-G., et al.: Second-order sliding mode control of 3d overhead cranes. In: 2013 International Conference on Control, Automation and Information Sciences (ICCAIS), pp. 341–346. IEEE (2013)
5. Zhang, M., Zhang, Y., Chen, H., Cheng, X.: Model-independent PD-SMC method with payload swing suppression for 3d overhead crane systems. Mech. Syst. Signal Process. **129**, 381–393 (2019)
6. Khatamianfar, A., Savkin, A.V.: A new tracking control approach for 3d overhead crane systems using model predictive control. In: 2014 European Control Conference (ECC), pp. 796–801. IEEE (2014)
7. Wang, X., Liu, J., Zhang, Y., Shi, B., Jiang, D., Peng, H.: A unified symplectic pseudospectral method for motion planning and tracking control of 3d underactuated overhead cranes. Int. J. Robust Nonlinear Control **29**(7), 2236–2253 (2019)
8. Giacomelli, M., Faroni, M., Gorni, D., Marini, A., Simoni, L., Visioli, A.: MPC-PID control of operator-in-the-loop overhead cranes: a practical approach. In: 2018 7th International Conference on Systems and Control (ICSC), pp. 321–326. IEEE (2018)
9. Wu, Y., Sun, N., Chen, H., Zhang, J., Fang, Y.: Nonlinear time-optimal trajectory planning for varying-rope-length overhead cranes. Assembly Automation (2018)
10. Nagai, S., Kaneshige, A., Ueki, S.: Three-dimensional obstacle avoidance online path-planning method for autonomous mobile overhead crane. In: 2011 IEEE International Conference on Mechatronics and Automation, pp. 1497–1502. IEEE (2011)
11. Lee, H.-H.: Motion planning for three-dimensional overhead cranes with high-speed load hoisting. Int. J. Control. **78**(12), 875–886 (2005)
12. Shen, C., Shi, Y., Buckham, B.: Trajectory tracking control of an autonomous underwater vehicle using Lyapunov-based model predictive control. IEEE Trans. Industr. Electron. **65**(7), 5796–5805 (2017)
13. Mahmood, M., Mhaskar, P.: Lyapunov-based model predictive control of stochastic nonlinear systems. Automatica **48**(9), 2271–2276 (2012)
14. Gong, P., Yan, Z., Zhang, W., Tang, J.: Lyapunov-based model predictive control trajectory tracking for an autonomous underwater vehicle with external disturbances. Ocean Eng. **232**, 109010 (2021)
15. Nguyen Manh, C., Nguyen, N.T., Bui Duy, N., Nguyen, T.L.: Adaptive fuzzy Lyapunov-based model predictive control for parallel platform driving simulators. Trans. Inst. Measur. Control. 01423312221122470 (2022)

Deep Learning-Based Object Tracking and Following for AGV Robot

Ngo Thanh Binh[1](✉), Bui Ngoc Dung[1], Luong Xuan Chieu[1], Ngo Long[1], Moeurn Soklin[1], Nguyen Danh Thanh[1], Hoang Xuan Tung[2], Nguyen Viet Dung[3], Nguyen Dinh Truong[4], and Luong Minh Hoang[5]

[1] University of Transport and Communications, No.3 Cau Giay Street, Lang Thuong Ward, Dong Da District, Hanoi, Vietnam
ngobinh74@utc.edu.vn
[2] VNU University of Engineering and Technology, Hanoi, Vietnam
[3] Hanoi University of Science and Technology, 1St Dai Co Viet Street, Hanoi, Vietnam
[4] Toshiba Software Development (Viet Nam) Co., Ltd., Kim Ma, Ba Dinh, Hanoi, Vietnam
[5] University of Bristol, Beacon House, Queens Road, Bristol BS8 1QU, UK

Abstract. This paper proposes a solution for the AGV (Autonomous Guided Vehicles) robot to effectively monitor a moving object using deep learning by enabling the robot to learn and recognize movement patterns. Using a model of a four-wheeled self-propelled robot vehicle, a highly adaptable and modifiable platform AGV was built. A customized development of the TensorFlowLite ESP32 module from the TensorFlow CoCo SSD model enables the ESP32-CAM camera module on the robot to self-identify objects and autonomously follow the human object in front. Using its built-in distance tracking algorithm, the robot can also detect and adjust its speed to safely follow the individual in front at an appropriate distance. It can function autonomously or manually via the local network. Even with a minimal configuration, the algorithm is appropriate for the automaton. The experimental findings demonstrate the method's precision and effectiveness including sensors, algorithms, and mapping technologies that enable the robot to identify obstacles and navigate around the AGV.

Keywords: Robot · AGV · tracking · following · deep learning

1 Introduction

Automated Guided Vehicles (AGV) are a type of Unmanned Ground Vehicle (UGV) designed with self-propelled capabilities for various applications, including autonomous driving, vehicle research, and robots with designated autonomous missions. The effectiveness of AGV applications depends on the robot's ability to navigate predetermined paths, evade obstacles, or follow a cargo support object such as a pallet or container [1]. To suit different terrains and purposes, AGVs may be equipped with various navigation technologies. Navigation solutions for AGV autonomous vehicles are developed based on signal processing that incorporates cameras, LiDAR [2, 3], and other supporting

T. D. L. Nguyen et al. (Eds.): ICISN 2023, LNNS 752, pp. 204–214, 2023.
https://doi.org/10.1007/978-981-99-4725-6_26

sensors, utilizing 3D cameras, multi-angle cameras, 3D object detection, or obstacle avoidance during movement [4, 5]. It is noteworthy that these robots or groups of robots are primarily designed to avoid obstacles instead of tracking objects. Currently, a number of commercial products begin to be marketed with tasks such as delivery, security, or rescue mission support [6], but these robots or groups of robots are all self-propelled that avoid obstacles instead of tracking objects.

The Automated Guided Vehicle (AGV) is widely regarded for its ability to detect and track objects accurately. Recent findings by the Ren research group [7] indicate that a graphical processing unit (GPU) that operates at a rate of 5 frames per second can achieve nearly real-time performance (FPS). However, object detection is subject to delays associated with the sequential division of region proposal and classification. To address this issue, Redmon et al. [8] proposed an innovative solution (YOLO) that unifies region proposal and classification, resulting in a comparable performance at a significantly higher frame rate of 30 FPS. The authors of [9] further enhanced the YOLO algorithm, leading to the development of YOLOv2. Tenguria et al. [10] report that YOLOv2 is being implemented in robotic platforms. Two Raspberry Pi microprocessors were employed to test Neural Networks for integrated processing, but they yielded abysmal performance.

In a recent study, a depth-based application was developed as reported in [11]. Microsoft Kinect was used to obtain depth data about a pair of legs through the YOLOv2 algorithm for image development. The study demonstrated successful execution of YOLOv2 on NVIDIA Jetson TX2 with satisfactory detection efficiency, even under variable (low to moderate) traffic conditions. Another study [12] incorporated the development of a surrounding map and positions of previously trained objects for identification by the neural network, allowing the robot to follow a given path. The authors used the YOLO algorithm in combination with a 2D laser sensor, odometers, and an RGB-D camera to detect objects. Furthermore, a camera with a depth sensor and higher computational capacity than the Microsoft Kinect was developed. The authors of [13] employed the YOLO algorithm for object identification and tracking in the NAO humanoid robot, which significantly aided the robot in real-time object recognition and tracking, as indicated by the testing results. In a similar vein, [14] utilized YOLO and SLAM algorithms in a Robot Operating System (ROS) application to detect and classify household objects and furniture for localization and mapping purposes. However, these studies all employ GPUs or powerful embedded computers for training and detection, leading to large, heavy, and costly robots.

This paper describes the development of an autonomous Automated Guided Vehicle (AGV) robot constructed with ESP32-CAM that can adaptively track the speed and direction of a human's movements. Using Tensor-FlowLite ESP32 and TensorFlowLite ESP32, respectively, the robot employs a detection and tracking algorithm based on TensorFlow CoCo SSD to recognize humans and monitor their movements. The resulting autonomous AGV robot could be utilized in hospitals, where it could provide nurses and patients with assistance without physical contact.

2 Proposed Robot System

2.1 AGV Robot Architecture

As depicted in Fig. 1, the autonomous guided vehicle (AGV) is constructed using fundamental electrical components such as ESP32-CAM, DC combination gear reducer V1, DEBO MOTODRIVER2/L298N module, DFRobot DFR0379 20W DC-DC Buck Converter, HC-SR04 module, servo, single LED, and buzzer. The system has a compact design and can be installed on our AGV with ease.

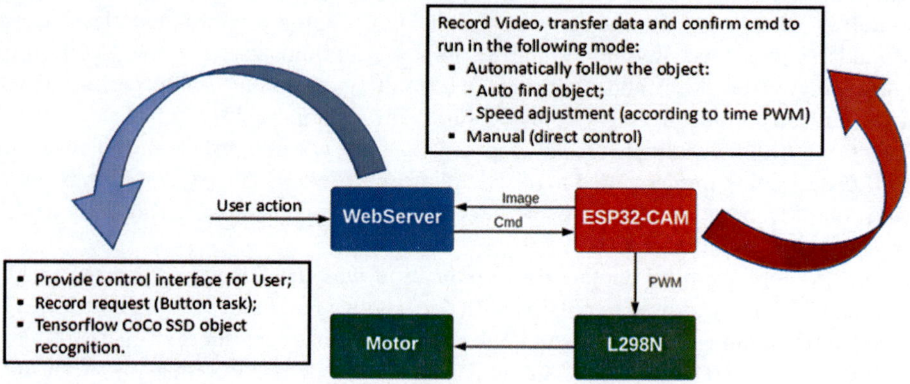

Fig. 1. Block diagram and main components of the designed system.

The robot's central processing unit is based on an OV2640 camera-equipped ESP32-CAM AI-Thinker. This integrated circuit features a primary processor that employs the ESP32 + Camera OV2640 module for image transmission and image processing via Wifi, Bluetooth, and IoT applications. The ESP32-CAM is a low-cost circuit, making it suitable for widespread use. In addition, it can serve as a web hosting server. When a local network website sends a request, the ESP32-CAM serves the request and sends the response back to the website. In this study, the ESP32-CAM is used as the web server, while any web browser or Android application can be used as the web client. When requesting access to a web file, a web client sends an HTTP GET request to the web server. When the server receives an HTTP GET request, it responds with the requested web page.

In this study, an ESP32-Cam was used to establish a STA mode Web server. The ESP32-Cam is connected to an existing Station (STA) WiFi network, which is generated by the wireless router. In STA mode, the ESP32-Cam obtains an IP address from the wireless router to which it is connected, allowing it to establish a Web server and serve Web pages to all devices connected to the existing WiFi network. In addition, the ESP32-Cam can serve as a central hub for one or more Access Points (APs), thereby establishing its own WiFi network. This mode of operation is referred to as a Soft Access Point (soft AP) because it does not interface with a wired network and the number of connected stations is limited. In contrast to AP mode, in which the ESP32-Cam generates a new WiFi network and assigns an SSID and IP address, STA mode was utilized for this study. HTML was used to program the Web Server interface buttons, allowing control of fundamental robot functions such as moving up, moving down, turning right, turning left, and stopping.

2.2 Human Detection and Tracking

The robotic system repeatedly captures images of its surroundings and uses tracking algorithms to locate and monitor a specified object. Based on the location of the object on the image plane, the system generates robot control signals. As shown in Fig. 2, the image plane is divided into five discrete regions, each of which corresponds to a specific command. Due to the varying perspective of the camera and the dynamic movements of the object, the object's position is constantly changing, necessitating the development of five distinct commands. These commands allow the robot to move forward, backward, to the left, to the right, or to stop. The accompanying illustration depicts the image plane and its corresponding segments, as well as an example tracked object. Regardless of the object's position on the image plane, the system algorithm guarantees effective object tracking.

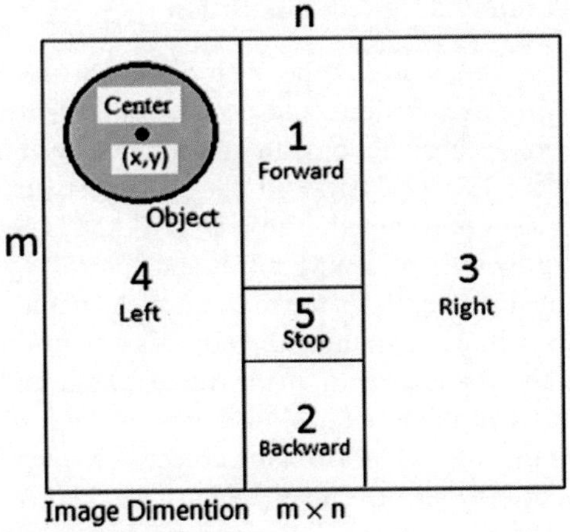

Fig. 2. Different segments of the image plane

During the initial phase of object tracking, an image is captured and the object region is identified. The subsequent step involves dividing the image into five segments based on its center, thereby facilitating the location of the target object and directing the robot accordingly. Following the calculation of the center coordinate, specific conditions are applied to determine the corresponding image segment. To elaborate, the robot is programmed to move forward if segment 1 detects the object's center. For segment 2, the robot moves in reverse, while for segments 3 and 4, it turns to the right and to the left, respectively. In segment 5, when the center is detected, the robot stops moving. Figure 3 depicts the overall process of controlling. The processing unit generates the control signals that direct the robot's four motors, which are responsible for locomotion. These signals are then transmitted to the data pins of the parallel port and fed to the motor driver circuits.

TensorFlow, an open-source machine learning library developed by Google, serves as the basis for the object detection solution in this study. The machine learning models supported by TensorFlow Lite for Microcontrollers are designed to run on hardware like microcontrollers and others with low memory requirements. The Arm Cortex M3's core runtime is only 16 KB in size, allowing it to run a variety of fundamental models without the need for an operating system, common C or C++ libraries, or dynamic memory allocation. In this study, machine learning models can be run locally on ESP32 devices by using the Tensor-FlowLite ESP32 library. This library enables the execution of TensorFlow machine learning models on microcontrollers, enabling the development of AI/ML applications utilizing deep learning and neural networks.

2.3 Software Algorithm and System Operation

The interface utilized in this study is the Web Server of the ESP32-CAM, which is responsible for receiving user actions and transmitting control commands (Cmd) to the ESP32-CAM based on those actions. In addition, the Web Server receives images (Images) captured by the ESP32-CAM and detects objects using a TensorFlow SSD-CoCo model that has been pre-trained. The ESP32-CAM is primarily accountable for controlling the actuator by sending PWM pulses and control signals to the L298N module. The L298N module then regulates the motor's speed based on the command output from the ESP32-CAM, while maintaining the object's size by comparing it to the object frame. In particular, if the object detector in the frame is large and the robot is close to the object, the program instructs the robot to slow down. In contrast, if the object detector is small in the frame and the robot is far from the object, the controller will increase motor speed to facilitate robot movement. This mechanism is known as Continuous Actions and is described in [15].

When pressing the 'Start Detect' button (Streaming Video & Run Model), if the robot detects that the object is a human, it will send data to the ESP32-CAM for processing. Similar to the Button Panel buttons include No Stop (do not stop), FrontLeft (forward tilt left), Front (forward), FrontRight (forward tilt right), Left (turn left), Stop (stop) Back), Right (turn right), LeftAfter (backward tilt left), Back (go backward) and RightAfter (go backward tilt right), the Web Server will send the data segment corresponding to the values. As for the command 'person', the system will check it, and if it is correct, it will transmit verbatim data about the vehicle's journey. The software interface is shown in Fig. 4.

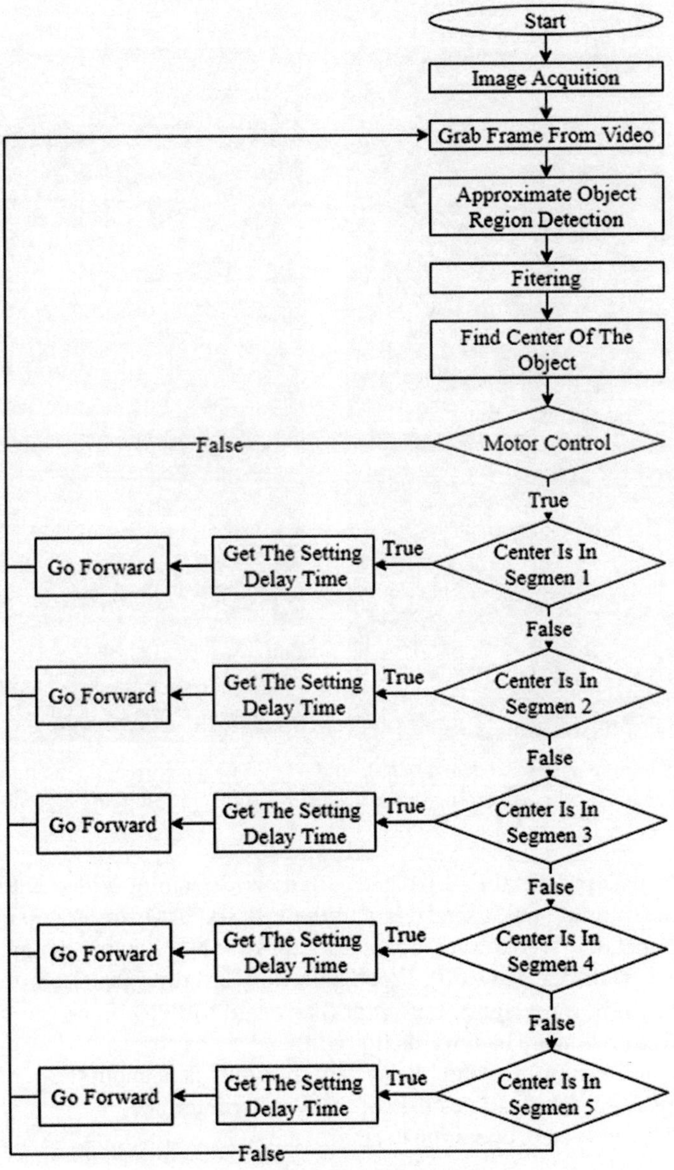

Fig. 3. Algorithm for object tracking and following by distance

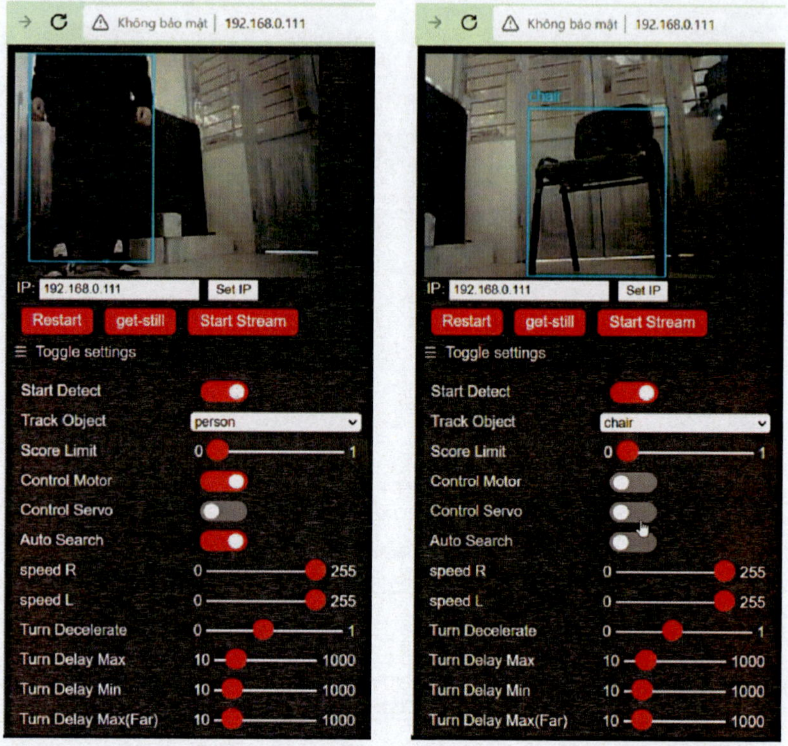

Fig. 4. Checking the system to detect people and other objects, correcting the impact parameters of the AGV robot's motors according to the object-tracking and following algorithm by distance.

When the user presses the 'Start Detect' button (Streaming Video & Run Model), if the robot detects a human, it sends data to the ESP32-CAM for processing. Similar to the Button Panel, the Web Server includes buttons for No Stop (do not stop), FrontLeft (forward tilt left), Front (forward), FrontRight (forward tilt right), Left (turn left), Stop (stop), Back, Right (turn right), LeftAfter (backward tilt left), Back (go backwards tilt right), and RightAfter (go backwards tilt right) and sends corresponding data segments. If the command 'person' is valid, the system transmits data about the vehicle's journey in its entirety. Figure 4 illustrates the software's user interface.

Algorithm and system operation to control the AGV on tracking and following object mode:

Input: image from camera.
Output: motors, led.

1) Powering the device will proceed to initialize the camera, configure the frame, configure controlled pins of the motors and led, connect to wifi and then broadcast wifi, initiate connections for the server and then turn off the led.
2) Connect browser (client) to esp32 (client and esp32 must be on the same wifi network).
3) When the connection is successful, the system will load the web page from the server (esp32 as the server). During the page load, the AI module will be configured and installed.
4) To start the control AGV, press the "Set IP" button for the client to connect to esp32.
5) If: The connection is successful.
6) Then: Display successful connection message.
7) If: changing flash indicator.
8) Then: Send flash value to esp32.
9) Esp32 receives the value and controls the led according to that value.
10) If: Change the parameters.
11) Then: Send new parameters to esp32. Esp32 separates data and updates parameters.
12) If "Get-sill" button is pressed.
13) Then: Send a photo request to esp32.
14) Esp32 takes pictures, encodes images.
15) Send the image to the client.
16) Client receives the image, decodes and displays it.
17) If: Click "Start Stream" button.
18) Then: Proceed to send a request to transmit video images to esp32.
19) Take pictures, encode, send to client.
20) Client receives the image, decodes and displays it.
21) If: Press the "Stop Stream" button.
22) Then: Stop image transmission.
23) If: Press "Start detect" button.
24) Then: Search for the specified object in the "Track Object" section.
25) Determine the center of the image.
26) Determine the center of bounding box.
27) Add Bounding box to the image.
28) If: "Control motor" button is high
29) Then: Compare the center of the bounding box with the center of the image along the x-axis.
30) Send motor control and running time signals to esp32.

31) Esp32 control motors.
32) If: "Control servo" button
33) Then: Compare the center of the bounding box with the center of the image along the Y axis.
34) Send control signal to esp32.
35) Esp32 outputs motor and servo motor control signals
36) If: The "Auto Search" button is high and the object is not found.
37) Then: Control AGV to rotate around (priority the last object appearing).
38) Send dynamic control and running time signals to esp32.
39) Esp32 control motors.
40) If: press the "reset" button.
41) Then: restart esp32.
42) If: Lost or disconnected.
43) Then: finish and wait for another client to connect.

3 Experiment and Result

The AGV robot model is trained using the Microsoft Common Objects in Context (MS CoCo) dataset, which supports a variety of computer vision tasks including object detection, segmentation, key-point detection, and captioning. The dataset consists of a vast collection of 328 thousand images, and the AGV robot model is constructed using the aforementioned hardware components and architecture. The algorithm of the Web Server is responsible for regulating the robot's response to the user's control commands. In order to function properly, the Web Server algorithm must initially load the Model CoCo SSD.

On the AGV robot, the Web Server and ESP32-CAM algorithms for direct control commands and object tracking mode have been successfully implemented. After loading the Model CoCo SSD, the AGV robot responds to Web server control inputs. Upon pressing the 'Start Detect' (Streaming Video & Run Model) button, the AGV robot scans and detects human objects using the servo motor. In auto mode, when the 'person' button is pressed to determine whether an object is a person, the system verifies its accuracy and initiates auto mode by transmitting frame data and the movement of the AGV robot. Figure 5 depicts how, upon detecting a human object, the robot sends frame and object data to the ESP32-CAM for processing and interactive control, using the Continuous Actions [15] mechanism based on Sequences of Images [16]. When the object is closer or further away, the robot responds with forward or reverse control, respectively. The robot also responds to control inputs by moving back and forth at variable speeds in 4.5 s, when the subject's movement speed varies rapidly from 50 cm to 200 cm away.

Figure 5 depicts an AGV robot following a human subject in automatic mode. The human object initially faces away from the robot, so the robot advances towards the object. As the human object rotates to the left, the robot detects that the object's center is in the left portion of the frame and rotates the subject accordingly to keep it within the frame. When the human object is in the center of the frame, the robot will stop. The design, production, and testing of modern AGV equipment is an urgent task in the present day. These AGVs are capable of performing both low- and high-risk tasks,

Fig. 5. Experiment with the robot's object tracking and following function

such as transporting goods in factories and working in hazardous environments or with infections. In the event of COVID-19-related lockdowns, the AGV robot developed by the research team can be deployed in hospitals, where its automatic object tracking capability can facilitate the delivery of medications and medical supplies to patients in isolated wards, thereby minimizing patient-to-staff contact.

4 Conclusion

This paper proposes a novel AGV robot control solution based on deep learning that is capable of object tracking and has been successfully implemented and tested on a small four-wheeled self-propelled robot car model. The proposed algorithm uses ESP32-CAM to identify objects on the robot's body, enabling autonomous self-propulsion to follow a human object. The robot's built-in distance tracking algorithm allows it to detect and match the speed of a moving human object. Such robots have the potential to be utilized in a variety of scenarios, such as the transportation of tools and objects and in environments where human intervention is hazardous. This technology's successful deployment represents a significant advancement in robotics and demonstrates the potential of deep learning-based algorithms in robotics applications.

Acknowledgment. We would like to thank the Ministry of Education and Training for sponsoring this research within the framework of the Ministry-level scientific research project, code B2023-GHA-02.

References

1. Nishimura, S., Itou, K., Kikuchi, T., Takemura, H., Mizoguchi, H.: A study of robotizing daily items for an autonomous carrying system-development of person following shopping

cart robot. In: 9th International Conference on Control, Automation, Robotics and Vision, ICARCV 2006, pp. 1–6. IEEE (2006)
2. Ma, F., et al.: Self-supervised Sparse-to-Dense: Self-supervised Depth Completion from LiDAR and Monocular Camera. Computer Vision and Pattern Recognition (cs.CV); Artificial Intelligence (cs.AI); Machine Learning (cs.LG); Robotics (cs.RO). arXiv:1807.00275 [cs.CV] (2018)
3. Feng, Z., et al.: Advancing self-supervised monocular depth learning with sparse LiDAR. In: CoRL - 2022 Conference on Robot Learning, Dec 14–18, 2022 – Auckland, NZ. arXiv:2109.09628v4 [cs.CV], 29 November 2021
4. Hu, H.-N., et al.: Joint Monocular 3D Vehicle Detection and Tracking. Computer Vision and Pattern Recognition (2019). arXiv:1811.10742v3 [cs.CV], 12 September 2019
5. Lei, T., et al.: Virtual-to-real deep reinforcement learning: continuous control of mobile robots for mapless navigation. intelligent robots and systems (IROS). In: 2017 IEEE/RSJ International (2017). https://doi.org/10.1109/IROS.2017.8202134
6. Smprobotics - Autonomous Mobile Robot and Unmanned Ground Vehicles. https://smprobotics.com/productsautonomousugv/
7. Ren, S., He, K., Girshick, R., Sun, J.: Faster R-CNN: towards real-time object detection with region proposal networks. Adv. Neural. Inf. Process. Syst. **28**, 91–99 (2015)
8. Redmon, J., Divvala, S., Girshick, R., Farhadi, A.: You only look once: unified, real-time object detection. In: Proceedings of the IEEE Conference on Computer Vision and Pattern Recognition, Las Vegas, NV, USA, pp. 779–788 (2016)
9. Redmon, J., Farhadi, A.: YOLOv3: an incremental improvement. CoRR, vol. abs/1804.0.2767, pp. 1–6 (2018)
10. Tenguria, R., Parkhedkar, S., Modak, N., Madan, Ṙ., Tondwalkar, A.: Design framework for general purpose object recognition on a robotic platform. In: 2017 International Conference on Communication and Signal (2017)
11. Lucian, A., Sandu, A., Orghidan, R., Moldovan, D.: Human leg detection from depth sensing. In: 2018 IEEE International Conference on Automation, Quality and Testing, Robotics (AQTR), Cluj-Napoca, Romania, pp. 1–5 (2018)
12. Bersan, D., Martins, R., Campos, M., Nascimento, E.R.: Semantic map augmentation for robot navigation: a learning approach based on visual and depth data. In: 2018 Latin American Robotic Symposium, 2018 Brazilian Symposium on Robotics (SBR) and 2018 Workshop on Robotics in Education (WRE), Pessoa, Brazil, pp. 45–50 (2018)
13. Zhao, X., Jia, H., Ni, Y.: A novel three-dimensional object detection with the modified You only look once method. Int. J. Adv. Rob. Syst. **15**(2), 1–13 (2018)
14. Maolanon, P., Sukvichai, K., Chayopitak, N., Takahashi, A.: Indoor room identify and mapping with virtual based SLAM using furnitures and household objects relationship based on CNNs. In: 2019 10th Int. Conf. of Information and Communication Technology for Embedded Systems (IC-ICTES), Bangkok, Thailand, pp. 1–6 (2019)
15. Duo, N., et al.: A deep reinforcement learning based Mapless navigation algorithm using continuous actions. In: 2019 International Conference on Robots & Intelligent System (ICRIS). https://doi.org/10.1109/ICRIS.2019.00025 (2019)
16. Magán E., et al.: Driver drowsiness detection by applying deep learning techniques to sequences of images. Appl. Sci. **12**, 1145 (2022)

Predict Risk Assessment in Supply Chain Networks with Machine Learning

Thuy Nguyen Thi Thu[1](✉), Thi-Lich Nghiem[1], and Dung Nguyen Duy Chi[2]

[1] ThuongMai University, Hanoi, Vietnam
{ingthuynguyenthithu,lichnt72}@tmu.edu.vn
[2] Ulsan National Institute of Science and Technology (UNIST), Ulsan, Korea

Abstract. Supply chain gradually becomes a core factor to operate and develop for businesses. Using machine learning, especially with neural networks, to assess the risk in supply chain network has been attracted many research and become potential approaches. Via machine learning particular to Bayesian neural network, risk evaluation in supply chain network can be performed effectively to support supply chain partners to assess, identify, monitor, and mitigate risks. In detail, by using reliability theory, supply chain network's risk is divided in alternative scales (from Very high risk to Very low risk). The Bayesian neural network allows to treat the weights and outputs as the variables in order to find their marginal distributions that best fit the data. By taking the advantage of Bayesian neural network in deep learning, the experiment in this paper shows a very high accuracy rate in supply chain risk prediction. This implicates the performance of using machine learning in supporting of managerial decision making in selecting suppliers.

Keywords: Supply chain networks · Risks · Machine learning · Bayesian neural network

1 Introduction

A supply chain is a system that includes not only businesses, suppliers, product distributors, but also logistics systems, retail systems and its customers. During the operation of the chain, distributors are required to increase the quality of products and services, so that distributors act as key players with the privilege of mastering the actual flow, and information flows in the supply chain. Supply chain gradually becomes a core factor to operate and develop for businesses. Supply chain is also understood as a chain of products and services that are closely linked together [4, 6]. Directly or indirectly, the "links" of the supply chain participate in the production and distribution system of products and services and are affected by many factors. In the big data era, data is considered a key factor for organizations and businesses. Capturing the data, using data to predict latent information is extremely important. Specially, through extracted information, early detection of various supply chain is essential for giving well-timed countermeasures to prevent supply chain disruptions. Machine learning (ML) can aid in the early detection of risks [13].

T. D. L. Nguyen et al. (Eds.): ICISN 2023, LNNS 752, pp. 215–223, 2023.
https://doi.org/10.1007/978-981-99-4725-6_27

The most advantage of machine learning is dealing with big data, especially to supply chain risk management area [2, 14]. Machine learning can identify and extract automatically the patterns inside data from multi-dimensional resource. The disadvantage of traditional methods such as linear regression or statistics is hardly to deal and analyze big data. This let scholar to use machine learning techniques which can handle non-linear problems or unstructured data. Moreover, machine learning techniques also are stronger than statistical ones in recognizing and predicting risk identification, assessment, etc. [7]. The availability of numerous supply chain dataset as well as computing power can be performed by using alternative machine learning techniques. Many publications in supply chain risk management (SCRM) shown the popularity of machine learning techniques in the field [2]. In supply chain risk management (SCRM), it includes many strategies to identify, assess, mitigate and monitor some unexpected issues, which usually having disadvantages, happening during supply chain. The decisions in SCRM are based on numerous data from variety data sources. Therefore, the use of machine learning techniques can be suitable for SCRM [2].

The paper has been organized as follows: general introduction about risk assessment in supply chain as well as machine learning in supply chain area is shown in Sect. 1. Section 2 shows the related works. Framework of using machine learning in supply chain management is shown in Sect. 3. In here, the definition of how to calculate the risk indexes for supplier, distributor, retailer, etc. also is introduced. An experimental case study with data is shown in Sect. 4. Implications and conclusion of using machine learning, in particular Bayesian network is introduced in the last section.

2 Related Work

As the significance of data in supply chain has grown, supply chain management researchers and practitioners have explored every avenue for improving data management throughout supply chain in order to produce better decisions. Risk evaluation in supply chain network is considered as one of vital tasks mentioning to the coordinated and collaborative efforts of all supply chain stakeholders to control risks in order to increase robustness and resilience, reduce supply chain vulnerabilities. Applying neural networks in supply chain risk management tend to increase dramatically via the many publications in the supply chain management area [13].

Banerjee et al. [1] proposed Pareto Optimization based on a genetic approach to estimate uncertainty of the model and compare the results with several state-of-art machine learning techniques such as LSTM, CNN, RNN. They handle the uncertainty of supplier workload distribution based on demand. From alternative demands, number of supplier for supply chain network will be produced.

Schroeder & Lodemann [13] proposed a framework to identify the risks in supply chain management practical use cases based on machine learning, including procurement, production, transportation and sales. His results showed that applicable examples are mainly related to early risks' detection in order to quickly resolve potential supply chain issues. Besides, integrating machine learning (such as integrating new data sources) gives to supply chain risk management identifying the added value based on his analyzing [13]. According to [12], machine learning makes more practical, more objectivity, more robustness research in supply chain management. In addition, applying Bayesian theorem in neural network to estimate the risk of supply chain and resilience problems is suitable approach for estimating supply chain. Based this approach, he can capture the uncertainty of the model [7]. Several semi-supervised machine learning techniques such as random forest, SVM, C5 have been used in order to evaluate and detect the fraud in smart supply chains [5]. Although many studies have shown some limitation of these approaches such as not offering any real-world examples, machine learning has positive effect on supply chain risk management. So, applying machine learning can become a bright future in supply chain management [12, 15].

3 Risk Assessment in Supply Chain Network Using Machine Learning

3.1 Risk Assessment Framework

Supply chain network risk management is a long-term strategy, to develop a complete risk management plan, businesses often rely on 4 basic steps [8] as Identify, Evaluate and analysis, Handle and Monitor risks.

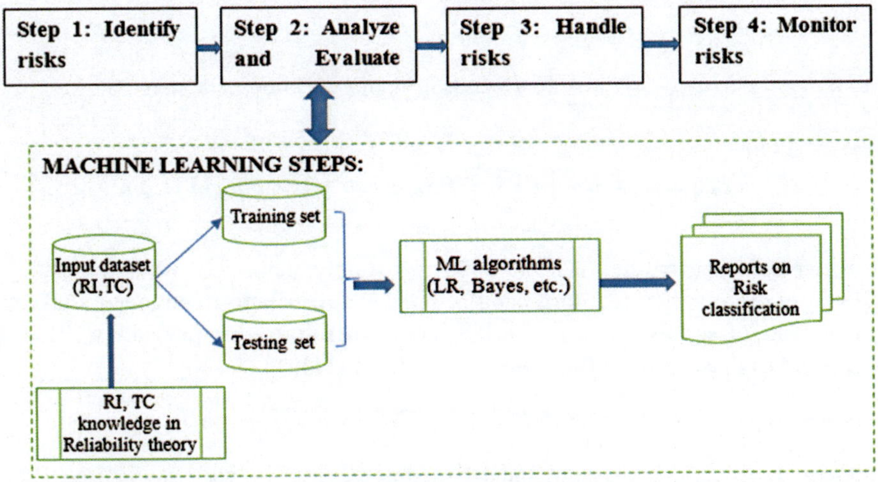

Fig. 1. Framework of using machine learning in Risk Assessment in supply chain network.

In Fig. 1, the sub-steps of machine learning are used to predict supply chain risks. Theses sub-steps will be used in step 2 to support users in analyzing and evaluating supply chain network risks. Detail of sub-steps can be seen as:

Sub-step 2.1. Inputting RI and TC: These variables are calculated basing on reliability theory. Also in this step, alternative risks are identified.
Sub-step 2.2. Data is divided into two sets of training and testing. The rate is 80:20 for training and testing sets.

Sub-step 2.3. Alternative machine learning algorithms can be used here to predict or classify the risk. In this paper, Bayesian neural network with deep learning is used to predict supply chain risks which are divided on 5 risk classes such as "Very High"; "High"; "Medium"; "Low" and "Very Low".
Sub-step 2.4. Reporting of predicting model can be created in order to send to users.

3.2 Reliability Theory and Risk Assessment in Supply Chain Network

If the supply chain can be seen as a system, there, the supplier, retailor, manufacturer, and distributor act as the components of that system. By using reliability analysis methodologies, the supply chain's performance can be evaluated via building alternative probability functions to assess the success of system. The success of supply chain system means as the ability of delivering product/service to the right customers. Therefore, the reliability assessment should be performed, which can measure the uncertainty of supply chain components. In general, a supply chain system model based on reliability can be seen as a summary of probability of all components in the system [8] (see Eq. 1).

$$Risk_{system} = P(C_1 \cup C_2 \cup \cdots \cup C_n) \tag{1}$$

where, C_i is the ith component in the system and it can be seen as success or failure one. The C_i is success or failure depending on probability of its sub-components success or failure.

$$P(C_{isuccess}) = \prod_{j=1}^{m} P\left(S_{ij}\right) \; if \; S_{ij} \; is \; sub - component \; success \tag{2}$$

$$P(C_{ifailure}) = \prod_{j=1}^{m} P\left(\tilde{S}_{ij}\right) \; if \; S_{ij} \; is \; sub - component \; failure \tag{3}$$

If we consider that the risk index (RI) for supply chain network includes the total risks from the supplier, retailor, manufacturer, and distributor mentioned above as the components of supply chain network system. Therefore, this risk index (RI) can be calculated as (based on the formulas derived from [11]:

$$RI_{system} = w_1 RI_S + w_2 RI_D + w_3 RI_M + w_4 RI_R \tag{4}$$

where w_i (i = 1,2,3,4) is weight and satisfied as $w_1 + w_2 + w_3 + w_4 = 1$;

And RI_S is Risk Index of Supplier; RI_D is Risk Index of Distributor; RI_M is Risk Index of Manufacturer; and RI_R is Risk Index of Retailer.

In detail, the Risk Index of Supplier, Distributor, Manufacturer, and Retailer formulas are shown as mathematical forms which are based from Neureuther & Kenyon [11] as follows:

$$RI_S = \sum_{i=1}^{n} \alpha S_{ij} \beta S_{ij} (1 - (1 - \prod_{j=1}^{m} P(\tilde{S}_{ij})) \tag{5}$$

$$RI_D = \sum_{i=1}^{n} \alpha D_{ij} \beta D_{ij} (1 - (1 - \prod_{j=1}^{m} P(\tilde{D}_{ij})) \tag{6}$$

$$RI_M = \sum_{i=1}^{n} \alpha M_{ij} \beta M_{ij}(1 - (1 - \prod_{j=1}^{m} P(\tilde{M}_{ij})) \tag{7}$$

$$RI_R = \sum_{i=1}^{n} \alpha R_{ij} \beta R_{ij}(1 - (1 - \prod_{j=1}^{m} P(\tilde{R}_{ij})) \tag{8}$$

where αS_{ij} is the consequence to the supply chain if the ith supplier fails, βS_{ij} is the percentage of value added to the product by the ith supplier, and $P(\tilde{S}_{ij})$ denotes the marginal probability that the ith supplier fails for jth demand.

Similarity for the rest denotes of αD_{ij}, αM_{ij}, αR_{ij}, βD_{ij}, βM_{ij}, βR_{ij}, $P(\tilde{D}_{ij})$, $P(\tilde{M}_{ij})$, and $P(\tilde{R}_{ij})$.

The cost of resource across a supply chain network as well as the lead-time should be concerned with an optimal choice in supply chain system. As mentioned above, supply chain is composed of n components (C_i) where each component contains many number of resource options (sub-components). The sub-components here contain their own cost and processing lead-time [9]. Therefore, total cost can be calculated as:

$$TC = \theta \sum_{i=1}^{n} (\mu_i \sum_{j=1}^{m} Cost_{ij} y_{ij}) \tag{9}$$

where $Cost_{ij}$ is the cost of the jth resource for the component ith, μ_i is the average demand per unit time at the component ith, θ is the period of interest, which is depended on the unit time, and y_{ij} is a binary variable denoting whether the ith component is a participant for the jth resource option.

The performance of supply chain system is categorized into alternative classes. The states used in the analysis are divided more detail as "Very high risk", "High risk", "Medium risk", "Low risk", and "Very low risk". Each of these states are defined by the consequence (α) of a failure, the percentage of value added to the final product (β) by the sub-item, by the manufacturer or percentage purchased by the customer (retailer), and by the probability of failure $P(\tilde{S}_{ij})$ of the supplier, distributor, manufacturer, or retailer within the structure.

In general, the success probability of each supplier, distributor, manufacturer, or retailer within the supply chain network system will range from 0.5 to 1.0 [11]. Probabilities of less than 0.5 show that it is more likely to fail than to succeed. In that case, the supply chain system should replace to others of supplier, distributor, manufacturer, or retailer.

4 Experimental Case Study

4.1 Data

The dataset is derived from [1]. The first version was created in March 2019 whereas the second one was created in April 2019 for academic purpose to test alternative machine learning techniques. This dataset is generated to evaluate Risk Index and Total supply chain cost by using MATLAB Simulink with probabilistic risk assessment model and a Linear Programming models. The generated dataset is used to deal with classification problems related to Supply Chain Risk Management (SCRM) such as Non-linear Autoregressive and Deep Neural Networks. In this paper, data is used with Bayesian Network which describes the probability of Risk Index and Total Cost.

The dataset includes two files of time series data. The training set has 649999 samples whereas the test set has about 150000 ones. The training data is generated in different time from 2016 to August 2018 whereas the test data is from 2016 to 2017. The inputs including RIs in the formula (5), (6), (7), (8), and TC in formula (9) are already simulated via MATLAB Simulink. The outputs for data are classified into 5 categories according to the objective function of Z as: $Z = w_1\ TC_n + w_2\ TRI_n$, where TC_n is normalized total cost and TRI_n is normalized total Risk Index, and w_1, w_2 are the weights and $w_1 + w_2 = 1$. The classification is ordered like likert scale of as "Very high risk", "High risk", "Medium risk", "Low risk", and "Very low risk"according to the labels of 0, 1, 2, 3, 4 respectively.

By analyzing dataset, it founds that the missing data for column RI Distributor1 and Total Cost are quite high. Therefore, preprocessing data is performed by filling missing values with appropriate columns' average values.

4.2 Algorithm and Evaluation Metric

The detailed machine learning using Bayesian neural network can be described as follows:

Step 1: Build the hyper-parameters and architecture of the model
Step 2: Apply LSTM model
Step 3: Use the Bayesian optimization:
In this step, in order to enhance the performance and capture the confidence and uncertainty of the model, a probability distribution to sample the weights and the biases instead of deterministic weights are performed. These weights can be computed based on the input feature standard deviation and the input feature mean at the special time on the position of the layer in the LMST model.
Then, the distribution parameters are optimized by finding the objective function and the hyper-parameters. If the result is good, next to step 4 Else back to the Step 1
Step 4: Predict the values Step 5: Evaluate the model

The predictive performance (sharpness in statistics) can be assessed by treating the estimator $\hat{y}$ as the prediction. To evaluate the performance of the model's prediction, we use MSE to measure the mean square of errors in the prediction and is calculated as average squared of the difference between predicted value and actual value, called prediction error. MSE is a risk function, corresponding to the expected value of the squared error loss and is defined as follows:

$$MSE = \frac{1}{n}\sum_{i=1}^{n}(y_i - \hat{y}_i)^2 \tag{10}$$

where $\hat{y}_i$ and y_i are the observed and predicted values at time step i, n is the length of the sample data.

4.3 Result

The model was trained for 10 epochs to minimize the mean absolute error loss function with a batch size of 128. We used the Adam optimization algorithm, with $\alpha = 1 \times 10^{-4}$ for epochs 1 to 10. L_2 regularization with $\lambda = 1 \times 10^{-5}$ was used in all convolutional layers to decrease overfitting. These hyper parameters were determined heuristically iteratively. Only one hyper parameter was changed at a time, while the others remained constant. Because the training, validation, and test sets were recorded during separate sessions, they were not drawn at random from the entire dataset. Figure 2 depicts the minimization of the loss function during the training process.

When applying Bayesian in deep learning model, we obtained the results on MSE value, 0.0025. This shows the accuracy for classification is nearly 100% (see Fig. 2).

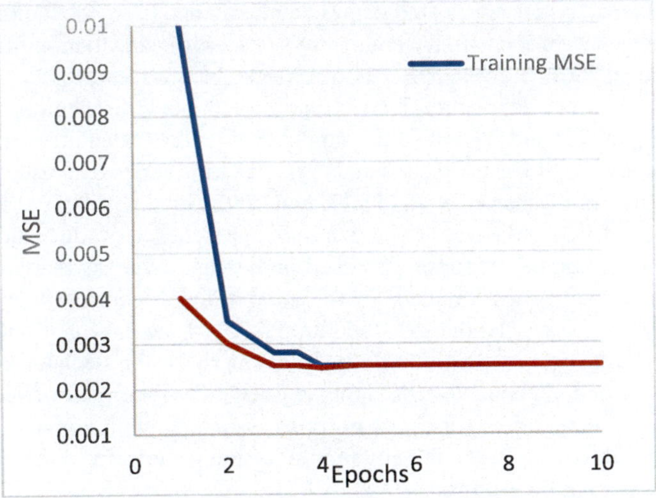

Fig. 2. MSE results for training and validation dataset

5 Implications and Conclusions

In supply chain network, according to [3], after creating a supply base from potential suppliers, a collection of supplier chain network's information is performed. From this step, a process of evaluating supply chain network as well as the evaluation of supply chain risks should be performed [11].

Therefore, if we can evaluate the alternative supply chain risks before choosing suppliers or distributors, it will help companies control the supply process in order to save managerial cost. This contributes to infer a reduction of products from saving managerial cost.

The process of risk evaluation is build in order to create the supporting reports for supply chain selection decision. In this period, the enterprise should conduct the interviews to assess the potential vulnerability to them. According to [11], the interviews can be divided into three categories as financial, technological, and operational. These categories can be seen as the sub-components in the supply chain network risk evaluations using in above equations. The goal of interviews is to evaluate the probability of success/fail for a supply chain network. Therefore, the interviews should be used for rendering of hidden failure probabilities. Depending on the failure probabilities of high or low scale, the enterprise can have right action such as having contingency plan, changing other suppliers, or splitting the demand to among suppliers, etc. instead of attempting to lower the failure probability. Therefore, in the supply chain risk management process, the defining risks are very important. The more well-defined risks, the more effective the following steps will be. Based on reliability theory, supply chain risk can then be defined as the product of the consequences of a specific incident and the probability of its occurrence. Mathematically, supply chain partners such as suppliers, distributors, manufacturers, and retailers can be seen as core components in supply chain networks. All supply chain activities are related to these components. From the internal or external risk activities such as problem in the enterprise's process or cancellation from customers, the mathematical formula for supply chain risks can be produced.

In this paper, we re-defined alternative risks of suppliers, distributors, etc. and Total Cost according to [11] and [9], where the supply chain risk is calculated with the weights of Risk Index and Total Cost (Eqs. (4) and (9)). The time series data used in the experiment is derived from Banerjee et al. [1]. They used the data with Mixed Integer Linear Programming (MILP) and Pareto Optimization approaches to produce alternative solutions of choosing number of suppliers based on demand whereas this paper used data with Bayesian neural network in deep learning to predict categorical supply chain risks such as "Very high risk", "High risk", "Medium risk", "Low risk", and "Very low risk". Through the utilization of predicting supply chain risks, the decision of managerial making is supported via choosing the right suppliers, distributors or retailers. The good understanding of categorical supply chain risks associated with a given suppliers, distributors or retailers can determine supplier selection criteria for companies to match their strategic plan in long-term.

To sum up, the enterprises can remain profitable and improve the assets if they have a good decision in choosing the right supply chain networks. The mathematical formulas for risk index calculation presented that supply chain architects should balance between coordination costs, the efficiency and risk of failure. This ensures for the supply chain can remain competitive and successful. The alternative levels of balance depend on enterprise's objectives, the market environment, etc.

Acknowledgements. This research is funded by Thuongmai University, Hanoi, Vietnam.

References

1. Banerjee, H., Ganapathy, V., Shenbagaraman, V.M.: Time Series Dataset for Risk Assessment in Supply Chain Networks. Mendeley Data, V2 (2020). https://doi.org/10.17632/gystn6d3r4.2
2. Baryannis, G., Dani, S., Antoniou, G.: Predicting supply chain risks using machine learning: the trade-off between performance and interpretability. Futur. Gener. Comput. Syst. **101**, 993–1004 (2019)

3. Beil, D.R.: Supplier selection. In: Cochran, J.J., Cox, L.A., Keskinocak, P., Kharoufeh, J.P., Smith, J.C. (eds.) Wiley Encyclopedia of Operations Research and Management Science. https://doi.org/10.1002/9780470400531.eorms0852. Accessed 27 Jan 2023
4. Molamohamadi, Z., BabaeeTirkolaee, E., Mirzazadeh, A., Weber, G.-W. (eds.): Logistics and Supply Chain Management. CCIS, vol. 1458. Springer, Cham (2021). https://doi.org/10.1007/978-3-030-89743-7
5. ConstanteNicolalde, F.-V., GuerraTerán, P., PérezMedina, J.-L.: Fraud prediction in smart supply chains using machine learning techniques. In: Botto-Tobar, M., ZambranoVizuete, M., TorresCarrión, P., MontesLeón, S., PizarroVásquez, G., Durakovic, B. (eds.) Applied Technologies. CCIS, vol. 1194, pp. 145–159. Springer, Cham (2020). https://doi.org/10.1007/978-3-030-42520-3_12
6. Harland, C.M., Knight, L.A.: Supply network strategy: role and competence requirements. Int. J. Oper. Prod. Manag. **21**(4), 476–489 (2001)
7. Hosseini, S., Ivanov, D.: Bayesian networks for supply chain risk, resilience and ripple effect analysis: a literature review. Expert Syst. Appl. **161**, 113649 (2020)
8. Kolarik, W.J.: Creating Quality Concepts, Systems, Strategies, and Tools. Mc Graw-Hill, New York (1995)
9. Mastrocinque, E., Yuce, B., Lambiase, A., Packianather, Michael S.: A multi-objective optimization for supply chain network using the bees algorithm. Int. J. Eng. Bus. Manage. **5**, 38 (2013). https://doi.org/10.5772/56754
10. Musa, S.N.: Supply Chain Risk Management: Identification, Evaluation and Mitigation Techniques (2012). http://www.diva-portal.org/smash/get/diva2:535627/fulltext01, last assessed 27/01/2023
11. Neureuther, B.D., Kenyon, G.: Mitigating supply chain vulnerability. J. Mark. Channels **16**(3), 245–263 (2009)
12. Ni, D., Xiao, Z., Lim, M.K.: A systematic review of the research trends of machine learning in supply chain management. Int. J. Mach. Learn. Cybern. **11**(7), 1463–1482 (2019). https://doi.org/10.1007/s13042-019-01050-0
13. Schroeder, M., Lodemann, S.: A systematic investigation of the integration of machine learning into supply chain risk management. Logistics **5**(3), 62 (2021)
14. Tirkolaee, E.B., Sadeghi, S., Mooseloo, F.M., Vandchali, H.R., Aeini, S.: Application of machine learning in supply chain management: a comprehensive overview of the main areas. Math. Probl. Eng. **2021**, 1–14 (2021)
15. Wang, D., Zhang, Y.: Implications for sustainability in supply chain management and the circular economy using machine learning model. Inf. Syst. E-Bus. Manage. 1–13 (2020). https://doi.org/10.1007/s10257-020-00477-1

Odoo: A Highly Customizable ERP Solution for Vietnamese Businesses

Cong Doan Truong(✉), Thao Vi Nguyen, and Anh Binh Le

International School, Vietnam National University, Hanoi, Vietnam
{tcdoan,20070011,19071012}@vnu.edu.vn

Abstract. In an era when digitization is on the rise, Vietnamese businesses are rapidly adopting technology to centralize business processes, support in system optimization, lessen time to solve problems, and scale up operations. The referred technology, or a software solution to be more specific, is called ERP (Enterprise resource planning). Among the top international ERP solution providers, Odoo has consistently become the standard software suite for many companies in Vietnam owing to its great adaptability, optimal interface, and inexpensive implementation and recurring costs. ERPViet is a prestigious Odoo solution supplier in Vietnam, not only do they provide implementation support, but they also develop several bespoke features like Overtime Management to meet the special requirements of Vietnamese firms. The goal of this paper is to decode how a new app such as Overtime Management is created by looking at the architecture, module structure, and the method to construct and install a new customized module on Odoo. The detailed procedure can be found in our Github repository.

Keywords: ERP · Odoo · Vietnamese Businesses · HRM · Overtime Module

1 Introduction

ERP is the acronym for Enterprise Resource Planning. Although ERP originated in manufacturing, it has rapidly evolved to cover a wide range of other functions and sectors. The ERP system can simply be described as an integrated information system servicing all aspects of the business. It handles transactions, maintains records, provides real-time information, and facilitates planning and control [1].

According to a 2021 report from the Ministry of Planning and Investment, 97% of Vietnamese businesses are small and medium-sized enterprises (SMEs) [2]. These SMEs are versatile, allowing them to swiftly establish new specialized procedures in response to changing circumstances. According to the Kearney Global Service Location Index (GSLI) 2021 study, an assessment of the potential and appropriateness of a country to supply business services, Vietnam placed sixth in the world in software outsourcing sites [3]. This emphasizes the increasing need for outsourcing to support business process management in Vietnam. Furthermore, the diversity of regional cultures complicates company culture, which has a direct impact on business operations, particularly human resource management. All of the aforementioned factors necessitate

T. D. L. Nguyen et al. (Eds.): ICISN 2023, LNNS 752, pp. 224–232, 2023.
https://doi.org/10.1007/978-981-99-4725-6_28

the use of a customized ERP solution to ensure seamless operation and administration for Vietnamese companies.

To meet this demand, various ERP products from Vietnam have emerged, such as ERPViet, WeUp ERP, Atoom ERP, and others, all primarily based on the Odoo platform. Odoo, which was first released in 2005, has steadily risen to become one of the top ERP softwares widely used across the globe [4]. Due to its fully integrated, configurable, and open-source properties, more and more businesses are adopting Odoo, and most ERP solution providers in Vietnam have selected Odoo as the foundation of their service. Above all, customizability has enabled many Odoo users to tailor-make new features to serve specific needs, contributing to the variety of the Odoo appstore, which in turn helps to expand the platform even further.

This paper will highlight the most notable feature of Odoo: customizability, by examining the process of establishing a new Odoo module. To provide a thorough understanding of the concept, we will firstly go through the architecture of the platform and, more specifically, the structure of a module. A summary of the important steps to create and install a new app on Odoo will be specified, and to exemplify, we will introduce a basic module developed by ERPViet that addresses the issue of managing overtime in Vietnam.

2 Methodology

To efficiently employ all accessible modules from Odoo for the desired usage or construct a new one, it is essential to dig into the system, from backend to frontend elements. This part will concentrate on breaking down and decoding factors such as architecture, structure, and the requirements for developing a module with Odoo.

2.1 Odoo Architecture

Odoo system design focuses on the work of enabling application development by splitting the labor into multiple levels of components and functionalities [5], as illustrated in Fig. 1.

The Data tier is the lowest-level layer, responsible for data storage and persistence. For which, Odoo relies on a PostgreSQL server. While not as well-known as Microsoft SQL Server or MySQL, PostgreSQL is an enterprise-class database server with numerous sophisticated capabilities. In reality, PostgreSQL compares favorably to significantly more costly databases like Oracle Database [6].

The Logic tier, handled by the Odoo server, in charge of all interactions with the data layer. As a general rule, the low-level database should only be accessed by this layer as the only way to ensure security access control and data consistency. At the core of the Odoo server lies the Object-Relational Mapping (ORM) engine for this interface. The ORM provides the application programming interface (API) used by the addon modules to interact with the data.

The Presentation tier handles the task of presenting data and interacting with the user. The client interacts with the ORM API by using ORM API methods via remote procedure calls (RPCs) to read, write, verify, or execute any other activity. These are sent to the Odoo server for processing, and then the results are sent back to the client for further handling.

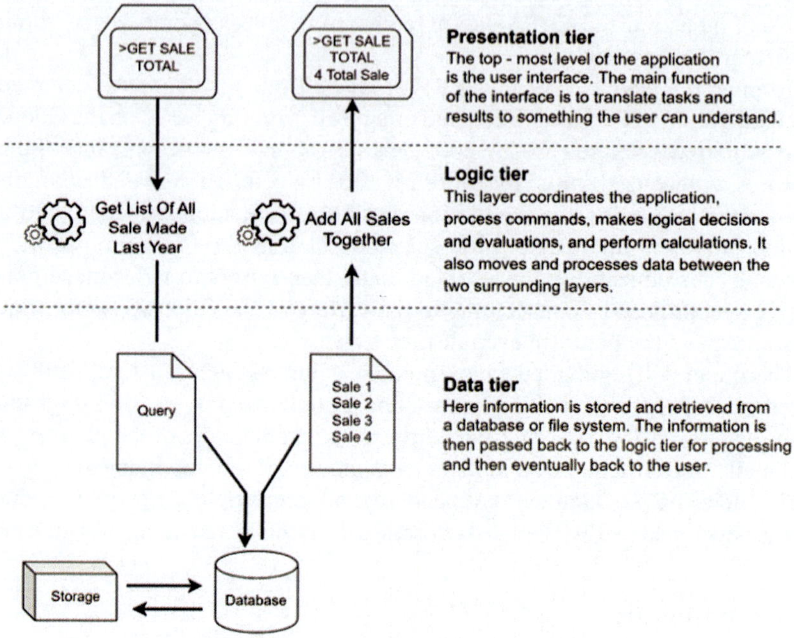

Fig. 1. An overview of the three-layer architecture

2.2 Odoo Structure

Odoo is built on top of Model-View-Controller (MVC) architecture, in which database tables are created with PostgreSQL, general codings are done with Python and views are created through XML. HTML templates are also used to build its view on the website, as depicted in Fig. 2.

Just as in Fig. 3, each individual folder storing the Odoo module usually contains a number of files such as business objects, object views, data files, web controllers, though they are not mandatory [7]. Some modules may just add data like configuring employee's data for a specific department, whereas others may only add business objects. Business objects (for example, an invoice) are defined as Python classes, and are automatically persisted by Odoo based on their settings. The ORM layer will automatically map the fields declared in these classes to the corresponding database columns [8]. Object views dictate how records should be shown to end users. Because they are described in XML, they may be modified independently of the models that they represent [9]. The data files, on the other hand, declare the model data in the form of XML or CSV files. They support the purpose of generating views or reports, configuration data (modules parametrization, security rules), demonstration data, and so on [10]. Static web data consists of all images, CSS or JavaScript files used by the web interface, and on top of that, the controllers will handle requests from web browsers.

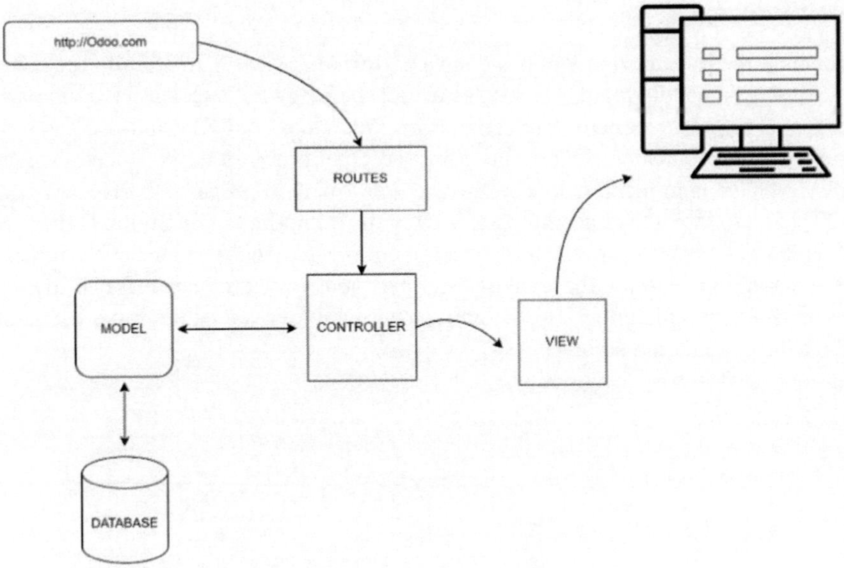

Fig. 2. A basic MVC architecture

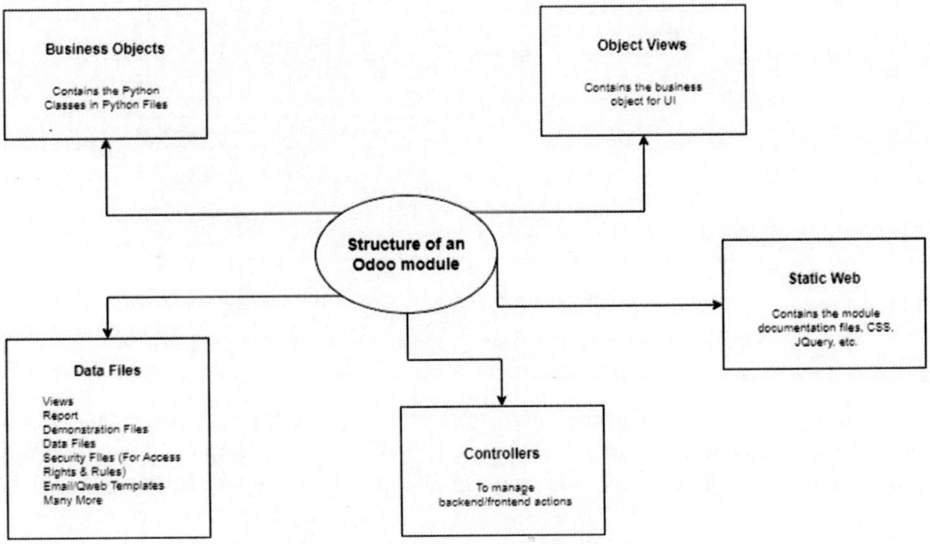

Fig. 3. The structure of an Odoo module

3 Results

To showcase the relevance of Odoo, we have performed research to identify the competitive advantages that distinguish Odoo as the most befitting ERP solution for Vietnamese businesses. The most recent research comparing Odoo to other ERP solutions has yielded favorable results for Odoo. Odoo outperformed Openbravo in most aspects like functionality, service, and technical, according to a assessment prepared by Gómez-Llanez in 2020 [11]. On top of that, according to a 2021 study conducted by Simon Huber, Odoo matches the precise criteria of Microsoft Dynamics NAV and provides even more [12]. Some outstanding features that make Odoo pertinent in Vietnam are flexibility, easy-to-use platform, simple integration, highly customizable, reasonable price, Vietnamese language supported, and so on (Fig. 4).

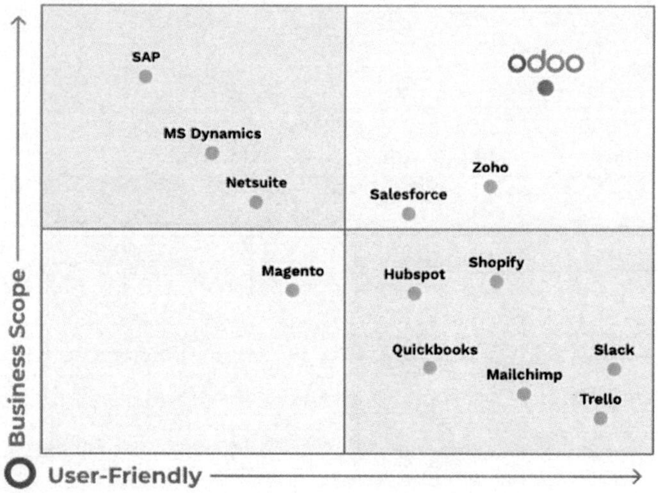

Fig. 4. Market position statistics created by the Odoo company

Among those, we'd like to emphasize the end-to-end user customization capabilities. With the provided packages, any firms may easily tailor an ERP application to their own needs on the Odoo platform by following a few simple steps that will be discussed below.

3.1 Odoo Customizability

Odoo, as previously stated, is a three-tiered multitenant architecture. HTML5, JavaScript, and CSS make up the presentation layer. The logic tier is written entirely in Python, whereas the data tier only supports PostgreSQL as an RDBMS. Odoo development can be done at either of these levels, depending on the scope. It is important to first prepare the adequate versions of Python and PostgreSQL in order to create and operate the application.

Aside from the elements mentioned in part 2.2, a folder containing components of a module must include two important files which are usually named __init__.py and __manifest__.py. All of the Python files that will be utilized must be imported into the __init.py file, while __manifest__.py acts as a directory defining the module name, version, the author name, description, company, category, etc.

Detailed guidance on how to create each component belonging to the new module can be found in our Github repository. After successfully design and specify all entailed files, following six steps in Fig. 5 would lead to the new feature being installed.

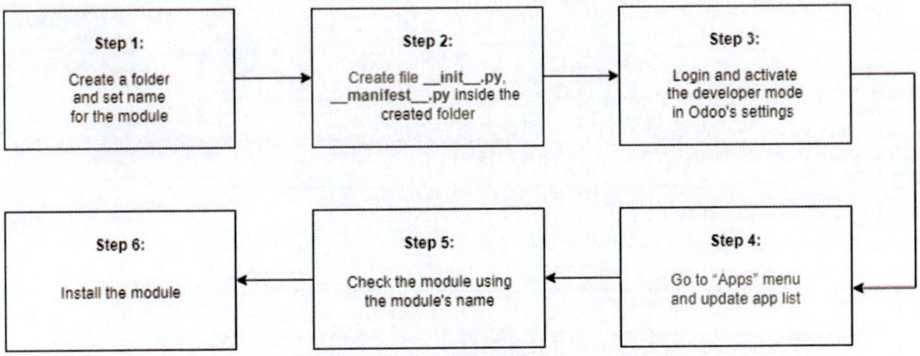

Fig. 5. Steps to install a new Odoo module

3.2 HRM Overtime Module

The open-source development methodology of Odoo has enabled thousands of developers and business professionals to create one of the biggest ecosystems of fully connected business apps. In fact, it has become the trusted foundation for many Vietnamese ERP systems, and one of the most popular examples is ERPViet.

Recent years saw a trend in working overtime as more and more employees want to increase their income [13]. Under that circumstances, firms are implementing numerous productivity strategies and monitoring work hour practices to stay up with the changing workplace culture. There is no method for requesting overtime in Odoo by default, thus with the view to providing Vietnamese businesses with an all-rounded ERP solution, ERPViet has added the Overtime Function into their app collection.

What distinguishes the overtime management procedure in Vietnam is that Vietnamese Labor Law specifically stipulates the number of overtime hours that employees may work - to which all employers must comply. For example, 200 h is the maximum amount of overtime per year. To guarantee that no company ever violate the law, as a result, this module has the capability of automatically verifying and notifying businesses about the limit accordingly (Fig. 6).

Employees and management can both communicate via this Overtime module to ensure a smooth work flow. To begin with, employees can put in an overtime request using the default registration form as shown in Fig. 7.

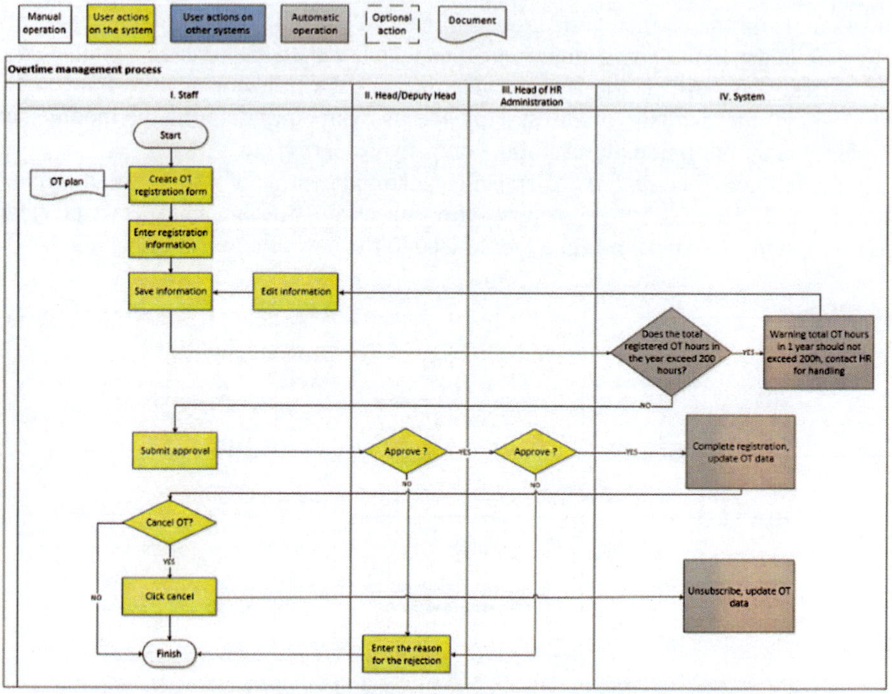

Fig. 6. Overtime proposal management process developed by ERPViet

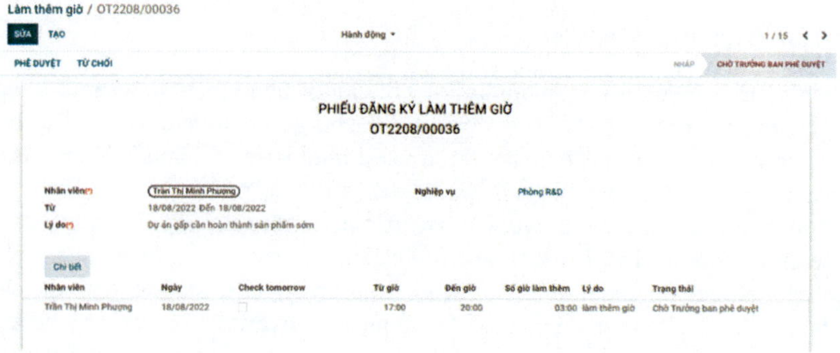

Fig. 7. Overtime registration form in Overtime management module developed by ERPViet

The system will then automatically verify and issue a warning if the total overtime hours surpass the prescribed limit. Once authenticated, a notification will be forwarded to the management for approval (Fig. 8). The system will automatically aggregate and determine the employee overtime regime (working overtime on weekdays, holidays, on public holidays, etc.) and update the data on payroll reports. Whether the request is allowed or denied, the employee should receive an immediate response along with sound explanations.

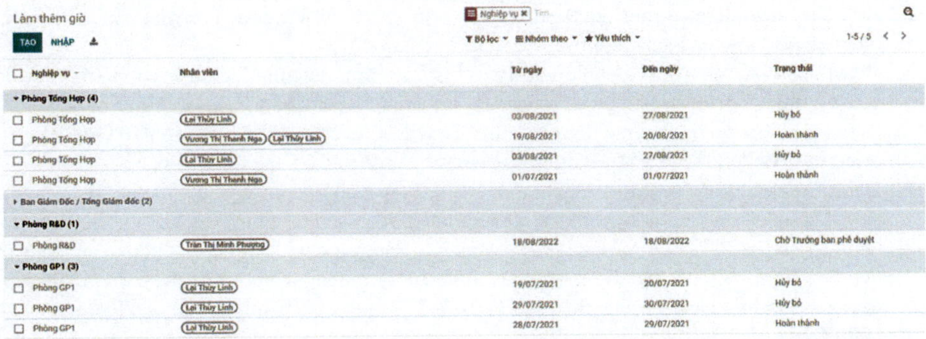

Fig. 8. All overtime requests on display in Overtime management module developed by ERPViet

4 Conclusion

ERP is an indiapensable part of all companies in this era of technology development, and Odoo is at the forefront by possessing a modular architecture, allowing developers to quickly and easily create useful solutions for their clients. Not to mention, Odoo also provides developers with a powerful API, comprehensive reference manual, and tools for quickly constructing new modules. With the presented resources, Vietnamese enterprises such as ERPViet have been able to swiftly and easily build exclusive, integrated, well-thought-out solutions such as the Overtime Management app to serve a specific group of consumers with minimal effort.

Acknowledgement. This research is funded by International School, Vietnam National University, Hanoi (VNUIS) under project number CS.2023–05. We would like to acknowledge the assistance of four students (Tran Viet Long, Dang Thuy Ngan, Nguyen Ngoc Thien, and Pham Hai Long).

References

1. ERP: The implementation cycle - 1st edition - Stephen Harwood - Routledge. https://www.routledge.com/ERP-The-Implementation-Cycle/Harwood/p/book/9780750652070. Accessed 05 Jan 2023
2. Ministry of planning and investment portal. https://www.mpi.gov.vn/en/Pages/tinbai.aspx?idTin=49802&idcm=133. Accessed 28 Jan 2023
3. Read @Kearney: toward a global network of digital hubs, Kearney. https://www.kearney.com/digital/article/-/insights/the-2021-kearney-global-services-location-index. Accessed 28 Jan 2023
4. Odoo - Wikipedia. https://en.wikipedia.org/wiki/Odoo. Accessed 29 Jan 2023
5. Reis, D.: Odoo 11 Development Essentials - Third Edition. Packt Publishing (2018)
6. Martins, P., Tomé, P., Wanzeller, C., Sá, F., Abbasi, M.: Comparing Oracle and PostgreSQL, performance and optimization. In: Rocha, Á., Adeli, H., Dzemyda, G., Moreira, F., Ramalho Correia, A.M. (eds.) WorldCIST 2021. AISC, vol. 1366, pp. 481–490. Springer, Cham (2021). https://doi.org/10.1007/978-3-030-72651-5_46

7. Building a Module—Odoo 16.0 documentation. https://www.odoo.com/documentation/16.0/developer/howtos/backend.html. Accessed 05 Jan 2023
8. ORM API—Odoo 16.0 documentation. https://www.odoo.com/documentation/16.0/developer/reference/backend/orm.html#reference-orm. Accessed 05 Jan 2023
9. Views—Odoo 16.0 documentation. https://www.odoo.com/documentation/16.0/developer/reference/backend/views.html#reference-views. Accessed 05 Jan 2023
10. Data Files—Odoo 16.0 documentation. https://www.odoo.com/documentation/16.0/developer/reference/backend/data.html#reference-data. Accessed 05 Jan 2023
11. Gómez-Llanez, C.Y., Diaz-Leal, N.R., Angarita-Sanguino, C.R.: A comparative analysis of the ERP tools, Odoo and Openbravo, for business management. Aibi Revista de Investigación, Administración e Ingeniería **8**(3), 145–153 (Sep.2020). https://doi.org/10.15649/2346030X.789
12. Huber, S.: ERP software system comparison between Odoo and Microsoft Dynamics NAV (2021). http://www.theseus.fi/handle/10024/497869. Accessed 12 Jan 2023
13. VnExpress, Workers' dilemma as Vietnam considers increase in overtime cap - VnExpress International. VnExpress International – Latest news, business, travel and analysis from Vietnam. https://e.vnexpress.net/news/economy/workers-dilemma-as-vietnam-considers-increase-in-overtime-cap-4418344.html. Accessed 17 Jan 2023

Optimal Pressure Regulation in Water Distribution Systems Based Mathematical Program with Vanishing Constraints

Pham Duc Dai[1](✉) and Dang Khoa Nguyen[2]

[1] Thuyloi University, 175 Tayson, Dong Da, Hanoi, Vietnam
daipd@tlu.edu.vn

[2] VNU-International School, 144 Xuan Thuy, Cau Giay, Hanoi, Vietnam
khoand@vnuis.edu.vn

http://www.tlu.edu.vn

Abstract. Optimal pressure control to water leakage reduction can be accomplished by controlling pressure reducing valves (PRVs) installed in water distribution systems (WDSs). The optimal pressure control can be casted into a nonlinear program (NLP) where the model of PRVs is of important for proper operation of the systems. In this paper, at first, we reformulated the mathematical model of PRVs by using vanishing constraints which is suitable for the use in practice, then we applied an efficient relaxation approach for solving the mathematical program with vanishing constraints (MPVCs). The proposed relaxation approach has strong convergence. The application of MPVCs for optimal pressure management has been evaluated on one WDS in Vietnam showing that the MPVCs results in highly accurate solutions and it is suitable for the use in the decision support system for optimal pressure control.

Keywords: Water distribution system · Leakage reduction · Mathematical program with vanishing constraints · Pressure reducing valves

1 Introduction

Water distribution systems (WDSs) are now coping with the problems of water loss which reaches a large part and considered as a non- revenue water. In India, the non-revenue water (NRW) is high and reaches from 5% to about 70% of the total drinking water production[1]. In WDSs, water leakage flows [2,3] can be controlled by pressure reducing valves (PRVs) [4–6]. The optimal pressure management in practice can be transformed into a nonlinear programing problem (NLP) where PRV pressure settings are decision variables while state variables are flows and nodal heads for links and nodes, respectively. Many solution approaches from modeling to solving the NLP have been proposed in the literature. Genetic algorithm (GA) have been extensively employed for addressing the optimization problems for optimal pressure control [7,8]. A reinforcement learning combined with a

T. D. L. Nguyen et al. (Eds.): ICISN 2023, LNNS 752, pp. 233–239, 2023.
https://doi.org/10.1007/978-981-99-4725-6_29

greedy algorithm for control of pressure in a WDS was used in [9] to control pressure of the WDS. About mathematical programming, in [4], the SQP algorithm (sequential quadratic program) was applied to solve the NLP for optimal pressure management problem where either excessive pressure or water leakage amount is minimized. The same optimization model is applied to regulate pressure in domestic area meters (DMAs) with boundary and internal PRVs. The sequential convex program method (SCP) aiming to improve the quality of the NLP solution [10] was employed to address the NLP for the problem of optimal pressure management to improve the solution quality. In this work, the constraints for modeling of PRV operations are reformulated by vanishing constraints. This formulation is suitable since in the check valve mode, the model equation of PRV in the active mode must be vanished. This formulation has not been considered so far and the NLP formulated for pressure control becomes a mathematical program with vanishing constraints (MPVCs). We applied a regulation approach with strong convergence property to solve the MPVCs. Numerical simulations have been used to verify the proposed regulation approach.

2 Problem Formulation for Optimal Pressure Management

We formulate the NLP for a general WDS with N_n nodes, N_p links, N_{prv} PRVs, and N_r reservoirs. The defined objective function aims to minimize the surplus pressure at all demand nodes while ensuring that the hydraulic model constraints as well operational constraints are satisfied [11].

2.1 Objective Function

$$\min F = \sum_{i=1}^{N_n} \sum_{k=1}^{T} (H_{i,k} - H_{i,k}^{L}) \tag{1}$$

where $H_{i,k}$ and Hi, k^L are nodal head and its minimum allowable value, respectively. T is the whole time horizon.

2.2 Hydraulic Model Equality and Inequality for the Optimization Problem

Model equations for node i at scenario k

$$\sum_{j,k} Q_{j,i,k} - \sum_{j,k} Q_{i,j,k} - d_{i,k} - l_{i,k} = 0 \tag{2}$$

$k = 1, ..., T$; $i = 1, ..., N_n$ where $d_{i,k}$ is the demand at node i and at time interval k; $Q_{i,j,k}$ is the flow coming out from node i at time interval k,while $Q_{j,i,k}$ and $l_{i,k}$ are the flows coming into and the leakage flows at node i, respectively.

Hydraulic equations for pipes The energy equations for pipes ij

$$-H_{i,k}+H_{j,k}+\Delta H_{i,j,k}=0; i,j=1,...,N_P \tag{3}$$

In the equation, $\Delta H_{i,j,k}$ represent the head loss over a pipe from node i to node j is calculated by the Hazen- Williams equation

$$\Delta H_{i,j,k}=\frac{10.67L_{i,j}}{D_{i,j}^{4.87}}\left(\frac{Q_{i,j,k}}{C_{i,j}}\right)^{1.852} \tag{4}$$

Hydraulic Model Equalities for PRVs

$$\begin{aligned}\left(R_{i,j}Q_{i,j,k}^2-H_{i,k}+H_{j,k}\right)Q_{i,j,k}&\leq 0\\ Q_{i,j,k}&\geq 0\end{aligned} \tag{5}$$

The principle operation of a PRV is that when PRV operates in active mode, we have from the constraint in (5) that $Q_{i,j,k}\geq 0$ and $R_{i,j}Q_{i,j,k}^2-H_{i,k}+H_{j,k}\leq 0$. In the case as a PRV operates in the check valve mode the constraint $R_{i,j}Q_{i,j,k}^2-H_{i,k}+H_{j,k}\leq 0$ is vanished as soon as $Q_{i,j,k}=0$. The PRV model constraint formulated in such the way can be classified into the kind of vanishing constraint [12].

Operational Constraints: We impose the lower and upper bound constraints on flow and head quantities to ensure that the WDS works properly: $H^L\leq H_{i,k}\leq H^U$, $Q^L\leq Q_{i,j,k}\leq Q^U$. The reservoir i is assumed with a constant head $(\overline{H}_i)$, i.e., $H_{i,k}=\overline{H}_i$. Since the constraints for PRVs are kind of vanishing ones (5), the formulated optimization for optimal pressure management is a mathematical program with vanishing constraints (MPVCs) [12]. In general, the MPVC can be written as

$$\begin{aligned}&\min F(x)\\ &H_i(x)=0, i=1,\ldots,N_n\\ &G_i(x)\leq 0, j=1,\ldots,N_p\\ \varphi_i\left(m_i(x),n_i(x)\right)=m_i(x)n_i(x)&\leq 0, i=1,\ldots,N_{prv}\end{aligned} \tag{6}$$

where $H_i\left(x\right)$ and $G_i\left(x\right)$ stand for constraints for junctions and pipe lines, respectively; φ is denoted as constraints for PRVs; $m_i\left(x\right)$ and $n_i\left(x\right)$ stand for $R_{i,j}Q_{i,j,k}^2-H_{i,k}+H_{j,k}$ and $Q_{i,j,k}$, respectively. We will present the reformulation approach in [12,13] in the next section for addressing the MPVCs efficiently.

3 An Efficient Regularization Scheme for the MPVCs

Solving the MPVCs directly is complicated. For a sake of simplification, we assign $z=R_{i,j}Q_{i,j,k}^2-H_{i,k}+H_{j,k}$ and $y=Q_{i,j,k}$. We develop a function $\varphi:\ \mathbb{R}^2\rightarrow\mathbb{R}$ with the following properties [12]: $\varphi(z,y)=0\Leftrightarrow y\geq 0, zy\leq 0$. For this purpose, we adapt the idea from [13] for developing an efficient regularized function with strong convergence as applied to solving the MPCCs. This regularization scheme has not been used so far for optimal pressure control. To this end, the following function $\varphi : \mathbb{R}^2 \rightarrow \mathbb{R}$ is considered

$$\varphi(z, y) = \begin{cases} zy, & \text{if } z + y \geq 0 \\ -\frac{1}{2}\left(z^2 + y^2\right), & \text{if } z + y < 0 \end{cases} \tag{7}$$

The above function has the following properties (a) φ: $\varphi(z, y) = 0$ if and only if $z \geq 0, y \geq 0, zy = 0$. (b) φ is continuous and differentiable (c) φ with

$$\varphi(z, y) = \begin{cases} > 0, & \text{if } z > 0 \text{ and } y > 0 \\ < 0, & \text{if } z < 0 \text{ or } y < 0 \end{cases}$$

From this function, the regularized function $\Phi(\cdot; t) : \mathbb{R}^2 \rightarrow \mathbb{R}$ is defined as bellows:

$$\Phi(z, y; t) := \begin{cases} z(y - t), & \text{if } z + y \geq t \\ -\frac{1}{2}\left(z^2 + (y - t)^2\right), & \text{if } z + y < t \end{cases}$$

for small and positive value of t, i.e., $t \rightarrow 0$, $\Phi(z, y; t) \rightarrow \varphi(z, y)$. For numerical computation, we replace vanishing constraints with such the regularized and smooth function ($\Phi(z, y; t)$) and, as a result, the MPVCs now becomes a NLP. To the end, instead of solving the MPVCs, we address a sequence of the following $NLP(t)$ formulated with gradually decreased values of t.

$$\begin{aligned} &\min F(x) \\ &G_i(x) = 0, i = 1, \ldots, N_n \\ &H_i(x) \leq 0, j = 1, \ldots, N_p \\ &\Phi(m_i(x), n_i(x); t) \leq 0, i = 1, \ldots, N_{prv} \end{aligned} \tag{8}$$

4 Case Studies

The efficacy of our proposed solution approach is evaluated with the problem of optimal pressure regulation for a WDS in Thainguyen City, in Vietnam as depicted in Fig. 1. The WDS studied in [14] is regulated to ensure the mnimum pressure values at demand nodes of 17.00 [14]. The MPVCs (8) is formulated and solved with a sequence of t, namely $t = 0.1t_0$ and with $t_0 = 1$. More over, an IPOPT solver [15] is employed to address the NLP efficiently. The optimal PRV pressure settings are given in Fig. 2. It is seen that as demand patterns varied, the PRV settings change accordingly to regulate the pressure. In addition, as seen in Fig. 3, the average pressure of the WDS as we applied control of PRVs (i.e., the blue line) is significantly reduced as compared with the case where pressure management is not accomplished (i.e., the red line). In the low demand pattern periods (i.e., from 0 to 8:00am), high pressure are observed in the system (i.e., without PRV control), however, as PRVs are used, much more excessive pressure are absorbed. It is well known that, as pressure is kept under control and at appropriate levels, the probability of pipe break and new leak is avoided.

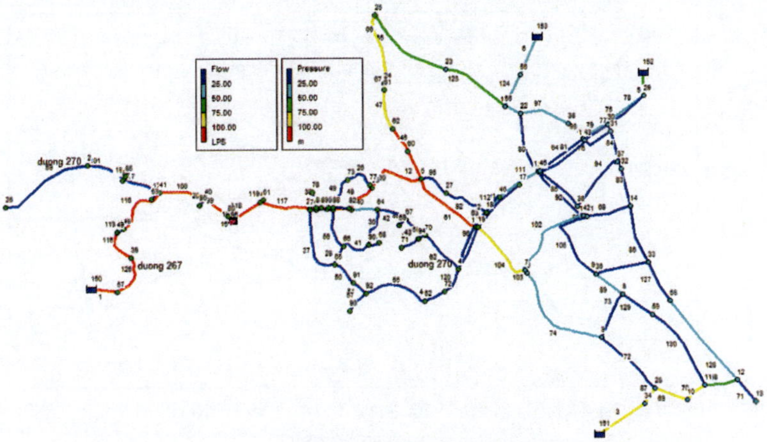

Fig. 1. Water distribution Network in Thainguyen City

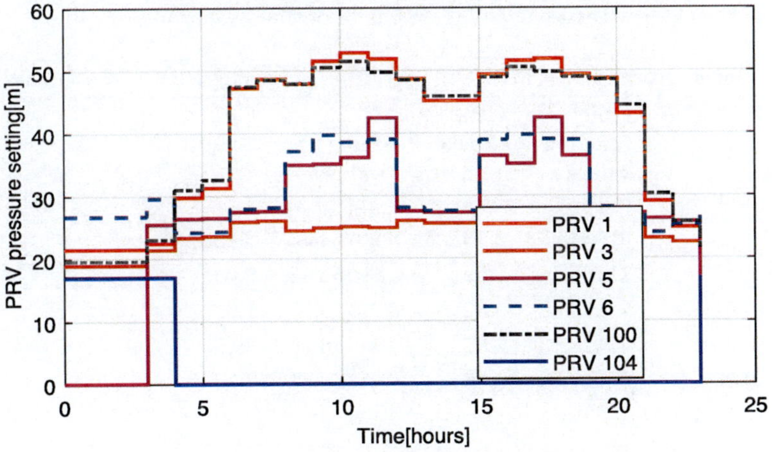

Fig. 2. Optimal PRV pressure settings

Now, we accomplish comparisons of solutions obtained from solving the MPVCs (i.e., formulated with our new PRV model) with the ones obtained from solving the NLP formulated with a PRV model in [14] using complementarity constraints. The results are given in Table 1. It can be recognized that using both PRV models, the same optimal solutions are achieved at the small values of t. However, with our PRV model, IPOPT took much less the computation time.

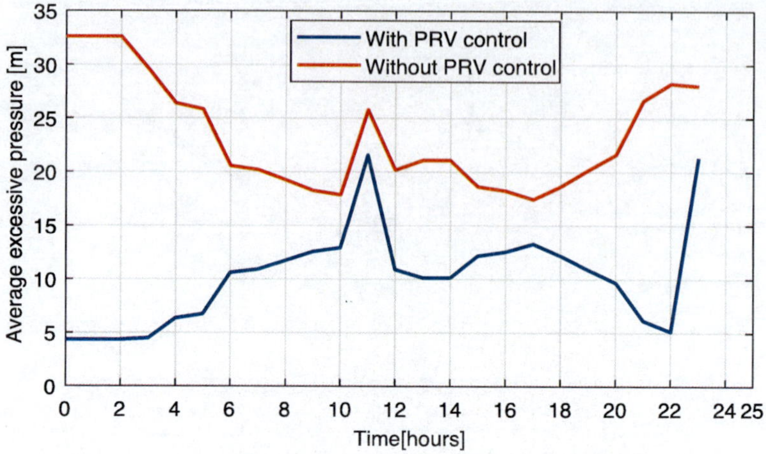

Fig. 3. Average pressure in the WDS. (Color figure online)

Table 1. Comparisons of optimal solutions

Our approach using MPVCs			Approach using PRV model in [14]	
Value of t_k	Objective function values [m]	CPU time[s]	Objective function values [m]	CPU time [s]
1	21074.57	2.18	20754.13	3.86
0.1	21185.14	0.52	20754.13	2.63
0.01	21173.82	0.32	20754.13	1.03
0.001	21128.83	2.07	21146.66	3.15
0.0001	21147.85	0.56	21166.61	2.16

4.1 Conclusions

This paper has proposed MPVCs approach to address the optimal pressure valve setting. The efficient regularized function is proposed to convert the MPVCs to a NLP, which can be efficiently solved by existing standard NLP solvers. The approach has been applied to determine hourly pressure settings for valves showing the potential reduction of excessive pressure.

References

1. Rajakumar, A.G., Cornelio, A.A., Mohan Kumar, M.S.: Leak management in district metered areas with internal-pressure reducing valves. Urban Water J. **17**(8), 714–722 (2020)
2. Ulanicki, B., Bounds, P.L.M., Rance, J.P., Reynolds, L.: Open and closed loop pressure control for leakage reduction. Urban Water **2**(2), 105–114 (2000)

3. Taha, A.-W., Sharma, S., Lupoja, R., Fadhl, A.-N., Haidera, M., Kennedy, M.: Assessment of water losses in distribution networks: methods, applications, uncertainties, and implications in intermittent supply. Resour. Conserv. Recycl. **152**, 104515 (2020)
4. Vairavamoorthy, K., Lumbers, J.: Leakage reduction in water distribution systems: optimal valve control. J. Hydraul. Eng. **124**(11), 1146–1154 (1998)
5. Pham Duc Dai and Pu Li: Optimal pressure regulation in water distribution systems based on an extended model for pressure reducing valves. Water Resour. Manage **30**(3), 1239–1254 (2016)
6. Dai, P.D.: A new mathematical program with complementarity constraints for optimal localization of pressure reducing valves in water distribution systems. Appl Water Sci **11**(9), 1–16 (2021). https://doi.org/10.1007/s13201-021-01480-8
7. Savić, D.A., Walters, G.A.: Integration of a model for hydraulic analysis of water distribution networks with an evolution program for pressure regulation. Comput.-Aid. Civil Infrastruct. Eng. **11**(2), 87–97 (1996)
8. Araujo, L.S., Ramos, H., Coelho, S.T.: Pressure control for leakage minimisation in water distribution systems management. Water Resour. Manage **20**(1), 133–149 (2006)
9. Mosetlhe, T.C., Hamam, Y., Du, S., Monacelli, E.: Appraising the impact of pressure control on leakage flow in water distribution networks. Water **13**(19), 2617 (2021)
10. Wright, R., Stoianov, I., Parpas, P., Henderson, K., King, J.: Adaptive water distribution networks with dynamically reconfigurable topology. J. Hydroinformat. **16**(6), 1280–1301 (2014)
11. Germanopoulos, G., Jowitt, P.W.: Leakage reduction by excess pressure minimization in a water supply network. In: ICE Proceedings, vol. 87, pp. 195–214. Ice Virtual Library (1989)
12. Achtziger, W., Hoheisel, T., Kanzow, C.: A smoothing-regularization approach to mathematical programs with vanishing constraints. Comput. Optim. Appl. **55**(3), 733–767 (2013)
13. Hoheisel, T., Kanzow, C., Schwartz, A.: Mathematical programs with vanishing constraints: a new regularization approach with strong convergence properties. Optimization **61**(6), 619–636 (2012)
14. Pham Duc Dai: Optimal pressure management in water distribution systems using an accurate pressure reducing valve model based complementarity constraints. Water **13**(6), 825 (2021)
15. Wächter, A., Biegler, L.T.: On the implementation of an interior-point filter line-search algorithm for large-scale nonlinear programming. Math. Program. **106**(1), 25–57 (2006)

Blockchain and Federated Learning Based Integrated Approach for Agricultural Internet of Things

Vikram Puri[1](✉), Vijender Kumar Solanki[2], and Gloria Jeanette Rincón Aponte[3]

[1] Duy Tan University, Da Nang, Vietnam
purivikram@duytan.edu.vn
[2] Department of Computer Science and Engineering, CMR Institute of Technology, Hyderabad, TS, India
[3] Universidad Cooperativa de Colombia, Bogota, Colombia
gloriaj.rincon@campusucc.edu.co

Abstract. The agriculture industry has undergone considerable changes in recent decades due to technological improvements and the introduction of new technologies such as the Internet of Things (IoT), artificial intelligence (AI), and secure communication protocols. Even in harsh weather, it is now possible to grow plants. Knowledge transfer continues to be a significant barrier to the adoption of AI technologies, specifically in the agriculture sector. The concept of "federated learning" (FL) was developed to safeguard data from various users or devices. FL is adequate for protecting data, but there are still several issues to be concerned about, including single-point failure, model inversion attacks, and data and model poisoning. For this study, a blockchain-enabled, FL-integrated framework is suggested as a solution to these problems. Additionally, the experimental setup and the Ethereum network are used to validate the suggested framework and the blockchain transactions.

Keywords: Federated Learning · Blockchain Technology · Agricultural Internet of Things · IoT · Privacy

1 Introduction

Smart-Agriculture (SA) is being enthusiastically embraced because of the rapid integration of modern technologies such as the Internet of Things (IoT), Artificial Intelligence (AI), big data, and centralized and decentralized communication with traditional agriculture [1]. The main objective of the SA system is to provide an intelligent, trustworthy and reliable system for the farmers. Low-cost sensors and gadgets are combined into "smart agriculture" to improve both the qualitative and quantitative productivity of agriculture. AI models are also integrated

T. D. L. Nguyen et al. (Eds.): ICISN 2023, LNNS 752, pp. 240–246, 2023.
https://doi.org/10.1007/978-981-99-4725-6_30

to help farmers make better decisions about their crops [2]. SA ensures the delivery of fresh goods while also giving farmers financial rewards [3]. AI has been used in a variety of industries recently, including the identification of plant diseases, crop recommendations based on atmospheric conditions, determining the quality of crops, and livestock output forecasting [4]. Although these devices use encryption to transfer data to servers or IoT devices, due to the enormous volume of data generated, third parties, such as hackers, may still misuse or improperly manage this data. Some researchers have suggested solutions to resolve these problems in recent years. For IoT-enabled agriculture, Ferrag [5] categorized the vulnerabilities into five areas, including user privacy, device authenticity, secrecy, dependability, and integrative characteristics, and described a four-tier green-IoT architecture for SA. Additionally, qualitative comparison also highlighted security and privacy in green IoT technologies and examined blockchain's significance in SA. Similarly, [6][7], the authors explored the security threat in the SA IoT systems. Authors [8] suggested a method to strengthen the security in IoT systems by cluster key management in order to increase the security in the SA. In addition, a web interface has been set up to gather acknowledgments from the devices and provide notifications when a signal is detected. An algorithm [9] is developed to optimize the deployment of security mechanisms in order to mitigate cyber threats in the agriculturally equipped IoT system. The above-mentioned studies consider different techniques and algorithms to secure IoT devices for agriculture, but it is still vulnerable at the collection and storage point. A smart irrigation system with cloud computing and IoT integration is suggested in [10]. The system is restricted to sensing the soil moisture and humidity before applying a variety of machine-learning models to forecast water consumption for irrigation. The single point of failure is a challenge in cloud computing deployment. Data can fall into the hands of a third party and be readily modified if the central point server is compromised or if any malicious malware is inserted. To overcome these issues, federated learning (FL) is introduced. By decentralizing information from the primary server to end users, FL provides a way to protect user privacy while enabling AI gains for heterogeneous and sensitive data-rich domains. There are some frameworks [11] [12] considered FL in the SA systems. These frameworks' main goal is to make IoT systems for agriculture more secure. Data privacy is also a goal, along with security. In this study, Blockchain integrated FL approach is proposed to overcome the above the mentioned issues. The main contributions of the study are:

1. Integrate FL with the blockchain technology to store the models' parameters in a decentralized way.
2. Deploy experiment setup to validate the performance of the system.

The following are the other topics covered in the paper: The methodology of the proposed work is demonstrated in Sect. 2, the results of the experimental setup are evaluated in Sect. 3, and the study is wrapped up with future directions in Sect. 4.

2 Methodology

The proposed architecture of the Agricultural Internet of Things (A-IoT) system is categorized into three different layers as follows (see in Fig. 1):

1. A-IoT Device: The concept of "precision farming" refers to a set of methods and equipment that farmers can use to improve soil fertility and profitability by implementing a number of focused strategies. IoT devices with sensors aid in automating farming tasks like watering and rinsing while also monitoring farm productivity. Soil NPK sensor [13] is employed for A-IoT devices, allowing for the collection of soil temperature, humidity, and pH as well as nitrogen, phosphorus, and potassium content. Due to the parameters' resemblance to the suggested strategy, pre-collected dataset [14] is taken into consideration in this approach. The dataset contains the N, P, K values, temperature, humidity, PH, rain and labels.
2. Local Models: Individuals can use federated learning to develop a model without transmitting raw data to a centralized platform, preventing the acquisition of information that should be kept private [15]. There are two categories of models used in federated learning: 1) local model 2) global model. The global model is trained on the central server, while the local model is trained on IoT devices. When using a federated learning technique, only models can communicate with one another; data cannot be shared. Additionally, it supports data privacy. In this study, a machine learning model is considered for the various devices with the same dataset that is mentioned above.
3. Federated Learning and Blockchain Network: Even so, FL is changing the market in terms of protecting privacy and securing distributed ML models. However, it also brought forth other problems, including communication costing, system variability, and model credibility [16]. This study combines the FL and blockchain networks to address the aforementioned problems. The immutable nature of the blockchain network makes it difficult to modify the local and global model parameters stored there. It also manages the centralized servers' single point of failure.

Figure 2 represents the workflow of the proposed architecture of the A-IoT system. A-IoT devices download a global trained model for the number of devices in the first stage and register the global parameters in the blockchain network. These devices collect the environmental parameters as per the condition and train local models in the devices. After training the local models, parameters are

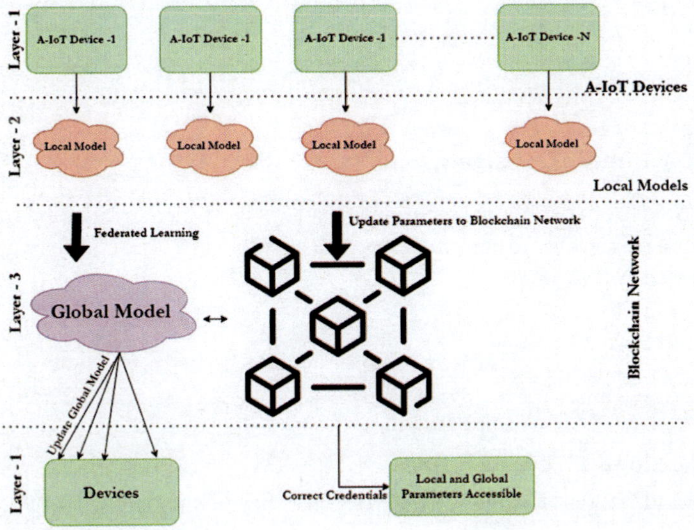

Fig. 1. Blockchain enabled Federated learning architecture for the A-IoT System

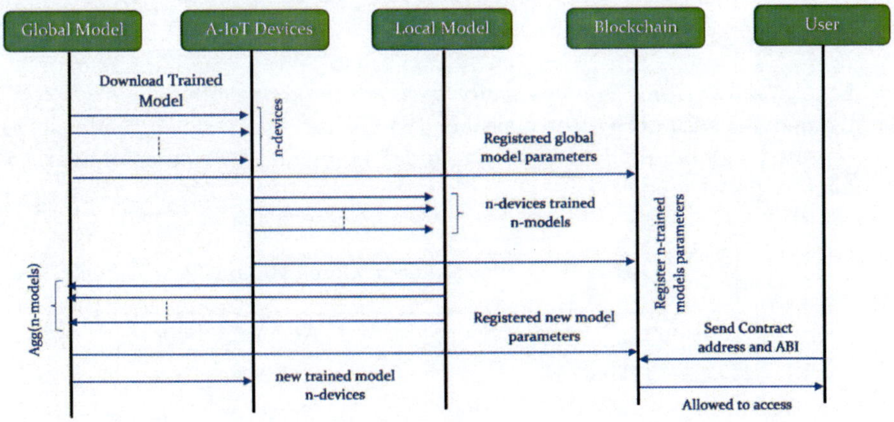

Fig. 2. Workflow of the Proposed Approach

registered in the blockchain network and shared with all models with the global model. The fundamental idea behind the FL is to combine models developed on many devices in order to create a new, more inclusive "average" model or global model. Once more, this global model is used with devices that have additional capabilities. New parameters are similarly added to the blockchain network. If there any user enters the blockchain network, the contract address and ABI are verified at the initial level [17].

3 Results and Evaluation

In this section, experimental setup and evaluation of the results are discussed.

3.1 Experimental Testbed

The proposed approach's experimental setup is shown in Table 1. The following two machines are taken into consideration: 1) The FL learning strategy 2) Blockchain Network.

Table 1. Experimental Testbed

Machine	Configuration	Role
Windows 11	512 GB SSD, 8GB RAM	FL learning approach
Windows 11	Ganache, REMIX Solidity	Blockchain Network

3.2 Analysis

1. Federated Learning: For this study, we considered one as a global model which trained the support vector classifier (SVC) and two as local models. Table 2 represents global model training rounds. There are ten rounds of aggregation for the global model.

Table 2. Global Model Training Rounds

Round	Accuracy Score	Round	Accuracy Score
1	95.73	6	95.62
2	95.91	7	95.83
3	96.02	8	96.86
4	95.47	9	95.29
5	95.31	10	95.35

2. Blockchain Technology: For this study, Ethereum network is considered to deploy this approach through the help of smart contract. Smart contracts, in contrast to typical software, execute on the network itself rather than on any local or server machine. Ganache is also considered in this study to work as dummy Ethereum public blockchain network. The Ethereum network is used because it allows for the development of smart contracts and makes it simple to test them on ganache. In other cases, test tokens need to request first.

Table 3. Time to transmit data to the blockchain network at different intervals

S.No	Transaction Number	Time(seconds)
1	1st	0.27
2	5th	0.28
3	10th	0.42
4	15th	0.43
5	20th	0.34

Table 4. Time to receive data from the blockchain network at different intervals

S.No	Transaction Number	Time(seconds)
1	1st	0.48
2	5th	0.40
3	10th	0.50
4	15th	0.49
5	20th	0.49

The time for sending and receiving data from the blockchain network are shown in Table 3 and Table 4, respectively. In Table 3, transaction at 1st, 10th, 15th and 20th, the data transmission times are 0.48, 0.50, 0.49, and 0.49 s, respectively. The variation in seconds is extremely tiny, but the variation in the 5th transaction is fairly significant when compared to other transactions. The local machine or compiler processing time could be the cause of the time lag. Similarly in Table 4.

4 Conclusion

Ever since farm technology was made commercially available, there has been a boom in enthusiasm for agricultural data. However, due to how erratic this information is, investigators are worried about their integrity because it is highly likely that someone else may have altered the accuracy of the report at distinct positions along the data stream A blockchain-integrated FL strategy is suggested for the A-IoT ecosystem in this paper. Real-time setup and separate analysis of the results based on the blockchain network and federated learning are utilized to validate the proposed framework. Single-point failure and data privacy can be eliminated with the adoption of the blockchain. The FL method is verified using the model accuracy score obtained via various rounds, and the blockchain network is verified using the amount of time required to send and receive data through the device. The study took into account the Ethereum network, which provides smart contracts to put the proposed strategy into practice but has very high transaction costs. The polygon or IOTA networks will eventually take the place of the Ethereum network in order to address this issue. Additionally this methodology can be implemented into hospitals ecosystem.

References

1. Sinha, B.B., Dhanalakshmi, R.: Recent advancements and challenges of internet of things in smart agriculture: a survey. Future Gener. Comput. Syst. **126**, 169–184 (2022)
2. Kumar, P., Gupta, G.P., Tripathi, R.: PEFL: deep privacy-encoding-based federated learning framework for smart agriculture. IEEE Micro **42**(1), 33–40 (2021)
3. Kumar, R., Mishra, R., Gupta, H.P., Dutta, T.: Smart sensing for agriculture: applications, advancements, and challenges. IEEE Consum. Electron. Mag. **10**(4), 51–56 (2021)
4. Thakur, P.S., Khanna, P., Sheorey, T., Ojha, A.: Trends in vision-based machine learning techniques for plant disease identification: a systematic review. Expert Syst. Appl., 118117 (2022)
5. Ferrag, M.A., Shu, L., Yang, X., Derhab, A., Maglaras, L.: Security and privacy for green IoT-based agriculture: review, blockchain solutions, and challenges. IEEE Access **8**, 32031–32053 (2020)
6. Demestichas, K., Peppes, N., Alexakis, T.: Survey on security threats in agricultural IoT and smart farming. Sensors **20**(22), 6458 (2020)
7. Kassim, M.R.M.: IoT applications in smart agriculture: issues and challenges. In: 2020 IEEE Conference on Open Systems (ICOS), pp. 19–24. IEEE (2020)
8. Anand, S., Sharma, A.: AgroKy: an approach for enhancing security services in precision agriculture. Measur.: Sens. **24**, 100449 (2022)
9. Shaaban, A.M., Chlup, S., El-Araby, N., Schmittner, C.: Towards optimized security attributes for IoT devices in smart agriculture based on the IEC 62443 security standard. Appl. Sci. **12**(11), 5653 (2022)
10. Phasinam, K., et al.: Application of IoT and cloud computing in automation of agriculture irrigation. J. Food Qual. **2022**, 8285969 (2022)
11. Kumar, P., Gupta, G.P., Tripathi, R.: PEFL: deep privacy-encoding-based federated learning framework for smart agriculture. IEEE Micro **42**(1), 33–40 (2021)
12. Friha, O., Ferrag, M.A., Shu, L., Maglaras, L., Choo, K.K.R., Nafaa, M.: FELIDS: federated learning-based intrusion detection system for agricultural internet of things. J. Parallel Distrib. Comput. **165**, 17–31 (2022)
13. NPK Sensor. https://www.jxctiot.com/product1/product195.html. Accessed 10 Jan 2023
14. NPK Dataset. https://www.kaggle.com/datasets/atharvaingle/crop-recommendation-dataset. Accessed 10 Jan 2023
15. Hanzely, F., Richtárik, P.: Federated learning of a mixture of global and local models. arXiv preprint arXiv:2002.05516 (2020)
16. Li, D., et al.: Blockchain for federated learning toward secure distributed machine learning systems: a systemic survey. Soft. Comput. **26**(9), 4423–4440 (2022)
17. Puri, V., Kataria, A., Sharma, V.: Artificial intelligence-powered decentralized framework for Internet of Things in Healthcare 4.0. Trans. Emerg. Telecommun. Technol., e4245 (2021)

Personal Federated Learning via Momentum Target with Self-Improvement

T-Binh Nguyen[1], H-Khoi Do[1], M-Duong Nguyen[2], and T-Hoa Nguyen[1(✉)]

[1] School of Electrical and Electronic Engineering, Hanoi University of Science and Technology, Hanoi, Vietnam
hoa.nguyentien@hust.edu.vn

[2] Department of Information Convergence Engineering, Pusan National University, Busan 46241, Republic of Korea

Abstract. Federated learning (FL) is a new artificial intelligence concept that enables Internet-of-Things (IoT) devices to learn a collaborative model without sending the raw data to centralized nodes for processing. Despite numerous advantages, Federated Learning faces huge challenges from model overfitting due to the lack of data and statistical diversity among clients. To address these challenges, we propose a novel personalized federated learning method via momentum adaptation, the so-called pFLTI. Specifically, pFLTI generates the target model by adapting from the temporal ensemble of the meta-learner, i.e., the momentum network. This momentum network and its task-specific adaptations enjoy a favorable generalization performance, enabling self-improving of the meta-learner through knowlodels to find the across task relationsedge distillation. Moreover, we found that perturbing parameters of the meta-learner, e.g., dropout, further stabilize this self-improving process by preventing fast convergence of the distillation loss during meta-training. Our experimental results demonstrate that our algorithm is awesome.

Keywords: Non-IID · Personalized Federated Learning · Semantic Communication

1 Introduction

Due to the high bandwidth cost and risk of privacy leakage, careless data transmission and aggregation have increasingly become intolerable with the rapid expansion of data and the strong privacy protection policy. Federated learning [16] has recently been presented as a replacement for the old largely centralized learning paradigm [19,20] while protecting data privacy.

One of the major issues in Federated Learning is data heterogeneity, which occurs when data in clients is distributed non-identically and separately (Non-IID). In this circumstance, the vanilla Federated Learning method, FedAvg cite,

T. D. L. Nguyen et al. (Eds.): ICISN 2023, LNNS 752, pp. 247–253, 2023.
https://doi.org/10.1007/978-981-99-4725-6_31

leads to drifting local models and catastrophically forgets global information, resulting in reduced performance and delayed convergence.

To address the problem of data heterogeneity, we offer a unique methodology for personalized Federated Learning that improves Federated Learning performance through a self-improving method, nicknamed personalized Federated Learning via momentum Target with self-Improving (pFLTI). The main contributions of our work can be summarized as follows:

- We formulate a new personalized Federated Learning optimization problem aiming at finding an initial model that can generalize well with individual users with just a few steps of gradient-descent update.
- We exploit the momentum network that operates as the coordinator for the specific-task learner to improve the meta-generalization performance.

2 Related Works

In recent years, several insightful review papers on federated learning have been published in the literature as a result of the quickly rising research interest in Federated Learning. The authors in [24] provide a comprehensive overview of Federated Learning and its application; Otherwise, result in [8] has been provided in-depth analyses of advancements and difficulties in Federated Learning and the promised solution. Reference [22] on the other hand, gives analyses of threats and further privacy preservation strategies in Federated Learning. There have also been summaries of Federated Learning applications to wireless networks [7], mobile devices [15], healthcare [23] and IoT [21]. However, federated learning also has some challenges such as: device heterogeneity, data heterogeneity due to non-IID distribution of data which can significantly reduce the performance [11].

Our research focuses on addressing the problem of data heterogeneity; we use momentum adaptation for personalized federated learning, which can perform good generalization on each local user.

3 Proposed Algorithms

In the Federated Learning setting, the clients' model are agnostic (i.e., do not have access to other clients' knowledge or data distribution). Hence, it is difficult for clients to leverage other clients' knowledge to improve their local training. Thus, we aim to design a framework, where the client can improve their performance via the self-improving process. To be more specific, we design dual-independently trained models, coined task-specific model and momentum-adapted model, respectively. The first model performs as the main model, where learn the task on the specific client. This task-specific model acts the same as the inner model as demonstrated in Per-FedAvg [3] or clients' model in conventional Federated Learning settings [1,2,9,14,16–18]. Meanwhile, the second model, the so-called momentum-adapted model, performs a variant step from the main gradient trajectory.

3.1 System Architecture

We start by giving a high-level overview of the method. The algorithm, dubbed personalized Federated Learning via momentum Target with self-Improvement (pFLTI), is demonstrated as in Algorithm 1. It follows a common pattern used by meta-learning algorithms: use an additional perturbed model, the so-called momentum model, to self-verify the inner adaptation process of the local user.

There are two main procedures in our algorithm: on the server and on each clients. At the server, we first utilize the temporal elements of the network, i.e., the meta-model ϕ and the momentum-model θ. For each round of communication, the server chooses a fraction of users and send its current model to these clients. Each client applies two distinct updates, dubbed task-specific adapt and momentum adapt. Consequently, the client uses the momentum adapt as an advisor to guide the task-specific model to become more generalized. Finally, the meta update is applied on the clients' models to find the across task relationship among clients.

3.2 Meta Update with a Hybrid Loss

Self-improving Process via Knowledge Distillation To update meta-model, we use a hybrid loss function, which includes: the MAML udpate [4] and the knowledge transfer via knowledge distillation [6]. Formally, let $\mathcal{S}, \mathcal{Q} = (\mathcal{S}_i, \mathcal{Q}_i | i = \{1, \cdots, N\})$ be the given clients' datasets with support-query split (i.e., the support and query dataset on client i is randomly sampled from the client's datapool). Let θ_u be the task-specific model generated by our target generation procedure. We have a hybrid loss optimization problem for our meta update process as follows:

$$\mathcal{L}_u^{\text{tot}} = \frac{1}{U} \sum_{u=1}^{U} (1 - \lambda) \cdot \mathcal{L}(\theta_{u,\text{drop}}, \mathcal{D}_u^q) + \lambda \cdot \mathcal{L}_{\text{KD}}(\theta_{u,\text{drop}}, \theta_{u,\text{mom}}, \mathcal{D}_u^q) \tag{1}$$

3.3 Self-improving Momentum Model via Slow Update

We use the meta-momentum model's network θ_{mom} to generate target models in a compute-efficient manner. Specifically, we compute the exponential moving average of the meta-model parameter θ after each meta-model training iteration as follows:

$$\theta_{\text{mom}} = \zeta \cdot \theta_{\text{mom}} + (1 - \zeta) \cdot \theta, \tag{2}$$

where $\zeta \in [0, 1]$ is the momentum coefficient. The θmom meta-model can adapt better than the θ meta-model itself, and the loss landscape has flatter minima, which might be a signal for understanding generalization improvement. As a result, the following adaption function exists:

$$\widetilde{\theta}_{u,\text{mom}} = \theta_{u,\text{mom}} - \eta \nabla_{\theta_{u,\text{mom}}} \mathcal{L}(f_{\theta_{u,\text{mom}}}; \mathcal{D}_u^s) \tag{3}$$

Algorithm 1. pFL with Self-Momentum Adaptation

Require: Run on the server

Initialize ϕ^0_{glob} for the global model.
Initialize $\phi^0_{\text{mom}} \leftarrow \theta$.
for each communication round $r = 1, 2, \ldots$ **do**
    Collect client models from clients.
    $\phi^r_{\text{glob}} = \frac{1}{U}\sum_{u=1}^{U} \phi_u$.
    $\phi^r_{\text{mom}} \leftarrow \eta \cdot \phi^r_{\text{mom}} + (1-\eta) \cdot \phi^r_{\text{glob}}$.
    Broadcast global model ϕ^r_{glob} and momentum model ϕ^r_{mom} to every users.
    Run on clients.
end for

Require: Run on clients $u \in \mathcal{U}$

Set $\theta_u = \phi^r$, $\theta_{u,\text{mom}} = \phi^r_{\text{mom}}$
Sample a support set $\mathcal{D}^s_u = (\mathbf{X}^s_u, \mathbf{Y}^s_u)$ and a query set $\mathcal{D}^q_u = (\mathbf{X}^q_u, \mathbf{Y}^q_u)$ from $\mathcal{D}_u$
for $t : 1$ to τ **do**
    Compute the stochastic gradient $\nabla_{\theta_u}\mathcal{L}(f_{\theta_u}; \mathcal{D}^s_u)$ using data set $\mathcal{D}^s_u$.
    Calculate the user-specific parameters θ_u by the inner-loop adapt and dropout,

$$\widetilde{\theta}_{u,\text{drop}} = \theta_u - \eta\nabla_{\theta_u}\mathcal{L}(f_{\theta_u}; \mathcal{D}^s_u)$$

    Calculate the momentum parameters $\theta_{u,\text{mom}}$ by the inner-loop adapt,

$$\widetilde{\theta}_{u,\text{mom}} = \theta_{u,\text{mom}} - \eta\nabla_{\theta_{u,\text{mom}}}\mathcal{L}(f_{\theta_{u,\text{mom}}}; \mathcal{D}^s_u)$$

    Calculate the momentum loss $\mathcal{L}^{\text{tot}}_u(\theta_u)$ by the inner-loop adapt,

$$\mathcal{L}^{\text{tot}}_u(\theta_u, \mathcal{D}^q_u) = (1-\lambda) \cdot \mathcal{L}(\theta_{u,\text{drop}}, \mathcal{D}^q_u) + \lambda \cdot \mathcal{L}_{\text{KD}}(\theta_{u,\text{drop}}, \theta_{u,\text{mom}}, \mathcal{D}^q_u)$$

    Calculate the meta-based parameters $\phi^{L-l}_{u,cr}$ by the inner-loop adap,

$$\phi_u = \theta_u - \beta\left[\mathbb{I} - \eta\nabla^2_{\widetilde{\theta}_u}\mathcal{L}(f_{\widetilde{\theta}_u}; \mathcal{D}^s_u)\right]\nabla_{\widetilde{\theta}_u}\mathcal{L}^{\text{tot}}_u(f_{\widetilde{\theta}_u}, \mathcal{D}^q_u)$$

end for
Update model ϕ_u of client u to the server for the aggregation.

4 Experimental Evaluation

4.1 Experimental Setup

Dataset. We use two multi-class categorisation benchmark datasets in our evaluations, including **1)** Rainbow-MNIST [13], **2)** CIFAR-10 [10].

Settings. We apply two DNNs for our experimental evaluations, including LeNet-5 [12] for Rainbow-MNIST, and ResNet-9 [5] for CIFAR-10. It is worth noting that we do not apply pre-trained models on both LeNet-5 and ResNet-18 because the pre-trained models can help pFLTI approach the global minima easily, which can reduce the generalization of the evaluation.

4.2 Experimental Result

We used state-of-the-art Federated Learning training algorithms, FedAvg [16], Per-FedAvg [3] to compare against pFLTI. We reveal our experiments on MNIST with LeNet-5 and CIFAR-10 on ResNet-9 on Figs. 1a, 1b, respectively. As we can see from the two figures, pFLTI outperforms Per-FedAvg and FedAvg on both data. This phenomenon is held due to the additional of the momentum model as coordinator for the specific-training task on every client, which leads the clients' models to be more deterministic.

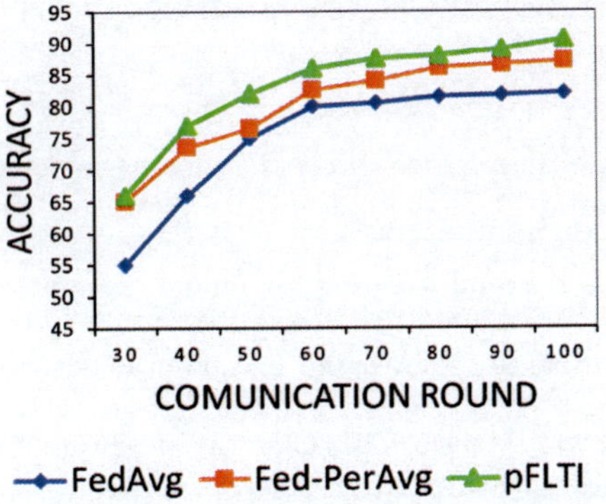

(a) Evaluation on Rainbow-MNIST

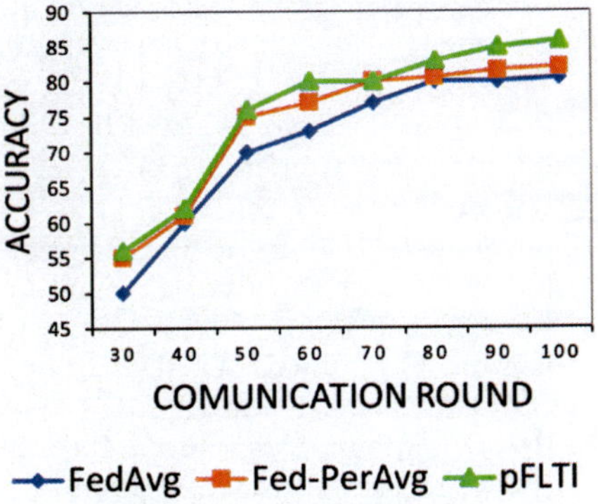

(b) Evaluation on CIFAR

Fig. 1. Evaluation of pFLTI on various dataset.

To be more specific, the momentum model is utilized with the dropout function. Due to the reduction in parameters via the dropout function, the demonstration function of the momentum network becomes less convex, which thus improves the generalization of the momentum network. Thus the momentum network acts as a guide for the specific-training task network (which is more prone to noise and related to the data representation). The client's model can realize the universal trajectory towards the global optima.

4.3 Conclusions

In this research, we propose pFLTI, a simple yet effective strategy for enhancing customized Federated Learning. Our central idea is to use a momentum network to efficiently produce target models and then use that knowledge to self-improve the distributed client. Our findings show that pFLTI considerably increases the performance of meta-learning approaches in a variety of applications.

References

1. Acar, D.A.E., Zhao, Y., Matas, R., Mattina, M., Whatmough, P., Saligrama, V.: Federated learning based on dynamic regularization. In: International Conference on Learning Representations, December 2021
2. Dinh, C.T., Vu, T.T., Tran, N.H., Dao, M.N., Zhang, H.: FedU: a unified framework for federated multi-task learning with Laplacian regularization, February 2021
3. Fallah, A., Mokhtari, A., Ozdaglar, A.: Personalized federated learning with theoretical guarantees: a model-agnostic meta-learning approach. In: Larochelle, H., Ranzato, M., Hadsell, R., Balcan, M., Lin, H. (eds.) Advances in Neural Information Processing Systems, vol. 33, pp. 3557–3568. Curran Associates, Inc. (2020). https://proceedings.neurips.cc/paper/2020/file/24389bfe4fe2eba8bf9aa9203a44cdad-Paper.pdf
4. Finn, C., Abbeel, P., Levine, S.: Model-agnostic meta-learning for fast adaptation of deep networks. In: Precup, D., Teh, Y.W. (eds.) Proceedings of the 34th International Conference on Machine Learning. Proceedings of Machine Learning Research, vol. 70, pp. 1126–1135. PMLR, 06–11 August 2017. http://proceedings.mlr.press/v70/finn17a/finn17a.pdf
5. He, K., Zhang, X., Ren, S., Sun, J.: Deep residual learning for image recognition, December 2015
6. Hinton, G., Vinyals, O., Dean, J.: Distilling the knowledge in a neural network. arXiv:1503.02531, March 2015
7. Hosseinalipour, S., Brinton, C.G., Aggarwal, V., Dai, H., Chiang, M.: From federated to fog learning: distributed machine learning over heterogeneous wireless networks. IEEE Commun. Mag. **58**(12), 41–47 (2020). https://doi.org/10.1109/MCOM.001.2000410
8. Kairouz, P., et al.: Advances and open problems in federated learning. Found. Trends Mach. Learn. **14**(1–2), 1–210 (2021). https://doi.org/10.1561/2200000083
9. Karimireddy, S.P., Kale, S., Mohri, M., Reddi, S., Stich, S., Suresh, A.T.: SCAFFOLD: stochastic controlled averaging for federated learning. In: Proceedings of the 37th International Conference on Machine Learning, 13–18 July 2020
10. Krizhevsky, A.: Learning multiple layers of features from tiny images. Technical report, August 2009
11. Kulkarni, V., Kulkarni, M., Pant, A.: Survey of personalization techniques for federated learning. In: 2020 Fourth World Conference on Smart Trends in Systems, Security and Sustainability (WorldS4), pp. 794–797. IEEE (2020)
12. LeCun, Y., et al.: Backpropagation applied to handwritten zip code recognition. Neural Comput. **1**(4), 541–551 (1989). https://doi.org/10.1162/neco.1989.1.4.541
13. LeCun, Y., Cortes, C.: MNIST handwritten digit database (2010). http://yann.lecun.com/exdb/mnist/
14. Li, T., Sahu, A.K., Zaheer, M., Sanjabi, M., Talwalkar, A., Smith, V.: Federated optimization in heterogeneous networks, April 2020
15. Lim, W.Y.B., et al.: Federated learning in mobile edge networks: a comprehensive survey. IEEE Commun. Surv. Tutor. **22**(3), 2031–2063 (2020). https://doi.org/10.1109/COMST.2020.2986024
16. McMahan, H.B., Moore, E., Ramage, D., Hampson, S.: y Arcas. Communication-efficient learning of deep networks from decentralized data, B.A. (2017)
17. Nguyen, M.D., Lee, S.M., Pham, Q.V., Hoang, D.T., Nguyen, D.N., Hwang, W.J.: HCFL: a high compression approach for communication-efficient federated learning in very large scale IoT networks. IEEE Trans. Mob. Comput. 1–13, June 2022. https://doi.org/10.1109/TMC.2022.3190510

18. Nguyen, M.D., Pham, Q.V., Hoang, D.T., Tran-Thanh, L., Nguyen, D.N., Hwang, W.J.: Label driven knowledge distillation for federated learning with non-IID data (2022). https://doi.org/10.48550/ARXIV.2209.14520, arxiv.org:2209.14520
19. Nguyen, V.D., Bui, N.D., Do, H.K.: Skin lesion classification on imbalanced data using deep learning with soft attention. Sensors. **22**(19) (2022). https://doi.org/10.3390/s22197530, https://www.mdpi.com/1424-8220/22/19/7530
20. Nguyen, V.D., Luong, M.P., Do, H.K.: Endoscopic image classification using block-based color feature descriptors. In: 2022 RIVF International Conference on Computing and Communication Technologies (RIVF), pp. 214–219 (2022). https://doi.org/10.1109/RIVF55975.2022.10013868
21. Shi, Y., Yang, K., Jiang, T., Zhang, J., Letaief, K.B.: Communication-efficient edge AI: algorithms and systems (2020). https://doi.org/10.48550/ARXIV.2002.09668, https://arxiv.org/abs/2002.09668
22. Vepakomma, P., Swedish, T., Raskar, R., Gupta, O., Dubey, A.: No peek: a survey of private distributed deep learning (2018). https://doi.org/10.48550/ARXIV.1812.03288, arxiv.org:1812.03288
23. Xu, J., Glicksberg, B.S., Su, C., Walker, P., Bian, J., Wang, F.: Federated learning for healthcare informatics (2019). https://doi.org/10.48550/ARXIV.1911.06270, arxiv.org:1911.06270
24. Yang, Q., Liu, Y., Chen, T., Tong, Y.: Federated machine learning: concept and applications. ACM Trans. Intell. Syst. Technol. **10**(2) (2019). https://doi.org/10.1145/3298981

Adaptive Radial-Basis Function Neural Network Control of a Pneumatic Actuator

Van-Vuong Dinh[1,2], Bao-Long Pham[1], Viet-Thanh Nguyen[1], Minh-Duc Duong[1], and Quy-Thinh Dao[1(✉)]

[1] Hanoi University of Science and Technology, Hanoi, Vietnam
thinh.daoquy@hust.edu.vn

[2] Hanoi Vocational College of High Technology, Hanoi, Vietnam

Abstract. This paper explores an advanced adaptive controller for Pneumatic Artificial Muscles (PAMs). PAMs offer lightweight, simple, and safe operation advantages but are difficult to model and control due to the non-linearity and hysteresis caused by their physical manufacturing. An adaptive controller based on a neural approximation is proposed to address these challenges, incorporating Radial Basis Function algorithms and adapting to model parameter uncertainty. The efficiency of the control approach is confirmed through a multi-scenario experiment with high-accuracy results, promising future developments in intelligent control of PAMs.

Keywords: RBF neural network · Pneumatic artificial muscle · Adaptive control · Neural approximation

1 Introduction

Pneumatic artificial muscles (PAMs) have recently been used as actuators instead of electrical motors in robotics rehabilitation because of their natural characteristic. An emblematic PAM consists of an elastomeric inner bladder inside a braided fiber clasped at two ends. When supplying the air equivalent to increase the pressure, PAM will be compressed longitudinally, inflated radially, and return to its initial shape if the gas escapes. Due to the similarity with human muscle's behavior and many other advantages such as lightweight, easy to fabricate with low cost, and high gravimetric specific power, PAM has widely appeared in biorobotic [7,13], medical [3,4,9], and many other applications. Notwithstanding many advantages, there are still numerous considerable downsides to the PAM. The most serious problems are elastic hysteresis and difficulty recognizing an accurate model because of its nonlinearity and uncertainty. Various control strategies for PAM systems have been carefully considered for high-efficiency trajectory tracking. The classical Proportional-Integral-Derivative (PID) control approach and its variants were soon introduced in many

This research is funded by Hanoi University of Science and Technology (HUST) under project number T2022 - PC - 002.

T. D. L. Nguyen et al. (Eds.): ICISN 2023, LNNS 752, pp. 254–262, 2023.
https://doi.org/10.1007/978-981-99-4725-6_32

types of research. In [14], an unscented Kalman-filter-based PID control system only needs one pressure sensor. An enhanced nonlinear PID controller was reported in [1] to improve tracking effectiveness and reducing hysteresis issues. While bounteous intelligent control theories are widely used, many of them are applied to PAM configurations, such as fuzzy-PID control [2,5]. Nevertheless, these PID controllers can not deal with PAMs' cons effectively.

The application of artificial neural network theory has penetrated all fields, including the field of control; therefore, the limitations of previous control methods have been partially overcome thanks to its significant advantage, that is, no precise modeling is required [12]. Fuzzy control, neural networks, and genetic algorithms are frequently utilized in creating intelligent control systems that have proven their superiority over classical controllers [8,11]. Since 1965, when Zadeh published his works [15], fuzzy logic has received increased attention from researchers. This field did not truly take off until the 1980s in Japan, when several fuzzy logic-based solutions were implemented in control systems for subways, fans, air conditioners, washing machines, etc. The concept of combining fuzzy logic with conventional controllers for PAM has also been widely implemented and achieved good results, such as fuzzy-PID [2,5]. Initially, fuzzy logic systems were based primarily on the experience of experts, in which the system's parameters were selected empirically and permanently installed.

Therefore, people have utilized the concept of developing learning algorithms for fuzzy systems to enable the fuzzy system's parameters to be adaptively adjusted based on the signal samples provided. These systems are also known as neural networks with fuzzy logic. One of these systems that we take into consideration is the Radial Basis Function (RBF) neural network, which has been employed with great success in a range of applications and is regarded as an effective solution for resolving various control problems, including dynamic uncertainty due to its capability to approximatively reduce uncertainty in complex and unknown functions [10] of model-uncertain systems. In this study, we use indirect adaptive control methods. The estimated parameters are used to calculate required controller parameters combined with RBF neural network to estimate those uncertain parameters. In summary, this research has the following contributions:

- An indirect adaptive controller based on neural approximation is designed for an antagonistic configuration of PAMs
- RBF algorithms are used to approximate uncertainties in systems and calculate controller parameters.
- Results from experiments conducted under various circumstances show how useful the suggestion controller is in rehabilitation applications.

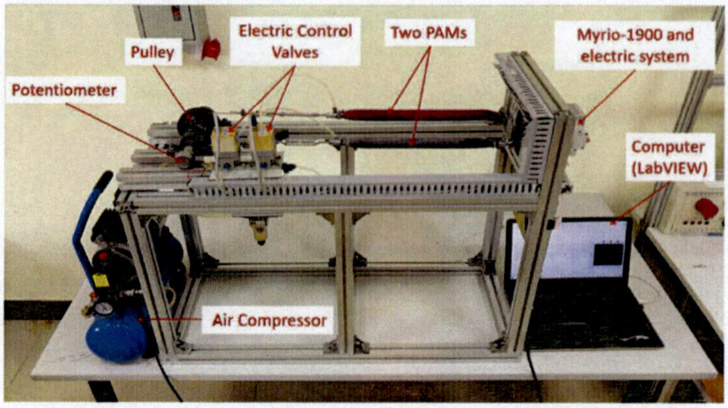

Fig. 1. Experiment Platform of PAM-Based Actuator.

2 System Modeling

A complete model of the pneumatic experimental platform is illustrated in Fig. 1. A pair of PAMs, which are arranged in an antagonistic setup, have 25 mm of diameter and 400 mm of nominal length. When there is a difference between the pressure of two PAMs, the pulley wheel rotates, and the angle will be recorded through a potentiometer (WDD35D8T). All that information will be recorded by an embedded controller (MyRIO1900 from National Instrument), which helps us to intervene in the system through the computer. Because of the similarity of research subjects, the mathematical model of the antagonistic actuator in the continuous time domain is inherited from the previous work [6].

$$\ddot{\theta}(t) = \kappa_2\dot{\theta}(t) + \kappa_1\theta(t) + \eta_1 \Delta P + \eta_0 \tag{1}$$

where η_i, κ_j are the parameters of the mathematical model. The values $\eta_0 = 205$, $\eta_1 = 18.01$, $\kappa_1 = -4.83$, $\kappa_2 = 7.35 \times 10^{-4}$ are specified through the identification process.

3 Control Design

This paper aims to offer an online adaptive controller based on RBF neural approximation as a method for improving control precision and adapting to parameter variations. The Lyapunov stability theory served as the basis for the adaptive law's creation, which will produce stable closed-loop systems. Figure 2 depicts the structure of the proposed closed-loop neural-based control system, where the control laws are generated by an RBF neural network. Output y is the angle measured by a potentiometer, and y_d is the desired signal. Due to the closed-loop control system, the tracking error provides feedback to the controller, the adaptive mechanism, and the RBF neural network. The uncertain $g(\boldsymbol{x})$ will be estimated by the adaptive cluster and then supplied to the controller. The control signal u is transmitted to two electrical valves, which alter the pressure P_1, P_2 of the two PAMs in an antagonistic arrangement.

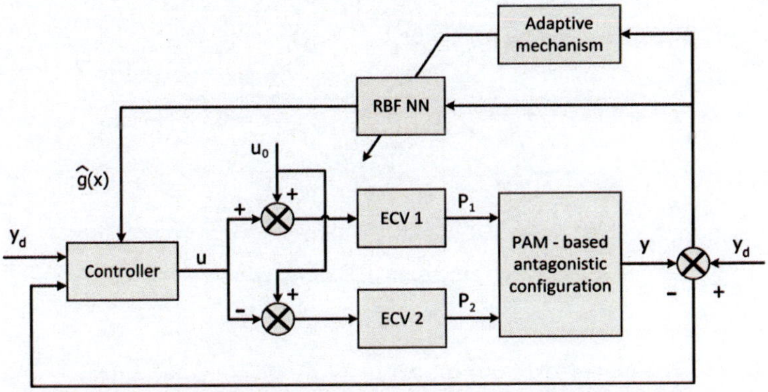

Fig. 2. RBF Neural network based control diagram.

Two PAMs in an antagonistic configuration is a SISO systems. According to (1), we represented the PAM system by the controllable canonical form to assist the building of the control algorithm:

$$\begin{cases} \ddot{x} = g(x, \dot{x}) + \eta_1 u \\ y = x \end{cases} \tag{2}$$

in which $g(x, \dot{x}) = \kappa_2 \dot{x}(t) + \kappa_1 x(t) + \eta_0$ is an unknown function and η_1 is a known element, $u \in R^n$ and $y \in R^n$ are the plant's input and output, respectively. $\boldsymbol{x} = [x, \dot{x}]^T \in R^n$ is the system's state vector. Presume that y_d is the ideal position signal. Let's define:

$$e = y_d - y, \quad \boldsymbol{E} = [e \ ; \ \dot{e}] \tag{3}$$

The control signal is:

$$u^* = \frac{1}{\eta_1}[-g(\boldsymbol{x}) + \ddot{y}_d + \boldsymbol{M}^T \boldsymbol{E}] \tag{4}$$

Substituting (4) into (2), we have:

$$\ddot{e} + m_d \dot{e} + m_p e = 0 \tag{5}$$

To ensure the polynomial $s^2 + m_d s + m_p = 0$ is Hurwitz, $\boldsymbol{M} = [m_p, m_d]^T$ is chosen so that all the roots of the polynomial function are in the open left-half complex plane. Then, system stability will be attained, and the plant output y will converge to the ideal output y_d asymptotically. We can deduce from (4) that the control law can not be realized if $g(\boldsymbol{x})$ is unknown.

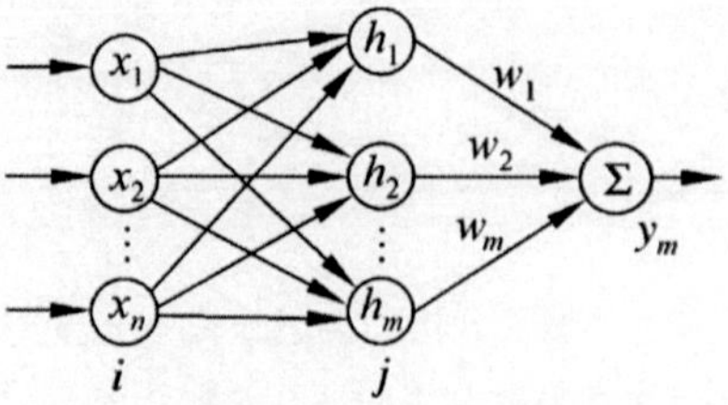

Fig. 3. Diagram of an RBF neural network.

3.1 RBF Neural Network Design

Figure 3 demonstrates the typical three-layer diagram of an RBF neural network. Specifically, $\boldsymbol{x} = [x_i]^T$ and $\boldsymbol{h} = [h_j]^T$ are the input-output vectors of the hidden layer, and each h_j represents a Gaussian function value for net j in the hidden layer:

$$h_j = exp(-\frac{||x - c_{ij}||^2}{b_j^2}) \tag{6}$$

where the neural network's Gaussian function's center j for the i_{th} input has coordinate value $c = [c_{ij}]$ $(n \times m)$ and the neural network's Gaussian function width j is b_j with $i = 1, 2, ..., n; j = 1, 2, ..., m$. Consequently, the RBF neural network's output is:
$y_m = \boldsymbol{W}^T \boldsymbol{h} = \sum_{j=1}^{m} W_j h_j$ where $\boldsymbol{W}$ is the weight value.

Since RBF neural networks can estimate uncertainties of the systems [10], with the input $\boldsymbol{x}' = [e \;\; \dot{e}]^T$ (e is the tracking error), the RBF algorithm for approximating $g(\boldsymbol{x})$ is as follows:

$$h_j = exp(-\frac{||x' - c_{ij}||^2}{b_j^2})$$
$$\hat{g}(\boldsymbol{x}) = \hat{\boldsymbol{W}}^T \boldsymbol{h} \tag{7}$$

in which $\hat{\boldsymbol{W}}$ is the approximate weight vector tuned by using the adaptive law and $\boldsymbol{h}$ is the output of the hidden layer.

3.2 Control Law Design

From the previous part, the unknown nonlinear function $g(\boldsymbol{x})$ has been represented by the RBF neural network. The control signal is:

$$u = \frac{1}{\eta_1}[-\hat{g}(\boldsymbol{x}) + \ddot{y}_d + \boldsymbol{M}^T \boldsymbol{E}] \tag{8}$$

where $\hat{g}(\boldsymbol{x})$ is the estimated parameter for $g(\boldsymbol{x})$.

3.3 Adaptive Law Design

In this subsection, the adaptive mechanism is chosen depending on Lyapunov stability theory to estimate $\hat{\boldsymbol{W}}$ for the RBF neural network's input. From (8) and (2), the overall system can be showed as:

$$\ddot{e} = -\boldsymbol{M}^T\boldsymbol{E} + [\hat{g}(\boldsymbol{x}) - g(\boldsymbol{x})]. \tag{9}$$

Let:

$$\boldsymbol{A} = \begin{bmatrix} 0 & 1 \\ -m_p & -m_d \end{bmatrix}, \boldsymbol{B} = \begin{bmatrix} 0 \\ 1 \end{bmatrix} \tag{10}$$

Now, (9) can be written into the vector form:

$$\dot{\boldsymbol{E}} = \boldsymbol{E} + \boldsymbol{B}[\hat{g}(\boldsymbol{x}) - g(\boldsymbol{x})]. \tag{11}$$

Define the optimal weight value as

$$\boldsymbol{W}^* = argmin[sup|\hat{g}(\boldsymbol{x}) - g(\boldsymbol{x})|]. \tag{12}$$

Define the error of the model as:

$$\delta = \hat{g}(\boldsymbol{x}|\,\boldsymbol{W}^*) - g(\boldsymbol{x}). \tag{13}$$

Then equation (11) becomes

$$\dot{\boldsymbol{E}} = \boldsymbol{A}\boldsymbol{E} + \boldsymbol{B}\{[\hat{g}(\boldsymbol{x}) - \hat{g}(\boldsymbol{x}|\,\boldsymbol{W}^*)] + \delta\} \tag{14}$$

substituting (7) into (14), we obtain:

$$\dot{\boldsymbol{E}} = \boldsymbol{A}\boldsymbol{E} + \boldsymbol{B}[(\hat{\boldsymbol{W}} - \boldsymbol{W}^*)^T\boldsymbol{h}(\boldsymbol{x}) + \delta] \tag{15}$$

The Lyapunov function to select is:

$$V = \frac{1}{2}\boldsymbol{E}^T\boldsymbol{P}\boldsymbol{E} + \frac{1}{2\mu}(\hat{\boldsymbol{W}} - \boldsymbol{W}^*)^T(\hat{\boldsymbol{W}} - \boldsymbol{W}^*) \tag{16}$$

with constant $\mu > 0$. $\hat{\boldsymbol{W}} - \boldsymbol{W}^*$ is the error of the estimation. The symmetric and matrix defined as positive $\boldsymbol{P}$ that complies with the following Lyapunov formula:

$$\boldsymbol{A}^T\boldsymbol{P} + \boldsymbol{P}\boldsymbol{A} = -\boldsymbol{Q} \tag{17}$$

where $\boldsymbol{Q}$ is a random choice 2×2 matrix defined as positive; $\boldsymbol{A}$ is given by (10).

Differentiating the Lyapunov function V in (16), we have:

$$\dot{V} = -\frac{1}{2}\boldsymbol{E}^T\boldsymbol{Q}\boldsymbol{E} + \boldsymbol{E}^T\boldsymbol{P}\boldsymbol{B}\delta + \frac{1}{\mu}(\hat{\boldsymbol{W}} - \boldsymbol{W}^*)^T[\dot{\hat{\boldsymbol{W}}} + \mu\boldsymbol{E}^T\boldsymbol{P}\boldsymbol{B}\boldsymbol{h}(\boldsymbol{x})] \tag{18}$$

Since $-\frac{1}{2}\boldsymbol{E}^T\boldsymbol{QE}$ is negative, choose the adaptation law as

$$\dot{\hat{\boldsymbol{W}}} = -\mu \boldsymbol{E}^T\boldsymbol{PBh(x)} \tag{19}$$

Then

$$\dot{V} = -\frac{1}{2}\boldsymbol{E}^T\boldsymbol{QE} + \boldsymbol{E}^T\boldsymbol{PB}\delta \tag{20}$$

Remark: If the neural network is designed effectively so that the modeling error $\delta \approx 0$, we get

$$\dot{V} = -\frac{1}{2}\boldsymbol{E}^T\boldsymbol{QE} < 0 \tag{21}$$

4 Experimental Results

Different scenarios are tested with different trajectories to see how well the suggested controller works. The control algorithm is implemented with sampling time is $T_s = 5$(ms). Following the experiment, the optimal set of parameters is given as follows. We choose the 2-5-1 structure for the neural network. Zero is assigned as the initial weight value., and the parameters of $\boldsymbol{c}_{ij}$ and b_i are designed as $[\boldsymbol{c}_{ij}] = \begin{bmatrix} 10 & 5 & 0 & -5 & -10 \\ 20 & 10 & 0 & -10 & -20 \end{bmatrix}$ and $b_i = 2$. For the control and adaptive law, we choose $\mathbf{Q} = \begin{bmatrix} 20 & 0 \\ 0 & 20 \end{bmatrix}$, $m_p = 40, m_d = 0.1$ and $\mu = 120$.

In this following part, sinusoidal and triangular signals with an amplitude of 40° and two frequencies of 0.2Hz and 0.5Hz are chosen as the reference trajectories. Figures 4 and 5 show the experiment results of two above types of signals. Tracking performance and tracking error are represented in the upper sub-figure and lower sub-figure, respectively.

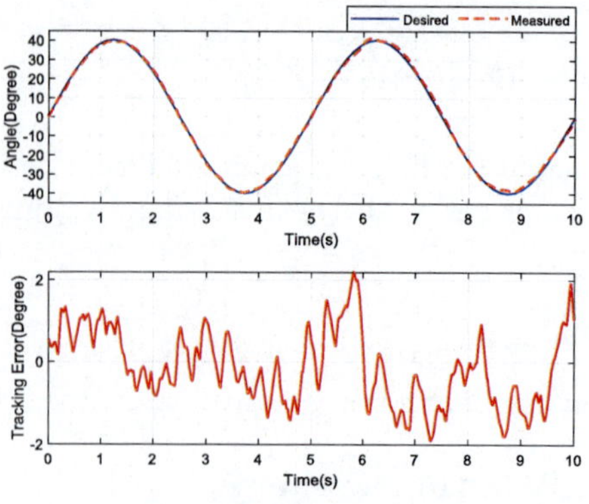

(a) 0.2 Hz

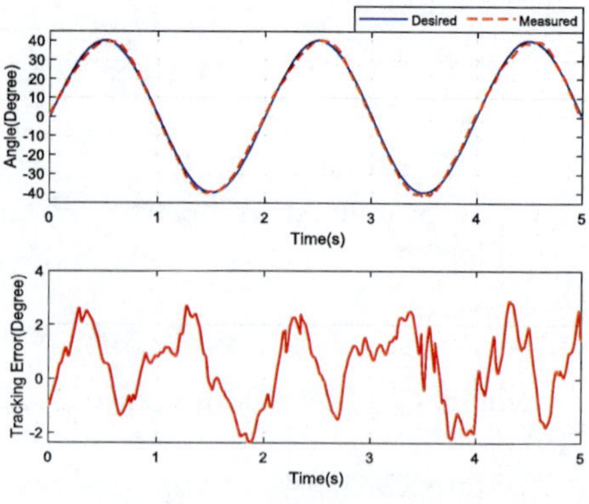

(b) 0.5Hz

Fig. 4. Experiment results when tracking sinusoidal trajectories.

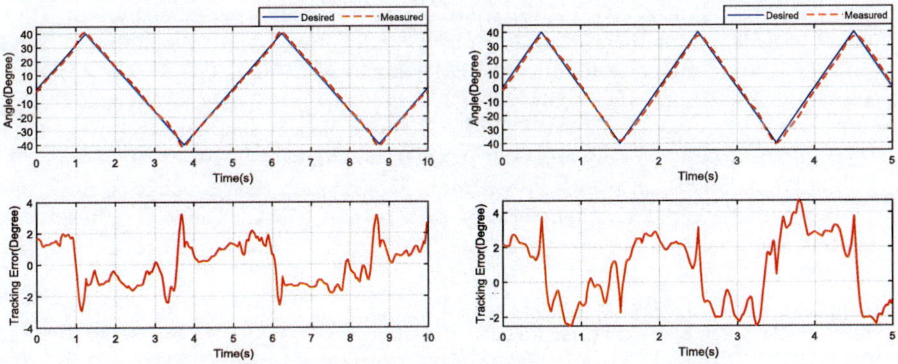

(a) 0.2 Hz (b) 0.5Hz

Fig. 5. Experiment results when tracking triangular trajectories.

It can be observed that in all experiment results, the adaptive RBF controller can track the desired signals. The control performance is slightly degraded when increasing the trajectory frequency, but still produces accurate results. In the case of tracking sinusoidal signals, the tracking errors only reach around $\pm 2.0^\circ$ with low frequency and less than 4.0° with higher frequency, which is a decent tracking accuracy. Next, in the case of triangle trajectories, the results show just slightly less precision. However, we have noticed that if we raise the frequency to more than 0.5Hz, it is very difficult to maintain the performance. This is mainly due to the hysteresis and large delay in characteristic of PAMs.

5 Conclusion and Discussion

This study offers an adaptive control method for an antagonistic configuration of dual PAMs based on RBF neural approximation. An RBF neural network is utilized in the design to approximate the control law. Besides, Lyapunov's stability theory has been applied to analyze the system's stability and the adaptive law. As a result, the proposed controller can produce control signals that can adapt to the diversity of the system's conditions and still achieve excellent tracking performance. For example, when tracking sinusoidal signals with 40° amplitude without a load, the tracking error always goes below 4.0° (10% of amplitude) regardless of frequency. In the future works of this paper, we will combine more advanced RBF algorithms with other control strategies to make use of the handy capabilities of neural networks and execute more practical experiments to serve the rehabilitation purpose.

References

1. Andrikopoulos, G., Nikolakopoulos, G., Manesis, S.: Advanced nonlinear PID-based antagonistic control for pneumatic muscle actuators. IEEE Trans. Ind. Electron. **61**(12), 6926–6937 (2014). https://doi.org/10.1109/TIE.2014.2316255

2. Anh, H.P.H., Ahn, K.K.: Hybrid control of a pneumatic artificial muscle (PAM) robot arm using an inverse NARX fuzzy model. Eng. Appl. Artif. Intell. **24**(4), 697–716 (2011). https://doi.org/10.1016/j.engappai.2010.11.007
3. Banala, S.K., Kim, S.H., Agrawal, S.K., Scholz, J.P.: Robot assisted gait training with active leg exoskeleton (ALEX). IEEE Trans. Neural Syst. Rehabil. Eng. **17**(1), 2–8 (2009). https://doi.org/10.1109/TNSRE.2008.2008280
4. Carvalho, A.D.D.R., Karanth P, N., Desai, V.: Characterization of pneumatic muscle actuators and their implementation on an elbow exoskeleton with a novel hinge design. Sens. Actuators Rep. **4**, 100109 (2022). https://doi.org/10.1016/j.snr.2022.100109
5. Chan, S., Lilly, J., Repperger, D., Berlin, J.: Fuzzy PD+I learning control for a pneumatic muscle. In: The 12th IEEE International Conference on Fuzzy Systems, 2003. FUZZ '03, vol. 1, pp. 278–283 (2003). https://doi.org/10.1109/FUZZ.2003.1209375
6. Dao, Q.T., Le Tri, T.K., Nguyen, V.A., Nguyen, M.L.: Discrete-time sliding mode control with power rate exponential reaching law of a pneumatic artificial muscle system. Control Theor. Technol. (2022). https://doi.org/10.1007/s11768-022-00117-8
7. Escobar, F., et al.: Simulation of control of a Scara robot actuated by pneumatic artificial muscles using RNAPM. J. Appl. Res. Technol. **12**(5), 939–946 (2014). https://doi.org/10.1016/S1665-6423(14)70600-5
8. Gupta, P., Sinha, N.K.: Intelligent control of robotic manipulators: experimental study using neural networks. Mechatronics **10**(1), 289–305 (2000). https://doi.org/10.1016/S0957-4158(99)00059-8
9. Kadota, K., Akai, M., Kawashima, K., Kagawa, T.: Development of Power-Assist Robot Arm using pneumatic rubbermuscles with a balloon sensor. In: RO-MAN 2009 - The 18th IEEE International Symposium on Robot and Human Interactive Communication, pp. 546–551 (2009). https://doi.org/10.1109/ROMAN.2009.5326335
10. Liu, J.: Radial Basis Function (RBF) Neural Network Control for Mechanical Systems. Springer, Berlin, Heidelberg (2013). https://doi.org/10.1007/978-3-642-34816-7
11. Naganna, G., Kumar, S.: Conventional and intelligent controllers for robotic manipulator. In: 2006 IEEE International Conference on Industrial Technology, pp. 424–428 (2006). https://doi.org/10.1109/ICIT.2006.372240
12. Passino, K.: Bridging the gap between conventional and intelligent control. IEEE Control Syst. Mag. **13**(3), 12–18 (1993). https://doi.org/10.1109/37.214940
13. Rezoug, A., Tondu, B., Hamerlain, M.: Experimental study of nonsingular terminal sliding mode controller for robot arm actuated by pneumatic artificial muscles. IFAC Proceedings Volumes **47**(3), 10113–10118 (2014). https://doi.org/10.3182/20140824-6-ZA-1003.00730, 19th IFAC World Congress
14. Yokoyama, K., Kogiso, K.: PID position control of McKibben pneumatic artificial muscle using only pressure feedback. In: 2018 Annual American Control Conference (ACC), pp. 3362–3367 (2018). https://doi.org/10.23919/ACC.2018.8431631
15. Zadeh, L.: Fuzzy sets. Inf. Control **8**(3), 338–353 (1965). https://doi.org/10.1016/S0019-9958(65)90241-X

A Novel Private Encryption Model in IoT Under Cloud Computing Domain

Suchitha Yadala[1], Chandra Shaker Reddy Pundru[2(✉)], and Vijender Kumar Solanki[3]

[1] Department of Computer Science and Engineering, VNR Vignana Jyothi Institute of Engineering and Technology, Hyderabad, TS, India

[2] School of Computer Science and Engineering, Lovely Professional University, Phagwara, Punjab, India
chandu.pundru@gmail.com

[3] Department of Computer Science and Engineering, CMR Institute of Technology, Hyderabad, India

Abstract. The presence of the Internet of Things (IoT) works with the assortment and diffusion of metropolitan information data. In any case, it can release clients' very own security data in brilliant urban areas. Many works have been done and different security systems identifying with the precautions of distributed computing have been implemented from copious points of view. Be that as it may, they don't propose a quantitative way to deal with dissecting and assessing protection and security in distributed computing frameworks. Accordingly, we propose another private data encryption strategy in IoT under distributed computing climate. Below IoT, as per the properties and securing time, protection data can be isolated into numerous subspaces. Because of the stream figure instrument, we plan an encryption framework model of data assortment. In the subspace, the security data is encoded and moved to the transfer hub. Subsequent to encoding, they are divided and rebuilt. At long last, we use stream figure and dual-key calculation to finish opportunity nondestructive change among plaintext and ciphertext to guarantee the trustworthiness of the encoded private data. Test outcomes illustrate that the presented strategy requires some investment in the encryption and unscrambling measure, which has superior ciphertext transformation yield impact and experiences fewer organization assaults in a similar encryption time. As far as computation cost, the proposed strategy diminishes by roughly 11%. Furthermore, it has greater security and works on the security and integrity of the protection data assortment procedure.

Keywords: Encryption · IoT · cloud computing · Data security

1 Introduction

Expanding amounts of information are being produced and put away from our regular routines, prompting an ascent in huge information examinations. Our examination centers around metropolitan information encryption advancements that are pointed toward utilizing the huge metropolitan informational indexes produced as individuals live, work,

T. D. L. Nguyen et al. (Eds.): ICISN 2023, LNNS 752, pp. 263–270, 2023.
https://doi.org/10.1007/978-981-99-4725-6_33

and move across a city [1]. It is as of now not the interconnection amid clients in the basic sense, however the association amid things. However, the improvement of the portable Internet is confronted with numerous issues, for example, in the confidential course of data assortment and diffusion security encryption, which it should have been tackled. With the fast turn of events and wide utilization of portable Internet innovation and correspondence innovation, interconnection causes the procurement and use of private data to turn out to be defter [2, 3]. Because of the versatile Internet, data assortment is handily attacked and assaulted by malignant information as far as dependability or countenances the danger of information being taken.

While network clients appreciate ideal and helpful customized administrations, they as a rule need to reveal individual protection data in return for relating administrations. In this manner, under the state of IoT, the assortment and transmission of data ought to be scrambled to guarantee the security of the data framework. Versatile Internet innovation and correspondence innovation have quickly evolved and broadly applied in private data encryption. The clients appreciate helpful and customized benefits however ordinarily need to unveil individual protection data as the cost for comparing administration simultaneously. Private information proprietors that give clients better nature of administration should comprehend the clients' touchy data, the security data, and the inconsistency between protection data exposure and the nature of administration [4, 5]. Protection data security has slowly turned into a typical worry of portable Internet clients and is likewise identified with the Internet's dire issues for chiefs. In the interim, the proprietor of protection information should comprehend clients' delicate data, which brings about the logical inconsistency between security data revelation and administration quality [6, 7].

Based on the security data encryption technique [8, 9] using obtaining time, this paper presents the division and revamping innovation to isolate the extended protection data into more modest data cuts and afterward coordinates them after the change. Moreover, versatile Internet hub validation is added to build the adequacy and unwavering quality of the wellspring of protection data [10]. This article is coordinated as pursues: Sect. 2 presents a Literature survey. Section 3 proposed model execution. Section 4 shows the investigational results and examination. There is an end in Sect. 5 .

2 Related Works

Most of the researchers had played out an esteemed conversation about the safety allied issues in Cloud processing frameworks presenting a subjective investigation and studies identified with the security challenges. They design security methodologies to create and send subjective security management structure on distributed computing frameworks. Mukhin, V et al. [11] designed a model and structure for protected distributed computing frameworks that recognize the security prerequisites, assaults, dangers and apprehensions related to the organization of the mists. They additionally recommended that cloud security isn't only a specialized issue, yet furthermore includes normalization, directing mode, laws and guidelines, and abundant diverse perspectives. Mahajan et al. [12] designed a Law-as-a-Service for programmed authorizing of legitimate arrangements to deal with inquiries for cloud specialist co-ops (CSPs) and their clients. The law-mindful super-peer goes about as a gatekeeper giving information coordination and insurance. R Sun et al. present a powerful multidimensional trust model in the premise of time-variation far reaching assessment multi-dimensional strategy [13]. In [14] the creators proposed conventional security management structure permitting suppliers of cloud information the board frameworks to characterize and authorize complex security approaches. They planned the structure to recognize and stop an enormous number of assaults characterized through an expressive arrangement portrayal language and to be effectively interfaced with different information management systems. In [15] the creators examined the security issues in a Cloud processing climate. They zeroed in on specialized security issues emerging from the utilization of cloud administrations. They talked about security dangers introduced in the cloud, for example, VM-Level assaults, segregation disappointment, the board interface compromise and consistency chances and their alleviation. In [16] the creators broke down weaknesses and security hazards explicit to distributed computing frameworks.

FElgendy et al. [17] presented a unique structure to divest heightened estimation endeavors from the wireless to the cloud. This skeleton used an upgraded design to choose the offloading decision intensely reliant upon crucial limits, explicitly, energy usage, CPU use, execution time, and memory use. Anyway, this work can't deal with endeavors equivalent. Karthikeyan et al. [18] introduced key trade methods dependent on got energy effectiveness in versatile distributed computing. Cloud suppliers should take solid safety efforts to secure the uprightness and protection of the medical services related information. Be that as it may, a large portion of the analysts don't at the same time focus on both uprightness and security for Health-CPS. In this manner, Xu et al. [19] proposed a protection saving information respectability check model by utilizing lightweight streaming verified information structures for Health-CPS. PCS Reddy et al. [20] introduced a multiuser asset allotment and calculation offloading model with information security to address the impediments of such gadgets. The calculation and radio assets were together considered for multiuser situations to ensure the productive usage of shared assets. Singhal A et al. [21] introduced a Secure Sensor-Cloud-Architecture for IoT applications to further develop network adaptability with effective information preparation and security. Above or existing works have been done and different security systems identifying with the precautions of distributed computing have been implemented from copious points of view. Be that as it may, they don't propose a quantitative way to deal with dissecting and assessing protection and security in distributed computing frameworks.

3 Proposed Methodology

Under cloud computing, data assortment, securing, and diffusion are more helpful. Encryption and security preparation in data assortment ordinarily incorporate three levels, to be specific, start to finish encryption, encryption among hubs, and connection encryption. The progressive system between information obtaining and transmission guarantees the security of data procurement generally. Connection encryption encodes the correspondence way between two correspondence hubs and gives security confirmation to information assortment and transmission. Before gathering and communicating join data, the hubs in sending and communicating should be encoded. This data encryption technique is generally utilized for the change of coordinated and offbeat lines. The encryption preparation of gadgets at the two closures of the connection ought to be completed simultaneously to work on the security of data obtaining. Notwithstanding, this encryption strategy requires the continuous transformation of encryption gadgets, expanding the danger of misfortune in information assortment. The proposed security model is addressed in Fig. 1.

With regards to distributed computing, encryption interaction of data assortment dependent on bedlam innovation ought to adhere to the essential working guidelines, to be specific, substitution rules and unscrambling rules. Each component in the plaintext data, like sound, archive, and picture, is planned and changed into a ciphertext component as per certain secret key standards. The ciphertext data has been adjusted and consolidated to supplant the plaintext data with ciphertext data. Under the state of distributed computing, it utilizes the tributary code to encode the arrangement dependent on the characters during the time spent on data encryption measure. In the change of plaintext and ciphertext, the bidirectional coupling picture cross sections and arbitrariness test strategies are utilized to check the produced succession esteems, to decide if the arrangement upsides of the tumultuous method match the stream figure. During the time spent changing ciphertext into plaintext, the first plaintext data can be recuperated by handling ciphertext signal as information esteem. The above overall process is displayed in Fig. 2.

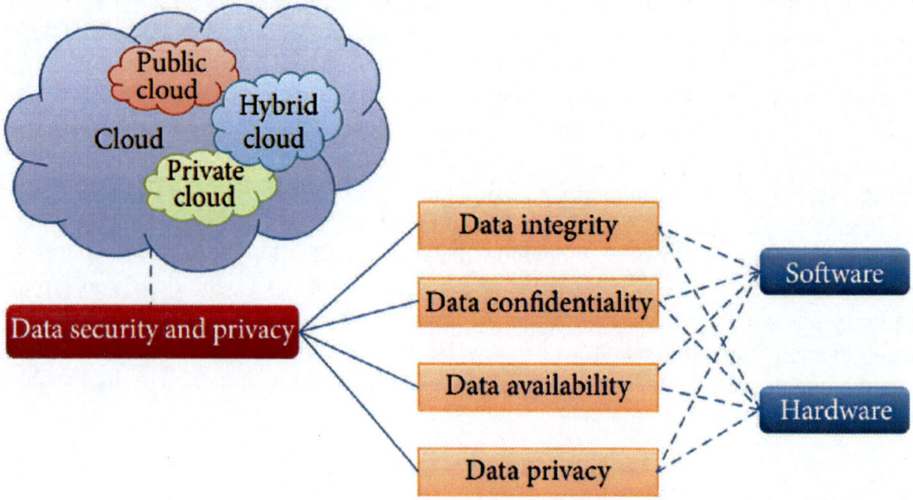

Fig. 1. Proposed Security Model

3.1 Integrity Optimization of Privacy Encryption

In view of the security data encryption of procurement time, we present division recombinant innovation, and the extended protection data is isolated into more modest cuts.

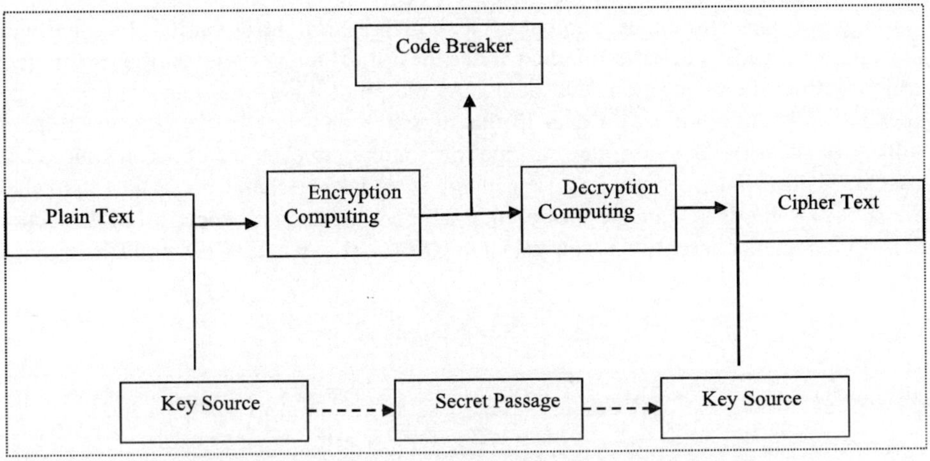

Fig. 2. Safe encryption information achievement method

The combination is performed after transformation. Versatile Internet hub validation is added to expand the legitimacy and dependability of security data sources. The particular interaction is depicted underneath.

Expect to be that η addresses the hubs of the versatile Internet. The organization hub arbitrarily chose Sη, J = |Sη |, utilizing h jump. One of the hubs η haphazardly isolates the recognized security data into κ patches. The organization hub η holds just one fix and sends the excess κ − 1 cuts of protection data to the remainder of the hubs of the portable Internet utilizing a scrambled technique. Assume that the Internet hub η sends dη 'κ' security data to the hub κ. The last total consequence of the protection data can be communicated as follows:

$$\mathrm{F} = \sum\nolimits_{k}^{N} d_{(n^1)} = 0 \tag{1}$$

$$r_{k^1} = 0 \tag{2}$$

$$\sum\nolimits_{k^1}^{N} r_{n^1} = \sum\nolimits_{k=1}^{N} \sum\nolimits_{n=1}^{N} d_{n^1 k^1} \tag{3}$$

Security assurance during the time spent protection data assortment is acknowledged by the security data recombination and total. The collected security data is at long last transferred to the sink hub, re-establishing the first protection data and accomplishing the motivation behind security assurance.

4 Results and Discussions

To confirm the exhaustive viability of the implemented strategy, we lead correlation tries different things with BKG, ACBE, and DSSE as far as security, trustworthiness, and computational intricacy. The security of protected data is frequently mirrored by the anti-aggressiveness in the assortment cycle, which is straightforwardly identified with the number of subkeys. Table 1 demonstrates the critical integer with various techniques which outline the encryption well being. At the point when the presented technique encodes and sends protection data in the subspace, it wants to demand the relating subkey of security data from the information source hub. The information source hub concludes if to approve as per its arrangement. It tends to be seen from Table1 that when the subkey value is 188 of the proposed technique, the secure execution of protection data procurement encryption is equivalent to BKG (701), ACBE (679), and DSSE (645).

Table 1. Key number contrast.

Iteration number	BKG	ACBE	DSSE	Implemented
8	641	652	561	121
16	667	623	620	145
24	702	679	652	178
32	1156	1657	956	241

Table 2 shows the real measure of scrambled security data information and extra overhead. PE% displays the extent of scrambled information. AO% presents the extra overhead. As per the examination of Table 2, when utilizing the technique in this study to scramble private data, the extra expense brought about by subregion division, data encryption communication time, and information source hub approval is about 60%, which implies that the additional expense is longer than the real season of protection data encryption activity. In any case, the BKG, ACBE, and DSSE techniques have more expenses, representing 85.47%, 84.79%, and 79.64%, individually. As far as the extent of real encoded private information, since the BKG strategy utilizes non-retransmission information bundles to refresh the key of the two players simultaneously and embraces the XOR calculation to finish symmetric encryption and decoding of security data, the genuine measure of scrambled protection information is enormous, coming to 38.72%.

Table 2. Calculation similarity.

Index	BKG	ACBE	DSSE	Implemented
PE %	39.63	35.63	32.46	14.73
AO %	83.34	83.69	78.32	58.31

At the point when the protection data is scrambled by the proposed technique, the security data is encoded in the subspace and sent to the transfer hub. At the point when the transfer hub needs to peruse the protection data, just the kid keys are created for certain information source hubs, and the aggregate sum of really encoded security information is just 14.72%. The trial results confirm that the implemented model in this study has little computational intricacy. BKG doesn't plan the insurance measure in the protection data assortment, which can prompt the deficiency of part of the encoded security data. Yet, in the new strategy, the data is scrambled by division and rebuilding, the extensive protection data is separated into more modest cuts. Through the incorporation of the reemergence after transformation, adequately ensures the protection of data uprightness during the time spent polymerization.

5 Conclusions

Since tumultuous calculation has no conspicuous balance and periodicity prerequisite, it is more compact in issue preparation, so it is truly reasonable for private data encryption. The disorder has rich progressive design qualities. This paper utilizes the benefits of mayhem innovation to work on the security of portable Internet clients' data assortment and yield. Then, at that point, we utilize stream figure and double key calculation to finish opportunity non-destructive change among plaintext and ciphertext to guarantee the trustworthiness of the encoded data. The test result demonstrates the predominance of the designed encryption innovation. Later on, more profound learning techniques will be applied to private data encryption. Furthermore, they will be functional to viable ventures.

References

1. Ma, H., Zhang, Z.: A new private information encryption method in internet of things under cloud computing environment. Wireless Commun. Mobile Comput. **2020**, 8810987 (2020)
2. Namasudra, S.: A secure cryptosystem using DNA cryptography and DNA steganography for the cloud-based IoT infrastructure. Comput. Electr. Eng. **104**, 108426 (2022)
3. Rathore, M.S., et al.: A novel trust-based security and privacy model for internet of vehicles using encryption and steganography. Comput. Electr. Eng. **102**, 108205 (2022)
4. Sujihelen, L., et al.: Node replication attack detection in distributed wireless sensor networks. Wireless Commun. Mobile Comput. **2022**, 7252791 (2022)
5. Ali, A., et al.: An industrial IoT-based blockchain-enabled secure searchable encryption approach for healthcare systems using neural network. Sensors **22**(2), 572 (2022)
6. Almaiah, M.A., Hajjej, F., Ali, A., Pasha, M.F., Almomani, O.: A novel hybrid trustworthy decentralized authentication and data preservation model for digital healthcare IoT based CPS. Sensors **22**(4), 1448 (2022)
7. Dhanalakshmi, R., et al.: Onboard pointing error detection and estimation of observation satellite data using extended Kalman filter. Comput. Intell. Neurosci. **2022**, 4340897 (2022)
8. Sharma, A., Singh, U.K.: Modelling of smart risk assessment approach for cloud computing environment using AI & supervised machine learning algorithms. Global Transitions Proc. **3**(1), 243–250 (2022)
9. Sabitha, R., Shukla, A.P., Mehbodniya, A., Shakkeera, L., Reddy, P.C.S.: A Fuzzy Trust Evaluation of Cloud Collaboration Outlier Detection in Wireless Sensor Networks. Adhoc Sens. Wireless Netw. **53**, 165–188 (2022)
10. Kumar, S., et al.: Novel method for safeguarding personal health record in cloud connection using deep learning models. Comput. Intell. Neurosci. **2022** (2022)

11. Mukhin, V., Volokyta, A.: Notice of violation of IEEE publication principles: security risk analysis for cloud computing systems. In: Proceedings of the 6th IEEE International Conference on Intelligent Data Acquisition and Advanced Computing Systems, vol. 2, pp. 737–742. IEEE (2011)
12. Mahajan, H.B., et al.: Integration of Healthcare 4.0 and blockchain into secure cloud-based electronic health records systems. Appl. Nanosci., 1–14 (2022) https://doi.org/10.1007/s13204-021-02164-0
13. Liu, L., Shafiq, M., Sonawane, V.R., Murthy, M.Y.B., Reddy, P.C.S., Kumar, K.C.: Spectrum trading and sharing in unmanned aerial vehicles based on distributed blockchain consortium system. Comput. Electr. Eng. **103**, 108255 (2022)
14. Jensen, M., Schwenk, J., Gruschka, N., Iacono, L.L.: On technical security issues in cloud computing. In: 2009 IEEE International Conference on Cloud Computing, pp. 109–116. IEEE (2009)
15. Ashok, K., Boddu, R., Syed, S.A., Sonawane, V.R., Dabhade, R.G., Reddy, P.C.S.: GAN base feedback analysis system for industrial IOT networks. Automatika, 1–9 (2022)
16. Kuldeep, G., Zhang, Q.: Multi-class privacy-preserving cloud computing based on compressive sensing for IoT. J. Inf. Secur. Appl. **66**, 103139 (2022)
17. Elgendy, I.A., Zhang, W.Z., Liu, C.Y., Hsu, C.H.: An efficient and secured framework for mobile cloud computing. IEEE Trans. Cloud Comput. **9**(1), 79–87 (2018)
18. Karthikeyan, B., Sasikala, T., Priya, S.B.: Key exchange techniques based on secured energy efficiency in mobile cloud computing. Appl. Math. Inf. Sci. **13**(6), 1039–1045 (2019)
19. Xu, J., Wei, L., Wu, W., Wang, A., Zhang, Y., Zhou, F.: Privacy-preserving data integrity verification by using lightweight streaming authenticated data structures for healthcare cyber–physical system. Futur. Gener. Comput. Syst. **108**, 1287–1296 (2020)
20. Reddy, P.C., Nachiyappan, S., Ramakrishna, V., Senthil, R., Sajid Anwer, M.D.: Hybrid model using scrum methodology for software development system. J. Nucl. Energy. Sci. Power Gener. Technol. **10**(9), 2 (2021)
21. Singhal, A., et al.: Minimization of latency using multitask scheduling in industrial autonomous systems. Wirel. Commun. Mob. Comput. **2022**, 1–10 (2022)

Development of an Autonomous Mobile Robot System for Hospital Logistics in Quarantine Zones

Tang Quoc Nam, Hoang Van Tien, Nguyen Anh Van, and Nguyen Dinh Quan(✉)

Le Quy Don Technical University, Hanoi, Vietnam
ndquan@lqdtu.edu.vn

Abstract. This paper presents the development of an autonomous mobile robot (AMR) system for hospital logistics in quarantine zones. The designed system includes a remote control station, five autonomous mobile robots Vibot-2, a wireless communication network and an observation system for reducing workload and avoiding the risk of infection for frontline health workers in quarantine zones. The overall architecture and hardware system are designed to meet all design requirements. The control software architecture of the robotic system is constructed in the robot operating system (ROS) framework. The navigation framework is developed based on the ROS Navigation Stack to move the robot safely and avoid the humans in its surrounding working area. In addition, the perception system is integrated a deep learning model for human detection and use the AprilTags to help the robots recognize and navigate to cart and charging stations. Finally, the low-cost solution based on the fusion of localization using AprilTags on the ceiling and the AMCL algorithm is developed for improving robustness and precision. The autonomous mobile robot system has been successfully built and deployed at COVID-19 field hospitals in Ha Nam, Bac Giang and Ho Chi Minh City in Vietnam in the year 2021.

Keywords: Autonomous Mobile Robots · Indoor Navigation · Indoor Localization

1 Introduction

Recently healthcare services have been growing steadily. It is necessary to make the workflow of the healthcare staffs better. In the past several decades, it can be seen that service mobile robots play significant role in increasing the productivity of logistic tasks in many hospitals such as delivery or collection [1]. Healthcare personnel can better serve the community by concentrating on other duties by streamlining such logistical chores.

Since the late 2019, the COVID-19 epidemic has spread to every continent. Healthcare personnel are overburdened by the epidemic and face a significant risk of infection. The use of robots during the pandemic has been advocated on a worldwide scale to enhance patient care and reduce the burden on the medical system. [2].

T. D. L. Nguyen et al. (Eds.): ICISN 2023, LNNS 752, pp. 271–281, 2023.
https://doi.org/10.1007/978-981-99-4725-6_34

There are various commercial systems that have been put in hospitals with great success and have demonstrated amazing performance [3–5]. HelpMate is a robot that can deliver lunch trays, laboratory supplies, pharmaceutical supplies, or patient data. [3]. In several hospitals and labs, TUG Smart Autonomous Mobile Robot are used to haul and carry commodities, materials, and medical supplies [4].

In this paper, the development of an AMR system, namely the Vibot-2, is introduced. The overall architecture and hardware system are designed to meet all technical requirements. Most of control modules are developed on the ROS framework [6]. We use ROS Navigation Stack as the base to implement our own navigation algorithm to navigate the robot safely, avoid static and dynamic obstacles (including human). In addition, a deep learning ANN-based model for detecting and tracking human is integrated in the robot control system. Finally, we develop a fusion scheme for localization using April-Tags [7–9] mounted on the ceiling together with the AMCL package [10] for improving robustness and precision in the robot pose estimation.

2 Problem Description and Design Requirements

Because of the COVID-19's impact on global public health, healthcare professionals are increasingly vulnerable to burnout owing to job stress. Robots can partially replace some activities in the healthcare industry, such as delivering medications and meals to patients and conducting remote video medical exams, protecting frontline healthcare workers from coronavirus exposure and reducing the need for medical staff.

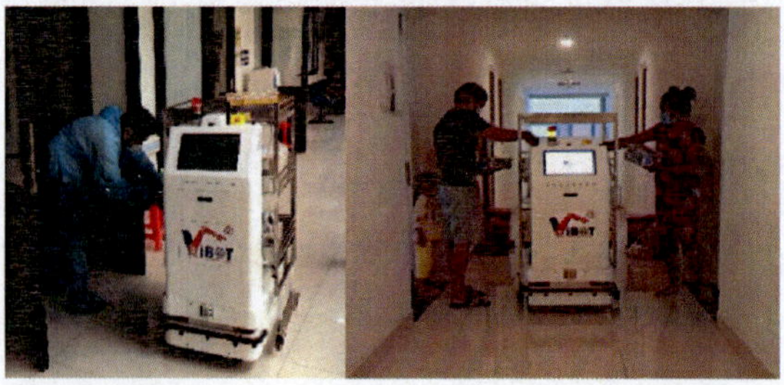

Fig. 1. Robot Vibot-2 delivered food for Covid-19 patients

In that context, the Vibot transportation medical robotic system (see Fig. 1) has been developed and manufactured to ensure some functions: *food delivery*, *medication delivery, garbage collection, remote examination* and *autonomous battery management*.

In order to meet the above functional requirements, the proposed robotic system should have met the following design requirements:

– Several robots need to work together on the same floor or on multiple floors, and can be controlled from a remote control station via a wireless local area network.

– Having an observation camera system allowing health workers to easily monitor the activities of the robots or manually control them.

– To perform the transportation tasks and remote examination, the robots should have the following features:

+ Autonomous capabilities: mapping, localization, navigation, obstacle avoidance. (including people in the hospital environment following social rules).

+ Perception: recognizing the working environment including patient rooms, charging station, cart types, people.

+ Power management and autonomously recharging: monitoring the battery level and automatically returning to charging when the battery is low.

3 System Design

3.1 Hardware System Architecture

In order to meet the above design requirements, the hardware system architecture of the Vibot system is designed as shown in the Fig. 2. Medical staff can control and monitor the robots from the remote control station and from smart mobile devices.

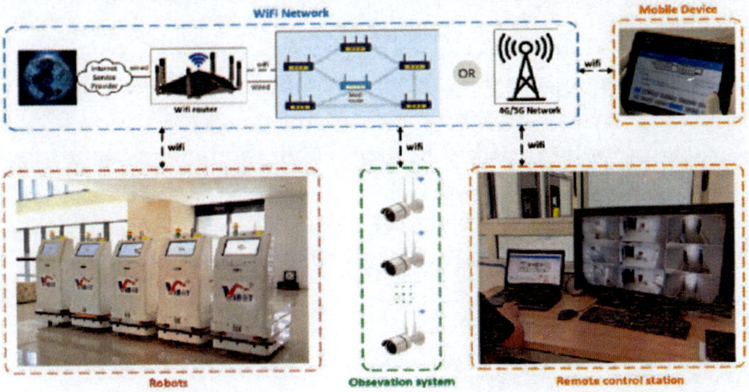

Fig. 2. Overview of the medical robot transport system Vibot-2

The observation system includes the wireless IP cameras mounted on the ceiling of the floors. The WiFi network provides a wireless connection between the remote control station with the robots and IP cameras.

Figure 3 shows the hardware structure of the Vibot-2 AMR. The robot has two 120W brushless DC servo motors driving two fixed wheels. There are four supported caster wheels at the four corners of the base frame. This combination allows the AMR to carry large loads (up to 120 kg) and move with max speed of 0.78 m/s. One lifting table is attached to the base to pick up the carts at the cart stations (see Fig. 3a).

For navigation and obstacle detection, we use two SICK TIM571 laser range finders (placed at two opposite corners of the robot base) that enable a fully 360^0 scanning range around the base. Ultrasonic sound sensors and several IR sensors together with two bumper sensors are used to increase the safety factor for the robot.

The Vibot-2 AMR has four cameras in total. The front camera (mounted below the display) is used to identify the carts. One rear camera is used to aid the robot in approaching cart stations or charging station. One camera is mounted on top of the

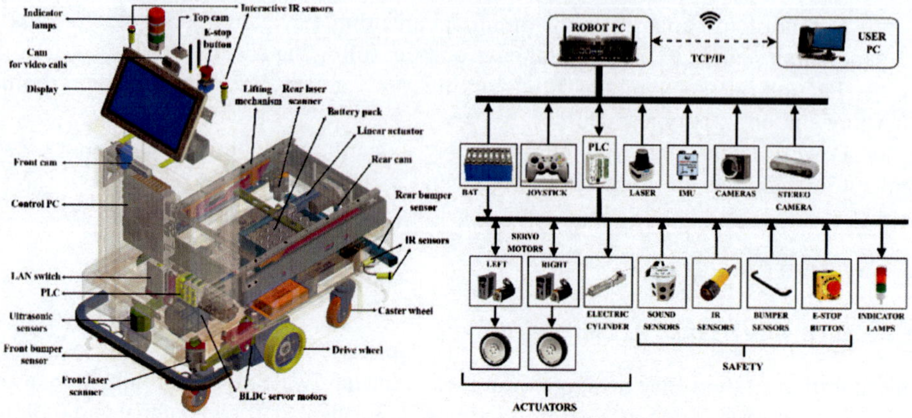

(a) Vibot-2 mechanical design (b) Hardware components

Fig. 3. Hardware architecture of Vibot-2 AMR

control cabin that is facing the ceiling to read some AprilTags for the re-localization process. A camera is used for remote video calls.

One embedded computer is used to manage all the main system functions such as navigation, localization, delivery and remotely examination tasks. All actuators in the robotic system are controlled directly by a PLC. The PLC gets specific commands being sent from the control PC and sends back information on robot states including: motor encoders, sensor states.

3.2 Software System Architecture

The software system of the Vibot-2 robots was designed according to the general architecture diagram (see Fig. 4), which consists of four layers: *organization, coordination, navigation* and *device manager*.

The first layer receives commands from the control station, organizing the tasks and sending to the coordination layer. At the second layer, the tasks will be divided into three separated control tasks: delivery tasks control, docking control and remote examination control. Then these tasks are planned and executed through the layers below.

For each control task at the coordination layer, the navigation layer conducts navigation control with specific functions: map building, localization, path planning, trajectory control, and obstacle avoidance. Finally, the third layer sends control commands to the device manager layer and receives a feedback information about the environment and robot states.

The last layer is responsible for processing different data sent from wheel encoders, range sensors, cameras, safety sensor and patient signal sensors. In addition, the device manager layer also directly performs motion control via the motors and indicates the robot states to users via the warning lamps.

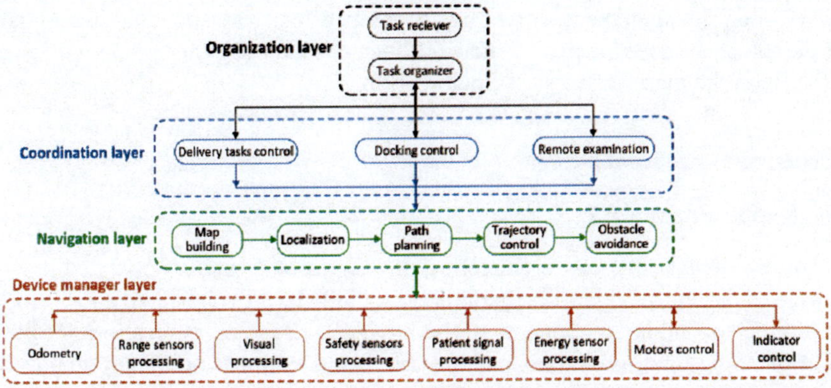

Fig. 4. Software architecture of Vibot-2

3.3 Communication System

The communication system of Vibot-2 (see Fig. 5) has two main tasks: managing video call in remotely examinations, and managing data transmission between robot control computer and the user remote control devices.

For video call in remotely examinations, we use the Jitsi platform [11] as the base framework to integrate video call function into our web-based GUI control. The connection can be made between a Vibot-2 robot and any PC or smart portable devices that runs internet, thus enables a flexible solution.

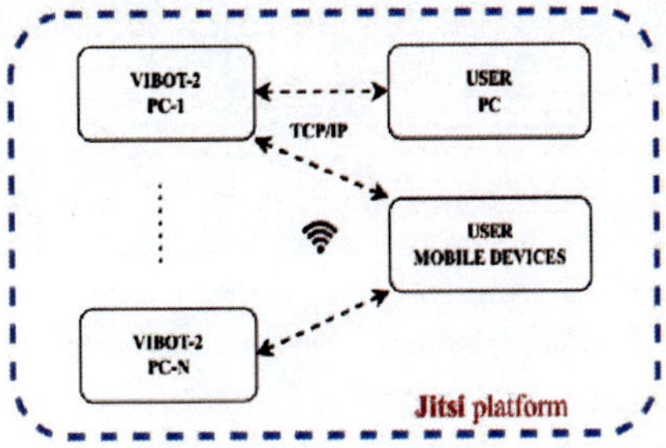

Fig. 5. Communication of Vibot-2 systems

For transmitting the control commands (from the remote user device to the robot control PC) and feedback of robot states (from the robot control PC to server or remote user device), we use a separate channel to transceive data packages. The real-time condition is strict for this data transmission function. The control commands are encoded with information on the specific tasks that one robot is assigned to do. The feedback data contains important information on robot states such as: battery level, current robot pose, connection status, task processing status, obstacle detection and emergency state.

The robotic management software is developed in a way such that it can manange multiple robots at the same time. A new robot can be added to the system with a new ID together with the map ID of its assigned working area.

3.4 Navigation System Design

a) ***Navigation Framework***

Figure 6 depicts the new architecture of our Vibot-2 robots' extanded navigation system. A traditional navigation scheme and a socially aware robot navigation framework make up the two main components of the navigation system. Perception, localization, motion planning, and motion control are the four functional building components on which the traditional navigational system is normally built. By extracting the socio-spatio-temporal features of people nearby the robots and developing socially aware motion planning, the socially aware navigation framework in the second section tries to distinguish humans from common obstacles. Specifically, Human detection and tracking block is used to detect and track people in the robot's field of view. Human group detection is used to identify groups of people and people interacting with objects. To move the robot smoothly and amiably while preserving the safety and comfort of the human in the robot's proximity, the "Social timed elastic band (STEB)" model [13] was developed using information about a person or a group of people, such as position, orientation, motion, and hand gestures.

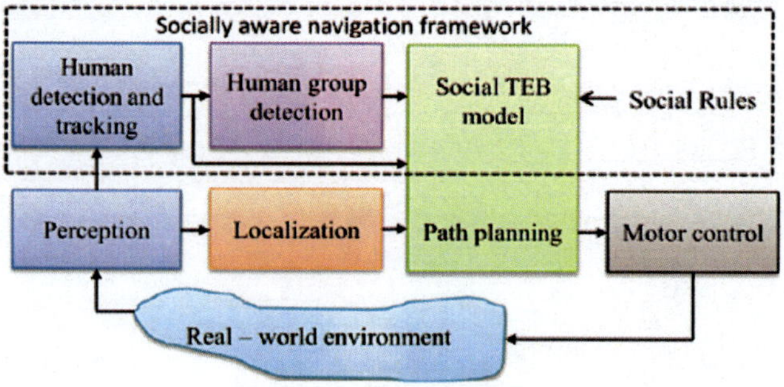

Fig. 6. The social-aware navigation system for robot Vibot-2

b) ***Perception System***

Acquiring information about its surroundings is one of the most crucial jobs for any sort of autonomous system. To estimate human poses, the Vibot-2 uses the front depth camera Intel Realsense D435i, while the rear monocular camera is used for recognizing and estimating the positions of the cart and charging stations with respect to the robot frame.

For human pose estimation, we use the MobileNet-SSD model [14] to detect people in each video frame, and the received results are the bounding boxes containing people.

The coordinates of the humans with respect to the robot frame are approximated by fusing the coordinates of those boxes in the depth camera's color picture with the point cloud it created. Those coordinates are then used as the inputs for the proposed socially aware navigation frameworks.

For cart station recognition, we use AprilTags to help the robots recognize and navigate to cart and charging stations.

c) ***Localization System***

The localization process of proposed method is depicted in Fig. 8. It can be divided into two stages: mapping and localization. Mapping the environment is performed based on the scan data from laser rangers and the GMapping ROS package [15], while localization implements the AMCL algorithm with four required inputs such as static grid map, initial pose, motion and measurement models.

For the Vibot-2 localization, the low-cost solution based on fusion of localization using AprilTags on the ceiling and the AMCL algorithm is proposed for fast global localization with the high accuracy, effectively solving kidnapped robot problem and reducing accumulating errors in the position tracking process (Fig. 7).

Fig. 7. The AprilTags is attached on the ceiling and the AprilTag bundle label the cart stations at the Bac Giang General Hospital in Bac Giang province, Vietnam

While the robot is moving, the top camera is used to detect AprilTags and the found tag IDs is combined with their predetermined global pose stored in the AprilTag database for calculating a robot pose in the world frame [16]. Then that pose is used as an initial pose for AMCL algorithm.

d) ***Path Planning***

For Vibot-2 AMR, the path planning system is developed based on the ROS Navigation Stack [17] with the setup diagram as shown in Fig. 9. Specifically, The STEB model is used to generate a viable path in the local map and a motion control order to move the robot safely and socially away from the hu-mans in its proximity. The A* algorithm [18] is used to calculate the shortest path in the global map.

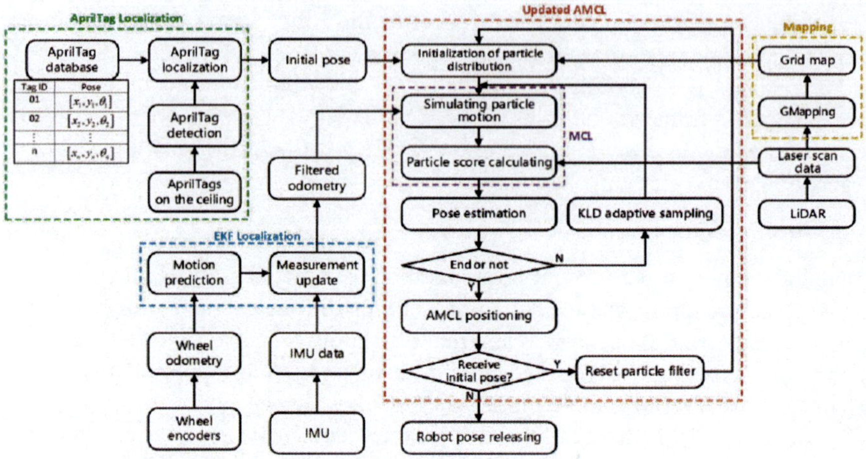

Fig. 8. The localization process of proposed method for Vibot-2

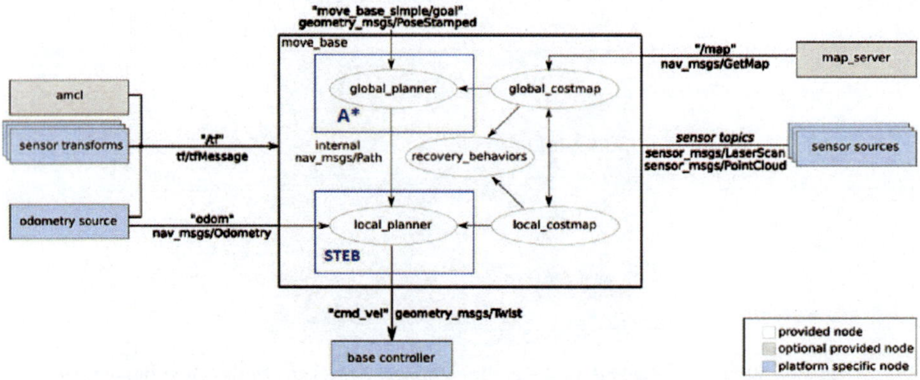

Fig. 9. Incorporating A* and STEB algorithm into the path planning system for Vibot-2

e) ***Motor Control***

There are two main motor control problems for the Vibot-2 AMR. The first one is controlling the motor speeds to meet the robot desired linear and angular velocities in navigation tasks. The second problem is control the motor speeds in docking tasks (to get or return carts at cart stations or to return to the charging station).

Figure 10 shows the motor control system for navigation tasks. To ensure smooth motion for the robot, one saturation block is added to the control scheme to manage the robot linear and angular velocities.

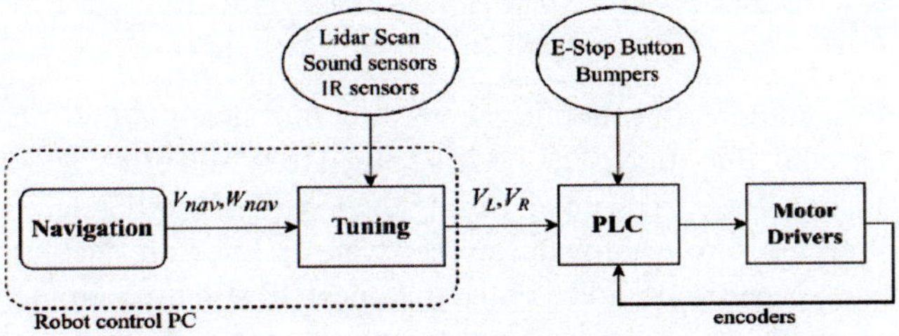

Fig. 10. Motor control scheme of the Vibot-2 AMR

4 Field Deployment and Evaluation

In 2021, when the covid-19 epidemic broke out in Vietnam, the number of infections increases rapidly. The Vibot-2 robot system had been fully deployed at the Bac Giang General Hospital in the North of Vietnam in June and July of 2021. The robot system consists of two robots working on the identical third and fourth floors of the Department of Infectious Diseases's building (see Fig. 11).

During deployed period, the robots had completely replaced healthcare workers in delivering food, medicine and supplies and helped to reduce workload and the risk of infection. The robot system worked in accordance with the design requirements and demonstrated its effectiveness in helping medical staff. The following link leads to videos showing the robot system in operation: https://www.youtube.com/@RRDGroup.

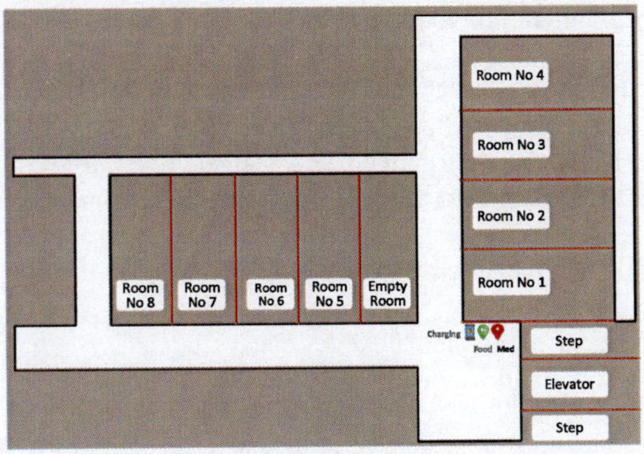

Fig. 11. The map of the third and fourth floors of the Bac Giang General Hospital, Vietnam

The implementation shows that the robotic system still has some limitations that need to be improved such as: the robots could not automatically move between floors using the elevator, and in heavy rain conditions leading to the wet floor, the robot's wheels slip, leading to incorrect positioning during operation.

5 Conclusion

The creation of an autonomous mobile robot system for hospital logistics in quarantine areas was discussed in this study. The AMRs helped to lessen effort and lower infection risk for front-line healthcare personnel during the Covid-19 epidemic.

Completed solutions from hardware and software architecture design to control modules have been developed and implemented that meet all strict requirements. The robots can deliver food, medicine and accessories to patient rooms in the quarantine areas.

The benefits of the robots in lowering effort and infection risk for medical workers have been demonstrated by experiments conducted in field hospitals during the Covid-19 epidemic. However, there are several issues that need to be improved in future works. Due to the limitation of the infrastructure of hospital building, the robots cannot move automatically between floors because of the inaccessibility to the elevator control. Besides, there are cases the robot still lost track of its location, especially under slippage condition.

References

1. Fragapane, G., Hvolby, H.-H., Sgarbossa, F., Strandhagen, J.O.: Autonomous mobile robots in hospital logistics. In: Lalic, B., Majstorovic, V., Marjanovic, U., von Cieminski, G., Romero, D. (eds.) APMS 2020. IAICT, vol. 591, pp. 672–679. Springer, Cham (2020). https://doi.org/10.1007/978-3-030-57993-7_76
2. Wang, X.V., Wang, L.: A literature survey of the robotic technologies during the COVID-19 pandemic. J. Manuf. Syst. **60**, 823–836 (2021)
3. Evans, J.M.: Help Mate: an autonomous mobile robot courier for hospitals. In: Proceedings of the International Conference on Intelligent Robots and Systems (IROS 1994), pp. 1695–1700 (1994)
4. Aethon Automates Intralogistics, TUG Smart Autonomous Mobile Robot. http://www.aethon.com/tug
5. Niechwiadowicz, K., Khan, Z.: Robot based logistics system for hospitals – survey. In: Proceedings of the IDT Workshop on Interesting Results in Computer Science and Engineering (2008)
6. Stanford Artificial Intelligence Laboratory et al., Robotic Operating System. https://www.ros.org
7. Olson, E.: AprilTag: a robust and flexible visual fiducial system. In: IEEE International Conference on Robotics and Automation (ICRA 2011), pp. 3400–3407 (2011)
8. Wang, J., Olson, E.: AprilTag 2: efficient and robust fiducial detection. In: International Conference on Intelligent Robots and Systems, pp. 4193–4198 (2016)
9. Kallwies, J., Forkel, B., Wuensche, H.-J.: Determining and Improving the Localization Accuracy of AprilTag Detection. In: IEEE International Conference on Robotics and Automation (ICRA 2020), pp. 8288–8294 (2020)
10. Fox, D.: Adapting the sample size in particle filters through KLD-sampling. Int. J. Robot. Res. **22**(12), 985–1003 (2003). https://doi.org/10.1177/0278364903022012001
11. Jitsi, Free video conferencing software for web and mobile. https://jitsi.org
12. Siegwart, R., Nourbakhsh, I.R., Scaramuzza, D.: Introduction to Autonomous Mobile Robots. The MIT Press, Cambridge (2011)
13. Hoang, V.B., Nguyen, V.H., Nguyen, L.A., Quang, T.D., Truong, X.T.: Social constraints-based socially aware navigation framework for mobile service robots. In: Seventh NAFOSTED Conference on Information and Computer Science (NICS 2020), pp. 84–89 (2020)

14. Chiu, Y.C., Tsai, C.Y., Ruan, M.D., Shen, G.Y., Lee, T.T.: Mobilenet-SSDv2: an improved object detection model for embedded systems. In: International Conference on System Science and Engineering (ICSSE 2020), pp. 1–5 (2020)
15. Grisetti, G., Stachniss, C., Burgard, W.: Improved techniques for grid mapping with Rao-Blackwellized particle filters. IEEE Trans. Rob. **23**(1), 34–46 (2007)
16. Hoang, V.T., Tang, Q.N., Truong, X.T., Nguyen, D.Q.: An indoor localization method for mobile robot using ceiling mounted ApriTags. J. Sci. Tech. **17**(05), 70–91 (2022)
17. Marder-Eppstein, E., Berger, E., Foote, T., Gerkey, B., Konolige, K.: The Office Marathon: robust navigation in an indoor office environment. In: IEEE International Conference on Robotics and Automation (ICRA 2010), pp. 300–307 (2010)
18. Duchon, F., Babinec, A., et al.: Path planning with modified a star algorithm for a mobile robot. Procedia Eng. **96**, 59–69 (2014)

Position Control for Series Elastic Actuator Robot Using Sliding Mode Control

Minh-Duc Duong(✉), Duc-Long Nguyen, and Van-Hung Nguyen

HaNoi University of Science and Technology, Hanoi, Vietnam
duc.duongminh@hust.edu.vn

Abstract. Series Elastic Actuator (SEA) has attracted much attentions in recent years because of its high-performance torque control. However, the flexibility of SEA makes it difficult to position control. In this paper, we propose sliding mode-based position control for a single-joint SEA robot system. Three reaching laws including constant rate, exponential, and power rate reaching laws are considered. The simulations show the effectiveness of the considered control algorithms and a comparison has been done among them.

Keywords: Sliding Mode Control · Series Elastic Actuator · Position Control · Flexible Structure

1 Introduction

A SEA can be described as a device that adds elastic elements between the actuator and the load [1]. By introducing an elastic element, the robot is passively flexible and can buffer external torques, ensuring its safety when in contact with humans and the environment. Because of its outstanding features, SEA has been used in many human-related applications such as human-friendly robots, rehabilitation, human-robot interaction, etc. [2–4].

In order to apply SEA actuators to a practical system, two control problems need to be solved. That is force control and position control. For force/impedance control of SEA actuators, various control techniques have been applied successfully such as force observer [5], robust control [6], sliding mode control [7], passive control [8], decoupling control [9], adaptive control [10].

For position control, the flexibility of SEA causes the vibration of the SEA robot system thus reducing the position precision. Thus, it is required to control the motor position precisely while suppressing the vibration as fast as possible. To overcome this vibration problem, several control approaches have been applied such as robust control [11], sliding mode control [12,13], and backstepping control [14].

In this paper, we develop a hierarchical sliding mode control for a SEA robot system with three fundamental reaching laws including constant rate, exponential, and power rate reaching laws. The simulation of three control laws is done and the comparative evaluation among them is also discussed in the paper.

T. D. L. Nguyen et al. (Eds.): ICISN 2023, LNNS 752, pp. 282–287, 2023.
https://doi.org/10.1007/978-981-99-4725-6_35

2 SEA Robot Model

For simplicity, let us consider a single-link SEA robot as shown in Fig. 1. In general, the mathematical model of the SEA robot can be described as follows:

$$D(q)\ddot{q} + C(q, \dot{q}) + G(q) = K(\theta - q) \tag{1}$$

$$J\ddot{\theta} + K(\theta - q) = u \tag{2}$$

For a single-link SEA robot, $D = ml_c^2, G = gl_c, C = 0$.

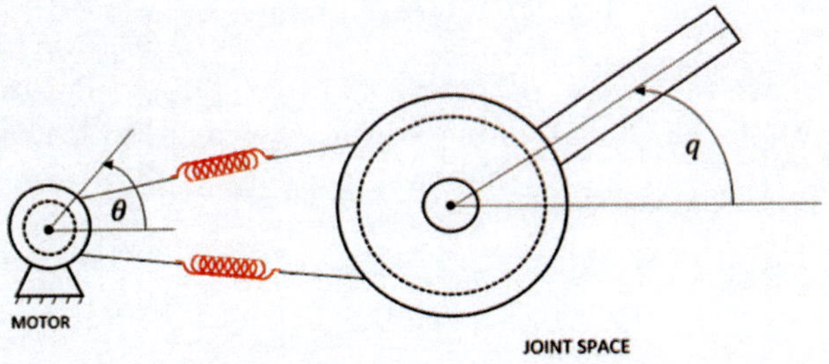

Fig. 1. A Single-Link Flexible Joint Robot SEA

Set $x_1 = q(t)$, $x_2 = \dot{q}(t)$, $x_3 = \theta(t)$, $x_4 = \dot{\theta}(t)$, then the state-space form of the SEA robot model is :

$$\begin{cases} \dot{x_1} = x_2 \\ \dot{x_2} = -\frac{K}{M}x_1 - \frac{G}{M}cos(x_1) + \frac{K}{M}x_3 \\ \dot{x_3} = x_4 \\ \dot{x_4} = \frac{K}{J}x_1 - \frac{K}{J}x_3 + \frac{1}{J}u(t) \end{cases} \tag{3}$$

3 Sliding Mode Controller Design

We define

$$\begin{cases} e_1(t) = q_d(t) - x_1 \\ e_2(t) = \theta_d(t) - x_3 \end{cases} \tag{4}$$

where $e_1(t)$, $e_2(t)$ are position error of robot joint and motor, respectively; and $q_d(t)$, $\theta_d(t)$ are corresponding desired position values.

Substituting (3) into (2):

$$\begin{cases} \dot{e_1}(t) = \dot{q_d}(t) - x_2 \\ \ddot{e_1}(t) = \ddot{q_d}(t) - (-\frac{K}{M}x_1 - \frac{G}{M}cos(x_1) + \frac{K}{M}x_3) \\ \dot{e_2}(t) = \dot{\theta}_d(t) - x_4 \\ \ddot{e_2}(t) = \ddot{\theta}_d(t) - (\frac{u(t)}{J} - \frac{k(x_3-x_1)}{J}) \end{cases} \tag{5}$$

Select the sliding surface as follow:

$$\begin{cases} s_1(t) = c_1e_1(t) + \dot{e_1}(t) \\ s_2(t) = c_2e_2(t) + \dot{e_2}(t) \end{cases} \tag{6}$$

where c_1 and c_2 are the design parameters of two sliding sides to ensure system stability. We propose a global sliding surface by a linear combination of two sliding surfaces $s_1(t)$ and $s_2(t)$ as follows:

$$\begin{aligned} S(t) &= \lambda_1 s_1(t) + \lambda_2 s_2(t) \\ &= \lambda_1(c_1e_1(t) + \dot{e_1}(t)) + \lambda_2(c_2e_2(t) + \dot{e_2}(t)) \end{aligned} \tag{7}$$

where λ_1, λ_2 are the constant coefficients of the sliding surface. Derivative of (7), one can obtain:

$$\begin{aligned} \dot{S}(t) &= \lambda_1(c_1\dot{e_1}(t) + \ddot{e_1}(t)) + \lambda_2(c_2\dot{e_2}(t) + \ddot{e_2}(t)) \\ &= \lambda_1[c_1(\dot{q_d}(t) - x_2) + \ddot{q_d}(t) - (-\frac{K}{M}x_1 - \frac{G}{M}cos(x_1) \\ &+ \frac{K}{M}x_3)] + \lambda_2[c_2(\dot{\theta}_d(t) - x_4) + \ddot{\theta}_d(t) \\ &- (\frac{u(t)}{J} - \frac{k(x_3 - x_1)}{J})] \end{aligned} \tag{8}$$

3.1 Constant Rate Reaching Law Sliding Mode Control

To establish this control law, the sliding mode is chosen as follows:

$$\dot{S} = -\epsilon.sat(S), \quad \epsilon > 0 \tag{9}$$

Then, the control signal $u(t)$ is calculated as:

$$\begin{aligned} u(t) = J.\frac{\lambda_1}{\lambda_2}.&[c_1(\dot{q_d}(t) - x_2) + \ddot{q_d}(t) + \frac{K}{M}x_1 + \frac{G}{M}cos(x_1) \\ &- \frac{K}{M}x_3)] + J[c_2(\dot{\theta}_d(t) - x_4) + \ddot{\theta}_d(t)] + K(x_3 - x_1) \\ &+ \epsilon.sat(S) \end{aligned} \tag{10}$$

3.2 Exponential Reaching Law Sliding Mode Control

To establish this control law, the sliding mode is chosen as follows:

$$\dot{S} = -\epsilon.sat(S) - kS, \quad \epsilon > 0, k > 0 \tag{11}$$

The control signal $u(t)$ then can be calculated as

$$\begin{aligned} u(t) = {} & J.\frac{\lambda_1}{\lambda_2}.[c_1(\dot{q}_d(t) - x_2) + \ddot{q}_d(t) + \frac{K}{M}x_1 + \frac{G}{M}cos(x_1) \\ & - \frac{K}{M}x_3)] + J[c_2(\dot{\theta}_d(t) - x_4) + \ddot{\theta}_d(t)] + K(x_3 - x_1) \\ & + \epsilon.sgn(S) + kS \end{aligned} \tag{12}$$

3.3 Power Rate Reaching Law Sliding Mode Control

To establish this control law, the sliding mode is chosen as follows:

$$\dot{S} = -k|S|^{\alpha}.sat(S), \quad k > 0, 1 > \alpha > 0 \tag{13}$$

The control signal $u(t)$ then can be calculated as

$$\begin{aligned} u(t) = {} & J.\frac{\lambda_1}{\lambda_2}.[c_1(\dot{q}_d(t) - x_2) + \ddot{q}_d(t) + \frac{K}{M}x_1 + \frac{G}{M}cos(x_1) \\ & - \frac{K}{M}x_3)] + J[c_2(\dot{\theta}_d(t) - x_4) + \ddot{\theta}_d(t)] + K(x_3 - x_1) \\ & + k|S|^{\alpha}.sat(S) \end{aligned} \tag{14}$$

It is noted that to reduce chattering, the saturation function $sat(s)$ instead of the sign function is used.

$$sat(s) = \begin{cases} 1, & s > \Delta \\ ks, & |s| \leq \Delta, \quad k = \dfrac{1}{\Delta} \\ -1, & s < -\Delta \end{cases}$$

4 Simulations

The parameters of the robot SEA model for simulation are as follows: $K = 25[N/m]; l_c = 0.3 \pm 0.1[m]; m = 1.53 \pm 0.1[kg]; J = 0.0218[kg.m^2]; g = 9.8[m/s^2]; d(t) = 10sin(t)$.

The simulation results for sliding mode controllers with three reaching law are shown in Fig. 2. For constant rate reaching law, it meets the requirement that the orbital approach control be placed for a finite period of time. However, the control input signal quality is not good and lasts a long time before switching to zero. This causes output position response with large deviations.

For exponential reaching law, $-kS$ component is added to the control input u resulting in exponentially faster arithmetic progression. This results is better

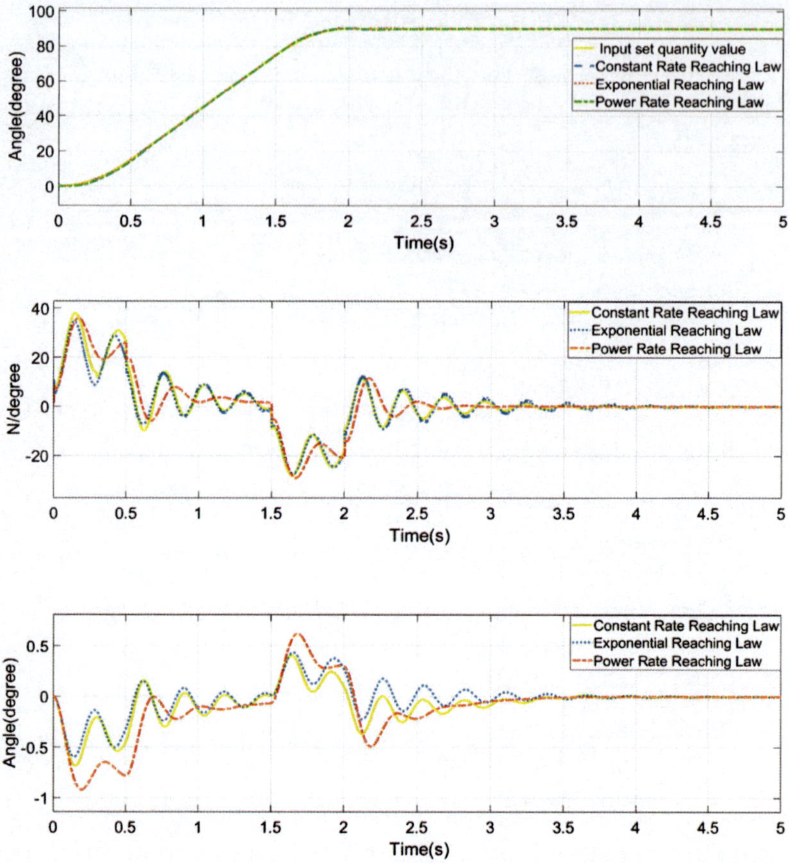

Fig. 2. Position response, control input and tracking error with three sliding mode controllers

than in the case the absence of the $-kS$ component. The derivation of the control input signal significantly reduces, and the output responses over a finite period of time compared to the constant rate reaching law.

For power rate reaching law, the output signal will move towards the orbit exponentially, which obviously creates a shorter response effect, however, due to moving towards the orbit too quickly, it will lead to large movements in a short period of time initially, which will then gradually become more stable and reduce noise significantly compared to the two above control types.

5 Conclusion

In this paper, sliding mode controllers with three reaching laws for SEA robots are evaluated. All of them can guarantee system stability and good tracking. In addition, the comparison among three sliding mode controllers is also done. The more complicated the controller is, the better the performance is. It is the trade-off between simplicity and good performance. In the near future, the comparison of other controllers will be done in order to find the best position controller for SEA robot.

References

1. Williamson, M.M.: Series elastic actuators Master Thesis, MIT (1995)
2. Laffranchi, M., et al.: Development and control of a series elastic actuator equipped with a semi active friction damper for human friendly robots. Robot. Auton. Syst. **62**, 1827–1836 (2014)
3. Yu, H., Huang, S., Chen, G., Pan, Y., Guo, Z.: Human-robot interaction control of rehabilitation robots with series elastic actuators. IEEE Trans. Ind. Electron. **31**, 1089–1100 (2015)
4. Li, X., Pan, Y., Chen, G., Yu, H.: Adaptive Human-Robot Interaction Control for Robots Driven by Series Elastic Actuators. IEEE Trans. Ind. Electron. **33**, 169–182 (2017)
5. Park, Y., Paine, N., Oh, S.: Development of force observer in series elastic actuator for dynamic control. IEEE Trans. Ind. Electron. **65**(3), 2398–2407 (2017)
6. E. Sariyildiz and H. Yu, *A robust force controller design for series elastic actuators,* IROS 2017, pp. 2206-2212
7. Sariyildiz, E., Yu, H., Nozaki, T., Murakami, T.: Robust force control of series elastic actuators using sliding mode control and disturbance observer. In: IECON 2016, pp. 619-624 (2016)
8. Kenanoglu, C.U., Patoglu, V.: Passivity of series elastic actuation under model reference force control during null impedance rendering. IEEE Trans. Haptics **15**(1), 51–56 (2022)
9. Lin, Y., Chen, Z., Yao, B.: Decoupled torque control of series elastic actuator with adaptive robust compensation of time-varying load-side dynamics. IEEE Trans. Ind. Electron. **67**(7), 5604–5614 (2019)
10. Losey, D.P., Erwin, A., McDonald, C.G., Sergi, F., O'Malley, M.K.: A time-domain approach to control of series elastic actuators: adaptive torque and passivity-based impedance control. IEEE/ASME Trans. Mechatron. **21**(4), 2085–2096 (2016)
11. Sariyildiz, E., Chen, G., Yu, H.: An acceleration-based robust motion controller design for a novel series elastic actuator. IEEE Trans. Ind. Electron. **63**(3), 1900–1910 (2015)
12. Sariyildiz, E., Wang, H., Yu, H.: A sliding mode controller design for the robust position control problem of series elastic actuators. In: ICRA 2017, pp. 3055-3061 (2017)
13. Sun, H.-J., Ye, J., Chen, G.: Trajectory tracking of series elastic actuators using terminal sliding mode control. In: CCDC 2021, pp. 189-194 (2021)
14. Zhao, W., Sun, L., Yin, W., Li, M., Liu, J.: Robust position control of series elastic actuator with backstepping based on disturbance observer. In: IEEE/ASME International Conference on Advanced Intelligent Mechatronics (AIM), vol. 2019, pp. 618–623 (2019)

Check for updates

Development of a High-Speed and Accurate Face Recognition System Based on FPGAs

Ha Xuan Nguyen[1,2](✉), Dong Nhu Hoang[2], and Tuan Minh Dang[2,3,4]

[1] Hanoi University of Science and Technology, No. 1 Dai Co Viet, Hanoi, Vietnam
ha.nguyenxuan@hust.edu.vn
[2] CMC Applied Technology Institute, CMC Corporation, 11 Duy Tan, Hanoi, Vietnam
[3] CMC University, CMC Corporation, 11 Duy Tan, Hanoi, Vietnam
[4] Posts and Telecommunication Institute of Technology, Ha Dong, Hanoi, Vietnam

Abstract. In this work, a high-speed and accurate face recognition system based on Field Programmable Gate Arrays (FPGAs) was completely developed. A complete pipeline that contains a sequence of processing steps, including preprocessing, face feature extraction, and matching, is proposed. For processing steps, lightweight deep neural models were developed and optimized so that they could be computationally accelerated on an FPGA. Besides the core processing pipeline, a database as well as a user application server were also developed to fully meet the requirements of readily commercialized applications. The experimental evaluation results show that our system has a very high accuracy based on the BLUFR benchmark, with a precision of 99.336%. Also, the system is very computationally efficient, as the computing time to recognize an ID in a dataset of 1000 IDs with 4000 images on the FPGA ZCU-104 is only 30.3 ms. For the critical case, the system can process 8 camera streams and simultaneously recognize a maximum of 80 IDs within a computing time of 342.3 ms for each ID. With its high-speed and accuracy characteristics, the developed system has a high potential for practical applications.

Keywords: Face Recognition · Deep Learning · FPGAs · Edge Processing

1 Introduction

The rapid development and advancement of deep learning and computing hardware have shown many advantages for the image processing problem. The face recognition issue has received much attention in recent years due to its very high potential for practical applications in access/check-in/check-out control systems, public security surveillance systems, and electronic commercial transactions [1]. Face recognition, like any other image processing problem, has three main modules: i) the preprocessing module, which includes processes such as face detection, anti-spoofing, alignment, and quality checking; ii) the face feature extraction module, which employs a deep neural network and typically yields face-feature vectors (FFVs); and iii) the face feature matching module, which calculates cosine distances between FFVs to obtain the final identification of a

T. D. L. Nguyen et al. (Eds.): ICISN 2023, LNNS 752, pp. 288–294, 2023.
https://doi.org/10.1007/978-981-99-4725-6_36

face. This processing pipeline requires a lot of computational effort, and as a result, high-performance computing hardware is needed. Thus, the need for developing and optimizing lightweight models that satisfy both high accuracy and computational efficiency is a current research trend [1]. Several works [2–5] use FPGAs to accelerate the computation of the processing pipeline. It has been demonstrated that the use of FPGAs can significantly accelerate computation while consuming less power. However, the obtained accuracy as well as the computational efficiency can be more improved. Thus, due to the limited hardware resources of FPGAs, the processing pipeline as well as deep neural network models for face recognition should continue to be optimized in such a way that they are lightweight and have high accuracy.

In this work, a high-speed and accurate face recognition system on FPGAs is developed. Compared to other works [2–5], our system has a complete processing pipeline that contains pre- and post-processing algorithms as well as deep convolutional neural models. These models are developed so that they are lightweight while maintaining high accuracy characteristics. For this reason, pretrained models with the backbone of Resnet [6] and Densebox [7] were fine-tuned using transfer learning techniques with a public dataset [8]. These models are then optimized and quantized so that they can be implemented and computationally accelerated on Xilinx FPGAs ZCU-104 [9]. The accuracy of the system is evaluated via the benchmark BLUFR [10], and the computational efficiency is verified via several critical scenarios. For a complete system, a database that contains data for recognizing people and a user application server were developed.

2 System Description

The overall processing pipeline of the design system is shown in Fig. 1. There are four main modules, each of which is responsible for a specific task. The first module is the face recognition pipeline, which contains a sequence of several processing steps. Video frames captured from cameras are streamed to the processing unit via the RTSP protocol. The video streams are decoded by the Video Codec Unit (VCU). The decoded frames are converted to a format suitable for deep learning models, including the color format (RGB) and the input size. After this step, a series of deep learning models are performed in sequence, including face detection, face tracking, face spoofing detection, face quality control, face alignment, and face feature vector extraction. Deep neural networks used for this processing are listed in Table 1. Most of these processes are implemented on the deep learning processing unit (DPU) of a FPGA, namely the ZCU-104. To meet the requirements of real-time applications, we can use the FPGA to accelerate the computation of deep learning models by using parallel computing. The second module is the database using SQLite3 DB, which contains the face feature vectors of the corresponding images of the person identification. This database is created by the third module, namely the server updater. This module also has full deep learning models for face detection and face feature vector extraction. The extracted face featured vectors are then updated in the Feature Vector DB. Inputs to this module are the face images of interested persons and the corresponding record of information such as their name, age, company, and email. These inputs can be imported from the so-called CIVAMS Web Server or from the image database. When a person is required to be queried, its face feature vectors are matched with feature vectors in the database to find the best-matched vector, which has a maximum cosine similarity larger than a threshold. This process is carried out by the ARM CPU of the system on a chip, the ZCU-104. The fourth module is the user application, which contains the CIVAMS Web Server and CIVAMS Application Servers and other display devices and actuators for user-specific applications.

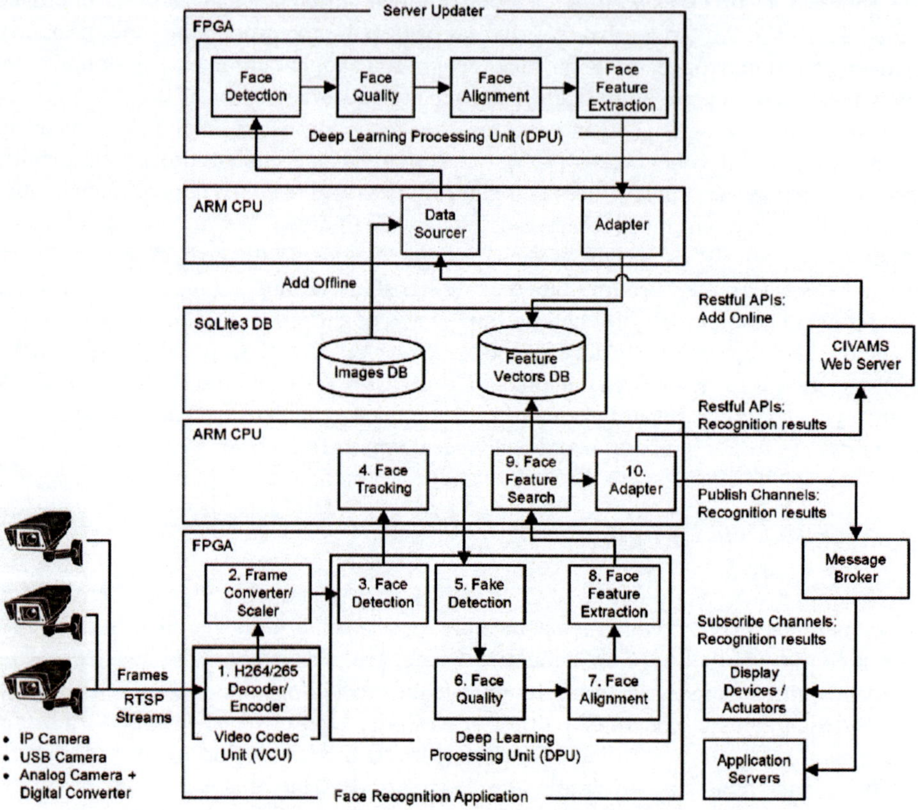

Fig. 1. Processing pipeline of the whole face recognition system based on FPGA ZCU-104 [9].

Figure 2 describes a block diagram of the development and deployment of a deep learning model on the FPGA ZCU-104 platform. First, a deep neural network is trained based on the so-called Caffe framework and a dataset. After the training and testing process, a weight file (*.caffemodel) and a network architecture file (*.prototext) are obtained. These files are then quantized and combined via the tools Decent and DNNC from the hardware provider. After this, we obtain corresponding files whose formats are compatible with the hardware. These files are then combined by the Xilinx SDK to make the execution file (*.elf), which is stored on the SD card. For the hardware platform creation, a design for the DPU IP is generated, which is then converted to hardware-defined format via the tool Vivado Design Suite. This file, together with other system files, is used to build the operating system Petalinux, resulting in a sysroot and bootable image.

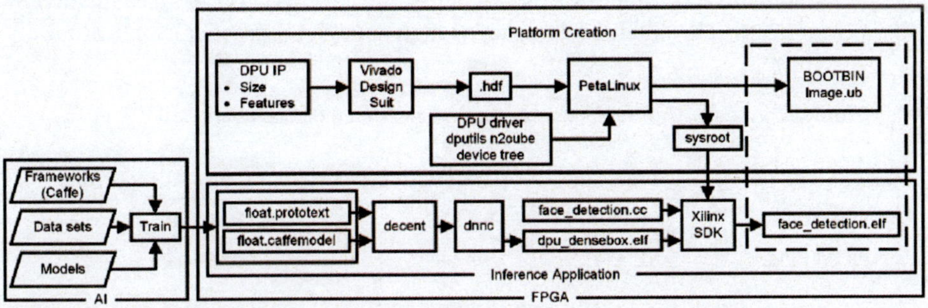

Fig. 2. Development and deployment pipeline of deep learning model on FPGA ZCU-104 [9].

Table 1. Technical detail of the developed models.

Model	Back-bone/specification
Face anti-spoofing	Resnet34 [6]
Face and landmark detection	Densebox 640x360 [7]
Face quality	Resnet10 [6]
Face feature extraction	Resnet36 [6]
Face alignment	Affine transformation

The detailed technical specifications of the FPGA ZCU from Xilinx can be found in [9]. This hardware is intended for the edge processing of deep neural networks. The hardware is a completed multi-process system on chip (MP-SOC), which can be used for image processing. The hardware features a Zynq® UltraScale + ™ MPSoC EV device with video codec and supports many common peripherals and interfaces for embedded vision use cases. The included ZU7EV device has a quad-core ARM® Cortex™-A53 applications processor, a dual-core Cortex-R5 real-time processor, a Mali™-400 MP2 graphics processing unit, a 4KP60 capable H.264/H.265 video codec, and 16 nm FinFET + programmable logic.

3 Results and Discussions

The face feature vector extraction model was developed based on the backbone of Resnet36 [6] with a pretrained model. This model was fine-tuned using a dataset, namely the MS-Celeb-1M [8], which contains a set of one million images of a total of 100,000 people. The fine-tuned model is then evaluated using a standard benchmark, namely the BLUFR [10], on a self-collected test dataset containing 6000 images of 1000 IDs. The results are shown in Table 2. Evaluation on two types of hardware, including the

GPU and the FPGA ZCU-104, was performed. The developed model has an impressive accuracy of 99.336% and a true positive rate of 99.244%, with a false acceptance rate of only 0.00104. Since the system has the pre-processing modules, challenging cases like hard ambient light, camera's viewing angle can be solved via the face quality and alignment process. The obtained accuracy is high enough for practical applications.

Table 2. Accuracy of the face recognition model on the benchmark BLUFR [10].

Model on hardware	Precision	True Positive Rate	False Accepts Rate
Fine-tuned GPU	98.060	98.350	0.00100
Fine-tuned FPGA	**99.336**	99.244	0.00104

Aside from accuracy, the system's computational efficiency is the most important factor. Evaluation results of the computational performance on FPGA ZCU-104 are shown in Table 3. The evaluation was performed on a test dataset containing 4000 images representing 1000 IDs. In this evaluation, the total processing time for the overall proposed pipeline is calculated under different scenarios. The number of camera streams and the number of IDs are varied to test the computing time for each processing step, including the detection time, the recognition time, the matching time, and as a result, the total processing time for one ID. Critical cases are considered. It is seen that for the fastest case, one ID on only one camera stream, the total processing time is only 30.3 ms. For the critical case, where a total of 80 IDs from 10 camera streams can be recognized in a total processing time of 342.3 ms for each ID. It can be concluded that the designed system has very high computational efficiency on the FPGA ZCU 104 since the overall processing pipeline was optimized and only lightweight models were developed. This shows a high potential for practical applications. Compared to other works [2–5], our system has higher both in the accuracy and computational efficiency.

Table 3. Computational performance evaluation results.

No. of IDs	No. of streams	FPS	Detection time (ms)	Recognition time (ms)	Matching time (ms)	Total time for 1 ID (ms)	Total no. of processed ID
1	1	59	12.7	10.3	7.3	**30.3**	**1**
	4	13	26.3	38.1	10.5	74.9	4
	8	7	68.9	141.2	12.3	223.4	8
7	1	58	16.3	10.5	7.5	34.3	7
	4	16	62.6	73.4	9.7	145.7	28
	8	9	136.5	193.5	12.2	342.2	56
10	1	57	18.5	10.9	7.3	36.7	10

(*continued*)

Table 3. (*continued*)

No. of IDs	No. of streams	FPS	Detection time (ms)	Recognition time (ms)	Matching time (ms)	Total time for 1 ID (ms)	Total no. of processed ID
	4	19	63.1	70.1	9.9	143	40
	8	9	138.2	192.2	11.8	**342.3**	**80**

For user applications, a webserver was developed to manage the results of the system. The graphical user interface is shown in Fig. 3. The web server contains both a back-end and a front-end. The back-end has a user database and modules to allow administration as well as let the normal user manage the results. This back-end can communicate with other functional systems via APIs. The front-end contains several tabs for statistics, dashboard, timeline, access control, device configuration, and import and export data tools.

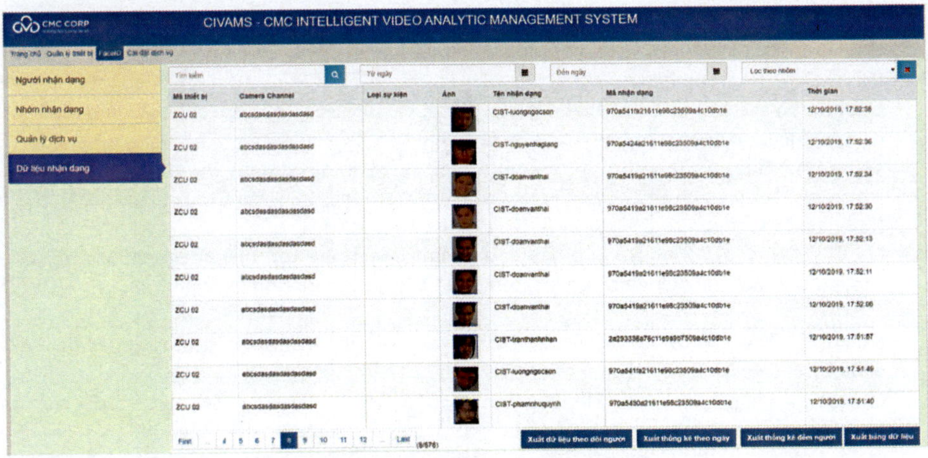

Fig. 3. User application webserver

4 Conclusions and Outlook

In this work, a complete, high-speed, and accurate face recognition system has been successfully developed. For an accurate system, the processing pipeline must contain a certain number of processing units, where the preprocessing is of the greatest importance. Our system is computationally efficient since only lightweight models were developed and optimized. Furthermore, the computation of these models can be accelerated using the FPGA ZCU-104. The use of ZCU-104 is efficient. On the one hand, it allows for very high computational performance, with a maximum processing ability of 8 camera streams for 80 IDs in only 342.3 ms. On the other hand, the used hardware has a very low power consumption, which is significant for practical applications where heat emission is a big issue. Our system outperforms others in term of the accuracy and computational efficiency. It can be concluded that the system is efficient and can be used for edge-processing applications.

In the future, the system will be extended in some directions. The goal is to develop data privacy protection using blockchain, backup, and system high availability.

Acknowledgement. This research is funded by the CMC Applied Technology Institute, CMC Corporation, Hanoi, Vietnam.

References

1. Wang, M., Deng, W.: Deep face recognition: a survey. Neurocomputing **429**, 215–244 (2021)
2. Qu, X., Wei, T., Peng, C., Du, P.: A fast face recognition system based on deep learning. In: 2018 11th International Symposium on Computational Intelligence and Design (ISCID), vol. 1, pp. 289–292 (2018)
3. Chen, C.: Design of image recognition system based on FPGA. In: 2022 7th International Conference on Intelligent Computing and Signal Processing (ICSP), pp. 1924–1928 (2022)
4. Selvi, S.S., Bharanidharan, D., Qadir, A., Pavan, K.R.: FPGA implementation of a face recognition system. In: 2021 IEEE International Conference on Electronics, Computing and Communication Technologies (CONECCT), pp. 1–5 (2021)
5. Wang, H., Cao, S., Xu, S.: A real-time face recognition system by efficient hardware-software co-design on FPGA SoCs. In: 2021 IEEE 3rd International Conference on Artificial Intelligence Circuits and Systems (AICAS), pp. 1–2 (2021)
6. He, K., Zhang, X., Ren, S., Sun, J.: Deep residual learning for image recognition. In: Proceedings of the IEEE Conference on Computer Vision and Pattern Recognition, pp. 770–778 (2016)
7. Huang, L., Yang, Y., Deng, Y., Yu, Y.: DenseBox: unifying landmark localization with end to end object detection. arXiv preprint arXiv:1509.04874 (2015)
8. Guo, Y., Zhang, L., Hu, Y., He, X., Gao, J.: MS-Celeb-1M: a dataset and benchmark for large-scale face recognition. In: Leibe, B., Matas, J., Sebe, N., Welling, M. (eds.) ECCV 2016. LNCS, vol. 9907, pp. 87–102. Springer, Cham (2016). https://doi.org/10.1007/978-3-319-46487-9_6
9. Zynq UltraScale+ MPSoC ZCU104 Evaluation Kit. https://www.xilinx.com/products/boards-and-kits/zcu104.html. Accessed 29 Nov 2020
10. Liao, S., Lei, Z., Yi, D., Li, S. Z.: A benchmark study of large-scale unconstrained face recognition. In: IEEE International Joint Conference on Biometrics, pp. 1–8 (2014)

Nonlinear Model Predictive Control with Neural Network for Dual-arm Robots

Hue Luu Thi[2], Chung Nguyen Van[1], and Tung Lam Nguyen[1](✉)

[1] Hanoi University of Science and Technology, Hanoi, Vietnam
lam.nguyentung@hust.edu.vn
[2] Electric Power University, Hanoi, Vietnam

Abstract. This study proposes a Lyapunov-based Model predictive control with an estimate disturbance radial neural network technique for an extremely complex system-a dual-arm robot. By using a radial neural network to estimate the disturbances, their effects on the system are reduced. This technique also ensures the cooperation between the two arms of the robot and guarantees stabilization through the Lyapunov inequality constraint. Simulations were performed to investigate controller qualities.

Keywords: Model predictive control · Lyapunov-based model predictive control · Radial basis function · Dual-arm robot

1 Introduction

Dual-arm (DA) robotic manipulator is a type of robot with two manipulators, which as human for two hands. Using two manipulators to cooperate together will be more efficient and flexible than using a single robot. Consequently, it can be used to replace humans when working in hazardous and dangerous environments as sea exploration, firefighting, etc. The research shows that there are some typical control methods such as PID controller, PD controller, impedance control, and simultaneous control of both force control and position control in the DA manipulator system. The coordination controller for the DA manipulator was designed based on a kinematic analysis of the system. A Coordinated motion of a DA robot under the PD gravity compensation controller based on inverse kinematics is performed [1]. Alternatively, coordinated operations between two manipulators by motion control at the level of acceleration are presented [2]. The reference velocity/acceleration of the object is designed according to the required task. Impedance control methods will be combined with other control methods to increase the efficiency of the controller. The impedance controller is designed using the Jacobian relationship [3]. The proposed impedance controller will be more efficient when adjusting the desired trajectory and desired impedance. The double impedance controller with two impedance loops is researched in [4], with the outer loop impedance between the object and the environment, and the

T. D. L. Nguyen et al. (Eds.): ICISN 2023, LNNS 752, pp. 295–304, 2023.
https://doi.org/10.1007/978-981-99-4725-6_37

inner loop impedance between the controller and the object. This controller has good controlling over force, position, and impedance control. Other studies focus on combined force and position control for DA robot manipulator coordination [5,6]. In [5] a force sensor is placed at the end-effector used for the master/slave controller combination, to coordinate control the object tracking the desired trajectory to avoid the object drifting and falling. The controller [6] monitors the desired force and compensates for trajectory deviation due to external disturbance. However, the system working is influenced by external noise, the sliding mode controller in [7] reacts quickly and strongly to the external noise. The controllers above, the limit of the control input, and the state variable of the actuator have not been considered when the controller design. They need to be included in the controller during the design process. Predictive control methods (MPC) can solve this problem. The nonlinear model predictive controller (NMPC) is presented in [8], the nonlinear constraints are approximated to economize time calculation. In addition, to reduce the computation time, Jason Bettega et al. [9] used the MPC with embedded reference dynamics method in the controller design. A robust multi-objective predictive controller is proposed in [10], which increases the feasible region without losing important properties such as system stability. With normal MPC controllers, the stability in the control closed loop has not been satisfactorily resolved. Solving this problem, the backstepping controller combined with the Lyapunov - based MPC controller presented in [11,12], the stability in the closed loop controller is analyzed and proved to be stable. The control Lyapunov-Barrier function (CLBF)- MPC controller for the nonlinear system is proposed in [13] with input constraints and ensures two requirements: stability and safe operation. To solve the above, a controller with a structure consisting of MPC based on Lyapunov and the SMC combined with a radial basis function neural network (RBFNN) to compensate for the disturbance is presented in this paper. The contributions in the paper are summarized as follows:

- An RBFNN is added to compensate the disturbance, which reduces the influence of noise on the system in the MPC of the nonlinear model.
- Simultaneously, the constraints of the state variable as well as the control input are considered in the LMPC controller design. The combination of LMPC and RBFNN ensures that the system is robust and stable, and the constraint conditions are guaranteed.

The paper has the following structure: Dynamics of the system present in Sect. 2. In sections three and Sect. 4, the controller is designed. Simulation results and conclusions denote in Sect. 5 and Sect. 6, respectively.

2 Dynamics of the System

The DA robot has two manipulators. The purpose is to cooperate in holding a rigid object, each arm with n degrees of freedom (DOF) is shown in Fig. 1. {O} is the base frame. $\{O_v\}$ represents the object frame, which has a center attached

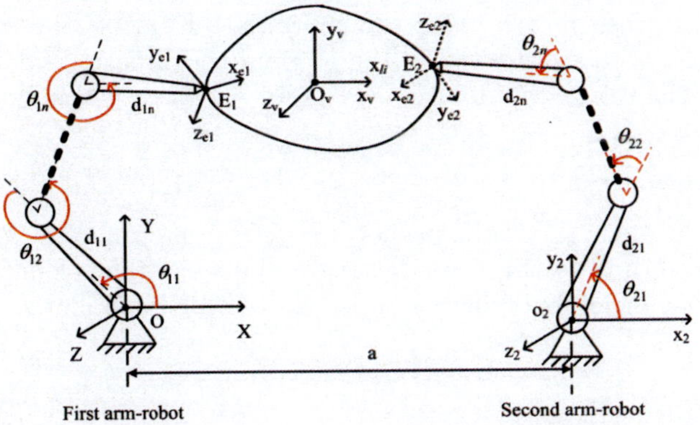

Fig. 1. Dual-arm robot and object frames

to the center of mass of the object, and $\{E_i\}$ $(i = 1 \div 2)$ is the ith end-effector frame. The dynamic equation of the ith robot with n DOF can be construct in the joint space as follows [14]:

$$\tau_i = H_i(q_i)\ddot{q}_i + C_i(q_i, \dot{q}_i)\dot{q}_i + G_i(q_i) + D_i + J_{0i}^T F_i \tag{1}$$

where $q_i = [\theta_{i1}, \theta_{i2}, ..., \theta_{in}]^T \in R^{n\times 1}$ represents the joints angle vector of the ith robot; $C_i(q_i, \dot{q}_i) \in R^{n\times n}$ is the Coriolis and centrifugal matrix. $H_i(q_i) \in R^{n\times n}$ is the inertia matrix, the $\dot{H}_i - 2C_i$ matrix is a skew-symmetric matrix; $G_i(q_i) \in R^{n\times 1}$ is a gravitational vector; J_{0i} is Jacobi matrix; D_i denotes bounded unknown disturbances; F_i is the 6×1 vector of forces/torque vector applied by the ith manipulator to hold object. According to (1), the dynamic of the DA robot can be written by:

$$H(q)\,\ddot{q} + C(q, \dot{q})\,\dot{q} + G(q) + D + J_B^T(q)\,F = \tau \tag{2}$$

where $q = [q_i]^T \in R^{2n\times 1}$; $\tau = [\tau_i]^T \in R^{2n\times 1}$; $F = [F_i]^T \in R^{12\times 1}$; $D = [D_i]^T$; $H(q) = [H(q_i)] \in R^{2n\times 2n}$; $C(q, \dot{q}) = [C(q_i, \dot{q}_i)] \in R^{2n\times 2n}$; $G(q) = [G_i(q_i)]^T \in R^{2n\times 1}$, $i = 1, 2$

The dynamics of the grasped object are represented by the motion equation: [14]

$$H_c\,\ddot{c} + C_c(c, \dot{c})\,\dot{c} + g_c = F_c, \tag{3}$$

where $c \in R^{6\times 1}$ is the vector of orientations and positions; $H_c \in R^{6\times 6}$ represents the inertia matrix; $C_c(c, \dot{c}) \in R^{6\times 6}$ is the Coriolis and centrifugal matrix and $g_c(c) \in R^{6\times 1}$ denotes the gravitational vector of the object; $F_c \in R^{6\times 1}$ is the total force/moment impact the object at the center in the base frame {O}.

The force/torque vector F_c at the center of the object can be described by force/torque vectors F affected on the object at the end-effectors as follows [15]:

$$F_c = T\,F \tag{4}$$

where T is the grasp matrix from the coordinate frame of the end-effectors to the object frame $\{O_v\}$ respected in the original coordinate frame {O}.

From (4), the relationship between F_c and F can be rewrite as:

$$F = T^+ F_c \tag{5}$$

where $T^+ = T^T T (T^T)^{-1}$ is the pseudo-inverse of grasp-matrix T.

Besides according to [14] the relationship between the velocity of the object and the velocity of joints of the ith robot is described as follows:

$$\dot{q} = A\,\dot{c};\ \ddot{q} = A\,\ddot{c} + B\dot{c} \tag{6}$$

where $A = \left[(A_i^{-1})^T \right]^T, B = \left[(-A_i^{-1} \dot{A}_i A_i^{-1})^T \right]^T$, $i = 1, 2$ with $A_i \in R^{6\times n}$ is the relationship matrix between object velocity and joints velocity.

3 The Controller for the Dual-arm Robot

3.1 The Controller Design

Firstly, the object reference velocity is defined as $\dot{c}_r = \dot{c}_d + \gamma e_p$, with γ is a positive constant gain, $e_p = c_d - c$ is the tracking errors. Where $c_d = [x_d, y_d, z_d, \psi_d, \varphi_d, \theta_d]^T$. So, the desired force effect on the object is determined through the object's reference model:

$$F_c{}^d = H_c \ddot{c}_r + C_c(c, \dot{c}_r)\dot{c}_r + g_c \tag{7}$$

The errors between the reference and the actual velocities of the object are calculated as follows:

$$s_0 = \dot{c} - \dot{c}_r = -\dot{e}_p - \gamma e_p \tag{8}$$

From (5), the desired forces effect at the ending points determines as follows:

$$F^d = T^+ F_c{}^d \tag{9}$$

Now, the robotic manipulator reference velocity is defined as $\dot{q}_r$. The errors between the reference and the actual velocity of the DA robot joints are calculated as:

$$s = \dot{q} - \dot{q}_r = A\, s_0 \tag{10}$$

The RBFNN is used to estimate the disturbances in (2) as follow [16]:

$$D = W^{*T} \phi(\bar{n}) + \varepsilon \tag{11}$$

where W^* is the optimal weight of RBFNN; $\phi(\bar{n})$ is Gaussian function; ε is approximate error. $\hat{D}$ is an estimate of D, which is determined:

$$\hat{D} = \hat{W}^T \phi(\bar{n}) \tag{12}$$

The error between the disturbances D and approximation value $\hat{D}$ can be proposed as:

$$\tilde{D} = D - \hat{D} = (W^* - \hat{W})\,\phi(\bar{n}) + \varepsilon = \tilde{W}\,\phi(\bar{n}) + \varepsilon \tag{13}$$

From Eqs. (2), (10) and (11), after a few calculations, the dynamic of the DA robot is rewritten as follows:

$$H(q)\,\dot{s} + C(q,\dot{q})\,s + f(q,\dot{q}_r,\ddot{q}_r) + W^{*T}\phi(\bar{n}) + \varepsilon = \tau - J_B^T(q)\,F \tag{14}$$

here $f(q,\dot{q}_r,\ddot{q}_r) = H(q)\,\ddot{q}_r + C(q,\dot{q}_r)\,\dot{q}_r + G(q)$. The hybrid position/force controller of the system is proposed as follows:

$$\tau = f(q,\dot{q}_r,\ddot{q}_r) + \hat{W}\phi(\bar{n}) + J_B^T F^d - K_s\,(A\,s_0) - K_1\,sign(s) \tag{15}$$

where K_s and K_1 are positive coefficients matrix

3.2 Weight Updates Law for Neural Networks

From Eqs. (3), (7) and (8), after some calculations, the dynamics of the object are rewritten following:

$$H_c\,\dot{s}_0 + C_c\,s_0 = F_c - F_c{}^d \tag{16}$$

From Eqs. (14) and (16), the candidate of the Lyapunov function is selected according to the Lyapunov stability principle.

$$V = \frac{1}{2}s^T H\,s + \frac{1}{2}tr\left(\tilde{W}^T\Gamma^{-1}\tilde{W}\right) + \frac{1}{2}s_0^T H_c s_0 \tag{17}$$

The derivation of the Eq. 17 is calculated as :

$$\dot{V} = s^T H\,\dot{s} + \frac{1}{2}s^T\,\dot{H}\,s + tr(\tilde{W}^T\Gamma^{-1}\dot{\tilde{W}}) + s_0^T H_z\,\dot{s}_0 + \frac{1}{2}s_0^T\dot{H}_c s_0 \tag{18}$$

Using (9), (10), (14), (16) with skew symmetrics $\dot{H} - 2C$ and $\dot{H}_c - 2C_c$, so $s^T\left(\dot{H} - 2C\right)s = 0$ and $s_0{}^T\left(\dot{H}_c - 2C_c\right)s_0 = 0$. The Eq. (18) becomes:

$$\dot{V} = -(A\,s_0)^T K_s A\,s_0 - s^T\,K_1\,sign(s) - (A\,s_0)^T\tilde{W}^T\phi(\bar{n}) + tr(\tilde{W}^T\,\Gamma^{-1}\dot{\tilde{W}}) \tag{19}$$

The stable condition $\dot{V} \le 0$, so the update law for weights of the neural network must be as follows:

$$\dot{\hat{W}} = -\Gamma\phi(\bar{x})\,(A\,s_0)^T. \tag{20}$$

With the control law Eq. (15) and the integrated adaptation law Eq. (20). The Lyapunov stability principle proves that the system is stable.

4 Lyapunov-Based MPC Controller

The desired trajectory for the object in three-dimensional space is chosen by: $z_d = [x_d,\, y_d,\, z_d,\, \psi_d,\, \varphi_d,\, \theta_d]^T$
Set $\chi = \left[z,\, \dot{z}\right]^T$, from (3) the dynamics of the object is rewritten as follows:

$$\dot{\chi} = \left[\dot{z},\, H_z{}^{-1}\left(F_z - C_z(z,\dot{z})\dot{z} + g_z\right)\right]^T = f(\chi, F_z). \tag{21}$$

where $F_z = [F_x,\, F_y,\, F_z,\, \tau_\psi,\, \tau_\varphi,\, \tau_\theta]$ is the total force and moment at the center of the object.

In the Eq. (2), τ is the input control vector, and $u = \tau$ controls the joint angles of the DA robot, thereby controlling the movement of the object. So, the MPC for this system can be established as follows:

$$\min_{\hat{u}\in\varpi(T)} J = \int\limits_0^{T_p} \left(\left\|\tilde{\chi}(\varpi)\right\|^2{}_Q + \left\|\hat{F}_z(\varpi)\right\|_{R1}^2 + \left\|\hat{u}(\varpi)\right\|_R^2 \right) d\varpi \tag{22}$$

s.t

$$\begin{aligned} \dot{\hat{\chi}}(\varpi) &= f\left(\hat{\chi}(\varpi), F_z(\varpi)\right). \\ \hat{\chi}(0) &= \chi(t_0). \\ \left|\hat{u}(\varpi)\right| &\le u_{\max} \\ \left|\hat{F}_z(\varpi)\right| &\le F_{z\,\max}. \\ \dot{V}\left(\hat{\chi}(\varpi),\, \hat{u}(\varpi)\right) &\le \dot{V}\left(\hat{\chi}(\varpi),\, h(\hat{\chi}(\varpi))\right). \end{aligned} \tag{23}$$

where $\hat{\chi}(\varpi)$ is the predicted state horizon of the object corresponding to the predictive controller $u(\varpi)$, $\hat{\chi}(0)$ is the initial value taken from the system model; $\tilde{\chi} = \hat{\chi} - \chi_d$ is the error state; $\varpi(T)$ and $T_p = N.T$ denotes respectively a family of feature functions with the sampling period T and the prediction horizon; $Q,\ R_1,\ R$ are the positive weighting matrices. $h(.)$ is the Lyapunov-based nonlinear controller and $V(.)$ represent respectively the Lyapunov function.

The constraints (23) ensure that the LMPC guarantees stable properties regardless of the length of the prediction horizon. The auxiliary controller $h(\chi)$ as follows:

$$h(\chi) = f(q, \dot{q}_r, \ddot{q}_r) + \hat{D} + J_B^T F^d - K_s\left(A s_0\right) - K_1 \,\mathrm{sign}(s) \tag{24}$$

The V_{LMPC} function is proposed as:

$$V_{LMPC} = \frac{1}{2} s^T H\, s + \frac{1}{2}\mathrm{tr}(\tilde{W}^T \Gamma^{-1} \tilde{W}) + \frac{1}{2} s_0{}^T H_c\, s_0 \tag{25}$$

Using update law (20) for $\dot{V}_{LMPC}$ therefore the constraint condition (23) is shortened as follows:

$$\begin{aligned} & s^T(\varpi)\tau(\varpi) - s^T(0) f(\varpi) - s^T(\varpi)\hat{D}(\varpi) - s_\varpi{}^T(\varpi) F_c{}^d(\varpi) \\ & \le -(A(\varpi)\, s_0(\varpi))^T K_s\left(A(\varpi)\, s_\varpi(\varpi)\right) - s^T(\varpi) K_1 \mathrm{sign}\left(s(\varpi)\right) \le 0, \forall t \in [0,\, T_p] \end{aligned}$$

5 Simulation for Dual-arm Robot

The purpose of this section is to make the DA system track the reference trajectory while ensuring the stabilization of the system and the cooperation between to arms. The desired trajectory of the robot are chosen as:

$$[x_r, \theta_r] = [0.42 + 0.24t^3 - 0.12t^4 + 0.016t^5, -0.24t^3 + 0.12t^4 - 0.016t^5] \quad (26)$$

$$y_r = \begin{cases} 1.352 + 0.5t^3 - 0.4t^4 + 0.08t^5 (t \leq 1.5) & (27a) \\ 1.982 - 2.7t + 4.5t^2 + 2.9t^3 + 0.8t^4 - 0.08t^5 (t > 1.5) & (27b) \end{cases}$$

Table 1 shows the parameters of this system.

Table 1. Parameters of the DA robot.

Class	d_{11}	d_{12}	d_{13}	d_{21}	d_{22}	d_{23}
Value	$1(m)$	$0.8(m)$	$0.6(m)$	$1(m)$	$0.8(m)$	$0.6(m)$
Class	m_{11}	m_{12}	m_{13}	m_{21}	m_{22}	m_{23}
Value	$1.5(kg)$	$1.2(kg)$	$1(kg)$	$1.5(kg)$	$1.2(kg)$	$1(kg)$
Class	g	L	M	L_v	L_d	
Value	$9.8(m/s^2)$	$1.086(m)$	$1(kg)$	$0.5(m)$	$0.6(m)$	

The initial state of the DA robot is: $[x_0, y_0, \theta_0] = [0.6, 1.48, 0]$, the sampling period $T = 0.01(s)$, the prediction horizon $N_p = 10$, the weights of matrices Q, R_1, R are chosen by: $Q = \text{diag}(50{,}40{,}60{,}40{,}50{,}40{,}50{,}50{,}45)$, $R_1 = 0.25.diag(eye(4))$ and $R = 0.25.diag(eye(6))$. The results for simulations are shown in Fig. 2, Fig. 3 which indicated the feasibility of the control method. The tracking trajectory of DA robot is shown in Fig. 2, and the control input is shown in Fig. 3. With the feasible trajectory, the control input is bounded and ensures the feasibility of the system.

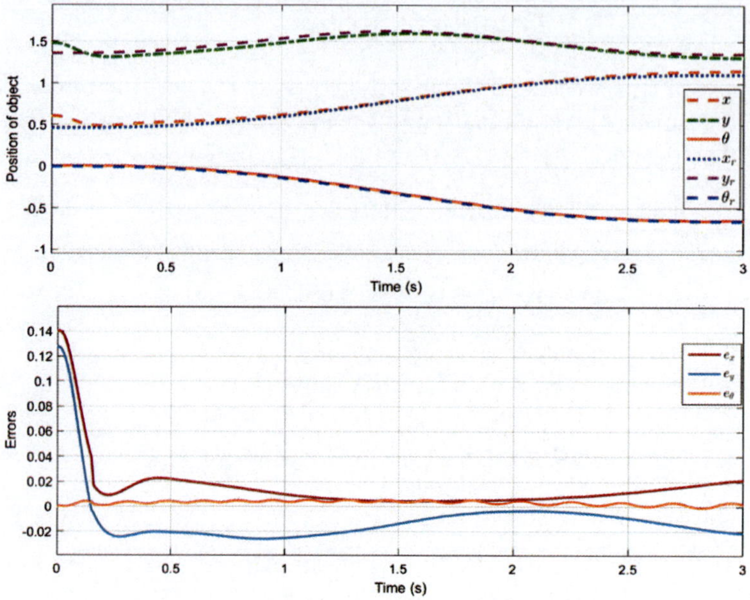

Fig. 2. The output x, y, θ tracking performance and the errors of x, y, θ.

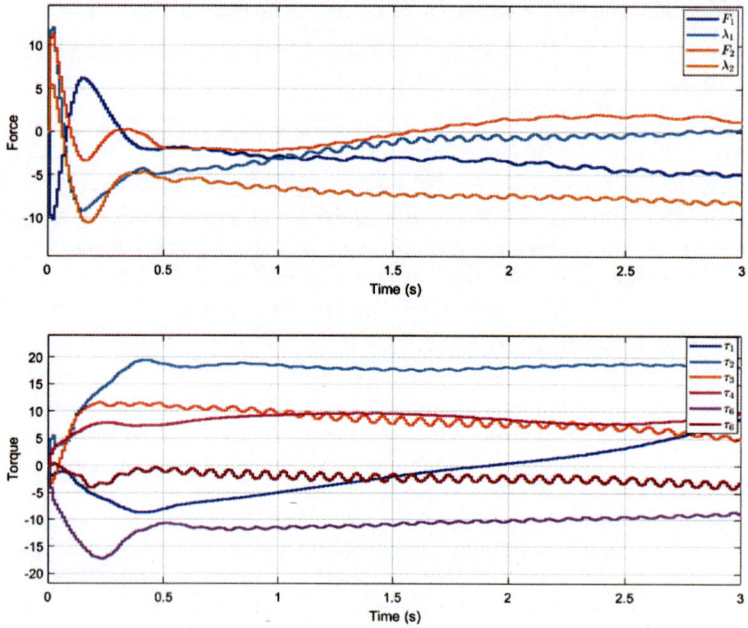

Fig. 3. The control signals: Torques and Forces.

6 Conclusion

This paper studies a new method to control DA robot systems. Although the complexity of the systems, the LMPC control method could ensure cooperation between the two arms of the robot. The LMPC technique is applied and guarantees closed-loop stability. Besides, a radial neural network estimates the disturbance and reduces the influence of disturbance on the system. The simulation was executed to show the advantages of this control technique.

Acknowledgements. This research is funded by Electric Power University (EPU) under project number DTKHCN.19/2022.

References

1. Zheng, N., Han, L., Xu, W.: A Dual-arm cooperative manipulator: modularized design and coordinated control. In: 2018 IEEE International Conference on Information and Automation (ICIA), pp. 945–950. IEEE (2018)
2. Liu, T., Lei, Y., Han, L., Xu, W., Zou, H.: Coordinated resolved motion control of dual-arm manipulators with closed chain. Int. J. Adv. Robot. Syst. **13**(3), 1–14 (2016)
3. Lee, J., Chang, P.H., Jamisola, R.S.: Relative impedance control for dual-arm robots performing asymmetric bimanual tasks. IEEE Trans. Industr. Electron. **61**(7), 3786–3796 (2013)
4. Sadeghian, H., Ficuciello, F., Villani, L., Keshmiri, M.: Global impedance control of dual-arm manipulation for safe interaction. IFAC Proc. Vol. **45**(22), 767–772 (2012)
5. Benali, K., Brethé, J.-F., Guérin, F., Gorka, M.: Dual arm robot manipulator for grasping boxes of different dimensions in a logistics warehouse. In: 2018 IEEE International Conference on Industrial Technology (ICIT), pp. 147–152. IEEE (2018)
6. Jinjun, D., Yahui, G., Ming, C., Xianzhong, D.: Symmetrical adaptive variable admittance control for position/force tracking of dual-arm cooperative manipulators with unknown trajectory deviations. Robot. Comput.-Integr. Manuf. **57**, 357–369 (2019)
7. Xu, F., Wang, J., Guo-dong, L.: Adaptive robust neural control of a two-manipulator system holding a rigid object with inaccurate base frame parameters. Front. Inf. Technol. Electron. Eng. **19**(11), 1316–1327 (2018)
8. de Freitas Virgilio Pereira, M., Kolmanovsky, I.V., Cesnik, C.E.S.: Nonlinear Model Predictive Control with aggregated constraints. Automatica **146**, 110649 (2022)
9. Bettega, J., Richiedei, D.: Trajectory tracking in an underactuated, non-minimum phase two-link multibody system through model predictive control with embedded reference dynamics. Mech. Mach. Theory **180**, 105165 (2023)
10. Oh, T.H., Kim, J.W., Son, S.H., Jeong, D.H., Lee, J.M.: Multi-strategy control to extend the feasibility region for robust model predictive control. J. Process Control **116**, 25–33 (2022)
11. Kim, Y., Oh, T.H., Park, T., Lee, J.M.: Backstepping control integrated with Lyapunov-based model predictive control. J. Process Control **73**, 137–146 (2019)
12. Gong, P., Yan, Z., Zhang, W., Tang, J.: Lyapunov-based model predictive control trajectory tracking for an autonomous underwater vehicle with external disturbances. Ocean Eng. **232**, 109010 (2021)

13. Zhe, W., Christofides, P.D.: Handling bounded and unbounded unsafe sets in Control Lyapunov-Barrier function-based model predictive control of nonlinear processes. Chem. Eng. Res. Des. **143**, 140–149 (2019)
14. Vikas Panwar, N.K., Sukavanam, N., Borm, J.-H.: Adaptive neural controller for cooperative multiple robot manipulator system manipulating a single rigid object. Appl. Soft Comput. **12**, 216–227 (2012)
15. Baigzadehnoe, B., Rahmani, Z., Khosravi, A., Rezaie, B.: On position/force tracking control problem of cooperative robot manipulators using adaptive fuzzy backstepping approach. ISA Trans. **70**, 432–446 (2017)
16. Song, Y.D., Huang, X., Jia, Z.J.: Dealing with the issues crucially related to the functionality and reliability of NN-associated control for nonlinear uncertain systems. IEEE Trans. Neural Netw. Learn. Syst. **28**(11), 2614–2625 (2016)

A Multi-layer Structured Surface Plasmon Resonance Sensor with Improved Sensitivity

Manish Jangid, Vijay Janyani(✉), and Ghanshyam Singh

Department of Electronics and Communication Engineering, Malaviya National Institute of Technology, Jaipur 302017, India
{2020rec9503,vjanyani.ece,gsingh.ece}@mnit.ac.in

Abstract. Surface plasmon resonance (SPR) is a vital, fast, and robust approach to exploring the real-time detection of molecular interactions. In recent decades, many innovative ways suggested making SPR-based sensors more sensitive and reliable. In order to make a Kretschmann-based SPR sensor more sensitive, many different versions of the sensor were investigated. We proposed a multi-layer structured (Glass-SF11/Ag-Au/ITO) SPR sensor, which is mathematically modelled and simulated via MATLAB environment. A numerical analysis was carried out to optimize the design parameters and characteristics of the proposed design. The results show that the sensor's sensitivity (for RI ranges of 1.33–1.37) is affected by various design parameters, including the choice of metal, the metal layer's thickness, and the supportive layer's thickness. The highest sensitivity of 183°/RIU is achieved for the proposed sensor, which is significantly higher than the conventional SPR sensors. These findings suggest a novel method to design sensitive plasmonic refractive index-based SPR sensors for analyte detection.

Keywords: SPR sensor · Sensitivity · Multi-layered · Numerical analysis · Plasmonics

1 Introduction

Nowadays, the advancements in sensor technologies have gained the attention of everyone. Their vast applications in every field make life better and easier. Surface plasmon resonance is one of the most viable technology that is used in various fields like bioscience, environmental science, chemical industries, etc. [1, 2]. Due to its real-time monitoring, label-free features, and sensitive responses, SPR sensors have caught the attention of researchers interested in analyzing biomolecular interactions [3]. SPR occurs when light hits a metal sheet at a particular angle, some of the light's intensity can interact with the freely moving electrons in the conductive metal sheet (plasmon) and reduce the amount of reflected light. Surface plasmons (SPs) are electromagnetic waves traveling on the boundary surface of a metal-dielectric layer. The excitation of SPR occurs in resonant mode when the incident light beam has the wave-vector same as plasmon [4]. The analogous condition of resonance occurring can be expressed as:

$$n_{prism} sin\theta_{reso} = \sqrt{\frac{\varepsilon_{diel}\varepsilon_{metal}}{\varepsilon_{diel} + \varepsilon_{metal}}} \tag{1}$$

T. D. L. Nguyen et al. (Eds.): ICISN 2023, LNNS 752, pp. 305–311, 2023.
https://doi.org/10.1007/978-981-99-4725-6_38

Here n_{prism} = refractive index of prims, θ_{reso} = incident light beam, ε_{diel} & ε_{metal} are the dielectric constants of dielectric media and metal sheet respectively. So, the reflectance constant can be calculated by using the following equation:

$$R_p = |r_p|^2 = \left| \frac{r_p + r_m exp(2ik_m d_m)}{1 + r_p r_m exp(2ik_m d_m)} \right|^2 \tag{2}$$

Here, r_p, d_m and r_m are prism-metal interface, metal thickness, and metal-sensing medium interface. With the help of (2), the SPR dip's resolution can be calculated by varying both the medium's thickness (D_k) and refractive index value (n).

In SPR detection, gold metal is commonly used. It exhibits aqueous solution chemical stability and surface chemistry that is adaptable to various surface functional processes [5]. The major goal of this study is to determine whether it is possible to protect a silver film while maintaining its optical qualities to reduce the amount of oxidation that occurs on silver.

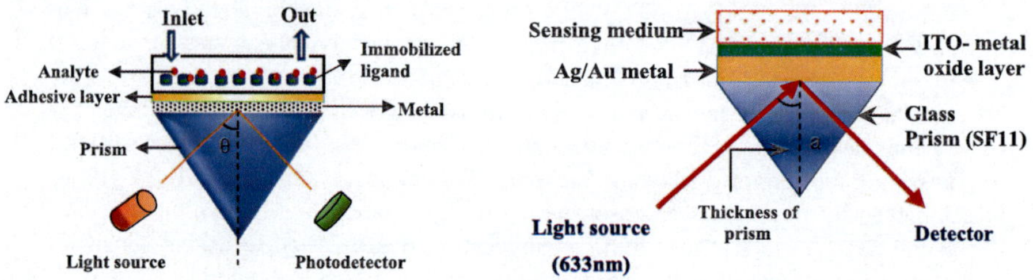

Fig. 1. **(a)** Standard configuration of Prism-based SPR biosensor [6]. **(b)** Proposed structure of multilayer SPR sensor.

The Silica (SiO_2) layers atop gold were the first deposits to be exploited. [7, 8] developed other oxide layers, name indium tin oxide (ITO) and tin oxide (SnO_2). They have relatively low sensitivity, however, due to their complicated structures. To make a biosensor work better, one can improve the sensing surface to detect biomolecular interaction. The Kretschmann configuration (in Fig. 1a), is generally preferred as the SPR configuration for its simplicity. We chose ITO as the protective layer atop of silver because it has tunable optical characteristics, superior conductivity and low infrared loss [9]. In this work, we propose an SPR architecture with minimal design complexity and a high level of sensitivity. Optimizing the sensor structure parameters, we achieved the highest sensitivity of 183°/RIU for a refractive index detection range of 1.33–1.37. The proposed sensor has broad applications in the chemical and healthcare industries.

2 Theory and Mathematical Modelling

Figure 1b shows a geometrical representation of the proposed multi-layer structured SPR sensor. Layer-1(L_1) is made of glass with a thickness (altitude of triangular prism) of 110 nm. The refractive index (RI) of SF11-glass is modelled by the Sellmeier equation

[10] as follows:

$$n(\lambda) = \sqrt{1 + \frac{A_1\lambda^2}{\lambda^2 - B_1} + \frac{A_2\lambda^2}{\lambda^2 - B_2} + \frac{A_3\lambda^2}{\lambda^2 - B_3}} \tag{4}$$

Here λ is the wavelength of incident light and Sellmeier coefficients A_1, A_2, A_3 & B_1, B_2, B_3 are calculated from [11]. A metallic layer (L_2) is deposited over a glass (SF11), whose dielectric constant and RI are calculated by the Drude formula [12] as follows:

$$n^2(\lambda) = \epsilon_m(\lambda) = 1 - \frac{\lambda^2\lambda_c}{\lambda_p^2(\lambda_c + i\lambda)} \tag{5}$$

Here, λ_c & λ_p are collision and plasma wavelengths, respectively. Next, a metal oxide-ITO (L_3) is deposited over L_2, whose RI is calculated as from [11]. Lastly, a sensing medium (L_4) of thickness 250 nm is deposited over the metal oxide layer. The sensing layer has a value of RI ranging from 1.33–1.37. Details of all layers with material type, refractive index value, and thickness are listed in Table 1.

Table 1. Designing parameters of each layer of the proposed SPR sensor at 633 nm.

Material	Thickness (nm)	RI values
Glass (SF11)	110	1.7786
Ag	50	0.1437+i3.8097
Au	50	0.31208+i3.1478
Ag (90%)+Au (10%)	50	0.21686+i3.6338
ITO	10	1.7824+i0.00324
Sensing Medium (SM)	250	1.33–1.37

2.1 Performance Key Parameters for SPR Sensor

The characteristic parameters of the designed sensor are evaluated using a MATLAB environment [13]. Since a measure of the sensor's ability to sense, sensitivity (S_{SPR}) is expressed as the ratio of the shift in resonance angle ($\Delta\theta_{Reso.} = \theta_2 - \theta_1$) to the shift in RI of the sensing layer seen in SPR reflectance spectra.

$$S_{SPR} = \frac{(\Delta\theta_{Reso.})}{\Delta n_s}(^\circ/RIU) \tag{6}$$

The difference in resonance angles at 50% reflection intensity is known as the Full Width at Half Maximum (FWHM). Furthermore, it represents the wavelength spreading of the SPR characterizing curve. D_A stands for "Detection accuracy," which is inversely proportional to the FWHM.

$$D_A = \frac{1}{FWHM}(1/^\circ) \tag{7}$$

A figure of merit (FM) is defined as a product of sensitivity to detection accuracy.

$$FM = S \times D_A(1/RIU) \tag{8}$$

3 Results and Discussion

To improve sensitivity, we looked at how all of the design parameters affected the loss spectrum. An angular interrogation technique is used to investigate the sensitivity of the proposed SPR sensor. Firstly, all design parameters of the SPR sensor, like the incoming light wavelength of 633 nm, material choice, Refractive index (RI), and thickness of prism, metal, metal oxide layer, and sensing media, are determined.

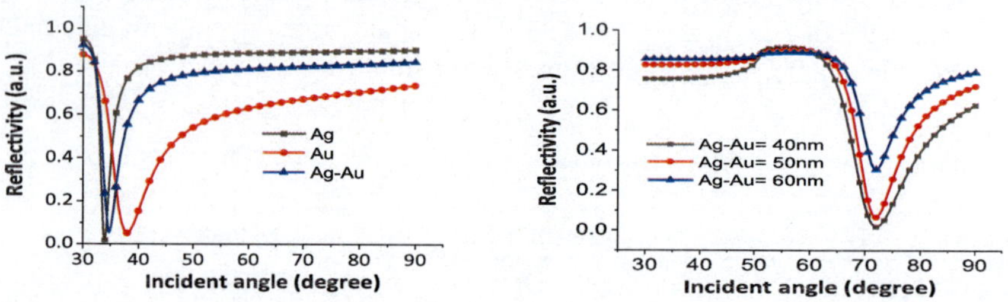

Fig. 2. **(a)** Change in SPR reflectivity for different metals (thickness = 50 nm) at an RI value of 1.00. **(b)** Change in SPR reflectivity concerning the different values of metal thickness for the proposed sensor.

The SPR reflectivity curve for common plasmonic metals silver and gold is depicted in Fig. 2a which outperforms the visible spectra (400–700 nm). The results show that silver-based SPR sensors are superior in performance to gold-based SPR sensors. But silver-based SPR sensors have oxidation problems. Gold-based SPR sensors resolve this problem with improved stability but they also have wider SPR spectra (red shaded in Fig. 2a). So, a combination of silver (90%) + gold (10%) can be used to enhance the sensor performance with maintaining a sharp dip in SPR reflection spectra as silver-based SPR sensor. There is a shift observed in SPR reflectivity as we change the (Ag-Au) metal thickness shown in Fig. 2b.

It can clearly show that the higher value of metal thickness reduces the sensor's performance. There is also an investigation carried out for finding the optimum value of the ITO layer. Figure 3a illustrates the changes that occur in the SPR reflectivity curve concerning different ITO layer thicknesses. The observations show that the sensor performance degrades with increasing ITO thickness. The optimum value of the ITO layer's thickness were obtained at 20 nm with sharp and deeper reflective characteristics for the proposed SPR sensor. The proposed sensor's performance was analyzed at optimum parametric values ($D_{Ag\text{-}Au} = 50$ nm, $D_{ITO} = 20$ nm) with different RI values of the sensing medium. The Fig. 3b shows the important characteristic i.e. sensitivity of the sensor. It is clear that as the value of RI increases from 1.33 to 1.37, the SPR reflectance spectra

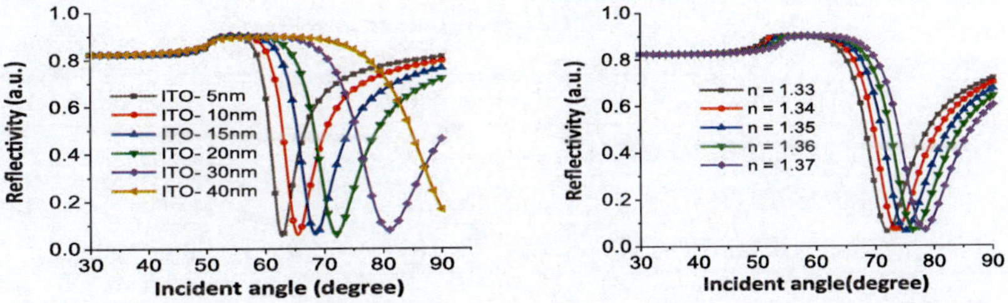

Fig. 3. **(a)** Change in SPR reflectivity concerning the different thickness values of the ITO layer for the proposed sensor. **(b)** Change in SPR reflectivity of the proposed sensor (SF11/Ag-Au/ITO/SM) for different refractive indexes (varying from 1.33–1.37) of the sensing medium.

move towards the redshift. The proposed SPR sensor exhibits the highest sensitivity of 183°/RIU. A high sensitivity value indicates that the sensor has good sensing ability with high resolution. In addition, the detection accuracy is also calculated to determine how the sensor can detect the minimum concentration of a substance in a sensing sample. All the performance parameters are calculated and shown in Table 2.

Table 2. Performance parameters of the proposed SPR Sensor

RI values (n)	Sensitivity (°/RIU)	θ_{SPR} (°)	Shift in θ_{SPR} (°)	Detection Accuracy (1/°)	Figure of Merit (1/*RIU*)	FWHM (°)	Reflectivity
1.34	160	73.51	1.6	0.0884	14.14	11.3	0.0627
1.35	150	75.08	1.57	0.085	13.345	11.7	0.063
1.36	150	76.67	1.59	0.0819	13.022	12.2	0.0655
1.37	183	78.52	1.83	0.079	15.01	12.6	0.0671

At last, we give a performance comparison of the proposed multilayer structured SPR sensor with available research work in a tabular form. Table 3 shows how well the proposed sensor outperforms compared to other sensors. It is simple in design and has achieved the highest sensitivity.

Table 3. Comparison of sensitivity value to previously published work of SPR sensor

S. No	Configuration	S (°/ RIU)	Research work
1	Glass/Ag/Sio_2/Zro_2/SM	98.04	[14]
2	Glass/Ag/dielectric/Au/SM	50	[15]

(*continued*)

Table 3. (*continued*)

S. No	Configuration	S (°/ RIU)	Research work
3	2SG2/Al/Si/BP/graphene	148.2	[16]
4	BK7/Au/Bi_2Te_3/SM	175	[17]
5	Glass/Ag-Au/ITO/SM	183	Proposed

4 Conclusions and Future Scope

An angular interrogation technique based on MATLAB modelling is used to examine the sensitivity of the proposed SPR sensor. It is shown that adding a protective layer (metal oxide) on the metal surface improves the performance and sensitivity of existing SPR sensors. Other performance parameters, like FOM and DA, also had significant values. Further research has been carried out on the impact of geometrical and optical parameters on the biosensor response, which aids in determining the performance characteristics of the designed sensor. Compared with available research with various multi-layer structured SPR sensors, the proposed sensor (SF11/Ag-Au/ITO/SM) demonstrated the highest sensitivity of 183°/RIU at 633 nm source wavelength using multi-layer, structured SPR sensors. The proposed sensor has the capability to detect analytes with high sensitivity and is simple to fabricate.

References

1. Cheng, S.F., Chau, L.K.: Colloidal gold-modified optical fiber for chemical and biochemical sensing. Anal. Chem. **75**(1), 16–21 (2002)
2. Jorgenson, R., Yee, S.: A fiber-optic chemical sensor based on surface plasmon resonance. Sens. Actuators B Chem. **12**(3), 213–220 (1993)
3. Hill, R.T.: Plasmonic biosensors. WIREs Nanomed. Nanobiotechnol. **7**(2), 152–168 (2014)
4. Mahmood, R., Johnson, M.B., Hillier, A.C.: Massive enhancement of optical transmission across a thin metal film via wave vector matching in grating-coupled surface plasmon resonance. Anal. Chem. **91**(13), 8350–8357 (2019)
5. Mrksich, M., Sigal, G.B., Whitesides, G.M.: Surface plasmon resonance permits in situ measurement of protein adsorption on self-assembled monolayers of alkanethiolates on gold. Langmuir **11**(11), 4383–4385 (1995)
6. Syed Nor, S.N., Rasanang, N.S., Karman, S., Zaman, W.S.W.K., Harun, S.W., Arof, H.: A review: surface plasmon resonance-based biosensor for early screening of SARS-CoV2 infection. IEEE Access **10**, 1228–1244 (2022)
7. Szunerits, S., Castel, X., Boukherroub, R.: Preparation of electrochemical and surface plasmon resonance active interfaces: deposition of indium tin oxide on silver thin films. J. Phys. Chem. C **112**(29), 10883–10888 (2008)
8. Szunerits, S., Shalabney, A., Boukherroub, R., Abdulhalim, I.: Dielectric coated plasmonic interfaces: their interest for sensitive sensing of analyte-ligand interactions. Rev. Anal. Chem. **31**(1) (2012)
9. Rafi, H.N., Kaysir, R., Islam, M.J.: Air-hole attributed performance of photonic crystal fiber-based SPR sensors. Sens. Bio-Sens. Res. **29**, 100364 (2020)

10. Ghatak, A., Thyagarajan, K.: An Introduction to Fiber Optics, 1st edn. Cambridge University Press, New Delhi (1998)
11. Villar, I.D., et al.: Generation of lossy mode resonances by deposition of high-refractive-index coatings on uncladded multimode optical fibers. J. Opt. **12**(9), 095503 (2010)
12. Homola, J.: On the sensitivity of surface plasmon resonance sensors with spectral interrogation. Sens. Actuators B Chem. **41**(1–3), 207–211 (1997)
13. Xiang, Y., Zhu, J., Wu, L., You, Q., Ruan, B., Dai, X.: Highly sensitive terahertz gas sensor based on surface plasmon resonance with graphene. IEEE Photonics J. **10**(1), 1–7 (2018)
14. Bao, S., Li, H.-J., Zheng, G.: Concentration sensor with multilayer thin filmoupled surface plasmon resonance. Optoelectron. Lett. **17**(5), 289–293 (2021). https://doi.org/10.1007/s11801-021-0088-4
15. Yadav, A., Sharan, P., Kumar, A.: Surface plasmonic resonance based five layered structure-biosensor for sugar level measurement in humans. Results Opt. **1**, 100002 (2020)
16. Su, M., et al.: Black phosphorus (BP)–graphene guided-wave surface plasmon resonance (GWSPR) biosensor. Nanophotonics **9**(14), 4265–4272 (2020)
17. Zhao, Y., Gan, S., Zhang, G., Dai, X.: High sensitivity refractive index sensor based on surface plasmon resonance with topological insulator. Results Phys. **14**, 102477 (2019)

Formation Control Scheme of Multiple Surface Vessels with Model Predictive Technique

Thanh Trung Cao, Manh Hung Vu, Van Chung Nguyen, The Anh Nguyen, and Phuong Nam Dao(✉)

School of Electrical and Electronic Engineering, Hanoi University of Science and Technology, Hanoi, Vietnam
nam.daophuong@hust.edu.vn

Abstract. This article studies formation control consideration of cooperative path following problem in a group of multiple Surface Vehicles (SVs). A proposed formation control protocol that contains optimal control problem in two subsystems of each SV with Model Predictive Control (MPC) and Approximate/Adaptive reinforcement Learning (ARL) Controller. The MPC is developed for nonlinear sub-system of SV with the tracking performance to be guaranteed by considering an appropriate optimization problem. Moreover, RL control design is carried out for time-varying sub-system by indirect method. Finally, the proposed control protocol is demonstrated by simulation result to show the effectiveness of this control protocol.

Keywords: Formation control · Surface Vehicles (SVs) · Model Predictive Control (MPC) · Approximate/Adaptive reinforcement Learning (ARL) Controller

1 Introduction

As a typical consideration of multiagent systems, the formation control problem multiple Surface Vehicles (SVs) has got much attention from many researchers [1–3]. Studies in this area are motivated by extending the consideration of each SV with kinematic and dynamic sub-systems, which implies the challenges in establishing the formation control. Additionally, due to the difficulties of the unification between trajectory tracking control problem and optimal control requirement, almost researches are focused on Lyapunov stability theory using conventional nonlinear control schemes for single robotic system as well as multiple robotics. Compared with previous references [1–3] implementing Back-stepping technique for multiple SVs, this article extends the consideration of optimal control problem as well as the unification with trajectory tracking control effectiveness. Moreover, unlike the related references only consider MPC as well as RL based optimal control [4–6], the advantages of the proposed control framework in this article are shown that optimal control performance is satisfied not only in kinematic sub-system with MPC but also in dynamic model by RL strategy. The

Supported by School of Electrical and Electronic Engineering, Hanoi University of Science and Technology.

T. D. L. Nguyen et al. (Eds.): ICISN 2023, LNNS 752, pp. 312–317, 2023.
https://doi.org/10.1007/978-981-99-4725-6_39

organization of this article is given as follows. The preliminary and problem statement are discussed in Sect. 2. In Sect. 3, the formation control structure is proposed with the combination between three control loops. In Sect. 4, the simulation implementations are given to illustrate.

2 Preliminary and Problem Statement

In this section, we considered the group of N surface vehicles. With each agent, after disregarding motion in heaven, roll, and pitch axis, the model of each surface vehicle with three degrees of freedom (3DOF) is taken into consideration like:

$$\dot{\eta_i} = J(\eta_i)\xi_i,\ M_i\dot{\xi_i} + C(\xi_i)\xi_i + D(\xi_i)\xi_i + g(\eta_i) = \tau_i + \Delta(\eta_i, \xi_i) \tag{1}$$

Where $\eta_i = [x_i, y_i, \psi_i]^T$ denoted the ith SV's the position of frame (x, y) and the heading angle ψ, and let $\xi_i = [u_i, v_i, r_i]^T$ be surge, sway and yaw linear velocities of ith SV. Moreover, the matrices $J(\eta_i)$, M_i, $C(\xi_i)$ and $D(\xi_i)$ in (1) described in [6,7]. The formation control strategy is to find distributed control design τ_i for obtaining the cooperative path following problem as shown in Fig. 1.

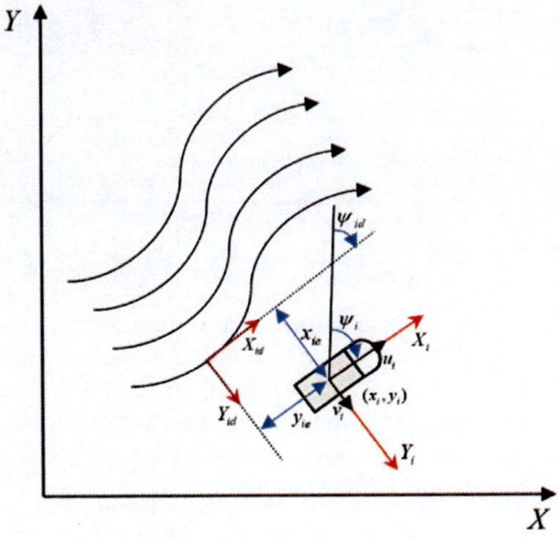

Fig. 1. Diagram shows the cooperative path following problem [3]

3 Formation Control

3.1 Cooperative

Considering the high-level protocol in Fig. 1 to archive the formation. According to (1), the kinematics model of surface vehicle i^{th} can be rewritten as:

$$\begin{cases} \dot{x_i} = u_i \cos(\psi_i) - v_i sin(\psi_i) \\ \dot{y_i} = u_i \sin(\psi_i) - v_i \cos(\psi_i) \\ \dot{\psi_i} = r_i \end{cases} \tag{2}$$

Assumption 1. $x_{id}(\theta_i), y_{id}(\theta_i)$ are the reference curve path of agent i^{th} defined by the path variable θ_i follow the curve of the river, which is defined by:

$$\theta_i = v_s - \omega_i \tag{3}$$

Where v_s is the reference speed and w_i is the cooperative variable. Therefore the tracking error x_{ie}, y_{ie} of agent i^{th} can be defined as follows:

$$\begin{cases} x_{ie} = (x_i - x_{id}) \cos(\psi_{id}) + (y_i - y_{id}) \sin(\psi_{id}) \\ y_{ie} = -(x_i - x_{id}) \sin(\psi_{id}) + (y_i - y_{id}) \cos(\psi_{id}) \end{cases} \tag{4}$$

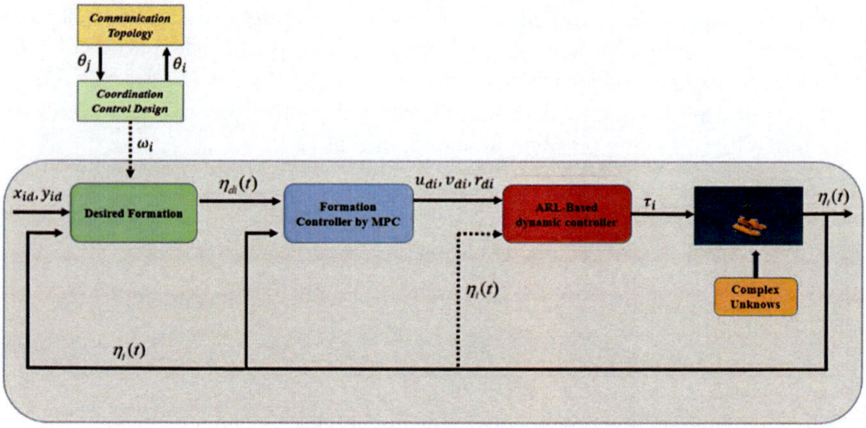

Fig. 2. Schematic representation of the system and control strategy

Based on the formation of neighbors, the error of the formation was proposed to archive the cooperative path following, the error is defined by:

$$e_i = \sum_{i \in \mathcal{N}_i} a_{ij}(\theta_i - \theta_j) \tag{5}$$

According to Graph Theory, the cooperative error can be rewritten as $e = \mathcal{L}\theta$ and the error can be calculated by:

$$\dot{e} = -\mathcal{L}\omega \tag{6}$$

In order to update the trajectories for each agent at each cycle, updating the cooperative variables is necessary, so an intermediate cooperative variable $\bar{\omega_i}$ is given in [3] and the law of updating the variable ω_i is defined as follows:

$$\begin{cases} \omega_i = -k_{\omega i} x_{ie} u_{id}^* + k_{\omega i} e_i + \bar{\omega_i} \\ \dot{\bar{\omega_i}} = -k_{\bar{\omega} i} \bar{\omega_i} - x_{ie} u_{id}^* + e_i \end{cases} \tag{7}$$

3.2 NMPC

In Fig. 2, consider a continuous time nonlinear model of the vessel following the form of (1) subject to the constraints:

$$\begin{aligned} \eta_i &\in \aleph \\ \xi_i &\in F \end{aligned} \tag{8}$$

Where $\eta \in \Re^n$ is the vector of position, $\xi \in \Re^n$ is the control input vector. $\aleph \in \Re^n$ and $F \in \Re^n$ are compact sets containing the origin in their interior points. The initial states have a value equal to 0 at time 0. In general terms, the scheme of NLMC is to predict the future position of each agent following the (2) by updating cooperative variable using [3] under some finite time prediction horizon. The NMPC algorithm has a periodicity, at the first sample, the control value is calculated to make the value of the cost function as small as possible while ensuring the stability of the system. In the following sample, the calculation is repeated. The desired trajectory of each agent can be calculated by using the cooperative variable by solving [3], the desired trajectory defined by: $\eta_{di} = (x_{id}(\theta_i), y_{id}(\theta_i), \psi_{id})^T$

Remark 1. The NMPC control input can be calculated by minimizing the cost function of the system inputs and states from the reference trajectories, the control input has been defined as: $\xi_{di} = [u_{di}, v_{di}, r_{di}]^T$.

The form of the finite horizon cost function is defined by:

$$\text{minimize} \quad J(\eta, \xi) = \int_{t_0}^{t_0+N_p} C(\eta(t), \xi(t))dt + G(\eta(N_p)) \tag{9}$$

$$\text{subject to:} \quad \dot{\hat{\eta}}(t) = f(\eta(t)) + J(\hat{\eta}(t))\hat{\xi}(t),\ \hat{\eta}(t) \in \aleph,\ \hat{\xi}(t) \in F$$

where: $C(\eta(t), \xi(t)) = \left\| e(t) \right\|_X^2 + \left\| \Delta\xi(t) \right\|_Y^2, G(\eta(N_p)) = \left\| \eta(T_p) - \eta_{id}(T_p) \right\|_Z^2$, with $e(t) = \eta(t) - \eta_{di}(t)$ and $\Delta\xi(t) = \xi(t) - \xi_{di}(t)$ are respectively predicted error trajectory and predicted change of control input made at time t. In Fig. 2, given a fleet of USVs $\mathcal{N} = \{1, 2, ..., N\}$, the distributed control law $\xi_{di} = [u_{di}, v_{di}, r_{di}]^T$ can be archived by solving (9). To guarantee the stability of each agent using control law $\xi_{di} = [u_{di}, v_{di}, r_{di}]^T$, it's necessary to use infinity prediction horizon by letting $N_p = \infty$, but the solution of the infinite horizon nonlinear optimization problem is not feasible. A nonlinear model predictive control (MPC) with finite horizon has been introduced in [4] to guarantee the tracking requirement.

4 Dynamic Control Design with ARL-Based for Each Surface Vehicle

The control design of sub-sytem in each SV (Fig. 2) is carried out using the dynamic equation (3) to be rewritten as:

$$M_i\dot{\xi}_i = \tau_i - C(\xi_i)\xi_i - D(\xi_i)\xi_i - g(\eta_i) \tag{10}$$

The objective is to design a controller with all signals are bounded and the coordination η_{di} is tracking to η_i. Each agent's primary tracking controller τ_i will be shown in two components. A model-based term is first created using the data related to the desired trajectory and the mathematics model. Second, each SV cascade control system's tracking problem is resolved, an RL-based optimal control is created for the transformed autonomous model. So we proposed the control law with two components as:

$$\tau_i = \tau_{di} + u_i \tag{11}$$

In the light of [6], we can approximate the Bellman function and the optimal controller by NN presented as follows:

$$\hat{V}(X_i) = \hat{W}_{ci}^T \Phi_i(X_i) \tag{12}$$

$$\hat{u}_i(X_i) = -\frac{1}{2} R_i^{-1} D_i^T(X_i) (\frac{\partial \Phi_i}{\partial x_i})^T \hat{W}_{ci} \tag{13}$$

Where the activation function $\Phi_i(X_i)$ is selected based on the type of system and the updated law for $\hat{W}_{ci}$ is given in [6].

Remark 2. It is worth noticing that the advantage of the proposed control design in this article is to guarantee optimal control problem in two control loops (Fig. 2) with MPC and RL techniques, which has not been mentioned in the previous researches [3].

5 Simulation Results

To show the effectiveness and feasibility of the control method, we consider a group of five SV agents with model parameters in each SV as follows:

$$M = \begin{bmatrix} 28.135 & 0 & 0 \\ 0 & 45.568 & 0 \\ 0 & 0 & 131.423 \end{bmatrix}, C = \begin{bmatrix} 0 & 0 & -45.568(v_2) \\ 0 & 0 & 28.135(v_1) \\ -8.7165(v_2) & -8.7165(v_1) & 0 \end{bmatrix} \tag{14}$$

In the outer loop of formation control system (Fig. 2), the reference velocity of each agent is chosen as: 0.6 m/s and the parameter of cooperative constants $k_{\omega i}$ as well as $k_{\bar{\omega} i}$ are the same value being 10. Additionally, the control weights of the MPC are selected as: $X = diag(50, 50, 25)$, $Y = diag(0.1, 0.1, 0.2)$, and $Z = 0$. The design parameters of sample time and the number of steps of prediction horizon are chosen $t = 0.2(s), N_p = 10$, respectively. On the other hand, the smooth activation function in ARL based dynamic controller (Fig. 2) is given for each SV as:

$$\Phi(X) = [X_1^2, X_1X_2, X_1X_3, X_2^2, X_2X_3, X_3^2, X_1^2X_7^2, X_2^2X_8^2, X_3^2X_9^2, X_1^2X_4^2, X_2^2X_5^2, X_3^2X_6^2]^T \tag{15}$$

As shown in Fig. 3, the 5 SVs group is moved and converged to desired formation under the proposed control system in Fig. 2. Thus, the simulation result further show the effectiveness of the presented formation controller.

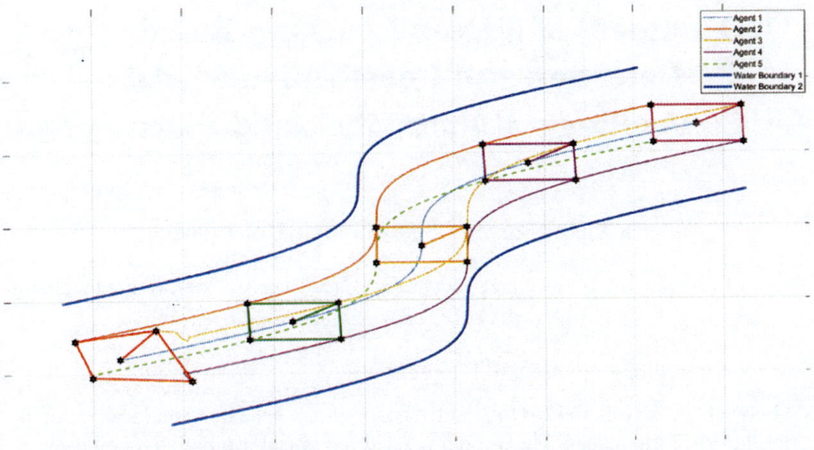

Fig. 3. Formation illustration of five agents

6 Conclusions

This article presented a formation control structure for multiple SVs with two optimal control loops to be considered. The kinematic sub-system and dynamic sub-system of each SV are investigated using nonlinear MPC strategy and ARL based dynamic controller, respectively. The unification between optimal control performance and trajectory tracking problem is described by simulation studies.

References

1. Dai, S.-L., Lu, K., Jin, X.: Fixed-time formation control of unicycle-type mobile robots with visibility and performance constraints. IEEE Trans. Ind. Electron. **68**(12), 12 615–12 625 (2020)
2. Dai, S.-L., Lu, K., Fu, J.: Adaptive finite-time tracking control of nonholonomic multirobot formation systems with limited field-of-view sensors. IEEE Trans. Cybernet. **52**, 10695–10708 (2021)
3. Huang, Y., Wu, D., Li, L., Feng, N.: Event-triggered cooperative path following control of multiple underactuated unmanned surface vehicles with complex unknowns and actuator saturation. Ocean Eng. **249**, 110740 (2022)
4. Dao, P.N., Nguyen, H.Q.: Robust model predictive kinematic tracking control with terminal region for wheeled robotic systems. Automatika časopis za automatiku, mjerenje, elektroniku, računarstvo i komunikacije **62**(3-4), 513–519 (2021)
5. Dao, P.N., Liu, Y.-C.: Adaptive reinforcement learning in control design for cooperating manipulator systems. Asian J. Control **24**, 1088–1103 (2022)
6. Pham, T.L., Dao, P.N., et al.: Disturbance observer-based adaptive reinforcement learning for perturbed uncertain surface vessels. ISA Trans. **130**, 277–292 (2022)
7. Vu, V.T., Tran, Q.H., Pham, T.L., Dao, P.N.: Online actor-critic reinforcement learning control for uncertain surface vessel systems with external disturbances. Int. J. Control Autom. Syst. **20**(3), 1029–1040 (2022)

Development of a Deep Learning-Based Object Detection and Localization Model for Controlling a Robotic Pick-and-Place System

Ha Xuan Nguyen(✉) and Phuc Hong Pham

Hanoi University of Science and Technology, No. 1 Dai Co Viet, Hanoi, Vietnam
ha.nguyenxuan@hust.edu.vn

Abstract. A new model for the multi-task of object detection and localization is developed in this paper. This model is used to control a robotic pick-and-place system for real-time applications. The model is developed from a deep neural network, namely RetinaFace with MobileNet as its backbone, with modifications to the output to allow simultaneous detection and localization of objects. A self-generated dataset was prepared for the training and testing processes. The error calibration of the camera is implemented. The robot's overall control algorithm is created. The results show that our model has a high accuracy of 97.4% for object detection and an error of less than 2.41 mm for object localization. The computation is also efficient, reaching 25 FPS with the very lightweight hardware of the Jetson Nano. Experiments with the pick-and-place method have a success rate of 100%.

Keywords: Object Detection · Deep Learning · Robotic Manipulation

1 Introduction

With the development of deep learning-based computer vision, there have been many applications of the robotic-vision-based pick-and-place task [1]. Unlike traditional methods, which mostly use pure image processing algorithms with the library OpenCV [2], deep learning-based methods allow for object detection and localization with higher accuracy and reliability [3]. For the pick-and-place tasks, the robotics system must overcome four steps: i) Object localization; ii) Object pose estimation; iii) Grasp estimation; and iv) Motion planning depending on the types of end-effectors, the pick-and-place strategies can be varied. There are two typical types of end-effectors: gripper-type and suction-based end-effectors. The use of grippers has many advantages when grasping complex-geometry objects, but it requires much more complicated image processing algorithms since the orientation of the object must be determined. For the tasks of object detection, localization, and pose estimation, RGB-D cameras are mostly used [4, 5]. RGB-D cameras are used for sensor fusion, which is then combined with the data processing algorithms (image or laser data). The three data processing tasks (localization, pose estimation, and grasp estimation) are performed separately using deep neural networks. A large amount of data is generated from RGB-D cameras, and these processing

T. D. L. Nguyen et al. (Eds.): ICISN 2023, LNNS 752, pp. 318–325, 2023.
https://doi.org/10.1007/978-981-99-4725-6_40

tasks require much computational effort. Thus, multi-task learning with lightweight deep networks is expected to reduce the complexity of the processing and achieve high computational efficiency 6.

In this work, a new approach to developing a multi-task deep neural network model is introduced. A model is developed based on the network architecture of the RetinaFace [7] and the MobileNet [8] backbone. The developed model allows for simultaneous detection and localization of objects. The model was trained and tested based on our self-generated dataset. The outputs of the model are used to determine control parameters for the robotic-based pick-and-place system. The invert kinematics, working space, and path planning issues of the robot are solved. Experiments are implemented based on the hardware Jetson Nano of Nvidia [9] to evaluate the accuracy, computational efficiency, and reliability of the system.

2 Implementation

2.1 System Description

The overall architecture of the design system is shown in Fig. 1. Core components of the system consist of a camera from Cisum [10], a 5-DOF Yahboom [11] robot, and a conveyor, all of which are controlled via an embedded computer of the Jetson Nano. The camera is used to collect data for training a deep neural network model and later to capture images for the task of object detection and localization. The conveyor supplies billets via the billet feeder and moves them to the region of interest, which is inside the working space of the robot. The Yahboom robot is used for the task of picking and placing objects in a predefined position. Depending on the type of billet, the robot will place them in the corresponding discharge bins.

The system is operated automatically via a control program, in which the processing pipeline is described in Fig. 2. First, images captured from the camera is sent to the embedded computer, which are the processed by the object classification and localization. After this process, objects in each image frame are detected with the coordinates of their key points. The transformation between the coordinates of the camera and the origin coordinates of the robot is performed to get object positions. The invert kinematics and path planning are solved in the second step, where the inputs are object positions, and the outputs are joint variable values of the robot. These parameters are sent to the microcontroller, which then sends them to the power driver, which drives the motor in each joint. The overall processing pipeline is run on the Jetson Nano, which is equipped with GPU-based computational acceleration.

2.2 Object Detection and Localization

Unlike conventional image processing methods using the OpenCV library, in this paper, a model based on a deep neural network is developed. The model allows for the detection and classification of objects having similar features like color and dimension. Furthermore, the model can localize not only the position but also the orientation of objects with complex and asymmetric boundaries. This localization makes robotic-based assembly

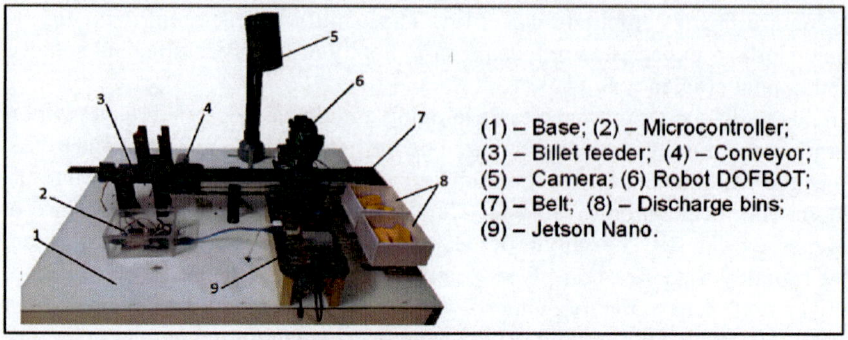

Fig. 1. Description of the robotic pick-and-place system.

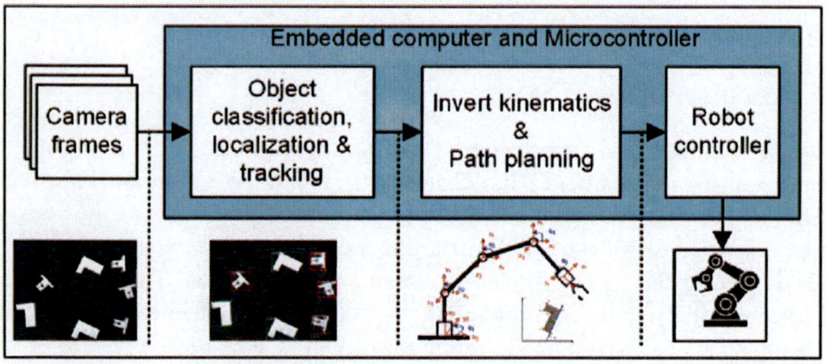

Fig. 2. Overall processing pipeline for the robotic pick-and-place system.

processes more feasible and reliable. In order to make the model lightweight and thus computationally efficient, multi-task learning is applied, which combines the detection, classification, and localization tasks in only one model. The RetinaFace with MobileNet backbone is selected for the developing model. Two types of objects, whose shapes are shown in Fig. 2, are used for the evaluation. The size of objects has a bounding dimension of 40 mm × 31 mm × 13 mm. The weight of the objects is quite small, so dynamic effects caused by the object are neglected. A dataset, which contains 3896 training images and 491 testing ones, is generated.

Objects in each image are labelled with their class type (class_id), the coordinates of two conners of the bounding box ((x_i, y_i), $i = 1, 2$) and coordinates of six keypoints ((p_{xj}, p_{yj}), $j = 1 \div 6$). The output of the RetinaFace network is modified to meet the requirement. In addition, for multi-task learning, the loss function is set according to Eq. (1), where $loss_{bbox}$, $loss_{cls}$, and $loss_{land}$ are loss regression for or bounding box prediction, landmark localization, and object classification, respectively. The parameters α, β, γ were empirically chosen ($\alpha = 2$, $\beta = 1$, $\gamma = 1$). The model was fine-tuned with in 250 epoches, bathsize of 32, learning rate of 0.0.1, and SGD optimizer.

$$loss = \alpha.loss_{bbox} + \beta.loss_{cls} + \gamma.loss_{land} \quad (1)$$

2.3 Invert Kinematics and Control

To determine the parameters of a joint (q_i, $i = 1 \div 5$), invert kinematic equations are established and solved. For the robotic-based pick-and-place task using a gripper, both the position and orientation of the end-effector (the gripper) must be pre-defined. Since the fourth link of the robot must always be perpendicular to the picking plane, the orientation of the gripper can be defined via the fifth joint parameter q_5. This parameter can be determined based on the orientation of picking objects lying on the working plane via the coordinates of their key points, according to Eq. (2). The picking position on the object is calculated from Eq. (3).

$$q_5 = arctan\left[(p_{y2} - p_{y3})/(p_{x2} - p_{x3})\right] \tag{2}$$

$$\begin{cases} x_E = (p_{x2} + p_{x3} + p_{x4} + p_{x5})/4 \\ y_E = (p_{y2} + p_{y3} + p_{y4} + p_{y5})/4 \end{cases} \tag{3}$$

Table 1. Dernavit-Hartenberg parameters.

Link	θ_i	d_i	a_i	α_i
1	q_1	l_1	0	$\pi/2$
2	q_2	0	l_2	0
3	q_3	0	l_3	0
4	q_4	0	l_4	0

The coordinate systems for the robot are drawn in Fig. 3a. Accordingly, the Dernavit-Hartenberg parameters are set as in Table 1. Based on these parameters, the invert kinematic equation is obtained using a homogeneous transformation matrix as in Eq. (4) and coordinate homogenization as in Eq. (5). Solving Eq. (5), the joint parameters are obtained according to Eqs. (6)–(9). The working space of the robot is defined by Eqs. (10) and (11) and is determined by geometric parameters and joint variable constraints.

The joint parameters are determined by the inverted kinematic problem, which takes into account the working space constraint and optimal path planning to avoid collision. This task is performed on the embedded computer, the Jetson Nano. The obtained parameters are then sent to the microcontroller Arduino, which uses a Proportional Integral Derivative controller (PID) to control the motors and actuate the joints to the desired positions. Parameters of the PID controller are adapted from the robot's manufacturer.

$$T_4^0 = T_1^0.T_2^1.T_3^2.T_4^3 = \begin{bmatrix} a_{11}(q) & a_{12}(q) & a_{13}(q) & a_{14}(q) \\ a_{21}(q) & a_{22}(q) & a_{23}(q) & a_{24}(q) \\ a_{31}(q) & a_{32}(q) & a_{33}(q) & a_{34}(q) \\ a_{41}(q) & a_{42}(q) & a_{43(q)} & a_{44}(q) \end{bmatrix} \tag{4}$$

$$a_{14}(q) = x_E;\ a_{24}(q) = y_E;\ a_{34}(q) = z_E; \tag{5}$$

$$q_1 = arctan(y_E/x_E) \tag{6}$$

$$q_2 = arctan\frac{z_E - l_1}{\sqrt{x_E^2 + y_E^2}} + arccos\frac{l_3^2 - l_2^2 - \left[(z_E - l_1)^2 + x_E^2 + y_E^2\right]}{-2.l_2.l_3} \tag{7}$$

$$q_3 = \pi - arccos\frac{(z_E - l_1)^2 + x_E^2 + y_E^2 - l_3^2 - l_2^2}{-2.l_2.l_3} \tag{8}$$

$$q_4 = \frac{\pi}{2} - q_3 + q_2 \tag{9}$$

$$r_{min} = \sqrt{l_3^2 - \left[(l_1 + l_2 + z) - \left(l_4 + l_g\right)\right]^2} \tag{10}$$

$$r_{max} = \sqrt{(l_2 + l_3)^2 - \left[\left(l_4 + l_g\right) - (l_1 + z)\right]^2} \tag{11}$$

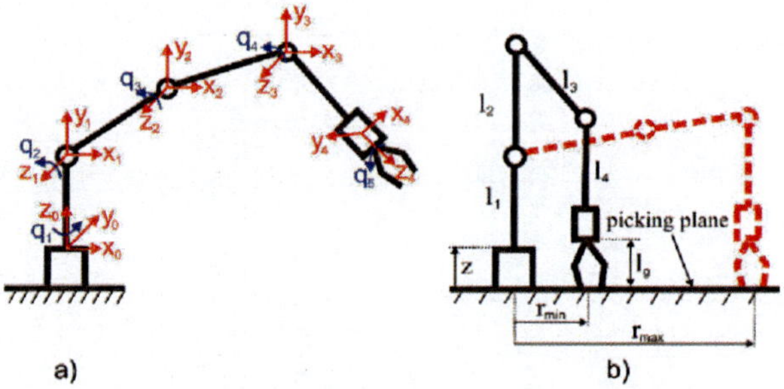

Fig. 3. Kinematics and working space.

3 Results and Discussions

The developed object detection and localization model was evaluated via a standard benchmark. For the object detection task, the benchmark average precision (AP) on several thresholds of the intersection over union (IoU) is used. Table 2 shows the evaluation results. It is evident that the accuracy of the detection task is quite high. The mean average precision reaches 97.4% for all thresholds. In this paper, only two classes of objects are considered. So, the shape of objects is quite easy to differentiate. This result can be extended to the detection of more complex classes of objects without any difficulty.

For the object localization task, the position of key points on objects must be predicted. According to Eq. (12), the accuracy is evaluated via the benchmark percentage of correct key points (PCK) over different error thresholds. If the prediction error of the key point is smaller than a threshold, this prediction is true; otherwise, it is false. Table 3 shows the results. It is seen that for an error threshold smaller than 3 mm, the model predicts the position of key points very well with a PCK of 92.99%. If the threshold is decreased to 2 mm, the prediction accuracy is reduced to 77.18%. The prediction accuracy can also be evaluated via the average error according to Eq. (13), where p_i is the ground truth, $\tilde{p}_i$ is the predicted key point, and N is the number of key points. The average deviation is found to be 1.1 mm. With this level of accuracy, this result can be applied to the task of robotic-based pick-and-place.

$$PCK = \frac{\sum correct\ keypoints}{\sum all\ keypoints} \quad (12)$$

$$e = \frac{1}{N}\sum_{i=1}^{N} |p_i - \tilde{p}_i| \quad (13)$$

Table 2. Accuracy of the object detection task using average precision (AP).

Class	AP50	AP_{60}	AP_{70}	AP_{80}	AP_{90}	AP_{95}	mAP
0	1	1	0.996	0.995	0.99	0.925	0.974
1	0.995	0.995	0.994	0.994	0.962	0.845	

Table 3. Evaluation of measurement error by PCK method.

Error (mm)	< 0.5	< 1.0	< 1.5	< 2.0	< 2.5	< 3.0
PCK (%)	11.47	35.85	60.89	77.18	86.8	92.99

In fact, localization errors can appear during the labeling process. Since a two-dimensional camera is used, images of objects that are not at the center of the viewpoint of the camera can be distorted, leading to the inaccuracy of the key point labeling. Thus, this issue is also evaluated and solved. The discrepancy between the coordinates of key points and those of ground truth is measured, and the error is consequently calculated according to Eq. (14). Fifty sampling points are measured for the calibration. Based on the calibration, the compensation for labeling error is obtained via a linear regression as in Eq. (15). The labeling error is shown in Table 4. It is visible that after the calibration process, the error is reduced from 4.53 mm to 1.31 mm. So, the total error of the localization of the key point in both the labeling process and the prediction is 2.41 mm. This value is acceptable for practical applications.

$$e_{x,y} = \frac{1}{N}\sum_{i=0}^{N} e^{i}_{x,y} \text{ and } e = \sqrt{e_x^2 + e_y^2} \quad (14)$$

$$Calib_x(x) = 0.0527x - 8.636;\ Calib_y(y) = 0.0296y - 3.0871 \quad (15)$$

The computational efficiency is also evaluated. The developed model is converted and quantized to run on the embedded computer of the Jetson Nano. The total processing

Table 4. Evaluation of measurement error by PCK method.

Error (mm)	e_x	e_y	e
Before calibration	3.133	3.273	4.53
After calibration	0.836	1.004	1.31

time is 50 ms (25FPS). The MobileNet network's lightweight backbone has the advantage of having a low computational time, allowing for real-time applications. With the vision system's accuracy improved, pick-and-place experiments on the entire system were carried out. It is achieved that all one hundred tests are 100% successful. This confirms the reliability of the developed system.

4 Conclusions and Outlook

In this work, a deep learning model for the multi-task of object detection and localization has been successfully developed. The model can be used for controlling robotic systems to perform pick-and-place tasks. The accuracy and the computational efficiency are of high quality and can thus be exploited for many real-time applications without requiring a high-performance computer. The approach used in this paper is simple and efficient, which has advantages compared to other approaches using stereo cameras or laser-based point clouds, which typically require complex image processing algorithms and thus lead to computational inefficiency. In the future, the deployment of our model for much more complex application scenarios is planned. Further optimization for computational efficiency and robustness under high-speed pick-and-place conditions will be implemented.

Acknowledgement. This research is funded by the Vietnam National Foundation for Science and Technology Development (NAFOSTED) under Grant number "107.01-2019.05".

References

1. Du, G., Wang, K., Lian, S., Zhao, K.: Vision-based robotic grasping from object localization, object pose estimation to grasp estimation for parallel grippers: a review. Artif. Intell. Rev. **54**(3), 1677–1734 (2021)
2. OpenCV library Homepage. https://opencv.org/. Accessed 09 Nov 2021
3. Sharma, V., Mir, R.N.: A comprehensive and systematic look up into deep learning based object detection techniques: a review. Comput. Sci. Rev. **38**, 100301 (2020)
4. Dhillon, A., Verma, G.K.: Convolutional neural network: a review of models, methodologies and applications to object detection. Progr. Artif. Intell. **9**(2), 85–112 (2019). https://doi.org/10.1007/s13748-019-00203-0
5. Gao, M., Jiang, J., Zou, G., John, V., Liu, Z.: RGB-D-based object recognition using multimodal convolutional neural networks: a survey. IEEE Access **7**, 43110–43136 (2019)
6. Crawshaw, M.: Multi-task learning with deep neural networks: a survey. arXiv preprint arXiv:2009.09796 (2020)

7. Deng, J., Guo, J., Zhou, Y., Yu, J., Kotsia, I., Zafeiriou, S.: RetinaFace: single-stage dense face localisation in the wild, arXiv:1905.00641v2 (2019)
8. Sandler, M., Howard, A., Zhu, M., Zhmoginov, A., Chen, L.C.: Mobilenetv2: inverted residuals and linear bottlenecks. In: Proceedings of the IEEE Conference on Computer Vision and Pattern Recognition, pp. 4510–4520 (2018)
9. Jetson Nano Developer Kit Homepage. https://developer.nvidia.com/embedded/jetson-nano-developer-kit
10. Cimsum Document Camera Scanner. https://www.cimfax.com/scanneren/index.html
11. Yahboom Robot Arm. http://www.yahboom.net/study/Dofbot-Jetson_nano

$\mathbf{H}_\infty$ Optimal Full-State Feedback Control for a Ball-Balancing Robot

Duc Cuong Vu[1], Thuy Hang Nguyen Thi[1,2], Dinh Dat Vu[3], Viet Phuong Pham[1], Danh Huy Nguyen[1], and Tung Lam Nguyen[1](✉)

[1] Hanoi University of Science and Technology, Hanoi, Vietnam
cuong.vd202313@sis.hust.edu.vn,
{phuong.phamviet,huy.nguyendanh,lam.nguyentung}@hust.edu.vn
[2] Thuy Loi University, Hanoi, Vietnam
hangntt@tlu.edu.vn
[3] Hung Yen University of Technology and Education, Hai Duong, Vietnam
vudinhdat@utehy.edu.vn

Abstract. A ball-balancing robot is a robot that can move and balance on a ball. This research comes from system modeling, finding the binding equations between state variables of the system through a differential equation. In order to construct the state-space model, Lagrange equations are used for each of the three planes. The system is linearized around the equilibrium position, to approximate the system to a linear system. This controller aims to minimize the influence of exogenous output, and in this paper, this control method is approached by linear matrix inequalities. Besides, the LQR controller is also presented to compare with the controller base on $\mathbf{H}_\infty$ method. For the xy plane, it does not affect the balance of the system, thus, the controller in this plane is not presented.

Keywords: Ball-balancing robot · $\mathbf{H}_\infty$ · Linear matrix inequalities · LQR

1 Introduction

As presented in [1], A robot that balances a ball called a ballbot consists of a ball, a body, and several motors to direct the movement of the ball through the wheels. The first version of the ballbot was made by Ralph Hollis at Carnegie Mellon University (CMU) in 2005 [1]. The ballbot from CMU was controlled by an inverse mouse-ball driven with four DC motors. The second ballbot from Tohoku Gakuin University (TGU) in 2008 [2]. Besides, there is another ballbot version from Amirkabir University of Technology [3], University of Adelaide [4] in 2009, and Aalborg University [5] in 2019, etc. In this paper, the ballbot version used is the ballbot model that was developed at Hanoi University of Science and Technology. The parameters of this ballbot version are shown in Table 1.

There are two directions to approach the modeling of the ballbot. One of them is to separate the 3D ballbot model into 3 independent planes, then the

T. D. L. Nguyen et al. (Eds.): ICISN 2023, LNNS 752, pp. 326–334, 2023.
https://doi.org/10.1007/978-981-99-4725-6_41

equation of motion of the ballbot for each plane can easily be found. Following this method, the model of the ballbot is approached easier. However, it eliminates the relationships between the planes when the object is rotates. The second is to model the ballbot without separating it into 3 planes, this requires knowledge of physics as well as mathematics. Therefore, this model implemented this way brings highly accurate results. This article repeats the planar ballbot's model equations of motion, it reduces the computational, as the $\mathbf{H}_\infty$ control which is the primary tool used in this study.

The crucial contribution of this paper is the successful application of the $\mathbf{H}_\infty$ in control theory for the ballbot balancing. Prior to that, many control theories were proposed to balance the ballbot. For example, the LQR/PID controller was applied to the first version of the ballbot and the second version ballbot [2,6]. Otherwise, more controllers such as Sliding Mode Control (SMC) and Model Predictive Control (MPC) are also used in [5], etc. For this system, the $\mathbf{H}_\infty$ control was used in [7], nevertheless, in [7], the $\mathbf{H}_\infty$ method was approached by transferring functions and applying to the mouse type ballbot. In this paper, the $\mathbf{H}_\infty$ control is approached by linear matrix inequalities (LMIs) for balancing the ballbot, like in [8] use for the wheeled inverted pendulum systems. One disadvantage of this method is that it requires knowledge of algebraic transformations related to linear matrix inequalities. And the LQR controller is presented to compare with the $\mathbf{H}_\infty$ controll.

2 Planar Model of a Ballbot

To understand how the ballbot works, a simplified planar model in 2D was derived. The ballbot's 3D model is divided into 3 separate planes as xy, yz, and zx to create the planar model. Evidently, the balancing of the ballbot is unaffected by its state. In this study, the states in the xy plane are unnecessary, therefore will be ignored. For zx/yz planes, the state-space model is found by solving Euler - Lagrange equation. The detailed description of each step is written in [9]. Thus, matrix form of the equation of motion of the ballbot in yz/zx plane as follow (Fig. 1):

$$\mathbf{M}(\mathbf{q})\ddot{\mathbf{q}} + \mathbf{C}(\mathbf{q}, \dot{\mathbf{q}}) + \mathbf{G}(\mathbf{q}) = \mathbf{f}_{NP} \tag{1}$$

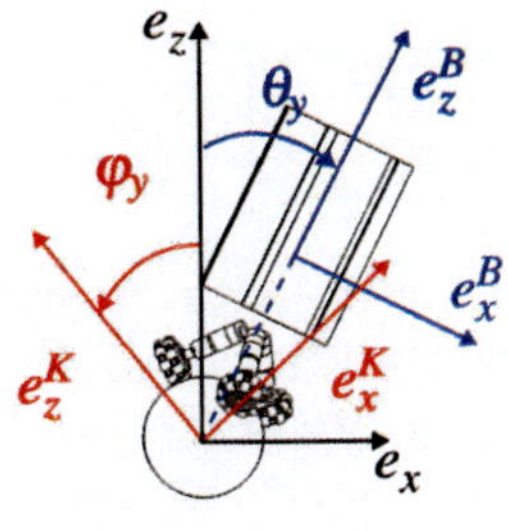

Fig. 1. Ballbot model on a plane

Table 1. Parameters of planar model of ballbot

Parameters	Symbols	Values
Mass of the ball	m_K	7.13 kg
Moment of inertia of the ball	J_K	0.041 kgm^2
Radius of the ball	r_K	0.12 m
Mass of the body	m_B	4.59 kg
Moment of inertia of the body	J_B	0.2 kgm^2
Distance form COM of body to COM of ball	l	0.5 m
Mass of the omniwheel	m_W	0.19 kg
Radius of the omniwheel	r_W	0.05 m
Moment of inertia of the omniwheel	J_W	2.3750e−04 kgm^2
Angle of the omniwheel	α	45°
Angle between motors in xy plane	β	120°
Gravity acceleration	g	9.81 m/s^2
Torque limit	T_{max}	5.5 Nm

where, $\mathbf{q} = \left[\varphi\ \theta\right]^T$, φ is the Euler angle of the ball, θ is the Euler angle of the body, the matrices containing masses and inertia $\mathbf{M}$, Coriolis forces $\mathbf{C}$, gravitational $\mathbf{G}$ and non - potential forces $\mathbf{f}_{NP}$ are defined below:

$$\mathbf{M} = \begin{bmatrix} m_t\, r_K^2 + J_K + \frac{r_K^2}{r_W^2} J_W & -\frac{-r_K}{r_W^2} r_t\, J_W + \gamma\, r_K \cos\theta \\ -\frac{-r_K}{r_W^2} r_t\, J_W + \gamma\, r_K \cos\theta & \frac{r_t^2}{r_W^2} J_W + J_B + m_B\, l^2 + m_W\, r_t^2 \end{bmatrix};$$

$$\mathbf{C} = \begin{bmatrix} -r_K\,\gamma\, \sin\theta\, \dot{\theta}^2 \\ 0 \end{bmatrix}; \quad G = \begin{bmatrix} 0 \\ -g\,\gamma\, \sin\theta \end{bmatrix}; \quad \mathbf{f}_{NP} = \begin{bmatrix} \frac{r_K}{r_W} T_x & \frac{r_K}{rW} T_x \end{bmatrix}^T$$

$$m_t = m_k + m_B + m_W; \quad r_t = r_K + r_W; \quad \gamma = l\, m_B + (r_K + r_W)\, m_W$$

Linearize the system by first-order Taylor expansion at equilibrium position with parameters of ballbot in Table 1, the matrices of the system are determined:

$$\mathbf{A} = \begin{bmatrix} 0 & 0 & 1 & 0 \\ 0 & 0 & 0 & 1 \\ 0 & -29.709 & 0 & 0 \\ 0 & 22.918 & 0 & 0 \end{bmatrix}; \quad \mathbf{B} = \begin{bmatrix} 0 \\ 0 \\ 18.39 \\ -5.53 \end{bmatrix}; \quad \mathbf{C} = \mathbf{I}_4; \quad \mathbf{D} = \begin{bmatrix} 0 \\ 0 \\ 0 \\ 0 \end{bmatrix}$$

where, I_4 is the unit matrix. The linearization of the system only approximates the system around the equilibrium position. With the above equations, it is obvious that this model is similar to an inverted pendulum.

Since the inputs in the above models are virtual torques, T_x, T_y, T_z, a transformation is needed to find the torques required for each motor T_1, T_2, T_3.

From [5], the relationships between the virtual and real torques of the omni-wheels are shown as below:

$$T_1 = \frac{1}{3}\left(T_z + \frac{2}{\cos\alpha}(T_z \cos\beta) - T_y \sin\beta\right) \tag{2}$$

$$T_2 = \frac{1}{3}\left(T_z + \frac{1}{\cos\alpha}(\sin\beta\,(-\sqrt{3}T_z + T_y) - \cos\beta\,(T_x + \sqrt{3}T_y))\right) \tag{3}$$

$$T_3 = \frac{1}{3}\left(T_z + \frac{1}{\cos\alpha}(\sin\beta\,(\sqrt{3}T_z + T_y) + \cos\beta\,(-T_x + \sqrt{3}T_y))\right) \tag{4}$$

3 LMIs for $\mathbf{H}_\infty$ Full-State Feedback Control

In [10], the model of a closed loop of an $\mathbf{H}_\infty$ problem was shown, where $\mathbf{w}$ is the exogenous, $\mathbf{z}$ is the control error, $\mathbf{u}$ is the control signal, $\mathbf{y}$ is the output of the system, the signals that can be measured by sensor, $\mathbf{P}$ is controlled, and $\mathbf{K}$ is the controller. Equations with $\mathbf{H}_\infty$ problem [10]:

$$\dot{\mathbf{x}} = \mathbf{A}\mathbf{x} + \mathbf{B_1}\mathbf{w} + \mathbf{B_2}\mathbf{u} \tag{5}$$

$$\mathbf{z} = \mathbf{C_1}\mathbf{x} + \mathbf{D_{11}}\mathbf{w} + \mathbf{D_{12}}\mathbf{u} \tag{6}$$

$$\mathbf{y} = \mathbf{C_2}\mathbf{x} + \mathbf{D_{21}}\mathbf{w} + \mathbf{D_{22}}\mathbf{u} \tag{7}$$

$$\mathbf{u} = \mathbf{K}\mathbf{y} \tag{8}$$

or can be written follow as:

$$\begin{bmatrix}\mathbf{Z}(s)\\ \mathbf{Y}(s)\end{bmatrix} = \mathbf{P}(s)\begin{bmatrix}\mathbf{W}(s)\\ \mathbf{U}(s)\end{bmatrix} = \begin{bmatrix}\mathbf{P_{11}}(s) & \mathbf{P_{12}}(s)\\ \mathbf{P_{21}}(s) & \mathbf{P_{22}}(s)\end{bmatrix}\begin{bmatrix}\mathbf{W}(s)\\ \mathbf{U}(s)\end{bmatrix} \tag{9}$$

where, $\mathbf{P} = \left[\begin{array}{c|cc}\mathbf{A} & \mathbf{B_1} & \mathbf{B_2}\\ \hline \mathbf{C_1} & \mathbf{D_{11}} & \mathbf{D_{12}}\\ \mathbf{C_2} & \mathbf{D_{21}} & \mathbf{D_{22}}\end{array}\right]$ and control matrix $\mathbf{K} = \left[\begin{array}{c|c}\mathbf{A_K} & \mathbf{B_K}\\ \hline \mathbf{C_K} & \mathbf{D_K}\end{array}\right]$. The controller is designed by choosing a matrix $\mathbf{K}$ to minimize [10]:

$$\|\mathbf{S}(\mathbf{P},\mathbf{K})\| = \|\mathbf{P_{11}} + \mathbf{P_{12}}\mathbf{K}(\mathbf{I} - \mathbf{P_{22}}\mathbf{K})^{-1}\mathbf{P_{21}}\| \tag{10}$$

To simplify the controller, remove the $\mathbf{A_K}$, $\mathbf{B_K}$, $\mathbf{C_K}$ matrices, thus, only the $\mathbf{D_K}$ matrix is found. Next, for full-state feedback, $\mathbf{B_1} = \mathbf{B}$, $\mathbf{C_1} = \mathbf{C_2} = \mathbf{C}$, $\mathbf{D_{12}} = \mathbf{D_{22}} = \mathbf{0}$ and since the noise is eliminated, $\mathbf{B_1}$, $\mathbf{D_{11}}$ and $\mathbf{D_{21}}$ are equal zero. Thus, the transfer function of this problem is:

$$\mathbf{S}(\mathbf{P},\mathbf{K}) = \left[\begin{array}{c|c}\mathbf{A} + \mathbf{B_2}\mathbf{D_K} & \mathbf{0}\\ \hline \mathbf{C_1} & \mathbf{0}\end{array}\right] \tag{11}$$

Lemma 1. Kalman-Yakubovich-Popov Lemma [11]: Let $\hat{G}(s) = \left[\begin{array}{c|c}\mathbf{A} & \mathbf{B}\\ \hline \mathbf{C} & \mathbf{D}\end{array}\right]$, inequality $\|\hat{G}(s)\| \leq \gamma$ satisfies, equivalent to existing a $\mathbf{X} > 0$ such that

$$\begin{bmatrix}\mathbf{A}^T\mathbf{X} + \mathbf{X}\mathbf{A} & \mathbf{X}\mathbf{B}\\ \mathbf{B}^T\mathbf{X} & -\gamma\mathbf{I}\end{bmatrix} + \frac{1}{\gamma}\begin{bmatrix}\mathbf{C}^T\\ \mathbf{D}^T\end{bmatrix}[\mathbf{C}\ \mathbf{D}] < 0 \tag{12}$$

Proof. Due to the inequality being strict, there are several existences $\epsilon > 0$ such that

$$\begin{bmatrix} \mathbf{A}^T\mathbf{X} + \mathbf{X}\mathbf{A} & \mathbf{X}\mathbf{B} \\ \mathbf{B}^T\mathbf{X} & -(\gamma - \epsilon)\mathbf{I} \end{bmatrix} + \frac{1}{\gamma}\begin{bmatrix} \mathbf{C}^T \\ \mathbf{D}^T \end{bmatrix}[\mathbf{C}\ \mathbf{D}] < 0 \tag{13}$$

equivalent to

$$\begin{bmatrix} \mathbf{u} \\ \mathbf{x} \end{bmatrix}^T \left[\begin{bmatrix} \mathbf{A}^T\mathbf{X} + X\mathbf{A} & \mathbf{X}\mathbf{B} \\ \mathbf{B}^T\mathbf{X} & -(\gamma - \epsilon)\mathbf{I} \end{bmatrix} + \frac{1}{\gamma}\begin{bmatrix} \mathbf{C}^T \\ \mathbf{D}^T \end{bmatrix}[\mathbf{C}\ \mathbf{D}]\right]\begin{bmatrix} \mathbf{u} \\ \mathbf{x} \end{bmatrix} < 0 \tag{14}$$

By algebraic transformations, and selecting Lyapunov function $V(\mathbf{x}) = \mathbf{x}^T\mathbf{X}\mathbf{x}$, (14) become

$$\dot{V}(\mathbf{x}(t)) - (\gamma - \epsilon)\|\mathbf{u}(t)\|^2 + \frac{1}{\gamma}\|\mathbf{y}(t)\|^2 < 0 \tag{15}$$

Integrating from 0 to infinity with $\mathbf{x}(0) = \mathbf{0}$; $\mathbf{x}(\infty) = \mathbf{0}$; $V(0) = 0$ ($\mathbf{A}$ is Hurwizt)

$$-(\gamma - \epsilon)\int_0^\infty \|\mathbf{u}(t)\|^2 dt + \frac{1}{\gamma}\int_0^\infty \|\mathbf{y}(t)\|^2 dt < 0 \tag{16}$$

$$-(\gamma - \epsilon)\|\mathbf{u}\|_{L_2}^2 + \frac{1}{\gamma}\|\mathbf{y}\|_{L_2}^2 < 0 \tag{17}$$

So,

$$\|G(s)\|_{\mathbf{H}_\infty} < \gamma$$

From close - loop state-space presented in Eq. (11) and KYP Lemma, the KYP condition of this system is:

$$\begin{bmatrix} \tilde{\mathbf{A}}^T\mathbf{X} + \mathbf{X}\tilde{\mathbf{A}} & \mathbf{0} \\ \mathbf{0} & -\gamma\mathbf{I} \end{bmatrix} + \frac{1}{\gamma}\begin{bmatrix} \mathbf{C_1}^T\mathbf{C_1} & \mathbf{0} \\ \mathbf{0} & \mathbf{0} \end{bmatrix} < 0 \tag{18}$$

where, $\tilde{\mathbf{A}} = \mathbf{A} + \mathbf{B_2}\,\mathbf{D_K}$

Theorem 1. Schur Complement: With any $\mathbf{M} \in \mathbb{S}^n$, $\mathbf{Q} \in \mathbb{S}^m$ and $\mathbf{R} \in \mathbb{R}^{n\times m}$

$$\begin{bmatrix} \mathbf{M} & \mathbf{R} \\ \mathbf{R}^T & \mathbf{Q} \end{bmatrix} < 0 \quad \text{equivalent} \quad \mathbf{Q} < 0\, and\, \mathbf{M} - \mathbf{R}\mathbf{Q}^{-1}\mathbf{R}^T < 0$$

In this problem, $\mathbf{Q} = -\gamma\mathbf{I}$, $\mathbf{M} = \begin{bmatrix} \tilde{\mathbf{A}}^T\mathbf{X} + \mathbf{X}\tilde{\mathbf{A}} & \mathbf{0} \\ \mathbf{0} & -\gamma I \end{bmatrix}$ and $\mathbf{R} = \begin{bmatrix} \mathbf{C_1}^T \\ \mathbf{0} \end{bmatrix}$

Thus, (18) satisfied if and only if:

$$\begin{bmatrix} \tilde{\mathbf{A}}^T\mathbf{X} + \mathbf{X}\tilde{\mathbf{A}} & \mathbf{0} & \mathbf{C_1}^T \\ \mathbf{0} & -\gamma\mathbf{I} & \mathbf{0} \\ \mathbf{C_1} & \mathbf{0} & -\gamma\mathbf{I} \end{bmatrix} < 0 \tag{19}$$

Lemma 2. Kalman-Yakubovich-Popov Dual: Let $\hat{G}(s) = \left[\begin{array}{c|c} \mathbf{A} & \mathbf{B} \\ \hline \mathbf{C} & \mathbf{D} \end{array}\right]$, inequality $\|\hat{G}(s)\| \le \gamma$ satisfies, equivalent to existing a $\mathbf{Y} > 0$ such that

$$\begin{bmatrix} \mathbf{Y}\mathbf{A}^T + \mathbf{A}\mathbf{Y} & \mathbf{B} & \mathbf{Y}\mathbf{C}^T \\ \mathbf{B}^T & -\gamma\mathbf{I} & \mathbf{D}^T \\ \mathbf{C}\mathbf{Y} & \mathbf{D} & -\gamma\mathbf{I} \end{bmatrix} < 0 \tag{20}$$

Proof. Let $\mathbf{X} = \mathbf{Y}^{-1}$, satisfy inequality (20) if and only if: $X > 0$ and

$$\begin{bmatrix} \mathbf{Y}^{-1} & \mathbf{0} & \mathbf{0} \\ \mathbf{0} & \mathbf{I} & \mathbf{0} \\ \mathbf{0} & \mathbf{0} & \mathbf{I} \end{bmatrix} \begin{bmatrix} \mathbf{Y}\,\mathbf{A}^T + \mathbf{A}\,\mathbf{Y} & \mathbf{B} & \mathbf{Y}\,\mathbf{C}^T \\ \mathbf{B}^T & -\gamma\mathbf{I} & \mathbf{D}^T \\ \mathbf{C}\,\mathbf{Y} & \mathbf{D} & -\gamma\mathbf{I} \end{bmatrix} \begin{bmatrix} \mathbf{Y}^{-1} & \mathbf{0} & \mathbf{0} \\ \mathbf{0} & \mathbf{I} & \mathbf{0} \\ \mathbf{0} & \mathbf{0} & \mathbf{I} \end{bmatrix} < 0 \tag{21}$$

deduce

$$\begin{bmatrix} \mathbf{A}^T\mathbf{X} + \mathbf{X}\mathbf{A} & \mathbf{X}\mathbf{B} & \mathbf{C}^T \\ \mathbf{B}^T\mathbf{X} & -\gamma\mathbf{I} & \mathbf{D}^T \\ \mathbf{C} & \mathbf{D} & -\gamma\mathbf{I} \end{bmatrix} < 0 \tag{22}$$

and, by the KYP lemma, it is equivalent $||G(s)|| < \gamma$

Now, apply this result to the ballbot model. There exists the controller $\mathbf{D_K}$ such that,

$$\left\| \mathbf{S}\left(\left[\begin{array}{c|cc} \mathbf{A} & \mathbf{0} & \mathbf{B_2} \\ \hline \mathbf{C_1} & \mathbf{0} & \mathbf{0} \\ \mathbf{I} & \mathbf{0} & \mathbf{0} \end{array}\right], \left[\begin{array}{c|c} \mathbf{0} & \mathbf{0} \\ \hline \mathbf{0} & \mathbf{D_K} \end{array}\right] \right) \right\| \le \gamma \tag{23}$$

when LMIs

$$\begin{bmatrix} \mathbf{X}\mathbf{A}^T + \mathbf{A}\mathbf{X} + \mathbf{Z}^T\mathbf{B_2}^T + \mathbf{B_2}\mathbf{Z} & \mathbf{0} & \mathbf{X}\mathbf{C_1}^T \\ \mathbf{0} & -\gamma I & \mathbf{0} \\ \mathbf{C_1}\mathbf{X} & \mathbf{0} & -\gamma\mathbf{I} \end{bmatrix} < 0 \tag{24}$$

must be satisfy with positive define matrices $\mathbf{X}$ and $\mathbf{Z}$.

Solve LMIs (24) with $\mathbf{X} > 0$ by LMI Solvers base on Minimization of a Linear Objective under LMI Constraints [12],

$$\mathbf{K} = \left[6.22\ 117.740\ 7.67\ 31.26\right] \tag{25}$$

4 Simulation

In this paper, an LQR controller has been designed to compare with the controller base on $\mathbf{H}_\infty$. The control signal of the LQR controller is calculated by multiplying the matrix $\mathbf{K}$ with the motion-feedback signal as Eq. (8). The matrix $\mathbf{K}$ is designed to get optimal gain matrix $\mathbf{K}$, by minimizing cost function:

$$Q = \frac{1}{2}\int_{0}^{\infty}\left(\mathbf{x}^T\mathbf{Q}\mathbf{x} + \mathbf{u}^T\mathbf{R}\mathbf{u}\right)\mathrm{d}t \tag{26}$$

After try some Q and R matrices, chose $Q = I_4$ and $R = 0.05$, then

$$\mathbf{K}_{LQR} = -\left[4.472\ 150.71\ 6.95\ 40.216\right] \tag{27}$$

Both of the controllers are full-state feedback and the control signal is generated by multiplying the state matrix by the matrix K as Eq. (8). The torques inputs are limited by T_{max} given in Table 1.

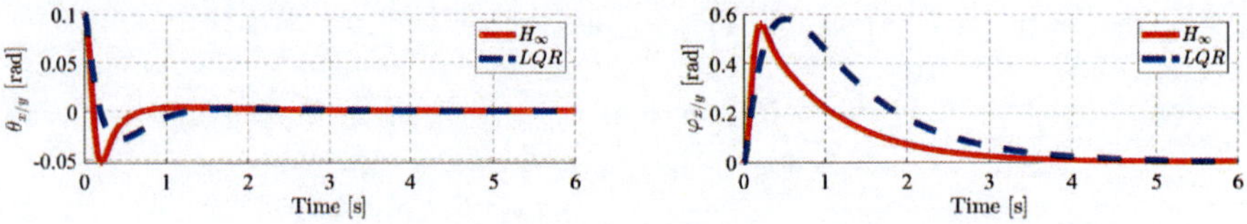

(a) Body's angle. (b) Ball's angle.

Fig. 2. Ball and body angle.

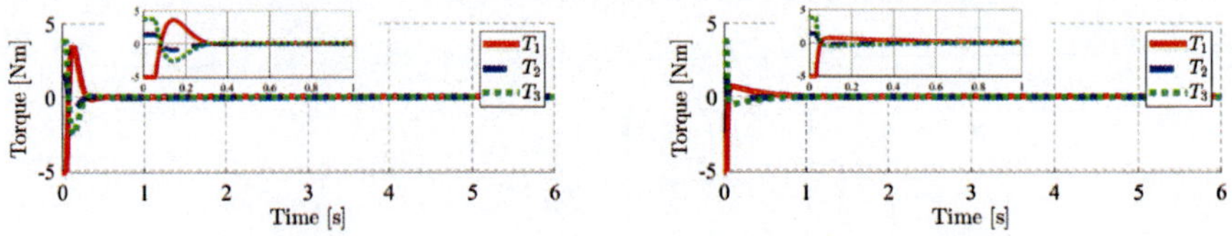

(a) $\mathbf{H}_\infty$ (b) LQR

Fig. 3. Torque input in $\mathbf{H}_\infty$ and LQR controller.

Simulation results with initial deflection $\varphi(0) = \theta(0) = 0.1$ are shown in Fig. 2. With the $\mathbf{H}_\infty$ controll, the system stabilized faster than the LQR controller, while ensuring that the torques are less than the maximum torques that the motor can provide (Fig. 3) (Figs. 4 and 5).

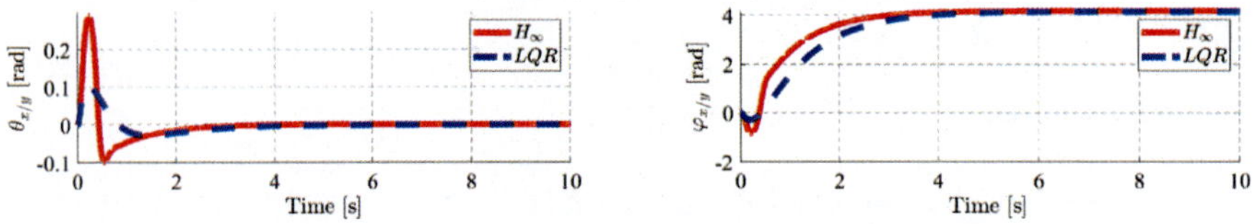

(a) Body's angle (b) Ball's angle

Fig. 4. Body and ball angle.

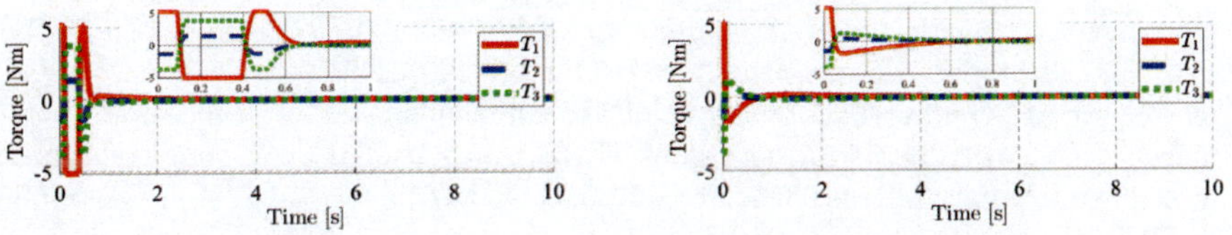

(a) $\mathbf{H}_\infty$ (b) LQR

Fig. 5. Torque input in $\mathbf{H}_\infty$ and LQR controller.

Similarly, with moving the ballbot to any position, here, the ballbot is balanced, and the ball roll to $(x, y) = (1, 1)(m)$ in the Oxy plane. Conversion between the position of the ballbot and the angles of the ball can be calculated through the perimeter of the ball. With the $\mathbf{H}_\infty$ controll, the ballbot reaches this position faster than the LQR controller but still maintains the torque limits. However, the overshoot of the angles of the body with $\mathbf{H}_\infty$ controll is higher than the overshoot of the angles of the body with the LQR controller.

At some time, the toques of the omniwheels reaches the saturation value. This is a limitation of the controller base on $\mathbf{H}_\infty$ method compared to the LQR controller because it cannot optimize input. However, its output response is better than that of the LQR.

5 Conclusion

The $\mathbf{H}_\infty$ based control has been utilized successfully thank to linear matrix inequalities. This controller demonstrates its effectiveness in comparison with LQR controller. The simulation results are presented to illustrate the control performances. A complete ballbot model will be introduced to fully demonstrate the dynamical properties of the ballbot in the future works.

Acknowledgments. This study was funded by Thuy Loi University.

References

1. Lauwers, T., Kantor, G., Hollis, R.: One is enough! In: Thrun, S., Brooks, R., Durrant-Whyte, H. (eds.) Robotics Research, pp. 327–336. Springer, Heidelberg (2007)
2. Kumagai, M., Hollis, R.: Development of a three-dimensional ball rotation sensing system using optical mouse sensors, pp. 5038–5043 (2011). https://doi.org/10.1109/ICRA.2011.5979899
3. Kordbacheh, S., Baghestan, K., Gheidary, S.S.: Modeling and robust control of ballbot robot with improved power transfer mechanism. In: 2018 6th RSI International Conference on Robotics and Mechatronics (IcRoM), pp. 353–358 (2018). https://doi.org/10.1109/ICRoM.2018.8657623
4. Fong, J., Uppill, S., Cazzolato, B.: 899: Ballbot (2009)

5. Jespersen, T.: Kugle - modelling and control of a ball-balancing robot. Ph.D. thesis, April 2019. https://doi.org/10.13140/RG.2.2.31490.73928
6. Kumagai, M., Ochiai, T.: Development of a robot balanced on a ball - application of passive motion to transport. In: 2009 IEEE International Conference on Robotics and Automation, pp. 4106–4111 (2009). https://doi.org/10.1109/ROBOT.2009.5152324
7. Boonto, S., Puychaisong, S.: Mouse type ballbot identification and control using a convex-concave optimization. J. Mar. Sci. Technol. **28**(5), 10 (2020)
8. Thanh, P.T., Nam, D.P., Van Tu, V., Huy, T.Q., Van Huong, N.: Robust control law using h-infinity for wheeled inverted pendulum systems. Int. J. Mech. Eng. Robot. Res. **8**(3), 483–487 (2019)
9. Fankhauser, P., Gwerder, C.: Modeling and control of a ballbot. Ph.D. thesis, June 2010
10. Gahinet, P., Apkarian, P.: A linear matrix inequality approach to h - infinity control. Int. J. Robust Nonlinear Control **4**(4), 421–448 (1994)
11. Yakubovich, V.A.: The solution of certain matrix inequalities in automatic control theory. Dokl. Akad. Nauk SSSR, 1304–1307 (1962)
12. Gahinet, P., Nemirovski, A.: The projective method for solving linear matrix inequalities. Math. Program. **77**, 163–190 (1997). https://doi.org/10.1007/BF02614434

Pathloss Modelling and Evaluation for A Wireless Underground Soil Moisture Sensor Network

Xuan Chinh Pham, Thi Phuong Thao Nguyen, and Minh Thuy Le(✉)

Hanoi University of Science and Technology, Hanoi, Vietnam
thuy.leminh@hust.edu.vn

Abstract. Wireless Underground Sensor Network (WUSN) is well known to monitor the soil quality for precision agriculture application. However, the uncertain property of the soil leads to many difficulties in designing a WUSN, one of them being the pathloss determination in different channels. In this study, we investigate the pathloss and the effect of soil moisture content on the pathloss in the Vietnam popular type of soil. From that foundation, we will construct a WUSN based on 920 MHz LoRa wireless technology. The experiment is implemented to validate underground wireless communication. The results show a good performance and stability of the network, where the connection between two underground sensor nodes could be extended up to 3 m with 5% soil moisture content.

Keywords: Pathloss modelling · Wireless underground sensor network · Soil moisture sensor · LoRa

1 Introduction

Wireless Sensor Network (WSN) plays an important role to collect data in the most of internet-of-thing (IoT) system applications. Since the WSN could be established without complicated physical connection, it can be installed with high density over a large area, therefore, improving the accuracy and the reliability of collected data. In recent year, WSN are widely used to monitor several parameters of soil environment from toxic substances for soil pollution warning to temperature and humidity on-demand irrigation system in precision agricultural applications [1, 2]. In such application, wireless moisture sensors are buried underground, and the measured data is collected via a suitable wireless technology to an above ground data hub. This kind of WSN namely wireless underground sensor network (WUSN) is usually used for any crops because no wiring is required [3–5]. Despite attracting much attention from both researchers and potential end-users, state-of-the-art WUSNs still has a main critical challenge of unreliable communication performance and limited lifetime. As the underground sensors located at different positions communicate to one another wirelessly through soil with varying composition, thus with high and variable pathloss compared to air. This is a significant challenge for underground wireless communication, especially since the heterogeneous

T. D. L. Nguyen et al. (Eds.): ICISN 2023, LNNS 752, pp. 335–345, 2023.
https://doi.org/10.1007/978-981-99-4725-6_42

environment is represented by different effective dielectric constants. Therefore, underground wave propagation must be characterized and calibrated for each type of soil. For these reasons, the wave propagation and link budget between underground to underground (UG2UG) sensor nodes, underground sensor node to aboveground data hub (UG2AG), and vice versal (aboveground hub to underground sensor node - AG2UG) must be analyzed for WUSN. In [6], the multiple ray channel has been studied due to the tunnel environments of the coal mine; two-ray signal propagation in soil medium have been modelling in [7], and an extended study with regard to the aboveground to underground channel and vice versal have been proposed in [8]. But the wave propagation characteristics is not the same for every soil medium due to its ambiguous profile such as soil moisture content (SMC), the ingredients and types of soil. Most of the studies only investigate the pathloss in case of low concentration of SMC, and therefore, cannot conduct an accurate result of the pathloss especially at high SMC value. In addition, the communication technology and sensor antenna is also a tremendous problem in constructing an efficient with low power consumption for a long service lifetime WUSN since different operating frequencies will lead to different pathloss. The communication technology also must satisfy the low power consumption to extend the battery lifetime of WUSN since the sensors are in-situ and buried in the ground. There have been few wireless technologies to deploying a WUSN, for example, narrow-band IoT [9], Sigfox [10], ... However, LoRa [11, 12] is one of the most favorable option to design a WUSN since it has low energy consumption, easy access with open source code and free license.

In this study, the channel modelling for underground sensor-to-sensor and underground sensor-to-aboveground communication hub will be addressed. Different parameters influences (such as soil moisture content, operating frequency, and soil ingredient regarding the linear correction for the Topp's equation) on the characteristic of wave propagation in different channel will be investigated. From that foundation, a WUSN will be proposed and discussed.

2 Underground Channel Pathloss Evaluation

The SMC is not uniform across the soil medium, and therefore, there is a need to place multiple sensors underground for temperature and humidity monitoring. However, for these sensor nodes and the command hub on the ground to be able to communicate with each other, it is compulsory to surpass the pathloss along the channeling path. Hence, in this section, the pathloss of different channel such as: UG2UG, AG2UG will be investigated, especially the effect of the soil permittivity since this have direct effects on the pathloss of different channel.

2.1 Relation Between SMC and Permittivity

Water in the soil could be divided into two different categories: (i) bounded water, which is water particles that got bounded soil particles, and cannot freely move in the medium and (ii) free water, which is available for plant root to absorb. Therefore, the soil medium is considered to be the combination of soil bulk, air, bounded water, and free water. Since the dielectric constant of air is 1, between 2.7 and 10 for common minerals in soils and rock [13], while water has the permittivity of 81 (depending on the temperature and frequency), so it is very obvious that the SMC is the main influence on the soil dielectric. Therefore, relation between the SMC and the soil permittivity have to be thoroughly studied in order to have an accurate pathloss model. Multiple studies have been studied in order to obtain the relationship between SMC and the soil permittivity, one of the most famous model is Topp's equation [14]. This shows the relationship between ϵ' and volumetric soil water content m_v:

$$\epsilon' = 3.03 + 9.3m_v + 146m_v^2 - 76.6m_v^3 \tag{1}$$

$$m_v = -5.3 \times 10^{-2} + 2.92 \times 10^{-2}\epsilon' - 5.5 \times 10^{-4}\epsilon'^2 + 4.3 \times 10^{-6}\epsilon'^3 \tag{2}$$

This model shows relatively better accuracy compared to other types of empirical model [15], but these traditional mixing formulas, since they only consider a very ideal model of the soil: the soil is the combination of soil bulk and free water only, they failed to show the complex dielectric behavior of the medium especially at GHz-scaled frequency. Therefore, a four-component dielectric mixing model is developed in [16] conducted based the data collected on five soil types, with a wide range of moisture conditions as well as a big frequency bandwidth (1.4 GHz to 18 GHz). This semiempirical has the following form:

$$\epsilon' = \left\{1 + \frac{\rho_b}{\rho_s}\left[(\epsilon_S)^{\alpha'} - 1\right] + [m_v]^{\beta'}\left[\epsilon'_{fw}\right]^{\alpha'} - m_v\right\}^{\frac{1}{\alpha'}} \tag{3}$$

$$\epsilon'' = [(m_v)^{\beta''}(\epsilon''_{fw\_low\_SMC})^{\alpha'}]^{\frac{1}{\alpha'}} \tag{4}$$

where m_v is the volumetric SMC usually have the saturated value of 60% [17], ρ_b is the bulk density, ρ_s is the specific density of the solid soil (2.66 g/cm^3), $\epsilon_S = (1.01 + 0.44\rho_s)^2 - 0.062$ (the dielectric constant of the soil solid), $\alpha' = 0.65$ $\beta' = 1.2748 - 0.519S - 0.512C$, $\beta" = 1.33797 - 0.063S - 0.166C$ are empirically determined constants, S and C is the percentage of sand and clay in the soil medium.

For Southeast Asia, since the temperature is swinging between 25 and 35° with high humidity and a lot of rain, which make the SMC relatively higher than other regions, especially the top surface of the soil after raining. Therefore the Eq. (3) and (4) need a simple linear adjustment to improve the correlation between SMC and permittivity at high level of water concentration [18], (3) (4) becomes:

$$\epsilon' = 1.15\left\{1 + \frac{\rho_b}{\rho_s}\left[(\epsilon_S)^{\alpha'} - 1\right] + [m_v]^{\beta'}\left[\epsilon'_{fw}\right]^{\alpha'} - m_v\right\}^{\frac{1}{\alpha'}} - 0.68 \tag{5}$$

$$\epsilon'' = [(m_v)^{\beta''}(\epsilon''_{fw\_high\_SMC\_})^{\alpha'}]^{\frac{1}{\alpha'}} \tag{6}$$

To ensure the performance of the WUSN, it is crucial to make sure that the link budget of the network could surpass the pathloss in the medium even in the worst-case scenario. "It is obvious that the permittivity (which is heavily dependent on the SMC according to (5) and (6) of the soil" has direct impact on the channeling pathloss, hence, we must evaluate the medium permittivity in this case, where the soil is fully saturated with water. According to [19], the porosity of the soil could be considered as the saturation of the SMC, and in addition, with Vietnam soil properties [20], we could see that most of the common soils of the region have the saturation value that is below 60% [21–25]. Figure 1 is the simulation conducted to evaluate the biggest permittivity based on (5), (6) with the parameter obtaining from [20] for acrisol (the most popular type of soil in Vietnam, could be found in more than half of Vietnam): $f = 920$ MHz, $\rho_b = 1.3–1.7$ g/cm^3, $\rho_s = 2.55–2.7$ g/cm^3, $S = 31\%$, $C = 52\%$ [20]. Therefore, we found the biggest permittivity value of the soil: $\epsilon_{v\_max} = 51.14 - 6.34j$, and this value will be used to calculate the biggest pathloss in different channel.

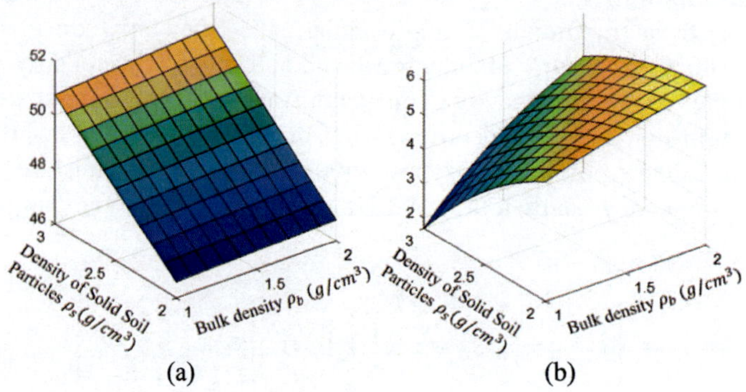

Fig. 1. Permittivity dependence on bulk density and solid particles density: a) the real part of the permittivity, b) the imaginary part of the permittivity

2.2 UG2UG Channel Model

Firstly, we investigate the UG2UG Line-of-sight (LOS) channel pathloss in order to deliver the suitable operation frequency for our WUSN. This channel has the longest propagation distance and therefore mostly depends on the soil permittivity. Therefore, if the channeling of the UG2UG LOS is ensured, then the other channels such as UG2AG, AG2UG are working as well. Assume a plane wave is transmitted in a homogenous space with no obstacle to a receiver at distance d. This transmission is called a LOS channel, and the pathloss of this channel according to [26] is:

$$P_L(dB) = 6 + 20\log(d) + 20\log(\beta) + 8.96\alpha d - 10\log(G_t G_r) \quad (7)$$

where G_t, G_r are the antenna gain of the transmitter and receiver, $\lambda = \frac{2\pi}{\beta}$ is the wavelength in the soil, α is the attenuation constant of the soil (Np/m), β is the phase constant (rad/m).

The UG2UG as in Fig. 2 is simulated based on (3)–(7), with the depth of sensor node is $h = 40$ cm, the antenna gains: $G_t = G_r = 1$ dBi, sand and clay ratio: $S = 0.31$ and

$C = 0.52$, soil moisture content is 60%, distance between underground nodes is $d = 2$ m. Figure 3a shows the simulation of how operation frequency effects on UG2UG channel pathloss (LOS channel). The increasing of operation frequency leads to the increasing of the pathloss, at $f = 1.5$ GHz, and $d = 2$ m, the pathloss is bigger than 150 dB. Therefore, the operation frequency for WUSN must less than $f = 1.5$ GHz:

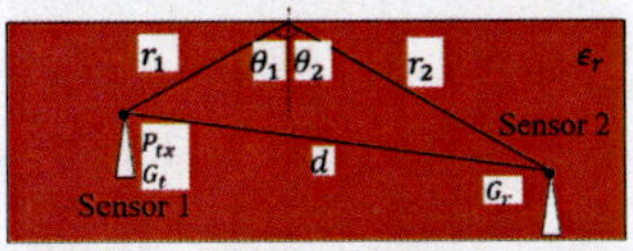

Fig. 2. Channel modeling of UG2UG

From this result, the frequency of $f = 920$ MHz is the most suitable since all our study and experiments are conducted in Vietnam.

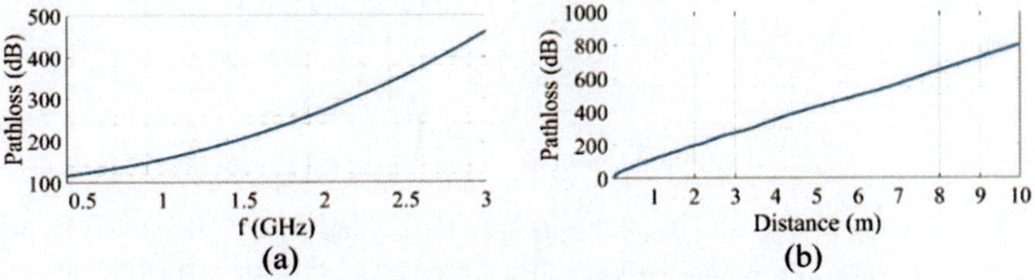

Fig. 3. Pathloss of UG2UG varies with: a) frequency, b) distance

We now evaluate the pathloss in UG2UG channel with the present of another indirect ray of signal transmitting in Fig. 2 to calculate the optimal distance between underground nodes for the WUSN. Assume that the soil has infinitive depth, therefore the reflection on the boundary between the soil and the air. This also means that there are only two-ray of channeling, one is the LOS channel, the other is the reflective path on the boundary between two regions. The reflective path could be the combination of two different path r_1 and r_2. The pathloss will be:

$$P_{2r}(dB) = P_L dB - 10\log\left[1 + \frac{\sqrt{G_{t1}G_{r1}}Rde^{-\alpha\Delta r}}{\sqrt{G_t G_r}(r_1 + r_2)}e^{-j\Delta\phi}\right]^2 \tag{8}$$

where G_{t1}, G_{r1}, G_t, G_r is the gain of the transmitter in the direction of r_1, the receiver in the direction of r_2, the transmitter and receiver in direction of d respectively, τ is the delayed time with respect to the LOS ray, $\Delta r = (r_1 + r_2) - d$ is the distance difference between two paths, and $\Delta\phi = \frac{2\pi\Delta r}{\lambda}$ is the phase difference, R is the reflective coefficient of the soil-air boundary that can be calculated using the following equation [26]:

$$R = \frac{\frac{1}{\epsilon_r}\cos\theta - \sqrt{\frac{1}{\epsilon_r} - \sin^2\theta}}{\frac{1}{\epsilon_r}\cos\theta + \sqrt{\frac{1}{\epsilon_r} - \sin^2\theta}}\text{(perpendicular polarized)} \tag{9}$$

$$R = \frac{\cos\theta - \sqrt{\frac{1}{\epsilon_r} - \sin^2\theta}}{\cos\theta + \sqrt{\frac{1}{\epsilon_r} - \sin^2\theta}} \text{(parallel polarized)} \quad (10)$$

$\theta_1 = \theta_2 = \theta$ denote the incident and reflected angles.

With the constructed model, the following simulation (Fig. 3b) shows the pathloss of this channel in the worst-case scenario. At $d > 2$ m, the pathloss is already surpassed 135 dB, there is no communication technology could be used to construct a stable WUSN in such conditions, which means that the distance between each underground node cannot be bigger than 2 m.

2.3 UG2AG (AG2UG) Channel Models

In this section, the pathloss between the underground node and the aboveground gateway is evaluated to find the suitable distance for the transmitting signal process from underground to the aboveground hub (UG2AG modeling). The model is the superposition of two parts (Fig. 4a): underground path d_1 and aboveground path d_2, the pathloss for this channel will be calculated by [8]:

$$P_{UG2AG}(dB) = P_u + P_a + 10\log\left(\frac{1}{Te^{-j2\pi\left(\frac{d_1}{\lambda}+\frac{d_2}{\lambda_0}\right)}}\right)^2 \quad (11)$$

with $P_u = 6 + \log d_1 + 20\log\beta + 8.69\alpha d_1 - 10\log G_u$ and $P_a = 20\log f + 20\log d_2 - 147.56 - 10\log G_a$. λ_0 is the wavelength in free-space, T is the refractive coefficient from soil to air and can be calculated with following equation [27]:

$$T = \frac{2\cos\theta_1}{\sqrt{\frac{1}{\epsilon_r}}\cos\theta_1 + \sqrt{1-\epsilon_r\sin^2\theta_1}} \text{(Perpendicular polarized)} \quad (12)$$

$$T = \frac{2\cos\theta_1}{}\cos\theta_1 + \sqrt{\frac{1}{\epsilon_r}} - \sin^2\theta_1 \text{(Parallel polarized)} \quad (13)$$

Since ϵ_r is much larger than 1, θ_1 will be bigger than the critical angle $\theta_c = \arcsin\sqrt{\left(\frac{1}{\epsilon_r}\right)}$, Therefore, $\theta_2 \approx 90°$, which mean that the aboveground signal will propagate along the soil surface.

Assuming that the aboveground node is place on the ground, the blue line in Fig. 5 is the simulated pathloss of UG2AG channel with respect to distance. In this channel, since the underground path of the transmitting signal is relatively short compared to UG2UG, therefore, the pathloss is significantly smaller, only 95 dBi at 4 m.

Similar to UG2AG modelling, we also investigate the pathloss for AG2UG channel (Fig. 4b) to reassure about the transmitting distance between each element in the WUSN. The pathloss in this case would be:

$$P_{AG2UG}(dB) = P_a + P_u + 10\log\left(\frac{1}{Te^{-j2\pi\left(\frac{d_1}{\lambda_0}+\frac{d_2}{\lambda}\right)}}\right)^2 \quad (14)$$

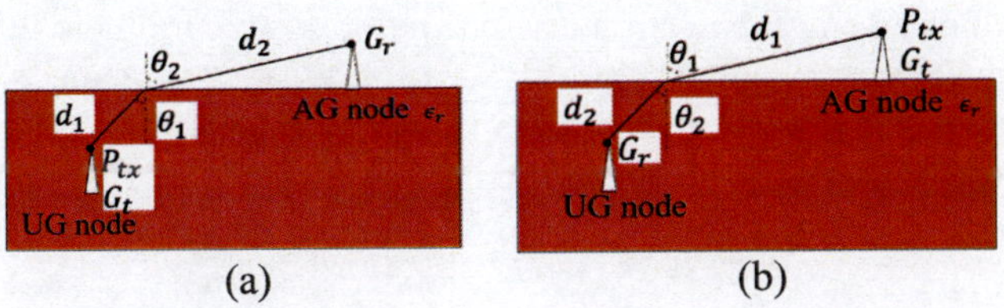

Fig. 4. Channel modeling of: a) UG2AG, b) AG2UG

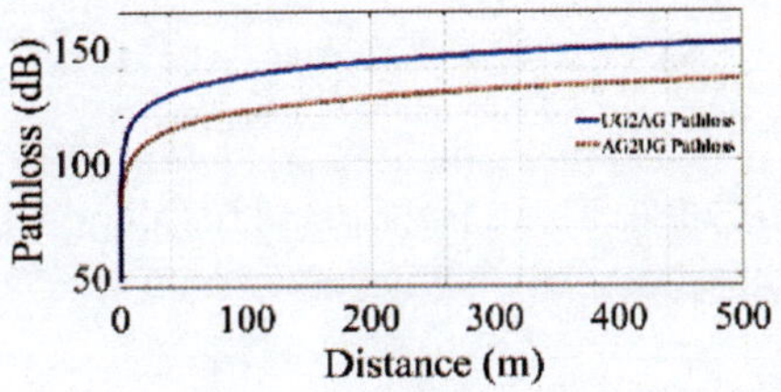

Fig. 5. Pathloss in UG2AG and AG2UG channel

where: $P_a = 20\log f + 20\log d_1 - 147.56 - 10\log G_a$; $P_u = 6 + \log d_2 + 20\log\beta + 8.69\alpha d_2 - 10\log G_u$ and T is the refractive coefficient from air to soil [27]:

$$T = \frac{2\cos\theta_1}{\sqrt{\epsilon_r}\cos\theta_1 + \sqrt{1 - \frac{1}{\epsilon_r}\sin^2\theta_1}} \text{(Perpendicular polarized)} \tag{15}$$

$$T = \frac{2\cos\theta_1}{\cos\theta_1 + \sqrt{\epsilon_r - \sin^2\theta_1}} \text{(Parallel polarized)} \tag{16}$$

Assuming the aboveground node is now placed 1 m higher than the ground level, the red line in Fig. 5 shows the pathloss of this channel with respect to distance. Similar to UG2AG channel, the underground signal transmitting path is also short, hence, less dependent on soil characteristics. However, according to (15) and (16), it can be seen that the transmission coefficient in AG2UG is bigger, the distance between two nodes can be extended up to 500 m (using LoRa communication technology which have received signal strength indication of RSSI = −146 dBm).

3 Underground Wireless Sensor Network Proposal

3.1 Chosen Wireless Technology for Underground Communication

As mentioned and evaluated in Sect. 2, the UG2UG channel is the one with most difficulties in communication and channeling signal since it is mostly dependent on the soil as a transmission media. Figure 3b shows that the pathloss of UG2UG with m_{v_max}, the pathloss is 135 dB at $d = 2$ m. With such pathloss and conditions, the technology must meet these conditions: long transmitting distance, small RSSI, small signal-to-noise ratio (SNR). LoRa with a range of received sensitivity RSSI from −120 dBm to −148 dBm is used for the longest communication distance purpose. According to [28], the noise in the soil holds the average value $P_{floor} = -103$ dBm. We now construct a stable WUSN, which can guarantee the performance even in worst case scenario by this link budget formula, suppose that the loss in cable is zero:

$$RSSI = P_{tx} + G_r + G_t - P_{pathtloss} \quad (17)$$

The smallest RSSI value that the receiving underground node is defined as in (18) to allow the two nodes in the soil capable of communicating with each other,

$$RSSI = P_{floor} + SNR = -123\,\text{dBm} \quad (18)$$

For UG2UG channel: at $d = 2$ m since $P_{pathloss} = -135$ dBm, smallest RSSI is –123 dBm, assuming the antenna gain is the same for all nodes. From (17), we derive:

$$P_{tx} = 12 - 2G \quad (19)$$

LoRa has the transmitting power in the range of $P_{tx} = 0 \div 20$ dBm. The underground node has small power consumption when the transmitting power is 8 dBm, with the antenna gain $G = 2$ dBi.

For UG2AG and AG2UG channel: suppose that $G_r = G_t = 0$ dBi, smallest RSSI is -123 dBm. From (17), we find the longest communication distance in two channels. From simulation in Sect. 2.3, the longest communication distance is 500 m:

$$P_{pathloss} = P_{txmax} - RSSI = 143\,\text{dBm} \quad (20)$$

3.2 The Proposed Underground Soil Moisture Sensor Network

The deployment monitoring area on farmland is 100 m^2 (10 m × 10 m). The distance between each node is 2 m, allowing reliable transmissibility. The longest communication between sensor nodes and gateway is 500 m. Therefore, the gateway is located 10 m away from sensor network. All sensor nodes are deployed in underground at 40–50 cm deep. For the network configuration in Fig. 6, firstly, sensor nodes will vote a cluster head node among themselves. Cluster head node is selected based on distance and the remaining energy. Cluster head node selection process happens on cycles to balance operation time between nodes. After that, every sensor node is provided a timeslot to communicate with the cluster head node. When a new node joins the network, it selects its cluster head node or when a sensor node is died, it does not affect to another nodes. After the network configuration step is done, sensor nodes begin to communicate to cluster head node with timeslots provided to them. 24 bytes of data communicating between nodes include: moisture percentage, GPS, battery percentage. With the payload of 256 bytes, LoRa (RFM95) can manage 10 nodes maximum.

In this section, we conduct an experiment to evaluate the performance of LoRa technology as well as to measure the pathloss using two sensors as a transmitting and a receiving device. Each sensor node includes three main parts: a microcontroller STM32F103C6T6 as a control unit, a RFM95 module as a transceiver module and an

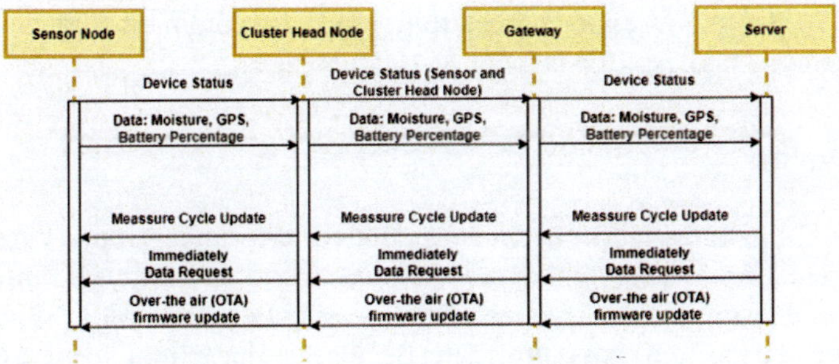

Fig. 6. Data flow in the proposed WUSN

antenna. Depending on the testing case: communication in soil environment (case 1-UG2UG) and communication in soil-air environment (case 2 – UG2AG), the node can be placed into the soil or on the ground. In both cases, the receiving node is buried in the soil at a depth of 40 cm and is not moved throughout the experiment. For the UG2UG case, the transmitting node is buried in the soil at a depth of 40 cm and is placed at different distances 1 m, 2 m or 3 m compared with the receiving node to do the measurement. For the UG2AG channel testing case, the transmitting node is placed on the ground at different distances to obtain the measured results. Besides, in each measurement, we sent 100 messages from transmitting node to the receiving node to evaluate the stability and reliability of the communication technology.

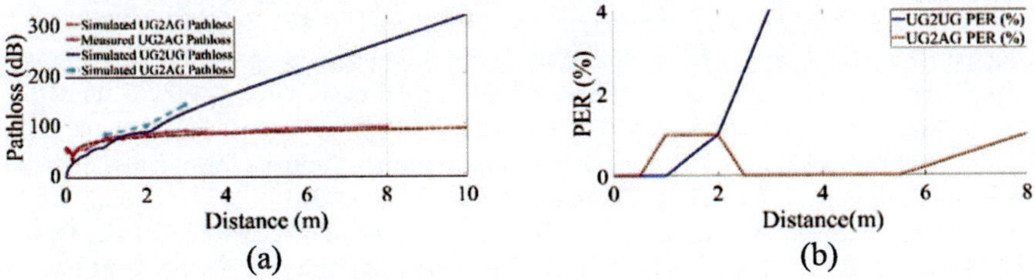

Fig. 7. Experiment results: a) simulated and measured pathloss results, b) PER

Figure 7 shows the simulated and measured results of the case 1 and case 2, respectively, at the field testing with SMC = 5%. The measured results show a good agreement with the simulated pathloss in Fig. 7a. The measured data is slightly bigger than the simulated data since we assume that the soil medium is homogenous (soil bulk and SMC is distributed uniformly across the field), and there was no scattering or complex reflection happened in our simulation. In case 1, the simulation result is highly dependent on the SMC content, where the connection between two nodes could even establish at 3 m, which can ensure that constructed WUSN can perform well in real-life condition. On contrary, case 2 shows less differences between the measurement result, which indicates that in this channel, the pathloss is less dependent on the soil properties since the travelling distance of the signal in the high-permittivity medium is way less than case 1. The

PER in Fig. 7b also shows a good compatibility of LoRa communication technology for WUSN, where the highest PER is 4 out of 100 messages.

4 Conclusion

In this paper, the pathloss model between underground-to-underground wireless sensors and underground to aboveground data Hud is discussed and evaluated in the North of Vietnam. Based on this pathloss model, a WUSN using Lora technology is proposed with the evaluation of measured packet error rate. Besides, we have implemented underground communication at a depth of 40 cm at different distances from the receiving unit. These results are potential for a WUSN in precision agriculture application.

References

1. Vuran, M.C., Salam, A., Wong, R., Irmak, S.: Internet of underground things in precision agriculture: architecture and technology aspects. Ad Hoc Netw. **81**, 160–173 (2018)
2. Nguyen, V.A., et al.: Realizing mobile air quality monitoring system: architectural concept and device prototype. In: 2021 26th IEEE Asia-Pacific Conference on Communications (APCC), pp. 115–120. IEEE (2021). https://doi.org/10.1109/APCC49754.2021.9609931
3. Fahrmeier, N., Goeppert, N., Goldscheider, N.: A novel probe for point injections in groundwater monitoring wells. Hydrogeol. J. **30**, 1021–1029 (2022)
4. Jiao, W., Wang, J., He, Y., Xi, X., Chen, X.: Detecting soil moisture levels using battery-free Wi-Fi tag. Preprint at http://arxiv.org/abs/2202.03275 (2022)
5. Pandey, G., Weber, R.J., Kumar, R.: Agricultural cyber-physical system: in-situ soil moisture and salinity estimation by dielectric mixing. IEEE Access **6**, 43179–43191 (2018)
6. Ranjan, A., Misra, P., Dwivedi, B., Sahu, H.B.: Studies on propagation characteristics of radio waves for wireless networks in underground coal mines. Wireless Pers. Commun. **97**(2), 2819–2832 (2017). https://doi.org/10.1007/s11277-017-4636-y
7. Akyildiz, I.F., Sun, Z., Vuran, M.C.: Signal propagation techniques for wireless underground communication networks. Phys. Commun. **2**, 167–183 (2009)
8. Xiaoya, H., Chao, G., Bingwen, W., Wei, X.: Channel modeling for wireless underground sensor networks. In: 2011 IEEE 35th Annual Computer Software and Applications Conference Workshops, pp. 249–254. IEEE (2011). https://doi.org/10.1109/COMPSACW.2011.46
9. Castellanos, G., Deruyck, M., Martens, L., Joseph, W.: System assessment of WUSN using NB-IoT UAV-aided networks in potato crops. IEEE Access **8**, 56823–56836 (2020)
10. Zhang, X., et al.: Thoreau: a subterranean wireless sensing network for agriculture and the environment. In: 2017 IEEE Conference on Computer Communications Workshops (INFOCOM WKSHPS), pp. 78–84. IEEE (2017). https://doi.org/10.1109/INFCOMW.2017.8116356
11. Ebi, C., Schaltegger, F., Rust, A., Blumensaat, F.: Synchronous LoRa mesh network to monitor processes in underground infrastructure. IEEE Access **7**, 57663–57677 (2019)
12. Froiz-Míguez, I., et al.: Design, implementation, and empirical validation of an IoT smart irrigation system for fog computing applications based on LoRa and LoRaWAN sensor nodes. Sensors **20**, 6865 (2020)
13. Takahashi, K., Igel, J., Preetz, H., Kuro, S.: Basics and application of ground-penetrating radar as a tool for monitoring irrigation process. In: Kumar, M. (ed.) Problems, Perspectives and Challenges of Agricultural Water Management. InTech (2012). https://doi.org/10.5772/29324

14. Topp, G.C., Davis, J.L., Annan, A.P.: Electromagnetic determination of soil water content: measurements in coaxial transmission lines. Water Resour. Res. **16**, 574–582 (1980)
15. Cui, F., Du, Y., Ni, J., Zhao, Z., Peng, S.: Effect of shallow-buried high-intensity mining on soil water content in Ningtiaota minefield. Water **13**, 361 (2021)
16. Dobson, M., Ulaby, F., Hallikainen, M., El-rayes, M.: Microwave dielectric behavior of wet soil-Part II: dielectric mixing models. IEEE Trans. Geosci. Remote Sens. **GE-23**, 35–46 (1985)
17. Alharthi, A., Lange, J.: Soil water saturation: dielectric determination. Water Resour. Res. **23**, 591–595 (1987)
18. Peplinski, N.R., Ulaby, F.T., Dobson, M.C.: Dielectric properties of soils in the 0.3-1.3-GHz range. IEEE Trans. Geosci. Remote Sensing **33**, 803–807 (1995)
19. Tarboton, D.G.: Rainfall-Runoff Processes. Civil and Environmental Engineering Faculty Publications (2003)
20. Spohrer, K., Herrmann, L., Ingwersen, J., Stahr, K.: Applicability of uni- and bimodal retention functions for water flow modeling in a tropical acrisol. Vadose Zone J. **5**, 48–58 (2006)
21. Dysli, M.: Characteristic coefficients of soils. Road Traffic **86**, 72–73 (2000)
22. Das, B.M.: Advanced Soil Mechanic. Taylor & Francis, Boca Raton (2008)
23. Hough, B.K.: Basic Soils Engineering. Ronald Press, New York (1969)
24. Terzaghi, K., Peck, R.B., Mesri, G.: Soil Mechanics in Engineering Practice. Wiley, Hoboken (1996)
25. Obzud, R., Trusty: The Hardening Soil Model - A Practical Guidebook Z. (2012)
26. Goldsmith, A.: Wireless Communications. Cambridge University Press, New York (2012)
27. Griffiths, D.J.: Introduction to Electrodynamics. Pearson, Boston (2013)
28. Li, L., Vuran, M.C., Akyildiz, I.F.: Characteristics of underground channel for wireless underground sensor networks. In: Proceedings of the MedHoc-Net 2007, Corfu, Greece, June 2007

Predicting Student Study Performance in a Business Intelligence System

Han Minh Phuong[1](✉), Pham Minh Hoan[2], Nguyen Trung Tuan[2], and Doan Trung Tung[3]

[1] Thuongmai University, Hanoi, Vietnam
phuonghm@tmu.edu.vn
[2] National Economics University, Hanoi, Vietnam
{hoanpm,tuannt}@neu.edu.vn
[3] University of Greenwich (Vietnam), Hanoi, Vietnam
tungdt27@fe.edu.vn

Abstract. Business Intelligence (BI) systems have been widely implemented in various industries to improve decision-making processes. In higher education institutions, BI systems have shown great potential in predicting student performance, which is crucial for academic and admission activities. This article begins by introducing the concept of BI and its applications in higher education institutions. A literature review is conducted to examine the existing studies on BI implementation in higher education institutions and predicting student performance. Based on the literature review, this article proposes the development of a BI system named to address the predicting student performance problems. The system includes an integrated data extraction module, a user-friendly interface for the school's managers, and several methods of data analysis based on machine learning and statistical probabilities. Binary Logistics Regression and Neural Network are used to forecast student performance. The results of the study suggest that the BI system can provide practical effects in academic and admission activities.

Keywords: business intelligence · predicting · education · framework · data warehouse

1 Introduction

The field of Business Intelligence (BI) is not novel; however, the advancement of technology in storing, managing, processing, and representing data has recently propelled BI into a trend. The ease of access, implementation, and use of BI has provided organizations and businesses with the opportunity to employ their own BI to optimize performance and profits.

The BI system leverages data visualization techniques to convert data into visual representations that enable users to easily comprehend and analyze data. Users can view reports that range from an overview to details and extract specific information fields promptly, regardless of time and place. Moreover, BI completely revolutionizes operational methodologies by shifting the focus from

T. D. L. Nguyen et al. (Eds.): ICISN 2023, LNNS 752, pp. 346–355, 2023.
https://doi.org/10.1007/978-981-99-4725-6_43

subjective experience-based operating skills to a data and information-based approach that synthesizes and accurately analyzes data.

Higher education institutions (HEIs), particularly autonomous and non-public higher education institutions, can be viewed as a particular type of enterprise that delivers training services to learners who are not influenced by the current competitive environment. The more training services aim at providing learners with good quality education that meets social requirements, the more prestige they will bring to educational institutions. Consequently, educational institutions will attract more learners, resulting in increased revenue generation.

This article expounds upon the construction of a BI system for a university and the application of BI to solve one of education's pressing issues: predicting student learning outcomes.

2 BI in HEIs

Extensive research has been conducted on the application of Business Intelligence (BI) systems in higher education institutions [1]. These studies demonstrate that BI systems play a pivotal role in managing big data resources, enhancing the admissions process, improving student retention and satisfaction, as well as generating significant financial reports that can minimize management costs of higher education institutions. Numerous studies have been conducted, with specific emphasis on the general framework [2], technology [3,4], and specific aspects of learning data mining in BI systems [5,6].

When implementing the BI system, higher education institutions benefit from improvements in training quality [7], increased retention rates and reduced dropout rates [8,9], enhanced student satisfaction, and improved enrollment rates [10]. Additionally, the BI system helps to reduce management costs and enable the rational allocation of institutional resources [11]. BI presents a comprehensive overview of university operations, offering valuable information for top management [12]. By providing efficient capabilities for creating and delivering various reports, such as annual performance reports [13], the system provides a clearer picture of the university's performance, allowing the management to closely monitor university operations. BI can support strategic vision by analyzing strategic planning and comparing educational quality outcomes across different units within and outside the university [14,17].

In the broader context of the field of education, Business Intelligence (BI) systems find numerous applications. For the scope of this paper, the authors aim to focus on the issue of predicting learning outcomes and leveraging BI systems to address this challenge.

3 Predicting Student Performance

The need to predict student performance is imperative in the realm of education, and it has been a subject of extensive research within the academic community. Forecasting techniques encompass a broad range, including predictions of student attrition, anticipated graduation results, and academic achievements in general or specific subjects. To address these prediction challenges, various approaches are employed, such as regression, machine learning, data mining, and neural networks [16].

The research team aims to predict student learning outcomes based on previous academic performance or current learning status. Typically, in the learning process, there are related or influencing subjects. This leads to the hypothesis that if the learning outcomes of previous subjects in the learning path are known, the learning outcomes of a particular subject can be predicted. Furthermore, adequate attendance can also be a factor that affects the results of that subject. Such predictions can help identify students who are likely to fail a subject, and provide early support to these students to improve learning quality and increase the passing rate. Multilinear regression models, as discussed in [18,19], are often employed for such forecasting.

Apart from the issue of predicting learning outcomes by subject, the research team is also concerned with forecasting student learning outcomes based on enrollment information. This forecast will assist the admissions department in providing better guidance to students regarding the majors they will study, thus enhancing the quality of education and lowering the dropout rate. Since the input data comprises both categorical and numerical forms and is not suitable for the regression model, the research team opted to employ the Neural network model for forecasting. Numerous studies have employed Neural networks in predicting learning outcomes from a variety of parameters beyond learning parameters [20–22]. Furthermore, the Neural network model has proven to be suitable for forecasting issues with varied data inputs, such as the problem of predicting students at risk of dropping out [23,24].

4 Buiding a BI System for HEI

The research team has developed a BI system for the Greenwich Vietnam university based on a proposed framework of four-layer in a previous publication [25]. The framework comprises the following layers:

Data Source Layer: This layer consists of data from various sources within the university. The data from the training department is extracted from the AP site (ap.greenwich.edu.vn) and includes information such as student scores, lecturer GPA (rate by students), class schedules, and attendance records. The admissions department provides data on student information prior to admission, while the student affairs department provides information on students participating in clubs, campus events, and OJT (on the job training) internships.

ETL Layer: The data undergoes extraction, transformation, and loading processes before it is uploaded to the data warehouse. The extraction module is implemented in Python with Jupyter Notebook, and performs normalization and data cleaning on the data, which comes from various sources, such as software, manual input, and spreadsheets.

Reporting Layer: This layer consists of dynamically generated visual reports, which provide information to different departments within the school, such as enrollment, training, and student collaboration.

Analytics Layer: This layer includes advanced data analytics tools that provide insights into the data, such as predictive modeling and machine learning algorithms.

Drawing from the proposed framework, the authors developed a BI system comprising an architecture described in Fig. 1.

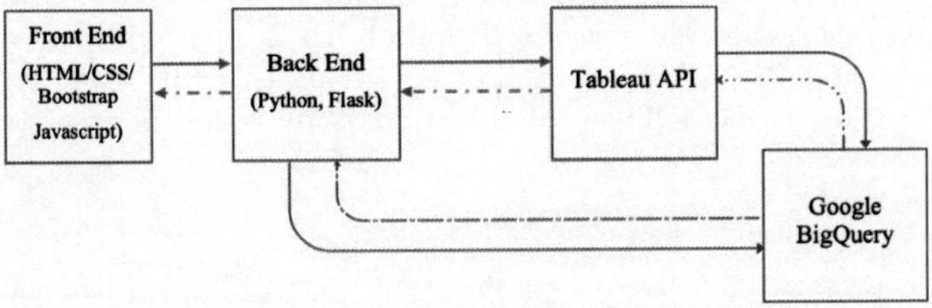

Fig. 1. BI system architecture

This architecture consists of 4 components:

- The front-end component employs HTML/CSS and JavaScript to render the web page user interface.
- The back-end component includes Flask, which is a Python-based web framework, as well as the forecasting and querying modules developed in Python.
- The reporting component includes Tableau API that is utilized to design visual reports.
- The data warehouse component is Google BigQuery - a cloud-based platform for data warehousing provided by Google.

The operation of the BI system is as follows: users access the system through the front-end layer and submit requests, which are then processed by the back-end layer. If the request is for visual reports, it is forwarded to the Tableau API component, which sends a data request to Google BigQuery. After receiving the data, a visual report is built using Tableau APIs and sent back to the back-end layer. The back-end layer then forwards the data and report images to the front-end layer for display. For querying or forecasting requests, the corresponding Python modules in the back-end layer are invoked to retrieve data from Google BigQuery for processing, and the results are returned to the front-end layer through Flask.

5 Applied BI System for Predicting Student Performance

5.1 Predicting Student Grade

The research team utilized the Binary Logistic Regression method to develop a predictive model for determining student success or failure in a subject based on their performance in previous subjects. Given that all variables are binary, with two possible values of 1 (failure) or 0 (pass), the authors chose this popular model in research to estimate the probability of an event occurring. Unlike multivariate regression, which estimates the value of the dependent variable Y based on independent variable X, Binary Logistic Regression estimates the probability of the event Y (i.e., probability of failure or passing) given the value of X. The dependent variable Y has two possible values of 0 (absence of the event) and 1 (occurrence of the event). The model evaluates the probability of the event occurring (Y = 1) when the predicted probability is greater than 0.5, and conversely, the probability of the event not occurring (Y = 0) when the predicted probability is less than 0.5. The probability function is described by the following formula:

$$Pi = P(Y = 1) = E(Y = 1 \mid X) = \frac{e^{B_0 + B_1 X_1 + B_2 X_2 + \cdots + B_n X_n}}{1 + e^{B_0 + B_1 X_1 + B_2 X_2 + \cdots + B_n X_n}} \tag{1}$$

Performing mathematical transformations, the Binary Logistic Regression equation is interpreted as follows:

$$\log_e \frac{P_i}{1 - P_i} = B_0 + B_1 X_1 + \cdots + B_n X_n \tag{2}$$

where:

- P(Y = 1) is the probability of the event (Y = 1) occurring
- B_0 is the intercept or constant term in the model
- B_1, B_2, ..., B_n are the coefficients associated with the independent variables X_1, X_2, ..., X_n

In order to assess the effectiveness of the regression model, the model will be evaluated using the Chi-squared test, the -2 Log Likelihood test, and the Goodness-of-fit test.

In order to evaluate the predictive performance of the binary logistic regression model, the authors conducted a test on the subject "Advance Programming" with the subject code "1651". The test dataset comprises of 496 students, which were extracted from the Data Warehouse component of the BI system described in Sect. 4 to ensure that all the relevant subjects have scores. The evaluation of the forecast results was based on the percentage of correct predictions. The program, which was written in Python, reads data from Google BigQuery and applies the calculations of binary logistic regression to estimate the model parameters and the regression equation.

The variables used in the forecasting process included the following:

- Avg1633: the average score of Website Design, an independent variable, and a quantitative variable
- Avg1649: the average score of Data Structure and Algorithm, an independent variable, and a quantitative variable
- Avg1644: the average score of Cloud Computing, an independent variable, and a quantitative variable

- IsPass: the final result of Advanced Programming, the dependent variable, and a binary variable

Based on the run of model for the data of 496 participating students, the authors derived the logistic regression equation:

$$log_e(\frac{p}{1-p}) = -3.016+0.268\times Avg1649+0.311\times Avg1644+0.266\times Avg1633 \quad (3)$$

and the probability equation:

$$p = \frac{e^{-3.016+0.268\times Avg1649+0.311\times Avg1644+0.266\times Avg1633}}{1+e^{-3.016+0.268\times Avg1649+0.311\times Avg1644+0.266\times Avg1633}} \quad (4)$$

The statistical tests of Chi-square, -2 Log likelihood and Hosmer and Lemeshow demonstrate that the model parameters are acceptable. By utilizing the probability equation stated above in the forecast, the authors found that out of 129 failed observations, 80 were mispredicted. The probability of correct prediction was calculated as 62%. Similarly, out of 366 passing observations, 346 were correctly predicted, resulting in a correct prediction rate of 94.5%. The overall average percentage of correct predictions was computed as $(61.2 + 94.3)/2 = 86.1\%$.

Based on the statistical tests, the model parameters were found to be acceptable and can be used to predict student grades.

5.2 Predicting Average Grades Based on Admission's Information

Admission information including the province, school, gender, math score, literature score, foreign language score, and the selected major are considered for this study. These information are stored in the Data Warehouse component of the BI system described in Sect. 4. The output information is the average score of the subjects that students have studied, categorized as either above average (score ≥ 6.5) or below average (score < 6.5). This analysis aims to assist the admission department in advising students on the choice of majors and to provide insight into the academic challenges that students may face. To achieve this objective, the research team will employ a neural network consisting of one input layer, three hidden layers, and one output layer as shown in the Fig. 2.

The input layer comprises six nodes that correspond to the following variables: the province of the student (represented as categorical data: 0, 1, 2, ...), the student's gender (represented as binary data: 0 or 1), and scores in math, literature, and foreign language (represented as real numbers between 0 and 10). The students' selected majors at Greenwich Vietnam are also included in the input layer and represented as categorical data (0, 1, 2).

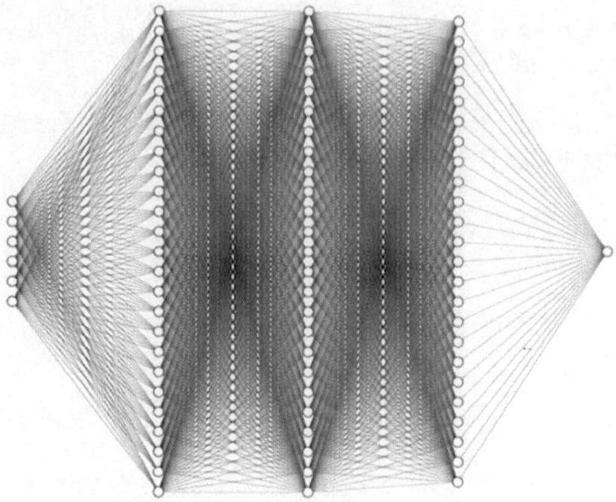

Fig. 2. Neural Network model for predicting

The output layer of the neural network model contains a single node that represents the average score of the subjects studied by the student after admission, which includes both the English stage score and the major stage score. This node has two categories:

- Category 0: represents students with an average GPA score below 6.5
- Category 1: represents students with an average GPA score above or equal to 6.5

The hidden layer of the neural network used in this study comprises three layers, each containing 25 nodes. The choice of the number of layers and nodes is based on the need to achieve accurate predictions without overfitting. The neural network is trained to predict learning outcomes based on six input variables:

- Province: where a student lives
- School: where a student studied in high school
- Gender: the gender of student (male or female)
- Math, Literature and Foreign Language scores: 3 most important scores of a student in the last year of high school.
- Major: the chosen Major to study in university.

The input layer uses a linear activation function and is initialized uniformly. The three hidden layers use the rectified linear unit (Relu) activation function and are initialized uniformly. The output layer has one node with a sigmoid activation function, also initialized uniformly. The learning rate is set to 0.005, the loss function used is the mean squared error (MSE), and the optimal function used is AdaMax.

The research team utilized the enrollment data from the years 2020 and 2021, and identified 797 students who both enrolled and obtained academic results at the institution. The dataset was then divided into a training set comprising 70%

of the data, and a test set consisting of the remaining 30%, which were randomly sampled. The accuracy of the prediction model was evaluated by determining the proportion of correctly predicted successful enrollments in both the training and test sets. The implementation of the model was conducted in Python, utilizing the Tensorflow library, and the program was executed on a Windows 11 machine equipped with 16 GB RAM and a CPU core i5 4.3 GHz. The loss values in Fig. 3 shows how the model is converged after 1000 epochs.

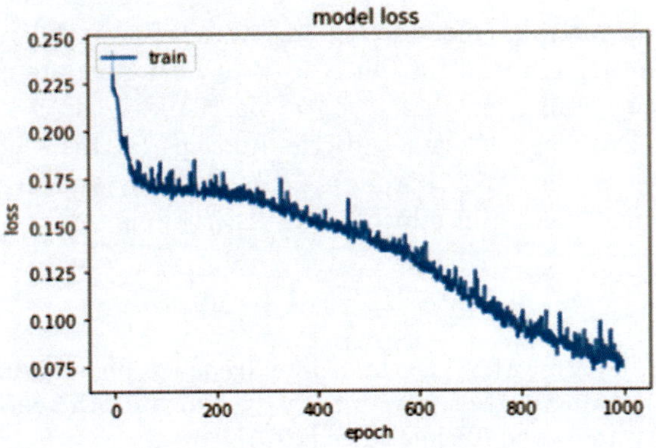

Fig. 3. Loss values of Neural Network model when training

Upon completion of training the model, its performance is evaluated by comparing actual learning outcomes with predicted outcomes. Since the model only classifies students into two categories, learned or unlearned, the model's outputs are rounded for evaluation. The training set was found to have a correct prediction rate of 0.93 while the test set had a correct prediction rate of 0.73.

For instance, if a student's input is [10. 0. 8.4 4. 6.6 0.] and the major parameter is changed from 0 (information technology) to 1 (business administration) or 2 (graphic design), the output results are 0.77, 0.05, and 0.38, respectively. This indicates that the student has chosen the correct major with a learning outcome of 0.77. Conversely, selecting majors 1 or 2 will lead to learning outcomes of 0.05 and 0.38, respectively.

Another example is a student with input [0. 0. 7.8 5.75 6.4 1.], which yields output results of [0.49, 0.18, 0], respectively. This indicates that the student should have chosen major 0 instead of major 1 to achieve better academic results.

6 Conclusion

In this study, we have developed and implemented a comprehensive business intelligence system, which is specifically designed for education and tested at the Hanoi campus of Greenwich Vietnam. The BI system is a complete solution that comprises an integrated data extraction module from multiple sources, including academic, admission, and student affairs data, which are uploaded to the Google BigQuery cloud service. The system includes a user-friendly website that provides an intuitive interface for school managers to generate and view reports and dashboards for the three school departments: academic, admission, and student affairs.

Furthermore, the BI system includes numerous methods of data analysis based on machine learning and statistical probabilities, leveraging the aggregated data warehouse. The research team has successfully tested two forecasting methods based on Binary Logistic regression and Neural network, which can have practical implications in academic and admission activities.

During the implementation of the BI system, the authors encountered challenges in extracting and synthesizing data, particularly data loss due to inadequate attention to storage. As a result, the findings of this study are somewhat limited. Nonetheless, the study highlights the necessity and urgency of having a business intelligence system in educational institutions.

References

1. Piedade, M.B., Santos, M.Y.: Business intelligence in higher education: enhancing the teaching-learning process with a SRM system. In: 5th Iberian Conference on Information Systems and Technologies (2010)
2. Zulkefli, N.A., et al.: A Business Intelligence framework for higher education institutions. J. Eng. Appl. Sci. **10**, 18070–18077 (2016)
3. Hamed, M., Mahmoud, T., Gómez, J.M., Kfouri, G.: Using data mining and business intelligence to develop decision support systems in Arabic higher education institutions. In: Marx Gómez, J., Aboujaoude, M.K., Feghali, K., Mahmoud, T. (eds.) Modernizing Academic Teaching and Research in Business and Economics. SPBE, pp. 71–84. Springer, Cham (2017). https://doi.org/10.1007/978-3-319-54419-9_4
4. Boulila, W., Al-kmali, M., Farid, M., Mugahed, H.: A business intelligence based solution to support academic affairs: case of Taibah University. Wirel. Netw., 1–8 (2018)
5. Villegas, W., et al.: Towards the Integration of Business Intelligence tools applied to educational data mining. In: 2018 IEEE World Engineering Education Conference (2018)
6. Valdiviezo-Diaz, P., Cordero, J., Reategui, R., Aguilar, J.: A business intelligence model for online tutoring process. In: IEEE Frontiers in Education Conference (FIE), El Paso, TX, USA (2015)
7. Baepler, P., Murdoch, C.J.: Academic analytics and data mining in higher education. Int. J. Scholarsh. Teach. Learn **4**(2) (2010)
8. Falakmasir, M.H.: Business Intelligence in eLearning (case study on the Iran University of Science and Technology dataset). In: The 2nd International Conference on Software Engineering and Data Mining (2010)
9. Villegas, W., Palacios-Pacheco, X., Lujan-Mora, S.: A business intelligence framework for analyzing educational data. Sustainability (2020)
10. Yellowfin. https://www.yellowfinbi.com. Accessed Aug 2021
11. Scholtz, B., Calitz, A., Haupt, R.: Business intelligence framework for sustainability information management in higher education. Int. J. Sustain. High. Educ. **19**, 266–290 (2018)

12. Mutanga, A.: A context-based business intelligence solution for South African higher education. J. Ind. Intell. Inf. **3**(2) (2015)
13. Chen, M.: Applying Business Intelligence in Higher Education Sector: Conceptual Models and Users Acceptance. University of Bedfordshire (2012)
14. Mihaela, et al.: Business intelligence systems in support of university strategy. In: Recent Researches in Educational Technologies (2011)
15. Khatibi, V., Keramati, A., Shirazi, F.: Deployment of a Business Intelligence model to evaluate Iranian national higher education. Soc. Sci. Humanit. Open **2**, 100056 (2020)
16. Alwarthan, S.A., Aslam, N., Khan, I.U.: Predicting student academic performance at higher education using data mining: a systematic review. Appl. Comput. Intell. Soft Comput. **2022** (2022)
17. Khatibi, V., Keramati, A., Shirazi, F.: Deployment of a BI model to evaluate Iranian National Higher Education. Soc. Sci. Humanit. Open (2020)
18. El Aissaoui, O., El Alami El Madani, Y., Oughdir, L., Dakkak, A., El Allioui, Y.: A multiple linear regression-based approach to predict student performance. In: AISC, vol. 1102
19. Kumar, A., Eldhose, K.K., Sridharan, R., Panicker, V.V.: Students' academic performance prediction using regression: a case study. In: ICSCAN 2020
20. Sikder, M.F., Uddin, M.J., Halder, S.: Predicting students yearly performance using neural network: a case study of BSMRSTU. In: 2016 5th International Conference on Informatics, Electronics and Vision (ICIEV), pp. 524–529 (2016). https://doi.org/10.1109/ICIEV.2016
21. Naser, A., Zaqout, S., Ghosh, I.A., Atallah, M., Eman, R.A.: Predicting student performance using artificial neural network: in the faculty of engineering and information technology. Int. J. Hybrid Inf. Technol. (2015)
22. Umar, M.A.: Student academic performance prediction using artificial neural networks: a case study. Int. J. Comput. Appl. **178**, 24–29 (2019). https://doi.org/10.5120/ijca2019919387
23. Niyogisubizo, J., Liao, L., Nziyumva, E., Murwanashyaka, E., Nshimyumukiza, P.C.: Predicting student's dropout in university classes using two-layer ensemble machine learning approach: a novel stacked generalization. Comput. Educ. Artif. Intell. **3**, 100066 (2022)
24. Alban, M., Mauricio, D.: Neural networks to predict dropout at the universities. Int. J. Mach. Learn. Comput. **9**(2), 149–153 (2019)
25. Han, M.P., Pham, M.H., Nguyen, T.T., Doan, T.T.: A proposed Business Intelligence framework for autonomous and non-public higher education institutions in Vietnam. In: ICISN 2022

Agent-Based Service Change Detection in IoT Environments

Tran Huu Tam[1], Cong Doan Truong[2], Nguyen Xuan Thu[3], Hoang Vu Hai[4], and Le Anh Ngoc[5,6](✉)

[1] FPT Corporation, Hanoi, Vietnam
tamth3@fpt.edu.vn
[2] International School, Vietnam National University, Hanoi, Vietnam
tcdoan@vnu.edu.vn
[3] People's Police University of Technology and Logistics, Hanoi, Vietnam
[4] Northern Kentucky University, Highland Heights, USA
hoanghl@nku.edu
[5] Swinburne Vietnam, FPT University, Hanoi, Vietnam
ngocla2@fe.edu.vn
[6] Swinburne University of Technology, Melbourne, Australia
nle@swin.edu.au

Abstract. The preferred way to establish diverse smart computing environments is by encapsulating functionalities as services and providing them through well-defined interfaces. However, such service environments are prone to constant changes and require proper change detection, which is essential for service providers and client applications. Current methods for detecting changes in services focus on structural and semantic changes, but they do not consider the actual behavior of a service by examining input and output values. This can result in the selection of inappropriate services for an autonomous replacement, as the behavior of the selected service may differ significantly from the replaced service. In this paper, we propose an architecture that captures and evaluates the behavior dimension of services to provide more reliable service replacements in IoT environments. We achieved this by using machine learning algorithms and a multi-agent architecture called EVA.

Keywords: Change Detection · Anomaly Detection · Service Evolution · IoT Services · Multi-agent Systems

1 Introduction

Service changes in IoT systems are inevitable and cannot be avoidable, this impact may incur huge losses in businesses if the services are not properly maintained leading to interruption of the binding between the service provider and service consumer. Changes due to updates may affect functional as well as non-functional properties. Thus detecting and communicating service changes play a vital role in service management [1,2].

T. D. L. Nguyen et al. (Eds.): ICISN 2023, LNNS 752, pp. 356–365, 2023.
https://doi.org/10.1007/978-981-99-4725-6_44

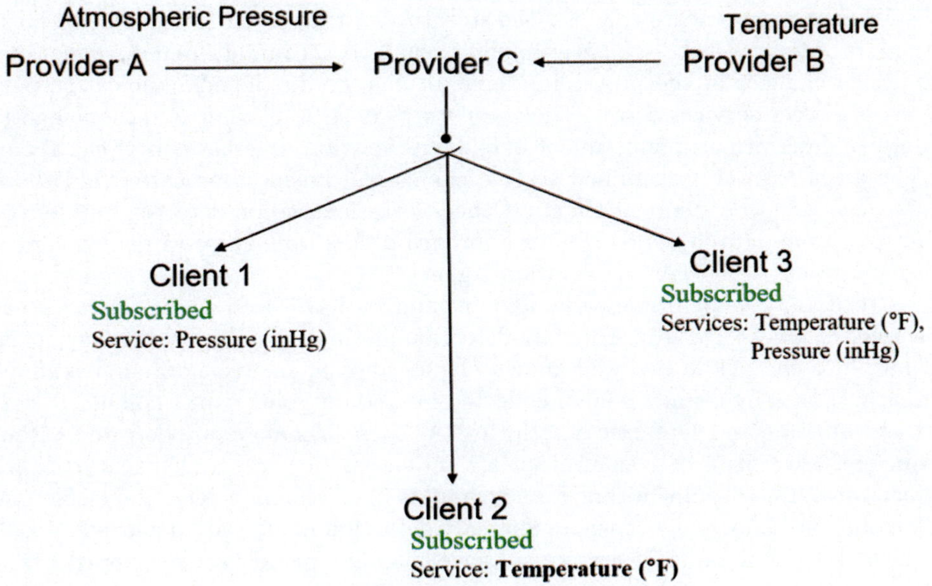

Fig. 1. Motivation Scenario [3]

Let's consider an example of an IoT scenario shown in Fig. 1. In this scenario, service provider C uses services provided by A and B to obtain information on atmospheric pressure and temperature, respectively. Additionally, service provider C has three clients, Client 1, Client 2, and Client 3, who subscribe to its services. In this case, service provider C provides the Temperature service in Fahrenheit to Clients 2 and 3 when they request the *getTempInF()* method. However, if service provider B updates its service to provide temperature information in Celsius instead of Fahrenheit, service provider C may not be notified of this update. This could cause a disruption in services provided to Clients 2 and 3. This type of change in service can be considered a behavioral change.

In this research work, we focus on behavior-based service change detection, an important tool for ensuring the stability and reliability of IoT systems. The tool helps to prevent service failures and outages by detecting changes in the behavior of services before they become a problem. The behavior-based service change detection system uses various methods to monitor the behavior of services, including log analysis, performance monitoring, network traffic analysis, and statistical analysis. The system compares the current behavior of a service with its expected behavior, and if there is a significant difference, the system raises an alarm to alert the other systems or end-users. This allows the interdependent services to quickly respond to changes in the behavior of services, and make any necessary changes to the system to ensure that it continues to operate effectively.

Existing approaches consider the structure of the service interface, and business process and take semantic annotations into account to detect changes or identify changes in the processing logic, or changes in the communication patterns between services. For instance, the works in [4, 9, 13] compared the semantic service description of subsequent versions to extract differences between them. The authors in [14] examined service changes in business processes and share the results with dependent clients. Other works focused on detecting outliers or anomalies in data [5, 6, 10, 17]. Our approach differs from these works by using a smart agent named EVA (Evolution Agent) [12].

To date, several frameworks and techniques have been proposed to detect service changes [8, 11, 13]. Anomaly detection methods usually classify a system's state into one of the two well-known classes anomalous and normal. We adapt this approach by using a pool of collected in- and outgoing values combined with their semantic and syntactic description. As soon as changes are detected within one of these dimensions, an appropriate action will be executed. In this work, we focus on IoT environments and detect changes in an anomaly detection fashion by introducing a smart agent where anomaly detection models are deployed at each client depending on their service subscription. The models are developed based on machine learning that can be learned and updated automatically detecting anomalies efficiently which reduces a part of the manual effort in the continuous provision of services to the end-users. We leverage this to introduce a new change detection architecture that does not only examine the interface description but inspects input and output to recognize changes in the service behavior.

The primary focus of this paper is to provide a solution that addresses the following research inquiries: (i) How to identify alterations in service behavior within an IoT setting? (ii) What are effective techniques for detecting changes while considering the limitations of IoT devices with limited resources? (iii) How can the reader recognize the association between multi-agents and their environment, as well as their interactions with others?

The remainder of this paper is organized as follows. Section 2 describes our approach. Section 3 presents some experiments and their evaluations. Finally, the main findings of this paper are summarized in Sect. 4.

2 Proposed Architecture

Our proposed architecture involves an intelligent agent to detect behavioral changes in an anomaly-detection manner. This is done by deploying anomaly detection models at each client depending on their service subscription. These models can be automatically learned and updated using machine learning, reducing the need for manual effort in providing continuous services to end users. The models are typically trained offline to recognize normal data behavior, and when presented with new data online, they can recognize anomalies as they occur. The key points of our proposed architecture can be described as follows:

2.1 Anomaly Detection

Anomaly detection is a technique used to identify unusual patterns or behavior that deviate from the expected behavior. This technique is helpful in various domains, such as finance, healthcare, cyber security, and marketing, to detect fraud, medical conditions, security threats, and unusual customer behavior. To date, there are many machine learning methods such as supervised, unsupervised, and semi-supervised are applied to detect changes in service systems [15,16].

Supervised methods or classification methods require a labeled training set that contains both normal and anomalous samples to build the predictive model. In theory, supervised methods have better detection rates than semi-supervised and unsupervised methods because they have access to more information [7,17]. However, some technical issues exist, which can make these methods appear less accurate than expected. The first issue is the shortage of a training dataset, which can be a challenge, and the training datasets usually contain some noise that results in higher false alarm rates. For these reasons, we will mainly focus on supervised learning algorithms for our implementations.

Supervised learning algorithms, such as Decision Trees, Naive Bayes, and K-Nearest Neighbours, are used to classify anomalies. Artificial Neural Networks (ANN), such as Feed-forward neural networks, Elman, and Jordan recurrent neural networks, are used to predict the data and classify the anomalies based on the Mean Square Error. Unsupervised learning, such as K-Means, and semi-supervised learning, such as One-Class SVM (OC-SVM), are used for the clustering and classification of anomalies, respectively. The significant characteristics of these algorithms can be summarized in references [6,7].

For solving the motivating problem, we propose a multi-agent plugin that monitors interface descriptions and captures service behaviors. In addition, the agents make use of unsupervised machine learning algorithms and a service registry framework equipped with reasoning capabilities.

In order to manage IoT devices efficiently, there is a need for resource-efficient mechanisms. To address this, the concept of having an externally hosted agent manage IoT devices is being considered. The proposed solution is to equip every service with an agent called EVA (Evolution Agent) that can be deployed on a powerful gateway node or edge computing. The EVA requires memory and storage to process incoming update messages and is comprised of components for analysis, evolution analytics, and evolution coordination. These components can be seen in the Fig. 2.

2.2 EVA

The IoT devices only need to run the Smart Update component, which allows the EVA to access and manage them. In the context of IoT environments, cloud or local computers are used to cater to a growing number of clients in the network. One EVA can manage several IoT services belonging to one provider, and

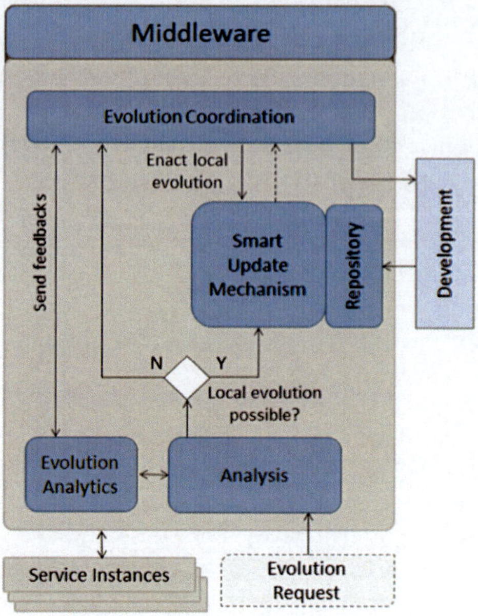

Fig. 2. EVA Architecture [12]

communication between clients and EVAs is done via REST. Additional information about EVA's detailed information architecture is provided in the referenced sources. Further details on the architecture of EVA can be found in references [1,2,12].

Figure 3 shows a service provider delivering temperature service of three different cities to three other clients 1, 2, and 3 respectively, depending on their subscription of services, for example. The temperature services are provided via REST service in Fahrenheit values. The agent EVAs are deployed to both clients and servers. In our scenario, EVAs are the only module the developer has to integrate. Assuming an application is running on different devices, e.g., a Raspberry, each is equipped with its own agent. The agents collect the in- and output sent between the application and the invoked services. Depending on the number of agents and the message sizes, the responsible agent will ask its agent intermittently for a subset of their collected data. This enables the agent to analyze the typical interaction between the application and its connected services and generate a representative model through machine learning algorithms.

2.3 Service Architecture

In order to support the models on constraint devices, the performance of each algorithm is analyzed with benchmarking and accuracy. The efficient algorithms are chosen and then deployed on agents at the clients based on their service subscriptions. When the deployed models detect normal values as anomalies,

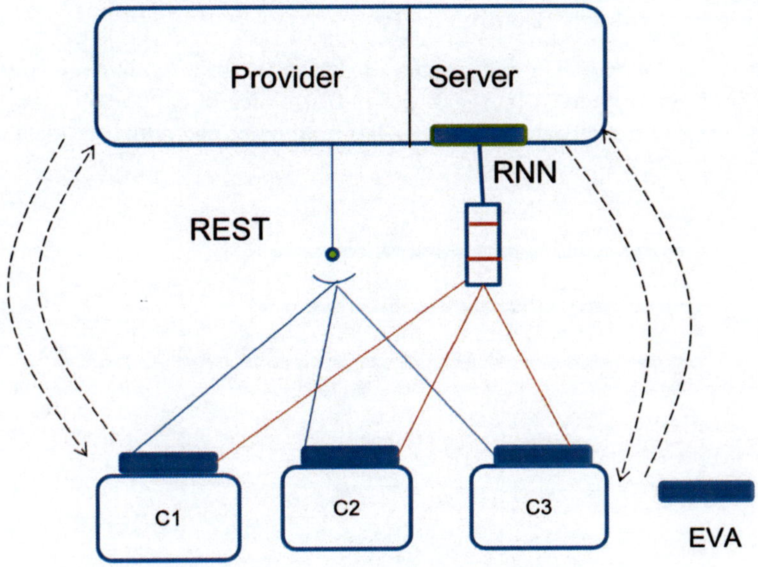

Fig. 3. Proposed Service Architecture

the value or specific data instance is sent to the server where other efficient models are employed. This is usually done to avoid false alarms. The model on the server checks the data instance for an anomaly. If it detects an anomaly, the model deployed at the clients performs well. If it detects as a normal value instead of an anomaly, then it replies back to the concerned agents to update the model. The models are then updated offline with new data and deployed back with updated models.

3 Experiments and Results

In this section, the implementation and evaluation of various machine-learning algorithms are examined. The classical machine learning algorithms like Decision Tree, Naïve Bayes, K-Nearest Neighbour, K-Means, and One-Class Support Vector Machine are implemented using a WEKA in Java. WEKA is a data mining software that has collections of machine learning algorithms for data mining tasks. The models based on Neural Networks are developed using ENCOG in Java. ENCOG is an advanced machine learning framework that supports a variety of advanced algorithms, as well as support tools to normalize and process data. Regarding datasets, we use available raw temperature sensor data collected from National Centres for Environmental Information (NCEI). The weather data that is considered contains 8 features, and 19,780 instances and covers 60 years period (1960–2022). The experiments in this section have been extended and adapted from the work [3,17].

3.1 Benchmarks on Raspberry Pi

To deploy efficient models on a number of devices including memory-constrained devices, it is very essential to check the performance of each algorithm through benchmarking and comparing through bar graphs to see which models are best suited to support IoT devices.

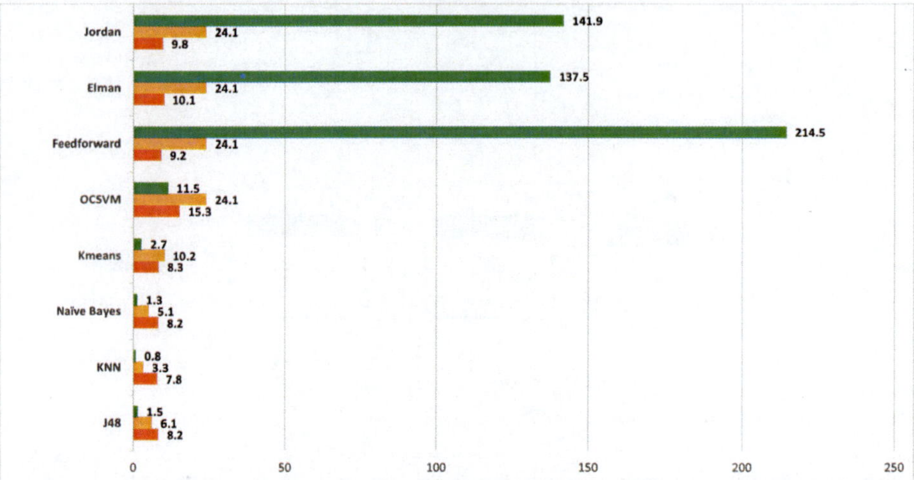

	J48	KNN	Naïve Bayes	Kmeans	OCSVM	Feedforward	Elman	Jordan
Runtime (seconds)	1.5	0.8	1.3	2.7	11.5	214.5	137.5	141.9
CPU Usage (%)	6.1	3.3	5.1	10.2	24.1	24.1	24.1	24.1
Heap Memory (Mb)	8.2	7.8	8.2	8.3	15.3	9.2	10.1	9.8

Runtime (seconds) CPU Usage (%) Heap Memory (Mb)

Fig. 4. Benchmark - Raspberry Pi, Quad-core 64-bit, 1.5 GHz

The bench-marking results of the algorithms on Raspberry Pi showed in Fig. 4. During training the models, the important performance characteristics such as heap memory usage, CPU usage, and runtime are considered for comparison and evaluation.

3.2 Algorithm Performance

The machine learning frameworks and temperature datasets were used to evaluate the performance of various algorithms. We found that J48, Naïve Bayes, KNN, and K-Means are classical machine learning algorithms that are easy to implement, understand, and have fast training and prediction speeds with low memory usage. The accuracy of these algorithms is high. However, K-Means accuracy cannot be easily calculated. Although these algorithms have high accuracy, they did not perform well in detecting contextual and collective anomalies. OCSVM had the highest accuracy of 96.38 percent and partially detected contextual anomalies. Figure 5 provides an overview of the algorithm's performance.

Although neural networks have slower training and prediction speeds than classical machine learning algorithms, they require less memory. As the number of features and data instances increases, neural networks become more complex. Feedforward neural networks, with their static memory, are not able to predict contextual anomalies well. However, Elman and Jordan recurrent neural networks have specialized contextual neurons that enable them to construct dynamic or short-term memory, effectively detecting contextual anomalies. Among the neural networks, Jordan has the highest accuracy in detecting anomalies and a relatively high training speed, converging to a small error in a low number of iterations. For IoT-constrained devices, J48, Naïve Bayes, and KNN algorithms are suitable for detecting point anomalies, with KNN being the best option considering high accuracy and low resource consumption. OCSVM can be used when there is a trade-off between accuracy and memory usage. For devices with high memory, Jordan and Elman recurrent neural networks are suitable.

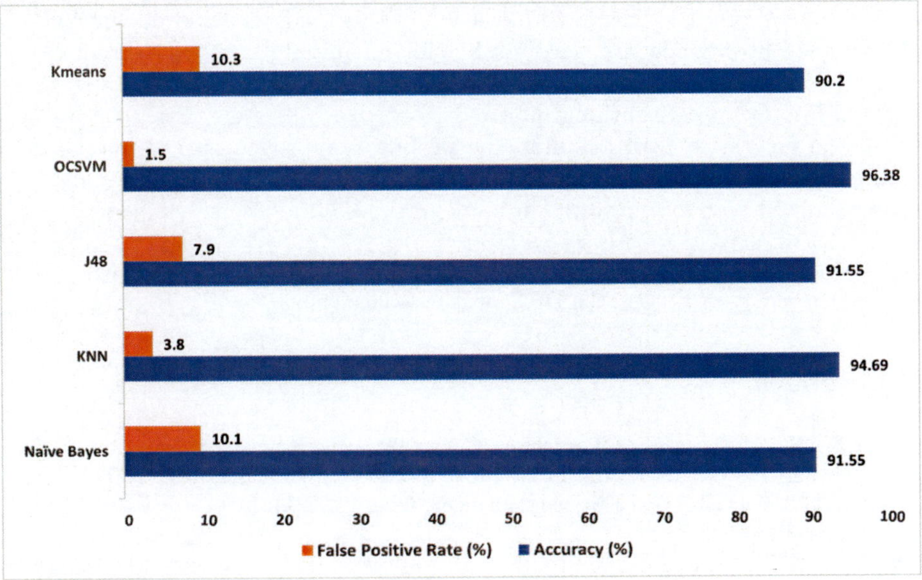

Fig. 5. Performance of Algorithms: Kmeans, J48, KNN, Naive Bayes and OCSVM

3.3 REST Service Architecture

We have implemented a REST service architecture that provides appropriate models for resource-constrained IoT devices. The implementation can be described in the following steps:

- The trained models "OCSVM" are deployed on the client side. Each client is provided with temperature data from different cities.
- The model receives the data pre-processes it and detects anomalies if any.

- The processed data is then sent to the server, where a trained Recurrent Neural Network (RNN) is employed via AMQP, an advanced messaging queuing protocol used for messaging services between a provider and various clients.
- If the model on the client side does not perform well, it automatically sends a request to the provider via REST to update its model. The updated model is then deployed automatically, making the system more efficient.

4 Conclusions and Future Works

In this paper, we presented an approach to detect anomalies efficiently in the streamed sensor data by deploying agents that automatically take care of these changes. To detect abnormalities in the sensor data, we have to use anomaly detection methods based on machine learning algorithms. The performances of several algorithms are analyzed through benchmarks, training, and test results. The performance is measured to satisfy the requirements of IoT-constrained devices. Therefore, the primary innovation of this study is achieved through the verifying agents, which are accountable for examining the behavior of client-server communication. This introduces a novel aspect that has not been incorporated into other conventional methods.

In the future, we will combine the proposed approach with other machine learning algorithms to realize a complete real-time service-oriented architecture for automated payment systems in banks or in Vietnamese stock systems [18].

Acknowledgements. This work is partly supported by the Distributed Systems Group, University of Kassel, the DFG project 280611965.

References

1. Tran, H.T., Nguyen, V.T., Phan, C.V.: Towards service co-evolution in SOA environments: a survey. In: Vinh, P.C., Rakib, A. (eds.) ICCASA/ICTCC -2020. LNICST, vol. 343, pp. 233–254. Springer, Cham (2021). https://doi.org/10.1007/978-3-030-67101-3_19
2. Tran, H.T., Baraki, H., Geihs, K.: An approach towards a service co-evolution in the internet of things. In: Giaffreda, R., et al. (eds.) IoT360 2014. LNICST, vol. 150, pp. 273–280. Springer, Cham (2015). https://doi.org/10.1007/978-3-319-19656-5_39
3. Tran, H.T.: Towards Service Co-evolution in the Internet of Things. Dissertation (2021)
4. Fokaefs, M., Stroulia, E.: Using WADL specifications to develop and maintain REST client applications. In: 2015 IEEE International Conference on Web Services (ICWS), pp. 81–88 (2015)
5. Amer, M., Goldstein M., Abdennadher, S.: Enhancing one-class support vector machines for unsupervised anomaly detection. In: Proceedings of the ACM SIGKDD Workshop on Outlier Detection and Description, pp. 8–15. ACM (2013)
6. Chandola, V., Banerjee, A., Kumar, V.: Anomaly detection: a survey. ACM Comput. Surv. (CSUR) **41**(3), 15 (2009)

7. Omar, S., Ngadi, A., Jebur, H.H.: Machine learning techniques for anomaly detection: an overview. Int. J. Comput. Appl. **79**(2) (2013)
8. Tran, H.T., Baraki, H., Kuppili, R., Taherkordi, A., Geihs, K.: A notification management architecture for service co-evolution in the internet of things. In: 2016 IEEE 10th International Symposium on the Maintenance and Evolution of Service-Oriented and Cloud-Based Environments (MESOCA), pp. 9–15. IEEE (2016)
9. Stavropoulos, T.G., Andreadis, S., Riga, M., Kontopoulos, E., Mitzias, P., Kompatsiaris, I.: A framework for measuring semantic drift in ontologies. In: 1st International Workshop on Semantic Change Evolving Semantics (SuCCESS 2016). CEUR Workshop Proceedings, Leipzig, Germany (2016)
10. Ghafoori, Z., Rajasegarar, S., Erfani, S.M., Karunasekera, S., Leckie, C.A.: Unsupervised parameter estimation for one-class support vector machines. In: Bailey, J., Khan, L., Washio, T., Dobbie, G., Huang, J.Z., Wang, R. (eds.) PAKDD 2016. LNCS (LNAI), vol. 9652, pp. 183–195. Springer, Cham (2016). https://doi.org/10.1007/978-3-319-31750-2_15
11. Jahl, A., Tran, H.T., Baraki, H., Geihs, K.: WIP: behavior-based service change detection. In: The Proceedings of IEEE International Conference on Smart Computing (SMARTCOMP), pp. 267–269. IEEE (2018)
12. Tran, H.T., Baraki, H., Geihs, K.: Service co-evolution in the internet of things. In: The Endorsed Transactions on Cloud Systems, pp. 1–15, EAI 2015 (2015)
13. Tran, H.T., Jahl, A., Geihs, K., Kuppili, R., Nguyen, X.T., Huynh, T.T.B.: DECOM: a framework to support evolution of IoT services. In: The Proceedings of the Ninth International Symposium on Information and Communication Technology, pp. 389–396. ACM (2018)
14. Jahl, A., Baraki, H., Tran, H.T., Kuppili, R., Geihs, K.: Lifting low-level workflow changes through user-defined graph-rule-based patterns. In: Chen, L.Y., Reiser, H.P. (eds.) DAIS 2017. LNCS, vol. 10320, pp. 115–128. Springer, Cham (2017). https://doi.org/10.1007/978-3-319-59665-5_8
15. Liao, H.-J., Lin, C.-H.R., Lin, Y.-C., Tung, K.-Y.: Intrusion detection system: a comprehensive review. J. Netw. Comput. Appl. **36**(1), 16–24 (2013)
16. Rana, P., Pahuja, D., Gautam, R.: A critical review on outlier detection techniques. Int. J. Sci. Res. (IJSR). **3** (2014)
17. Cholkar, S.: An Efficient Anomaly Detection Mechanism for Streamed Sensor Data in IoT Environments. Master thesis, the University of Kassel (2018)
18. Truong, C.-D., Tran, D.-Q., Nguyen, V.-D., Tran, H.-T., Hoang, T.-D.: Predicting vietnamese stock market using the variants of LSTM architecture. In: Cong Vinh, P., Huu Nhan, N. (eds.) ICTCC 2021. LNICSSITE, vol. 408, pp. 129–137. Springer, Cham (2021). https://doi.org/10.1007/978-3-030-92942-8_11

Development of a Human Daily Action Recognition System for Smart-Building Applications

Ha Xuan Nguyen[1,2](✉), Dong Nhu Hoang[2], Hoang Viet Bui[2], and Tuan Minh Dang[2,3,4]

[1] Hanoi University of Science and Technology, No. 1 Dai Co Viet, Hanoi, Vietnam
ha.nguyenxuan@hust.edu.vn
[2] CMC Applied Technology Institute, CMC Corporation, 11 Duy Tan, Hanoi, Vietnam
[3] CMC University, CMC Corporation, CMC Corporation, 11 Duy Tan, Hanoi, Vietnam
[4] Posts and Telecommunication Institute of Technology, Ha Dong, Hanoi, Vietnam

Abstract. In this work, a daily action recognition system for smart-building applications is developed. The system consists of a processing pipeline to perform tasks of human detection, pose estimation, and action class classification. The Yolov7-Pose was used for the human detection and pose estimation task, while a trained model based on the CRT-GC method was used for the action classification. The prediction of the start-to-finish duration of an action in a sequence video is performed via the sliding window method. For the trained model and the evaluation, a self-generated dataset of six classes of daily actions with challenging conditions was created. The evaluation results show that the Yolov7-Pose outperforms others in terms of accuracy, robustness, and computational efficiency. The pose estimation reaches an AP_{50} of 89.1%, and the action recognition has an mAP50 of 85.6%, in which the highest accuracy reaches 95.7%. The total computing time for the overall processing pipeline is 14ms. The obtained results indicate that there is a high potential for practical applications.

Keywords: Action Recognition · Pose Estimation · Object Detection

1 Introduction

With the rapid development of image processing techniques, computing hardware, and cameras, there have been many investigations into human daily action recognition [1–4]. The recognition of these activities allows for many practical applications in smart buildings, including the security surveillance, abnormal activity of elderly people or children, and safety monitoring [3]. There have been numerous datasets and methods for action recognition [3]. The datasets can be based on the 2D or 3D images, interest point, trajectory, depth, pose, motion, and shape. Also, the processing techniques can vary from traditional machine learning techniques, for example, graph-based, to deep learning techniques, such as convolutional neural networks [3].

T. D. L. Nguyen et al. (Eds.): ICISN 2023, LNNS 752, pp. 366–373, 2023.
https://doi.org/10.1007/978-981-99-4725-6_45

The majority of published datasets for the pose dataset are three-dimensional key-points captured from RGB-D cameras or combined position sensor systems [5]. The use of 3D or stereo cameras for action recognition is, in fact, inconvenient since complex infrastructure is required. Also, the field of view of 3D cameras is very limited, which reduces the potential applications of the system. Thus, in practical applications, 2D cameras are often used. As a result, datasets and processing algorithms for this type of camera are needed. The existing public datasets still lack challenging conditions as they appear in real-life applications, which should be further considered. Typical processing steps of action recognition are human detection, human pose estimation, and action classification. This sequence of consecutive processing steps can be computationally inefficient. This issue should also be improved.

In this work, a human daily action recognition system for smart-building applications is developed. Unlike traditional methods, the Yolov7-Pose [6] is used to detect humans while also estimating their corresponding pose. For the action classification, a model based on the CTR-GC method [7] is trained. The training and testing datasets are collected from NTU [5] and our self-built testbed. This dataset consists of many videos of six classes of actions, including standing, sitting, walking, standing up, sitting down, and falling. The overall processing pipeline is developed. The prediction of the start-to-end duration of an action in the video sequence is performed using the concept of the sliding window [8]. Accuracy, robustness against challenging conditions, and computational efficiency are all evaluated. The contribution of this work is thus a new system that allows for efficiently and accurately recognizing human actions in challenging conditions.

2 System Description and Method

The overall processing pipeline for the daily action recognition system is shown in Fig. 1, which consists of a sequency of processing steps. First, image frames captured from the camera system are pushed into the human detection, tracking, and pose estimation modules. A pretrained model of Yolov7-Pose [6], a branch of Yolov7 official implement, was used for the detection and pose estimation tasks, and the so-called SORT [9] method was used for the tracking task. After this processing step, each detected and tracked person in the video frames is assigned a tracking ID, which corresponds to a sequence of poses. As depicted in Fig. 1. There are N video frames and M persons that are detected and tracked. The sequence number of each ID is continuously pushed into the buffering system according to the rule First-In-First-Out (FIFO). Second, the data sampling module acquires data from the buffering system and pushes it to the action recognition module. For each ID, a given number of consecutive series of poses are sampled using the so-called "sliding window method" [8]. The use of a sliding window allows us to find the start-to-finish duration of an action since, in fact, an action can appear at any time in the sequence of videos. Without the sliding window, the consecutive series of poses of each action can be incorrectly acquired or overlapped with other actions, which reduces the accuracy of the action prediction module. The action recognition module employs the Channel-wise Topology Refinement Graph Convolution (CTR-GC) method [7]. This model outputs the prediction of each action in a set of six interesting actions with corresponding confidence. In this work, six classes of daily activities, including standing up, sitting down, falling, standing, sitting, and walking, are considered.

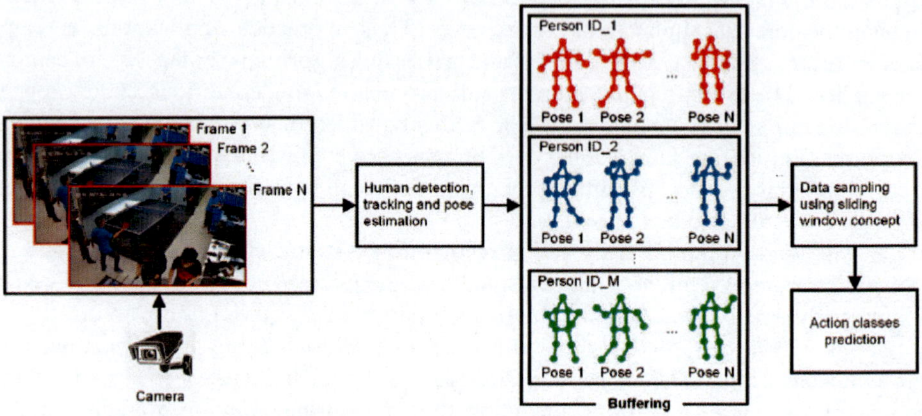

Fig. 1. Processing pipeline of the whole activity recognition system.

Details of the sliding window method are illustrated in Fig. 2. In the illustration, a sliding window has three sliders. These sliders have different lengths (number of poses), which were empirically chosen. The window is moved to a new location with a stride. The sliders "2" of the sliding window "2" contain start-to-finish poses of the appearing action very well, as can be seen. In contrast, all sliders in the sliding window "1" cannot hold the entire duration of the displayed action. In the sliding window "2", the slider "1" is too short while the slider "3" is too long, which may overlap with other actions.

The choice of the lengths of each slider as well as the stride is very important to correctly capture the start-to-finish duration of an action. Depending on the type of action, the duration lengths can be different, which leads to the fact that the lengths of the sliders and stride must be empirically chosen. Thus, an empirical model is proposed. Based on the collected dataset of the six classes of actions, histograms of duration for all classes of actions are plotted. From these histograms, the lengths of the three sliders are empirically found to be 25, 40, and 55 for each. The stride length is 15 and the buffering length is 55.

All three sliders of each sliding window are pushed to the trained CTR-GC model for action recognition. The prediction model is implemented with a batch size of three to improve the computational efficiency. After the prediction step, an action class with the highest mean confidence of three predictions is voted to be recognized.

The CTR-GC [7] backbone was trained for the action model using our collected daset. Unlike many published works, where 3D pose is used, in our work, only 2D pose is used. The use of 2D pose is so meaningful that, in most practical applications, 2D cameras are often used. The use of 2D cameras can expand the field of view and does not require complex processing algorithms like 3D cameras. Table 1 shows statistics from our self-generated dataset. The dataset was created using videos from NTU [5] as well as our own self-created videos on two different testbeds. The videos are recorded

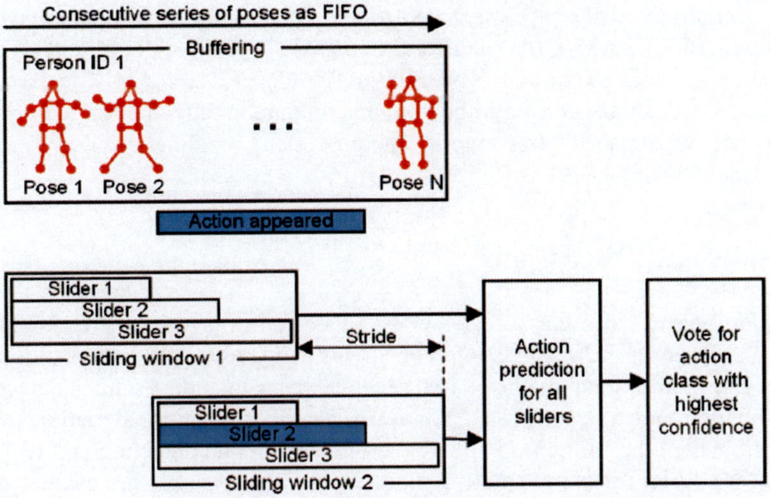

Fig. 2. Illustration of the sliding window for detecting start-to-finish duration of an action [8].

with full-HD resolution of 1920x1080 and 15FPS. In our testbeds, each testbed has four cameras, which are arranged at different viewpoints and heights. Each testbed is setup in a different location to add variety to the dataset. The number of videos for each class of action is listed in Table 1. Depending on the type of action, the video takes 2–5 s to complete.

Table 1. Statistics of pose dataset for fine-tuning the CTR-GC model.

Type of actions	From NTU [5]	From testbed1	From testbed2	Total videos
Falling	1122	1942	1278	4342
Sitting	886	1168	2953	5007
Standing	2893	508	980	4381
Sitting down	888	962	1380	3230
Standing up	871	912	1372	3155
Walking	3567	530	1027	5124

The generated videos are then fed into the Yolov7-Pose model, which detects and estimates the pose. It should be noted that the pose is 2D and has a skeleton of 17 keypoints that are fitted to the COCO dataset format [10]. After this step, a pose dataset is obtained and further used for the training process of the CTR-GC model. This model is trained from scratch, where the dataset obtained from NTU and testbed1 is used for the training and that of testbed2 is used for the testing. The trained model is customized so that the input must align with the MS-COCO dataset format and the output has six classes of actions.

The overall processing pipeline is deployed on a workstation computer having configuration of 16GB RAM, CPU Intel(R) Core (TM) i7–10700 CPU @ 2.90GHz, and a computation accelerated card GPU NVIDIA RTX 3060 12GB RAM. The used models are converted to suitable format to be able be computationally accelerated on Nvida's hardware via the TensorRT framework. The processing pipeline is coded by C++ and also the OpenCV library is used.

3 Results and Discussions

The use of Yolov7-Pose has the advantage of detecting people and estimating their corresponding poses in a single inference. This will reduce the computational time significantly. In traditional methods, two separate steps, including human detection and pose estimation, must be performed. An evaluation of the traditional methods was also performed, where the Yolov5s [11] was used for the human detection and the HRNet-w48 [12] was used for the pose estimation. The two approaches are evaluated on the MS-COCO dataset, which has 5000 images of different types of objects. The evaluation results are shown in Table 2. The object detection model is evaluated based on the benchmark average precision (AP) with a threshold of intersection over union (IoU) ranging from 0.5 to 0.95. Similarly, in the pose estimation, the benchmark object keypoint similarity (OKS) with a threshold is used.

The results show that, for the pose estimation task, with a threshold of 0.5, the discrepancy between the two approaches is very small (89.2 vs. 90.4). In contrast, at thresholds from 0.5 to 0.95, the mAP of HRNet-w48 is significantly higher. This is due to the fact that the image input size of the two approaches differs. With the HRNetw48, the image size is only 384x288 while that of the Yolov7-Pose is 640x640. This leads to the error in the Yolo7-Pose becoming higher. However, for the object detection task, the Yolov7-Pose outperforms the Yolov5s.

It is noticed that the results in Table 2 are obtained based on the evaluation of the MS-COCO dataset, in which objects and their poses are quite easy to detect and estimate. In practical applications, there are several challenging conditions that must be overcome. Thus, the two approaches must be evaluated based on challenging data to verify their robustness and accuracy. In our self-generated dataset from testbeds 1 and 2, challenging conditions are added. As illustrated in Fig. 3, images with very poor lighting conditions (infra-red mode) of humans with partial or occluded appearances are included. Furthermore, scenarios about different shooting angles, including from far to close, diagonal angles, hard-to-see, and the variety of ways to generate actions, are considered. It is demonstrably demonstrated that using Yolov7-Pose with the challenging dataset is far more robust than HRNet-w48. As can be seen in Fig. 4, the HRNet-w48 fails to estimate the pose, whereas the Yolov7-Pose performs much better. One key advantage of the Yolov7-Pose is that it is very computationally efficient and requires only 10 ms for estimating all poses in the image, whereas HRNet-w48 needs 20 ms for a single pose. If the image contains several poses, the computational time will be longer. Compared to other works [13–15], the use of Yolovv7-Pose has the best performance in terms of accuracy and computational efficiency, especially for challenging conditions like occluded and partial bodies.

Table 2. Comparison of average precision of the Yolov7-Pose and Yolov5s with HRNetw48.

Tasks	AP_{50} (%)	mAP_{50}-P_{90}(%)
Object detection with Yolov7-Pose	**93.8**	72.6
Pose estimation with Yolov7-Pose	89.1	68.0
Pose estimation with Yolov5 and HRNet w48	90.4	**76.3**

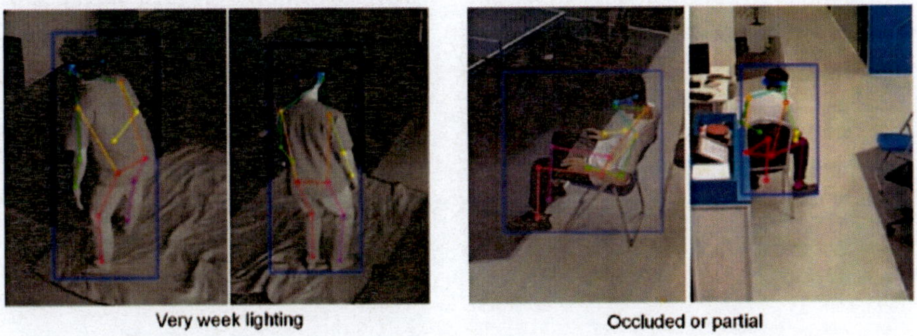

Fig. 3. Illustration of the challenging of the self-generated dataset.

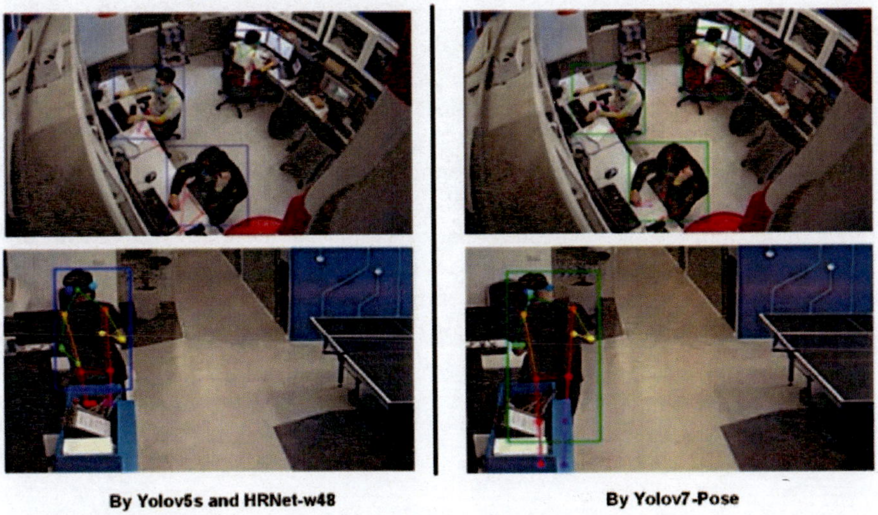

Fig. 4. Performance comparison between Yolov5s-HRNet-w48 and Yolov7-Pose.

Table 3 shows the evaluation results for the action recognition model CTR-GC. The model is evaluated based on the cross-view method. That means all videos of testbed 1 are used for the training process, and all videos of testbed 2 are used for the testing process. The percentage of true positives divided by the total number of tests is used to calculate accuracy. It can be seen that the standing action has the highest accuracy of 95.7%, which is followed by the falling action with an accuracy of 89.5%. Other actions have lower accuracy. This can be explained by the fact that standing is a very visible action. As a result, it is easier to recognize. For the falling action, our dataset contains more data about this action, including the volume as well as the variety, than other classes. Thus, the accuracy of this action is higher and more robust. The reason for this unbalanced data in this class is that we aim to detect the falling action of elderly people in smart-building applications. The average accuracy of all classes is 85.6%. The computation time of the model is only 4ms. With these results, our system can be used for practical applications.

Table 3. Accuracy of the action recognition model CTR-GC

Action	Falling	Sitting	Standing	Standing up	Sitting down	Walking
Accuracy (%)	89.5	90.6	95.7	77.8	80.9	78.9

4 Conclusions and Outlook

In this work, a daily action recognition system for smart-building applications has been successfully developed. The obtained results have high potential for practical applications. The use of Yolov7-Pose has advantages in terms of accuracy, robustness, and computational efficiency. The use of HRNet-w48 has less robustness and requires much more computational effort. Thus, Yolov7-Pose should be the best choice. The datasets that have been made public still lack challenging conditions as well as variety properties. Thus, they are hard to use for practical applications. A self-generating dataset containing challenging conditions and real-life scenarios should therefore be created. The sliding window parameter has a significant impact on accuracy. Depending on the interested action classes, these parameters must be optimized in an empirical way.

In the future, a thorough evaluation of the pose estimation model on challenging datasets and optimization of the sliding window are intended to be investigated.

Acknowledgement. This research is funded by the CMC Applied Technology Institute, CMC Corporation, Hanoi, Vietnam.

References

1. Özyer, T., Ak, D.S., Alhajj, R.: Human action recognition approaches with video datasets—a survey. Knowl. Based Syst. **222**, 106995 (2021)
2. Le, V.-T., Tran-Trung, K., Hoang, V.T.: A comprehensive review of recent deep learning techniques for human activity recognition. Comput. Intell. Neurosci. **2022**, 1–17 (2022). https://doi.org/10.1155/2022/8323962

3. Pareek, P., Thakkar, A.: A survey on video-based human action recognition: recent updates, datasets, challenges, and applications. Artif. Intell. Rev. **54**(3), 2259–2322 (2021)
4. Ren, B., Liu, M., Ding, R., Liu, H.: A survey on 3d skeleton-based action recognition using learning method. arXiv preprint arXiv:2002.05907 (2020)
5. Shahroudy, A., Liu, J., Ng, T.T., Wang, G.: NTU RGB+D: a large scale dataset for 3d human activity analysis. In: Proceedings of the IEEE conference on computer vision and pattern recognition, pp. 1010–1019 (2016)
6. Maji, D., Nagori, S., Mathew, M., Poddar, D.: YOLO-pose: enhancing YOLO for multi person pose estimation using object keypoint similarity Loss. In: Proceedings of the IEEE/CVF Conference on Computer Vision and Pattern Recognition, pp. 2637–2646 (2002)
7. Chen, Y., Zhang, Z., Yuan, C., Li, B., Deng, Y., Hu, W.: Channel-wise topology refinement graph convolution for skeleton-based action recognition. In: Proceedings of the IEEE/CVF International Conference on Computer Vision, pp. 13359–13368 (2021)
8. Ma, C., Li, W., Cao, J., Du, J., Li, Q., Gravina, R.: Adaptive sliding window-based activity recognition for assisted livings. Inf. Fusion **53**, 55–65 (2020)
9. Bewley, A., Ge, Z., Ott, L., Ramos, F., & Upcroft, B.: Simple online and realtime tracking. In: 2016 IEEE international conference on image processing (ICIP), pp. 3464–3468 (2016)
10. Lin, T.-Y., et al.: Microsoft coco: common objects in context. In: Fleet, D., Pajdla, T., Schiele, B., Tuytelaars, T. (eds.) Computer Vision – ECCV 2014. LNCS, vol. 8693, pp. 740–755. Springer, Cham (2014). https://doi.org/10.1007/978-3-319-10602-1_48
11. Yolov5s. https://github.com/ultralytics/yolov5
12. Sun, K., Xiao, B., Liu, D., Wang, J.: Deep high-resolution representation learning for human pose estimation. In: Proceedings of the IEEE/CVF Conference on Computer Vision and Pattern Recognition, pp. 5693–5703 (2019)
13. Huang, J., Zhu, Z., Huang, G., Du, D.: AID: pushing the performance boundary of human pose estimation with information dropping augmentation. arXiv preprint arXiv:2008.07139 (2020)
14. Cai, Y., et al.: Learning delicate local representations for multi-person pose estimation. In: Vedaldi, A., Bischof, H., Brox, T., Frahm, J.-M. (eds.) Computer Vision – ECCV 2020. LNCS, vol. 12348, pp. 455–472. Springer, Cham (2020). https://doi.org/10.1007/978-3-030-58580-8_27
15. Artacho, B., Savakis, A.: Omnipose: a multi-scale framework for multi-person pose estimation. arXiv preprint arXiv:2103.10180 (2021)

Check for updates

Analytical Constrains for Performance Improvement of the Integration INS/GNSS into Navigation System

Nguyen Trung Tan[1](✉), Nguyen Thi Dieu Linh[2], and Bùi Minh Tín[1]

[1] Faculty of Radio Electronics, Le Quy Don Technical University, Ha Noi, Vietnam
trungtan68@gmail.com

[2] Science and Technology Department, Hanoi University of Industry , Ha Noi, Vietnam

Abstract. The integration of Inertial Navigation System (INS) into Global Navigation Satellite System (GNSS) utilizing Inertial Measurement Unit (IMU) has become increasingly common in Mobile Mapping Systems (MMS) and navigation. It enables the accurate determination of the location, velocity, and attitude of mobile entites in a seamless manner. Besides, thanks to advantages such as compact light weight structures, low cost and energy consumption, the Micro-Electro-Mechanical System (MEMS) IMU and GPS transceivers have been an active research area. However, the quality of the small-cost INS/GPS systems remains low, specially in GNSS-noise and without-GNSS environments. To improve the system performance, this study applies analytical constraints, consisting of non-holonomic constraints and zero-velocity updates, to the data unification, such as the Enhanced Kalman Filter. Experiments and data analysis are used to validate the benefits of our proposal.

Keywords: INS/GPS · INS/GNSS · IMU · Kalman Filter

1 Introduction

The combination between the inertial navigation system with inertial measurement unit and global positioning system is commonly utilized in navigation applications and mobile mapping system to determine state in terms of the location, velocities and direction of mobile entities. INS offers good points such as high autonomy, large sampling rate for measurement, and short-term precision. Neverthless, the precision quickly decreases over time when an external entry is not added, especially when the effective-cost IMU is used. While the long-term position and velocity information can be provided extractly by GPS, it still exists several limitations for navigation applications when used individually, including the small rate for sampling, affecting by environment, and the absence of orientation determination with one antenna. Thus, intergrating INS into GPS has emerged as an effective method to improve the system performance [1]. INS can operate in the environment without GPS. However, their applicability is affected by the inertial-sensor cost and the unavailability duration of GPS signal. Either tactical-grade or better inertial system allows to provide both the location precision and sustainability when the GPS signal

T. D. L. Nguyen et al. (Eds.): ICISN 2023, LNNS 752, pp. 374–383, 2023.
https://doi.org/10.1007/978-981-99-4725-6_46

is congested for a long time [2]. High-end and high-cost INS can offer real-time position precision of less than 3 m during a one-minute GPS gap. Nevertheless, the prohibitive cost of sophisticated inertial sensors makes them unsuitable as the primary navigation module for road vehicles. Thus, the Micro-Electro-Mechanical System (MEMS) uses strap-down inertial sensors as the additive component to GPS to seamlessly conduct the vehicle navigation. However, their position precision decreases rapidly over time when GPS signals are not available. It can be seen that the sustainability of the integrated INS-GPS scheme applying MEMS inertial technique in without-GPS conditions is limited. Nevertheless, MEMS inertial sensor technology has rapidly developed, providing potential navigation solutions in terms of precision and expense for the road traffic [3]. This paper proposes applying analytic constraints to the Extended Kalman Filter (EKF) to give the performance improvement of a cheap INS-GNSS system in hostile environments with the goal of bounding errors during GNSS outages. Analytic constraints involve the usage of physical conditions and moving-platform theory to improve the INS-GPS system without complementary sensors. Non-holonomic constraints (NHC) were first proposed by Dissanayake et al., in [4] for cheap, strap-down INS in road applications. The NHC essence assumes that in the road vehicle, velocities in directions perpendicular to the movement direction are zero. The Zero Velocity Update (ZUPT) is based on the observation that when a vehicle stops, velocities in all directions are zero.

The following parts of the paper is organnized by: Sect. 2 presents the fundamental INS merchainzation, Sect. 3 introduces the principle of analystic constrains. Section 4 is about INS/GNSS integration with analystic constrains. The experiment and discussion are provided in Sect. 5.

2 Fundamental of INS-GNSS Integration

2.1 INS Mechanization

IMU provides the angular rate and specific force in the body frame, which is rigidly attached to and defined within the vehicle carrying the navigation system. The INS mechanization processes the IMU's raw measurements in the body frame to determine navigation paremeters such as the position, velocity, and attitude in the navigation frame. Figure 1 depicts the INS mechanization in the local-level frame. Equation (1) provides dynamic equations for the position, velocity, and attitude.

$$\begin{bmatrix} \dot{r}^l \\ \dot{v}^l \\ \dot{R}^l \end{bmatrix} = \begin{bmatrix} D^{-1}v^l \\ R_b^l f^b - (2\Omega_{ie}^l + \Omega_{el}^l)v^l + g^l \\ R_b^l(\Omega_{ib}^b - \Omega_{il}^b) \end{bmatrix} \tag{1}$$

where $\dot{r}^l$ is the time derivation of the position in the local-level frame. Let $\dot{v}^l$ and $\dot{R}^l$ denote the time derivative of the of the velocity and attitude, respectively. f^b is the vector of applied forces sensed by accelerometers. Ω_{ib}^b presents the angular velocity of the body frame related to the inertial frame and parameters in the body frame; R_b^l is the transformation matrix from the body frame to the local level frame. Let Ω_{ie}^l and Ω_{el}^l present the rotation rate of the earth with respect to the inertial frame and of the navigation frame with respect to the earth, respectively; g^l is the normal gravity in the local level frame. D^{-1} is defined as follows.

$$D^{-1} = \begin{bmatrix} 0 & \frac{1}{M+h} & 0 \\ \frac{1}{(N+h)\cos\phi} & 0 & 0 \\ 0 & 0 & 1 \end{bmatrix} \tag{2}$$

where N presents the curvature radius in vertical, M is the meridian radius of curvature, h is the ellipsoid height, the latitude is denoted by ϕ.

An INS mechanization algorithm's performance is often inadequate due to inertial sensor biases and fixed-step integration errors, leading to quick divergence of navigation parameters. The software for navigation must address these errors to fix estimated parameters. It is possible to determine the dynamic error model for navigation parameters consisting of the position, velocity, and attitude in the Kalman Filter (KF) through linearizing INS mechanization equations and ignoring trivial terms in the linear model.

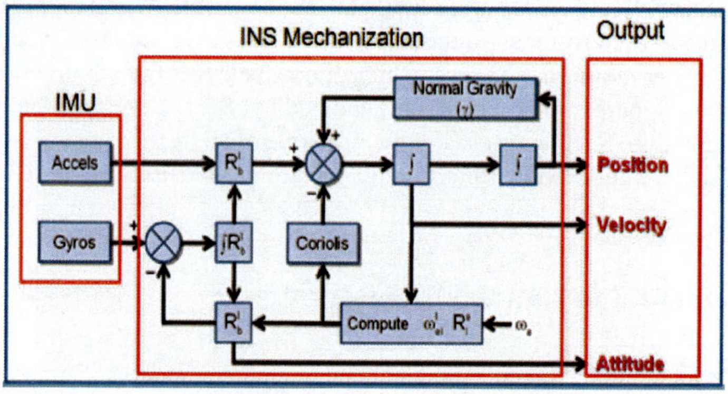

Fig. 1. Architecture of INS mechanization

2.2 INS-GNSS Integration

In this research, Loosely Coupled (LC) is utilized thanks to its simpility in processing the data. In the traditional LC INS-GPS model, the GPS processor computes location fixes and velocities at the local-level frame. Next, it transmits updated measurements to major EKF. Based on the comparison between the navigation methods given by INS and those given by GPS processor, states of navigation are estimated perfectly as shown in Fig. 2. LC offers the advantage of being straightforward to implement without advanced knowledge of processing GPS measurements. However, its major drawback is that INS can only perform measurement updates when more than four satellites are in view [5].

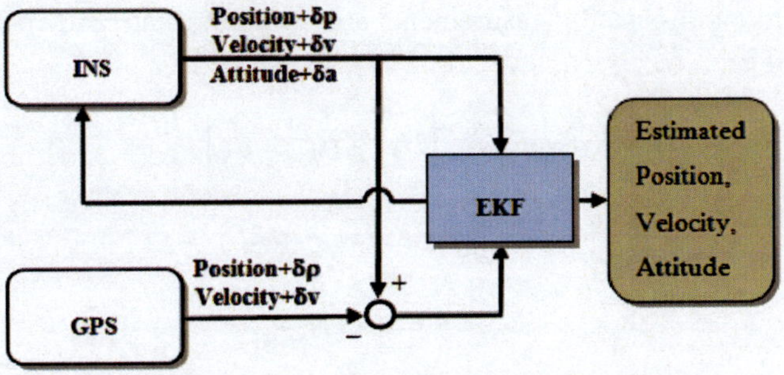

Fig. 2. Loosely coupled INS/GPS scheme

2.3 Estimation Algorithms

The estimation algorithm plays an important part in INS-GPS system to provide the best solution. EKF is known as a popular method thanks to its simple in the local-level frame. In order to use EKF, theoretical models are firstly formed. Based on the INS error model, the system model is given as follows [1]

$$x_k = \Phi_{k-1;k} x_{k-1} + w_k \tag{3}$$

In which $x = \left[\delta R \delta V \delta \psi b_a b_g s_a s_g\right]^T_{21\times1}$ denotes state vector with components compose the location, velocity and attitude errors, biases and scale factor of accelerometers and gyroscopes; $\Phi_{k-1;k}$ is the state changing matrix from *k* – *1* to *k*, w_k is referred to as the noise of the system. The measure model is given by GPS measurement as follows.

$$z_k = H_k x_k + n_k \tag{4}$$

where z_k and H_k present the measured vector and mapping matrix, respectively. n_k is measured noise at time *k*.

Using the system model presented in (3) and the state and covariance from time *k-1*, we can derive the state and corresponding covariance values at time *k* as follows.

$$\hat{x}_k^- = \Phi_{k-1;k} \hat{x}_{k-1} \tag{5}$$

$$P_k^- = \Phi_{k-1;k} P_{k-1} \Phi_{k-1;k}^T + Q_k \tag{6}$$

When additional assisted measurements are available, state and covariance values will be updated by

$$K_k = P_k^- H_k^T \left[H_k P_k^- H_k^T + R_k \right]^{-1} \tag{7}$$

$$\hat{x}_k = \hat{x}_k^- + K_k \left(z_k - H \hat{x}_k^- \right) \tag{8}$$

$$P_k = P_k^- - K_k H_k P_k^- \tag{9}$$

Here, $\hat{x}_k^-$, P_k^- represents the predicted state and covariance at time k, $\hat{x}_{k-1}, P_{k-1}$ are the estimated state and covariance at time $k - 1$, and $\hat{x}_k, P_k$ are the estimated state and covariance at time k.

3 INS-GNSS Integration with Analytics Constrain

3.1 Non-holonomic Constrain

Dissanayake, et al. [4] explained that if the vehicles stay in a land platform, the vehicle's velocity in the plane perpendicular to the direction of travel is close to 0. It is referred to as restrictions for the navigation of road traffic. To implement this constraint, components of the velocity in y and z dimensions of the body frame are assumed to be zero, as given in (10) and Fig. 3.

$$\begin{cases} v_y^b = 0 \\ v_z^b = 0 \end{cases} \tag{10}$$

where (b) is the payload of frame.

Concerning EKF, INS detects the velocity vector then transformers it into the body frame.

$$v^b = C_n^b v^n \tag{11}$$

The measurement expression for EKF is given by

$$\delta z = \begin{bmatrix} v_y^b - 0 \\ v_z^b - 0 \end{bmatrix} = \begin{bmatrix} 0\,1\,0 \\ 0\,0\,1 \end{bmatrix} C_n^b \delta v^n + \begin{bmatrix} \varepsilon_{vy} \\ \varepsilon_{vz} \end{bmatrix} \tag{12}$$

where ε_{vy} and ε_{vz} are velocity noise in y and z dimensions, respectively.

NHC is an analytical correction that can be utilized for any land-based INS to provide the precision improvement of the navigation without complementary sensors. Nevertheless, when it is inapplicable the vehicle's behavior, NHC will introduce more noise into the system. In general, when GPS is available, it is more reliable than NHC, especially under open-sky conditions. Hence, in our proposal, NHC is only used when the GPS signal is interrupted. Additionally, the update interval of NHC is variable and depends on the IMU performance, a higher quality IMU can allow for a longer update interval of NHC.

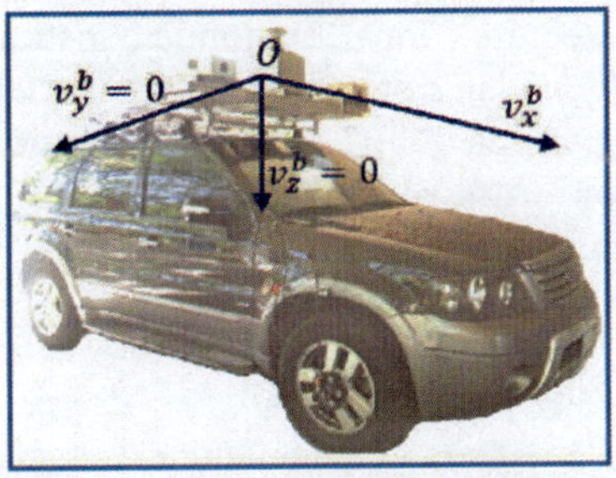

Fig. 3. Non-holonomic constraints.

3.2 ZUPT/ZIHR Update

Zero Velocity Updates (ZUPT) involves occasional short stops of the system to estimate errors and limit the increase of inertial sensor errors. When vehicles come to a complete stop, the output of the velocity in all directions is zero. To account for this constraint, the equation for measure update of ZUPT is given by

$$\delta z = \begin{bmatrix} \hat{v}_N^l & -0 \\ \hat{v}_E^l & -0 \\ \hat{v}_D^l & -0 \end{bmatrix} = \begin{bmatrix} 1\,0\,0 \\ 0\,1\,0 \\ 0\,0\,1 \end{bmatrix} \delta v^n + \begin{bmatrix} n_{vx} \\ n_{vy} \\ n_{vz} \end{bmatrix} \tag{13}$$

where $\hat{v}_x^n$, $\hat{v}_y^n$, $\hat{v}_z^n$ are north, east, and down parts of detected velocity vector of INS in the navigation frame, n_{vi} is the velocity noise in the direction i.

3.3 Integration Architecture

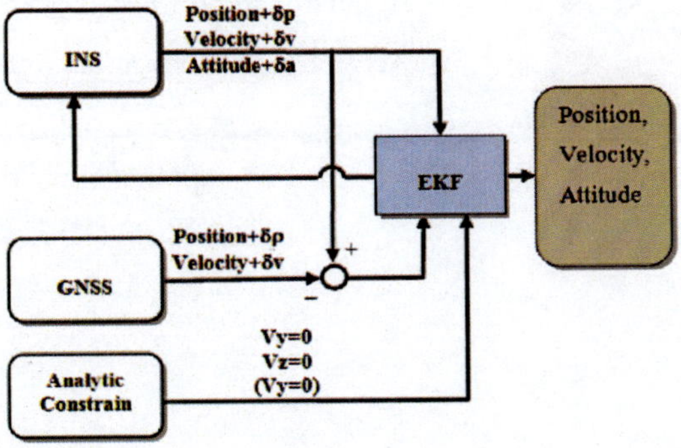

Fig. 4. Non-Holonomic constraint and ZUPT with velocity constrains

In new integration architecture, Non-Holonomic constraint and ZUPT with velocity constrains are considered by updates of measurement in the EKF given in Fig. 4. NHC is updated with given interval set. ZUPT is estimated automatically and activated by the velocity of the direction of travel. When this velocity is lower than a predefined threshold, the ZUPT activation is conducted.

4 Experiment and Discussion

For the field test, two integrated INS-GNSS navigation systems were established. The benchmark consists of a high-end tactical-grade IMU, SPAN-LCI from NovAtel, a dual-frequency geodetic-grade GNSS receiver, ProPak V3 also from NovAtel, and a distance measurement instrument (DMI). The testing system, on the other hand, utilized MEMS IMU, the STIM300 from Sensonor, as specified in Table 1. Systems are equipped on a mobile van to collect data and demonstrate the outperformance of our proposal.

Test data sets were gathered in both urban and suburban areas of Hanoi, Vietnam under various environmental conditions. The benchmark trajectory was established using the benchmark system and its IMU raw measurements as well as raw GPS carrier phase measurements that were processed in differential mode with commercial software, Inertial Explorer (NovAtel). The system utilized sensor fusion in TC smoothing mode and was aided by DMI. Overall, the kinematic positioning precision of the benchmark system was deemed sufficient with an extract level of less than 10 cm.

Table 1. Specifications for system testing.

<table>
<tr><th>Physical characteristics</th><th colspan="2">IMU performance</th></tr>
<tr><td rowspan="4">
</td><td>Gyro bias instability (deg/h)</td><td>0.5</td></tr>
<tr><td>Angular random walk (deg/√h)</td><td>0.15</td></tr>
<tr><td>Accelerometer bias instability (mg)</td><td>0.05</td></tr>
<tr><td>Velocity of random walk (m/s/√h)</td><td>0.07</td></tr>
</table>

Two scenarios were implemented for testing, INS/GNSS and INS/GNSS with analytic restrictions. Figure 5 depicts the complete trajectory of the test. An area of interest was extracted for performance analysis as shown in Fig. 6. The analysis of the position and orientation are compared to the benchmark data. Figure 7, 8, and Table 2 present the performance of two integration strategies regarding the Root Mean Square Error (RMSE) of position and attitude. The analysis indicates generally that the integrated INS-GNSS can overcome the issue of GNSS in GNSS-hostile environment such as in the urban area or through the tunnels, where GNSS signal is interfered or blocked. Besides, with the analytic restrictions such as NHC and ZUPT, the system obtains significantly better performance in comparíon with the traditional INS/GNSS for both location and attitude as shown in Fig. 6, 7 and Table 2.

Fig. 5. Test trajectory

Fig. 6. Trajectory of extracted area

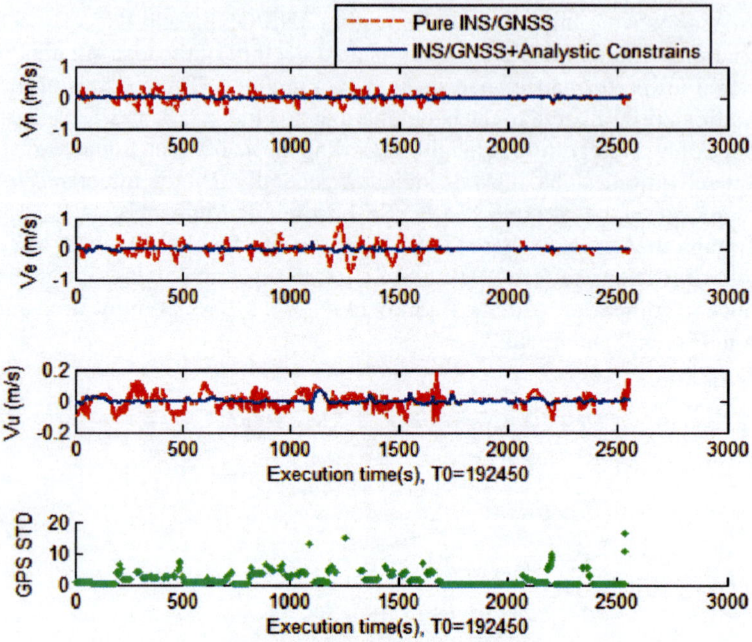

Fig. 7. Positional errors

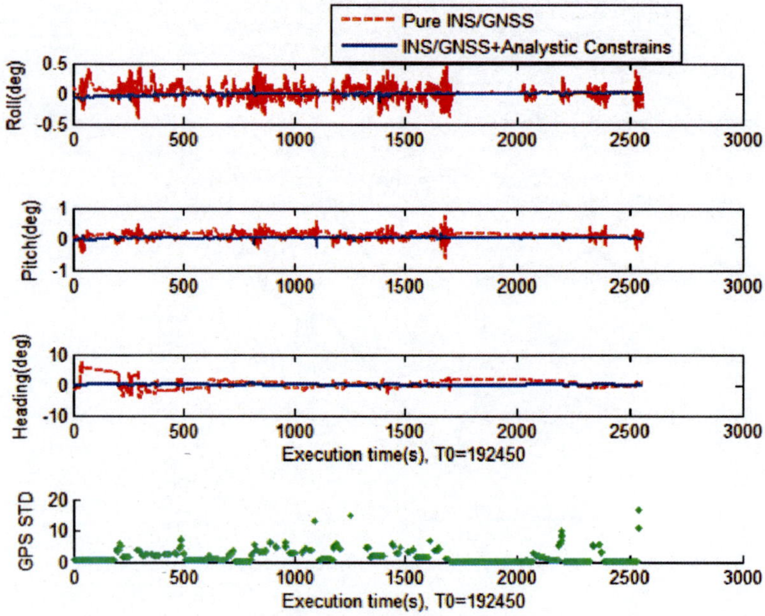

Fig. 8. Attitude errors

Table 2. RMSE of EKF and RTS

RMSE	INS-GNSS	INS-GNSS + Analystic Constrains
East (m)	2.4935	0.6834
North (m)	2.9559	0.6681
Up (m)	1.6486	0.6296
3D (m)	4.2039	1.1445
Roll (°)	0.1358	0.023
Pitch (°)	0.1772	0.0273
Heading (°)	0.0408	0.0128

5 Conclusions

Our study conduct the analysis and evaluation of the system performance for the integrated INS-GNSS system with analytical constrains consisting of non-holonomic constrain and 0-velocity update.

Experiment results revealed that our proposal results in a substantial improvement in the performance of the traditional INS/GNSS regarding to the location and attitude. Additionally, results demonstrated outperforms of the analytical constraints, as it allows to enhance system performance without requiring complementary sensors.

Error model of analytical constrains will be further evaluated in our future study. Stop statute detection strategies will be addressed to active ZUPT automatically.

References

1. Chiang, K.-W., Duong, T.T., Liao, J.-K.: The performance analysis of a real-time integrated INS/GPS vehicle navigation system with abnormal gps measurement elimination. Sensors **13**, 10599–10622 (2013)
2. Titterton, D.H., Weston, J.L.: Strapdown Inertial Navigation Technology, 2nd edn. American Institute of Aeronautics and Astronautics: Reston, VA, USA (2004)
3. Rogers, R.M.: Applied Mathematics in Integrated Navigation Systems, 3rd edn. American Institute of Aeronautics and Astronautics: Reston, VA, USA (2007)
4. Dissanayake, G., Sukkarieh, S., Nebot, E., Durrant-Whyte, H.: The aiding of a low cost, strapdown inertial unit using modeling constraints in land vehicle applications. IEEE Trans. Robot. Automation **17**, 731–747 (2001)
5. Wendel, J., Trommer, G.F.: Tightly coupled GPS/INS integration for missile application. Aerosp. Sci. Technol. **8**, 627–634 (2004)

Check for updates

Fault Analysis Approach of Physical Machines in Cloud Infrastructure

Thanh-Khiet Bui(✉)

Thu Dau Mot University, Binh Duong, Vietnam
khietbt@tdmu.edu.vn

Abstract. The large-scale cloud computing environment has raised great challenges for fault analysis in infrastructure. The openness of cloud computing makes it challenging to assess the state of the infrastructure, which affects the data center's continuing availability. In addition, the fault detection in which prior-knowledge of faults has not been defined yet can make mistakes since using supervise learning. To address this issues, the fault analysis model of physical machines in cloud infrastructure is proposed by three components, i.e., abnormal score, fault detection, and ranking suspicious metrics. The proposed model is validated by using a current Google cluster trace dataset.

Keywords: Fault analysis · Physical machines · Infrastructure cloud

1 Introduction

Cloud computing services are considered as resource power of Internet to e-commerce, social networks, computing scientific which are accessed by million users. However, continued availability of cloud-based services which minimize customer effect is critical. According to a research by Tellme Networks, identifying faults accounts for 75% of failure recovery time, and finding problems in time may save 65% of failures [1]. The abstraction layer built to provide services on top of the physical layer is affected by any hardware failure at the physical layer. Therefore, analyzing hardware faults is essential to ensure the reliability of cloud-based services. The data-center like as engine of cloud-based services while its data is fuel. During system operation, enormous amounts of data are produced and saved in the form of logs pertaining to various errors and events. In this study, the fault analysis model of physical machines (PMs) in cloud infrastructure is proposed to deal with unknown faults which have not been defined yet.

The approach is designed by combining three components.

- The abnormal score component showing ability of determining system status from monitoring data is formulated by applying one-class classification. The abnormal score is calculated by One-class Support Vector Machine's decision boundary value [2];
- The fault detection component tracks fluctuation of abnormal score values based on exponentially weighted moving average (EWMA). It can detect abrupt changes without domain knowledge. Also, its low-computation overhead responses to real-time requirement of cloud-based services;

T. D. L. Nguyen et al. (Eds.): ICISN 2023, LNNS 752, pp. 384–391, 2023.
https://doi.org/10.1007/978-981-99-4725-6_47

- After fault detection phase, suspicious metrics relating faults are ranked to locate root faults. Ranking suspicious metrics is abstracted to a feature selection problem. It is solved by combining Relief and REF-RF to locate suspicious metrics. A Google cluster trace dataset with workload and scheduler events from the Borg cluster management system over the course of a month in a cluster of more than 12,000 nodes serves as the basis for the validation of our model [3].

The remainder of the paper is structured as follows: Sect. 2 reviews the relevant literature. The proposed fault analysis approach is introduced in Sect. 3. The evaluation is shown in Sect. 4. In Sect. 5, we wrap up our study and suggest additional research.

2 Related Work

Fault detection problems are approached by machine learning including clustering, classification, and hybrid models. Clustering model divide data into groups of similar objects. This method allows for the detection of anomalies such as EM Clustering, k-Means, k-Medoids, etc. Classification model assigns a new data sample to a class based on the trained data set. When faults are found, the data is split into two classes: safe and fault. Support Vector Machine (SVM), Neural Networks, Classification Tree, Fuzzy Logic, etc. are a few of the methods now in use. Hybrid model combines different models and algorithms to improve algorithm accuracy. Others are Cascading supervised techniques. For example, decision tree in combination with Naive Bayesian or Support Vector Machine; Combining supervised and unsupervised techniques. Researchers can merge the modified version of already existing algorithms. Most models focus on the modeling of normal data to be able to identify faults. Lin and colleagues have proposed an abnormal discovery mechanism proposed for IaaS Cloud Computing [2]. Extraction of performance data by Global locality pre-serves projection algorithm and then uses the local outlier factor algorithm to identify abnormalities. Studies reveal that the Global locality preserving projection method and the Locality preserving projection algorithm are superior to the PCA algorithm and other anomalous recognition systems, respectively. The RUBiS standard is used in experiments to develop a cloud computing platform using OpenStack and Xen, and 50 mistakes are then introduced into OpenStack to test the system. Then, they collect system performance data in real time and analyze them the F-Measure parameter is used to measure the accuracy of the algorithm. Next, Lin and his colleagues applied an efficient data-driven method of identifying local outilier (LOF) algorithms to identify abnormal performance and find out the Metric causes anomaly of the performance of the cloud platform at runtime execution [3]. The empirical results show that data-based anomalies are better than previous ones designed for cloud computing platforms with precision, recall, and F-measure. Doelitzscher and colleagues identify the abuse of cloud cases, the study presents an abnormal identification system for IaaS systems. [4]. Based on the analysis of cloud user behavior, neural networks are used to analyze and understand the normal using behavior of cloud clients, then abnormal identifiers can be derived from cloud security issues caused by a virtual machine crossed threshold Experimental validation of pro-posed system efficiency. Since the system is based on labeled data, unattended ma-chine learning approaches such as data clustering, apriori algorithms or self-organizing maps are also considered for validation.

3 Fault Analysis Approach

3.1 System Architecture

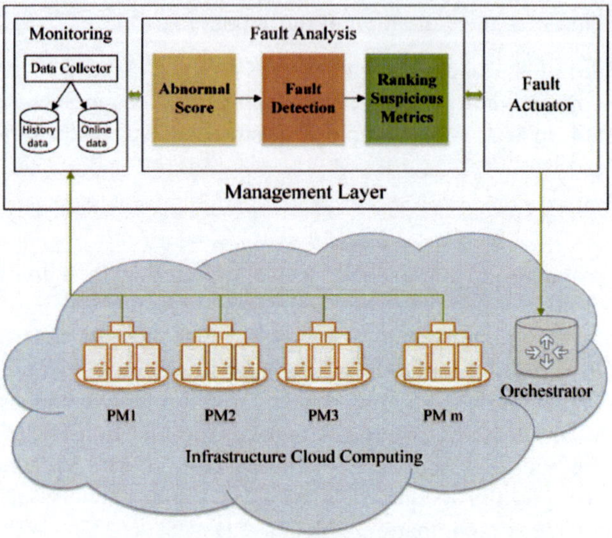

Fig. 1. System System architecture.

As shown in Fig. 1, the system architecture is divided into two layers, i.e., infrastructure layer with many PMs, and fault management layer. The PM agents gather monitoring data and send them to the Monitoring component. The system metrics are monitored through the interfaces of operation system or third party library. The collected data will be handled by normalizing, removing data, and correcting outliers. Following that, the collected data is stored in both old and new databases. Finally, the historical data with labeled faults will be exploited to discover suspicious metrics concerning faults, while the online data will be used to detect faults. The virtual machine migration technologies can be used in fault actuator component to avoid the influence of faults.

3.2 Abnormal Score

Scholkopf et al.'s One-class Support Vector Machine (OCSVM) is a method for drawing a clear line between the origin and the normal data set with the maximum margin [2]. Taking into account the training dataset $X = [x_1, x_2, \ldots, x_N] \in \mathcal{R}^{N \times M}$ which denotes the normal dataset. The primary objective of OCSVM is denoted as following.

$$\min_{w,\xi,\rho} \frac{1}{2}\|w\|^2 + \frac{1}{N\upsilon}\sum_{i=1}^{N}\xi_i - \rho \tag{1}$$

$$\textit{subject to}\, w.\Phi(x_i) \geq \rho - \xi_i, \xi_i \geq 0$$

where $\upsilon \in (0, 1]$ is a regularization parameter, N is the number of the data points, and ξ_i is nonzero slack variable for point x_i that displays the penalty of objection function. ρ and w are the target variables which determine the decisive boundary. The decisive boundary can be expressed as as:

$$f(x) = w.\Phi(x) - \rho \tag{2}$$

where $\mathcal{R}^{\mathrm{M}}$ and Φ refers to a type of mapping. The training dataset is not necessarily linearly separable in original space for the classification task. On the other hand, the original training dataset can be linearly segregated in high-dimensional space using the Φ. According to [4], the RBF kernel function can approximate the majority of kernel functions when the parameter is properly chosen. Because it is simple to turn σ properly with only one parameter, the RBF kernel $\mathrm{K}(x, y) = exp\left(-\frac{\|x-y\|^2}{2\sigma^2}\right)$ is chosen in this study.

To address the optimal problem in Eq. (1), the Lagrangian multipliers are used and the Lagrangian equation is described as following.

$$\begin{aligned} L(w, \rho, \xi, \alpha, \gamma) = \frac{1}{2}\|W\|^2 + \frac{1}{N\nu}\sum_{i=1}^{N}(\lambda_i\xi_i - \rho) \\ - \sum_{i=1}^{N}\alpha_i[(w \cdot \Phi(x_i)) - \rho + \xi_i] - \sum_{i=1}^{N}\gamma_i\xi_i \end{aligned} \tag{3}$$

where $(i = 1, \ldots, N)$ and $(\alpha_i, \gamma_i > 0)$ respectively. The partial derivatives of the Lagrangian equation are set to zero with respect to the variables w, ρ and, ξ. Following that, w and can be designated.

$$w = \sum_{i=1}^{N}\alpha_i\Phi(x_i) \tag{4}$$

$$\sum_{i=1}^{N}\alpha_i = 1 \tag{5}$$

$$\alpha_i = \frac{\lambda_i}{N\nu_i} - \gamma_i v\grave{a}\, \alpha_i \le \frac{\lambda_i}{N\nu} \tag{6}$$

When Eq. (4) - Eq. (6) are substituted into Eq. (3) and its dual form, the following results:

$$\begin{aligned} \min_{\alpha}\frac{1}{2}\sum_{i,j=1}^{N}\alpha_i\alpha_j\mathrm{K}(x_i, x_j) \\ subjec\ to \sum_{i=1}^{N}\alpha_i = 1, 0 \le \alpha_i \le \frac{\lambda_i}{N\nu} \end{aligned} \tag{7}$$

where $\alpha = [\alpha_1, \alpha_2, \ldots, \alpha_N]^T$.

The data points may be categorized into three groups based on the Kuhn–Tucker conditions:

(i) the data points locate within the boundary when $\alpha_i = 0$;
(ii) the data points are on the boundary with $0 < \alpha_i < \frac{1}{N\upsilon}$ and $\xi_i = 0$;
(iii) the data points fall outside the boundary with $\alpha_i = \frac{1}{N\upsilon}$.

The data points are called support vectors when $\alpha_i > 0$. Solve the Eq. (7) to get α and then ρ can be calculated as following.

$$\rho = \frac{1}{n_s}\sum_{i=1}^{n_s}\sum_{j}^{N}\alpha_j \mathrm{K}\left(x_j, x_i\right) \tag{8}$$

where N_{sv} is the quantity of support vectors with $\xi i = 0$ and $0 < \alpha_j < \frac{\psi i}{N\upsilon}$. The Eq. (2) can be calculated as following.

$$f(x) = \sum_{i=1}^{N}\alpha_i \mathrm{K}\left(x_j, x_i\right) - \rho \tag{9}$$

The following decision boundary in Eq. (10) calculates the abnormal score where an abnormal data point corresponds to a bigger score.

$$g(x) = \frac{f_{max} - f(x_i)}{f_{max}} \tag{10}$$

where f_{max} represents the greatest distance possible between training data points and the decision boundary.

3.3 Fault Detection

In a typical circumstance, the anomalous score $g(x)$ in Eq. (10) should remain steady; otherwise, it will change significantly when errors are detected. The EWMA chart is used in [5] similarly to identify changes in aberrant score values. The graph is suitable for spotting a slight change in the process. EWMA also costs less to compute than methods for sudden change detection. Additionally, EWMA does not require expertise in fault identification. The abnormal score values $g(x)$ are measured with EWMA chart denoted as following:

$$D_i = \tau g_i + (1 - \tau)D_{i-1} \tag{11}$$

where g_i is the anomalous score, D_i is the EWMA statistic, and $0 < \tau < 1$ is the smoothing constant. Initial deciding values are used to determine the initial value of EWMA D_0. The following formula is used to determine the upper $UCL(x_i)$ and lower $LCL(x_i)$ control limits.

$$UCL(x_i) = \mu_z + L_z\sigma_z$$

$$LCL(x_i) = \mu_z - L_z\sigma_z \tag{12}$$

where L_p is the unit normal distribution's percentage point, the initial data points of x are used to estimate the mean $\mu_p = \mu_x$ and standard deviation $\sigma_p = \sigma_x^2\left(\frac{\lambda}{2-\lambda}\right)$.

3.4 Ranking Suspicious Metrics

The suspicious metrics relate to causes of shifting system status from normal to fault. The more dramatically a measure changes before and after a fault is identified, the more likely it is to trigger a change in system status. Finding suspicious metrics associated with the root cause of the reported error has been abstracted in this work as a feature selection problem. The ranking suspicious metric problem is denoted as following.

$$\mathcal{L} = \mathcal{F}(Normal, Fault) \tag{14}$$

where $\mathcal{L}$ stands for set of suspicious metrics, *Fault* for fault dataset, *Normal* for normal dataset, and $\mathcal{F}$ for feature selection method. The methods choose an optimal feature set with great precision, but at the cost of a lot of overhead. To deal with this issues, we present a two-phase feature selection approach that includes a filter method, Relief [6], for removing redundant features and a wrapper method, RFE-RF [7], for ranking suspicious metrics. To rank the suspicious metrics associated to faults, the Recursive Feature Elimination (RFE) approach [8] is used, which applies the Random Forest (RF) algorithm on each iteration. RF is a decision tree extension that builds several classifiers in order to classify data with high accuracy. The appropriateness of the RF algorithm has been demonstrated in circumstances of defects in unbalanced training datasets [9]. The Root Mean Square Error (RMSE), which is determined as follows, is used to assess the ranking suspicious metric model.

$$RMSE = \sqrt{\frac{\sum_{i=1}^{N} (Y_{Predicted} - Y_{Actual})^2}{N}} \tag{15}$$

4 Evaluation

Google has published a tracking log data set of 12453 nodes that activity logs for these nodes for about 29 days, every 5 min [3]. Thanks to the data was shared by Sırbu [9], we obtain a set of labeled dataset including metrics, i.e., fail tasks, running task, started tasks, CPU load, MEMORY load, CPI (Cycles per Instruction), MAI(Memory Access per Instruction). As shown in Fig. 2, the abnormal score model is trained with 25-day data trace and 4-day data trace for fault detection and ranking suspicious metrics. Figure 3 shows the result of abnormal score of 4-day trace.

This model is applied to detect faults in 4-day trace by using EWMA chart to $g(x)$ in Eq. (9). The accuracy of EWMA method is benchmarked with threshold method $g(x) > 1.4$ shown in Fig. 4. Suspicious metrics for 4-day trace are ranked by using RFE-RF after labeling 4-day trace. The result of Fig. 5 shows that CPU is a suspicious metric in 4-day trace.

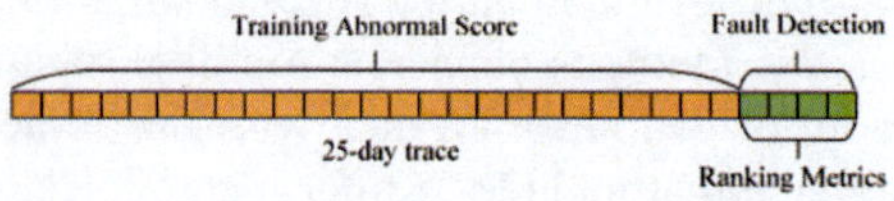

Fig. 2. Trace days

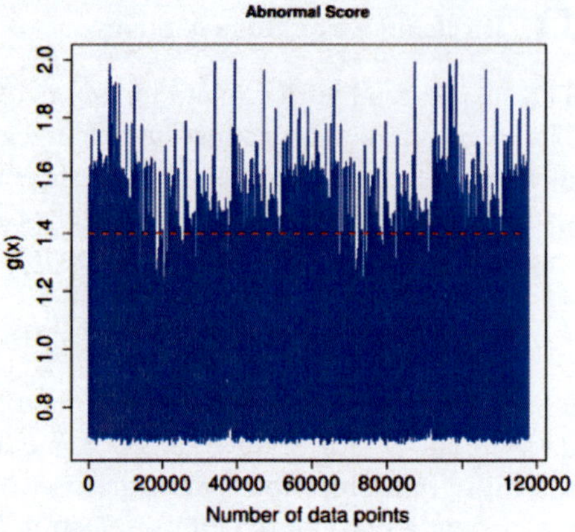

Fig. 3. Abnormal score of 4-day trace.

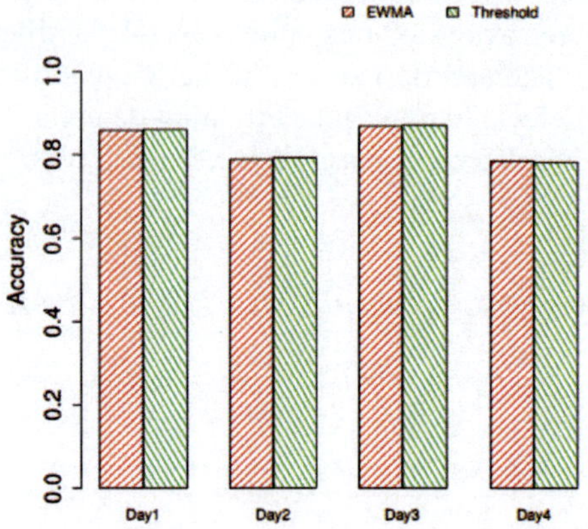

Fig. 4. Fault detection of 4-day trace.

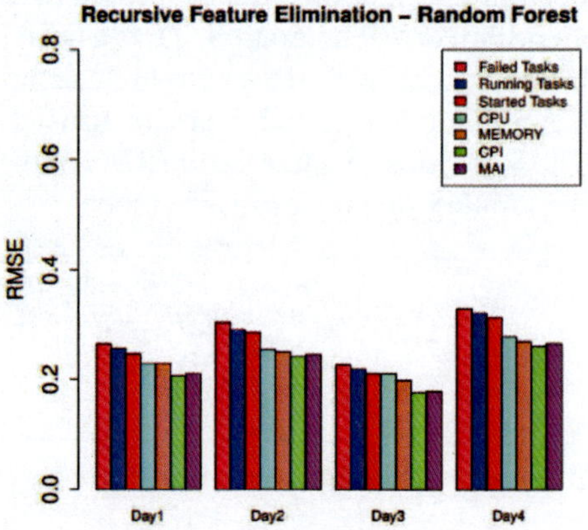

Fig. 5. RMSE Ranking suspicious metrics of 4-day trace.

5 Conclusion

In this study, we introduced a model of fault analysis for physical machine of cloud infrastructure. We proposed an abnormal score model to determine system status trained by normal data with OCSVM. The online faults are detected by using abnormal score with EWMA control charts without domain knowledge. After labeling data fault in fault detection component, suspicious metrics are ranked to locate the root fault with combining Relief and REF-RF. The evaluation results demonstrate that it can effectively detect faults of Google cluster trace dataset. In the future work, we plan to integrate some locating fault methods.

References

1. Oppenheimer, D., Ganapathi, A., Patterson, D.A.: Why do internet services fail, and what can be done about it? In: USENIX Symposium on Internet Technologies and Systems, vol. 67, Seattle, WA (2003)
2. Sch¨ olkopf, B., Platt, J.C., Shawe-Taylor, J., Smola, A.J., Williamson, R.C.: Estimating the support of a high-dimensional distribution. Neural Comput. **13**(7), 1443–1471 (2001)
3. Verma, A., Pedrosa, L., Korupolu, M., Oppenheimer, D., Tune, E., Wilkes, J.: Large-scale cluster management at google with borg. In: Proceedings of the Tenth European Conference on Computer Systems, pp. 1–17 (2015)
4. Keerthi, S.S., Lin, C.-J.: Asymptotic behaviors of support vector machines with gaussian kernel. Neural Comput. **15**(7), 1667–1689 (2003)
5. Bui, K.T., Van Vo, L., Nguyen, C.M., Pham, T.V., Tran, H.C.: A fault detection and diagnosis approach for multi-tier application in cloud computing. J. Commun. Networks **22**(5), 399–414 (2020)
6. Kira, K., Rendell, L.A.: A practical approach to feature selection. In: Machine Learning Proceedings 1992, 1992, pp. 249–256. Elsevier (1992)
7. Darst, B.F., Malecki, K.C., Engelman, C.D.: Using recursive feature elimination in random forest to account for correlated variables in high dimensional data. BMC Genet. **19**(1), 1–6 (2018)
8. Guyon, I., Weston, J., Barnhill, S., Vapnik, V.: Gene selection for cancer classification using support vector machines. Mach. Learn. **46**(1), 389–422 (2002)
9. Sırbu, A., Babaoglu, O.: Towards data-driven autonomics in data centers. In: 2015 International Conference on Cloud and Autonomic. IEEE (2015)

Imaged Ultrasonic Scattering Object Using Beamforming Strategy Along with Frequency Compounding

Luong Thi Theu[1], Tran Quang Huy[2](✉), and Tran Duc-Tan[3]

[1] Hoa Binh University, Nam Tu Liem, My Dinh 2, Ha Noi, Vietnam
lttheu@daihochoabinh.edu.vn
[2] Faculty of Physics, Hanoi Pedagogical University, Hanoi, Vietnam
tranquanghuy@hpu2.edu.vn
[3] Faculty of Electrical and Electronic Engineering, Phenikaa University, Hanoi, Vietnam
tan.tranduc@phenikaa-uni.edu.vn

Abstract. Ultrasonic imaging is commonly known for its popular reconstruction methods such as B-mode, which is widely used in commercial ultrasound devices. However, B-mode measurement is still limited in quality and detailed information about the imaged object. Recently, the method of tomographic ultrasound has become interested due to the strong development of hardware and software for devices. This paper proposes the application of frequency compounding technique in beamformed DBIM tomographic imaging. Beamformed approach using several probe transmitting elements simultaneously to offer a narrow beam which has the ability to minimize the noise effect has been applied for DBIM-based density imaging. When applying frequency compounding, it promises to improve the convergence speed, spatial resolution of imaged objects, noise reduction, and can create images of high-contrast objects. The numerical simulation results show that the image recovery performance is significantly increased and the noise is considerably improved when applying the proposed method.

Keywords: Ultrasonic tomography · inverse scattering · DBIM · beamforming · frequency compounding

1 Introduction

Ultrasonic imaging and tomography have many potential applications such as medicine, geophysics, underwater acoustics and non-destructive testing. Commercial scanners frequently employ B-mode images, which are shown in gray scale and give a map of the brightness or echo amplitude in relation to position within the scanning region. Recently, a promising trend in ultrasound imaging is tomography. Tomographic techniques have the ability to quantify and analyze the scattered field based on the interaction of the ultrasound pulse when compared to the tissue through which it is spreading. Using tomography techniques, we can obtain higher contrast images which are useful for early diagnosis.

T. D. L. Nguyen et al. (Eds.): ICISN 2023, LNNS 752, pp. 392–399, 2023.
https://doi.org/10.1007/978-981-99-4725-6_48

Attempts to determine the speed of sound crossing the shell and the volume of tissue (scatter area) have often been made in ultrasonic tomography. Some commercial tomography equipments have recently hit the market. The inverse scatter is constrained by high computation and efficiency, which is the cause of this. Early methods made use of the projection theory, which is frequently employed in nuclear and X-ray tomography [1, 2]. These ray-based techniques, however, did not work well with the diffraction characteristics of ultrasonic propagation.

The Born approximation method is the foundation for the majority of ultrasonic tomography research [3–6, 12]. The dielectric constant distribution's two-dimensional image was recovered with the distorted Born iterative method (DBIM) [7]. Then, with enhanced standard parameter selection, the DBIM was deal with the ultrasound tomography [8]. The inverse problem based on linear measurements is solved using the DBIM approach and the Tikhonov norm [9]. The advantage of this method is a quick convergence speed. However, its downside is that noise might affect it. In DBIM, each loop updates the Green function. The beamforming technique is used to construct a narrow beam in combination with the DBIM to restore density information of the target function in order to overcome one of the DBIM method's shortcomings, namely dealing with noise. With the advantages of improving the speed of convergence, resolution and being able to create high-contrast object images by frequency compounding technique [4–6], this paper proposes the application of frequency compounding technique in beamformed DBIM tomographic imaging. When applying frequency compounding, it promises to improve the convergence speed, spatial resolution of imaged objects, noise reduction, and can create images of high-contrast objects.

2 Beamformed DBIM with Frequency Compounding

The reconstructing object, discretized into side-h pixels that are N × N squares, and positioned at the 2-D space's origin, makes up the region of interest (ROI). There are N_t and N_r transmitters and receivers, respectively. The target function is calculated using the circular scatter area from Fig. 1 by

$$\mathrm{O(r)} = \begin{cases} \omega^2\left(\frac{1}{c_1^2} - \frac{1}{c_2^2}\right) - \rho^{\frac{1}{2}}(r)\nabla^2\rho^{-\frac{1}{2}}(r) & ifr \leq R \\ 0 & ifr > R \end{cases} \tag{1}$$

where c_1 and c_2 represent the sound speed in the target and the background medium, respectively, f represents the ultrasonic frequency, ω represents the angular frequency ($\omega = 2f$), ρ stands for ambient density, and R represents the object's radius.

To gather the scattered data, we set up a measuring configuration for transducers. Assumed that N_t transmitters and N_r receivers are present. The N_t transmitters are positioned all around the object at various angles in order to get comprehensive information about it. Following is the technique for doing ultrasonic signal receiving and transmission: All of the receivers (N_r) will simultaneously receive ultrasonic signals from the first three transmitters (the other transmitters are dormant). A collection of measurements (i.e., 1 × N_r measurements) is used to identify the initial placements of the transmitters. Next that, the activity is started by the following three transmitters and all receivers pick

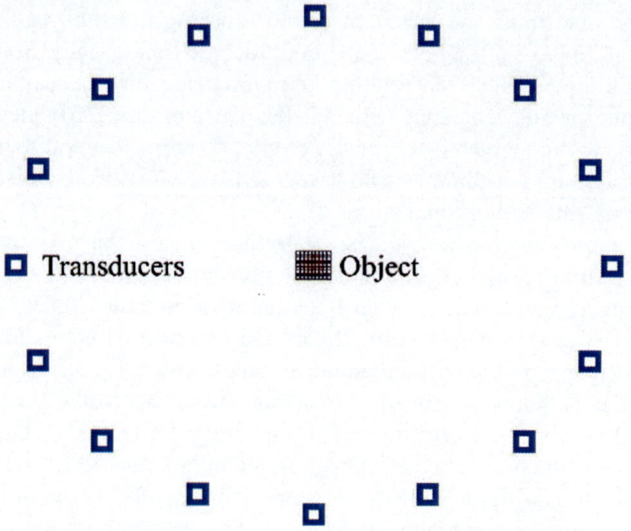

Fig. 1. Configuration of the tomographic imaging system

up a scattering signal in the second places of the following three transmitters. This results in a second set of measurements, or $2 \times N_r$ measurements. The final three transmitters, the same process is followed (N_t). At the conclusion of the measurement procedure, we have $N_t/3$ sets of obtained values (i.e., $N_t \times N_r/3$ measurements). In order to get sufficient information about the object at various angles around it, $N_t/3$ sets of obtained results are merged.

Given by is the inhomogeneous wave equation.

$$(\nabla^2 + k_0^2(r))p(r) = -O(r)p(r), \tag{2}$$

where $p(r)$ is the total pressure field and $k_0 = \omega/c_0$ is the reference medium's wavenumber, which in this case is water.

In terms of the Green's function, one can obtain the scattering pressure by solving (2) as follows:

$$p^{sc}(r) = p(r) - p^{inc}(r) = \int\int O(r)p(r)G(|r - r'|), \tag{3}$$

in which G is the Green's function in free space and $p^{inc}(r)$ is the incidence pressure. Using the *sinc* basis and *delta* functions, Eq. (3) can be resolved by the moment method [10].

A $N^2 \times 1$ vector can be used to represent the pressure in grid points.

$$p = p^{inc} + C^*D(O^*)p^*, \tag{4}$$

and a scalar value can likewise be derived for the scattered pressure.

$$p^{sc} = B_i^*D(O^*)p^* \tag{5}$$

where $\overline{C}$ is a $N^2 \times 1$ matrix made up of Green's coefficient across all the pixels in the mesh area, B_i^* is a $1 \times N^2$ vector transformed from a matrix generated by Green's coefficient $G_0(r,r')$ from every pixel to the ith receiver, and D(.) is an operator making a diagonal matrix out of a vector. In [10], there has detailed computations of B_i^* and C^*.

On condition that transmitters (N_t) and receivers (N_r) are utilized, then (5) can be used to produce the scattering pressure signal vector of size $(N_t/3)N_r \times 1$.

$$p^{sc} = B_i^* D(O^*) p^* = MO^* \tag{6}$$

where the matrix $M = BD(p^*)$ has a dimension of $(N_t/3)N_r \times N^2$.

As it satisfies the same differential equation as the pressure field, the forward solver is used to estimate the Green's function for any reference environment. The object function at step k + 1 is changed by from an initial value O_0^* and the associated referenced background at time step k.

$$O_{k+1}^* = O_k^* + \Delta O_k \tag{7}$$

where ΔO is the object function's update, which may be inferred from (6) as follows:

$$\Delta p^{*sc} = M \Delta O^* \tag{8}$$

Obviously, there is an iterative process in DBIM in order to estimate the object function O^*.

The matrix M is known to be ill-conditioned, therefore tiny measurement inaccuracies in the surface measure have a big impact on how the reconstruction turns out. The detectors are positioned outside of the meshed region, which results in a weakly diagonal and ill-conditioned inverse solver matrix M. As a result, it is frequently converted into a least-squares problem:

$$\min \| \Delta p^{*sc} - M \Delta O^* \|_2 \tag{9}$$

where the Euclidean vector space norm is denoted by the symbol $\| \|_2$.

Tikhonov regularization [9] can be used to estimate ΔO^* at the time step k in conventional ways [11]:

$$\Delta O_k^* = \min \| \Delta p^{*sc} - M \Delta O^* \|_2^2 + \gamma \| \Delta O^* \|_2^2 \tag{10}$$

where γ is the regularization parameter and Δp^{*sc} is the discrepancy between the predicted and obtained scattered fields.

The image recovery process is split into two stages: The first stage recovers the object with low frequency in the first few iterations. This stage needs to use a low frequency to satisfy the Born approximation condition; Stage 2 recovers the object with high frequency in the remaining iterations. Using high frequency improves the convergence speed, resolution and noise reduction in the recovered object.

3 Results of Numerical Simulation

Simulation parameters: Transmitter frequencies $f_1 = 0.5$ Hz, $f_2 = 1$ Hz, sound contrast 10%, object diameter 7.3 mm, transducer-to-object distances 100 mm, background environment's sound-speed 1540 m/s, number of iterations N = 4 (number of iterations using f_1: $N_{f1} = 2$, and the one using f_2: $N_{f2} = 2$), number of transmitters $N_t = 15$, and of receivers $N_r = 15$, noise 10%.

The input for the DBIM approach is Fig. 2, an initial image with sound contrast along with density information (20 pixels in total), the output of the DBIM method is the restored image when using beamformed DBIM and beamformed DBIM with frequency compounding. The restored results of the methods of beamformed DBIM and beamformed DBIM with frequency compounding are shown in Fig. 3. By visual observation, we note this, with the two methods, noise takes place quite large, especially background noise in the first and second iterations. As for the beamformed DBIM with frequency compounding, the noise is significantly reduced in the third and fourth iterations. Moreover, when compared to the original beamformed DBIM method, the object convergence by the beamformed DBIM with frequency compounding is quite good, as the recovered target function when using the advised method is near to the objective function's initialization.

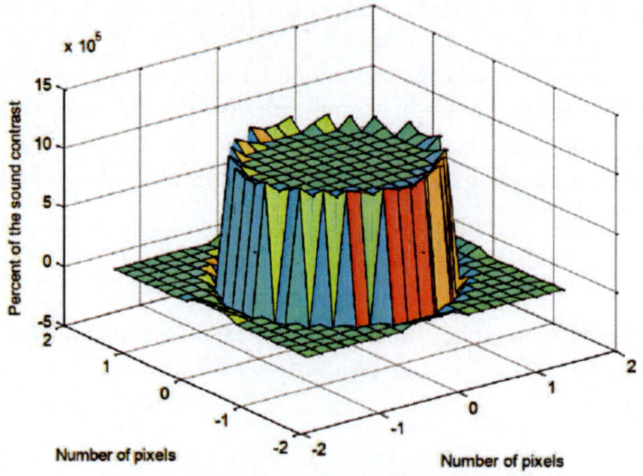

Fig. 2. Ideal initialization image (N = 20)

In order to compare the imaging quality on a quantitative basis of the beamformed DBIM and beamformed DBIM with frequency compounding, image recovery performance is calculated and shown in Fig. 4 with varying numbers of pixels, equivalent to N = 15, 20, 25 and 30. All four cases show that the image recovery performance of beamformed DBIM with frequency compounding is significantly increased in comparison with beamformed DBIM, especially right from the second iteration (at frequency hopping). The beamformed DBIM with frequency compounding is shown that the estimation of object convergence has been improved, and it can effectively overcome noise.

Methods	**Beamformed DBIM**	**Beamformed DBIM with frequency compounding**
The first iteration	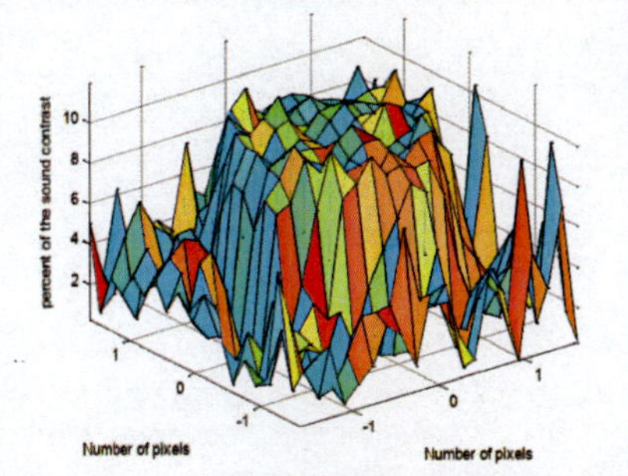	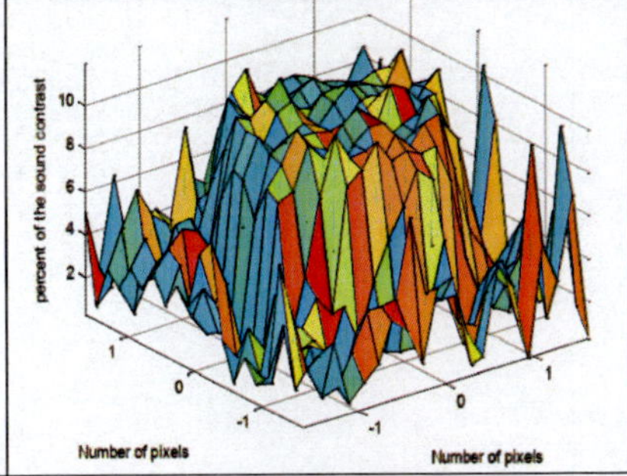
The second iteration	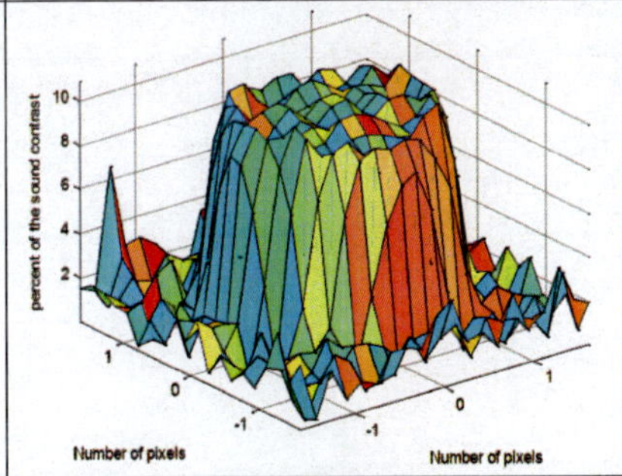	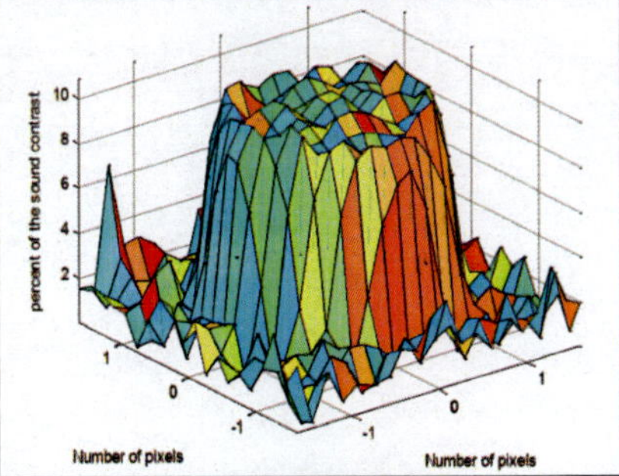
The third iteration	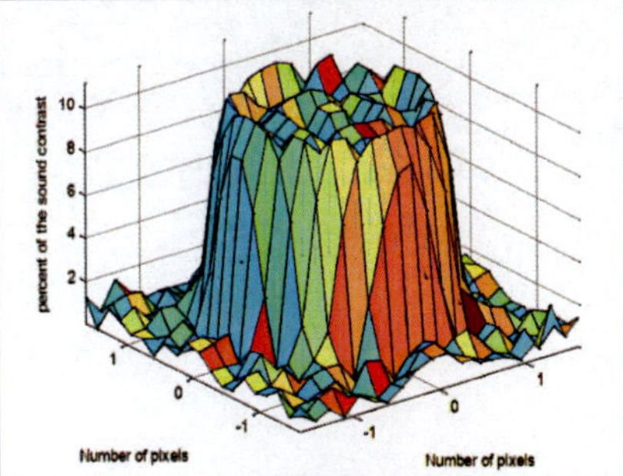	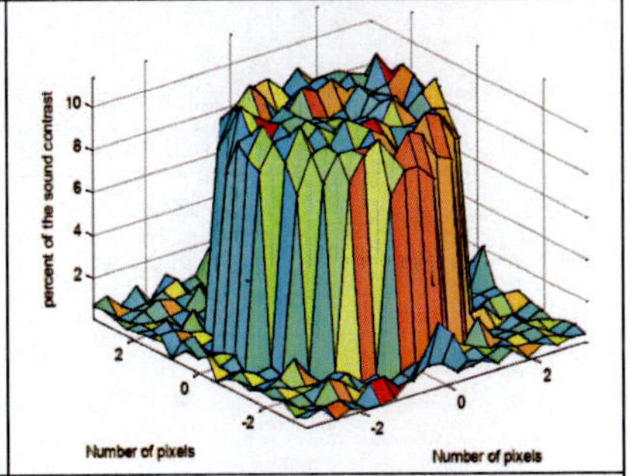
The fourth iteration	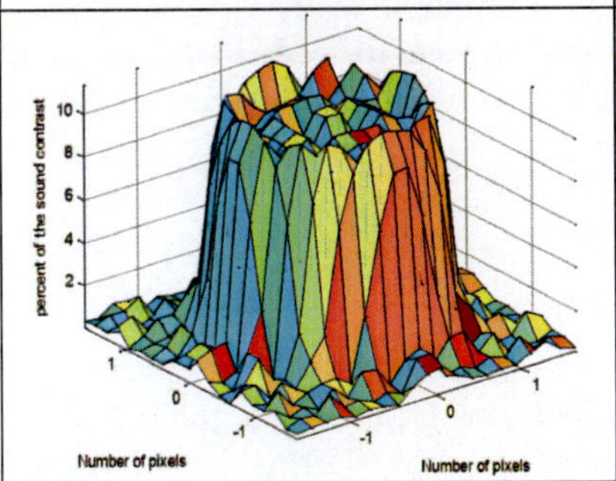	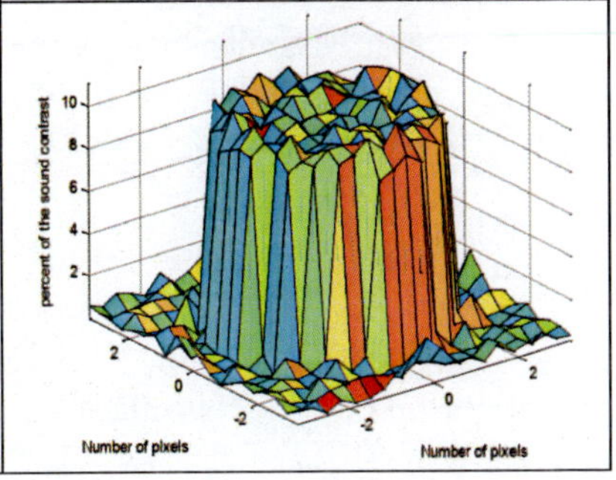

Fig. 3. Recovery images by methods of beamformed DBIM and beamformed DBIM with frequency compounding after 4 loops (N = 20)

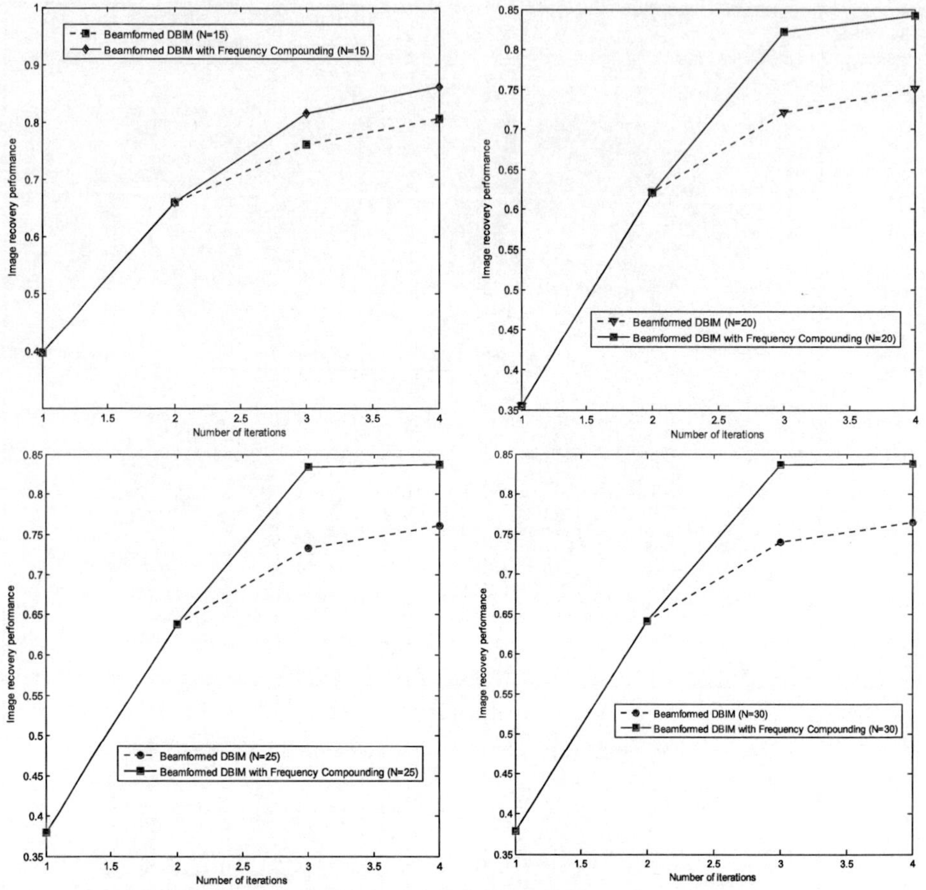

Fig. 4. Image recovery performance comparison between beamformed DBIM and beamformed DBIM with frequency compounding

After four iterations, when comparing with the beamformed DBIM, the image recovery performance using the beamformed DBIM with frequency compounding is increased by 36.72% with the case of $N = 20$.

4 Conclusions

In this work, we have applied the beamformed DBIM with frequency compounding (using a beam of three transmitters simultaneously) to restore the imaging object. The inherent drawbacks of the noise-prone beamformed DBIM can be partially avoided by the suggested method. The findings indicate that image recovery performance is significantly increased and noise has been notably reduced. Prior to imaging in practice, the beamformed DBIM with frequency compounding should be established using experimental data.

References

1. Greenleaf, J., Johnson, S., Samayoa, W., Duck, F.: Algebraic reconstruction of spatial distributions of acoustic velocities in tissue from their time-of-flight profiles. Acoustical Holography **6**, 71–90 (1975)
2. Greenleaf, J., Johnson, S., Lee, S., Herman, G., Wood, E.: Algebraic reconstruction of spatial distributions of acoustic absorption within tissue from their two-dimensional acoustic projections. Acoustical Holography **5**, 591–603 (1974)
3. Devaney, A.J.: Inversion formula for inverse scattering within the Born approximation. Opt. Lett. **7**, 111–112 (1982)
4. Huy, T.Q., et al.: An efficient procedure of multi-frequency use for image reconstruction in ultrasound tomography. In: Intelligent Computing in Engineering, pp. 921–929. Springer, Singapore (2020)
5. Huy, T.Q., et al.: The efficiency of applying compressed sampling and multi-resolution into ultrasound tomography. Ingeniería Solidaria **15**(3), 1–16 (2019)
6. Huy, T.Q., et al.: Tomographic density imaging using modified DF–DBIM approach. Biomed. Eng. Lett. 9(4), 449–465 (2019)
7. Chew, W.C., Wang, Y.M.: Reconstruction of two-dimensional permittivity distribution using the distorted born iterative method. IEEE Trans. Med. Imaging **9**, 218–225 (1990)
8. Lavarello, R., Oelze, M.: A study on the reconstruction of moderate contrast targets using the distorted Born iterative method. IEEE Trans. Ultrasonic, Ferroelectric, Frequency Control **55**, 112–124 (2008)
9. Golub, G.H., Hansen, P.C., O'Leary, D.P.: Tikhonov regularization and total least squares. J. Acoust. Soc. Am. **21**, 185–194 (1999)
10. Tracy, M.L., Johson, S.A.: Inverse scattering solutions by asinc basis, multiple sources, moment method–Part II: Numerical evaluations, Ultrasonic Imaging **5,** 376–392 (1983)
11. Lavarello, R., Oelze, M.: A study on the reconstruction of moderate contrast targets using the distorted born iterative method. IEEE Trans. Ultrasonic Ferroelectric Frequency Control **55**, 112–124 (2008)
12. Quang-Huy, T., Duc-Tan, T.: Sound contrast imaging using uniform ring configuration of transducers with reconstruction. In: 2015 International Conference on Advanced Technologies for Communications (ATC), Ho Chi Minh City, Vietnam, pp. 149–153 (2015). https://doi.org/10.1109/ATC.2015.7388308

Check for updates

Multiple Target Activity Recognition by Combining YOLOv5 with LSTM Network

Anh Tu Nguyen(✉) and Huy Anh Bui

Faculty of Mechanical Engineering, Hanoi University of Industry, Hanoi 159999, Vietnam
{tuna,buihuyanh}@haui.edu.vn

Abstract. Human detection plays an important role in several fields (such as autonomous mobile robots, bio-medical applications, military applications, etc.) and has received considerable attention from researchers in recent years. Especially, human gesture recognition provides information to predict human behavior for collision avoidance of the robots. The present paper proposes an approach in which deep-learning method and machine-learning method are integrated to classify activities and movements for multiple human targets. The proposed recognition process involves three sequential steps, namely the YOLOv5 model for detecting targets, the Media Pipe for drawing the skeleton, and the LSTM network for recognizing activities. The proposed method is examined through different scenarios. In the case of detecting the target with YOLOv5, the experimental results show that the loss accuracy always maintains below 5% for both training and the validation processes, and the mean Average Precision (mAP) of the designed YOLOv5 model is always higher than 99% for all consideration case studies. Furthermore, the activity recognition performance of the proposed method also successfully detects and tracks the behavior of the defined target inspected via three case studies: sitting, standing, and hands up. The experimental results prove the stability and precision of the method and point out that this approach can be applied to further studies and applications.

Keywords: Deep learning · Human Activity Recognition · YOLOv5 · LSTM network

1 Introduction

In recent years, the technologies of computer vision and artificial intelligence have experienced the remarkable achievement of human activity recognition (HAR) [1–4]. However, in actual contexts, human movements are seamless, with constant transitions between tasks of varying duration and a wide variety of dynamic movements.. The action of several targets is then more sophisticated and difficult to classify. Hence, it is noticeable that recognizing multiple target activities could be seen as an outstanding challenge to address. In [5], a semi-supervised recurrent convolutional attention model is proposed. To be more detailed, a semi-supervised framework is designed to extract and preserve diverse latent patterns of activities. Furthermore, the independence of multi-modalities of

T. D. L. Nguyen et al. (Eds.): ICISN 2023, LNNS 752, pp. 400–408, 2023.
https://doi.org/10.1007/978-981-99-4725-6_49

sensory data and the ability to identify salient regions indicative of human activities from inputs are exploited. Domingo *et al.* [6] integrates the LSTM with different data sources such as: features, object detection and skeleton concept to improve the performance of HAR process. In [7], a novel generalized Temporal Sliding Long Short-term Memory.

(TS-LSTM) network is constructed to handle the difficulties of skeletal feature representation and the modelling of temporal dynamics in order to detect human activities comprised of poses. The suggested networks are comprised of numerous TS-LSTM networks with varying hyper-parameters, allowing them to capture diverse temporal dynamics of activities. In [8], a hybrid network based on dense connection and weighted feature aggregation is built up to utilize the multimodal sensor data on the human body for activity detection. The effectiveness of the proposed model is demonstrated based on two benchmark datasets: Opportunity and UniMiB-SHAR.

Inspired by previous research works, in this paper, a real-time activity recognition system for multiple human targets is proposed. The key contributions of this article are summarized as:

- Utilize the YOLOv5 model [9, 10] to detect and investigate the possibility of human tracking and body recognition on different human targets. The designed model is evaluated against several indoor datasets under different types of transformation video frames.
- Apply the Media Pipe Human Pose Detection [11, 12] to establish the skeleton structure on the whole body of the defined targets. This structure displays the coordinates of 33 landmarks in total. Hence, in real-time video, the tracking possibility of the defined target is ensured precisely.
- Develop a long short term memory (LSTM) network [13] to recognition and classify automatically the activities of the human targets via the skeleton behavior. Moreover, only the image frames of the predefined targets are obtained and used in the video sequence. Hence, the data size is optimized during the operation process.

The remainder of this paper is as follows: The next section presents the detail description and the state-of-art of the applied techniques of the proposed method. In Sect. 3, the experiments result with the proposed approach obtained is reported. Finally, Sect. 4 notes the advantages of the designed system, provides an overall discussion, and suggests future developments.

2 Brief Overview of the Proposed Method

The proposed method is built on a series of integrated deep learning and machine learning models, which are implemented in the following order: First step - human target recognition by YOLOv5, second step - drawing the skeleton on the detected target, and third step - classifying target behavior using the enhanced LSTM model. The sequential structure of the proposed method is indicated as in Fig. 1.

Fig. 1. The sequential structure of the proposed method.

2.1 Human Detection with YOLOv5

To prepare the data training, the input dataset is established based on images collaborated with a specific target. A camera with a 16-megapixel resolution is set up to capture 100 photos at different side view for each target.

At the next stage, the input images are transmitted to YOLOv5 model. To ensure the precision of the human detection process, the YOLOv5 network model is established based of the following three primary components: The backbone consists of the Cross Stage Partial Network, the neck consists of the PANet and the head is made up of the Yolo layer. Figure 2 presents the structure of the YOLOv5 model in the proposed method.

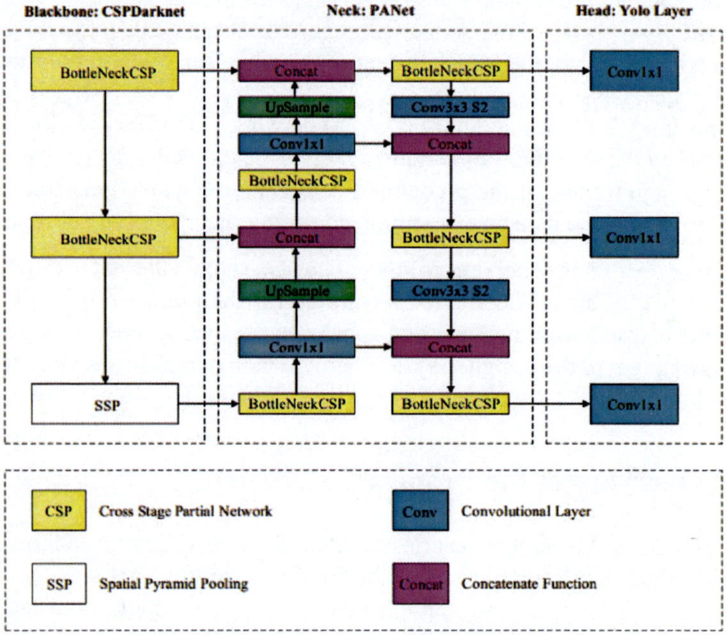

Fig. 2. The structure of the YOLOv5 model.

2.2 Display the Skeleton Concept on the Defined Target

After applying YOLOv5 to detect the correct target, the proposed system will draw the skeleton structure with 33 3D landmarks background segmentation mask on the whole body of the target by utilizing the Media Pipe Human Pose Detection and Tracking. In particular, a machine-learning pipeline that is composed of various stages is built from two models. The first model is a detector-tracker machine learning model that locates the correct target region-of-interest (ROI) inside the image. After applied the detector-tracker model over the image area, the second one - pose landmark model then functions on the cropped image region and accomplish exact key point localization of 33 3D coordinates via regression. As a result, the skeleton is display along the body of the defined target. Figure 3 illustrates the different landmarks on the skeleton.

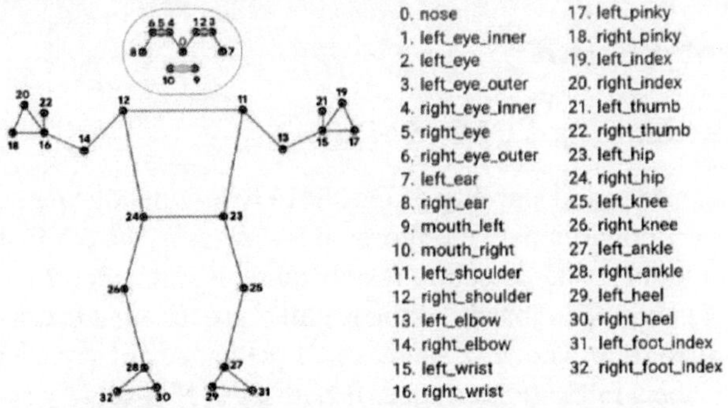

Fig. 3. 33 pose landmarks on the target body [11].

2.3 Activity Recognition Based on the LSTM Network

After rendering only the skeleton on the allowed target, the data will be recorded as frames of a 6-min video at 60fps. After that, the videos are included into the improved LSTM network. In this paper, we developed the LSTM model in the following way in order to increase the overall effectiveness of the training process: three LSTM layers and one dense layer at the final; the number of trainable params is at about 97353 (Fig. 4).

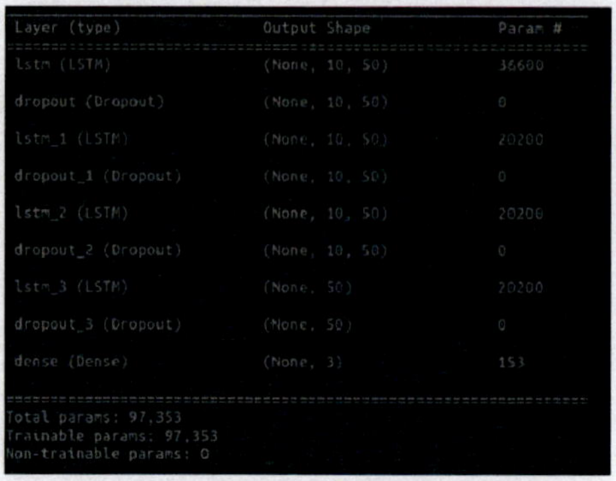

Layer (type)	Output Shape	Param #
lstm (LSTM)	(None, 10, 50)	10600
dropout (Dropout)	(None, 10, 50)	0
lstm_1 (LSTM)	(None, 10, 50)	20200
dropout_1 (Dropout)	(None, 10, 50)	0
lstm_2 (LSTM)	(None, 10, 50)	20200
dropout_2 (Dropout)	(None, 10, 50)	0
lstm_3 (LSTM)	(None, 50)	20200
dropout_3 (Dropout)	(None, 50)	0
dense (Dense)	(None, 3)	153

Total params: 97,353
Trainable params: 97,353
Non-trainable params: 0

Fig. 4. The structure of the enhanced LSTM network.

3 Experimental Result

3.1 Detection Result of the YOLOv5 Model

To examine the accuracy and stability of the YOLOv5 model during target recognition, the output results are continuously monitored in the form of the confusion matrix, accurate measurement graphs, and detection results on real-time video frames. Figure 5 and Fig. 6 illustrate the confusion matrix of the training process and the correlogram which is a group of 2d histograms showing each axis of the data against each other axis. In the correlogram, the labels in the input images are in XYWH space.

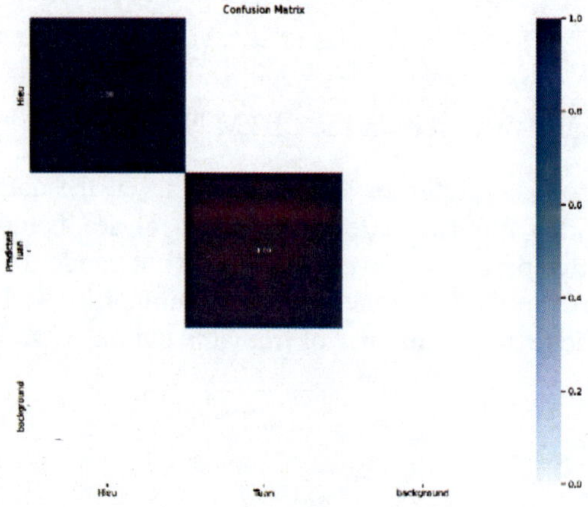

Fig. 5. The confusion matrix of the designed model.

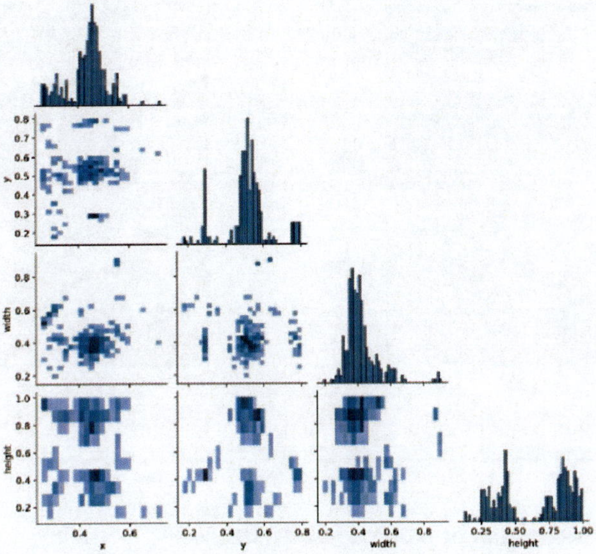

Fig. 6. The correlogram of the designed model.

Figure 7 and Fig. 8 indicates the general results of the designed model. It is evident that the loss accuracy during both training process and validation process always maintain below 5%. Moreover, the precision, the recall, and the mean Average Precision (mAP) of the designed YOLOv5 model are at about 99%, 99.6% and 99.5% respectively. Hence, the reliability of the recognition model has been fully demonstrated.

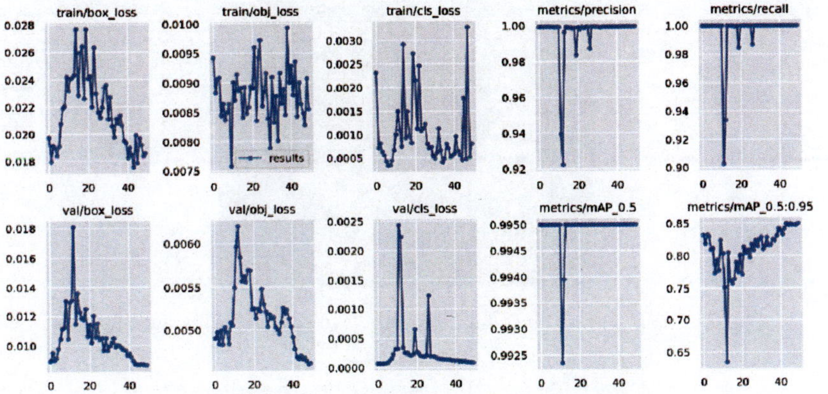

Fig. 7. The measurement of the designed model

(a) (b)

Fig. 8. The detection result of the designed model with real-time frames: a) Normal frames; b) Affine transformation frames

3.2 Activity Recognition of the Proposed Method

The activity recognition performance of the proposed method is inspected via different case studies. To be more specific, we divided the action of the target into three groups: sitting, standing and hands up. The proposed method only detects and tracks the behavior of the defined target. Otherwise, if it is not a predefined target, the model will skip identification and ignore defining behavior. As a result, the procedure retains frame-by-frame stability and saves a large amount of stored data. Figures 9, 10 and 11 shows the real-time activity recognition on multiple targets under different scenarios.

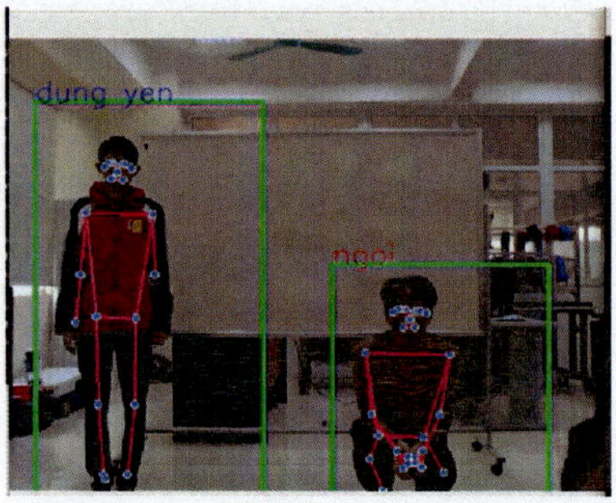

Fig. 9. The action result of multiple targets in case of sitting and standing.

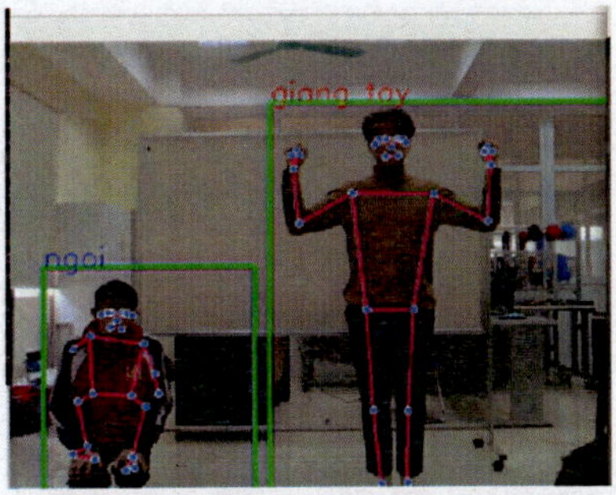

Fig. 10. The action result of multiple targets in case of sitting and hands up.

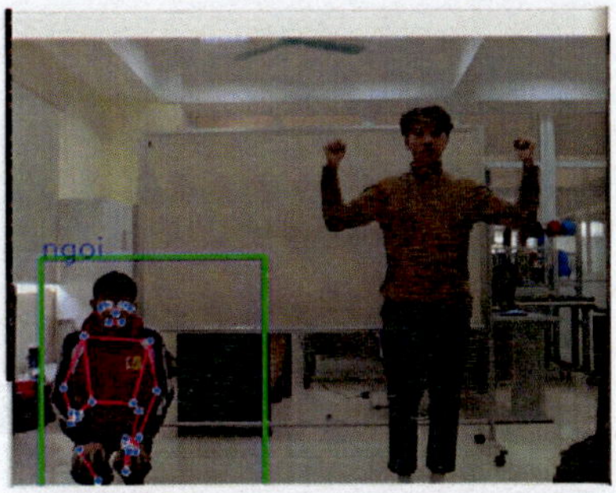

Fig. 11. The action result in case only one target is predefined and detected.

4 Conclusion

This paper presents an integrated deep-learning and machine-learning approach to classify activities and movements for multiple human targets. The proposed method is primarily built on three sequential components (the YOLOv5 model, the media pipe, and the LSTM network) to detect and recognize the right target's action. Unlike the majority of previous work, the data is captured as discrete recordings for each given activity, but as a continuous stream in which transitions between activities may occur at any moment and for an unbounded period. Besides, only the predefined targets are detected and conducted the behavior classification. Thus, necessary video frames are collected and stored. It could save a large amount of data during working process. The experimental results indicate the stability and the precision under several circumstances. Our future projects will concentrate on the potential of deeper convolution networks with better image feature extraction capabilities and the integration of our action recognition model in a residential environment with a flexible camera system. Furthermore, we are working on deploying this system with real-time tracking in military areas where specific human actions must be monitored and tracked.

References

1. Tripathi, R.K., Jalal, A.S., Agrawal, S.C.: Suspicious human activity recognition: a review. Artif. Intell. Rev. **50**(2), 283–339 (2017). https://doi.org/10.1007/s10462-017-9545-7
2. Li, R., Li, H., Shi, W.: Human activity recognition based on LPA. Multimed. Tools Appl. **79**(41–42), 31069–31086 (2020). https://doi.org/10.1007/s11042-020-09150-8
3. Beddiar, D.R., Nini, B., Sabokrou, M., Hadid, A.: Vision-based human activity recognition: a survey. Multimed. Tools Appl. **79**(41–42), 30509–30555 (2020). https://doi.org/10.1007/s11042-020-09004-3
4. Thomas, B., Lu, M.L., Jha, R., Bertrand, J.: Machine learning for detection and risk assessment of lifting action. IEEE Trans. Hum.-Mach. Syst. **52**(6), 1196–1204 (2022)
5. Chen, K., Yao, L., Zhang, D., Wang, X., Chang, X., Nie, F.: A semisupervised recurrent convolutional attention model for human activity recognition. IEEE Trans. Neural Networks Learn. Syst. **31**(5), 1747–1756 (2019)
6. Domingo, J.D., Gómez-García-Bermejo, J., Zalama, E.: Improving human activity recognition integrating LSTM with different data sources: features, object detection and skeleton tracking. IEEE Access **10**, 68213-68230 (2022)
7. Lee, I., Kim, D., Lee, S.: 3-d human behavior understanding using generalized TS-LSTM networks. IEEE Trans. Multimedia **23**, 415–428 (2021)
8. Lv, T., Wang, X., Jin, L., Xiao, Y., Song, M.: A hybrid network based on dense connection and weighted feature aggregation for human activity recognition. IEEE Access **8**, 68320–68332 (2020)
9. Du, X., Song, L., Lv, Y., Qiu, S.: A lightweight military target detection algorithm based on improved YOLOv5. Electronics **11**(20), 3263 (2022)
10. Ahmad, T., Cavazza, M., Matsuo, Y., Prendinger, H.: Detecting human actions in drone images using YOLOv5 and stochastic gradient boosting. Sensors **22**(18), 7020 (2022)
11. Lugaresi, C., et al.: Mediapipe: a framework for perceiving and processing reality. In: Third Workshop on Computer Vision for AR/VR at IEEE Computer Vision and Pattern Recognition (CVPR), vol. 2019, June 2019
12. Zhang, F., et al.: Mediapipe hands: on-device real-time hand tracking. arXiv preprint arXiv:2006.10214 (2020)
13. Shu, X., Zhang, L., Sun, Y., Tang, J.: Host–parasite: graph LSTM-in-LSTM for group activity recognition. IEEE Trans. Neural Networks Learn. Syst. **32**(2), 663–674 (2021)

An Analysis of the Effectiveness of Cascaded and CAM-Assisted Bloom Filters for Data Filtering

Quang-Manh Duong, Xuan-Uoc Dao, Hai-Duong Nguyen, Ngoc-Huong-Thao Tran, Ngoc-Hai Le, and Quang-Kien Trinh(✉)

Le Quy Don Technical University, Hanoi, Vietnam
kien.trinh@lqdtu.edu.vn

Abstract. This work proposes some optimization solutions and design methodologies for cascaded Bloom Filter design developed from the standard Bloom filter architecture presented in our previous study. These optimization solutions are all based on features extracted from the input data after passing through the first Bloom filter layer, which is used for the next Bloom layers in the entire filtering process. In addition, a solution using a small capacity of CAM instead of the Bloom filter in the last layer is also considered and evaluated in comparison with the solution using the pure Bloom filters. The CAM-assisted filter design could almost suppress the false positive rate with a trade-off in a small false negative rate.

Keywords: Bloom Filter · cascaded Bloom Filter · Bloom with CAM

1 Introduction

The Bloom filter, invented by Howard Bloom [1], is a relatively simple data structure for checking the membership of a dedicated set. The case when an object is not in the set but the result of checking is "exists" is called a "False Positive". The Bloom filter allows a non-zero probability of False Positives; however, it never gives a "False Negative" probability, which means, an element is not in the set but the answer is "yes". Bloom filter is a simple data structure that is efficient in terms of time and memory in querying the members of a set but accompanied by the trade-off on the correctness of the answer for the existence of an element in the given set. In many applications, the potential resource and energy savings and fast look-up offered by the Bloom filter outweigh the disadvantage of the small False-Positive probability. Bloom filters are found to be applied in areas such as database applications [2], spell-checking [3], and file processing [4]. Bloom filters with different usage schemes are present in numerous modern applications including Web caching [5], network routing [6], prefix matching [7], network security [8], etc.

In this work, we conduct a systematic study on how the impact of design topologies and the possibility of applying the feature extraction in the cascaded Bloom filter. We

T. D. L. Nguyen et al. (Eds.): ICISN 2023, LNNS 752, pp. 409–417, 2023.
https://doi.org/10.1007/978-981-99-4725-6_50

also consider combining the Bloom filter with Content-Address Memory (CAM) for enhancing the filtering performance with an acceptably small false negative rate.

The main contents of the paper are listed below:

- We proposed the novel design methodology for the cascaded Bloom filter structure based on the "feature extraction" of the input data;
- We investigated different variants of cascaded Bloom filter structures based on extracted features for improving the error rate of the data filtering;
- We proposed a novel CAM-assisted Bloom filter design that offers a very low false positive rate with an acceptable trade-off in the False Negative Rate.

The remaining sections are structured as follows. Section 2 describes the principle of operation of the standard Bloom filter. Section 3 proposes architectures of the cascaded Bloom Filters. Section 4 presents the architecture of the filter based on CAM and some comparisons with the cascaded Bloom filters. Section 5 gives some conclusions about the proposed architectures.

2 Premilitary

2.1 The Standard Bloom Filters

The Bloom filter represents the elements $m_1, m_2, ..., m_n$ of the set S by a vector of m bits, initialized with the values $\{0\}$. To insert the element m_i of the set S into the bit array, the Bloom filter hashes this element using a combination of k hash functions $h_1, h_2, \ldots, h_k$, where the hashed values range from 0 to $m - 1$, corresponding to the address range of memory cells. Thus, k hash functions for each element m_i will give k hashed values, which are respectively the addresses of k memory cells where the binary bits are set to $\{1\}$. This described process is called "adding", and the other is called "querying", in which the element c_i is checked if it belongs to S or not. Both the adding and querying operations are performed relatively simply using some hash functions. No "*False Negative*" occur with the standard Bloom filters, which means they never miss an element that exists in their pattern set. The False Positive Rate p is approximated according to the following formula:

$$p \approx \left(1 - e^{-kn/m}\right)^k \tag{1}$$

Given m and n, the optimal number of hash functions k_{opt} to obtain the minimum False Positive Rate p is:

$$k_{opt} = \frac{m}{n} ln2 \approx \frac{9m}{13n} \approx 0.6931 \frac{m}{n} \tag{2}$$

2.2 Related Works

A lot of studies on this topic focus on optimizing the Bloom filter with the reduced False Positive Rate or different design variants to extend the filter functionality. Yu Hua et. al. present a relatively complex filter architecture model called "Parallel Bloom Filters" (PBF) [9] using the auxiliary hash table for adding and querying elements, which can have multiple attributes. That proposed PBF, despite its low false positive rate, has disadvantages in terms of space occupation and a large number of used hash functions. Michael Paynter et. al. [10] propose the multiple-phase hashing pipelined Bloom filter architecture, in which the set of k hash functions is divided into different groups and each hash function is performed in a single phase. The execution of the next phase is only needed when a match in the hash value occurs in the previous phase. Therefore, the energy consumption for hashing operations is significantly saved. However, with the cascaded hashing, no improvement in FPR in their study was reported, while reducing FPR is our main concern in this work. Yuhua Chen et. al. use hardware implementation with the combination of TCAM and Bloom Filter [11]. Their proposed scheme shows a speed improvement, but due to memory capacity limitations, it can only be used for small data sets.

3 The Proposed Design of the Cascaded Bloom Filters

In some cases of data filtering using a single Bloom filter, the False Positive Rate (FPR) may be higher than required, so solutions to reduce the FPR should be considered experimentally. One of the possible solutions is the addition of Bloom filters behind the first filter, creating a cascaded Bloom filter structure. In this paper, we propose several different designs for the cascading Bloom filter to save the total memory space and achieve an FPR ratio approaching the theoretical value. With a fixed width of input data evaluated, we applied the "brute force" algorithm to get all possible input patterns sent to the first Bloom filter (BF_0), creating "extracted features" of the input data corresponding to this first filtering layer. Then, the output data of BF_0 is passed to the next filter structure, which can include one (BF_1) or more Bloom filters. The topologies of the next filter structure could be extended for larger values of input data width and will be presented in this section. In the actual application, the extracted features are predefined on the software platform (Python) to save time and hardware resources (FPGA).

3.1 The Feature Extraction of Input Data at the First BF Layer

To scan all possible values of the input data, we use multiple computation loops, each of which accumulates a finite number of FP elements. We generate the set of pattern elements randomly with variable parameters, then this set is fixed and used to evaluate all proposed Bloom filter structures later. With input data of 1 million randomized generated samples (1,000,000 input elements) in a single loop, the entire set of FP elements corresponding to the first Bloom filter is accumulated after about 300 iterations.

Feature Extraction Optimization

Several design topologies of the first Bloom layer were evaluated to minimize the number of FPs and depicted in Fig. 1. With the same fixed amount of memory space (1.0 Mb), we consider the options of using a single Bloom filter, a double Bloom filter, or up to a triple Bloom filter. When using the double Bloom filter, the two topologies examined are serial and parallel combinations. The optimal value of k corresponding to a fixed capacity of 1.0 Mb is 8, which is calculated by Formula (2). From the presented above results, the design topology using a single Bloom filter (in Fig. 1a) is the best solution with the number of accumulated FPs being about 52,214 elements.

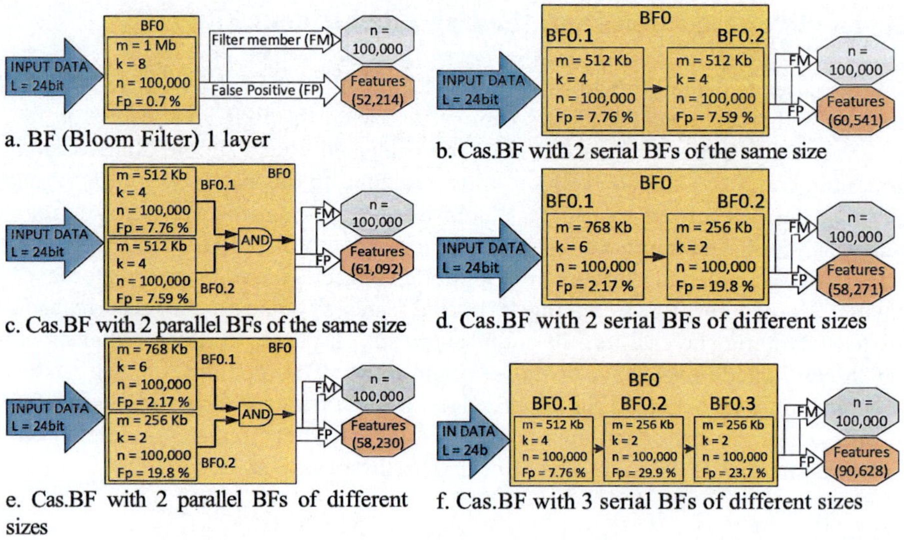

Fig. 1. Different structures for feature extraction after the first filter layer and the corresponding results of False Positive Rates.

The algorithm of the feature extraction from the input data is presented in Fig. 2.

```
Algorithm 1: Feature extraction from the input data
constant L = 24   # Number of bits of the input element
constant N = 1.000.000  # Number of elements of an input data set in a loop
constant Max_loop = 300  # Maximum number of loops
constant k = 8  # Number of hash function
function Get_features(featuresList)
    for i in Max_loop:
        for m in range(N):  # generate random input dataset
            element = random(0, 2^L - 1)  # get a random value
            InputSet_i. append(element) # add element to the input dataset
        end for
        Querying_BF(InputSet_i, k, QueryingList_i)   # data query
        for x_j in QueryingList_i:      # add a new feature
            if x_j = yes and InputSet_i(j) in featuresList = False:
                featuresList. append(InputSet_i(j))
            end if
        end for
    end for
    return featuresList
end function
```

Fig. 2. Feature extraction algorithm used in the proposed cascaded Bloom filter

3.2 The BF Structures Based on Extracted Features and Forward Serial Bloom Filters

Different from a random cascading BF, in our proposed design, the last BF layer of Forward Serial BFs is based on the features extracted by the first BF layer, which has been presented in the previous subsection.

Given the assumed size (1,375 Mb), we split the cascaded BF into 2 filter layers. The first filter layer (BF_0) has a size of 1.0 Mb. This filter layer is designed to minimize the features (FPs) obtained. The last BF layer is 384 Kb in size. We have randomly tested many BFs with different structures and sets of hash functions to determine the most suitable filter. For each BF structure, we tried and selected the most appropriate set of hash functions. Some of the designs with the best results are listed in Table 1.

Table 1. The number of false positives for each BF structure

Type of the BF structures	Amount of FPs
Single-layer BF with $k = 5$	9,751
Single-layer BF with $k = 3$	8,010
Cas.BF with 2 parallel BFs of different sizes and different k	12,089
Cas.BF with 3 parallel BFs of the same size and the same k	19,687

(continued)

Table 1. (*continued*)

Type of the BF structures	Amount of FPs
Cas.BF with 3 serial BFs of the same size and different k	18,178
Cas.BF with 2 serial BFs with different sizes and different k	12,089

The working principle of the serial-cascaded filter: An input element ($L = 24$ bits) is fed to query at filter BF_0, if it is evaluated as not in BF_0, it will be blocked at BF_0. If it is evaluated as belonging to BF_0, it is allowed to pass through, then queried at BF_1. If it is evaluated as not in BF_1, it will be blocked at BF_1. If it is evaluated as belonging to BF_1, it is passed through BF_1 and evaluated as a member of the cascaded BF. Elements blocked at BF_0 or BF_1 are evaluated as not members of the cascaded BF.

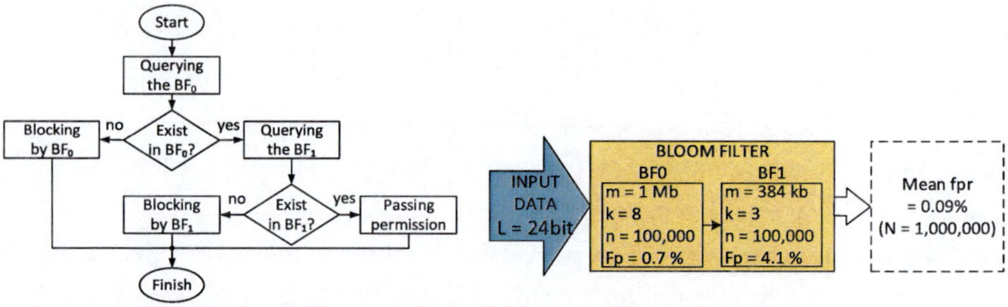

a. The working principle of the Forward Serial BF

b. Forward Serial BF structure

Fig. 3. The working principle of the Forward Serial Bloom filter.

The diagram of the cascaded filter is depicted in Fig. 3a. If a false positive of BF_0 (a member of features) is queried at BF_1, there are two possibilities. In the first case, it is blocked at BF_1, then the cascaded BF works correctly. In the second case, it is evaluated as belonging to BF_1, i.e., it is a false positive of BF_1 and also a false positive of the cascaded BF. An element is false positive for the forward-serial filter if and only if it is false positive for both filter stages BF_0 and BF_1.

With a total memory space of 1.35 Mb, the last Bloom filter layer is 384 Kb in size. According to our research results in Table 1, the optimal final Bloom filter layer is a single-layer variant with $k = 3$ (Fig. 3b). The average FPR of Forward Serial BFs is 0.094% for a random input data set of 1,000,000 elements.

4 The Proposed Design of the Filter Based on CAM

CAM is a data storage device, that stores a value or keyword in RAM, allows comparing input elements with all stored data in RAM, and gives the addresses of matched data. In our research, we designed the CAM as a filter.

From the feature of CAM that stores keywords, we proposed a design of the filter based on CAM. The filter consists of four parts: main CAM, auxiliary CAM, flag bit, and auxiliary flag bit. Main CAM stores keywords of length W bits and has *D_main* address bits. The flag bit at each main CAM address represents the empty or full status of the address. If it is {0}, the keyword whose length is W bits will be stored at address {*D_main*} of the main CAM. Otherwise, store that keyword at the address {*D_aux*} of auxiliary CAM and set the auxiliary flag bit to {1}. Therefore, auxiliary CAM also stores keywords of length W bits but has *D_aux* address bits that can be equal or different *D_main*. The diagram and the results of filtering using the CAM-based filter is presented in Fig. 4.

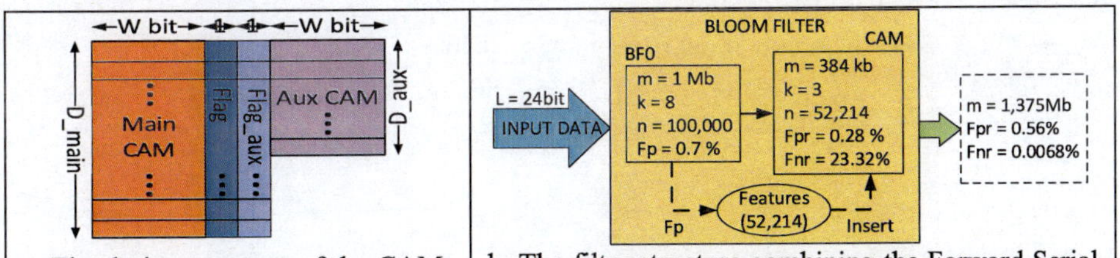

a. The design structure of the CAM-based filter

b. The filter structure combining the Forward Serial Bloom filter and the inverse CAM-based filter

Fig. 4. The design of the filter based on CAM

Using a CAM-based filter to store all the members of the data set requires a relatively large amount of memory and is inefficient in terms of resources. Therefore, we propose a filter structure combining the forward BF and the inverse CAM-based filter. The inverse CAM-based filter is defined as the CAM-based filter storing the feature of input data in Subsect. 3.1. Its working principle follows the operation of the CAM-based filter. The difference here is that we use inverse CAM, so when querying the member of the CAM, the elements that are evaluated as belonging to the CAM will be blocked, and the evaluated elements that do not belong to the CAM will be allowed to pass through. The inverse CAM-based filter has the same memory size as the designs in Subsect.3.2 and can store the maximum number of members in the feature, we consider giving the parameters of the filter, specifically: $W = 8$ bits, $D_{main} = 15$ bits, $D_{aux} = 13$ bits, $k = 3$.

After querying at the inverse CAM-based filter, the design has both *FPR* and small *FNR*. It should be noted that CAM-based filter design gives a small *FPR*, it's only 0.28%, while the BF with parameters as suggested gives $FPR = 3.55\%$ (according to (1)). The CAM-based filter can significantly reduce false positive rates, but at the expense of having a relatively high *FNR*. This is a special feature of CAM-based filters that needs more research in the future.

The evaluation of the effectiveness of the proposed data filter structures is based on the experimentally obtained *FPR* and *FNR* values, which are given in Table 2. Compared to the standard Bloom filter structure, our first proposed design based on pure serial Bloom filters using accumulated FPs provides a small improvement in FPR, and neither of these architectures allows the occurrence of False Negatives. Using the same extracted features as the first proposed design, the architecture using CAM in the latter filtering layer provides less accuracy than the rest of the filter structures in Table 2, in addition, there are disadvantages due to the presence of FNs. The complexity of the proposed designs is not much different, but soon when implementing them on hardware we will have more detailed evaluations.

Table 2. Comparison of different Bloom filter structures

Type of the Bloom Filter structure	False Positive Rate	False Negative Rate
Standard Bloom filter structure	0.1%	0%
Structure of Forward Serial Bloom filters	0.094%	0%
Inverse CAM-based filter with Serial Bloom filter structure	0.56%	0.0068%

5 Conclusion

We provided several solutions for data filtering problems using different topologies based on Bloom filters. As the key idea, we design and optimize the secondary filter layers based on the features extracted from the first filter layer to reduce the error rate. We proposed to use two designs for the secondary filter. The first design is a serial-connected or parallel-connected Bloom filter, whereas, in the second design, a CAM is employed instead of the Bloom filter. The evaluation results show that with the first design, the serial-connected Bloom filters are beneficial in terms of energy consumption, but the FPR is slightly lower than the single Bloom filter at the iso-area (0.094% vs 0.1%). The standalone CAM-based filter gives much better FPR than the single Bloom filters (0.28% vs 3.55%), but the second design based on an inverse CAM filter with a Serial Bloom filter comes with a relatively high FPR (0.56%) and a very small FNR (0.0068%). In the future, we will push forward on further enhancing the filter design and deployment for practical applications. Also, the filter will be seriously considered for implementation on hardware.

References

1. Bloom, B.H.: Space/time trade-offs in hash coding with allowable errors. Commun. ACM **13**(7), 422–426 (1970)
2. Mullin, J.K.: Optimal semijoins for distributed database systems. IEEE Trans. Software Eng. **16**(5), 558–560 (1990)
3. McIlroy, M.: Development of a spelling list. IEEE Trans. Commun. **30**(1), 91–99 (1982)
4. Gremillion, L.L.: Designing a bloom filter for differential file access. Commun. ACM **25**(9), 600–604 (1982)
5. Fan, L., et al.: Summary cache: a scalable wide-area web cache sharing protocol. IEEE/ACM Trans. Networking **8**(3), 281–293 (2000)

6. Jiang, P., et al.: Design of a multiple bloom filter for distributed navigation routing. IEEE Trans. Syst. Man Cybern. Syst. **44**(2), 254–260 (2013)
7. Dharmapurikar, S., Krishnamurthy, P., Taylor, D.E.: Longest prefix matching using bloom filters. In: Proceedings of the 2003 Conference on Applications, Technologies, Architectures, and Protocols for Computer Communications (2003)
8. Geravand, S., Ahmadi, M.: Bloom filter applications in network security: a state-of-the-art survey. Comput. Netw. **57**(18), 4047–4064 (2013)
9. Hua, Y., Xiao, B.: A multi-attribute data structure with parallel bloom filters for network services. In: Robert, Y., Parashar, M., Badrinath, R., Prasanna, V.K. (eds.) High Performance Computing - HiPC 2006. HiPC 2006. LNCS, vol. 4297. Springer, Berlin (2006). https://doi.org/10.1007/11945918_30
10. Paynter, M., Kocak, T.: Fully pipelined bloom filter architecture. IEEE Commun. Lett. **12**(11), 855–857 (2008)
11. Chen, Y., Oguntoyinbo, O.: Power-efficient packet classification using cascaded bloom filter and off-the-shelf ternary CAM for WDM networks. Comput. Commun. **32**(2), 349–356 (2009)

Detection of Fence Climbing Behavior in Surveillance Videos Using YOLO V4

Pham Thi-Ngoc-Diem[1(✉)], Chau Si-Quych-Di[1], Duong Quang-Thien[1], Tran Hoang-Le-Chi[2], Nguyen Thanh-Hai[1], and Tran Thanh-Dien[1]

[1] College of Information and Communication Technology, Can Tho University, Can Tho, Vietnam
{ptndiem,nthai.cit,ttdien}@ctu.edu.vn

[2] FPT University, FPT Polytechnic, Can Tho, Vietnam
chithl@fpt.edu.vn

Abstract. Nowadays, with the development of technology and the Internet, most households have surveillance cameras to observe everything around the house. Therefore, detecting abnormal human behaviors using videos generated by surveillance cameras has attracted much recent research. This paper focuses on applying the YOLO v4 to build the model detecting abnormal human behaviors, especially detecting fence climbing behaviors. Experimental results on the dataset, including 5340 images extracted from videos, showed that the model obtained the IoU measure of 71% and F1-score measure of 87%.

Keywords: abnormal human behavior · climbing fence behavior detection · YOLO v4 · Deep learning models

1 Introduction

Behavior detection refers to detecting individuals with certain intentions by automatically observing their behaviors and activities through images or videos such as running, jumping, standing, sitting, etc. In recent years, home invasion, burglary breaking into, and stealing property from a house have been increasing in many cases. The thieves can enter the house by breaking locks or climbing fences or walls. To improve home security, the houses' owners usually deploy surveillance cameras. However, current surveillance cameras do not support detecting a stranger breaking into the house yet. Therefore, it is necessary to have a solution for detecting activities breaking into the house using surveillance cameras. This helps keep the owner informed in case incidents happened to prevent burglars from committing crimes.

In this study, two one-stage object detection models, including YOLO (You Only Look Once) v4 [1] and Single Shot Detector (SSD) [5] are adopted to build models for detecting fence climbing behavior in surveillance videos. Experiments were conducted to compare the performance of these two models. Experimental results on the dataset, including 5430 images extracted from videos, showed

T. D. L. Nguyen et al. (Eds.): ICISN 2023, LNNS 752, pp. 418–425, 2023.
https://doi.org/10.1007/978-981-99-4725-6_51

that the YOLO V4 is better than SSD. With YOLO v4, the IoU and F1-score measurements are 71% and 87%, respectively.

This paper is organized as follows. Section 2 presents the related work. Section 3 introduces the proposed architecture, including the dataset used and the hyper-parameters of the models. Section 4 presents and explains the experimental results. The final Section is our conclusion.

2 Related Work

In recent years, with the frequent occurrence of scrimmages, riots, stampedes, or burglary raids, video surveillance equipment has been widely used in public places such as train stations, streets, parks, banks, hospitals, and schools. Besides, with the emergence and widespread application of machine learning algorithms in many fields, surveillance videos have been exploited for many different purposes, including automatically detecting abnormal human behaviors. Many methods have been used, including automatically detecting abnormal human behaviors. The authors [2,13] have proposed a general architecture for detecting and recognizing abnormal human behaviors. [13] has focused on detecting abnormal human behavior in the house using a combination of the Gaussian model for background image segmentation of each frame in video and FCM (fuzzy C-means clustering) to detect outliers in the data samples. [7] has suggested a method to learn anomalous behaviors in videos by finding regions of interest from spacetime information, as opposed to full-frame learning. The proposed model was trained and tested on the UCF dataset with an accuracy of 99.25%. Using a combination of OpenPose and YOLO, [4] has presented a model for detecting human behaviors, including walking, standing, sitting, and falling, on a dataset of 400 images. Experimental results showed that the accuracy in the training process of the model reached 95%. Meanwhile, [12] has presented a CNN-based abnormal behavior detection method such as walking, jogging, fighting, kicking, and punching.

[8] has used the SSD algorithm for recognizing human behaviors. SSD's average speed is 0.146s/frame, and the average accuracy on different datasets is 82.8%. To identify and detect anomalous human behavior, [9] used and improved the ResNet model. This improved model was trained on the ImageNet dataset and gave a higher accuracy than the ResNet-50 model of 2.8% when it was tested on the UTI dataset.

The authors [6] also used a deep learning network based on the YOLO v3 model to recognize human walking behavior on a dataset of 1198 samples; the achieved accuracy is 80.20%. The authors [11] used SVM model to detect 6 types of human behavior, including walking, jogging, running, boxing, waving, and clapping, on a dataset of 2391 videos. While [14] used a combination of CNN and SVM to recognize human movement. The research [3] used a combination of centroid features with SVM to detect walking and fence-climbing behavior. This work experimented on a dataset that consists of 15 videos for the training data and 50 videos for testing data. The accuracy of the model is over 90%. Also involved in detecting fence climbing behavior, [15] proposed a model consisting of extending the star frame model combined with the HMM training method to analyze a sequence containing many actions.

Although many studies have applied machine learning models and deep learning algorithms on different datasets to detect human behavior, most of these studies have focused on behaviors such as walking, standing, and running [4,6,9,11,14]. There are few studies on fence climbing behavior [3,15]. On the other hand, the dataset used to detect fence climbing behavior of these studies is subjective [3] or only applies to two types of iron fences [15]. Studies that detect fence climbing behavior on various fences and walls and random datasets are rather limited.

3 Method

3.1 YOLO V4 and SSD for Human Fence Climbing Behavior Detection

The overall architecture of the system detecting human fence climbing behavior is described in Fig. 1. The collected images were trained using YOLO v4 and SSD on MobileNet V2 [10]. The default values of hyper-parameters such as Learning rate, Input image size, etc. for each model are used to train described in **Table 1**.

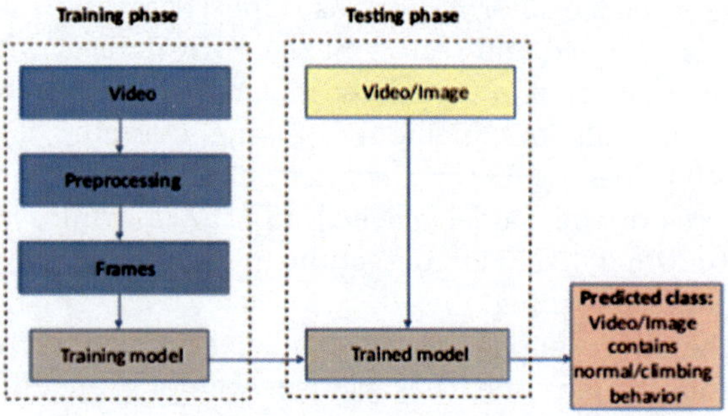

Fig. 1. Overall system architecture flow

The input for the training phase of the model is videos containing fence climbing behavior and other behaviors (running, walking, sitting, etc.). These videos are pre-processed by extracting frames and removing inappropriate frames. The resulting frames are labeled and then used to train and build the model.

In the testing phase, the videos or unlabelled images are the input of the trained model and the output is the predictive values (**climbing** or **normal**) that indicate whether a video or an image contains a fence climbing behavior with a bounding box.

Table 1. Description of Hyper-parameters and complexity of YOLO v4 and SSD MobileNet V2

Type	YOLO v4	SSD MobileNet V2
Learning rate	0.001	0.79999998
Input image	416×416	320×320
Number of classes	2 (Climbing and Normal)	2 (Climbing and Normal)
Decay	0.0005	0.996999979
Momentum	0.949	0.899999976
Model Complexity (#Parameters)	60×10^6	3.4×10^6

3.2 Dataset

Videos for training and testing the model were collected from the Internet (Youtube, Web, and Social networks) and were recorded by our research team. The videos are recorded both during the day and at night using a camera that is 4 m from the fence. These videos' resolution is 1280×720, and the frame rate is 30 fps. Depending on the specific situation, each video is from 10 s to 12 min in length. Videos for training are split into an image sequence (frames) using OpenCV and Python. Blurred images or images that do not include people are then removed. The rest of the images are labeled using LabelImg[1] tool. In this study, **climbing** or **normal** are two classes corresponding to fence climbing behavior and other behaviors. The details of the size of the dataset used for training are described in Table 2.

The original dataset contains 4882 images, of which there are 2497 images with fence climbing behavior. The ratio between the images that include the climbing behavior (51%) and the images that have the other behaviors (49%) is balanced.

Table 2. Description of dataset

Type of Images	Samples
Images containing fence climbing behavior	2497
Images containing other behavior	2385
Rotated images	231
Flipped images	227
Total of images	**5340**

For the augmentation dataset, 230 images were randomly selected from 2497 fence climbing images of the original dataset. From these 230 images, we create 231 images by random rotation and 227 images by flipping the original horizontally, vertically, and both horizontally and vertically. As a result, the augmented dataset contains 5340 images (original images and augmented images).

[1] https://github.com/tzutalin/labelImg.

4 Experimental Results

4.1 Model Evaluation

This section presents our experiments to evaluate the proposed model. The experiments were performed on the Google Colab environment. Training is repeated five times for every network on the same dataset of 4882 original images. The dataset was divided into two subsets to perform training and testing tasks (80% for training and 20% for testing). For every training, the number of iterations is set to 6000, and the data for training were randomly divided into a training set (70%) and a validation set (30%). Precision, Recall, F1-Score, Loss avg, and IoU (Intersection over Union) are metrics used to evaluate the performance of two architectures. Precision reflects the percentage of correct detections among the total number of detections, while recall represents the percentage of correct detections compared to the real positive cases. F1-score is one of the most important evaluation metrics and is the weighted average of Precision and Recall as shown in Eq. 1.

$$\text{F1-score} = 2 * \frac{Recall * Precision}{Recall + Precision} \tag{1}$$

A detection is considered true (True Positive) if the IoU is greater than a given threshold (0.6 in this work) and false (False Positive) if it is lower than a given threshold. IoU is an evaluation metric used to measure the accuracy of an object detector on a particular dataset.

4.2 Experimental Results

Two networks, YOLO v4 and SSD MobileNet V2, were used to train the data and build the model to detect fence-climbing behavior. Training results of two models (Precision, Recall, F1-Score, Loss average, IoU) are shown in Table 3.

Table 3. Comparison of YOLO v4 and SSD MobileNet V2

Metric	YOLO v4 (original dataset)	SSD (original dataset)	YOLO v4 (augmented dataset)
IoU	0.70 ± 0.01	0.56 ± 0.06	0.71 ± 0.003
Precision	0.85 ± 0.01	0.84 ± 0.01	0.86 ± 0.01
Recall	0.88 ± 0.01	0.67 ± 0.02	0.88 ± 0.01
F1-score	0.87 ± 0.01	0.74 ± 0.01	0.87 ± 0.00

The results in Table 3 show that YOLO v4 outperforms SSD. YOLO v4 produces better detection bounding boxes with high confidence scores as shown in Fig. 2 and Fig. 3. In Fig. 2, many objects are missed by SSD object detector. The average precision, average recall, and average F1-score of YOLO v4 are also higher than the corresponding values of the SSD model.

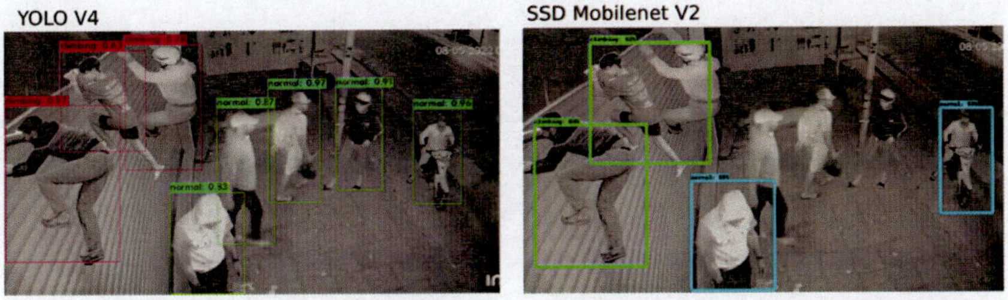

Fig. 2. Fence climbing behavior detection in Video

Fig. 3. Fence climbing behavior detection in Image using YOLO v4

4.3 Experiments with Augmented Dataset

As described in 3.2, the augmentation dataset was created from the original dataset by applying image rotation and flipping on the original image. The new dataset consists of 4882 original and 458 new images (increasing by approximately 9.4%). The training was done with YOLO v4 on an augmented dataset of 5340 images. This training is performed 5 times, where the augmented images are only in the training set and not in the validation set. Fig. 4 shows Loss and mAP graph during training of YOLO v4 model. The training results are illustrated in Table 3.

Compared with the result from the original dataset, the performance of the model with the augmented dataset improves a little. The IoU and precision improve by 1%. The other measurements are almost the same. This can be concluded that the collected dataset is reliable, and the augmented images contribute to the improvement of the model. However, the improvement is insignificant.

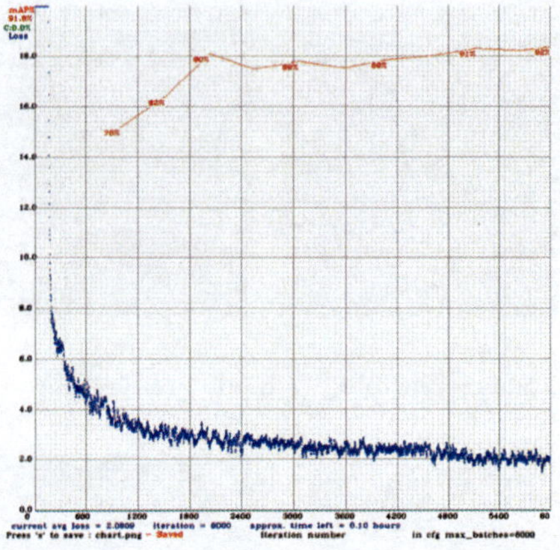

Fig. 4. Loss and mAP graph

5 Conclusion and Future Work

In this paper, we have investigated employing two networks, YOLO v4 and SSD MobileNet V2, to detect fence climbing behavior. The experiment on the dataset of 4882 images shows that YOLO v4 outperforms. With the use of the image augmentation method, YOLO v4 obtained an average IoU of over 70%, an average precision, an average recall, and an average F1-score of greater than 86%. Although the experimental results of the detection model achieved acceptable accuracy for detecting fence climbing activity, we expect that it could be improved more. Therefore, in future work, we will collect more images and augment the dataset to improve the performance of the fence climbing behavior detection model. In addition, we will study other abnormal activities that can be considered as a breaking into the house. Moreover, we will also try to adjust the model's hyper-parameters for better object detection.

Acknowledgement. This study is funded in part by Can Tho University, Code: T2022-03

References

1. Bochkovskiy, A., Wang, C., Liao, H.M.: Yolov4: optimal speed and accuracy of object detection. CoRR abs/2004.10934 (2020). https://arxiv.org/abs/2004.10934
2. Hu, Y.: Design and implementation of abnormal behavior detection based on deep intelligent analysis algorithms in massive video surveillance. Grid Comput. **18**, 227–237 (2020). https://doi.org/10.1007/s10723-020-09506-2
3. Kolekar, M.H., Bharti, N., Patil, P.N.: Detection of fence climbing using activity recognition by support vector machine classifier. In: 2016 IEEE Region 10 Conference (TENCON), pp. 398–402 (2016). https://doi.org/10.1109/TENCON.2016.7848029

4. Lina, W., Ding, J.: Behavior detection method of openpose combined with yolo network. In: 2020 International Conference on Communications, Information System and Computer Engineering (CISCE), pp. 326–330 (2020). https://doi.org/10.1109/CISCE50729.2020.00072
5. Liu, W., et al.: SSD: single shot multibox detector. In: Leibe, B., Matas, J., Sebe, N., Welling, M. (eds.) ECCV 2016. LNCS, vol. 9905, pp. 21–37. Springer, Cham (2016). https://doi.org/10.1007/978-3-319-46448-0_2
6. Lu, J., Yan, W.Q., Nguyen, M.: Human behaviour recognition using deep learning. In: 2018 15th IEEE International Conference on Advanced Video and Signal Based Surveillance (AVSS), pp. 1–6 (2018). https://doi.org/10.1109/AVSS.2018.8639413
7. Nasaruddin, N., Muchtar, K., Afdhal, A., Dwiyantoro, A.P.J.: Deep anomaly detection through visual attention in surveillance videos. Big Data **7**, 87 (2020). https://doi.org/10.1186/s40537-020-00365-y
8. Pan, H., Li, Y., Zhao, D.: Recognizing human behaviors from surveillance videos using the SSD algorithm. J. Supercomput. **77**(7), 6852–6870 (2021). https://doi.org/10.1007/s11227-020-03578-3
9. Qian, H., Zhou, X., Zheng, M.: Abnormal behavior detection and recognition method based on improved resnet model. Comput. Mater. Continua **65**, 2153–2167 (2020). https://doi.org/10.32604/cmc.2020.011843
10. Sandler, M., Howard, A.G., Zhu, M., Zhmoginov, A., Chen, L.: Inverted residuals and linear bottlenecks: mobile networks for classification, detection and segmentation. CoRR abs/1801.04381 (2018), http://arxiv.org/abs/1801.04381
11. Schuldt, C., Laptev, I., Caputo, B.: Recognizing human actions: a local SVM approach. In: Proceedings of the 17th International Conference on Pattern Recognition, 2004. ICPR 2004, vol. 3, pp. 32–36 (2004). https://doi.org/10.1109/ICPR.2004.1334462
12. Tay, N.C., Connie, T., Ong, T.S., Goh, K.O.M., Teh, P.S.: A robust abnormal behavior detection method using convolutional neural network. In: Computational Science and Technology. LNEE, vol. 481, pp. 37–47. Springer, Singapore (2019). https://doi.org/10.1007/978-981-13-2622-6_4
13. Wu, C., Cheng, Z.: A novel detection framework for detecting abnormal human behavior. Math. Probl. Eng. **2020,** 1–9 (2020). https://doi.org/10.1155/2020/6625695
14. Xu, H., Li, L., Fang, M., Zhang, F.: Movement human actions recognition based on machine learning. Int. J. Online Eng. (iJOE) **14**, 193 (2018). https://doi.org/10.3991/ijoe.v14i04.8513
15. Yu, E., Aggarwal, J.: Detection of fence climbing from monocular video. In: 18th International Conference on Pattern Recognition (ICPR 2006), vol. 1, pp. 375–378 (2006). https://doi.org/10.1109/ICPR.2006.440

Scalable Energy Efficiency Protection Model with Directed p-Cycles in Elastic Optical Network

Luong Van Hieu[1](✉) and Do Trung Kien[2]

[1] Faculty of Information Technology, Hanoi College for Electro-Mechanics, Hanoi, Vietnam
hieulv@codienhanoi.edu.vn

[2] Faculty of Information Technology, Hanoi National University of Education, Hanoi, Vietnam
kiendt@hnue.edu.vn

Abstract. The network protection problem is one of the important issues in optical network design and belongs to the NP-hard problem. The problem becomes even more difficult in the next-generation optical networks - Elastic Optical Networks (EONs). In EONs, the protection problem must take into consideration extra elements such as power consumption, spectrum allocation requirements, and format levels. Furthermore, several recent studies focus on solving the problem of power-efficient network protection in EONs. Almost all prior research focused on heuristics because of scalability concerns. In this paper, we propose an ILP model for large-scale optimization that can solve huge instances. Experiments were successfully conducted on the NSFNET and USANET networks, with very reasonable computing times.

Keywords: Column Generation · P-cycle Protection · Elastic Optical Network

1 Introduction

Elastic Optical Networks (EONs) are next-generation optical networks that are seen as an efficient solution for future optical transmission systems due to their flexibility in spectrum assignment and adaptive modulation formats [1]. Unlike traditional Wavelength-Division Multiplexing (WDM)-based networks, EONs are based on the Orthogonal Frequency Division Multiplexing (OFDM) [2], which supports flexible granularity, fractional data rates, and variable traffic by enabling the so-called sub-wavelength, super wavelength.

In optical networks, the failure of a network element (e.g., active components, a fiber cut) can affect a large amount of bandwidth in transmission and cause service interruptions to numerous end users and network services. Thus, survivability has become a fundamental problem in EONs design and has been widely investigated in various publications [1, 3].

Survivability is a critical issue in optical networks, and there are two approaches to restoration and protection [4]. This study is concerned with EON protection strategies. Many protection methods have been proposed in [5, 8]. The p-cycle protection scheme

T. D. L. Nguyen et al. (Eds.): ICISN 2023, LNNS 752, pp. 426–434, 2023.
https://doi.org/10.1007/978-981-99-4725-6_52

is a particularly attractive one due to p-cycles providing the restoration speed of rings and the resource efficiency of mesh networks [4, 10, 11]. The restoration speed of the p-cycle is because just two end nodes are involved in the case of a link failure to switch traffic to a pre-planned cycle. Another benefit of the p-cycle is that it can protect all on-cycle and straddling links.

We consider the problem of protection under the directed p-cycle protection scheme. Our model also includes practical parameters like the modulation format selection, taking the transmission distance into account. Our optimization goal is to reduce power consumption for protection. Numerous heuristic algorithms, such as [10], have been proposed. However, while heuristics can deal with difficult combinatorial problems more rapidly, they don't provide a way to determine how good the solution is or how far off it is from the best one.

For the exact solution of the power-efficient p-cycle protection problem, several authors investigated Integer Linear Programming (ILP) formulations with and without explicit modulation concerns. Early proposed formulations are not scalable and can only be applied on very small network topologies with a few nodes and links, e.g., [12]. To deal with cases involving a large number of variables, the authors in [16, 17] investigated a Column Generation (CG) technique that allows for reducing the number of variables (referred to as columns) in LP-based problem formulations [15]. However, their problem models mainly focus on spectrum allocation efficiency and apply to WDM networks. To the best of our knowledge, we are the first to propose a scalable exact method for solving the power-efficient protection problem in EON using directed p-cycles.

The paper is organized as follows. Section 2 describes the power-efficient p-cycle protection problem. The proposed optimization model is described in Sect. 3. The solution of the proposed model is presented in Sect. 4. Numerical results are discussed in Sect. 5. Finally, Sect. 6 is dedicated to the conclusion and future works.

2 Motivation and Problem Statement

2.1 Motivation

As previously stated, the goal of this study is to build a link protection p-cycle that is power-efficient. Many authors formulate the p-cycle protection problem as an ILP problem [10, 12], which is not scalable beyond very small network topologies. As a result, we propose developing a scalable exact method to solve the problem.

In EONs, optical devices such as BVT, OXC, and EDFA consume power and are calculated by [7]. Let e_m^{BVT}, e_v^{OXC} and e_l^{EDFA} are power consumption of BVT, OXC, and EDFA devices. These values depend on the number of FSs allocated to the connection. Hence, we also examine directed p-cycle protection in the context of asymmetric traffic in order to reduce power consumption in EON.

P-cycle protection strategies in EON are depicted in Fig. 1. There are two working paths (1-2-4-6) and (6-4-2-1) with bandwidth requirements of 300 Gbps and 150 Gbps, respectively. For simplicity, we assume that only QPSK modulation is used with a capacity of 25 Gbps per slot. As seen in Fig. 1(a), an undirected p-cycle allocates the same protection resource (twelve frequency slots) in both directions over the highest requested bit rate (i.e., the maximum amount of traffic) in each direction. However,

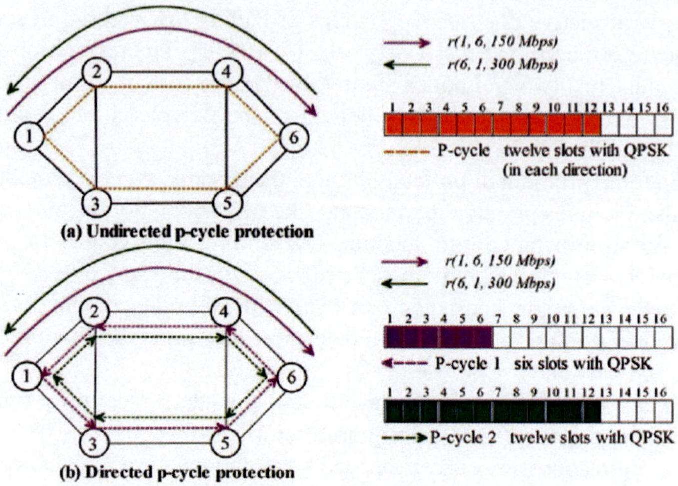

Fig. 1. Illustration of p-cycle protection techniques.

the directed p-cycle only allocates protection resources in one direction. As shown in Fig. 1(b), it may identify and supply distinct protection resources in two directions, with six frequency slots for p-cycle 1 and twelve frequency slots for p-cycle 2. We can see that the strategy in Fig. 1(b) employs fewer frequency slots, making it more suitable and helpful for power savings in EONs.

2.2 Problem Statement

Consider an EON, which is assumed by a directed graph $G = (V, L)$ with node set V and link set L. We assume that EONs support contiguous frequency slots (FS) with a spectral width of 12.5 GHz. The number of available FSs on each link is denoted by B. Let M is the set of available modulations (i.e., BPSK, QPSK, 8QAM, and 16QAM), indexed by m. The bandwidth provided to each slot at modulation level m is denoted TR_m. For each request $r \in R$, $\mathcal{SD}$ indicates the working route of request r. Let w_l be the overall required bandwidth for working traffic on link l that should be protected by p-cycles.

Formally, the problem of power-efficient p-cycle protection in elastic optical network is as follows: Given a set of asymmetric traffic demand matrix, find the set of directed p-cycles used for protection and assign spectra to each selected p-cycle such that the sum of the power consumed by all the protection paths is kept to a minimum.

3 Configuration Optimization Model

We propose a decomposition model relying on cycle configurations, where a cycle configuration $c \in C$ contains a cycle and a set of links that it can protect. A cycle configuration c is described by vectors $x = \left(x_l^c\right)_{l \in L}$ and $z = \left(z_{lm}^c\right)_{l \in L, m \in M}$, where:

- $x_l^c = 1$ if configuration c passes across link l, and 0 otherwise.
- $z_{lm}^c = 1$ if configuration c protect link l at modulation format m, and 0 otherwise.

The model uses the decision variable $n_c \in \mathbb{Z}_+$, where each n_c represents the number of occupied FSs of c on link l.

The objective function consists of minimizing the consumptive power across the entire network and is written as follows:

$$min \sum_{c \in C} \left(\sum_{m \in M} \sum_{l \in L} 2 \cdot e_m^{BVT} \cdot z_{lm}^c + \sum_{v \in V} \sum_{u \in N_v} \frac{x_{vu}^c}{B} \cdot e_v^{OXC} + \sum_{l \in L} \frac{x_l^c}{B} \cdot e_l^{EDFA} \right) \cdot n_c \quad (1)$$

Subject to:

$$\sum_{c \in C} \sum_{m \in M} z_{lm}^c \cdot TR_m \cdot n_c \geq w_l \qquad \forall l \quad (2)$$

$$\sum_{c \in C} x_l^c \cdot n_c \leq B \qquad \forall l \quad (3)$$

$$n_c \in \mathbb{Z}_+ \qquad \forall c$$

Constraint (2) ensures that overall the working capacity of each link is protected with 100% protection for single link failure, and constraint (3) guarantees that the requested bandwidth for constructing protection cycles does not surpass the link transport capability.

4 Solution Scheme

4.1 Generalities

A simple solution to the problem presented in (1)–(3) would necessitate enumerating all potential configurations, which would be impossible even with moderately sized instances. This section describes a CG-based solution process. The proposed CG algorithm's pseudocode is presented below.

CG algorithm with cycle configurations

```
Input: G = (V, L);
Output: set of selected cycles
Initialize set of cycle configurations (C)
Solve the Restricted Master Problem (RMP)
Solve the Pricing Problem (PP)
While (the reduced cost is negative) do
    Add the new cycle configuration to the set C
    Solve the RMP
    Solve the PP
Solve the Master Problem with set of cycle configurations C
```

In the column generation algorithm, the problem is split into a RMP (i.e., models (1)–(3) in Sect. 3) and a PP that generates a potential configuration. RMP and PP are solved alternately. First, initial configurations are generated and used to form the RMP. The RMP is solved, and the dual variables of the RMP are utilized to specify the PP's objective function. The PP is then solved in order to find a new configuration with a negative reduced cost. This new configuration is added to the current RMP and resolved iteratively. The process is repeated until the pricing problem's reduced cost is positive. Theoretically, in that instance, the optimal solution of the linear relaxation of the MP has been obtained.

Once we have found the optimal solution of the linear relaxation of the MP, we solve the ILP formulation of the final RMP to have the optimal solution of the MP.

4.2 Pricing Problem

Let $u_{lm}^{(2)}$ and $u_l^{(3)}$ be the dual prices come from constraints (2) and (3) of the RMP, respectively. The explicit formulation of the subproblem is given as follows:

$$min \sum_{m\in M}\sum_{l\in L} 2\cdot e_m^{BVT}\cdot z_l^m + \sum_{v\in V}\sum_{u\in N_v}\frac{x_{vu}}{B}\cdot e_v^{OXC} + \sum_{l\in L}\frac{x_l}{B}\cdot e_l^{EDFA} - \sum_{l\in L}\sum_{m\in M} u_{lm}^{(2)}\cdot z_l^m + \sum_{l\in L} u_l^{(3)}\cdot x_l \quad (4)$$

where $z_l^m = 1$ if link l is protected by c with modulation format m, and 0 otherwise; $x_l = 1$ If configuration c passes across link l, and 0 otherwise. For the set of constraints, including directed cycle generation and adaptive modulation selection, refer to [10].

5 Numerical Results

We use three realistic network topologies: Six-node (6-node, 16-(directed) link) in Fig. 2(a); NSFNET (14-node and 44-(directed) link) in Fig. 2(b); and USANET (28-node and 90-(directed) link) in Fig. 2(c). All computational results have been obtained by running the program on a Core i5 machine with 8GB of RAM, with the help of CPLEX (Version 12.6.0).

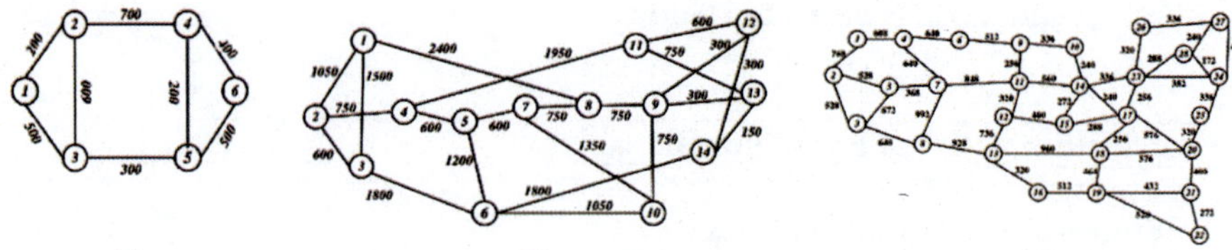

(a) (b) (c)

Fig. 2. Network topologies: (a) Six-node, (b) NSFNET, (c) USANET

Data Sets. In the NSFNET and USANET topologies, we consider 10 data sets. For the first data set (i.e., 01), the asymmetric traffic demand is randomly generated in {0, 12.5, 25, 37.5, 50, 62.5, 75, 87.5, 100 Gbps}. The following data sets correspond to increasing bandwidth such that $i \subseteq i+1$ where $i+1$ is constructed to be dependent on i by randomly deciding whether or not to add from 12.5 to 100 Gbps bandwidth demands for each pair of nodes. For each traffic instance, we provide the number of directed links with requests ($|\mathcal{SD}|$) and the total load.

5.1 Comparison of CG and ILP Algorithms

To assess the scalability and accuracy of the CG method, we compare it with the exact solution provided by the ILP model. The same objective function and majority constraints that are used to formulate CG models also apply to ILP models (Table 1).

Table 1. Comparative model/algorithm performance in six-node network

Data set	$\|\mathcal{SD}\|$	Total load (Tbps)	ILP		CG		$\frac{z^{CG}-z^*}{z^{CG}}$ (%)
			z^*(kW)	Excution time (s)	z^{CG}(kW)	Excution time (s)	
01	15	0.51	4.37	1,369	4.74	10	7.79
02	29	1.20	10.27	8,521	10.59	15	3.05
03	35	1.72	14.35	5,760	14.60	15	1.66
04	40	2.50	20.04	4,308	20.11	15	0.35
05	43	3.00	24.50	5,370	24.65	14	0.64

Tale 1 shows a comparison result and gives the achieved value of the objective function, the gap, and the execution time between the CG algorithm and the ILP model. Because the ILP model overflows the memory for higher traffic, we only implemented five data sets with low traffic on a six-node network. We can see that the proposed CG outperforms the ILP method significantly. It produced high-quality results that are quite similar to those of ILP. In terms of processing time, the CG approach is extremely quick. For example, in dataset 02, the solving process requires only 15 s for CG, but the solution in ILP takes more than 2 h. Despite the small optimality gap introduced, the large reduction in processing time in CG is quite attractive.

5.2 Performance of the CG Solution

We evaluate the efficiency of the CG algorithm on large data sets for the NSFNET and USANET networks, as shown in Tables 2 and 3. In these tables: Column z_{LP} denotes the LP optimal solution of the RMP, hence a lower bound on the optimal ILP solution, column z_{ILP} denotes the value of the integer solution, it is an ε-optimal solution, and column configurations represent the number of generated and selected configurations during the CG algorithm. The accuracy GAP evaluates the solution quality/accuracy and is defined as follows:

$$GAP = \frac{Z_{ILP} - Z_{LP}}{Z_{LP}} \times 100 \quad (5)$$

Table 2. Numerical results for the NSFNET network

Data set	$\|SD\|$	Total load (Tbps)	Optimal solution			Configurations		CPU time (s)
			z_{LP}(kW)	z_{ILP}(kW)	GAP(%)	Generated	Selected	
01	139	17.60	175.61	179.60	2.22	450	42	1,180
02	156	26.85	270.29	273.31	1.10	511	43	1,470
03	166	37.75	400.43	403.20	0.68	561	46	1,741
04	175	47.52	508.76	511.82	0.59	589	48	1,929
05	178	57.37	606.29	609.61	0.54	651	56	2,757
06	180	67.56	696.14	699.92	0.54	679	60	3,314
07	180	77.42	795.71	798.94	0.40	670	61	3,436
08	180	89.98	909.63	913.45	0.41	655	69	3,131
09	180	100.36	1,008.37	1,011.75	0.33	688	76	3,819
10	180	109.55	1,124.10	1,128.15	0.35	686	77	3,562

Table 3. Numerical results for the USANET backbone network

Data set	$\|SD\|$	Total load (Tbps)	Optimal solution			Configurations		CPU time (s)
			z_{LP}(kW)	z_{ILP}(kW)	GAP(%)	Generated	Selected	
01	102	21.79	171.57	178.53	3.89	589	76	3,758
02	146	44.95	353.43	361.03	2.08	758	80	5,223
03	163	66.61	522.23	528.73	1.23	796	82	4,779
04	170	83.86	656.81	663.11	0.95	868	90	6,399
05	175	101.54	795.60	801.97	0.79	899	93	7,199
06	177	118.53	927.28	934.79	0.80	890	97	6,613
07	179	136.22	1,065.46	1,071.76	0.58	907	103	7,292
08	179	152.58	1,193.30	1,200.25	0.57	967	105	9,388
09	180	172.20	1,345.37	1,352.96	0.56	950	112	9,003
10	180	191.33	1,477.47	1,484.51	0.47	937	114	8,331

We can observe that, in the column generation method, only a very small amount of configurations are generated while the actual number of configurations that need to be enumerated in the ILP method is very large. The gap ranges from ~ 0.3% to ~ 3.9%, meaning that the output solutions are always within a 10% accuracy. More remarkably, computing times for traffic instances on both NSFNET and USANET range between a few tens of minutes and a few hours. However, note that the results correspond to the large traffic instances solved optimally for 139 to 180 requests (resp. Total load from 17.60 to 109.55 Tbps) on the NSFNET network and 102 to 180 requests (resp. Total load from 21.79 to 191.33 Tbps) on USANET.

6 Conclusion

We considered the problem of power-efficient protection with the directed p-cycle in EON. We proposed a scalable and effective optimization model for a p-cycle protection scheme taking into consideration power-efficient. Through the experimental results, we have shown the effectiveness of our model in finding the solution within a reasonable time. Our planned future work includes improving the proposed problem's spectrum efficiency by employing the spectrum share-based techniques between the protective p-cycles, which would also result in large power savings in EONs.

References

1. Eshoul, A.E., Mouftah, H.T.: Survivability approaches using p-cycles in WDM mesh networks under static traffic. IEEE/ACM Trans. Netw. **17**(2), 671–683 (2009)
2. Sharma, D., Kumar, D.(Col) S.: An overview of elastic optical networks and its enabling technologies. Int. J. Eng. Technol. **9**(3), 1643–1649 (2017)
3. Wu, J., Liu, Y., Yu, C., Wu, Y.: Survivable routing and spectrum allocation algorithm based on p-cycle protection in elastic optical networks. Optik (Stuttg) **125**(16), 4446–4451 (2014)
4. Chen, X., Ji, F., Zhu, Z.: Service availability oriented p-cycle protection design in elastic optical networks. J. Opt. Commun. Netw. **6**(10), 901–910 (2014)
5. Chalasani, S., Rajaravivarma, V.: Survivability in optical networks. Proc. Annu. Southeast. Symp. Syst. Theory **2003**, 6–10 (2003)
6. Oliveira, H.M.N.S., Da Fonseca, N.L.S.: Algorithm for FIPP p-cycle path protection in flexgrid networks. In: 2014 IEEE Global Communications Conference GLOBECOM 2014, pp. 1278–1283 (2014)
7. Wang, C., Shen, G., Bose, S.K.: Distance adaptive dynamic routing and spectrum allocation in elastic optical networks with shared backup path protection. J. Light. Technol. **33**(14), 2955–2964 (2015)
8. López, J., Ye, Y., López, V., Jiménez, F., Duque, R., Krummrich, P.M.: On the energy efficiency of survivable optical transport networks with flexible-grid. Eur. Conf. Exhib. Opt. Commun. ECEOC **2012**, 8–10 (2012)
9. Davis, D.A.P., Vokkarane, V.M.: Resource survivability for multicast in elastic optical networks. In: Proceedings of 2016 17th International Telecommunications Network Strategy and Planning Symposium (Networks), pp. 199–206 (2016)
10. Ju, M., Zhou, F., Xiao, S., Zhu, Z.: Power-efficient protection with directed p-cycles for asymmetric traffic in elastic optical networks. J. Light. Technol. **34**(17), 4053–4065 (2016)
11. Ramaswami, G.S.R., Sivarajan, K.: Optical Networks: a Practical Perspective. Morgan Kaufmann, Burlington (2009)

12. Jaiswal, D.C., Asthana, R.: Power-efficient p-cycle protection with power conscious routing in elastic optical networks. In: Proceedings of 2018 International Conference on Current Trends towards Converging Technologies (ICCTCT), pp. 1–6, 2018
13. Hoang, H.A., Jaumard, B.: Design of p-cycles under a wavelength continuity assumption. In: 2011 IEEE International Conference on Communications (ICC) (2011)
14. Jaumard, B., Hoang, H.A., Kien, D.T.: Robust FIPP p-cycles against dual link failures. Telecommun. Syst. **56**(1), 157–168 (2014)
15. Chvatal, V.: Linear programming (1983)

Ranking E-learning Systems in Vietnamese K12 Market Based on Multiple Criteria

Ha Nguyen Thi Thu[1](✉), Linh Bui Khanh[2], and Trung Nguyen Xuan[3]

[1] Department of Greenwich, FPT University, Hanoi, Vietnam
hantt194@fe.edu.vn

[2] Electric Power University, Hanoi, Vietnam
linhbk@epu.edu.vn

[3] Vietnam Institute of Americas Studies, Hanoi, Vietnam
trungnguyenxuan@gmail.com

Abstract. Educational technology has become an indispensable tool in modern education at all school levels today. With the number of K12 students in Vietnam accounting for nearly 20% of the national population, the educational technology sector has attracted startups developing many new EdTech products and investors interested in this field. Not to mention international products that developed in Vietnam market, domestic products have also reached hundreds. The rapidly increasing use of educational technology leads to user choice problems. With an extremely large market, it is become too difficult for users of choice, difficult for management as well as to develop standards to guide the development and circulation of products in Vietnam. There have been quite a few methods of evaluating E-learning systems, but the ranking of E-learning systems has not been mentioned much, it is a suggestion for efforts to perfect the production of the companies and attract investors and users. In this study, a method of evaluating and ranking eLearning products is proposed based on a set of criteria such as information quality, system quality, scalability, and user satisfaction. With 174 eLearning training systems in the Vietnamese market for K12 students, data are collected and measured according to the criteria from the reputable systems, then an appropriate ranking formula used to measure the specific value of the product is also suggested for users as well as support for experts in evaluating educational technology products at a later in-depth stage.

Keywords: Online learning system · LCMS · EdTech · Digital Content · Platform

1 Introduction

The market for educational technology products in Vietnam is growing strongly and is ranked 3rd in Southeast Asia as well as 10th in attractive markets for investors in the world. According to the report of Tracxn Technologies in 2021, there are nearly 260 EdTech companies in Vietnam, most of which are start-ups that follow by the business-to-consumer model(B2C). EdTech startups in Vietnam are currently attracting the attention

T. D. L. Nguyen et al. (Eds.): ICISN 2023, LNNS 752, pp. 435–440, 2023.
https://doi.org/10.1007/978-981-99-4725-6_53

of internal and international venture capital funds. Startup companies bring many new types of business models to the online education and digital transformation market. In which, products such as Topica, Edumall, Forever Learning, Azota, Vietjack, etc. have attracted many international investors such as Do Ventures, Genesia Ventures, Chiba Dojo, Kaizen Private Equity, Spring Through Pte (Singapore)…. With a population of nearly 100 million people, according to the 2021 statistics, the number of K12 students about 23.5 million, accounting for more than 20% of the population of Vietnam. Products in this field are mainly lecture notes that associated with the curriculum content of the Ministry of Education and Training, in which a large product segment is for English learning and prepare for international English certificate as Ielts, Toefl,… Another part is exam systems, review, tutors, virtual schools, etc. It is estimated that by 2023 the Vietnam Edtech market will reach 3 billion USD.

With such a large and diverse product market, the quality of eLearning systems has received great interest from students, their parents, investors, and system developers [1, 10, 13, 14]. Therefore, the task of evaluating the quality of eLearning products needs to be considered in more detail from the point of view of users [9, 11], because the factor of user satisfaction is one of the important issues when developing an eLearning system, that ensures that it can match with user requirements and make investors want to pour capital [9]. Therefore, it is essential to develop effective methods for ranking eLearning systems, that helps user easier to choose [8, 15, 16].

There have been a number of previous studies dealing with the issue of evaluating an eLearning system [2, 3, 8, 10, 14–16]. The proposed methods are based on multiple criteria and create a model for measuring [10, 13]. Some methods used fuzzy or mathematical model, etc. [1, 4, 15, 16]. Most of them were developed from DeLone & McLean model [5, 6] then add more factors like interaction, support learners, quality assurance, learning tasks, publicity, and information to evaluate eLearning systems [5, 7, 12]. Others focus on users' perception [2, 11] of student expectations and experiences with eLearning systems in relation to academic success and course satisfaction. Some studies used the fuzzy axiom design method to evaluate eLearning websites and eLearning systems. They selected criteria of the user interface, interactivity, navigation, humanity, secure, complete, understandable content, and relevant information. [1, 4, 16].

Most of these studies often need a survey with end users like students or teachers to evaluate at least one or two criteria. So, it is difficult to arrange and rank for a large market of hundreds and thousands of products. On the other hand, sometimes it is a subjective perception by the evaluation users. Therefore, in this study, a set of criteria has been developed to rank including system quality, information quality, scalability, and user satisfaction. These criteria are able to measure by SEO criteria through Alexa, Neipatel, sea ranking systems to receive measuring values, then determine a set of weights for each criterion of eLearning products. In the next step, the study proposes the formula to calculate the ranking for eLearning products in the Vietnamese market.

The rest of this paper is organized as follows: Sect. 2 introduces the ranking method for K12 eLearning products in Vietnamese market. Section 3 is the ranking results, and final is the conclusion.

2 Methodology of Elearning Ranking

The DeLone and McLean model with the multi-criteria is a basis for the method of Elearning systems evaluation in this study. Elements of the model include the "systems quality" express the success of the technology; two factors: "organizational impacts" and "use, user satisfaction, individual impacts," measure the effectiveness of the system, and factor: "information quality" assesses the quality of information or lessons [7]. This study has proposed a measurement solution based on the available parameters according to the criteria and collected from the reliable website measurement systems in the world. Table 1 below summarizes the criteria and measurement variables and measurement methods.

Table 1. Criteria and measuring

Criteria	Meaning	Variables	Method of measuring
Systems quality (Tech)	The largest number of visits in 1 month	Tech1	Neilpatel or searanking measurement system
	Modern technology used in the system	Tech2	Identification of system features
Information quality (Infor)	Number of keywords on the system	Infor1	Neilpatel or searanking measurement system
Use, user satisfaction (U_sas)	An average number of visits within 6 months	U_sas1	Neilpatel or searanking measurement system
	Number of backlinks pointing to the system	U_sas2	Neilpatel or searanking measurement system
Scale ability (Scale)	Number of other countries accessing the system	Scale	Google Trends data

The method of ranking educational technology products for K12 on the Vietnamese market in this study focuses on summing up the values of the criteria measured in Table 1 above and is performed according to the following steps:

Step 1. Collect data on K12 products of the whole Vietnamese market and aggregate them into the official list after excluding products that are not built and developed by Vietnam.
Step 2. Use Neipatel or Searanking system to get product criteria data including Tech1, Info1, U_sas1, and U_sas2. In this study, data from November 2021 to March 2022 were used.
Step 3. Use the Google Trend system to get the Scaleability criterion score
Step 4. Human experts evaluate and assign scores to the Tech1 criteria of the systems.
Step 5. Summarize the criteria to calculate the weight of each product according to formula (1) below.
Step 6. Extract the product ranking in the order Z->A

The formula for calculating score for each product according to formula (1) as follows:

$$Score(Product_i) = \alpha * (Tech1_i + Infor1_i + U\_sas1_i + U\_sas2_i) + \beta * Tech2_i \quad (1)$$

In which:
$Score(Product_i)$: Score of the product i in the ranking list.
α, β: : the respective coefficients (weight of criteria).

3 Result

To evaluate online training systems, after collecting all products in the Vietnamese market, the study has grouped products into two main groups: (1). The platform includes tools for creating schools, searching for tutors, searching for schools, and reviewing,...; (2). The online training products include foreign language training content, lectures at all levels from elementary, middle, and high schools, exam preparation, etc. And then, only use group (2) for ranking. When collecting data for products for the K12 block market, the research aggregated 251 products. In which, there are 77 products is platforms and 174 products is E-learning systems.

Data from Neipatel were collected for each Elearning system with these criteria as Systems quality, Information quality, Scaleability, and user satisfaction. Table 2 below illustrates the value of the criteria of the Vietjack system.

Table 2. An illustrated product

Name	Vietjack
Score (Tech1)	10,790,615
Score (Tech2)	1000
Score (Infor1)	171,351
Score (U_sas1)	10,100,000
Score (U_sas2)	2,103,468
Score (Scale)	1,000,000
Score of Product	4,070,501,424

Table 4 below is an illustration of the top 20 products with the highest scores among the top K12 products in Vietnam.

With the above implementation, the research has ranked 174 Elearning products for K12 students in Vietnam.

Table 3. Top list 20 Vietnamese Edtech products with their score

No	Name of products	Website	Score
	vietjack	https://vietjack.com/	4,070,501,424.46
	vndoc	https://vndoc.com/	1,341,797,495.64
	download	https://download.vn	1,229,356,078.04
	doctailieu	https://doctailieu.com/	1,096,186,797.67
	Loigiaihay	https://loigiaihay.com/	670,733,422.80
	hoc247	https://hoc247.net/	403,290,558.50
	OLM	https://olm.vn	263,816,430.59
	123docz	https://123docz.net	180,247,667.66
	ejoy-english	https://ejoy-english.com/vi/	86,542,569.68
	hoidap247	https://hoidap247.com/	84,499,099.44
	Toploigiai	https://toploigiai.vn	80,851,515.78
	Vnedu	https://vnedu.vn/	78,240,091.60
	hoc24	https://hoc24.vn/	70,306,469.96
	Moon	https://moon.vn/	57,113,284.37
	stepup	https://stepup.edu.vn/	55,374,554.68
	Vungoi	https://vungoi.vn/	36,757,028.03
	tech24h	https://tech12h.com/	35,863,350.80
	Tech12h	https://tech12h.com/	35,851,828.64
	studytienganh	https://studytienganh.vn	32,527,924.12
	tuhoc365	https://tuhoc365.vn/	31,017,754.10

4 Conclusion

This study has come up with a solution to rank Elearning products for K12 students segmentation, this rating can suggest to users learning systems suitable to current conditions, as well as suggestions for investors to look at the potential Edtech market of Vietnam, choose the right investment products. In addition, the ranking is also a perspective for product developers based on the criteria to develop products that are suitable for users and easily attract investors. Education management agencies also better control products circulating on the market, thereby also having policies to manage and support education businesses today in Vietnam.

References

1. Garg, R., Sharma, R.K., Sharma, K.: Ranking and selection of commercial off-the-shelf using fuzzy distance based approach. Decis. Sci. Lett. 201–210 (2016). https://doi.org/10.5267/j.dsl.2015.12.004

2. Santoso, H.B., Hadi Putra, P.O., Farras Hendra, S.F.F.: Development & evaluation of e-learning module based on visual and global preferences using a user-centered design approach. Int. J. Emerg. Technol. Learn. **16**(15), 139 (2021). https://doi.org/10.3991/ijet.v16i15.24163
3. Effendy, F., Purwanti, E., Akbar, R.F.: Evaluation of e-learning: a case study of PsyCHE. In: International Conference on Mathematics, Computational Sciences and Statistics 2020, AIP Publishing (2021)
4. Garg, R., Kumar, R., Garg, S.: MADM-based parametric selection and ranking of e-learning websites using fuzzy COPRAS. IEEE Trans Educ. **62**(1), 11–18 (2019). https://doi.org/10.1109/te.2018.2814611
5. The DeLone and McLean model of information systems success: a ten-year update. J. Manag. Inf. Syst. **19**(4), 9–30 (2003). https://doi.org/10.1080/07421222.2003.11045748
6. Seta, H.B., Wati, T., Muliawati, A., Hidayanto, A.N.: E-learning success model: an extention of DeLone & McLean IS' success model. Indones. J. Electr. Eng. Inf. (IJEEI). **6**(3) (2018). https://doi.org/10.11591/ijeei.v6i3.505
7. Farid, S., Ahmad, R., Alam, M., Akbar, A., Chang, V.: A sustainable quality assessment model for the information delivery in e-learning systems. Inf. Discov. Deliv. **46**(1), 1–25 (2018). https://doi.org/10.1108/idd-11-2016-0047
8. Mastan, I.A., Sensuse, D.I., Suryono, R.R., Kautsarina, K.: Evaluation of distance learning system (e-learning): a systematic literature review. J. Teknoinfo. **16**(1), 132 (2022). https://doi.org/10.33365/jti.v16i1.1736
9. Mtebe, J.S., Raphael, C.: Key factors in learners' satisfaction with the e-learning system at the university of Dar es Salaam, Tanzania. Australas. J. Educ. Technol. **34**(4) (2018). https://doi.org/10.14742/ajet.2993
10. Al-Fraihat, D., Joy, M., Masa'deh, R., Sinclair J.: Evaluating E-learning systems success: an empirical study. Comput. Human. Behav. **102**, 67–86 (2020). https://doi.org/10.1016/j.chb.2019.08.004
11. Rajasekaran, V.A., Kumar, K.R., Susi, S., Mohan, Y.C., Raju, M., Hssain, M.W.: An evaluation of e-learning and user satisfaction. Int. J. Web-based Learn. Teach. Technol. **17**(2), 1–11 (2021). https://doi.org/10.4018/ijwltt.20220301.oa3
12. Information quality framework for e-learning systems. Knowl. Manag. E-learn. Int. J. 340–362 (2010). https://doi.org/10.34105/j.kmel.2010.02.025
13. Toan, P.N., Dang, T-T., Hong, L.T.T.: E-learning platform assessment and selection using two-stage multi-criteria decision-making approach with grey theory: a case study in Vietnam. Mathematics **9**(23), 3136 (2021). https://doi.org/10.3390/math9233136
14. Aslam, S.M., Jilani, A.K., Sultana, J., Almutairi, L.: Feature evaluation of emerging E-learning systems using machine learning: an extensive survey. IEEE Access **9**, 69573–69587 (2021). https://doi.org/10.1109/access.2021.3077663
15. Tao, R., Zhu, L., Wen, Q., Shi, Y., Feng, X.: The usability evaluation method of e-learning platform based on fuzzy comprehensive evaluation. In: Uden, L., Ting, I.H., Wang, K. (eds.) Knowledge Management in Organizations. KMO 2021. Communications in Computer and Information Science, vol. 1438, pp. 292–304. Springer, Cham (2021). https://doi.org/10.1007/978-3-030-81635-3_24
16. Garg, R., Sharma, R.K., Sharma, K., Garg, R.K.: MCDM based evaluation and ranking of commercial off-the-shelf using fuzzy based matrix method. Decis. Sci. Lett.117–136 (2017). https://doi.org/10.5267/j.dsl.2016.11.002

Evaluating the Improvement in Shear Wave Speed Estimation Affected by Reflections in Tissue

Nguyen Sy Hiep[1], Luong Quang Hai[2], Tran Duc Nghia[3], and Tran Duc Tan[4](✉)

[1] University of Information and Communication Technology, Thai Nguyen, Vietnam
Nshiep@ictu.edu.vn
[2] Le Quy Don Technical University, Hanoi, Vietnam
luonghai@mta.edu.vn
[3] Graduate University of Science and Technology, Hanoi, Vietnam
nghiatd@ioit.ac.vn
[4] Phenikaa University, Hanoi, Vietnam
tan.tranduc@phenikaa-uni.edu.vn

Abstract. Elastography has emerged as a promising technique for non-invasive clinical diagnosis in recent years. This method estimates tissue elasticity by analyzing the speed of shear waves captured by an ultrasonic Doppler transducer, and offers several advantages such as high reliability, low cost, and non-invasiveness. However, the propagation of sliding waves in diseased tissue generates reflected waves that can affect the accuracy of the results. To overcome this limitation, this study introduces a novel technique that employs a directional filter to eliminate reflected waves and increase the reliability of the elastography results.

Keywords: Shear wave · reflection · ultrasonic · directional filter

1 Introduction

Ultrasound elastography, which was introduced thirty years ago and has developed rapidly in recent years, is a technique for measuring the elasticity of tissue that is altered due to pathological factors. For example, a breast tumor has lower elasticity and viscosity than the healthy tissue surrounding it. Tissue elastography is applied to diagnose the liver, breast, thyroid, and prostate. Some ultrasound machines with the above features appear in machines such as the Hitachi Arietta 850, the SEQUOIA… with the specific problem mentioned in this paper. We use a vibrating needle to generate shear waves, then measure the propagation velocity in the tissue medium, thereby calculating the viscosity and elasticity of the tissue to be tested [1, 2]. The reflected wave plus the forward wave as the shear wave propagates in a heterogeneous medium in the diseased tissue. The data obtained is incorrect. To overcome this problem, We use a directional filter to remove noise [3, 4] (Fig. 1).

T. D. L. Nguyen et al. (Eds.): ICISN 2023, LNNS 752, pp. 441–451, 2023.
https://doi.org/10.1007/978-981-99-4725-6_54

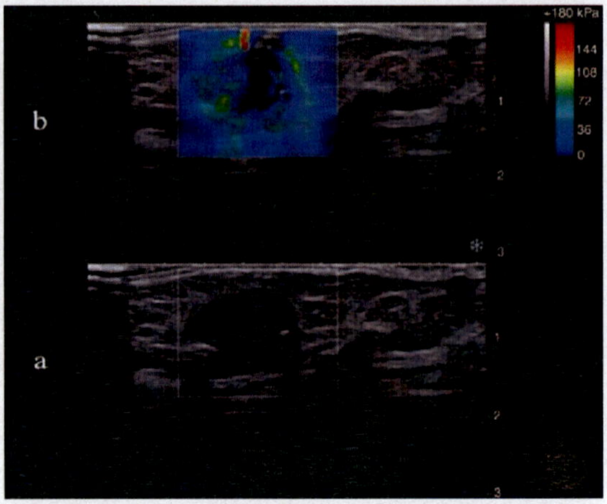

Fig. 1. (a) Greyscale ultrasound image and (b) Elastography image showing high peritumoural stiffness (mean kPa 148) [5].

1.1 Human Tissue Elasticity

To quantify tissue stiffness, a physical parameter known as Young's modulus is employed. This parameter is calculated by applying a uniform, external compression (or stress S) to diseased tissue, which generates a corresponding deformation (e) within the tissue. By measuring this deformation, the stiffness of the tissue can be evaluated and diagnosed (Fig. 2).

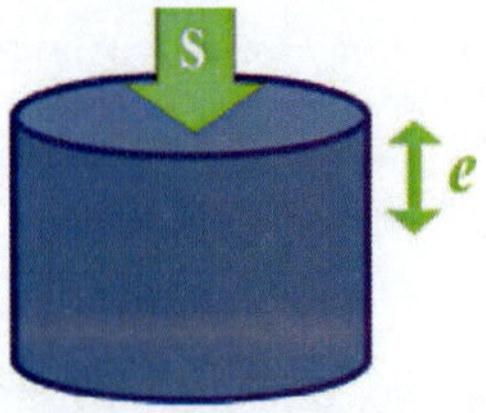

Fig. 2. Deformation of solid under an external stress.

$$E = \frac{s}{e} \quad (1)$$

s : External stress
e : Strain in side the tissue
E :Young Modulus

Hard tissue has a higher density than soft tissue, and diseased tissue is usually harder than healthy tissue. For example, the young modulus of breast adipose tissue has a value of 18–24 kPa, while the fibrous tissue has a value of 96–244 kPa. Normal liver has a value of 0.4–6 kPa while cirrhosis E has a value of 15–100 kPa. We know the density of tissues in the body is 1000 kg/m^3. Normal liver has a value of 0.4–6 kPa while cirrhosis of liver E has a value of 15-100 kPa. We know the density of tissues in the body is 1000 kg/m^3. While the density of tissue remains relatively constant in the body (1000 kg/m^3), tissue elasticity may differ significantly compared to tissue in a different pathological state, Compression waves propagate quickly in tissue (1500 m/s), while shear waves are much slower (1–10 m/s).

The correlation between elasticity and the velocity of propagation [6].

$$E = \rho c^2 \quad (2)$$

ρ: Density of tissues (1000kg/m^3)

c: Shear wave propagation speed

E: Elasticity

The measurement of shear wave velocity allows for the determination of tissue elasticity, given a known tissue density of 1000 kg/m^3 [6–8].

1.2 Determination of Elasticity and Viscosity of Tissues

In this study, we used a vibrating needle to produce a shear wave that is generated by tissue expansion and propagates perpendicularly to the source in the proximal direction. The speed of shear wave propagation is determined by the stiffness of the medium, with higher stiffness resulting in greater propagation speed (Fig. 3).

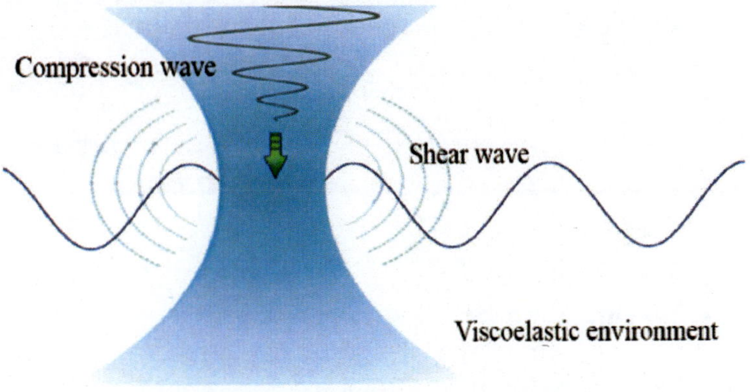

Fig. 3. The excitation source generates a shear wave.

Determine the speed of the shear wave at a position represented by the wave propagation equation. [6, 9, 10].

$$v_n(r) = \frac{1}{\sqrt{r - r_0}} A e^{-a(r-r_0)} \cos[\omega n \Delta t - k_s(r - r_0) - \phi] \quad (3)$$

Acoustic heterogeneity and tissue boundaries can cause the shear wave to undergo reflection or refraction as it propagates through tissue, which may affect the accuracy of the shear wave velocity (SWV) measurement.

Equation of the reflected wave.

$$v_n(r) = \frac{1}{\sqrt{r_0 - r}} A e^{-a(r-r_0)} \cos[\omega n \Delta t + k_s(r_0 - r) - \phi] \quad (4)$$

r	spatial coordinate	ϕ	Initial phase
r_0	coordinate of source	k_s	Wave number
A	amplitude of source	ω	Angular frequency
Δt	time step	α	the attenuation coefficient
n	the number of time step	v	velocity

According to the kelvin-voigt model, to calculate the elasticity E is also to calculate the complex shear modulus [11–13].

$$\mu = \mu_1 - i\omega\eta \quad (5)$$

For accurate location analysis, it is important to determine the elasticity and viscosity of the medium at that specific location.

Equation for viscosity and elasticity in space [6]

$$\eta = -\frac{2\rho\omega k_s \alpha}{(k_s^2 + \alpha^2)^2} \quad (6)$$

$$\mu_1 = \frac{\rho\omega^2 (k_s^2 - \alpha^2)}{(k_s^2 + \alpha^2)^2} \quad (7)$$

μ: Complex Shear Modulus

η: Viscosity of the medium

μ_1: Tissue elasticity

ρ: Density

2 Directional Filter

The reflected signal is in the opposite direction of the forward wave. The received signal is a mixed signal of the forward wave and the reflected wave. It is necessary to create a filter that removes the reflected signal. The received signal must be similar to the incoming wave signal [9] (Fig. 4).

The fourier-transformed input signal is then multiplied by the mask matrix Fig. 5, which will determine which signals are retained and which are discarded, represented by 0 (corresponds to dark blue) and 1(corresponds to yellow). With 1 for the signal to go through and 0 to keep it. The received signal is converted back to its original form. After

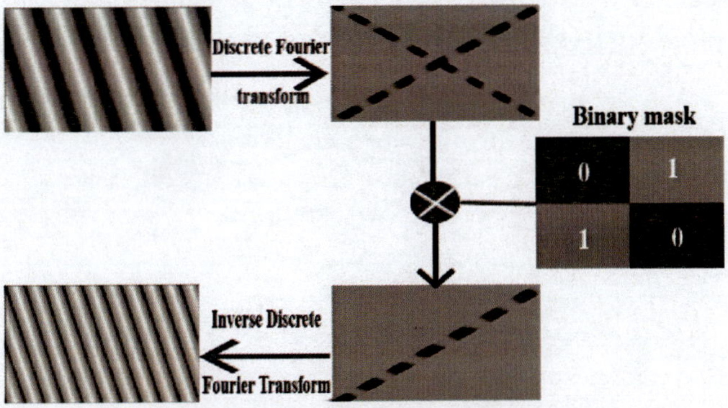

Fig. 4. Directional filter workflow.

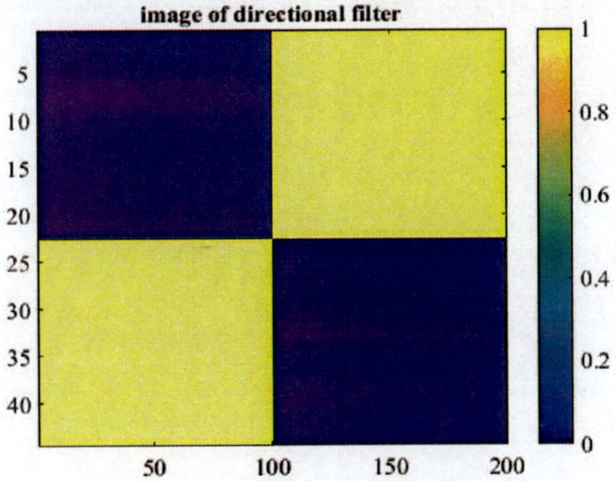

Fig. 5. Image of directional filter

passing through a filter with a mask design, it moves the reflected wave from right to left. We receive the signal that is close to the incoming signal when there is no reflected noise [14].

3 System Setup and Results

3.1 System Setup

This paper employs Matlab as a tool for simulating real-world scenarios (Fig. 6).

Details are shown in the Table 1 below.

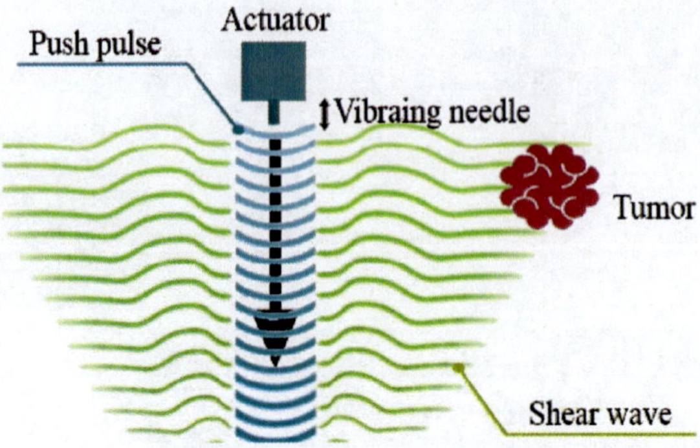

Fig. 6. Vibrating needle generate shear wave.

Table 1. Parameters used for simulation.

Frequency	100 Hz
Amplitude	0.0086
Normal tissue viscosity	650k Pas
Normal tissue elasticity	0.1 kPa
Elasticity of diseased tissue	800 kPa
Viscosity of diseased tissue	0.2 Pas
Density	1000 kg/m^3

3.2 Results

We evaluate the effect of the directional filter by comparing the results before and after using it.

Before Using Directional Filter
Observing Fig. 7(b) we can see that the reflected wave starts at a location of 20 in space and then propagates from the right to the left. Figure 8 We clearly see that reflected wave affect on forward wave Fig. 7(a).

The image's color represents the wave amplitude at a specific point in time and space. For example, dark blue corresponds to wave amplitude -0.8, yellow corresponds to wave amplitude 0.8 and green corresponds to wave amplitude 0.

The creation of a mix wave image shown in Fig. 10 involves the combination of the reflected wave image in Fig. 9(b) and the incident wave image in Fig. 9(a).

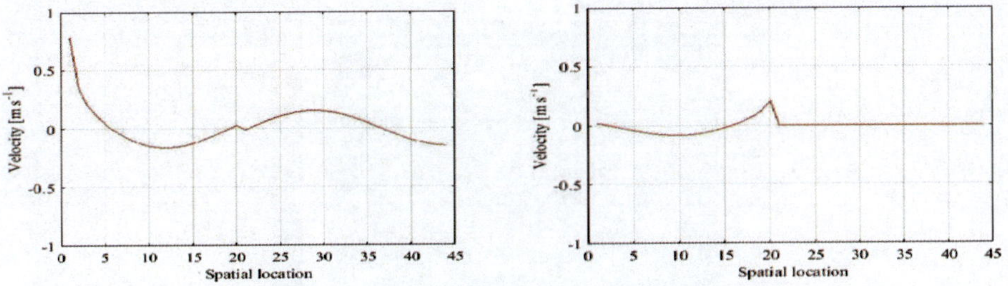

Fig. 7. (a) Incomming wave. (b) Reflected wave.

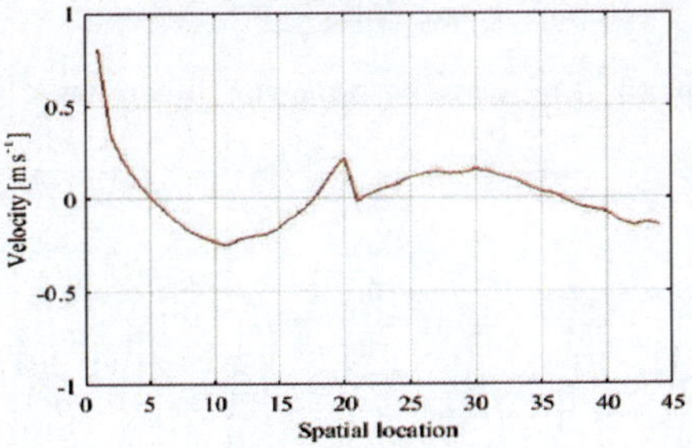

Fig. 8. Mix wave.

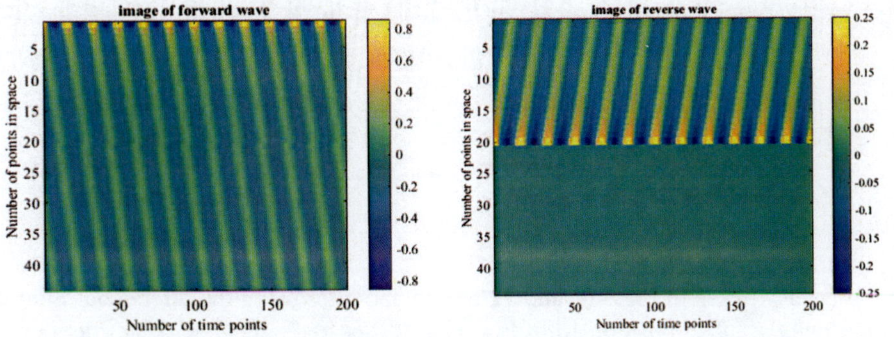

Fig. 9. (a) Image of incident wave. (b) Image of reflected wave.

Oscillation velocity graph at position 15 in Fig. 11. The received signal consists of an incoming wave signal and a reflected wave signal, so there is noise and a phase difference from the ideal waveform.

After Using Directional Filter

Signal After being passed through the directional filter, it will remove the data oriented from right to left. Only the data of the incoming wave is kept.

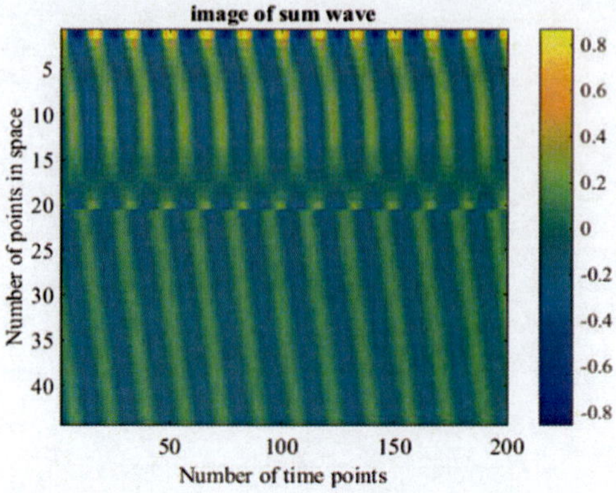

Fig. 10. Mix wave before using directional filter.

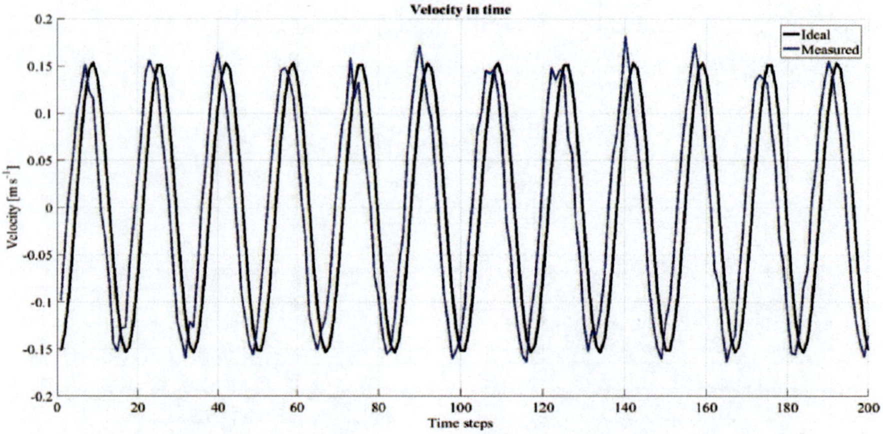

Fig. 11. The mix wave before using directional filter.

It is easy to see in Figs. 11 and 12 before and after using the directional filter that the velocity graph at position 15 of Fig. 12 is closer to the ideal velocity graph than in Fig. 11.

In Fig. 13, we observe that the blue line graph represents the ideal case, the dashed line represents the velocity graph prior to being passed through the directional filter, and the red line represents the velocity graph after it has been filtered. The red line is almost identical to the ideal velocity graph (blue line).

As depicted in Figs. 7(b) and 9(b), the velocity graph begins at position 20 in space, starts at 0, ends at 45, and decreases with increasing space. The amplitude of the reflected wave is greatest at position 20, as demonstrated in the aforementioned figures, and gradually decreases until it reaches position 0. This implies that the resulting wave will be affected by the reflected wave, which is strongest at point 20 and gradually weakens as it approaches point zero, as illustrated in Fig. 13. Additionally, Figs. 7(b) and 9(b)

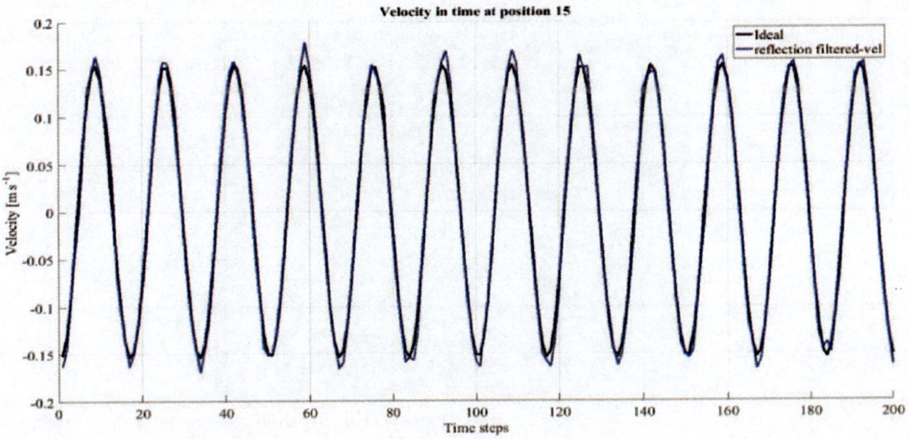

Fig. 12. The mix wave after using directional filter.

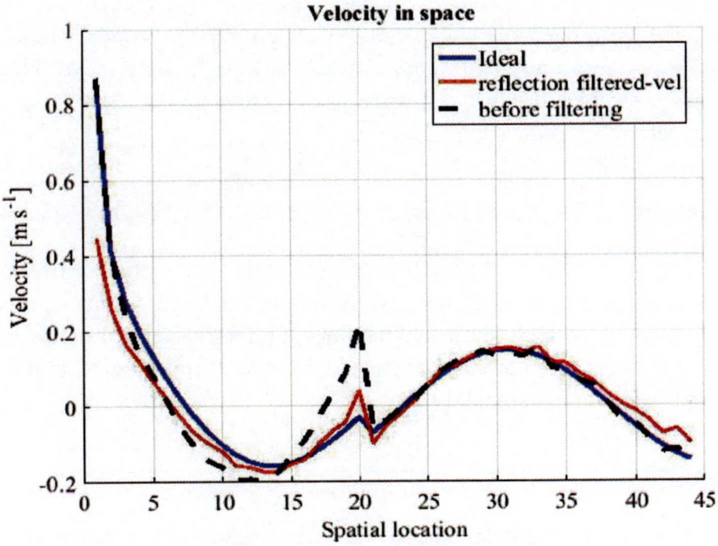

Fig. 13. Velocity in space.

indicate that the reflected wave has an amplitude of zero, as the potential does not affect the outcome of the composite wave, as shown in Fig. 13.

From positions 0 to 5, the effect of the reflected wave on the composite wave is minimal, and the composite wave graph closely follows the ideal graph. Beyond position 20 (positions 25–45), the graph of the composite wave, before and after filtering, is consistent with the ideal velocity graph. In conclusion, applying directional filtering to the object from positions 5 to 25 would yield the same results as filtering from positions 0 to 45 (Fig. 14).

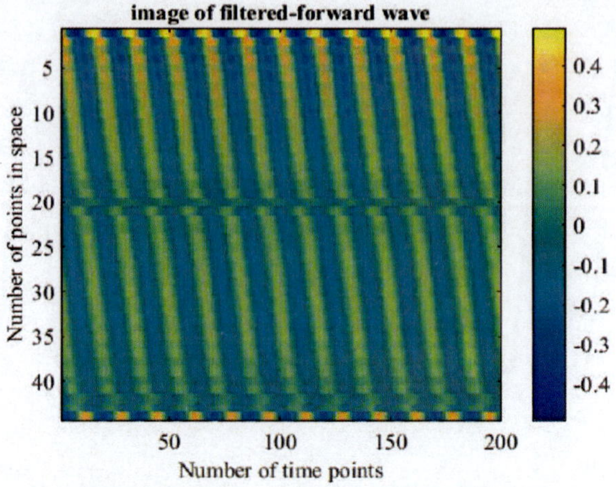

Fig. 14. Image of forward wave after using directional filter.

By using a directional filter, we can effectively eliminate and mitigate the impact of the reflected wave, resulting in an improved image of the forward wave. This improved image closely resembles the forward wave prior to filtering, leading to an increase in diagnostic accuracy.

4 Conclusion

This paper explores the effect of the reflected wave on the signal obtained in elastography. We propose the use of a directional filter to reduce its impact. Our results, obtained through simulation in Matlab, show that the filter is effective in minimizing the reflected wave. In future work, we aim to improve the filter to accommodate multi-directional signals, thereby enhancing signal quality.

References

1. Bercoff, J., Tanter, M., Fink, M.: Supersonic Shear imaging: a new technique for soft tissue elasticity mapping. IEEE Trans. Ultrason. Ferroelect. Freq. Control. (2014)
2. Sigrist, R.M.S., Liau, J., Kaffas, A.E., Chammas, M.C., Willmann, J.K.: Ultrasound elastography: review of techniques and clinical applications. Theranostics (2017)
3. Song, P., Manduca, A., Zhao, H., Urban, M.W., Greenleaf, J.F., Chen, S.: Fast shear compounding using robust two-dimensional shear wave speed calculation and multi-directional filtering. Ultrasound. Med. Biol. (2014)
4. Deffieux, T., Gennisson, J.L., Bercoff, J., Tanter, M.: On the effects of reflected waves in transient shear wave elastography. IEEE Trans. Ultrason. Ferroelect. Freq. Control **58**(10), 2032–2035 (2011). https://doi.org/10.1109/TUFFC.2011.2052
5. Evans, A.: Differentiating benign from malignant solid breast masses: value of shear wave elastography according to lesion stiffness combined with greyscale ultrasound according to BI-RADS classification. Br. J. Cancer **107**(2), 224–229 (2012). https://doi.org/10.1038/bjc.2012.253

6. Luong, Q.-H., Nguyen, M.-C., Tran-Duc, T.: Complex shear mod-ulus estimation using extended Kalman filter. Tạp chí khoa học& kỹ thuật s´ô179 (2016)
7. Carlson, L.C., Feltovich, H., Palmeri, M.L., Dahl, J.J., Munoz del Rio, A., Hall, T.J.: Estimation of shear wave speed in the human uterine cervix. Ultrasound Obstet. Gynecol. **43**, 452–458 (2014)
8. Dalong Liu, E.S.E.: Viscoelastic property measurement in thin tissue constructs using ultrasound. IEEE Trans. Ultrason. Ferroelect. Freq. Control. **2008** (2008)
9. Hao, N.T., Thuy-Nga, T., Dinh-Long, V., Duc-Tan, T., Linh-Trung, N.: 2D shear wave imaging using maximum likelihood ensemble filter. In: International Conference on Green and Human Information Technology (ICGHIT 2013) (2013)
10. Walker, W.F., Fernandez, F.J., Negron, L.A.: A method for imaging viscoelastic pa-rameters with acoustic radiation force. Phys. Med. Biol **45**(1437–1447), 2000 (2000)
11. Yuan, H., Guzina, B.B., Chen, S., Kinnickb, R.R., Fatemi, M.: Estimation of the complex shear modulus in tissue-mimicking materials from optical vibrometry measurements. Inv. Probl. Sci. Eng. (2011)
12. Callejas, A., Gomez, A., Faris, I.H., Melchor, J., Rus, G.: Kelvin–Voigt parameters reconstruction of cervical tissue-mimicking phantoms using torsional wave elastography. Sensors. **19**, 3281 (2019). https://doi.org/10.3390/s19153281
13. Huy, T.Q., et al.:2D shear wave imaging in Gaussian noise and reflection media. VNU J. Sci. Math. Phys. **37**(4) (2021)
14. song, S., Le, N.M., Huang, Z.: Quantitative shear-wave optical coherence elastography with a programmable phased array ultrasound as the wave source. Opt. Lett. **40**, 5007 (2015)

An Approach to Extract Information from Academic Transcripts of HUST

Nguyen Quang Hieu[1], Nguyen Le Quy Duong[2], Le Quang Hoa[1(✉)], and Nguyen Quang Dat[3]

[1] School of Applied Math. and Informatics, Hanoi University of Science and technology, Hanoi, Vietnam
hieu.nq185351@sis.hust.edu.vn, hoa.lequang1@hust.edu.vn

[2] School of Information and Communications Technology, Hanoi University of Science and Technology, Hanoi, Vietnam
duong.nlq210242@sis.hust.edu.vn

[3] HUS High School for Gifted Students, Hanoi University of Science, Vietnam National University, Hanoi, Vietnam
nguyenquangdat@hus.edu.vn

Abstract. In many Vietnamese schools, grades are still being inputted into the database manually, which is not only inefficient but also prone to human error. Thus, the automation of this process is highly necessary, which can only be achieved if we can extract information from academic transcripts. In this paper, we test our improved CRNN model in extracting information from 126 transcripts, with 1008 vertical lines, 3859 horizontal lines, and 2139 handwritten test scores. Then, this model is compared to the Baseline model. The results show that our model significantly outperforms the Baseline model with an accuracy of 99.6% in recognizing vertical lines, 100% in recognizing horizontal lines, and 96.11% in recognizing handwritten test scores.

Keywords: Academic transcript · Image Processing · CRNN · CTC · Digit string recognition

1 Introduction

At Hanoi University of Science & Technology, after every exam, scores are recorded in academic transcripts and then transferred to the school's database by teachers. Until now, this process has been done manually, which is time-consuming for the teachers and may lead to accidental mistakes such as the scores inputted being incorrect or the scores being assigned to the wrong students.

Currently, machine learning methods have been applied to automate these processes (see [1–3, 11–17]). It also helps to free up manpower. By utilizing Image-Processing Techniques and Deep Learning, we can automate this procedure with a system that can extract necessary data.

T. D. L. Nguyen et al. (Eds.): ICISN 2023, LNNS 752, pp. 452–460, 2023.
https://doi.org/10.1007/978-981-99-4725-6_55

This paper contains five sections. The introduction is the first section. The second section consists of prior research conducted by various authors on image processing research techniques. The third part is the method studied in this paper, including our proposed method. The fourth part is our results based on real data. And the last part is conclusion and acknowledgment.

2 Related Works

Digit recognition is extremely useful. This is especially the case for schools where the process of inputting grades into database is still being done manually. In such schools, the assistance of digit recognition can significantly increase accuracy and reduce the time allotted to this process.

In 2018, Liu et al. [4] proposed a hybrid CNN-LSTM algorithm for identifying CO2 welding defects. The algorithm is evaluated using 500 images of molten pools from the Robotics and Welding Technology Laboratory at Guilin University of Aerospace Technology. The results from CNN-LSTM algorithm were considered to be better than those of other basic algorithms (CNN, LSTM, CNN-3), with an accuracy of 85%, 88%, and 80% for 32×32, 64×64, and 128×128 images, respectively.

In 2019, Rehman et al. [5] utilized a hybrid CNN-LSTM model to analyze opinions in people's online comments. This model outperforms conventional machine learning techniques on the IMDB movie reviews dataset and the Amazon movie reviews dataset, with a precision of 91%, recall of 86%, an F-measure of 88%, and an accuracy of 91%.

In 2020, Yang et al. [6] compared the performance of CNN-LSTM model with analytical analysis and FEA algorithm in detecting the natural frequency of six types of beams. The goal was to assess the first, second, and third-order frequencies of each beam. Author's model was concluded to be superior in both the test on robustness, with 96.6%, 93.7%, and 95.1% accuracy, respectively, and the test on extrapolability, with 95.4%, 92%, and 92.5% accuracy, respectively.

In 2019, Sivaram et al. [7] proposed a CNN-RNN combination model to detect facial landmarks. The proposed model outperforms existing methods, such as CFAN, SDM, TCDN, and RCPR, on the FaceWarehouse database, with 99% precision, 99% recall, 99% F1-Score, 98.65% Accuracy, and 98.65% AUC or ROC.

In 2018, Xiangyang et al. [8] utilized a hybrid CNN-LSTM model to tackle the problem of text classification. With the Chinese news dataset (proposed by the Sogou Lab), the model proved superior to traditional KNN, SVM, CNN, and LSTM, with a precision of 90.13% under the CBOW model and 90.68% under the Skip-Gram model.

In 2017, Yin et al. [9] used CNN-RNN and C3D hybrid networks to detect emotions in videos from the AFEW 6.0 database. The objective was to assign emotion from 7 categories, namely anger, disgust, fear, happiness, sadness, surprise, and neutral, to each video in the test dataset. It was found that with 1 CNN-RNN model and 3 C3D models, an accuracy of 59.02% was achieved, surpassing the baseline accuracy of 40.47% and last year's highest accuracy of 53.8%.

In 2017, Zhan et el. [10] introduced a new RNN-CTC model to recognize digit sequences in three datasets, namely CVL HDS, ORAND-CAR (include CAR-A and CAR-B), and G-Captcha. Even though the proposed model only achieved a recognition rate of 27.07% for the CVL dataset, the model outperformed other state-of-the-art methods on the CAR-A, CAR-B, and G-Captcha datasets, with recognition rates of 89.75%, 91.14%, and 95.15%, respectively, due to the absence of segmentation.

3 Methodology

3.1 Convolutional Neural Networks

A convolutional neural network can be successfully applied for most computer vision problems. Its characteristics make it more effective than other conventional methods. Since its inception, CNN has witnessed numerous optimizations.

However, when it comes to deeper networks, a degradation problem arises. To tackle this, He et al. proposed a deep residual learning framework, ResNet. The basic structure of this network is using shortcut connections. Shortcut connections are those that traverse multiple layers. With this type of connection, we can overcome the problem of vanishing gradients and construct more extensive networks; in other words, better feature representations can be acquired. In practice, shortcut connections can be adjusted on a case-by-case basis, depending on each specific problem.

In our proposed model, we design a 10-layer residual network that doesn't have any global pooling layers. To prevent divergence, we avoid employing CNN that is excessively profound. In addition, we maximize the use of residual learning to enhance gradient propagation.

3.2 Recurrent Neural Networks

A Recurrent neural network is an architecture where multiple connections between neurons create a recursive cycle. Self-connection brings the advantage of utilizing contextual data when making a correspondence between sequences of input and output. Nevertheless, for conventional RNN, the range of accessible context is restricted in practice. due to the vanishing gradient problem. Applying a memory structure to RNN, which produces a cell that is known as a long short-term memory (LSTM), is one method that can be utilized to solve the problem. It has been demonstrated that the LSTM variant of RNN is capable of resolving some of the issues that are inherently present in conventional RNN, in addition to learning how to resolve issues of long-term dependency. At this time, LSTM has developed into one of the RNNs that is utilized the most frequently.

Regarding the sequence labeling problem, it is helpful to have access to context in both the future and the past. However, the normal LSTM solely considers information from the past and pays no attention to the context of the future. The creation of a second LSTM that processes input in the other direction is one alternate solution. This type of LSTM is known as a bidirectional LSTM, and its abbreviation is Bi-LSTM. Every training sequence is conveyed both forward and backward to two distinct LSTM layers, both of which are connected to the same output layer by the Bi-LSTM algorithm. This structure provides the output layer with full context for all points in the input sequence, both in the past and in the future. This context is supplied throughout the entire time period.

3.3 Handwritten Digit String Recognition with CTC

Sequence characters recognition is a common problem of OCR. In this paper, we proposed an approach to recognize handwritten digit string. After having features extracted by a convolutional neural network, the main idea is to use an output connectionist temporary classification layer to get the final predicted results after using a recurrent neural network to recognize sequence information in images. This is done after the convolutional neural network has been used to extract features.

The input image is one-dimensional tensor (after resizing 40×100x1). For feature extraction, a backbone network is constructed with convolutional, max-pooling layers, and residual network. After every convolution layer, we performed batch normalization to prevent internal covariance shift. Output of feature extraction block are fed as a sequence into labeling block. To avoid vanishing gradient problem, we use two Bi-LSTM layers instead of traditional RNN layer. Finally, a fully connected layer is used to reduce the dimension to 13, before passing CTC layer. The CTC layer serves primarily two functions: the first of these is to decode the output, and the second is to calculate the loss. The full architecture is shown in Fig. 1

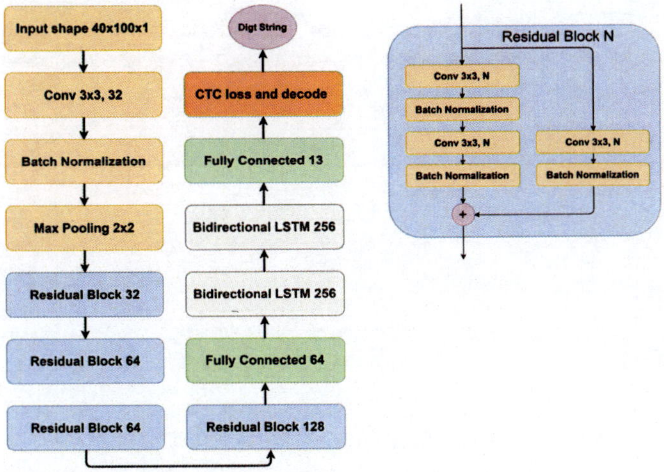

Fig. 1. Our proposed architecture model

3.4 Proposed Method

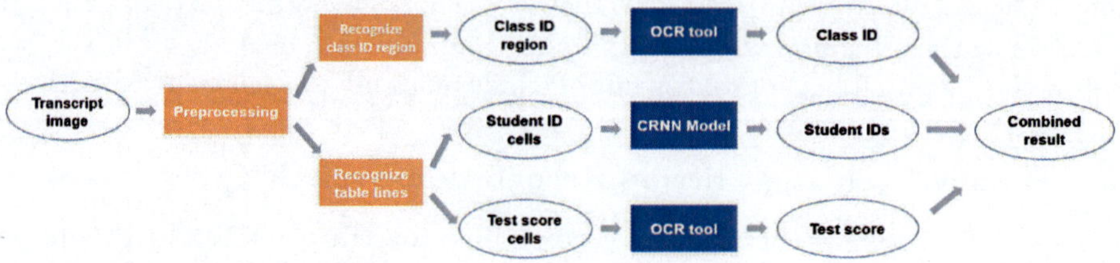

Fig. 2. Our proposed method flowchart

The first step of our method is image preprocessing. Transcript images are binarize, removing noises by Gaussian filter. We deskew the images by Projection profile algorithm. For class ID recognition, we use Template matching followed by OCR tools. To recognize and calculate coordinates of lines in transcript, horizontal and vertical masks generated by morphological operations are fed into Hough transformation. After having full coordinates of lines, cells of student IDs and test scores are cropped. For student IDs, are sequence of printed digits, can easily recognized by available OCR tools. In our method, we use Tesseract-OCR, which is inherently very accurate in recognizing printed characters. For test scores, we use our Handwritten digits recognition model with CTC decode. Finally, student IDs and test scores are combined (Figs. 2 and 3).

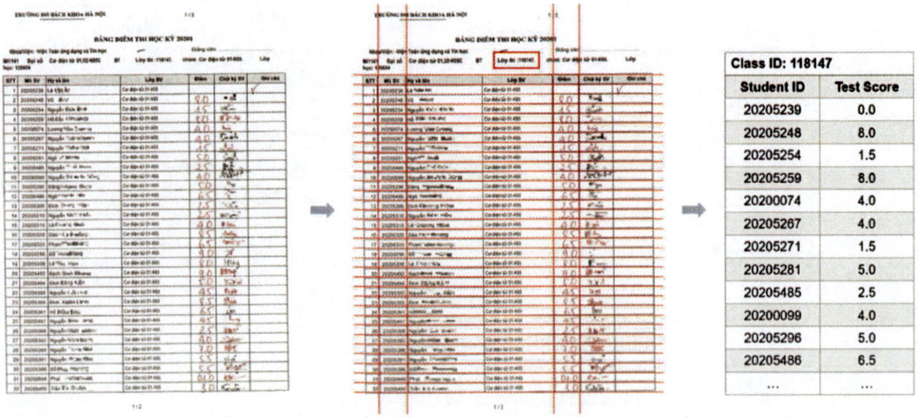

Fig. 3. Result of automatic score-inputting system

4 Experiment and Results

4.1 Evaluation of Image-preprocessing

By using a dataset consisting of images of 126 academic transcripts with 1008 vertical lines and 3859 horizontal lines, the results of the Baseline model and my improved model in detecting lines are shown below (Fig. 4):

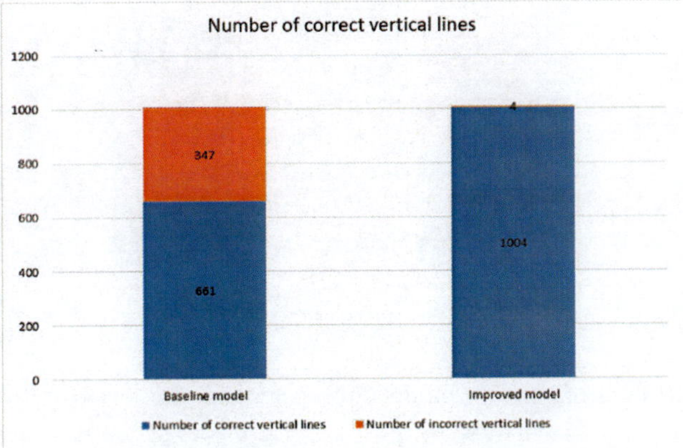

Fig. 4. Results of the two models in detecting vertical lines

The Baseline model achieved an accuracy of 65.57% for vertical lines, whereas the improved model had an accuracy of 99.6% (Fig. 5).

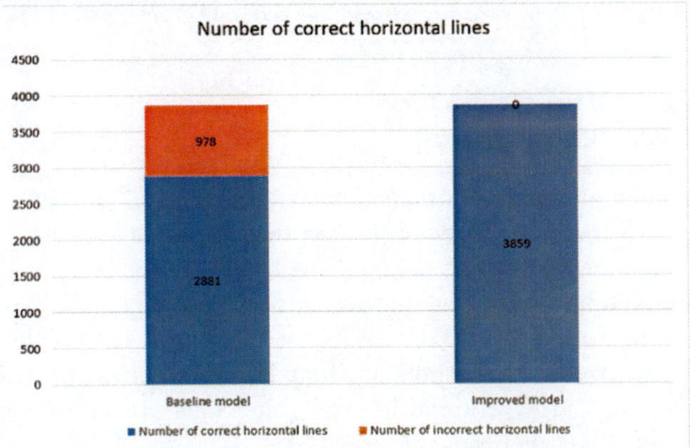

Fig. 5. Results of the two models in detecting horizontal lines

The Baseline model achieved an accuracy of 74.65% for horizontal lines, whereas the improved model had an absolute accuracy of 100%.

4.2 Evaluation of Models in Recognizing Handwritten Test Scores

By using an extracted dataset with 2139 handwritten test scores, the results of the CNN Baseline model and the my CRNN model are shown below (Fig. 6):

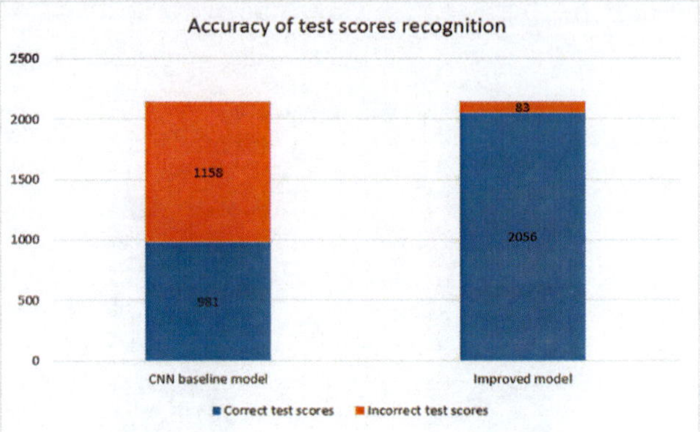

Fig. 6. Results of the two models in recognizing handwritten test scores

The Baseline model achieved an accuracy of approximately 45.86%, whereas my improved CRNN model had an accuracy of 96.11%.

4.3 Evaluation of Automatic Score-Inputting System

To evaluate the accuracy of the entire score-inputting system, we tested it on a dataset of 75 scanned academic transcripts with 162 images of size 1653×2337.

A) Evaluation of Baseline Model
Results of the Baseline model:

- The model was able to accurately detect lines in 92 images and recognize class IDs of 51 images (20 academic transcripts), achieving an accuracy of 31.4%.
- Among 3596 student IDs, the model correctly recognized 2201 IDs, achieving an accuracy of 61.2% (The majority of images in which the lines were accurately detected all had their student IDs recognized by the model).
- Among 3596 test scores, the model was able to accurately recognize 1532 test scores, achieving an accuracy of 42.6%.

B) Evaluation of Improved Model

Results of the improved model:

- In 22 images, the model misidentified 1 vertical line. However, these misidentified lines didn't affect the columns of data that needed to be recognized. Horizontal lines, on the other hand, were all accurately detected. In all 162 images, the model correctly recognized the class IDs with an accuracy of 100%.
- Among 3596 student IDs, the model correctly recognized 3481 IDs, achieving an accuracy of 96.8%.
- Among 3596 test scores, the model was able to accurately recognize 3451 test scores, achieving an accuracy of 95.9%.

5 Conclusion

In this research paper, we have introduced a new approach to the case of handwritten test scores into the computer. When using additional auxiliary features on the printout such as horizontal and vertical lines on the A4 paper, we have achieved very good results in clearly separating handwritten letters and numbers, thereby increasing adds precision to reading handwritten letters and numbers into numeric data.

In the future, we will put more research on some related issues, in order to further increase the accuracy:

- Identify several records of the same person.
- Identify both letters and numbers at the same time (points are written in both numbers and words in the one handwritten paper)

Acknowledgment. We would like to extend our gratitude to the researchers at Hanoi University of Science and Technology (Hanoi, Vietnam), who were of great assistance to us in the process of collecting data and testing the application of the model in real-world scenarios.

References

1. Duc, V.M., Thang, T.N.: Text spotting in Vietnamese documents. In: Anh, N.L., Koh, SJ., Nguyen, T.D.L., Lloret, J., Nguyen, T.T. (eds.) Intelligent Systems and Networks. LNNS, pp. 141–150, vol. 471. Springer, Singapore (2022). https://doi.org/10.1007/978-981-19-3394-3_17
2. Anh, H.T., Tuan, T.A., Long, H.P., Ha, L.H., Thang, T.N.: Multi deep learning model for building footprint extraction from high resolution remote sensing image. Multi Deep Learning Model for Building Footprint Extraction from High Resolution Remote Sensing Image. In: Anh, N.L., Koh, SJ., Nguyen, T.D.L., Lloret, J., Nguyen, T.T. (eds.) Intelligent Systems and Networks. LNNS, pp. 246–252, vol. 471. Springer, Singapore (2022). https://doi.org/10.1007/978-981-19-3394-3_29
3. Long, H.P., Dung, D.L., Anh, T.T., Thang, T.N.: Improving Pareto Front Learning via Multi-Sample Hypernetworks, https://doi.org/10.48550/arXiv.2212.01130

4. Liu, T., Bao, J., Wang, J., Zhang, Y.: A Hybrid CNN-LSTM algorithm for online defect recognition of CO2 welding. Sensors **18**(12), 4369 (2018). https://doi.org/10.3390/s18124369
5. Rehman, A.U., Malik, A.K., Raza, B., Ali, W.: A hybrid CNN-LSTM model for improving accuracy of movie reviews sentiment analysis. Multimed. Tools Appl. **78**(18), 26597–26613 (2019). https://doi.org/10.1007/s11042-019-07788-7
6. Yang, R., Singh, S.K., Tavakkoli, M., Amiri, N., Yang, Y., Karami, M.A., Rai, R.: CNN-LSTM deep learning architecture for computer vision-based modal frequency detection. Mech. Syst. Signal Process. **144**, 106885 (2020). https://doi.org/10.1016/j.ymssp.2020.106885
7. Sivaram, M., Porkodi, V., Mohammed, A.S., Manikandan, V.: Detection of accurate facial detection using hybrid deep convolutional recurrent neural network, Lebanese French University, Iraqi Kurdistan. ICTACT J. Soft Comput. **9**(2), 1844–1850 (2019). https://doi.org/10.21917/ijsc.2019.0256
8. She, X. and Zhang, D.: Text classification based on Hybrid CNN-LSTM hybrid model. In: 2018 11th International Symposium on Computational Intelligence and Design (ISCID), pp. 185–189, (2018). https://doi.org/10.1109/ISCID.2018.10144
9. Fan, Y., Lu, X., Li, D., Liu, Y.: Video-based emotion recognition using CNN-RNN and C3D hybrid networks, Conference: International Conference on Multimodal (2017). https://doi.org/10.1145/2993148.2997632
10. Zhan, H., Wang, Q., Lu, Y.: Handwritten digit string recognition by combination of residual network and RNN-CTC. In: Liu, D., Xie, S., Li, Y., Zhao, D., El-Alfy, E.S. (eds.) Neural Information Processing. ICONIP 2017. LNCS, vol. 10639, pp. 583–591. Springer, Cham (2017). https://doi.org/10.1007/978-3-319-70136-3_62
11. Williams, Ronald, J.; Hinton, Geoffrey, E.; Rumelhart, David, E: Learning representations by back-propagating errors. Nature. **323**(6088), 533–536, (1986). https://doi.org/10.1038/323533a0
12. Long, C.K., Trung, H.Q., Thang, T.N., Dong, N.T., Hai, P.V.: A knowledge graph approach for the detection of digital human profiles in big data, J. Sci. Technol. Issue Inf. Commun. Technol. **19** (2021). https://doi.org/10.31130/ict-ud.2021.118
13. Tung, N.X., Dat, N.Q., Thang, T.N., Solanki, V.K., Anh, N.T.N: Analysis of temperature-sensitive on short-term electricity load forecasting. In: 2020 IEEE-HYDCON, 20132699 (2020). https://doi.org/10.1109/HYDCON48903.2020.9242910
14. Hai, P.V., Hung, N.Q., et al.: A proposal model using deep learning model integrated with knowledge graph for monitoring human behavior in forest protection. TELKOMNIKA Telecommun. Comput. Electron. Control. **20**(6), 1276–1287 (2022). http://doi.org/10.12928/telkomnika.v20i6.24087
15. Hai, P.V., Thanh, D.H., Moore, P.: Hierarchical pooling in graph neural net-works to enhance classification performance in large datasets. Sensors. **21**, 6070 (2021). https://doi.org/10.3390/s21186070
16. Long, C.K., Hai, P.V., et al.: A novel fuzzy knowledge graph pairs approach in decision making. Multimed. Tools Appl. **81**, 26505–26534 (2022). https://doi.org/10.1007/s11042-022-13067-9
17. Nguyen, T., et al.: A Probabilistic framework for pruning transformers via a finite admixture of keys. In: The 2023 IEEE International Conference on Acoustics, Speech and Signal Processing (ICASSP 2023). https://doi.org/10.48550/arXiv.2110.08678

Framework of Infotainment Services for Public Vehicles

Hye-Been Nam, Joong-Hwa Jung, Dong-Kyu Choi, and Seok-Joo Koh(✉)

School of Computer Science and Engineering, Kyungpook National University, 80 Daehakro, Bukgu, Daegu 41566, Republic of Korea
sjkoh@knu.ac.kr

Abstract. The existing works on in-vehicle infotainment (IVI) services have so far focused on only private vehicles, such as car. This paper proposes a framework of infotainment services for public vehicles (PVIS), such as bus or train. In the proposed PVIS scheme, an agent is employed to support a large number of users. From testbed experimentation, we see that the proposed PVIS scheme can provide scalable services for large number of users in public vehicle.

Keywords: Infotainment Services · Public Vehicles · PVIS · Framework

1 Introduction

The industries on infotainment services for vehicles, as known as in-vehicle infotainment (IVI) services, have been growing rapidly [1]. It is envisioned that a variety of infotainment (or multimedia) devices and services will be newly developed for personal and public vehicles in the future. Such devices include navigations, cameras, speakers, headrest displays, air-conditioners, thermometers and heated seats, and lights [2, 3].

The IEC TC100 have so far developed a set of standards on Configurable Car infotainment Services (CCIS) [4]. The CCIS standards have been basically designed for personal users, such as car owner. In the meantime, there is also a crucial need to provide a variety of infotainment services for public vehicles (PVIS) such as bus or train.

PVIS services have different requirements and features from CCIS services. For user type, CCIS is targeted for one or two users (such as car owner), whereas PVIS is for a large number of guests or passengers within the public vehicle. For device type, CCIS deals with the personal devices (property or belonging) in the car, whereas PVIS will be targeted for a variety of public devices that are contained in a public vehicle.

This paper proposes a framework of PVIS. It is noted that there are many different features and requirements between CCIS and PVIS. Accordingly, we need to study how to effectively provide a variety of PVIS services. In particular, PVIS needs to employ a set of agents to effectively manage a large number of users or devices.

This paper is organized as follows. Section 2 compares the differences between private and public vehicles. Section 3 proposes a framework for PVIS. Section 4 discusses preliminary experimentation results for proposed PVIS scheme. Section 5 concludes this paper with future works.

T. D. L. Nguyen et al. (Eds.): ICISN 2023, LNNS 752, pp. 461–469, 2023.
https://doi.org/10.1007/978-981-99-4725-6_56

2 Infotainment Services for Private and Public Vehicles

Table 1 shows that the comparison of private vehicles and public vehicles.

Table 1. Comparison of infotainement services for private vehicles and public vehicles

Category	Private Vehicle (CCIS)	Public Vehicle (PVIS)
Users	private users (owner)	public users (passengers)
Examples	car, van	bus, train, tram, subway
Number of users	(usually) less than 20	20 (bus) ~ 1,000 (train)
Device type	personal devices (properties, belongings)	public devices (shared by many users)
Service duration	(usually) Long-Term	(usually) Short-Term
Security/Privacy	Moderate	Crucial

Public vehicles for PVIS (e.g., bus, train) have different requirements and features from private vehicles (e.g., car, van). A private vehicle is usually for a small number of users, whereas a public vehicle is targeted for a large number of guests or passengers. For device type, a private vehicle deals with the personal devices (property or belonging), whereas a public vehicle is targeted for a variety of public devices that can be shared by many users in the public vehicle. Some public services can be provisioned by interworking with the external networks, as shown in bus information service.

For service duration, the private vehicle usually provides long-term services, whereas the public vehicle tends to provide short-term services during which guests stay within the public vehicle. Security or privacy requirements are also strictly applied to the public vehicle, compared to the private vehicle.

3 PVIS Framework

3.1 Functional Entities

The PVIS functional entities can be classified into the five types: content provider, PVIS master, PVIS agent, PVIS device, and passenger device.

Content provider represents an external server or entity to provide multimedia infotainment services for PVIS passengers, such as multimedia on-line game or over-the-top (OTT) services. Content providers may deploy their contents as PVIS services with an appropriate negotiation. Such contents include a variety of applications, such as game, utility program, or media files for OTT service. For this purpose, a content provider may give an interworking function with the PVIS system for enhanced PVIS service provisioning.

PVIS master performs overall management and control for PVIS system and services. In initialization, PVIS master needs to identify a set of PVIS functional entities within the public vehicle, such as PVIS agents, PVIS devices and passenger devices. For service duration, PVIS master monitors these PVIS functional entities. PVIS master is also responsible for contents delivery from content provider to many passengers in the public vehicle.

PVIS agent is additionally employed for large-scale public vehicles, such as train, to provide scalable and effective PVIS services between PVIS master and a large number of passengers. It is expected that a PVIS agent is employed for each carriage in a large-scale public vehicle. The PVIS agent is responsible for management of PVIS devices in its carriage. PVIS agent is also responsible for service provisioning to PVIS passengers who are stating in its carriage. For this purpose, PVIS agent may temporally store multimedia contents for passengers during communication between PVIS master and passengers.

PVIS device represents a device that is attached and dedicated to the public vehicle, such as air condition, speaker, display, lights, sensors. The PVIS devices are used for a variety of PVIS services. Each PVIS device needs to be controlled and managed by PVIS master or agent. PVIS device supports the interaction of users with the PVIS agent or PVIS master.

Passenger device is a user device for PVIS services, such as smart phone. That is, a passenger uses PVIS services via the passenger device. By using such a passenger device, a passenger can request PVIS services to the PVIS device, PVIS agent, and further to PVIS master.

3.2 System Configuration by Network Environment

For description of PVIS services, the two types of public vehicles are considered: small-scale public vehicle (e.g., bus) and large-scale public vehicle (e.g., train).

It is expected that the communications between functional entities within a public vehicle (e. g., PVIS master, PVIS device, passenger device) are performed by using wireless personal area network (WPAN) technology, such as Bluetooth or Wi-Fi. In the meantime, the communications between a content provider outside the public vehicle and the PVIS master within the public vehicle can be done by using a mobile communication technology, such as 5G or 6G.

Figure 1 shows a system configuration in small-scale public vehicle environment, such as a bus. The small-scale public vehicle consists of a PVIS master and a set of PVIS devices and passenger devices. For device management, all PVIS devices and passenger devices need to be registered with the PVIS master. Passengers can control PVIS devices and enjoy PVIS service through the associated passenger devices.

Figure 2 shows a system configuration in large-scale public vehicle environment such as train. In this case, a PVIS agent is employed for each carriage in the public vehicle, additionally to the case of small-scale public vehicle environment. For PVIS service provisioning, all PVIS agents need to be registered with PVIS master, and all PVIS devices and passenger devices within the vehicle need to be registered with PVIS master or PVIS agents.

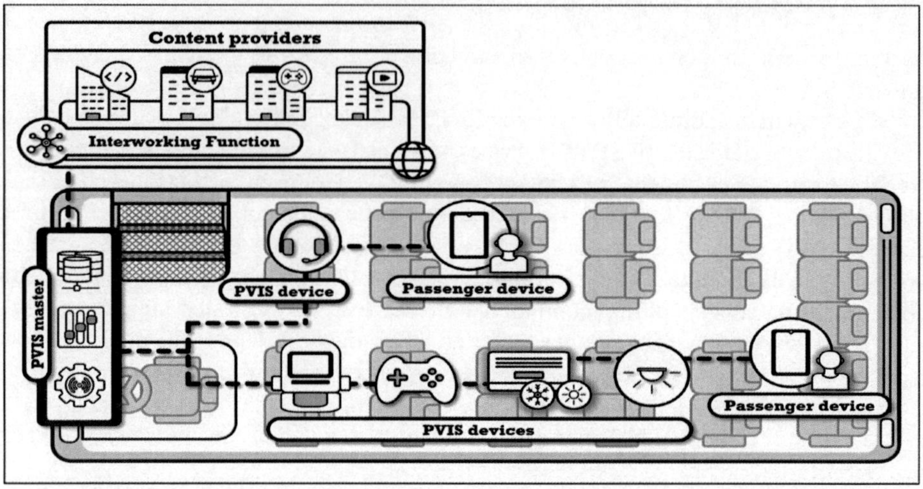

Fig. 1. System configuration for small-scale PVIS environment (e.g., bus)

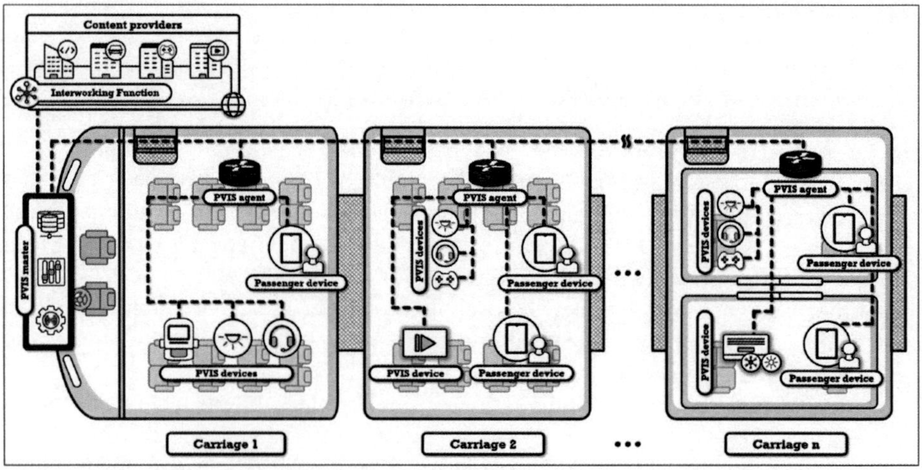

Fig. 2. System configuration for large-scale PVIS environment (e.g., train)

3.3 PVIS Operations in Small-Scale Public Vehicle

In small-scale public vehicle, each passenger uses device management services through PVIS master. Figure 3 shows the operation flows for PVIS service in small-scale public vehicles. For service provisioning, all PVIS devices are registered with PVIS master.

When a passenger is connected to the vehicle, its passenger device performs the registration process to retrieve the access token for PVIS system (step 1). In this process, OAuth (Open Authorization) which is an open standard and a protocol that allows users to grant access to their resources can be employed [5]. With the access token the passenger can query the available PVIS device list from PVIS master (step 2 and 3). Based on the list, the passenger device can request the PVIS device control to PVIS master (step 4). PVIS master will verify the request with the access token in the request message and it activates the corresponding PVIS device (step 5). The activated PVIS device sends the own status to the PVIS master (step 6), and then PVIS master will notify the status change on the PVIS device (step 7).

After the device is ready to be controlled, the passenger can play the content through the prepared PVIS device. Passenger can request the list of deployed content to PVIS master with the access token retrieved on registration (step 8). The PVIS master responds with the contents list deployed for system provisioning (step 9). Passenger selects the content and sends PVIS master the request message so as to play the content with the prepared PVIS device information (step 10). If the PVIS master has only content metadata, the PVIS master will download the content from the content provider specified in the metadata (step 11) and deliver them to PVIS device (step 12).

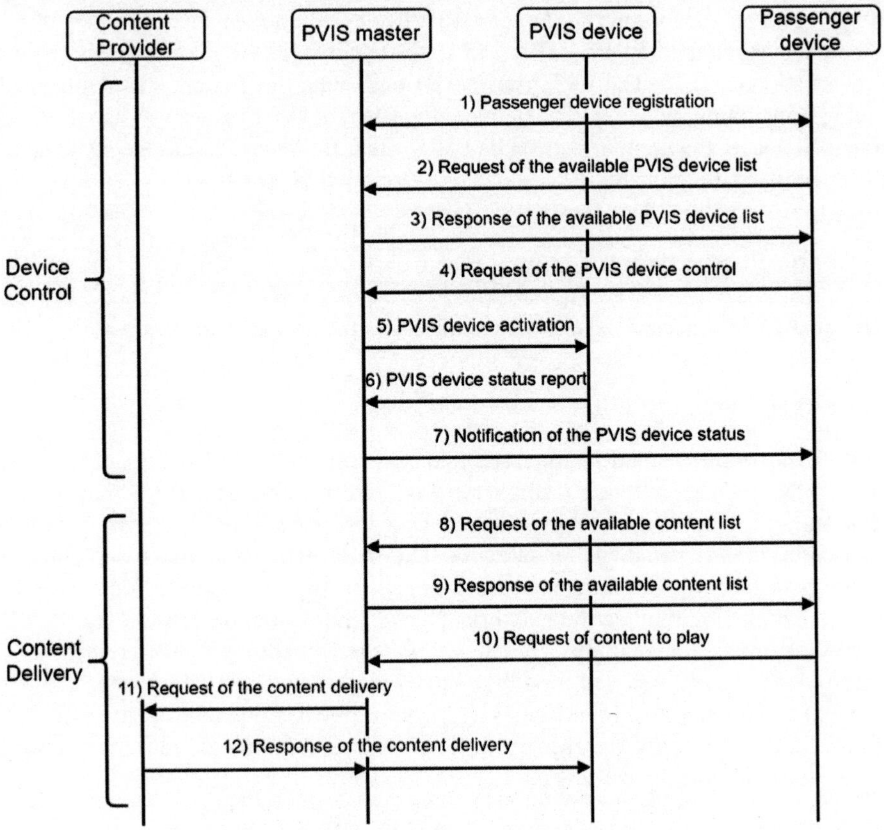

Fig. 3. PVIS operations for small-scale public vehicle

3.4 PVIS Operations in Large-Scale Public Vehicle

In large-scale public vehicle, a PVIS agent is employed for each carriage in the public vehicle. Figure 4 shows the operation flows for content delivery through a PVIS agent. The content delivery is performed by the PVIS agent. For service provisioning, all PVIS agents are registered with PVIS master. In addition, all PVIS devices and passenger devices are also registered with PVIS master via the associated PVIS agent. For content delivery service, a passenger device first obtains the list of multimedia contents available in the public vehicle from PVIS master via PVIS agent. Based on the available content list, the passenger will request the content to PVIS agent. Then, the PVIS agent will try to download the content from PVIS master or the content provider.

As we can see in the figure, the operation flows for large-scale public vehicle are almost the same with those for small-scale public vehicle. In this case, the PVIS agent will take a role of PVIS master for management of the devices and users within the associated carriage (step 1). After the device is ready to be controlled, the passenger can play the content through the prepared PVIS device. Passenger can request the list of deployed content to PVIS master via PVIS agent with the access token retrieved on registration (step 8). The PVIS master responds with the contents list deployed for system provisioning, which will be delivered to PVIS device by the PVIS agent (step 9). Passenger selects the content and sends PVIS agent the request message so as to play the content with the prepared PVIS device information (step 10).

4 Tetsbed Experimentation

In this section, we discuss the experimental results for performance evaluation.

4.1 Testbed Configuration

Figure 5 shows the testbed configuration for experimentation of the proposed agent-based PVIS scheme, in which *Raspberry pi* is used to implement PVIS master, PVIS agent, and *mock* passenger device. The *Linksys Velop Homekit* Router is used to set up a secure wireless personal area network. They communicate with the content server running a web server implemented using *gin-gonic* [6]. Communication utilizes the SKT-5G mobile communication networks. To facilitate communication over the SKT-5G network, the system employs a *phone-as-modem* methodology, called tethering. The access point (AP) is employed for a high-speed mesh network. PVIS master and PVIS agent will use ethernet for connection with AP, while the passenger devices use the Wi-Fi communication. The mock passenger devices are used to simulate multiple passenger devices through multiple processes.

4.2 Experimentation Results

Figure 6 shows a preliminary experimentation result to compare the existing centralized In-Vehicle Infotainment (C-IVI) scheme without PVIS agents and the proposed agent-based PVIS scheme (A-IVI), in which the master load means the volume of data sent

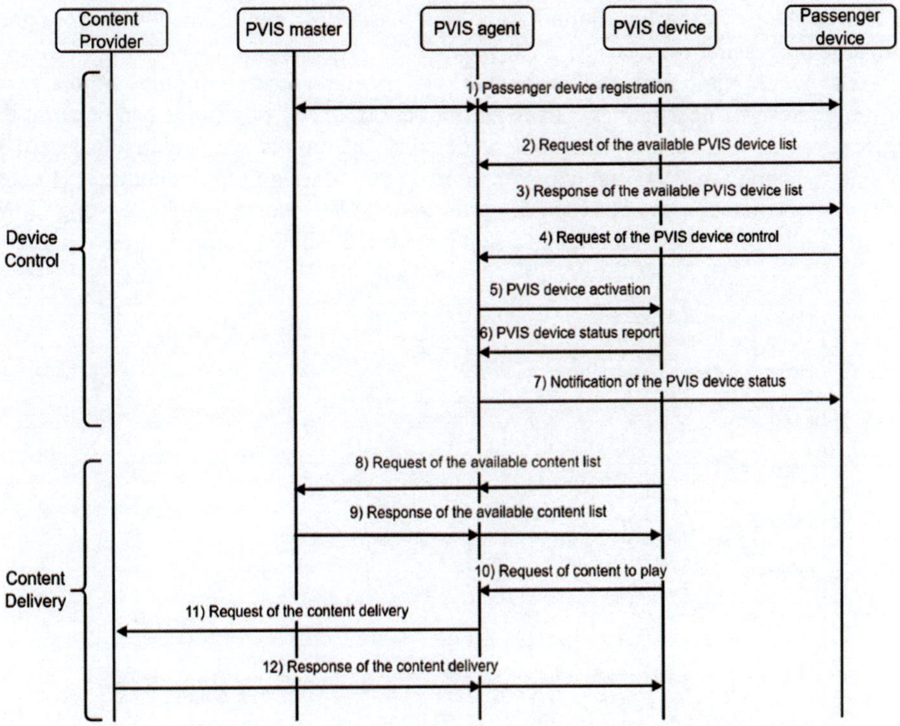

Fig. 4. PVIS operations for large-scale public vehicle

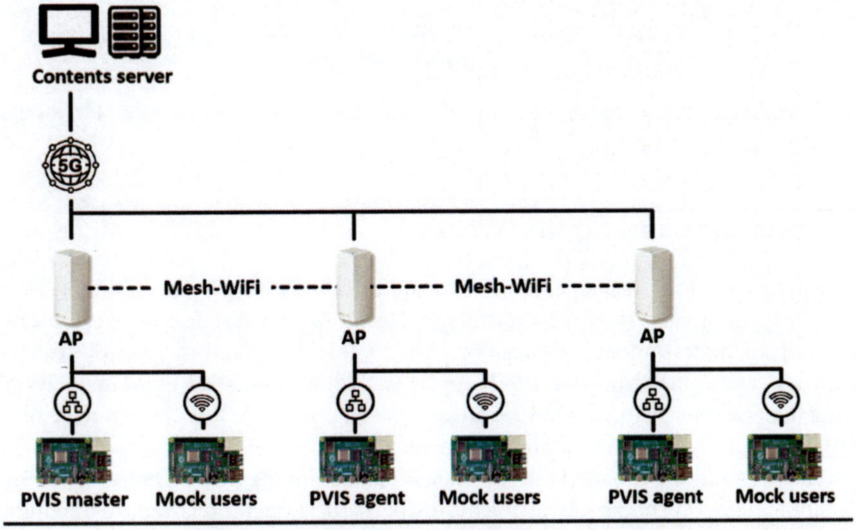

Fig. 5. Testbed configuration for agent-based PVIS scheme

to or from the PVIS master. In the figure, high master load represents low performance and large processing time.

As shown in the figure, as the number of users increases, the master load increases for the two candidate schemes. However, the proposed A-IVI scheme can perform the registration, device control and content delivery operations with relatively less data volume by using the PVIS agents, compared to the existing C-IVI scheme. It is noted that the performance gaps between the proposed A-IVI scheme and the existing C-IVI scheme get larger, as the numbers of users increase, as shown in the figures.

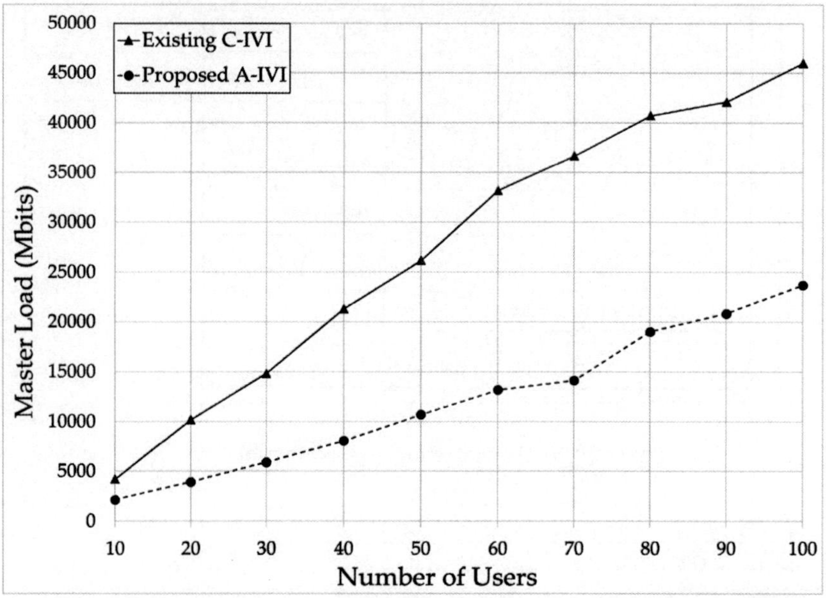

Fig. 6. Master loads for existing centralized scheme and proposed agent-based PVIS scheme

5 Conclusions and Future Works

In this paper, we took a deeper look into the differences between public and private vehicle infotainment service environments. Based on the discussion, we presented a framework of infotainment services for public vehicles, which include the functional entities and operation flows for PVIS services. From the testbed experimentation, we see that the proposed PVIS scheme with agents can provide better performance than the existing centralized scheme without any agent.

In the future, more detailed requirements for functional entities need to be identified in the implementation perspective, and the testbed experimentation needs to be conducted for validation of the PVIS services.

Acknowledgement. This work was supported by the Research and Development Programs of Korean Government (NRF2021R1I1A3057509). This work was also supported by the Technology

Innovation Program (20015107) funded By the Ministry of Trade, Industry & Energy (MOTIE, Korea).

References

1. Macario, G., Torchiano, M., Violante, M.: An in-vehicle infotainment software architecture based on google android. In: 2009 IEEE International Symposium on Industrial Embedded Systems, pp. 257–260 (2009)
2. Coppoal, R., Morisio, M.: Connected car: technologies, issues, future trends. ACM Comput. Surv. **49**, 1–36 (2016)
3. Choi, D.-K., et al.: IoT-based resource control for in-vehicle infotainment services: design and experimentation. Sensors **19**, 620–638 (2019)
4. IEC 63246–1: Configurable Car Infotainment Services (CCIS) – Part 1: General. IEC TC100. Published (2021)
5. Hardt, D.: The OAuth 2.0 Authorization Framework. Internet Engineering Task Force, Request for Comments RFC 6749 (2012). https://doi.org/10.17487/RFC6749
6. Gin Web Framework: Gin Web Framework. https://gin-gonic.com/. Accessed 19 Feb 2023

Check for updates

The Integration of Global Navigation Satellite System Kinematic Positioning and Inertial Measurement Unit for Highly Dynamic Surveying and Mapping Applications

Thi Dieu Linh Nguyen[1](✉), Trung Tan Nguyen[2], Xuan Thuc Kieu[1], Manh Kha Hoang[1], and Quang Bach Tran[3]

[1] Hanoi University of Industry, Hanoi, Vietnam
nguyen.linh@haui.edu.vn
[2] Le Quy Don University, Hanoi, Vietnam
[3] University of Economics - Technology for Industries, Hanoi, Vietnam

Abstract. Global Navigation Satellite System with Realtime Kinematic Positioning (GNSS RTK) is now widely applied for land-based surveying to provide positioning solution at centimeter level. However, for highly dynamic surveying and mapping applications such as UAV photogrammetry, hydrographical surveying and mobile mapping that require a high frequency and continuous navigation solution, GNSS RTK only is insufficient. To overcome this issue, we propose a system that integrates GNSS RTK and Inertial measurement Unit (IMU) to provide navigation solution including position, velocity, and attitude. For this scheme, Extended Kalman Filter is used for data fusion. The conducted field test indicated that the proposed system and solution is enabled to provide navigation solution of frequency up to 50Hz with positional accuracy of centimeter in open sky view and decimeter in GNSS hostile environment.

Keywords: GPS · GNSS RTK · IMU · Kalman Filter · Integration

1 Introduction

For collecting geo-spatial data, mobile mapping system (MMS) is widely used today. Various types of MMSs have been developed for different applications; land-based MMSs use land vehicles to carry the MMS [2]. Airborne MMSs refers to MMSs whose mapping sensors are mounted on airplanes or Unmanned Aviation Vehicles (UAVs) [1]. With the development of direct geo-referencing (DG) systems, the transformation parameters in an MMS can now be directly determined by using a DG system. DG is time-variable position and orientation parameters for a mobile mapping system in a certain reference frame [2]. Current technologies for this purpose can be satellite positioning with Global Positioning System (GPS) and the Inertial Navigation System (INS) using an Inertial Measuring Unit (IMU). Even though either GPS or IMU could determine both position and orientation in principle, they are usually integrated in such a way that INS and the

T. D. L. Nguyen et al. (Eds.): ICISN 2023, LNNS 752, pp. 470–477, 2023.
https://doi.org/10.1007/978-981-99-4725-6_57

GNSS receiver are the major orientation and position sensors, respectively [3]. Tuan Li [5] proposed a tightly-coupled integration of multiple GNSS single frequency RTK and MEMS-IMU to increase the positioning performance. The positioning accuracy can be archived at centimeter level in this scheme. However, the tightly coupled with GNSS signal intervention is not easy for certain equipment. Sudha [6] introduced a dual-frequency PPP GNSS and MEMS-IMU integration for continuous navigation in obstructed environments. The horizontal accuracy of this scheme can reach to centimeter level. This disadvantage of this strategy is that it is only suitable for post-processing application.

Considering the high cost and large size, high quality IMUs are restricted to commercial MMSs. The Micro-Electro-Mechanical System (MEMS) IMU with its small size, low cost, and low power consumption is now widely applied, particularly for UAV and portable MMSs where the payload is limited. Nevertheless, their performance is still poor for specific applications. Park and Gao [4] demonstrated that the performance of MEMS-based inertial sensors is not yet quite acceptable for land vehicle application for longer periods of GPS signal outrage. In the 10-s GPS position measurements updates of loosely coupled schemes, the maximum position error is about 60 m. However, for the large-scale mapping and precise navigation, decimeter to centimeter level of accuracy is required.

The aim of this study is to design an integrated scheme that uses GNSS RTK technology and the low-cost MEMS IMU with embedded Extended Kalman Filter for real-time solution.

2 Integration Strategies

2.1 System Design

The proposal of the integrated IMU and GNSS scheme is illustrated in Fig. 1. In order to to attain the navigation solutions such as the position, velocity and attitude in the navigation frame, the INS mechanization is used to process outputs of IMU, angular rates and specific forces are sensed by gyroscopes and accelerometers, respetively. The pseudo-range and carrier phase, the raw measurements from the Rover and Base GNSS receivers are pre-processed in the ambiguity resolution to achieve high accuracy positions. EKF plays the role of an estimator for the data fusion to obtain estimated navigation solutions.

2.2 System Model

The system and measurement models are required for data fusion by the usage of an estimator such as EKF. Thanks to the high sampling rate and seamless output, INS is utilized to design the system model. In order to form the measurement model, GPS measurements and other external aids are applied.

The scheme of proposed system is the discrete form of the INS mechanization whose state vector given by:

$$x_{21\times 1} = [r^l v^l r_b^l b_g b_a s_g s_a]^T \quad (1)$$

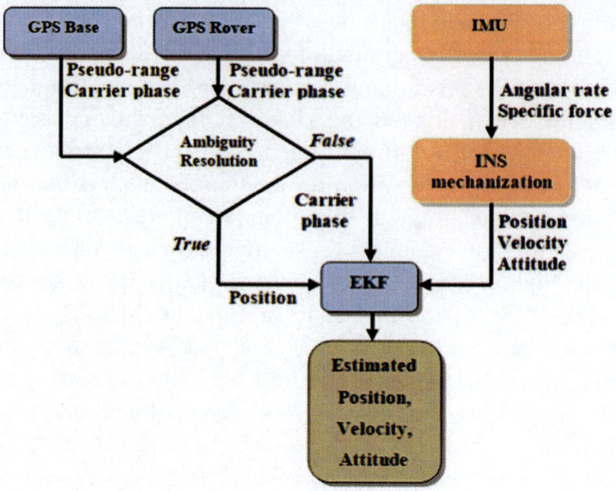

Fig. 1. The integrated scheme.

where r^l, v^l, and r_b^l represent the position, velocity, and attitude of the system in the local level frame; b_g, b_a, s_g, and s_a are the biases and scale factors of the gyroscopes and accelerometers, respectively.

For the system model is designed following Huang and Chiang (2009) as below:

$$\dot{x} = Fx + Gu \tag{2}$$

where

$$x = \begin{bmatrix} \delta r^c \\ \delta v^c \\ \psi \end{bmatrix}; F = \begin{bmatrix} F_{11} & F_{12} & 0 \\ F_{21} & F_{22} & F_{23} \\ 0 & 0 & F_{33} \end{bmatrix}; G = \begin{bmatrix} 0 & 0 \\ C_b^n & 0 \\ 0 & C_b^n \end{bmatrix}; u = \begin{bmatrix} \delta f^b \\ \delta \omega_{ib}^b \end{bmatrix} \tag{3}$$

The measurement parameters are position and velocity that are provided by the GPS receiver. For EKF, the measurement model is given by:

$$z = \begin{bmatrix} r_{INS}^e - r_{GPS}^e \\ v_I^e - v_{GPS}^e \end{bmatrix} = \begin{bmatrix} H_r & 0 \\ 0 & H_v \end{bmatrix} \begin{bmatrix} \delta r^e \\ \delta v^e \end{bmatrix} + \begin{bmatrix} \varepsilon_r \\ \varepsilon_v \end{bmatrix} \tag{4}$$

where $H_r = H_v = I_{3\times 3}$ denotes the mapping matrice, δr^e denotes the position error vector, δv^e is the velocity vector defined in ECEF (or e-frame). Let ε_r and ε_v represent the position and velocity noise, respectively. Likely to the noise of the system, the measured noise vector $\varepsilon = [\varepsilon_r, \varepsilon_v]^T$ can not be estimated directly, it can be accounted in the EKF by a representation of the measurement noise model R, where R is modelizated based on the GPS position and uncertain velocity.

After some transformations, the measured model can be represented in a required form of EKF as follows:

$$z_k = H_k x_k + \varepsilon_k \tag{5}$$

Since the system and measurement model are successfully established, the extended Kalman Filter is applied for estimation to provide navigation solution including position, velocity and attitude of the system. The procedure of the EKF can be found in the Huang and Chiang [3].

2.3 Hardware Design

The hardware design of the GNSS RTK/IMU system consists of five main components including Microcontroller Unit (MCU), GNSS module, IMU, communication port and power supply module. Microcontroller unit utilizes ARM-32 bit LPC1768FBD100 to receive the command from user, the signal and data from sensor such as GNSS receiver and IMU, synchronization and process the integration data. GNSS module is the dual frequency, multi-channel GNSS signal Ublox Neo-F9P module. IMU is the 6-degree freedom Xsens-MTi-3. The system uses 3–5 VDC-2A power supply. The diagram of the designed system is shown in Fig. 2, main board and the enclosure are depicted in Fig. 3.

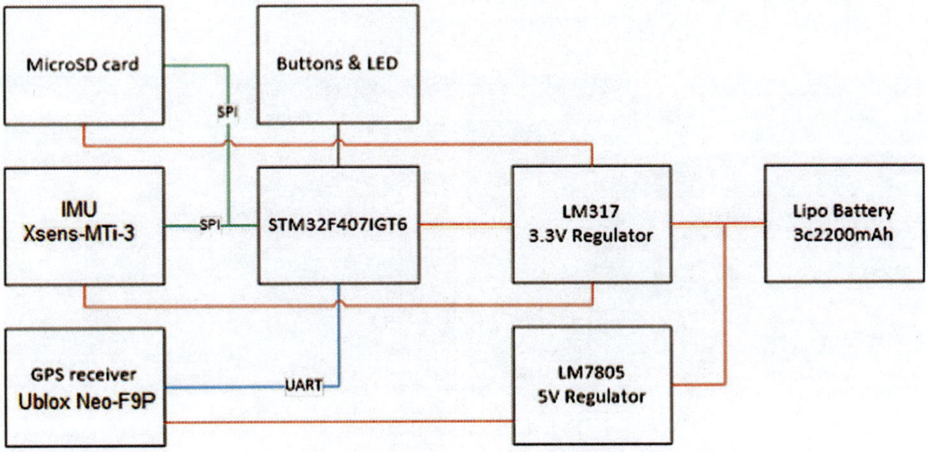

Fig. 2. Diagram of the system design.

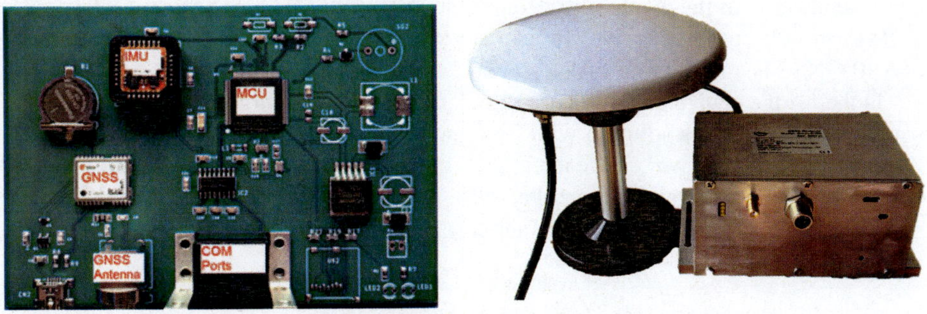

Fig. 3. Main board and the enclosure of the system.

3 Experiment and Discussion

For the test, three systems are utilized to investigate the proposed method. The first one is the dual frequency RTK GNSS receiver, Leica viva GS16 that connect with VNGEONET CORS network for getting RTK fixed solution. The second system is the single frequency GNSS receiver, Ublox Neo-M6T EVK with single point positioning solution (SSP). The third system is the integration of the GNSS RTK receiver, Ublox NEO-M8P and Xsens-MTi-3. Three systems are setup on a car for data collection (Fig. 4). To evaluate accuracy, ground control points with geodetic surveying are built along the testing trajectory for reference. The coordinate of the control points is measured using total station, connect with the third order geodetic national control network, the accuracy is guaranteed at level of millimeter.

The set of testing data is collected in two environment scenarios that are the open sky view and the bridge areas in Hanoi, Vietnam. The testing trajectory is shown in Fig. 5. For visual analysis, the enlargement of two typical scenarios including the open sky view and bridge view as illustrated in Fig. 6 and Fig. 7. The numerical analysis is given in Table 1 and Table 2.

Fig. 4. Testing platform.

The test results indicated that in the open sky, all three systems can provide continuosly solution with the availability from 95 to 99%. Both RTK GNSS and RTK GNSS + IMU provide centimeter level of accuracy with the RTK fixed solution. The accuracy of SSP GNSS is about 2.4 m for mean and 1.5 m for standard naviation.

In the bridge environment, most of time, GNSS RTK did not provide the RTK fixed solution. SPP provides solution with 72%, however, its accuracy is low, about 5 m for mean and maximum error can reach to 20.6 m. In contrast, with the integration of RTK GNSS and IMU, the availability still at 99%. Although it can not provide the RTK fixed solution, the positioning accuracy is about 0.2 m for mean and 0.34 m for the standard deviation.

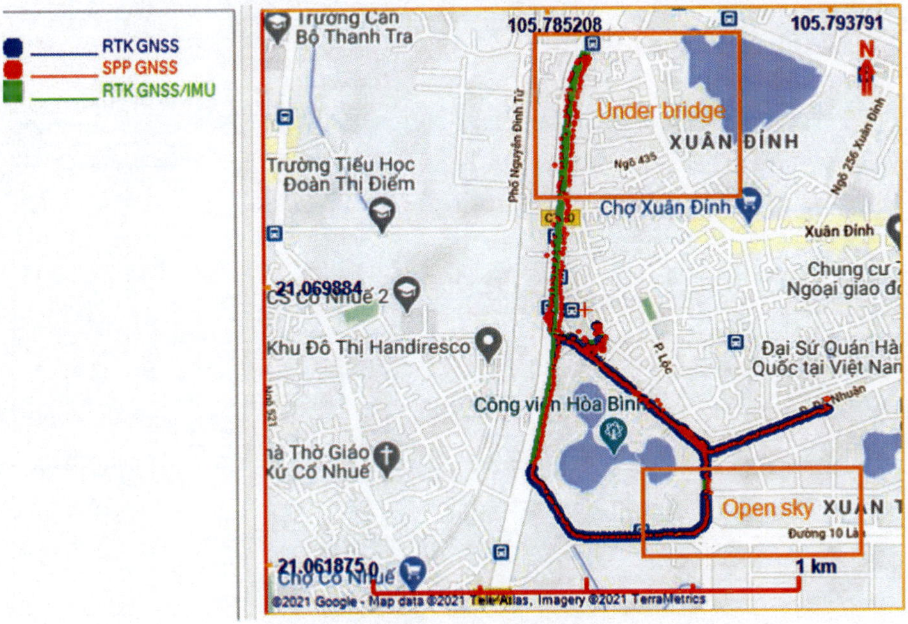

Fig. 5. Testing trajectory.

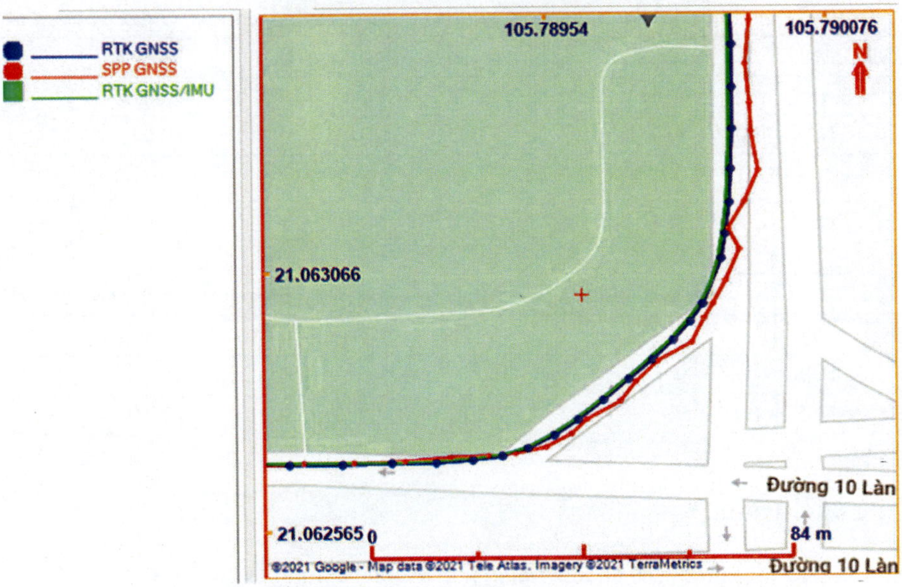

Fig. 6. Enlargement of scenario open sky view area

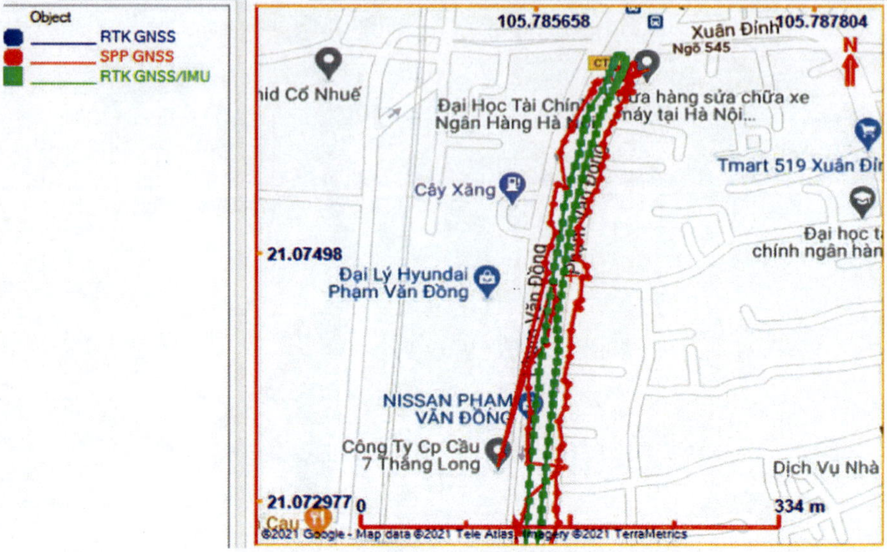

Fig. 7. Enlargement of under bridge area

Table 1. Numerial result in the open sky view area

	Availability (%)	Min(m)	Max(m)	Mean(m)	Std. Deviation(m)
SSP GNSS	99	0.450	9.610	2.400	1.560
RTK GNSS	95	0.002	0.720	0.030	0.026
RTK GNSS + IMU	99	0.003	0.810	0.040	0.035

Table 2. Numerial result in the bridge area

	Availability (%)	Min(m)	Max(m)	Mean(m)	Std. Deviation(m)
SSP GNSS	72	0.120	20.600	4.650	5.560
RTK GNSS	0	–	–	–	–
RTK GNSS + IMU	99	0.015	1.530	0.240	0.340

4 Conclusions

This research has proposed an integrated scheme that combines GNSS RTK and IMU. A compact board is designed with embedded Extended Kalman Filter in MCU for real-time solution output.

A field test with two environmental scenarios were implemented to investigate the performance of the proposed method compared to the GNSS SSP and GNSS RTK solution.

The results from test indicated that the GNSS RTK/IMU integration achieves the outperformed availability compared to GNSS RTK and is much better than SSP GNSS in the accuracy. In addition, the proposed scheme not only provides the position, but also the attitude including roll, pitch and heading for many applications.

Acknowledgments. This research is funded by Hanoi University of Industry under grant number 17–2022 RD/HD-ĐHCN for Nguyen Thi Dieu Linh.

References

1. Bossler, J.D., Schmidlay, R.W.: Airborne integrated mapping system promises large-scale mapping advancements. GIS World **10**(6), 46–48 (1997). Goad, C.C. The Ohio State University Mapping System: The Positioning Component. Proceedings of the 47th Annual Meeting, Williamsburg, VA, June 10–12, 1991, 121–124
2. El-Sheimy, N.: The Development of VISAT-A Mobile Survey System for GIS Applications. Ph.D. Dissertation, Depertment of Geomatics Engineering, University of Calgary, Calgary, AB, Canada (1996). Li, D., Zhong, S. D., He, X., Zheng, H. A mobile mapping system based on GPS, GIS and multisensor. Int. Arch. Photogrammetry and Remote Sensing 1999, 32, 1.3.1–1.3.5.
3. Huang, Y.W., Chiang, K.W.: Performance analysis of low cost INS/GPS POS systems for land based MMS utilizing LC and TC integration. In: ION GNSS 2009 Meeting, Savannah, Georgia, USA, September 22–25 (2009)
4. Park, M., Gao, Y.: Error and performance analysis of MEMS-based inertial sensors with a low-cost GPS receiver. Sensors **8**(4), 2240–2261 (2008)
5. Li, T., Zhang, H., Gao, Z., Chen, Q., Niu, X.: High-accuracy positioning in urban environments using single-frequency Multi-GNSS RTK/MEMS-IMU integration. Remote Sens. **2018**(10), 205 (2018). https://doi.org/10.3390/rs10020205
6. Vana, S., Bisnath, S.: Enhancing navigation in difficult environments with low-cost, dual-frequency GNSS PPP and MEMS IMU. In: Freymueller, J.T., Sánchez, L. (eds.) Beyond 100: The Next Century in Geodesy: Proceedings of the IAG General Assembly, Montreal, Canada, July 8-18, 2019, pp. 143–150. Springer International Publishing, Cham (2023). https://doi.org/10.1007/1345_2020_118

Efficiency Evaluation of Hanning Window-based Filter on Human Skin Disease Diagnosis

My N. Nguyen, Phuong H. D. Bui, Kiet Q. Nguyen, and Hai T. Nguyen(✉)

Can Tho University, Can Tho, Vietnam
nnmy@cit.ctu.edu.vn, {bdhphuong,nthai.cit}@ctu.edu.vn

Abstract. Skin diseases, one of the common human diseases, could be life-threatening if not diagnosed and treated early. This study proposes a skin disease detection model based on some image processing techniques and deep learning architectures. First, we deploy a data pre-processing procedure to convert the input images to Hue-Saturation-Value (HSV) color space and remove their unnecessary information with a Hanning Window-based filter. After applying the Hanning Window-based filter, we downsize the image to 64×64 before fetching it into the learning model. Next, we train the Convolutional Neural Network (CNN) model on the processed image dataset and the original image dataset to compare the effectiveness of two approaches. The experimental results show that using HSV color space and Hanning Window-based filter can improve the performance in diagnosing six out of eight considered types of skin diseases.

Keywords: Skin disease · Lesion · Hanning window · HSV color space

1 Introduction

The skin is the largest organ in the human body and serves as the first protective layer of defense against harmful environmental elements like bacteria and viruses. Nowadays, more and more people suffer from health problems, particularly skin diseases, one of the common human diseases, which may reduce self-confidence and affect human health. In the treatment activities, the initial diagnosis is an important step that assists doctors in understanding the patient's condition and making an accurate treatment plan. The correctness of disease diagnosis is a decisive factor in the success of the patient's treatment. Diagnosing skin diseases seen with the naked eye is easier and quicker than diagnosing internal organ problems that need the support of medical imaging techniques, e.g., Computed Tomography, Ultrasound, X-ray, etc. However, it requires doctors with solid expertise and years of experience in diagnosing skin lesions to make an accurate diagnosis. In addition, the overload of diagnostic and treatment activities for numerous patients challenges healthcare facilities. Therefore, in the strongly developing

T. D. L. Nguyen et al. (Eds.): ICISN 2023, LNNS 752, pp. 478–487, 2023.
https://doi.org/10.1007/978-981-99-4725-6_58

4.0 technology revolution, it would be necessary to apply deep learning in the medical field to solve the difficulties mentioned above, especially in detecting and classifying skin diseases based on dermatology images.

Deep learning has recently shown extraordinary potential in the studies of skin lesion detection and classification on dermatology images. In this research, we propose a skin disease detection model based on some image processing techniques and deep learning architectures using the images from the International Skin Imaging Collaboration (ISIC) 2019 dataset to detect abnormal skin lesions and classify them into eight different skin lesion categories, encompassing Actinic keratosis, Basal cell carcinoma, Benign keratosis lesion, Dermatofibroma, Melanoma, Melanocytic nevus, Squamous cell carcinoma, and Vascular lesion. First, a data pre-processing process is deployed to convert the input images to HSV color space and remove their unnecessary information with a Hanning Window-based filter. After applying the Hanning Window filter, we downsize the image to 64×64 before fetching it into the model. Next, we train the CNN model on the processed image dataset and the original image dataset to compare the effectiveness of the two approaches on human skin abnormality classification. We can see from the experimental results that the model trained on the processed image dataset, which used HSV color space and Hanning Window filter, gives better classification results on six out of eight skin disease types.

The rest of this paper is organized as follows. First, Sect. 2 briefly presents related work on skin lesion detection and classification studies. Then, Sect. 3 proposes a skin disease detection model based on some image processing techniques and deep learning architectures using the images from the ISIC 2019 dataset. Next, Sect. 4 presents experimental results obtained by training the model on the processed image dataset and the original image dataset, then evaluating and comparing the effectiveness. Finally, the conclusion is discussed in Sect. 5.

2 Related Work

Recently deep learning algorithms have been applied in many skin lesion detection and classification studies, especially CNN-based approach studies. In the work [1], Manzoor et al. presented a skin lesion detection method, applied on the ISIC and Mendeley datasets, to classify six common skin diseases, comprising Basal Cell Carcinoma, Actinic Keratosis, Seborrheic Keratosis, Nevus, Squamous Cell Carcinoma, and Melanoma. The method used a deep CNN for segmentation, extracted the in-depth features using AlexNet, GLCM features and ABCD rules, and then used a Support Vector Machine (SVM) for classification. A deep learning-based framework developed by Naeem et al. [2] was introduced for skin cancer classification. The framework, called skin cancer detection classifier network (SCD Net), combined Vgg16 with CNN applying on the ISIC 2019 dataset to classify the dermoscopy images into the four major skin cancer classes, such as Melanoma, Melanocytic Nevi, Basal Cell Carcinoma, and Benign Keratosis. Cano et al. [3] proposed a model using fine-tuning and data augmentation based on a CNN architecture, namely NASNet, to recognize 8 skin diseases, e.g., Actinic Keratosis, Squamous cell carcinoma, Benign keratosis, Dermatofibroma, Melanoma, etc. The model, trained on ISIC 2019 dataset, applied CNN architecture initialized with weight from ImageNet, fine-tuned to discriminate among skin lesion types and then 10-fold cross-validation. The work, presented by Nguyen et al. [4], leveraged processing techniques, called HS-UNET-ID, based on MorphologyEx (blackhat), thresholding, performing semantic segmentation with UNET, and several CNN architectures on human skin disease diagnosis tasks.

The HAM10000 dataset, a large collection of dermatoscopic skin lesion images taken from Kaggle, was used in many studies to train and evaluate their skin disease detection and classification methods. In the work [5], Polat and Koc proposed a deep learning method to classify skin diseases from dermatology images taken from the data HAM10000 dataset. The work applied two different models, encompassing alone CNN model and the combination of CNN and one-versus-all called OVA model. Jain et al. [6] developed a skin lesion classification method for skin cancer identification by taking the HAM10000 dataset. The method presented a comparative analysis of six transfer learning nets, comprising VGG19, InceptionV3, InceptionResNetV2, ResNet50, Xception, and MobileNet, which were used for feature selection and classification. In addition, an interesting work in [7] proposed a skin disease detection and classification method based on an ensemble deep learning model called the Predictive Ensemble Deep CNN Classifier (RF-DCNN), which combined two deep learning techniques, Random Forest classifier and CNN, to categorize skin diseases into seven classes from the HAM10000 dataset. Aldhyani et al. [8] developed a dynamic kernel deep-learning-based CNN model, where the ReLU activation function was used in the first three layers and leakyReLU was used in the last two layers of the CNN to improve the skin lesion classification of seven classes from the HAM10000 dataset. Another work [9] presented a method using the HAM10000 dataset to classify skin diseases through deep learning based on MobileNet V2 and Long Short Term Memory (LSTM). In the method, MobileNet V2 was used to classify skin disease categories, and LSTM was used to maintain the state information of the features over the iterations. Melanoma, the fifth most common skin cancer lesion, is regarded as the most serious skin cancer type due to its possibility of spreading to other body parts. In a study [10], Bhimavarapu and Battineni developed a melanoma detection framework, called the fuzzy-based GrabCut-stacked CNN (GC-SCNN) model, based on fuzzy logic and stacked CNNs for feature extraction combining with SVMs for lesion segmentation.

Although numerous studies based on deep learning algorithms have been proposed to diagnose skin diseases based on dermatology images, attempts on the Hanning Window filter in medical image processing have yet to be fully investigated for skin disease diagnosis. Therefore, our study has attempted Hanning Window-based filter and evaluated its effects on accuracy in skin disease classification.

3 Methods

3.1 Data Description

In this study, we use the dataset combining BCN20000 with HAM10000 datasets called the ISIC 2019 dataset, which consists of 25,331 images downloaded from[1], including eight types of skin anomalies, including Melanoma, Melanocytic nevus, Basal cell carcinoma, Actinic keratosis, Benign keratosis lesion (solar lentigo/seborrheic keratosis/lichen planus-like keratosis), Dermatofibroma, Vascular lesion, and Squamous cell carcinoma.

3.2 Data Pre-processing

The pre-processing procedure consists of four steps. First, image is converted to HSV color space. Next, the ratio between injury area and entire image is estimated. Then, Hanning window with parameter determined by the ratio of previous step is leveraged. This step is to shadow the periphery of the image, which almost contains no detail about the disease. Then, we downsize the image to 64×64 before fetching it into the model.

Converting Image to HSV Color Space. The HSV color space is very convenient for handling colors because each color will have a Hue (H) value in the range [0,360]. In addition, Value (V) and Saturation (S) indicate the brightness and colorfulness of color. This color space closely aligns with the human visual perception. Unlike HSV, RGB color space decomposes a color into three channels of Red (R), Green (G) and Blue (B), which is electronic display-oriented. The detection of skin lesions was mainly based on the color of the lesion. During the study of the dataset, we noticed that the injured skin areas are often darker than the normal skin, so we used the HSV color space to facilitate the training process later. Since the injured areas are darker than normal skin, it is very useful to use the Value (V) channel in the HSV color space. The V channel can separate the dark and light of the image to help the training process achieve high results. Figure 1 exhibits the difference between 2 color spaces, RGB and HSV.

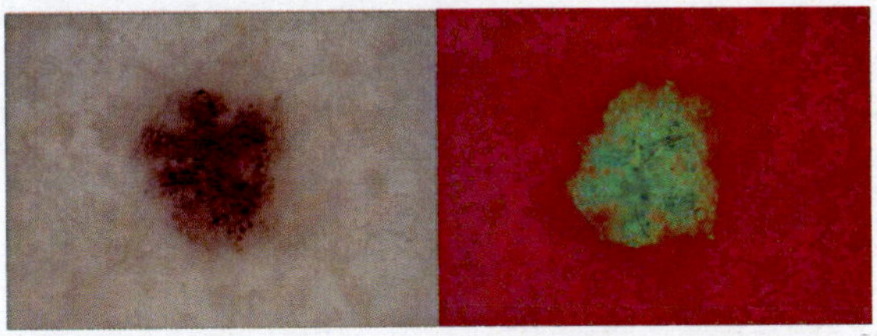

Fig. 1. Difference between RGB (left) and HSV (right) color spaces.

[1] https://www.kaggle.com/datasets/andrewmvd/isic-2019.

Estimating the Ratio of Dark Areas (lesions) in the Image. In the dataset, the injured area usually appears at the center of image and occupies only a small area compared to the whole image (except for a few images with lesions scattered throughout the image). So, we estimate the area of disease on the image and use it as the parameter for next step. Algorithm 1 exhibits the way for injured area estimation. In addition, Fig. 2 illustrates the result of converting from an image with only V channel to a binary image.

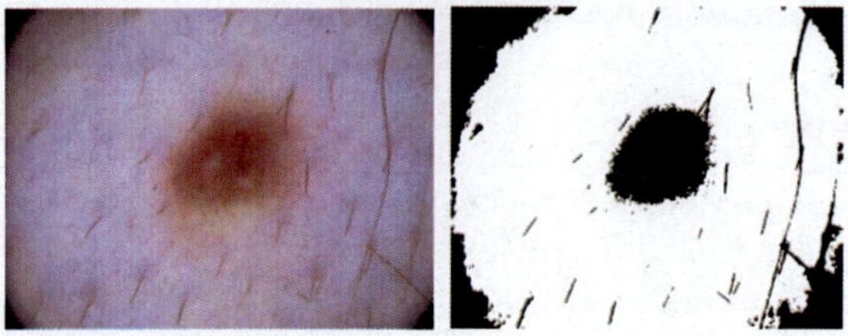

Fig. 2. An illustration of converting to a binary image.

Algorithm 1. The process of estimating the ratio of lesion area

Input: Pixels of an image, *Output*: Ratio of injured area
Converting the images from RGB to HSV color space.
Using K-means clustering technique with k=2.
Converting image from V channel (HSV) to binary image.
Estimating lesion area ratio = $\frac{\text{number of black pixels in binary image}}{\text{total number of pixels}}$
for all the pixels in the binary image from left to right and top to bottom **do**
  if any pixel value == 0 **then**
    counted as one and accumulated.
  end if
end for

Applying Hanning Window to the HSV Color Space Image. The training process's input data (image) only sometimes contains enough useful information. Besides, some pixels are not useful for the training process. This affects the training time as well as the results of the model. Therefore, we apply Hanning Window [11] to filter out unnecessary pixels in the image, which will be able to shorten the training time and give better results. The result at this step depends a lot on the step of calculating the ratio of the area of the lesion on the image. The main idea of applying this Hanning Window is based on the ratio of the lesion area calculated in the previous step. It can determine the image's degree of invasion (removing unnecessary components).

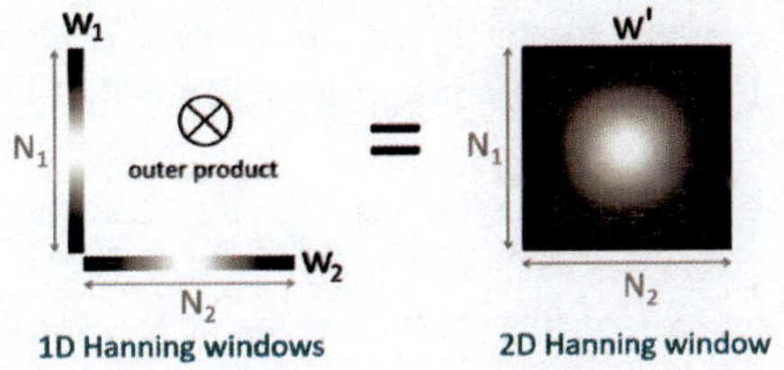

Fig. 3. The creation of 2D Hanning window.

When applying Hanning Window to the image, the image will be blackened from the edge (in the corners of the image) to the center. Because the image is two-dimensional, so the adopted Hanning window will also be two-dimensional. For the case of one-dimensional Hanning window, which corresponds to a line of pixels, the degree of blackening of the nth pixel $w[n]$ depends on the parameters as shown in Eq. 1. Two-dimensional Hanning window is derived from the outer product of two one-dimensional ones (Eq. 2). This derivation is illustrated in Fig. 3.

$$w[n] = \alpha_0 - (1 - \alpha_0) * cos\frac{2\pi n}{N}, 0 \leq n \leq N \tag{1}$$

$$w'[n_1, n_2] = w_1[n_1] \otimes w_2[n_2] \tag{2}$$

where N is the number of pixels, and α_0 is a real number that determines the strength of invasiveness.

The images contain eight disease types, so the injured areas on the skin can be different. In the Hanning Window function formula, the value of α_0 is fixed (α_0=0.5), so when applied, the invasiveness of the function can lose important information for images with large lesion areas. Therefore, we use the Hanning Window function with the parameter α_0 changing based on the ratio of the shadow area. The larger the parameter α_0, the less invasive the image will be, inferring the larger the disease area ratio, the larger α_0 and vice versa. To automatically change α_0 in the Hanning Window function, follow these steps. First, we tried values of α_0 on each specific image based on the ratio of lesion area. Then, we used the Linear Regression algorithm to predict the value of α_0. Next step, we apply Hanning Window to the image.

Various values of α_0 on each specific image We have attempted different values by applying Hanning Window to each image with the parameter α_0 based on the ratio of the area of the lesion and directly observing the results after applying. After choosing the appropriate parameter α_0 based on the ratio of the area of the lesion, the pairs (x, y) will be obtained, where x is temporarily called the ratio of the area of the lesion and y is temporarily called the value α_0.

Using Linear Regression algorithm to predict the value of α_0 Linear Regression is a method to predict a dependent variable (y) based on one or more independent variables (x) as Eq. 3.

$$y = \beta_0 + \beta_1 x + \epsilon \tag{3}$$

where y is the dependent variable (predicted value), x is an independent variable (value used for learning), ϵ deviation to compensate for errors (the smaller ϵ, the greater the association between x and y and vice versa), β_0 is the estimated value of y, β_1 is coefficient of the relationship between x and y. After obtaining the data pair (x, y), we predict the y value (parameter α_0). Then, we receive a weight file, which predicts the parameter α_0 value based on the newly arrived lesion area ratio. Figure 4 illustrates the results of the model training process using the Linear Regression algorithm.

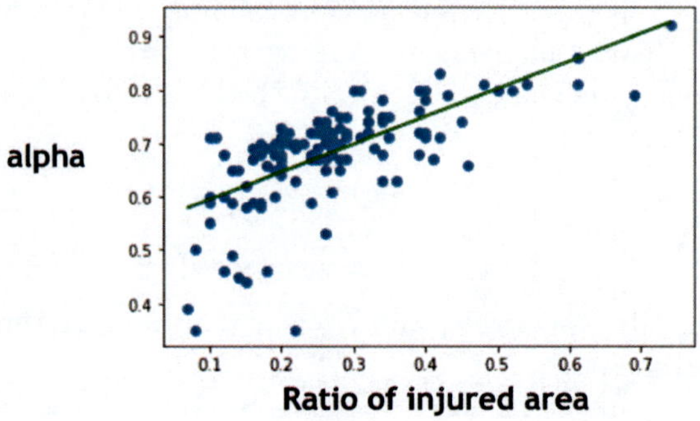

Fig. 4. A Linear Regression with a green line to illustrate the relationship between α_0 and ratio of the injured area. (Color figure online)

Applying Hanning Window to the Image. In this step, we will apply the Hanning Window to shadow the unnecessary peripheral region. After that, the image will be included in the training model. This filtering helps to reduce the training time and improve the model's performance. The difference between the original image and the image after filtering (in HSV color space) will be shown in Fig. 5.

To better understand the processing in this step, Algorithm 2 explains the processing in more detail after the ratio of the percentage of the area of the lesion. After performing Hanning Window, we downsize the image to 64×64 before fetching it into the learning model.

3.3 Convolutional Neural Network for Skin Diseases Classification

The architecture of the CNN of skin disease classification is presented in Table 1. Before training, the input image goes through a pre-processing process as presented in Sect. 3.2. The first and the second convolution layers use numbers of

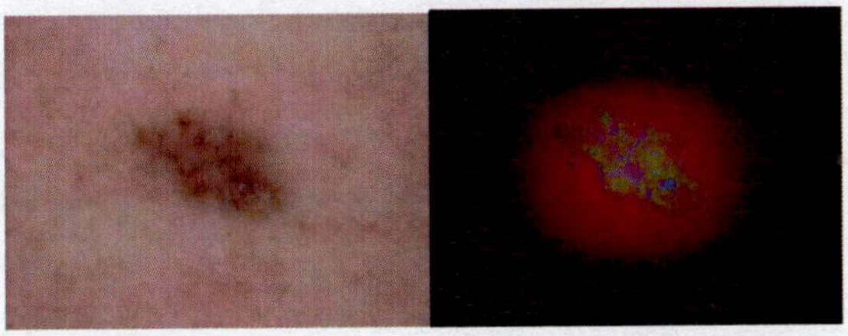

Fig. 5. The difference between the original image (left) and the applied Hanning Window image (right).

Algorithm 2. Detailed description of the filtering with Hanning Window

Input: an image, *Output*: the image filtered with Hanning Window
Calculating lesion area ratio ($\frac{\text{The number of black pixels in binary image}}{\text{total number of pixels}}$)
Taking the lesion area scale value into the model, then get a new parameter α_0 value of each image.
Taking the received parameter α_0 and put it into the Hanning Window formula to apply to the image.
The result is an image that has gone through the filtering process (Unimportant information is blackened in the image).

Table 1. Hyper-parameters of the convolutional neural networks.

Number of convolutional layers	2
Filter sizes and numbers for each convolution layer	$32 \times 3 \times 3$, $64 \times 3 \times 3$
Activation function	ReLU
Optimizer function	Adam
Number of Max-pooling layers	2
Number of classes Dense	4

filters 32 and 64 of 3×3, respectively, and a ReLU activation function follows each convolution. Furthermore, max-pooling layers of size 2×2 are used after each convolution layer. In addition, the model uses the fully connected layer (dense) using the Softmax function. Besides, we use a dropout rate of 0.2 for the neural layer, and the network runs to 100 epochs.

4 Evaluation

The dataset is randomly divided into three subsets: training set (18,238 images - occupying 72%), validation set (2,026 images - occupying 8%), and testing set (5,067 images - occupying 20%). The validation set is to select the best model, which then applies to the testing set. We evaluate the efficiency of the filtered dataset (Dataset 2), which used HSV color space and Hanning Window filter, and the original dataset (Dataset 1) with 8 types of skin abnormalities.

Table 2. Accuracy in percentages on each lesion type with the dataset filtered by Hanning Window (Dataset 2) and the original dataset (Dataset 1). The bold reveals that the performance of the processed images is better.

Type of injury	Dataset 1	Dataset 2
Actinic keratosis (AK)	28.365	24.531
Basal cell carcinoma (BCC)	31.468	**35.688**
Benign keratosis lesion (BKL)	16.867	**20.382**
Dermatofibroma (DF)	23.294	**31.636**
Melanoma (MEL)	32.441	**42.336**
Melanocytic nevus (NV)	92.482	**94.579**
Squamous cell carcinoma (SCC)	28.374	22.335
Vascular lesion (VASC)	35.436	**45.422**
Training Time (in minutes)	114	41

Table 2 shows the results of the model training process on each type of anomaly in accuracy on the testing set. The image, after processing, can give better classification results on 6 types of skin diseases, comprising Basal cell carcinoma (BCC), Benign keratosis lesion (BKL), Dermatofibroma (DF), Melanoma (MEL), Melanocytic nevus (NV), and Vascular lesion (VASC). And as we can see, many diagnostic results improve quite deeply, such as on MEL and VASC improving up to 10%, DF is better up to 8%. However, we also have lower results on 2 diseases, including Actinic keratosis (AK) (with more than 3% reduction) and Squamous cell carcinoma (SCC) (reduction of 6%). We can see a superiority for NV with an accuracy rate of more than 90%, while there is a great challenge for disease diagnoses such as AK, SCC, and BKL, with these diseases having a correct prediction rate of less than 30%.

We also see that the result of the training time on the dataset filtered by Hanning Window also converges nearly three times faster than that on the original dataset, with 41 min in comparison to 114 min of the latter.

5 Conclusion

In this study, we evaluated the effectiveness of using HSV color space in combination with the Hanning Window filter in diagnosing skin diseases with deep learning models. From the results, we see that the CNN model trained on the processed image dataset using HSV color space and Hanning Window filter gives better classification results on 6 out of 8 considered types of skin diseases. At the same time, we also clearly see the time benefits from learning the filtered images when the CNN model can converge in less time compared to the original images. However, we also noticed that the diagnostic performances are greatly different. We are still facing difficulties with the diagnosis of many skin diseases.

With 8 diseases considered, only Melanocytic nevus reached over 90% in accuracy. Future studies may focus on more complex models to improve accuracy. However, the study also brings to attention that filters like the Hanning Window can be useful to improve accuracy in diagnosing skin diseases. In addition, the efficiency evaluation of the regression algorithms when using the Hanning Window should also be noted, which can help improve the performance of this filter.

References

1. Manzoor, K., et al.: A lightweight approach for skin lesion detection through optimal features fusion. Comput. Mater. Continua. **70**(1), 1617–1630 (2022). https://doi.org/10.32604/cmc.2022.018621
2. Naeem, A., Anees, T., Fiza, M., Naqvi, R.A., Lee, S.W.: SCDNet: a deep learning-based framework for the multiclassification of skin cancer using dermoscopy images. Sensors. **22**(15), 5652 (2022). https://doi.org/10.3390/s22155652
3. Cano, E., Mendoza-Avilés, J., Areiza, M., Guerra, N., Mendoza-Valdés, J.L., Rovetto, C.A.: Multi skin lesions classification using fine-tuning and data-augmentation applying NASNet. PeerJ Comput. Sci. **7**, e371 (2021). https://doi.org/10.7717/peerj-cs.371
4. Nguyen, H.T., et al.: HS-UNET-ID: an approach for human skin classification integrating between UNET and improved dense convolutional network. Int. J. Imaging Syst. Technol. **32**(6), 1832–1845 (2022). https://doi.org/10.1002/ima.22776
5. Polat, K., K.O.K.: Detection of skin diseases from dermoscopy image using the combination of convolutional neural network and one-versus-all. J. Artif. Intell. Syst. **2**(1), 80–97 (2020). https://doi.org/10.33969/ais.2020.21006
6. Jain, S., Singhania, U., Tripathy, B., Nasr, E.A., Aboudaif, M.K., Kamrani, A.K.: Deep learning-based transfer learning for classification of skin cancer. Sensors. **21**(23), 8142 (2021). https://doi.org/10.3390/s21238142
7. Kalaivani, A., Karpagavalli, S.: Detection and classification of skin diseases with ensembles of deep learning networks in medical imaging. Int. J. Health Sci. 13624–13637 (2022). https://doi.org/10.53730/ijhs.v6ns1.8402
8. Aldhyani, T.H.H., Verma, A., Al-Adhaileh, M.H., Koundal, D.: Multi-class skin lesion classification using a lightweight dynamic kernel deep-learning-based convolutional neural network. Diagnostics. **12**(9), 2048 (2022). https://doi.org/10.3390
9. Srinivasu, P.N., SivaSai, J.G., Ijaz, M.F., Bhoi, A.K., Kim, W., Kang, J.J.: Classification of skin disease using deep learning neural networks with MobileNet v2 and LSTM. Sensors. **21**(8), 2852 (2021). https://doi.org/10.3390/s21082852
10. Bhimavarapu, U., Battineni, G.: Skin lesion analysis for melanoma detection using the novel deep learning model fuzzy GC-SCNN. Healthcare. **10**(5), 962 (2022). https://doi.org/10.3390/healthcare10050962
11. Kabe, A.M., Sako, B.H.: Analysis of continuous and discrete time signals. In: Structural Dynamics Fundamentals and Advanced Applications, pp. 271–427. Elsevier (2020). https://doi.org/10.1016/b978-0-12-821615-6.00005-8

Continuous Deep Learning Based on Knowledge Transfer in Edge Computing

Wenquan Jin[1], Minh Quang Hoang[4], Luong Trung Kien[2,3], and Le Anh Ngoc[2,4](✉)

[1] Yanbian University, Yanji 133002, China
jinwq@ybu.edu.cn
[2] FPT University, Hanoi, Vietnam
{kienlt6,ngocla2}@fe.edu.vn
[3] University of Sciences and Technologies of Hanoi, Hanoi, Vietnam
luong-trung.kien.d22.ict@st.usth.edu.vn
[4] Swinburne University of Technology, Melbourne, Australia
nle@swin.edu.au

Abstract. Edge computing enables real-time intelligent services to be provided near the environment where the data is generated. However, providing intelligent services requires sufficient computing ability that is limited by edge devices. Knowledge transfer is an approach that transfers the trained model from an edge device to another edge device. In this paper, we propose a continuous deep-learning approach based on knowledge transfer to reduce the training epochs of a single-edge device without sharing the data in the edge computing environment. For training deep learning continuously in the network edge, each edge device includes a deep learning model and local dataset to fine-tune the reached model. The fine-tuning is completed, then, the updated model is transferred to the next edge device. For experimenting with the proposed approach, edge computing is comprised of multiple edge devices that are emulated by the virtual machines to operate the deep learning model and simulate the network communication. The deep learning model is developed for classification based on Convolutional Neural Network (CNN). As expected, the prediction accuracy is improved in multiple iterations of transferring in the edge computing environment.

Keywords: Edge Computing · Deep Learning · Knowledge Transfer · Fine Tuning · Web Services

1 Introduction

Intelligent devices provide an efficient lifestyle by supporting personalized and autonomous services close to the environment such as smart buildings, smart campuses, smart farms, etc. Edge computing is a paradigm that brings computing ability to the environment where the data is generated and applied. Edge devices are deployed in the environment to handle the data based on low latency due to less network distance. However, the devices are constrained that are equipped with limited battery and computing resources including processor and storage [1–4]. Therefore, deriving the inference or

T. D. L. Nguyen et al. (Eds.): ICISN 2023, LNNS 752, pp. 488–495, 2023.
https://doi.org/10.1007/978-981-99-4725-6_59

prediction model based on the big size of data in the edge device is not sufficient. Mostly, the model is trained in a high-performance computing machine such as a cloud server and deployed to the edge device to provide services [5–7]. Nevertheless, for keeping the personal data in the environment and providing personalized intelligent services near to the user, edge devices are required to derive the inference or prediction model by themselves.

For enabling deep learning on the edge device, we propose a knowledge transfer approach to provide a continuous training process in edge computing. The proposed approach reduces the training epochs of a single-edge device without sharing the data in the edge computing environment. The edge device includes a deep learning model and local dataset to re-train the arrived model that is delivered by another edge device once the training is finished using the local dataset. The re-train process is performed by fine-tuning which improves the performance of the model by updating the weights of the deep learning. The edge device is implemented by the virtual machine that operates the deep learning model and simulates the network communication. The Fashion-MNIST and MNIST (Handwriting) datasets are used for experimenting the proposed continuous deep learning base on Convolutional Neural Network (CNN). Each dataset is comprised of 10 classes and 60,000 examples for training and 10,000 examples for testing. For the implementation, the training set is separated by 5 datasets and deployed in the edge devices respectively. The prediction accuracy is achieved to be test when the model transferred approximately 27 and 48 times in the continuous deep learning scenario.

The rest of the paper is structured as follows. Section 2 introduces the related works including existing knowledge transfer approaches and intelligent edge computing. Section 3 presents the proposed continuous deep learning approach in the edge computing architecture. Section 4 presents the implementation details and result of the prosed knowledge transfer process through the 5 virtual machines. Section 5 presents the experimental results that are collected over the iterations of knowledge transfer and fine-tuning on the edge devices. Finally, we conclude our paper in Sect. 6.

2 Related Works

Knowledge transfer enables parameters of the trained deep learning model to be re-trained in different platforms with different data. In the constrained environment, the devices transfer the models to continue the train for overcoming the battery and storage limitations [8]. Joyjit et al. [9] proposed an anomaly prediction model based on a Recurrent Neural Network (RNN) and transfer the model to an experimental environment for testing with unseen data. Sufian et al. [10] proposed an intelligent edge computing using a pre-trained CNN model with small data based on transferring the model between edge devices. Akram et al. [11] proposed a photovoltaic detection module based on model transfer and fine-tuning to improve accuracy. Li et al. [12] proposed an image classification approach based on transferring CNN for solving the small data problem.

Based on deep learning in the edge computing environment, the edge nodes are deployed to provide intelligent services by operating the inference model on each node. Le et al. [9] proposed an offloading strategy to enhance the performance of the deep learning processing in edge computing that overcomes the limitation of the computing ability of the edge device. For providing real-time inference service on the network edge, En Li et al. [13] proposed an on-demand deep learning model based on an agent that provides the inference model to be deployed in the edge computing environment. Vipul et al. [14] proposed a fine-tuned deep learning model based on MobileNet to handle complex networks in resource constraint environments for recognizing images on a chest dataset. Therefore, for collaborating edge nodes to work together, integrating the knowledge transfer technique with edge computing enables operating intelligent services on the constrained edge devices in the network edge.

3 Proposed Continuous Deep Learning in Edge Computing Architecture

Continuous deep learning is performed by multiple edge devices that are connected by network in the network edge. As shown in Fig. 1, the edge devices transfer the knowledge to others which is operated multiple time to improve the deep leaning model.

Each edge device includes data and deep learning model that is processed in the device and derives updated model. For updating the model, the fine-tuning approach is applied that updates the weights of the model based on the local data. Once the model is updated then the edge device transfers the model to another edge device that is pre-configured for the continuous updating scenario.

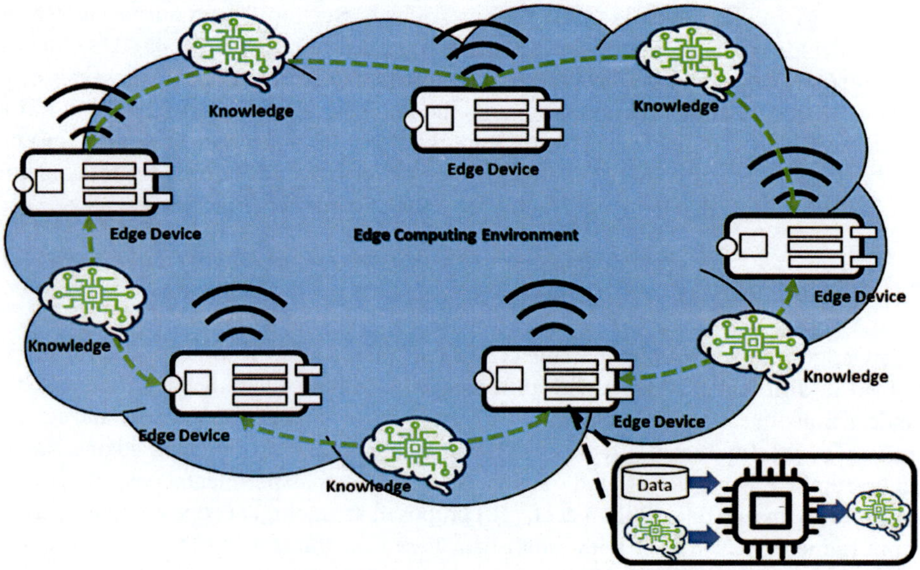

Fig. 1. Knowledge transfer in edge computing architecture.

Figure 2 shows the proposed edge computing platform architecture that is comprised of deep learning engine, knowledge transfer service provider and repositories. The deep learning engine is used for restoring the deep learning model from the model repository and fine tuning the model. The knowledge transfer service provider is used for providing the services of model transfer and continuous training based on web services. The platform includes data and model repositories respectively. In the data repository, the local dataset is stored for updating the model based on fine tuning. In the model repository, the model is stored that is transferred from other edge device through the model transfer service.

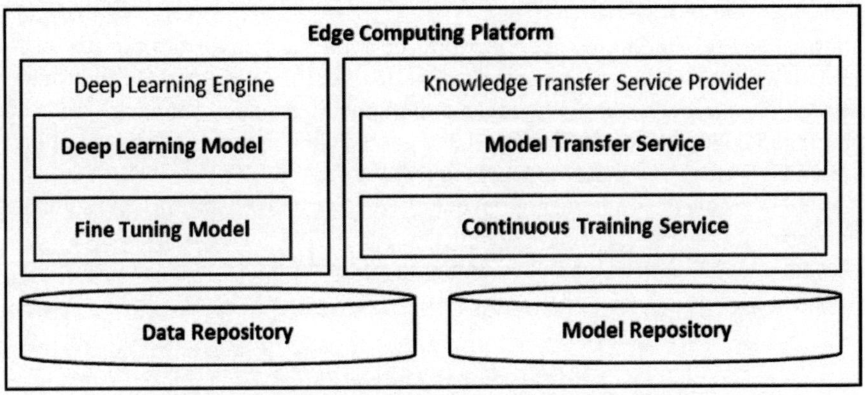

Fig. 2. Proposed edge computing platform architecture.

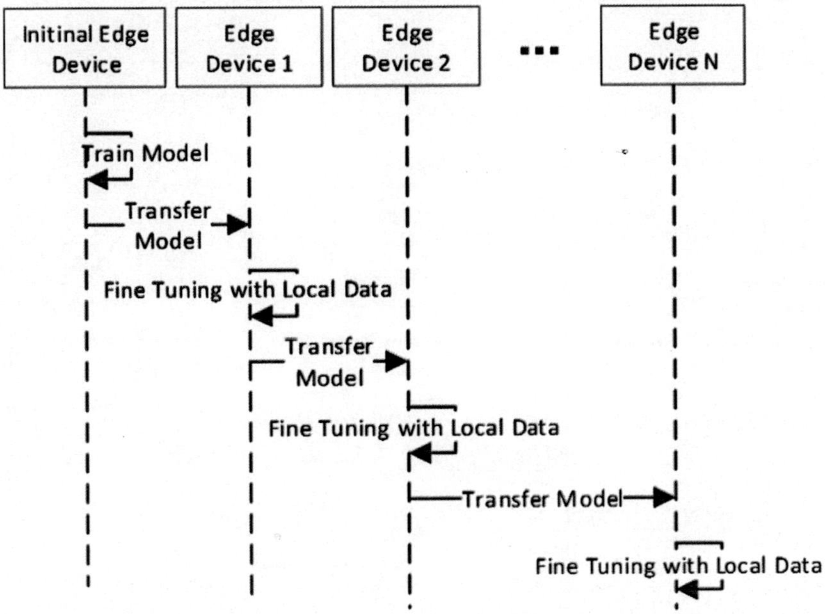

Fig. 3. Continuous deep learning scenario based on edge devices.

Figure 3 shows the continuous deep learning scenario based on the interactions between edge devices. In the initial state, a deep learning model is trained with a dataset to derive a inference model that is the pre-trained model. The model is comprised of fixed weights in the deep neural network. The initial edge device transfers the model to another edge device then the model is fine tuned with the local dataset. For the continuous deep leaning, the process is repeated until the configured scenario is finished.

4 Implementation of Knowledge Transfer in Edge Computing

For implementing continuous deep learning base on the knowledge transfer between multiple edge devices, the virtual machine with Ubuntu OS is operated to provide the functions of edge computing including communication and computing. In the edge computing, 5 edge devices are deployed and each device operates a Python application that includes web server and deep learning solution.

Table 1 presents the implementation details. The virtual machine is configured using Virtual Box and operates a latest Ubuntu OS. The virtual machine is emulated with 4 processors and 8 GB memory. The Python application is developed using Flask and TensorFlow libraries.

Table 1. Implementation details.

Virtual Machine	Operating System	Processor Count	Memory	Library
Virtual Box	Ubuntu 22.04	4	8GB	Flask 2.2.2, TensorFlow 2.10

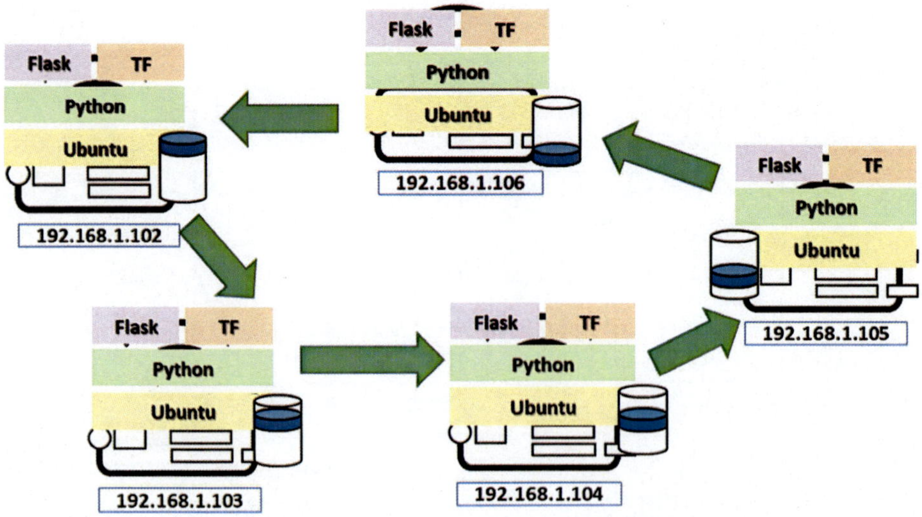

Fig. 4. Network architecture of edge computing environment.

Figure 4 shows the network architecture of the edge computing implementation that operates 5 edge devices with different network address. Each device is configured to transfer the model to the next edge device. The edge device includes the local dataset that is part of test dataset respectively.

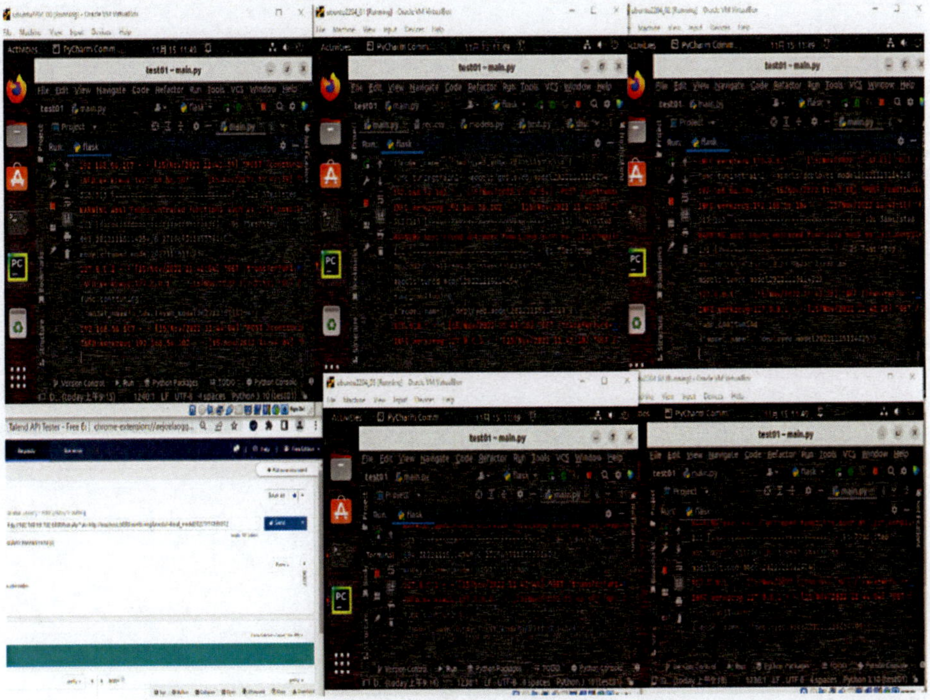

Fig. 5. Implementation result of edge devices.

Figure 5 shows the implementation result of edge devices. The Talend API tester is a web client that requests to the initial edge device to transfer the model to the next edge device. The edge devices are operated in the virtual machine based on Ubuntu OS.

5 Experiment Result

For experimenting the proposed the continuous deep learning, 5 edge devices are deployed by emulating the edge computing. Each device is deployed the different part of the Fashion MNIST and MNIST dataset. Once the model is transferred to a device, the device fine-tunes the model with the local host and predicts with a test data. The prediction result is the confidence that is recorded for each device 100 times.

As shown Fig. 6, the confidence results are combined and order by recorded time. The experimental result presents that the confidence is reached the best after 6 iterations by 27 times transfer approximately.

Figure 7 shows the experimental result of the MNIST-based continuous deep learning transfer using CNN. In this experiment, the CNN model is comprised using simpler architecture that is performed in faster operation. However, the overall continuous deep

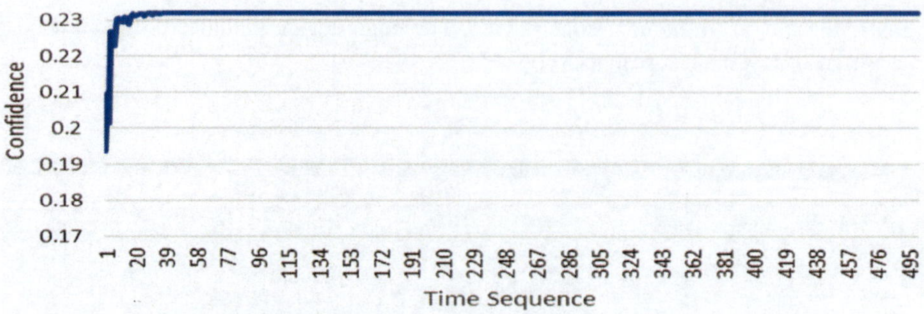

Fig. 6. Performance fine-tuning model in edge devices using Fashion-MNIST.

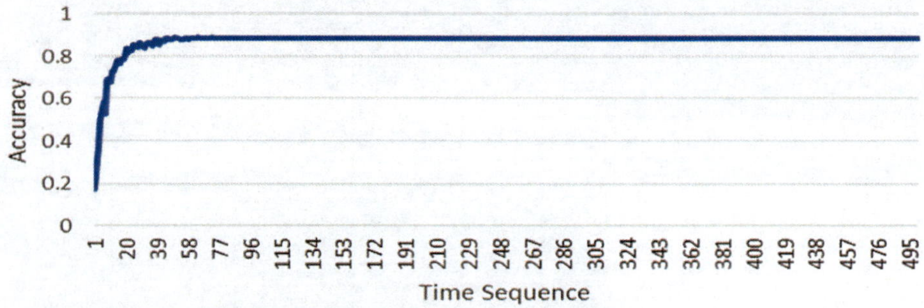

Fig. 7. Performance fine-tuning model in edge devices using MNIST.

learning transfer is achieved best accuracy in more iterations although the data set is easier to trained. The experimental result presents that the accuracy is reached the best after 10 iterations by 48 times transfer approximately.

6 Conclusions

In this paper, we proposed a knowledge transfer approach to provide continuous training process in the edge computing. In the edge computing environment, the edge device includes deep learning model and local dataset to re-train the arrived model that is delivered by another edge device once the training is finished using the local dataset. The re-train process is performed by fine tuning that improves performance of the model by updating the weights of the deep learning. For the implementation, the virtual machine is used to emulate edge devices that transfer the model to another emulated edge device. Over 100 iterations in the experiment, the collected prediction accuracy is improved by transferring the CNN model to fine-tune in each edge device. Based on the knowledge transfer, the proposed continuous deep learning reduces the training epochs of a single-edge device without sharing the data in the edge computing environment.

References

1. Wang, S., et al.: Adaptive federated learning in resource constrained edge computing systems. IEEE J. Sel. Areas Commun. **37**(6), 1205–1221 (2019)
2. Chatterjee, B., Cao, N., Raychowdhury, A., Sen, S.: Context-aware intelligence in resource-constrained IoT nodes: opportunities and challenges. IEEE Des. Test **36**(2), 7–40 (2019)
3. Jang, I., Kim, H., Lee, D., Son, Y.S., Kim, S.: Knowledge transfer for on-device deep reinforcement learning in resource constrained edge computing systems. IEEE Access **8**, 146588–146597 (2020)
4. Wang, F., Zhang, M., Wang, X., Ma, X., Liu, J.: Deep learning for edge computing applications: a state-of-the-art survey. IEEE Access **8**, 58322–58336 (2020)
5. Xu, R., Jin, W., Hong, Y., Kim, D.H.: Intelligent optimization mechanism based on an objective function for efficient home appliances control in an embedded edge platform. Electronics **10**(12), 1460 (2021)
6. Jin, W., Solanki, V.K., Le, A.N., Kim, D.: Real-time inference approach based on gateway-centric edge computing for intelligent services. In: Tran, D.-T., Jeon, G., Nguyen, T.D.L., Lu, J., Xuan, T.-D. (eds.) ICISN 2021. LNNS, vol. 243, pp. 355–361. Springer, Singapore (2021). https://doi.org/10.1007/978-981-16-2094-2_44
7. Jin, W., Xu, R., Lim, S., Park, D.H., Park, C., Kim, D.: Dynamic inference approach based on rules engine in intelligent edge computing for building environment control. Sensors **21**(2), 630 (2021)
8. Sharma, R., Biookaghazadeh, S., Li, B., Zhao, M.: Are existing knowledge transfer techniques effective for deep learning with edge devices?. In: 2018 IEEE International conference on edge computing (EDGE), pp. 42–49. IEEE (2018, July)
9. Li, H., Ota, K., Dong, M.: Learning IoT in edge: deep learning for the Internet of Things with edge computing. IEEE Netw. **32**(1), 96–101 (2018)
10. Sufian, A., You, C., Dong, M.: A deep transfer learning-based edge computing method for home health monitoring. In: 2021 55th Annual Conference on Information Sciences and Systems (CISS), pp. 1–6. IEEE (2021, March)
11. Akram, M.W., Li, G., Jin, Y., Chen, X., Zhu, C., Ahmad, A.: Automatic detection of photovoltaic module defects in infrared images with isolated and develop-model transfer deep learning. Sol. Energy **198**, 175–186 (2020)
12. Li, J., et al.: Autonomous Martian rock image classification based on transfer deep learning methods. Earth Sci. Inf. **13**(3), 951–963 (2020). https://doi.org/10.1007/s12145-019-00433-9
13. Li, E., Zhou, Z., Chen, X.: Edge intelligence: on-demand deep learning model co-inference with device-edge synergy. In: Proceedings of the 2018 Workshop on Mobile Edge Communications, pp. 31–36 (2018, August)
14. Singh, V.K., Kolekar, M.H.: Deep learning empowered COVID-19 diagnosis using chest CT scan images for collaborative edge-cloud computing platform. Multimedia Tools Appl. **81**(1), 3–30 (2021). https://doi.org/10.1007/s11042-021-11158-7

Detection of Abnormalities in Mammograms by Thresholding Based on Wavelet Transform and Morphological Operation

Yen Thi Hoang Hua[1,2](✉), Giang Hong Nguyen[3], and Liet Van Dang[1,2]

[1] Department of Physics and Computer Science, Faculty of Physics and Engineering Physics, University of Science, Ho Chi Minh City, Vietnam
hthyen@hcmus.edu.vn

[2] Vietnam National University, Ho Chi Minh City, Vietnam

[3] Department of General Education, Cao Thang Technical College, Ho Chi Minh City, Vietnam

Abstract. The most frequent type of cancer among women is breast cancer. Patients with breast cancer have a considerably higher chance of survival when they receive early identification and treatment. The greatest technique for early breast cancer detection is screening mammography analysis. Because cancerous tissues and glands are close to the illumination, the results of analysis by radiologists are frequently limited; therefore, using computer analysis is more convenient. Image thresholding is one of the simplest methods for separating tumors; however, the results are only good if the mammograms have high contrast and the threshold value is correctly chosen. In this article, we present an approach for increasing image contrast, while preserving tumor edges by combining the stationary wavelet transform with morphological transforms. The tumor is then extracted using the adaptive multi-threshold. By using performance evaluation criteria for segmentation methods, such as accuracy, sensitivity, specificity, and Dice similarity coefficient, the results are compared to the ground truth data, demonstrating the method's accuracy.

Keywords: Breast cancer · Mammography · Image enhancement · Image segmentation · Thresholding

1 Introduction

According to the World Health Organization (WHO), 685000 individuals died from breast cancer in 2020, and 2.3 million women received a diagnosis [1]. Screening is the primary method for finding breast cancer, and early diagnosis of the disease is crucial to lowering the mortality rate. There are a number of breast cancer screening methods available, including mammography, PET, MRI, ultrasonography, etc. Mammography is usually regarded as one of the most reliable and cost-effective methods. A low-dose x-ray is used in a mammography to study the tissues inside the breast, which results in images with weak contrast and noise contamination. Small tumors in low contrast mammography images closely resemble normal glandular tissue, making early identification of breast

T. D. L. Nguyen et al. (Eds.): ICISN 2023, LNNS 752, pp. 496–506, 2023.
https://doi.org/10.1007/978-981-99-4725-6_60

cancer challenging. Because there is only a slight variation in attenuation between lesions and normal tissues in mammography, abnormalities are difficult to detect; nevertheless, increasing detection accuracy may help.

Each image processing approach that has been presented for the purpose of detecting suspicious lesions has particular challenges. Anitha et al. [2] studies the use of dual-stage adaptive thresholding in a novel computer-aided method for mass detection. Sheba and Raj [3], Thawkar and Ingolikar [4] propose a method that employs a global thresholding and morphological procedures to identify and segment regions of interest (ROI) in mammograms. Nayak et al. [5] offer an undecimated wavelet transform and an adaptive thresholding approach for mass detection. Shanmugavadivu et al. [6] focus on mass detection using a wavelet and region-growing combination for efficient mass segmentation. Mencattini et al. [7] dealt with the problem of dyadic wavelet-based enhancing and denoising of mammographic images …

It has been demonstrated that image enhancement techniques are beneficial in improving segmentation efficiency and that gray level feature-based detection is a robust mass detection methodology. In this paper, we present another approach for increasing image contrast while preserving tumor edges by combining the stationary wavelet transform with morphological transforms and then extracting tumors using the multi-threshold technique.

2 Proposed Methodology

The proposed methodology consists of following steps as display in Fig. 1.

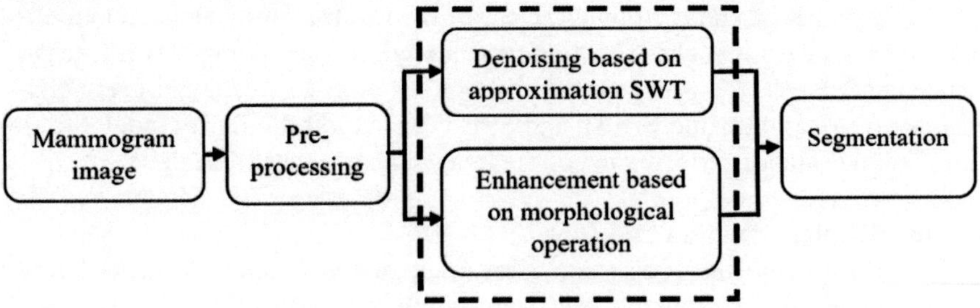

Fig. 1. Flowchart of the proposed method

2.1 Image Preprocessing

Preprocessing a mammogram is crucial since it improves the image's quality by reducing off noise.

Noise removal: The proposed technique employs a median filter on mammographic images.

Label and artifact removal: Noise, labels, and artifacts in mammograms can all have an impact on the detection algorithm's results. These must be eliminated by performing global thresholding and choosing the largest blob, which is the breast portion [3].

Pectoral muscle removal: The pectoral muscle is substantially more intense and almost equal in density to the dense breast tissue. This has an impact on detection and might result in false positive findings, eliminating it enhances the segmentation outcomes.

Figure 2 shows the steps of the preprocessing step:

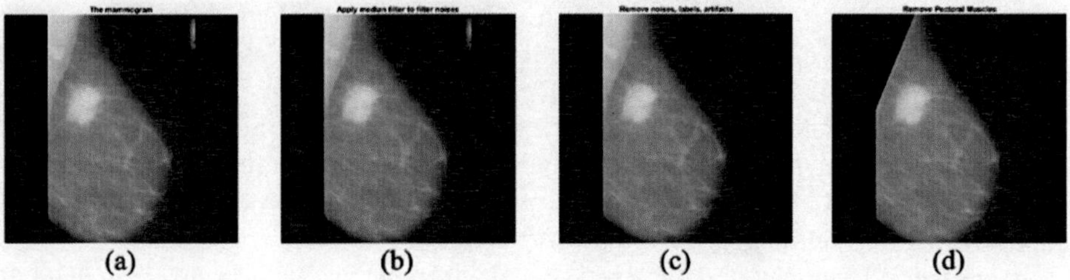

Fig. 2. The preprocessing steps with a mammogram: (a). The original mammogram, (b). The noise filtered image, (c). Image after labels and artifacts removal, (d). Image after pectoral muscle removal.

2.2 Image Enhancement

Appropriate image enhancement produces good results for both manual and automatic detection and segmentation process. For the best image enhancement, the approximation component of the Stationary Wavelet Transform (SWT) at the appropriate level is then combined with the top and bottom hat transforms with structuring element in multiscale image regions of each scale. The performance criteria including EME (Measure of Enhancement), PSNR (Peak Signal to Noise Ratio), AMBE (Absolute Mean Brightness Error), MSE (Mean Squared Error), and SSIM (Structural Similarity Index Metric) [8, 9] are used to evaluate performance of the image enhancement technique.

Selecting Wavelet Function and Level

In order to overcome the translational invariant of the Discrete Wavelet Transform (DWT), the SWT based on the "à trous" algorithm is designed [10]. Because the detail components of the wavelet transform at level 1 and level 2 contain the first and second highest frequencies that can be considered as noises, so their approximation components are used as the image denoising and taken into consideration to determine which level provides the best smoothing image [11]. Table 1 displays the performance metric parameters of some wavelet functions at levels 1 and 2.

According to the results of Table 1, approximation level 1 of the bior 2.2 wavelet provides the good smoothed image is a reasonable choice. This is in line with theory: (i) the family of biorthogonal wavelet functions has the property of linear phase, which is required for image reconstruction, (ii) the level 1 detail components contain the highest frequencies, which are considered noises, while level 2 detail components contain lower

Table 1. Image quality assessment measures for the mammogram mdb028.

Wavelet	Level	EME	PSNR	AMBE	MSE	SSI
Db1	SWT 1	1.7757396	44.327983	0.11197567	2.40039158	0.995624211
	SWT 2	2.0663913	38.40232	0.3413401	9.3939524	0.9815035
Db2	SWT 1	1.8298917	44.766658	0.093148232	2.16977406	0.996041477
	SWT 2	1.9752749	39.643882	0.2753096	7.0581713	0.985821
Sym2	SWT 1	1.8298917	44.766658	0.093148232	2.16977406	0.996041477
	SWT 2	1.9752749	39.643882	0.2753096	7.0581713	0.985821
Bior2.2	SWT 1	1.8427194	44.772863	0.093111992	2.16667652	0.996037402
	SWT 2	1.9752749	39.64553	0.2752113	7.0554924	0.9858387

frequencies, which contain noises as well as the image details. *So, the SWT and the approximation 1 of the wavelet function bior2.2 are used in this article.*

Image Enhancement Using Morphological Operations

The key to morphological operations is choosing an appropriate structural element (SE) that fits the target objects and applying it to an input image to get an identical-sized output image. By eliminating the bottom-hat transform and adding the top-hat transform from the original image, image enhancement employing morphological operations increases the contrast between the bright and dim portions of the image:

$$M = A + TH - BH \tag{1}$$

where, TH stands for the top-hat transform, BH for the bottom-hat transform, A for the pre-processed image, and M for the image enhancement.

This process increases the contrast of the image while preserving the sharpness of the objects. The formula (1) employs two morphological operators, opening and closing, with the key element being a structuring element whose shape and size are selected to correspond the objects in the image. Because breast tumors frequently vary in size, they cannot be removed with a single SE [12]. To solve this problem, multiple SEs, denoted by SE*i* (*i* = 1, 2, …, n) are used in the opening and closing operators. The challenge is to select n such that the set of elements SEi has a radius from 1 to n in an effort to the opening and closing operators are their maximum.

The top-hat transform and the bottom-hat transform are given by:

$$TH = A - \arg\max\,(A \circ SEi) \tag{2}$$

$$BH = \arg\max\,(A \cdot SEi) \tag{3}$$

where o denotes the opening operator and · denotes the closing operator.

To investigate this problem, four sets of disk structuring elements with radii: {i = 1,2,…,5}, {i = 1,2,…,10}, {i = 1,2,…,15}, {i = 1,2,…,20} are applied to three mammograms listed in column 1 of Table 2 to determine the best group SEi for image enhancement. Table 2 displays the performance metric parameters of three enhanced images using morphological operations. These parameters demonstrate that sets of disk structuring elements with radii of {i = 1,2,...,5} are used to improve image quality for better results when other sets are used. *Therefore, in this article, we chose a set of disk structuring elements ranging in radius from 1 to 5.*

Table 2. Image quality assessment measures of three mammograms.

Image	SE	EME	PSNR	AMBE	MSE	SSI
Mdb001	Radii = {1:5}	2.57176382	44.601493	0.27897739	2.2538815	0.9887205
	Radii = {1:10}	3.00042399	39.421408	0.52066612	7.4291582	0.9797591
	Radii = {1:15}	3.17266113	37.29894	0.66407776	12.111178	0.9756158
	Radii = {1:20}	3.34168922	35.688969	0.81604862	17.546215	0.9729755
Mdb028	Radii = {1:5}	2.72167354	42.606943	0.46389198	3.5676813	0.9818847
	Radii = {1:10}	3.76080562	37.475423	0.84363365	11.628885	0.9684931
	Radii = {1:15}	4.10489722	35.592998	1.0039444	17.938269	0.9634873
	Radii = {1:20}	4.35420057	34.214822	1.13767052	24.637588	0.9604657
Mdb184	Radii = {1:5}	2.2660079	41.853984	0.5294838	4.2430868	0.9777419
	Radii = {1:10}	2.57081089	36.963502	0.96099758	13.083689	0.9622582
	Radii = {1:15}	2.91616301	35.100033	1.18824577	20.094494	0.9558881
	Radii = {1:20}	3.25691423	33.795807	1.35653973	27.133115	0.9524215

Image Enhancement Using Proposed Algorithm

To improve image quality, the proposed algorithm combines the approximation component of SWT with morphological transforms using structuring elements ranging in radius. This algorithm enhances image contrast while preserving tumor edges.

The performance metric parameters such as HE (Histogram Equalization), CLAHE (Contrast Limited Adaptive Histogram Equalization), CS (Contrast Stretching) are used to evaluate image enhancement results obtained by proposed algorithms [13]. The results show that the proposed method does outperform the conventional methods for image enhancement (Table 3).

2.3 Image Segmentation

The segmentation of the mammographic image is crucial to mass detection so that the system can produce a more accurate classification result. Brightness and gray value are two of the most prevalent gray level characteristics for all types of lesions [2]. This article employs multi-threshold, and the key to multi-threshold is determining the number of thresholds required for the image. The histogram of an image with many

Table 3. Performance metrics of four methods: HE, CLAHE, CS and the proposed method.

Images	Methods	EME	PSNR	AMBE	MSE	SSI
Mdb028	HE	0.500907	5.728871	130.3593	17385.74	0.292273
	CLAHE	9.667995	20.96972	12.958	520.1253	0.525145
	CS	1.523161	17.33057	21.02715	1202.328	0.954807
	Proposed method	2.327568	45.66062	0.319806	1.766113	0.990487
Mdb184	HE	0.433217	6.666472	109.6438	14009.84	0.346121
	CLAHE	10.14356	24.08611	6.117743	253.7869	0.542756
	CS	1.911411	24.45955	9.508773	232.8763	0.993844
	Proposed method	2.260895	44.84125	0.364849	2.132825	0.988543
Mdb001	HE	0	3.295922	169.9121	30443.05	0.171233
	CLAHE	7.548709	24.67213	7.458947	221.7518	0.442996
	CS	1.819231	24.75758	6.626002	217.4315	0.992668
	Proposed method	2.4578	47.845	0.1918	1.0679	0.9943

objects, such as a mammogram, is multimodal and the Otsu approach can be used to identify numerous thresholds. To determine the number of thresholds, the histogram of the image is smoothed and its peaks are automatically counted by Matlab's *findpeaks* function; which is equal to the number of peaks minus 1. The image and the number of thresholds are then entered into Matlab's *multithresh* function to determine threshold values. Because tumors in the mammogram have the highest brightness in the enhanced mammogram, only the highest image segmentation threshold is used to extract tumors (Fig. 3).

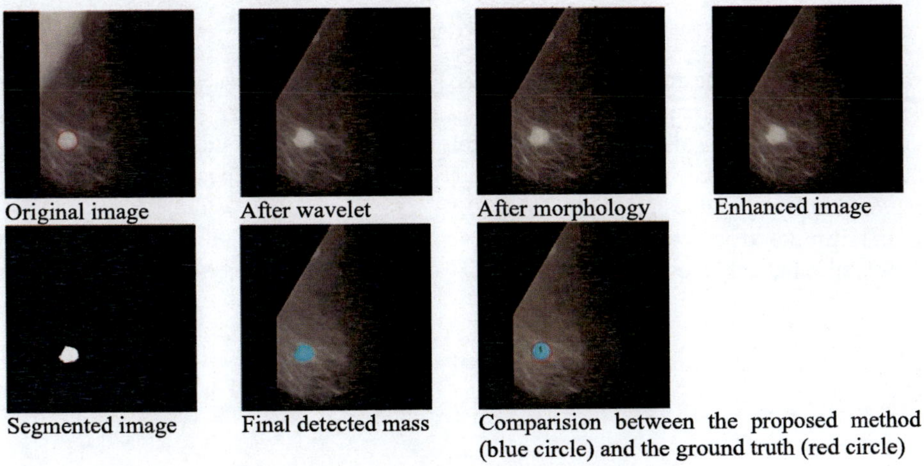

Fig. 3. Breast region segmentation and detection results.

3 Results

The mammogram images are taken from the 322 digitized mammograms in the mini-MIAS database, which include 202 normal and 120 abnormal images. The images were padded to a 8-bit word of each pixel in 1024 × 1024 size in Portable Gray Map (pgm) format. It also includes, if any anomalities are present, the radiologist's centroid and radius marks on the area. The Fig. 4 demonstrates the outcomes of the method's testing on 10 images of lesions from the mini-MIAS database.

The algorithm performs well in case of asymetry type lesions (mdb081, mdb083, mdb104, mdb111) because of significant gray level feature. In the case of circumcribed masses (mdb015, mdb028), owing to their nearly oval shapes, the proposed method achieves good sensitivity. However, in spiculated and architectural distortion masses (mdb117, mdb178, mdb184, mdb198), because they are vary in size and pattern, many lesions have low intensity, so, the segmentation method based on gray level approach relatively difficult, the proposed method has limited effect and these types of tests have a high rate of false positives.

Performance Evaluation Metrics for the Segmentation Methodology

Accuracy, sensitivity, specificity, and Dice similarity coefficient are the performance metrics utilized to assess the suggested algorithm.

$$\text{Accuracy} = \frac{\text{TP} + \text{TN}}{\text{TP} + \text{TN} + \text{FP} + \text{FN}} \tag{4}$$

$$\text{Sensitivity} = \frac{\text{TP}}{\text{TP} + \text{FN}} \tag{5}$$

$$\text{Specificity} = \frac{\text{TP}}{\text{TP} + \text{FP}} \tag{6}$$

$$\text{Dice} = \frac{2(\text{A} \cap \text{B})}{\text{A} + \text{B}} \tag{7}$$

where A and B are the amount of pixels in the ground truth and the segmentation regions, True Positive (TP) pixels that are defined as mass, True Negative (TN) pixels that are defined as not mass, False Positive (FP) pixels that are located outside of the mass region that is defined as mass, and False Negative (FN) pixels that are located inside the mass region that is defined as not mass (Table 4).

The proposed detection algorithm achieves a good accuracy of 98.88% and specificity of 99.86%, but achieves the worst sensitivity of 59.03%.

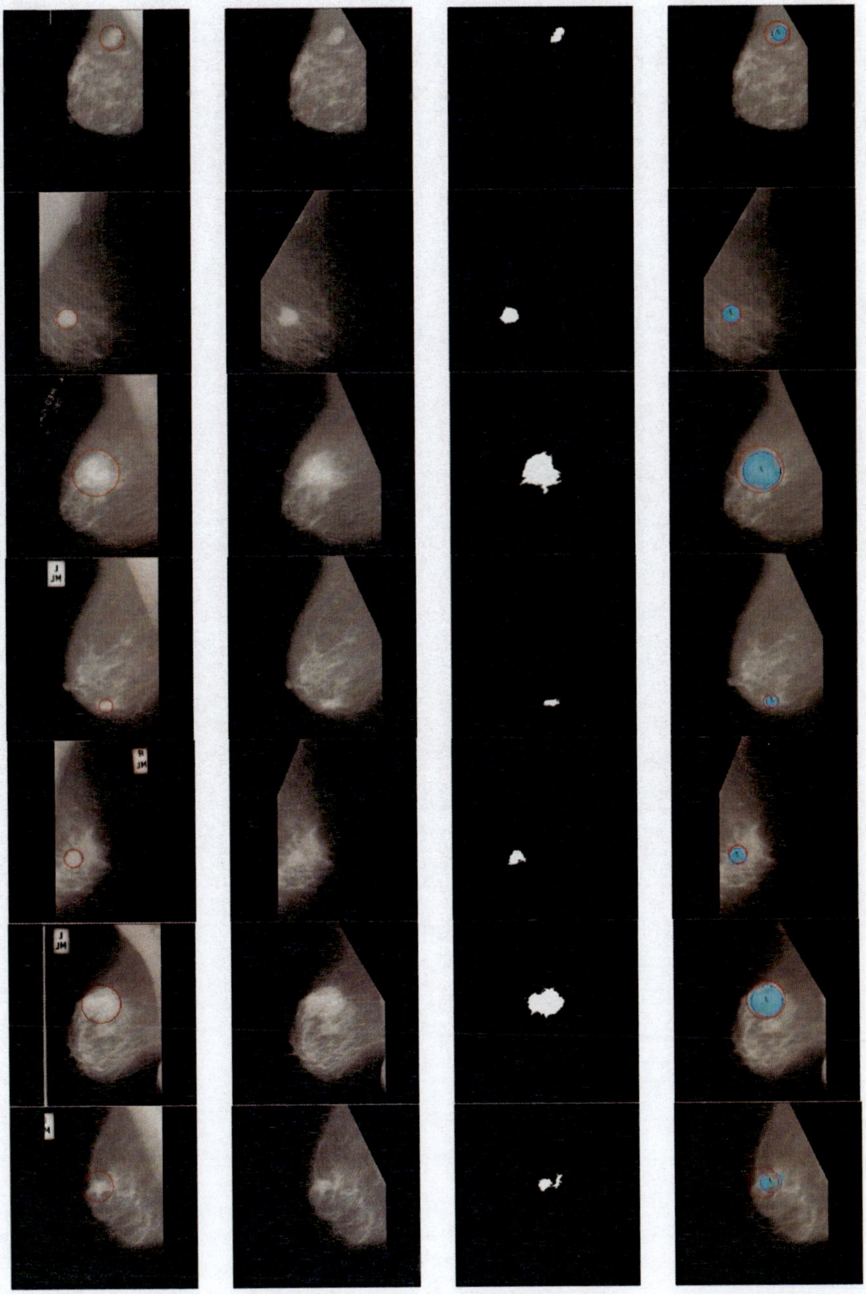

Fig. 4. Mammogram images and their suspicious regions detection by the proposed method for images: mdb015, mdb028, mdb081, mdb083, mdb104, mdb111, mdb117, mdb178, mdb184, mdb198. a). Original image with ground truth (the red circled region contains the lesion), b). Enhanced image by the proposed enhancement method, c). Final detected tumor mass by adaptive thresholding method, d). Segmented output image (the ground truth with red circle and the proposed method with blue circle). (Color figure online)

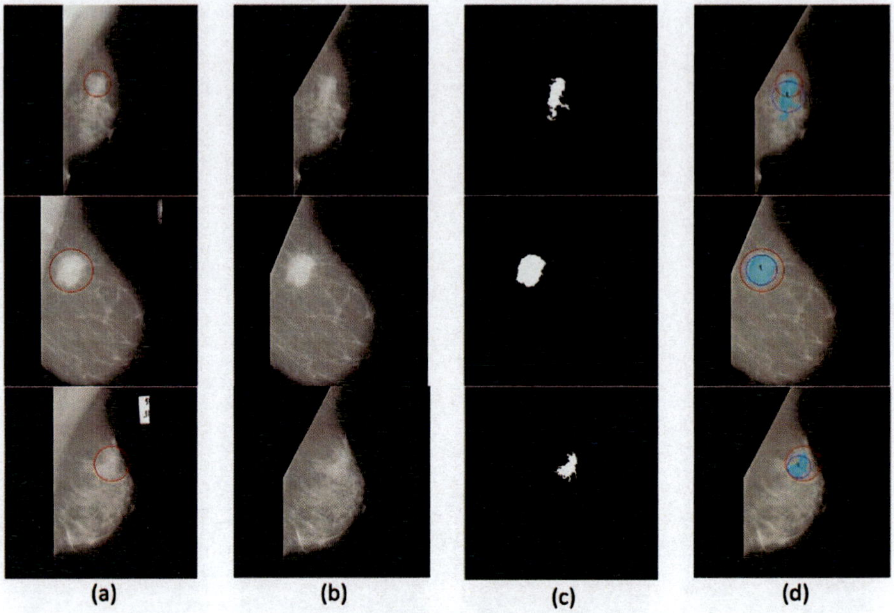

Fig. 4. (*continued*)

Table 4. Values of the performance measures.

Images	Accuracy (%)	Sensitivity (%)	Specificity (%)	Dice (%)
mdb015	99.10	34.64	100.00	51.51
mdb028	99.75	73.44	100.00	84.69
mdb081	98.03	61.84	100.00	76.42
mdb083	99.92	80.32	100.00	80.36
mdb104	99.88	84.45	100.00	84.47
mdb111	98.89	67.71	100.00	80.75
mdb117	98.64	41.89	99.87	56.70
mdb178	98.03	53.58	98.70	44.55
mdb184	98.02	49.35	100.00	66.08
mdb198	98.52	43.12	100.00	60.26
Average	98.88	59.03	99.86	68.58

4 Conclusion

The issue of enhancing mammographic images for the best segmentation is considered in this work. The pectoral muscle's triangle region, which has a high intensity and may be recognized as a mass or abnormality, necessitates the removal of artifacts as well as the pectoral muscle. The algorithm's sensitivity is increased by allowing for less distortion when noise and artifacts are suppressed by SWT. A morphological filter can enhance objects with shapes that are similar to a circle or an oval and is very beneficial for locating suspicious areas. The analysis' results indicate that:

i. The approximation component on level 1 of the wavelet 'bior' combined with the top-hat and bottom-hat transforms with the disk structure element in the radius range from 1 to 5 different directions for best contrast by quality measurement operators like EME, PSNR, AMBE, MSE, SSI values.
ii. The number of thresholds to choose for extracting the tumors to opt for a fine segmentation should be based on the distribution of the peaks of the enhanced image histogram.

The proposed methodology gives medical professionals another perspective and can be utilized to focus radiologists' attention to these areas, reducing the likelihood that they will overlook a region that is positively malignant. This effort may represent an adequate technical strategy for increasing the image quality in diagnostics.

Acknowledgement. This research is funded by University of Science, VNU-HCM under grant number T2022-05.

References

1. Sung, H., et al.: Global cancer statistics 2020: GLOBOCAN estimates of incidence and mortality worldwide for 36 cancers in 185 countries. Cancer J. Clin. **71**(3), 209–249 (2021)
2. Anitha, J., DineshPeter, J., AlexPandian, I.: A dual stage adaptive thresholding (DuSAT) for automatic mass detection in mammograms. Comput. Methods Progr. Biomed. **138**, 93–104 (2017). https://doi.org/10.1016/j.cmpb.2016.10.026
3. Sheba, K.U., GladstonRaj, S.: An approach for automatic lesion detection in mammograms. Cogent Eng. **5**(1), 1444320 (2018). https://doi.org/10.1080/23311916.2018.1444320
4. Thawkar, S., Ingolikar, R.: Segmentation of masses in digital mammograms using optimal global thresholding with Otsu's method. IJCST, **5**(3) (2014)
5. Nayak, A., Ghost, D.K., Ari, S.: Suspicious lesion detection in mammograms using undecimated wavelet transform and adaptive thresholding. IEEE (2013)
6. Shanmugavadivu, P., et al.: Wavelet transformation-based detection of masses in digital mammograms. Int. J. Res. Eng. Technol. **3** (2014)
7. Mencattini, A., Salmeri, M., Lojacono, R., Frigerio, M., Caselli, F.: Mammographic images enhancement and denoising for breast cancer detection using dyadic wavelet processing. IEEE Trans. Instrum. Measure. **57**(7), 1422–1430 (2008). https://doi.org/10.1109/TIM.2007.915470
8. Sheba, K., Gladston Raj, S.: Objective quality assessment of image enhancement methods in digital mammography-A comparative study. Signal Image Process. Int. J. (SIPIJ). **7**(4) (2016)
9. Moradmand, H., et al.: Comparing the performance of image enhancement methods to detect microcalcification clusters in digital mammography. Iranian J. Cancer Prevent. **5**(2) (2012)
10. Zhang, X., Li, D.: [Agrave] Trous wavelet decomposition applied to image edge detection. Annal. GIS **7**(2), 119–123 (2001). https://doi.org/10.1080/10824000109480563
11. Arya, D.P., Mini, M.G.: Mammographic image enhancement based on SWT and high boost filtering. Int. J. Comput. Theory Eng. **7**(5), 374–378 (2015). https://doi.org/10.7763/IJCTE.2015.V7.988

12. Zhang, X., et al.: A hybrid image filtering method for computer-aided detection of microcalcification clusters in mammograms. J. Med. Eng. **2013**, 1–8 (2013). https://doi.org/10.1155/2013/615254
13. Gupta, S., Porwal, R.: Appropriate contrast enhancement measures for brain and breast cancer images. Int. J. Biomed. Imaging **2016**, 1–8 (2016). https://doi.org/10.1155/2016/4710842

Determine the Relation Between Height-Weight-BMI and the Horizontal Range of the Ball When Doing a Throw-In

Nguyen Phan Kien(✉), Huynh Nguyen Cong, Hoang Pham Viet, Linh Nguyen Hoang, Nhung Dinh Thi, and Tran Anh Vu

Hanoi University of Science and Technology, Hanoi, Vietnam
{Nhung.dinhthi,Vu.trananh}@hust.edu.vn

Abstract. There have been many studies that have shown that throwing margins have techniques to make the throw-in more effective. Therefore, this paper is based on the results of studies that determine the angle of the throw and the optimal velocity for conducting the pitching experiment, data processing, and assessment of the effect of height - weight - BMI on long-range, focusing on Asian people. The study was practiced with 70 members with different indicators, throwing the right technique based on experimental conditions equivalent to reality on the field. The results showed that weight had the greatest effect on the distance of the ball. The result shows that in the selection of male players to perform the throw-in situation, the following factors are prioritized: height over 1.6m, weight from 65–78 kg, and having a BMI classified normally or overweight. In the future, we hope to do more research to get throwing a throw-in situation with higher difficulty and use other factors to support athletes to achieve the best results from the proposed system showed that the hand acceleration and punch forces correlated strongly with an average acceleration of 28.3 m/s^2 (without rotation) and 29.9 m/s^2 (with rotation) producing an average force of 107.5 N and 139.9 N, respectively. These results show that the punching velocity had a great impact on the punching forces. The experiments also proved that the system can use to monitor the force, acceleration of the punch, and also a posture of practitioners when doing punches.

Keywords: Thrown-in · Height · Weight · BMI

1 Introduction

Throwing is a very important part of football, but to throw to the highest efficiency, not all players know the technique. There has been a lot of research in the world about throwing boundaries. For example, in research of Cerrah, it shows no difference between throw-in velocity and the distance in two cases such as a run-and-run pitcher and a standing thrower [1]. Levendusky used 12 players to survey the classic throw-in. The result reports an average bounce speed of 18.3 m/s and the throw angle of 29° would be the most optimal [2]. Other research analyzed the time the fly reached the target, the spin of the ball, the time from drop to the first drop on the ground, and the corners when releasing the ball

T. D. L. Nguyen et al. (Eds.): ICISN 2023, LNNS 752, pp. 507–515, 2023.
https://doi.org/10.1007/978-981-99-4725-6_61

resulting in a throw angle of 28.1° (Messier) [3]. Kollath, E., and Schwartz, on the other hand, produced a result of 33.0 ± 6.9° [4]. Of all the studies of biomechanics references to throwing, most studies have been done on throwing speed, pitch drop height from hand, and throw angle [1–3].

The football throw-in is one of the activities in a football match that the players have to use their hands to throw the ball to restart the match [5]. The players have to face the field of playing, using two hands to keep the ball before throw and thrown-in by two hands. This rule creates difficulties for the players when doing thrown-in such as limiting the distance of the ball when thrown-in. The throw-in technique can divide into three styles such as standing throw with both feet parallel, standing thrown with feet staggered, and run-up thrown-in. If the distance of the throw-in ball is long so the players can create more chances to make pressure on the other side, so they normally try to make the best effort to throw the ball as long as possible. However, to have a long thrown-in ball, it needs some key events such as (a) retraction of the ball behind the head by trunk extension, shoulder flexion, and elbow flexion, (b) maximum ball retraction, (c) trunk, and shoulder forward extension to bring the ball forward, (d) elbow extension to ball release [5]. All key events are used to analyze the ball throw-in techniques. Some research analyzed the throw-in motion based on the accuracy of the throw-in [6] and they give some important factors to the throw-in such as the ball velocity at the release [8], the key motions related to trunk flexion, shoulder, and elbow extension [5], the optimum release angle of the ball, height of the released ball or large backspin of the released ball [7]. To increase the velocity of the ball, players have to increase their hands' velocity which is based on the combination of shoulder and elbow flexing/extension by analyzing these joint motions [8].

In conclusion, research on throw-in ball techniques has been done for a long time and it is related to the angle of ball release, the height of the ball, and the combination of shoulder, and elbow motion which creates a higher velocity of the release ball. Although most studies have been completed, most studies have not yet been included in the analysis of the effects of height, weight, and BMI on long ranges. And that is the goal of the research we are aiming for. For this study, we will use experimental methods and statistical analysis, focusing on Asian standards in order to support Vietnamese athletic and football players.

2 Materials and Methods

Modeling the pitching process is a point problem with the following factors [5, 6] (Fig. 1).

In the graph above, the symbols are interpreted: h is the ball's height out of hand, and H is the maximum height of the ball. Vo can be called Throw velocity. The throw angle is symbolized by θ. L is the long range of the ball and g is the gravity acceleration, equal to 9.81 m/s^2. From the analysis of kinematics and the use of classical mechanical equations, the connections between long range and throwing angle are obtained, specifically as follows:

$$y = h - \frac{gx^2}{2v_0^2.cos^2\theta} + x.tan\theta \quad (1)$$

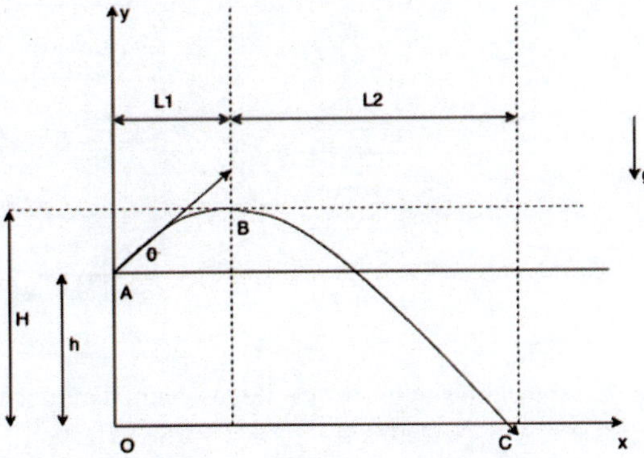

Fig. 1. Physical properties of the throw-in model

$$L = v_{ox}.t = v_0.cos\theta t = v_0\cos\theta\left(\frac{v_0 sin\theta}{g} + \sqrt{\frac{v_0^2 sin^2\theta}{g^2} + \frac{2h}{g}}\right) \quad (2)$$

Apply the results of previous studies, when performing the pitching movement: the thrower is instructed to throw at an angle of about 30°, in two cases there is a momentum run off about 2 m and stand still, throwing with the strongest possible force and low or non-swirling spin [7, 8]. It is necessary to make throws when the weather conditions are good, without affecting the results, ignoring the wind and air friction.

To determine the angle of the shot taken, we followed the steps, specifically as follows. Drawing two points on the ball's trajectory is the first step [11]. Then, join and measure the two sides of the right angle. The line is drawn on the computer, then it is calculated using the tangent theorem. Figure 2 describes the whole process, we stop each video in specific position of the ball, measure the length of a and b, then calculate the ratio a/b to determine the angle (Fig. 3).

Fig. 2. Angle determination procedure

Determine the Optimal Throwing Angle for Each Person Using Matlab

For the calculation of those parameters, the input of the program will be parameters that can be easily determined and measured, from which the missing variables can be calculated, and plotted related graphs [10]. To be able to do that, we will calculate based

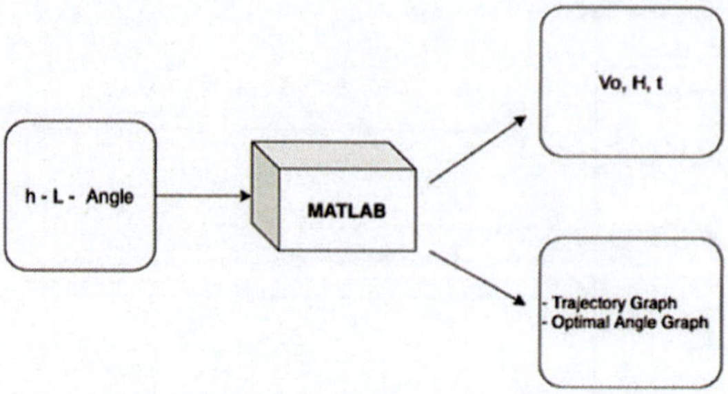

Fig. 3. Algorithms to determine the problem parameters.

on Eqs. 1, and 2, mentioned in the previous section, combined with the plot feature in Matlab to draw graphs.

Analyze the Effect of Height, Weight, and BMI

Height, weight, and BMI factors [9] affect the range in the Eq. (1). Height affects the height of the ball that flies h. Weight affects the initial throw velocity, Vo. BMI, including height, and weight, should also impact h and Vo. For BMI, we only evaluate adults over 20 years of age according to Asian classification standards [9]. For children under 20, we will evaluate other studies. Denote that all members' gender is male.

After throwing the following elements, height, weight, and long-range are recorded. The following chart of sample ratio statistics (Fig. 4):

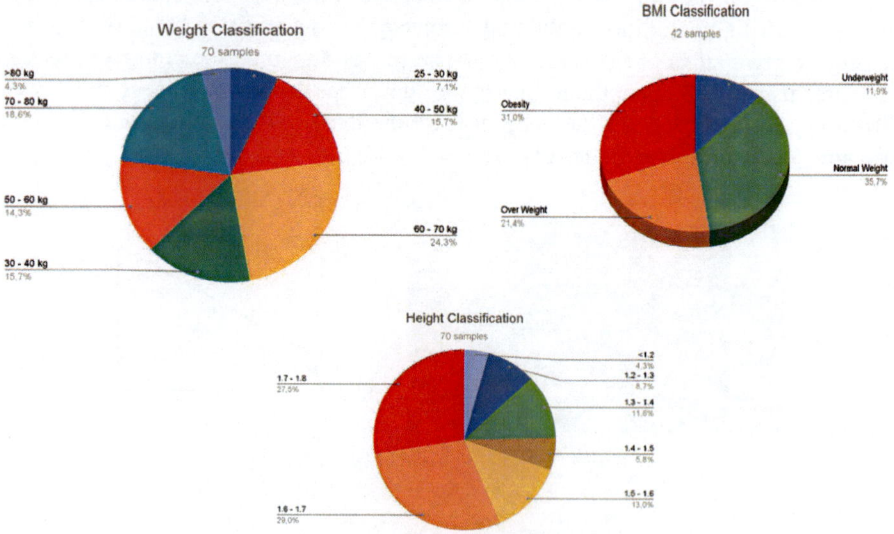

Fig. 4. Weight-Height-BMI sample statistics

The distribution of height, weight, and BMI is relatively uniform which is advantageous to evaluate the results.

3 Results

Optimal Angle for Each Person (Table 1)

Table 1. Matlab calculated results for 4 specific cases

	Measurement parameters			Calculating parameters		
Case	L(m)	h (m)	θ (Deg)	Vo (m/s)	t (s)	H (m)
1	7	2.2	20	7.57	0.98	2.54
2	10	2.1	30	9.11	1.28	3.16
3	12	2.25	40	9.88	1.59	4.31
4	6	2.05	60	7.53	1.59	4.22

As shown above, if we consider the initial throw velocity, gravitational acceleration, and height to be constant, the range can be expressed as a function of the angle θ. The following are a few graphs analyzing the effect of pitch angle on the range for a given initial velocity, altitude, and gravity (Fig. 5).

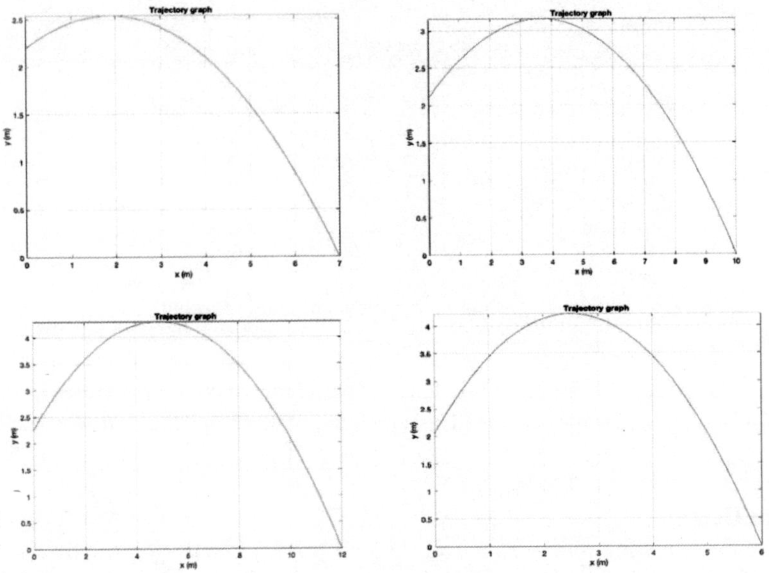

Fig. 5. Optimum angle of four given cases

All the results are depicted in the two diagrams (Fig. 6).

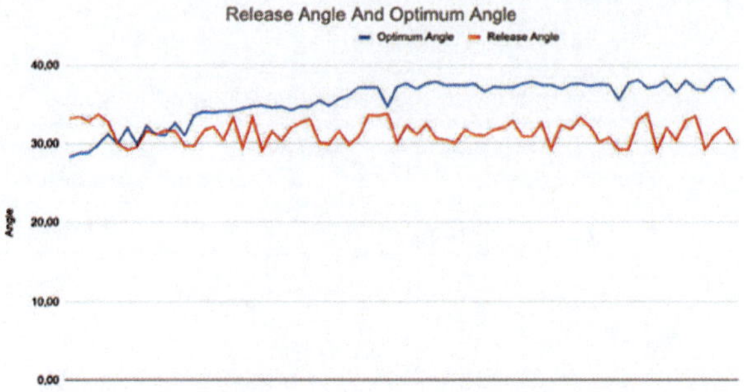

Fig. 6. Release and optimal angle for 70 samples

The application of this section is to calculate the exact angle of the throw for each specific person (Fig. 7).

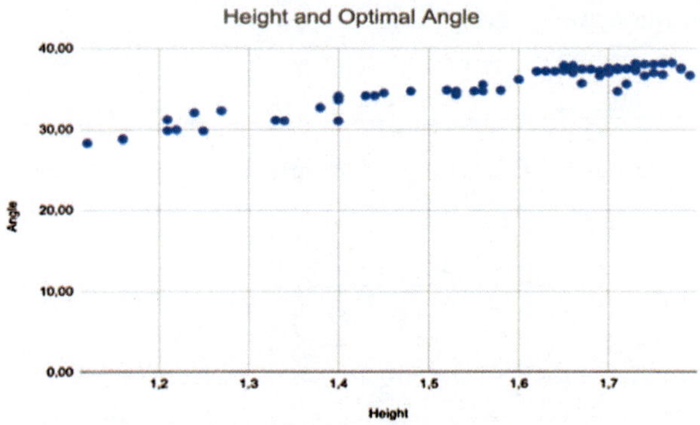

Fig. 7. Height and optimal angle relationship

As for the effect of height, we see that tall people have a larger optimal throw angle, about 37–38°, while people who are 1.2–1.5 m tall have an optimal throw angle between 30–34.

Effect of Height

In two images showing that people over 1.6m in height have the longest throw distance, usually, these people are men and adults, and the result of each throw of each person with a height of fewer than 1.6m has a fairly significant difference (Fig. 8).

In general, whether running momentum or not running momentum, height influences the long range in the pitch. A larger height will be thrown farther, but if the height is the

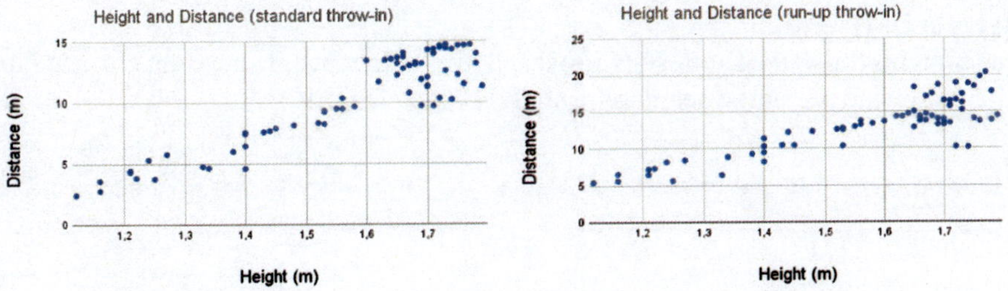

Fig. 8. Results of experiment in Standard Throw-in and Run-up Throw-in

same, it is necessary to consider the weight. For the measured results, a person with a height of >1.6 m will throw farther.

Effects of Weight

Weight has a relatively pronounced effect on long range when pitching the ball. At less than 70 kg, it will throw further the larger weight and vice versa. When the weight is greater than 70 kg, the range is slightly reduced but still high. For the members who have done it, those who weigh between 65–78 kg will throw the furthest (Fig. 9).

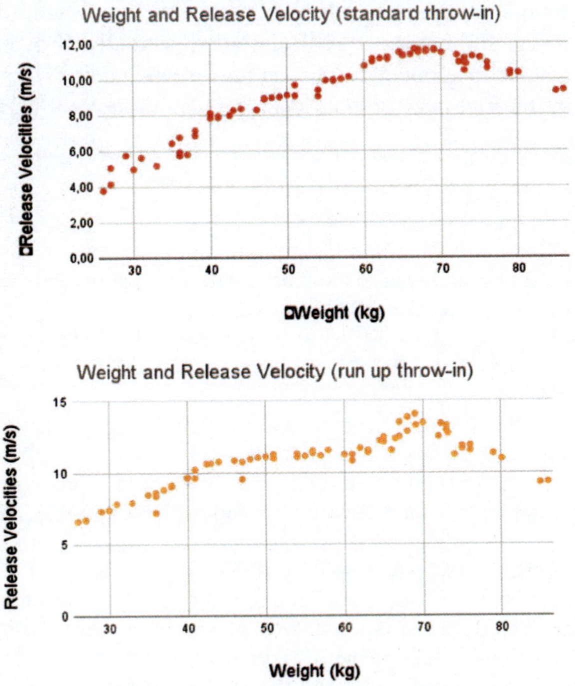

Fig. 9. Weight and distance in two cases

Effects of BMI

Underweights throw the shortest. Normal and overweight people have the greatest range, while those who are obese seem to be slightly lower (Fig. 10).

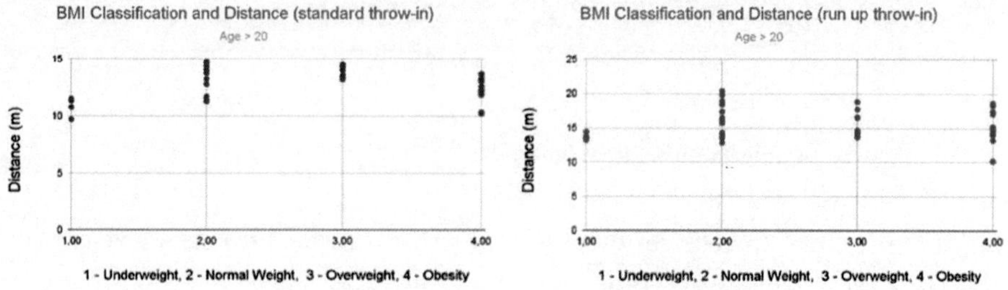

Fig. 10. BMI-Distance relation in two cases

4 Conclusion

The results of the analysis of the data showed that for the selection of male players to perform the throw-in situation, the following factors are prioritized: height over 1.6m, weight from 65–78kg, and having a BMI classified normally or overweight (by Asian standards). This result can support the chosen player when doing throw-ins in a football match, not only in the Vietnam football team but also in football games in universities and schools.

References

1. Cerrah, A.: Relationship between isokinetic strength parameters and soccer throw-in performance. Res. Q. Exerc. Sport **82**, A10–A10 (2011)
2. Levendusky, T.A., Clinger, C.D., Miller, R.E., Armstrong, C.W.: Soccer throw in kinematics. In: Terauds, J., Barham, J.N. (eds.) Biomechanics in Sports 2, California:Del Mar), pp. 258–268 (1985)
3. Messier, S.A.: Mechanics of Translation and rotation during conventional and handspring soccer throw- ins. Int. J. Sports Biomech. **2**, 301–315 (1986)
4. Kollath, E., Schwirtz, A.: Biomechanical analysis of the soccer throw-in. In: Reilly, T., Lees, A., Davids, K., Murphy, W.J. (eds.) Science and Football, (pp. 460–467, E & FN Spon, London (1988)
5. Lees. The biomechanics of football skills (2013)
6. Barraza, L.C.H., Yeow, C.: An investigation of full body kinematics for static and dynamic throw-in in proficient and non-proficient soccer players when they tried to hit a specific target. In: Colloud, F., Domalain, M., Monnet, T. (2015)
7. Chang, J.: The biomechanical analysis of the selected soccer throw-in techniques. Asian J. Phys. Educ. **2** (1979)
8. Cerrah, H.: The evaluation of ground reaction forces during two different soccer throw-in techniques: a preliminary study. In: Proceedings of 30th International Conference of Biomechanics in Sports, Melbourne, 2012, pp. 299–302

9. U. S. CDC. About Child & Teen BMI
10. . Nunome, H.: Elite Soccer referees. In: Football Biomechanics, pp. 3–4. Cenveo Publisher Services, London (2018)
11. Asai, T.: Direct free kicks. In: Football Biomechanics, p. 56. Cenveo Publisher Services, London (2018)

Reinforcement Control for Planar Robot Based on Neural Network and Extended State Observer

Duy Nguyen Trung[1], Thien Nguyen Van[1], Hai Xuan Le[2(✉)], Dung Do Manh[2], and Duy Hoang[2]

[1] Faculty of Information Technology, Hanoi University of Industry, Hanoi, Vietnam
nguyenthien@haui.edu.vn

[2] Faculty of Applied Sciences, Intenational School, Vietnam National University, Hanoi, Vietnam
hailx@vnu.edu.vn, {dungdm,hoangduy}@vnuis.edu.vn

Abstract. Based on the cooperation of neural network and extended state observer (ESO), in this paper, an approach for reinforcement control will be presented for Planar robot. By using the sliding surface as a state variable, the nominal system in quadratic form will be converted to first-order where the total uncertain component is estimated and remove by ESO. Then, a reinforcement algorithm will be added in collaboration to determine the nearly optimal solution of Hamilton-Jacobi-Bellman (HJB) equation. During the determination of the control signal, only one neural network is applied to reduce the computational complexity while still achieving the desired requirements. The simulation results of the algorithm will be examined on the Planar Robot with two degrees of freedom, thereby confirming the effectiveness of proposed control strategy.

Keywords: Reinforcement control · optimal control · dynamic programming · extended state observer · Planar robot

1 Introduction

In order to improve the productivity of manipulator systems in industry, the purpose of controller design is to control and adjust the robot's performance so that they can complete the tasks efficiently and optimally [1]. The problem of manipulator control is generally still a difficult problem because their mathematical models often have the nonlinear Euler-Lagrange form [2].

To address the issue of controlling industrial robotic manipulator, the approaches based on Lyapunov stability theory show the advantage when they ensure that the control errors is always in a neighborhood of the origin. Typical algorithms for this approach include the sliding mode controller [3,4]. By synthesizing the sliding surface from the system state variables, the controller can be designed so that not only can it guarantee tracking quality of system's states, but also control the speed of each joints. However, the dependence of control

T. D. L. Nguyen et al. (Eds.): ICISN 2023, LNNS 752, pp. 516–525, 2023.
https://doi.org/10.1007/978-981-99-4725-6_62

signal on system model is a limitation of sliding mode controller because the requirement to know the exact the robot model is difficult to fulfilled in practice. Therefore, to minimize the influence of uncertain parameters, the observers [5–7] can be used to estimate and remove their effects. In addition, adaptive control methods with purpose of steering system's states to desired trajectory in various operation situations where the neural networks are used as an effective tool to determined control signal were also studied [8,9] and gave good results. Thus, the applicability of neural networks in controller design for manipulator has been confirmed.

The requirement for manipulator control nowadays is not only to force the states to a given trajectory, but also to perform the tracking process optimally. This problem can be converted to a minimization problem for suitable cost function and solved by optimization method [10] through finding the solution of the HJB equation [11]. However, determining the direct analytic solution of the HJB equation is generally arduous, in many cases, it is almost impossible to find [12]. Therefore, numerical methods for approximating the solution of the HJB equation are preferred, where the minimization of the cost function and the design of the control signal is performed by the neural network [13]. Reinforcement control strategy designed for robotic manipulators are usually designed in Actor-Critic form [14–16] with two neural networks. The first plays the role of an approximation of optimal control evaluation function and the other is used to approximate optimal control law [16]. Two neural networks structure will bring convenience to controller development and management, but this also increase computational complexity and is a factor hindering the applicability of this method in practice. Therefore, reinforcement learning-based control strategy using only one neural network in whole design process [17] shows more efficiency in reducing the computational complexity.

Inspired from aforementioned works, in this paper, a cooperation of neural network and extended states observer based reinforcement control for Planar robot will be presented with two main contributions. First, an uncertain observer is designed thereby converted the nominal second order non-linear model of Planar robot to a first order system. Then, the single neural network optimal controller will be applied and the effectiveness of proposed cooperation will be verified through simulation results.

2 Problem Formulation

2.1 Dynamic Model of Planar Robot

The Euler-Lagrange form of 2 DoF robotic Planar model can be expressed as [5]

$$M(\underline{q})\ddot{\underline{q}} + C\left(\underline{q}, \dot{\underline{q}}\right)\dot{\underline{q}} + G\left(\underline{q}\right) = \underline{\tau} + \underline{\tau}_e. \tag{1}$$

where $\underline{q} \in \mathbb{R}^2$ denotes the state vector of joint variables, $\dot{\underline{q}}$ and $\ddot{\underline{q}}$ is accordingly first and second-order derivation of $\underline{q}$. Symmetry positive definite matrix $M(\underline{q}) \in \mathbb{R}^{2\times 2}$ [11] is symbolised for the inertia matrix, $C\left(\underline{q}, \dot{\underline{q}}\right) \in \mathbb{R}^{2\times 2}$ depicts all the centripetal-coriolis factors and vector $G(\underline{q}) \in \mathbb{R}^2$ is the gravity component. The control signal vector is represented by τ and the external disturbances are described by τ_e. For brief and convenience, this paper will use these notations M, C, G instead of $M(\underline{q}), C(\underline{q}, \dot{\underline{q}}), G(\underline{q})$, respectively.

2.2 Planar Robot Control Statement

The objective of reinforcement control for Planar robot is to determine the control signal so that the states of system track a given trajectory while minimizing the cost function. To design this cost function, based on [13], first, the tracking error of the system $\underline{e}_1 = [e_{1_1}; e_{1_2}]$ is defined by $\underline{e}_1 = \underline{r} - \underline{q}$ and the sliding surface $\underline{s} = [s_1; s_2]$ is determined by the formula

$$\underline{s} = \dot{\underline{e}}_1 + \lambda_1 \underline{e}_1 \tag{2}$$

with λ_1 is a positive definite symmetry parameter matrix with suitable dimension. From here, we get the dynamic model of the sliding variable

$$\begin{aligned} M_e \dot{\underline{s}} &= - C_e \underline{s} - \tau + \underline{\xi} \\ \underline{\xi} &= - \Delta M \dot{\underline{s}} - \Delta C \underline{s} + M(\ddot{\underline{r}} + \lambda_1 \dot{\underline{e}}_1) + C(\dot{\underline{r}} + \lambda_1 \underline{e}_1) + G - \tau_e \end{aligned} \tag{3}$$

where $\underline{\xi} = [\xi_1; \xi_2]$ is total uncertain function, M_e is a symmetric positive definite matrix that only depend on $\underline{e}_1$ and $\dot{\underline{e}}_1$, $\Delta M = M - M_e$, $\Delta C = C - C_e$. The control signal can be designed include two components $\underline{\tau} = \hat{\underline{\xi}} - \underline{u}$ where $\hat{\underline{\xi}}$ is the estimated value of $\underline{\xi}$ and $\underline{u}$ is the optimal control signal for system (4)

$$M_e \dot{\underline{s}} = -C_e \underline{s} + \underline{u} \tag{4}$$

Define the new state variable as $\underline{x} = \left[\underline{e}_1^T, \underline{s}^T\right]^T$, form (2) we have

$$\dot{\underline{x}} = F\underline{x} + B\underline{u} + B\left(\underline{\xi} - \hat{\underline{\xi}}\right) \tag{5}$$

in which $F = \begin{bmatrix} -\lambda 1 & I_2 \\ \Theta_2 & -M_e^{-1} C_e \end{bmatrix}$, $B = \begin{bmatrix} \Theta_2 \\ M_e^{-1} \end{bmatrix}$ with Θ_2 and I_2 are zero matrix and identity matrix dimension of 2×2. If the impact of $\underline{\xi}$ is removed or at least is minimized to a sufficiently small value by $\hat{\underline{\xi}}$, the system (5) becomes

$$\dot{\underline{x}} = F\underline{x} + B\underline{u} \tag{6}$$

The problem of reinforcement control for Planar Robot now can be separated into two smaller problems, the first is to determine the estimated component $\hat{\underline{\xi}}$ of the total uncertain $\underline{\xi}$ and the other objective is formulating optimal input signal $\underline{u}$ for nonlinear system (6). So, the main objective of this paper is to develop a control regime that solves each aforementioned problem.

3 Reinforcement Control Based on Neural Network and Extended State Observer

In this section, a reinforcement control strategy for Planar Robot is proposed. First, we will develop an observer for the objective of estimate the total uncertain component $\underline{\xi}$, then an optimal state feedback controller is applied to minimize the cost function while ensuring closed-loop tracking quality.

3.1 Total Uncertain Observer Design

An adaptive observer is constructed in this part to approximate the term $\underline{\xi}$, from (3) we have

$$\dot{\underline{s}} = -M_e^{-1}C_e\underline{s} - M_e^{-1}\tau + M_e^{-1}\underline{\xi} = -M_e^{-1}C_e\underline{s} - M_e^{-1}\tau + \underline{\eta} \tag{7}$$

with $\underline{\eta} = M_e^{-1}\underline{\xi}$ and instead of determining $\underline{\xi}$ directly, we will estimate $\underline{\eta}$ by $\hat{\underline{\eta}}$, then the estimated vector $\hat{\underline{\xi}}$ is calculated by

$$\hat{\underline{\xi}} = M_e\hat{\underline{\eta}} \tag{8}$$

Assuming that in a sufficiently small time interval, $\underline{\eta}$ is locally modeled by a first-order Taylor expansion, which mean $\ddot{\underline{\eta}} = \underline{0}$ and denote $\underline{z}_0 = \underline{s}$, $\underline{z}_1 = \underline{\eta}$, $\underline{z}_2 = \dot{\underline{\eta}}$, we have

$$\begin{cases} \dot{\underline{z}}_0 = -M_e^{-1}C_e\underline{z}_0 - M_e^{-1}\tau + \underline{z}_1 \\ \dot{\underline{z}}_1 = \underline{z}_2 \\ \dot{\underline{z}}_2 = \underline{0} \end{cases} \tag{9}$$

Define the observer values of $\underline{z}_0$, $\underline{z}_1$ and $\underline{z}_1$ are $\hat{\underline{z}}_0$, $\hat{\underline{z}}_1$ and $\hat{\underline{z}}_2$ respectively, the structure of adaptive observer is designed as

$$\begin{cases} \dot{\hat{\underline{z}}}_0 = -M_e^{-1}C_e\hat{\underline{z}}_0 - M_e^{-1}\tau + \hat{\underline{z}}_1 + \Upsilon_0\left(\underline{z}_0 - \hat{\underline{z}}_0\right) \\ \dot{\hat{\underline{z}}}_1 = \hat{\underline{z}}_2 + \Upsilon_1\left(\underline{z}_0 - \hat{\underline{z}}_0\right) \\ \dot{\hat{\underline{z}}}_2 = \Upsilon_2\left(\underline{z}_0 - \hat{\underline{z}}_0\right) \end{cases} \tag{10}$$

where Υ_0, Υ_1 and Υ_2 are parameter matrices with appropriate dimension. Subtract (10) from (9) and represent the result in matrix form we obtain

$$\dot{\underline{Z}} = A_z\underline{Z} \text{ with } A_z = \begin{bmatrix} -\Upsilon_0 - M_e^{-1}C_e & I_2 & \Theta_2 \\ -\Upsilon_1 & \Theta_2 & I_2 \\ -\Upsilon_2 & \Theta_2 & \Theta_2 \end{bmatrix} \tag{11}$$

in which $\underline{Z} = \mathrm{col}\left[\tilde{\underline{z}}_0, \tilde{\underline{z}}_1, \tilde{\underline{z}}_2\right]$ is the vector of all observer errors where $\tilde{\underline{z}}_i = \underline{z}_i - \hat{\underline{z}}_i$, $i = 0, 1, 2$ are the observer errors of $\underline{z}_i$. From (11) we can see that if Υ_0, Υ_1 and Υ_2 are designed to make A_z Hurwitz, the system (11) becomes stable and $\underline{Z}$ will converge to the origin which mean $\hat{\underline{z}}_i \to \underline{z}_i$ and we obtain that $\hat{\underline{\eta}} = \hat{\underline{z}}_1$ is the approximation value of $\underline{\eta}$. Then, the total uncertain component $\underline{\xi}$ is estimated by $\hat{\underline{\xi}}$ through (8).

Remark 1. The design procedure of proposed observer requires matrices M_e and C_e instead of M and C, therefore it is an adaptive observer when it can retain efficiency under the change of uncertain parameters in system's model.

Remark 2. The accuracy of the observer can be improved if we increase the degree of Taylor expansion used to approximate $\underline{\eta}$ and extend the state of the observer accordingly. However, this will also increase the computational complexity and limit the applicability of the observer in practice.

3.2 Optimal Controller Design Using a Single Neural Network

In this part, an optimal controller based on the dynamic programming principle will be designed for system (6) to minimized the cost function

$$J = \int_{0}^{\infty} \left(\underline{x}^T Q \underline{x} + \underline{u}^T R \underline{u}\right) dt \tag{12}$$

where $Q \in \mathbb{R}^{4\times4}$ and $R \in \mathbb{R}^{2\times2}$ are positive definite symmetry parameter matrices. On the other hand, with the design that M_e and C_e depend only on $\underline{e}_1$ and $\underline{\dot{e}}_1$, the matrices function F and B will depend only on $\underline{x}$ (because $\underline{\dot{e}}_1$ can be calculated form $\underline{e}_1$ and $\underline{s}$ through (2)). Therefore, system (6) is autonomous and based on [15], the optimal input signal can be archived by solving the HJB equation

$$H\left(\underline{x}, \underline{u}, V_x\right) = V_x^T \left(F\underline{x} + B\underline{u}\right) + \underline{x}^T Q \underline{x} + \underline{u}^T R \underline{u} \tag{13}$$

in which

$$V(\underline{x}(t)) = \int_{t}^{\infty} \left(\underline{x}^T Q \underline{x} + \underline{u}^T R \underline{u}\right) dt \tag{14}$$

is the evaluation function of control signal and $V_x = \partial V / \partial \underline{x}$. The optimal control signal evaluation function is defined through [15] and denoted by $V^*(\underline{x}(t))$. Hence, the optimal Hamiltonian function is obtained by

$$H\left(\underline{x}, \underline{u}, V_x^*\right) = \left(V_x^*\right)^T \left(F\underline{x} + B\underline{u}\right) + \underline{x}^T Q \underline{x} + \underline{u}^T R \underline{u} \tag{15}$$

and HJB equation becomes $\min H\left(\underline{x}, \underline{u}, V_x^*\right) = 0$. The optimal control signal $\underline{u}^*$ then will be determined according to the stationary condition [15], we get

$$\underline{u}^* = -\frac{1}{2} R^{-1} B^T V_x^* \tag{16}$$

Thus, it can be seen that, if the optimal evaluation function V^* is determined, we can completely calculate the optimal input signal $\underline{u}^*$. However, solving HJB equation to find analytical solution is significantly complicated, in many cases it is almost impossible to find [12]. Therefore, approximate V^* is a suitable approach and the neural network is a useful tool to solve this problem.

First, the ideal evaluation function $V^*(\underline{x})$ can be approximated by a neural network [17] as follows

$$V^*(\underline{x}) = W^T \Psi(\underline{x}) + \zeta \tag{17}$$

in which $W \in \mathbb{R}^m$ is the weight of neural network, vector $\Psi(\underline{x}) : \mathbb{R}^n \to \mathbb{R}^m$ is the activation mapping with m is the number of neurons in hidden layer and ζ is neural network's approximated error. Then, HJB equation is rewritten as

$$H(\underline{x}, \underline{u}, W) = W^T \Psi_x (F\underline{x} + B\underline{u}) + \underline{x}^T Q \underline{x} + \underline{u}^T R \underline{u} + \zeta_x^T (F\underline{x} + B\underline{u}) = 0 \tag{18}$$

where $\Psi_x = \partial \Psi / \partial \underline{x}$ and $\zeta_x = \partial \zeta / \partial \underline{x}$

Next, substituting the optimal control signal calculated by (16) into (18) and by representing the ideal evaluation function $V^*(\underline{x})$ through the neural network (17) we obtained

$$H(\underline{x}, W) = W^T \Psi_x A \underline{x} + \underline{x}^T Q \underline{x} - \frac{1}{4} W^T \Psi_x B_c \Psi_x^T W + \zeta_{hjb} = 0 \tag{19}$$

where $B_c = BR^{-1}B^T$ and ζ_{hjb} is the residual error calculated by

$$\zeta_{hjb} = \zeta_x^T A \underline{x} - \frac{1}{2} W^T \Psi_x B_c \zeta_x - \frac{1}{4} \zeta_x^T B_c \zeta_x \tag{20}$$

The control problem now is to determine the ideal weight W that satisfies the Eq. (19). For the dynamic programming principle, the ideal value of W will be estimated gradually by updating approximate weight $\hat{W}$ during the operation of the system by minimizing function H. At this point, the evaluation function $V(\underline{x})$ will be calculated according to the formula

$$\hat{V}(\underline{x}) = \hat{W}^T \Psi(\underline{x}) \tag{21}$$

When using $\hat{V}(\underline{x})$ to design the controller, from (16) we obtain the approximate optimal control signal.

$$\underline{\hat{u}} = -\frac{1}{2} R^{-1} B^T \Psi_x^T \hat{W} \tag{22}$$

and the HJB equation now becomes

$$H\left(\underline{x}, \underline{\hat{u}}, \hat{W}\right) = \hat{W}^T \Psi_x A \underline{x} + \underline{x}^T Q \underline{x} + \underline{\hat{u}}^T R \underline{\hat{u}} = \hat{\zeta} \tag{23}$$

where $\hat{\zeta}$ is the Hamilton error generated by the approximation of neural network and the approximate optimal control signal. Minimizing the function H is equivalent to minimizing $\hat{\zeta}$. To do this, a weight update rule for $\hat{W}$ will be constructed to minimize $\Phi = \frac{1}{2} \hat{\zeta}^T \hat{\zeta}$ whereas ensuring the stabilization of closed-loop system. According to [10], the proposed updating law for $\hat{W}$ is as follows

$$\dot{\hat{W}} = \begin{cases} \dot{\hat{W}}_1, & \underline{x}^T (A\underline{x} + B\underline{\hat{u}}) \leq 0 \\ \dot{\hat{W}}_1 + W_c, & \underline{x}^T (A\underline{x} + B\underline{\hat{u}}) > 0 \end{cases} \tag{24}$$

where $\dot{\hat{W}}_1$ is designed based on the modified Levenberg-Marquardt algorithm [10] and determined by

$$\dot{\hat{W}} = -\alpha \frac{\sigma}{(\sigma^T \sigma + 1)^2} \left(\sigma^T \hat{W} + \underline{x}^T Q \underline{x} + \underline{u}^T R \underline{u} \right) \tag{25}$$

in which $\sigma = \Psi_x (A\underline{x} + B\hat{\underline{u}})$ and W_c is the extra rule to keep the closed-loop system to be stable. It is computed by

$$W_c = \frac{1}{2} \beta \Psi_x B_c \underline{x} \tag{26}$$

These coefficients α and β are chosen as two positive constants. The stability of closed-loop system when using the controller (22) with updating rule (25) has been demonstrated in [17], thereby ensuring that the estimated weight $\hat{W}$ is asymptotically stable at the ideal weight W while guaranteeing that the tracking errors are bounded in a neighborhood of the origin. It can be seen that in the optimal estimator's design procedure, only one neural network is used, so the computational complexity has been significantly reduced.

4 Numerical Examples

This section will validate the performance of the proposed control strategy on Planar robot [16] with two state variables which are rotating angels denoted by q_1 and q_2. For the design of uncertain observer, these matrices M_e, C_e are defined by

$$M_e = \begin{bmatrix} \gamma_1 + 2\gamma_2 \cos e_{1_2} & \gamma_3 + \gamma_2 \cos e_{1_2} \\ \gamma_3 + \gamma_2 \cos e_{1_2} & \gamma_3 \end{bmatrix}$$

$$C_e = \begin{bmatrix} -\gamma_2 \sin e_{1_2} \dot{e}_{1_2} & -\gamma_2 \sin e_{1_2} (\dot{e}_{1_1} + \dot{e}_{1_2}) \\ \gamma_2 \sin e_{1_2} \dot{e}_{1_1} & 0 \end{bmatrix}$$

and the observer's parameters is chosen as

$$\Upsilon_0 = 3W_0 - M_e^{-1} C_e,\ \Upsilon_1 = 3W_0^2,\ \Upsilon_2 = W_0^3,\ W_0 = diag([40; 40])$$

For the optimal controller design, the matrices of cost function are defined according to [13] and the controller parameters are $\alpha = 5$, $\beta = 1$, the activation function $\Psi(\underline{x})$ is selected as

$$\Psi(\underline{x}) = \left[x_1^2, x_1 x_2, x_2^2, x_3^2, x_3^2 \cos q_2, x_3 x_4, x_3 x_4 \cos q_2, x_4^2 \right]^T$$

The initial condition of $\hat{W}$ given by

$$\hat{W} = [2; 1; 3; 2.5; 1; 1; 1; 0.5]^T$$

The PE condition is satisfied by combining a high-frequency explorational noise with the control input for the first 6 s and canceled after the convergence of parameters of neural network. The controller (22) will be utilized to steer the joints of robot to a cyclic desired trajectory $\underline{r} = [r_1; r_2]$

The simulation results obtained through Fig. 1 and Fig. 2 show that the angle of both joint quickly track the references, thereby confirming the effectiveness of proposed control strategy. The performance of the extended state observer is demonstrated in Fig. 3. We can see that the estimator give an excellent estimation to each elements of the total uncertain term even though the Taylor expansion used for design process was just the first-order.

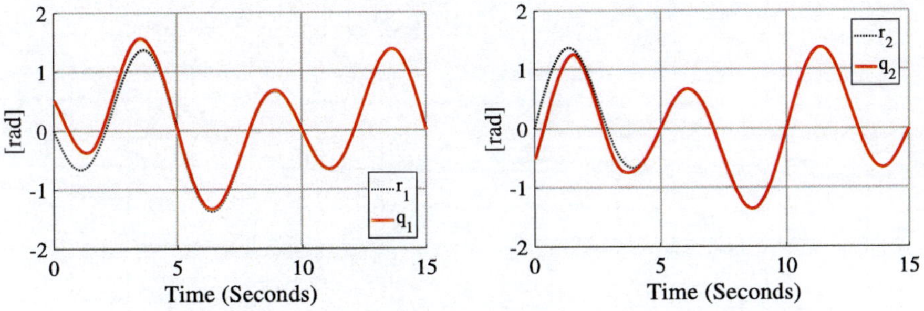

Fig. 1. Output response of link 1 (left) and link 2 (right)

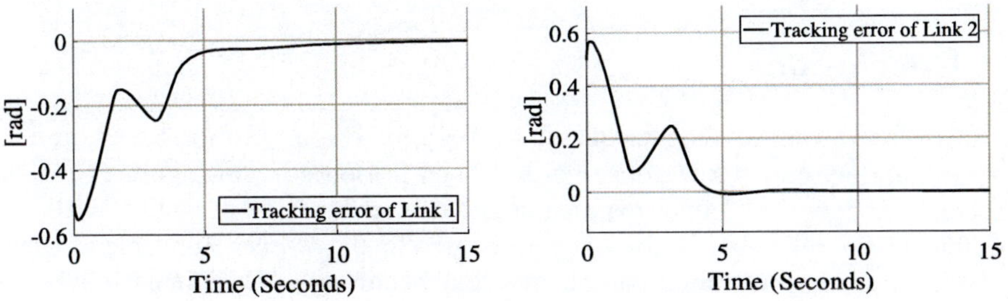

Fig. 2. Tracking error of link 1 (left) and link 2 (right)

The convergences of each elements of sliding surface are shown in Fig. 4a and the convergence of the weights of neural network is shown in Fig. 4b with $\hat{W} = [w_1; w_2; ...; w_8]$. It can be observed that with the updating rule (25), after a short amount of time, the approximate weights $\hat{W}$ of neural network approach the ideal weight and stable around this value thereby showing that the optimal controller can perform effectively in cooperation with the ESO.

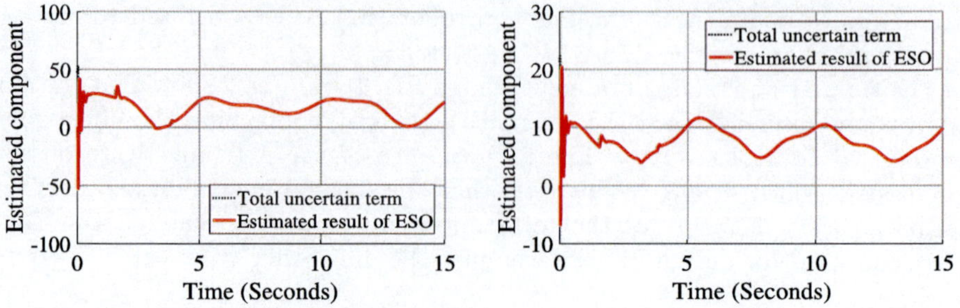

Fig. 3. Estimated result of ξ_1 (left) and ξ_2 (right)

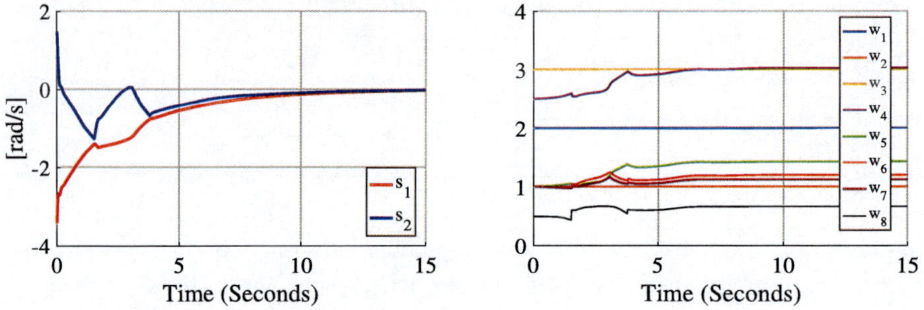

Fig. 4. a, The convergence to origin of each elements of sliding surface (left) b, The convergence to origin of the weight of the neural network (right)

5 Conclusions

In this work, a reinforcement control strategy for Planar Robot based on neural network and extended state observer has been presented. Throughout the design process, the degree of Taylor expansion for the construction of ESO as well as the number of neural network for optimal controller have been minimized thereby reducing the computational complexity. The accuracy of ESO in estimating and removing the influence of the total uncertain component in cooperation with the effectiveness of neural network in approximating the optimal solution of the HJB equation has brought an excellent control quality for Planar robot, which has been verified via the simulation results.

References

1. Blatnický, M., Dizo, J., Gerlici, J., Sága, M., Lack, T., Kuba, E.: Design of a robotic manipulator for handling products of automotive industry. Int. J. Adv. Robot. Syst. **17**(1) (2020)
2. Li, H., Liu, C.L., Zhang, Y., Chen, Y.Y.: Adaptive neural networks-based fixed-time fault-tolerant consensus tracking for uncertain multiple Euler-Lagrange systems. ISA Trans. **129**, 102–113 (2021)

3. Rahmani, M., Komijani, H., Rahman, M.H.: New sliding mode control of 2-DOF robot manipulator based on extended grey wolf optimizer. Int. J. Control Autom. Syst. **18**(6), 1572–1580 (2020). https://doi.org/10.1007/s12555-019-0154-x
4. Shao, K., Tang, R., Xu, F., Wang, X., Zheng, J.: Adaptive sliding mode control for uncertain Euler-Lagrange systems with input saturation. J. Franklin Inst. **358**(16), 8356–8376 (2021)
5. Ramírez-Neria, M., Madonski, R., Luviano-Juárez, A., Gao, Z., Sira-Ramírez, H.: Design of ADRC for second-order mechanical systems without time-derivatives in the tracking controller. In: 2020 American Control Conference (ACC), (2020)
6. Ha, W., Back, J.: A disturbance observer-based Robust Tracking Controller for Uncertain Robot Manipulators. Int. J. Control Autom. Syst. **16**(2), 417–425 (2018). https://doi.org/10.1007/s12555-017-0188-x
7. Zhang, Z., Leibold, M., Wollherr, D.: Integral sliding-mode observer-based disturbance estimation for Euler-Lagrangian systems. IEEE Trans. Control Syst. Technol. **28**(6), 2377–2389 (2019)
8. Zhou, Q., Zha, S., Li, R., Lu, R., Wu, C.: Adaptive neural network tracking control for robotic manipulators with dead zone. IEEE Trans. Neural Netw. Learn. Syst. **30(**12), 3611 - 3620 (2018)
9. Jouila, A., Nouri, K.: An adaptive robust nonsingular fast terminal sliding mode controller based on wavelet neural network for a 2-DOF robotic arm. J. Franklin Inst. **357**(18), 13259–13282 (2020)
10. Ioannou, P., Fidan, B.: Advances in Design and Control, Adaptive Control Tutorizal. SIAM, PA (2006)
11. Tang, L., Liu, Y.J., Tong, S.: Adaptive neural control using reinforcement learning for a class of robot manipulator. Neural Comput. Appl. **25**, 135–141 (2014)
12. Lewis, F.L., Vrabie, D.: Reinforcement learning and adaptive dynamic programming for feedback control. IEEE Circuits Syst. Mag. **9**(3), 32–50 (2009)
13. Dupree, K., Patre, P.M., Wilcox, Z.D., Dixon, W.E.: Asymptotic optimal control of uncertain nonlinear Euler-Lagrange systems. Automatica **1**, 99–107 (2011)
14. Hu, Y., Si, B.: A reinforcement learning neural network for robotic manipulator control. Neural Comput. **30**(7), 1983–2004 (2018)
15. Vamvoudakis, K.G., Lewis, F.L.: Online actor-critic algorithm to solve the continuous-time infinite horizon optimal control problem. Automatica, **46**, 878–888, (2010)
16. Vu, V.T., Dao, P.N., Loc, P.T., Huy, T.Q.: Sliding Variable-based online adaptive reinforcement learning of uncertain/disturbed nonlinear mechanical systems. J. Control Autom. Electr. Syst. **32**(2), 281–290 (2021). https://doi.org/10.1007/s40313-020-00674-w
17. Luy, N.T.: Reinforecement learning-based optimal tracking control for wheeled mobile robot. In: Proceedings of IEEE International Conference on Cyber Technology in Automation, Control, and Intelligent Systems (CYBER), pp. 371-376, May 2012

Proposing a Semantic Tagging Model on Bilingual English Vietnamese Corpus

Huynh Quang Duc(✉)

Faculty Information Technology, Robotics and Artificial Intelligence, Binh Duong University, Thu Dau Mot, Binh Duong, Vietnam
hqduc@bdu.edu.vn

Abstract. In this article, we report on the results of the study on building a semantic tagging system using the English-Vietnamese bilingual corpus to create a lexical resource with lexical notes based on translation similarities, transfer of vocabulary, and classification schemes through bilingual connections. This tool plays an important role in building and developing natural languages processing systems such as automatic translation, text summary, text extraction, information retrieval, and question-answer automatic. In this context, our goal to see English as the source language is roughly translated into Vietnamese, we use an English-Vietnamese bilingual concept dictionary for the purpose of annotating the semantics of words. In our experiments, we used hand-annotated vocabulary sets, compared with the proposed model results, and the results were achieved based on an average vocabulary coverage of 83.05%, with accuracy reaching up to 82.40%. This test can be applied to build lexical la-bels for other languages than bilingual corpus whose source language is English.

Keywords: Semantic Tagging · Bilingual Corpus · Lexical Label

1 Introduction

- In this paper, we implement a semantic labeling system in Vietnamese based on English semantics that is translated similarly on the LLOCE - LLOCV (it is shortened following the name of the Dictionary about the sense) bilingual concept dictionary and perform training for the model by English -Vietnamese bilingual corpus. During more than two decades of rapid growth in computer science, semantic resources and semantic labeling tools have been developed such as EuroWordNet [1], and USAS [2], these two systems have contributed greatly to the research and constructions of the NLP systems. Semantic-labeled developmental applications have been developed and achieved high results in practice [3, 4]. Information technology applications have also been researched and developed over time [5–7].
- Although, the semantic labeling tools implemented in English with the monolithic elements are much developed, as well as done in bilingual English - French; English - Japanese; English - Chinese. However, the implementation of bilingual English - Vietnamese has not received much attention. Recently, there have been studies of

T. D. L. Nguyen et al. (Eds.): ICISN 2023, LNNS 752, pp. 526–535, 2023.
https://doi.org/10.1007/978-981-99-4725-6_63

people who deceive science [8], plagiarism of people lacking scientific personality [9], analyzing linguistic semantics [10] has been reported, the USAS system has made great contributions to linguistic discovery and analysis. A system like USAS, which has been developed in many languages, will give science great benefits. Semantic label support systems implemented in languages such as Russian and Finnish [11, 12] initially achieved good results. Recently, the disambiguation of semantics in multi-language has been studied a lot, but mainly on semantic role labeling [23–25], so the de-ambiguation of languages has not yielded the same results desired, especially in Vietnamese. Therefore, the support of decision-making systems is not very effective.

- For the above reasons, we propose a solution for semantic labeling on English - Vietnamese bilinguals to help solve semantic ambiguity in Vietnamese, the research results can support decision-making systems to get more accurate results. We use the LLOCE-LLOCV semantic label set with 2449 label sets with more than 120000 bilingual sentence pairs as training corpus for the model, which is taken from the corpus that has been used in linguistic studies at the Center for Computational Linguistics (CLC), University of Natural Sciences, Ho Chi Minh City [13]. Observing the experimental results, we find that the results are very positive, which can be applied in practice to support semantic ambiguity reduction models for Vietnamese. The system efficiency is over 80%, we have built a bilingual corpus with semantic labels on Vietnamese over 100000 pairs of bilingual sentences.
- The remainder of the article is textured as follows: 2. Related work, 3. Approach method, 4. Experiment on the model, 5. Discussion and future development.

2 Related Work

In the development of semantic tagging systems, there have been studies with various methods performed on multiple languages such as GATE [14] and KIM [15] with the idea of building a labeling system based on ontologies to create a multilingual labeling tool. Freeling [16] also has a multilingual approach to tagging WordNet semantic entities and labels. Recently, Zhang and Rettinger [17] experimented with developing a labeling system based on multilingual texts on Wikipedia (wikization).

Studying semantic label sets, Paul Rayson and his colleagues developed a semantic analysis tool using LLOCE concept labels which are divided into 21 topics, of which 21 are divided into 232 different types of ideas [18]. In particular, the authors have relied on many different types of knowledge to determine semantic labels for words in sentences such as type words (POS tags), words with multiple words (MWEs), frequency dictionaries, and text scopes (the domain of discourse). We refer to the organization of the LLOCE conceptual point constructed by Mc Arthur in 1981 (Mc Arthur, 1981) with the organization of labels with additional elements to help increase the accuracy of semantic labels when assigning labels to text. For example: sex (sex) label with m/f (gender) or ± (proactive/passive) added to the label "chirpy" or "moody" with code "E4.1+" or "−E4.1".

In the study of semantic role labeling, the authors used heterogeneous data sources and language translation capabilities to compare semantics for the approaching model [19]. In this study, the authors presented approaches for learning multilingual phrase performances by manipulating coupled or uncoupled bilingual documents. The study hypothesizes that the multilingual aligning strategy is changeable, and thus, a model is created that trains for the alignment of the two languages so that it can be applied to the multilingual alignment serving the system [20]. Also with the multilingual approach, Yuan Chai et al. used the multilingual masked language model without the need for grammatical translation of multilingualism. The author has recognized the similarity between languages in terms of word constriction, word structure, and co-vocabulary to build the semantic structure for the language [21].

3 Approach Method

English-Vietnamese bilingual corpus after preprocessing steps including separating words in Vietnamese text, assigning word labels to Vietnamese and English, aligning English-Vietnamese bilingual, and identifying real words. Separating Vietnamese words using vnTokenizer 4.1.1, assigning word labels to English using Stanford-postagger tool 3.5.2, assigning magnetic type labels to Vietnamese using vnTagger 4.2.0, the same word as GIZA++v1.0.7. After the preprocessing steps, we have an English-Vietnamese bilingual corpus that has been lined up on the wording level, and in bilingual pairs only word labels for real words. These bilingual couples are input during the semantic tagging process through the following steps: Assigning base labels to nouns, filtering labels, calculating similarities, acknowledging errors, and assigning semantic labels. The output of this process is English-Vietnamese bilingual corpus with semantic labels on real words (nouns, verbs, adjectives) in English-Vietnamese sentence pairs.

3.1 Tokenization and Tagging POS (Part-Of-Speech) on Vietnamese

Vietnamese language materials must be tokenized from each sentence before conducting the labeling of category words to ensure greater accuracy. Because, specifically in the Vietnamese language, the definition of word boundaries is not implicitly understood by spaces like in English. For example, if we have the phrase "Tôi là giáo viên" then the word "giáo viên" if not separated, then by default "giáo viên" will be understood as two words, a word "giáo" and a word "viên". When the word is tokenized, the sentence above will become "Tôi là giáo_viên" and "giáo_viên" will be understood as a word.

The tagging POS in Vietnamese is done by the vnTagger tool with a set of 17-word tags. In particular, word labels for nouns include Np - Proper noun (proper noun), Nu - Unit noun (noun in unit), N - Common noun (common noun), and Nc - Classifier (noun type only)... Specifically, with labels for real words (nouns, verbs, adjectives) will start with the letters N, V, A (JJ for English words) in capital letters. Vietnamese language materials when being labeled with noun types will have the following form.

3.2 Tagging POS on English and Alignments for the Parallel Corpus

English is a language classified as a type of flexion language, the identification of word boundaries in English sentences is mainly white space or punctuation. In English words with two or more words are rarely focused on verbs, such as: carry out (make), take place (happen), calm down (calm), get in touch (contact), etc. The set of word labels used in Stanford-postagger tool: Penn Treebank tag set with 36 different word labels, including 4 labels for nouns, 6 verb labels, 3 adjective labels.

3.3 Tagging Foundational Labels

Baseline labeling is the first stage of the semantic tagging process to find all possible labels of actual words in a sentence, representing the degree of the meaning of the word. Then, using the method of collating labels through the word links between English words and Vietnamese words (from the results of alignment in GIZA++ in the preprocessing step) by the assignment (AND) between the two sets of mechanical labels Department of two words, the result of the assignment (AND) is a set of common labels. From this general set of labels, a reasonable semantic label will be selected for the actual word among the possible labels through its morphological, grammatical, and semantic similarity in the corpus. The system based on the link between words in two languages may include one-to-one links (this is the most expected link), one-to-many links, many-to-one links, and many-to-many links (Fig. 1).

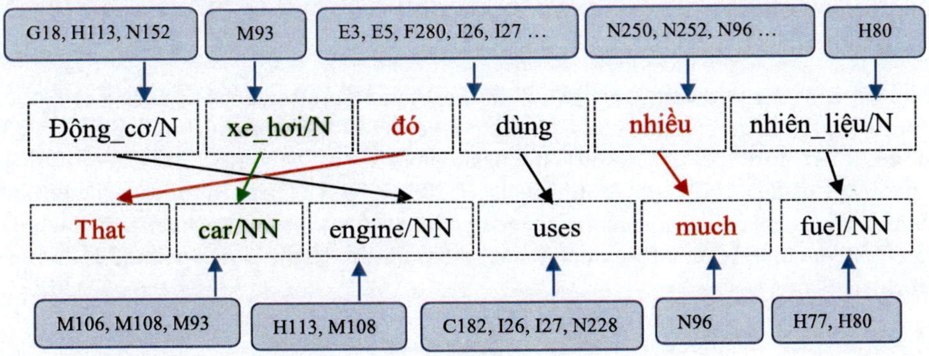

Fig. 1. A example for tagging foundational labels that the system used semantic labels of LLOCE – LLOCV

3.4 Implement and Labels

Case 1. The result of AND is one, the system will recognize this label as true and unique for both data sets in determining the sense Vietnamese and English. For example, in this case, we have the following bilingual sentence: “Chim có th bay   c” – “Bird can fly”, the word “Chim” has the basic label set {A32}, the word “bird” has the basic label set {A32, C4, C7}, the result of the AND is {A32} (in this case, the Vietnamese word has one label, the English word has many labels).

Case 2. The result of the AND is different from the 1 label, there are also two cases: zero or greater than or equal to 2.

Case Equals Zero. In this case, the system will recognize the error for the following reasons: The alignment of the error is incorrect, the word is not in the LLOCE or LLOCV label set, and the tagging POS is inaccurate.

Case Greater than or Equal to 2. In this case, the system has not been able to identify the label for the word, because after performing the AND there are 2 or more labels in the result label. For example, we have the following bilingual sentence:

i_tuy n/N bóng_ á/N ã dành/V c vài th ng_1 i/N 1 n/A.

The football/NN ({2}) team/NN ({1}) won/VBN ({4}) some ({6}) great/JJ ({8}) victories/NNS ({7}).

We consider the word "th ng_1 i" in the Vietnamese sentence and the word "victories" in the English sentences. The basic label set of the word "th ng_1 i" is {C283, J146, K107, K108, N130, N169}, the basic label set of "victories" is {C283, K108}, the result of the AND is a set of labels {C283, K108}. Then, in order to determine which labels are reasonable in the above two labels for Vietnamese and English words, the system will rely on information theory from Philip Resnik's viewpoint [22] through formula (1).

$$P(c) = \frac{x = -\sum_{w \in words(c)} count(w)}{N} \quad (1)$$

In particular, word count (c) is a set of words ordered according to the same principle as words c, and N is all the words in the training dataset. In the principle of information theory, the information meaning of class c in the training data set is determined by the formula (2).

$$IC(c) = -\log(P(c)) \quad (2)$$

Applied to the issue of counting the similarity of tags in the set of outcomes labels from the assignment (AND) to calculate the appropriate labels for the noun pairs in the bilingualism phrase in the prior example through the formula (3).

$$P(c_i) = \frac{\sum_{w \in words(c_i)} count(w)}{N} \quad (3)$$

c_i has the ith tag in the suite of result tags ($i \geq 2$), and w is the number of words in the training data set stored in a similar principle in c_i. N is the amounts of words in the training data set. Then, the system will choose the tag with the maximum information content in c_i ($i \geq 2$) tags with formula (4).

$$IC(c_i) = -\log(P(c_i)) \quad (4)$$

The Model will choose the tag by determining max (IC (c_i)).

Returning to the example in the bilingual sentence above, we need to choose the most appropriate tag in the set of tags {C283, K108} for two words "th ng_l i" and "victories".

According to Eq. (1), based on the 111,982 English sentence corpora, of which: 31,951 sentences in LLOCE, 60,032 sentences in EVC, and 19,999 sentences in machine-translated corpus, with a total of 1,348,968 words, we can calculate the semantic similarity of each label C283 and K108.

4 Experiment on Model

4.1 Prepare Corpus

The corpus is edited, then worded and labeled as the sample below:

Tr ng/Nn này ào_t o/V các/Nq em sau_này ph c_v /V quân_ i/Nn.
This college/NN prepares/VB boys/NNS for the army/NN.
ây là/V m t/Nq t_n c/Nn có nhi u/A ti m_n ng/Nn.
This is a country/NN of great/JJ potentiality/NN.

We prepare the corpus for the approach as follows: English and Vietnamese corpus, after being separated from and labeled with words, will be word aligned to create word links between the two languages English - Vietnamese serving the process of labeling and comparing labels between English and Vietnamese words translated in parallel. The alignment process is done by selecting the source language Vietnamese, and the target language English. The output corpus results are pairs of English-Vietnamese bilingual sentences with alignment indexes between English and Vietnamese sentences of the following form:

C u/Pp y/Pd i u_khi n/Vv m t/Nq chi c/Nc xe_h i/Nn /PU
NULL ({2}) He/PRP ({1}) drove/VBD ({3}) a/DT ({4}) car/NN ({5 6})./. ({7})
C u/Pp y/Pd b i/Vv qua/D sông/Nn. /PU
NULL ({2}) He/PRP ({1}) swam/VBD ({3 4}) the/DT ({}) river/NN ({5}). /. ({6})
Cô/Pp ta/Pp h c/Vv nh c/Nn. /PU
NULL ({}) She/PRP ({1 2}) studies/VBZ ({3}) music/NN ({4})./ ({5})

The above bilingual corpus will then go through the real word recognition unit, other words will be removed, leaving only the type words of real words (noun, verb, adjective) to serve the process of identifying word types and assigning labels is more efficient, in line with the requirements of the original problem (only assigning semantic labels to real words). The aligned bilingual corpus results after removing other words are as follows:

C u y i u_khi n/V m t/Nq chi c/Nc xe_h i/Nn.
NULL ({2}) He ({1}) drove/VBD ({3}) a ({4}) car/NN ({5 6}). ({7}).
C u y b i/V qua sông/Nn.
NULL ({2}) He ({1}) swam/VBD ({3 4}) the ({}) river/NN ({5}). ({6}).
Cô ta h c/V nh c/Nn.
NULL ({}) She ({1 2}) studies/VBZ ({3}) music/NN ({4})./ ({5})

4.2 System Architecture

Based on the LLOCE-LLOCV bilingual semantic label set, the English type of word set with the Penn Treebank tagset, and preprocessing steps described above. We combine with the package nltk 3.0.4 and vnTokenizer [26] to separate words and assign labels, jpype-py3 package integrates java into python supporting preprocessing and identifying English words with the stemming package 1.1.0. Along with the tools and components look up semantic vocabulary and identify the semantic shift of cross-language. The proposed model architecture is described through the basic steps as shown in Fig. 2.

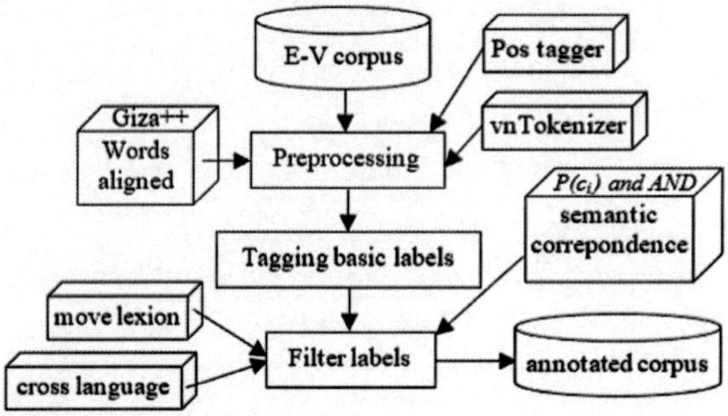

Fig. 2. The model of the semantic tagging system in Vietnamese - English bilingual corpus using LLOCE and LLOCV

4.3 Experiment

For evaluation, we use the labeling tools for the corresponding target language and check the percentage of actual words that are semantically labeled. The corpus that we used is listed in Table 1 below and the results of lexicon coverage.

Table 1. The outcomes that the approaching model performed regarding Lexicon coverage.

Resources	Pairs of bilingual sentences	Number of words	Tagged words	Lexicon coverage
LLOCE-LLOCV	31,951	344,180	278,945	**81.05%**
EVC	60,032	696,485	536,382	77.01%
MT	20,000	198,211	159,543	**80.49%**
Others	33,201	296,966	230,421	77.59%

Assessing the accuracy, currently, gold corpus (Language materials are labeled by linguists) on Vietnamese language materials is very limited. So, in practice, we carried out manual labeling through the criteria referenced by linguists at universities that have foreign languages and linguistics departments. The documents are based on the following criteria: first, consider which one of the 14 subjects of the LLOCE topic, followed by which group of 129 groups was divided into 14 topics, and finally, we considered semantic words in 2,449 topics that divided into 129 groups.

To ensure accuracy, we took 2,000 pairs of sentences in the corpus that have not been tested for vocabulary coverage above, then proceeded with experimental steps to measure vocabulary coverage (Lexicon Coverage - LC) and checked the results. Labeling accuracy (P - Precision), and coverage (R - Recall) are calculated using the following formula (5).

$$P = \frac{S \cap U}{S}, R = \frac{S \cap U}{U} \quad (5)$$

where S is the total number of tags lined up, and U is the total number of tags lined up exactly by hand. Experimental results are summarized in table 2 below.

Table 2. The outcomes that the proposed model performed about the Recall and the Precision

Manually tagged Labels	Tagged labels by model	Correct labels	Recall (%)	Precision (%)
17,362 (LLOCE-LLOCV)	18,060	15,011	**86.46%**	**83.12%**
17,009 (EVC)	18,002	14,342	**84.32%**	79.67%
18, 981 (MT)	18,201	15,330	80.76%	**84.23%**
18,442 (Others)	19,103	15,227	80.67%	82.57%

5 Discussion and Future Development

Through testing the experimental outcomes, some of the words are not labeled including aligning incorrectly, words not in LLOCE, proper nouns, the verbs as to be (is, are, was, were), to have (has, have, had) that also were not labeled because it acts as the auxiliaries. Some words that were separated from wrong words leading to wrong meanings were also not labeled. For example, in the bilingual sentences " ó là m t c n gió ông l nh" and "That is a cold east wind", the word " ông l nh" in the Vietnamese sentence that was separated to the word " ông_l nh", which leads to wrong meanings of the sentence, which also affects the quality of labeling.

In this article, we studied the feasibility of sense tagging studies on multiple languages by mapping a lexical and semantic structure on bilingualism. Specifically, we have tested the ability to translate semantic vocabulary from English into Vietnamese. Our experiments show that, if we have a high-quality English-Vietnamese bilingual corpus translated by language experts (gold corpus), assigning semantic labels with higher vocabulary coverage. In the future, we continue to pay more attention to the language labeled by experts, as a basis for evaluating that our algorithm is effective and applicable in several other languages as well as the translation of semantic vocabulary from one language to another to apply the effective automatic translation systems.

Acknowledgments. I would like to express my sincere thanks to Associate Professor Dr. Le Anh Cuong and colleagues at NLP-KD labs for their professional guidance and support in completing this research. We also thank our colleagues at the Faculty of Information Technology, Robotics, and Artificial Intelligence, Binh Duong University for helping us in processing training data and labeling word senses for all experiences and finishing this article.

References

1. Vossen, P. (ed): EuroWordNet: A Multilingual Database with Lexical Semantic Networks. Kluwer, Dordrecht (1998)
2. Rayson, P., Archer, D., Piao, S., McEnery, T.: The UCREL semantic analysis system. In: Proceedings of the Workshop on Beyon Named Entity Recognition Semantic labelling for NLP Tasks in Association with 4th International Conference on Language Resources and Evaluation (LREC 2004), Lisbon, Portugal, pp.7–12 (2004)
3. Klebanov, B.B., Leong, C.W., Dario Gutierrez, E., Shutova, E.: Semantic classifications for detection of verb metaphors. In: Proceedings of the 54th Annual Meeting of the Association for Computational Linguistics (2016). https://doi.org/10.18653/v1/P16-2017
4. Potts, A., Baker, P.: Does semantic tagging identify cultural change in British and American English. Int. J. Corpus Linguist. **17**(3), 295–324 (2012). https://doi.org/10.1075/ijcl.17.3.01pot
5. Hardmeier, C., Volk, M.: Using linguistic annotations in statistical machine translation of film subtitles. In: Jokinen, K., Bick, E. (eds.) NODALIDA 2009 Conference Proceedings, pp. 57–64 (2009)
6. Collings, D.G., Doherty, N., Luethy, M., Osborn. D.: Understanding and supporting the career implications of international assignments. J. Voc. Behav. **78**(3), 361–371 (2011). https://doi.org/10.1016/j.jvb.2011.03.010
7. Gacitua, R., Mazon, J.N., Cravero. A.: Using Semantic Web technologies in the development of data warehouses: a systematic mapping (2018). https://doi.org/10.1002/widm.1293
8. Markowitz, D., Hancock, J.T.: Linguistic Traces of a Scientific Fraud: The Case of Diederik Stapel. (2014). https://doi.org/10.1371/journal.pone.0105937
9. Kramer, A.D.I., Guillory, J.E., Hancock, J.T.: Experimental evidence of massive-scale emotional contagion through social networks. Psychol. Cogn. Sci. **121**, 8788–8790 (2014). https://doi.org/10.1073/pnas.1320040111
10. Balossi. G.: A Corpus Linguistic Approach to Literary Language and Characterization: Virginia Woolf's The Waves (2014). https://doi.org/10.1075/lal.18
11. Löfberg, L., Piao, S., Rayson, P., Juntunen, J-P.: A semantic tagger for the Finnish language. In: The Corpus Linguistics Conference 2005 at: Birmingham, UK (2015)
12. Sharoff, S., Babych, B., Rayson, P., Mudraya, O., Piao, S.: ASSIST: automated semantic assistance for translators. Demonstrations. In: 11th Conference of the European Chapter of the Association for Computational Linguisitcs (EACL 2006}, pp. 139–142 (2006)
13. http://www.clc.hcmus.edu.vn/?lang=en
14. Cunningham, H., Maynard, D., Bontcheva, K.: Text Processing with GATE. Gateway Press, California (2011). ISBN: 0956599311 9780956599315
15. Goranov, M.: KIM - semantic annotation platform. In: Proceedings of 2nd International Semantic Web Conference (ISWC2003), Florida, USA, pp. 834–849 (2003)
16. Padró, L., Stanilovsky, E.: FreeLing 3.0: towards wider multilinguality. In: Proceedings of the Language Resources and Evaluation Conference (LREC 2012). Istanbul, Turkey. May 2012

17. Zhang, L., Rettinger, A.; Semantic annotation, analysis and comparison: a multilingual and cross-lingual text analytics toolkit. In: Proceedings of the Demonstrations at the EACL 2014, Gothenburg, Sweden, pp. 13–16 (2014)
18. Mc Arthur, T.: Longman Lexcicon of Contemporary English. Longman, London (1981)
19. Conia, S., Bacciu, A., Navigli, R.: Unifying cross-lingual semantic role labeling with heterogeneous linguistic resources. In: Proceedings of the 2021 Conference of the North American Chapter of theAssociation for Computational Linguistics: Human Language Technologies, pp. 338–351, June 6–11, 2021. ©2021 Association for Computational Linguistics (2021)
20. Tien, C.-C., Steinert-Threlkeld, S.: Bilingual alignment transfers to multilingual alignment for unsupervised parallel text mining. Published at ACL 2022. arXiv:2104.07642 [cs.CL]
21. Chai, Y., Liang, Y., Duan. N.: Cross-lingual ability of multilingual masked language models: a study of language structure. ACL 2022. arXiv:2203.08430 [cs.CL]. 16 Mar 2022
22. Resnik, P.: Semantic similarity in a taxonomy: an information-based measure and its application to problems of ambiguity in natural language. J. Artif. Intell. Res. **11**(1999), 95–130 (1999)
23. Larionov, D., Shelmanov, A., Chistova, E., Smirnov. J.: Semantic role labeling with pretrained language models for known and unknown predicates. In: Proceedings of Recent Advances in Natural Language Processing, pp. 619–628, Varna, Bulgaria, 2–4 September 2019
24. Alam, M., Gangemi, A., Presutti, V., Reforgiato Recupero, D.: Semantic role labeling for knowledge graph extraction from text. Progr. Artif. Intell. **10**(3), 309–320 (2021). https://doi.org/10.1007/s13748-021-00241-7
25. Phuong, L.H.,et al.: Vietnamese Semantic Role Labelling. Articles, vol. 33, No. 2 (2017). Published Nov 21, 2017. https://doi.org/10.25073/2588-1086/vnucsce.166
26. https://github.com/undertheseanlp/word_tokenize

A Synthetic Crowd Generation Framework for Socially Aware Robot Navigation

Minh Hoang Dang[1], Viet-Binh Do[1], Tran Cong Tan[2], Lan Anh Nguyen[2], and Xuan-Tung Truong[2](✉)

[1] Institute of Information Technology, AMST, Hanoi, Vietnam
[2] Le Quy Don Technical University, Hanoi, Vietnam
xuantung.truong@gmail.com

Abstract. Socially aware robot navigation has gathered more and more interest from research communities due to its promising applications. Recent breakthroughs in Deep Reinforcement Learning (DRL) have opened many approaches to archive this task. However, due to the data-hungry characteristic of DRL-based methods, many promising proposed works have only trained on simulation, making real life applications still an open question. In this paper, we propose (i) a new Synthetic Crowd Generation (SCG) framework along with (ii) a world model for generating valid synthetic data. As a data-generating framework, SCG can be easily integrated into existing DRL-based navigation models without changing it. According to evaluations on simulation as well as real life data, our SCG has successfully boosted the published state-of-the-art navigation policy in terms of sample efficiency.

Keywords: Socially aware robot navigation · model-based reinforcement learning · synthetic data generation · sample efficiency

1 Introduction

Socially aware robot navigation is an active research area due to its applications and potential developments. Early approaches have proposed well-designed interaction models, which guide the robot to its goal while avoiding moving and static obstacles [2,3]. The key challenge is that they heavily relied on hand-crafted functions and cannot generalize to various situations in social crowds. Furthermore, these methods normally result in short-sighted navigation policies as well as unnatural behaviors. Recent approaches [4,8] utilize the power of DRL to archive promising results in terms of success rate, navigation time as well as socially acceptable trajectories. A recent popular one that can be named is SARL [5], which rethinks crowd-robot interactions with a self-attention mechanism to attain superior results. However, DRL-based approaches require a lot of data for training the policy. The lack of data results in not only underfitting but also biased models, which directly affect the robot's performance [7]. Many

T. D. L. Nguyen et al. (Eds.): ICISN 2023, LNNS 752, pp. 536–545, 2023.
https://doi.org/10.1007/978-981-99-4725-6_64

attempts have been made in [10] to collect training data in real life. Due to the complicated social navigation task, the cost of infrastructure and time as well as hidden dangerous situations, this task still remains challenging.

In recent years, synthetic data has emerged as one of the main engines for future machine learning based methods, since it is cheaper to obtain, faster to generate, and safer to collect. In the domain of socially aware robot navigation, existing work is diverse, ranging from high-fidelity 3D simulators that focus on raw sensor processing to simplified grid worlds that center a high level of navigation abstraction. In terms of high-fidelity synthetic data, a particular note is the Gazebo simulator [11], a general-purpose 3D simulator that robotics communities have widely used. Other popular choices are Unity and Unreal which are the underlying engine for AirSim [14], CARLA [6], and UnrealROX [13]. However, these environments center around raw sensor processing and neglect socially compliant factors. Besides, the more sensors are simulated, the bigger the cumulative sim-to-real gap explodes, which complicates transferring the trained policy to the real world. The 2D simulators like SocialGym [9], focus on agent-level information instead of sensor-level. By centering around only agents' position and velocity, it is easier to develop a navigation policy that encompasses human-human, human-robot interactions at the social-accepted level. Nevertheless, these works lack adaptation mediums to transfer from simulation to real life. Bridging the gap between synthetic data and real life applications, we design the Social Crowd Generation (SCG) framework. Our SCG incorporates a learnable world model, which enables it to learn the input high-level information to form the new close-real training samples. Therefore, on one hand, SCG provides DRL-based models with infinite data. On the other hand, it is capable of learning a variety of dynamic motions within social crowds.

In this paper, we proposed (i) the SCG framework, which is able to enhance existing navigation policy models in terms of sample efficiency, and collision-free navigation by providing a training phase with closed-real synthetic data. Also, we introduce (ii) a data generation model (which we call the world model) which is the heart of the framework. As the framework only changes training data, therefore, it is widely compatible with current and future DRL-based navigation models.

2 Proposed Method

2.1 Robot Navigation Task Formulation

The task of robots navigating through humans can be considered a sequence of decision-making based on observations. This can be formulated as a typical Markov Decision Process (MDP) [4,8]. An MDP is a tuple of (S, A, P, R, γ), where S is state space, A is a set of finite actions that the robot can take, P is the unknown state transition function, R is reward function and $\gamma \in (0, 1)$ is the discount factor. In n-agent (robot or human) scenario ($\mathbb{N} = \{1, \ldots, n\}$), the state vector for each agent is composed of an observable and hidden potion $s_i^t = [s_i^{t,o}, s_i^{t,h}]$. Observable states are position $p = [p_x, p_y]$, velocity $v = [v_x, v_y]$, and

Algorithm 1. Proposed DRL-based robot navigation using SCG

Input : Training data D
Output: Trained navigation policy π

Initialize policy π
Initialize empty memory D for training world model
Initialize empty memory D'
$D \underset{k}{\leftarrow} COLLECT(env)$ ▷ collect k episodes from *env*.
while not done **do**
  Initialize parameters θ for *env'*
  $\theta \leftarrow TRAIN\_WORLD\_MODEL(D)$ ▷ train world model, section 2.3 .
  $D' \leftarrow DATA\_GENERATION(env')$ ▷ generate data from *env'*.
  $\pi \leftarrow TRAIN\_RL(D')$ ▷ update policy using world model.
end while

Algorithm 2. Generate training sample from *env'*

Input : A random part of real training sample d_i
Output: A completed synthetic training sample d_i'

$d_i = [d_i^0 : d_i^r] \leftarrow D$ ▷ get a random part of a sample from real data.
$g_i^0 = d_i^r$
while not end episode **do**
  $g_i^{t+1} = env'(g_i^t)$ ▷ using world model to generate synthetic data.
end while
$d_i' = [d_i^0 : d_i^r, g_i^1 : g_i^{end}] \rightarrow D'$ ▷ join as a completed synthetic training sample.

radius r which is denoted as $s^{t,o} = [p_x^t, p_y^t, v_x^t, v_y^t, r]$. Hidden (or self-aware-only) states are the goal position $g_i = [g_x, g_y]$ preferred speed v_{pref}, and orientation Ψ, $s^{t,h} = [g_x, g_y, v_{pref}, \Psi]$. At every step t, the robot is aware of other observable states as well as its self-state, which is denoted as joined state $s^{jn} = [s^h_{robot}, s^o_{1:n}]$. We assume that the robot archives desired velocity instantly after taking action a_t. The optimal policy, $\pi^* : s^{jn} \longmapsto a_t$ is developed by maximized expected return $\mathbb{E}(s^{jn})$ [4]:

$$\pi^* = \underset{a_t}{argmax} R(s^{jn}, a^t) + \gamma^{\Delta t.v_{pref}} \int_{s^{jn}_{t+\Delta t}} P(s_t^{jn}, a_t, s_{t+\Delta t}^{jn}) V^*(s_{t+\Delta t}^{jn}) ds_{t+\Delta t}^{jn} \quad (1)$$

where V^* is the optimal value function:

$$V^*(s_{t+\Delta t}^{jn}) = \sum_{t'=t}^{T} \gamma^{t'.v_{pref}} R_{t'}(s_{t'}^{jn}, \pi^*(s_{t'}^{jn})) \quad (2)$$

2.2 Proposed Framework

In general, the policy π is trained by data, sampled from a given environment *env* (which can be real life or simulation). Inspired by model-based reinforcement

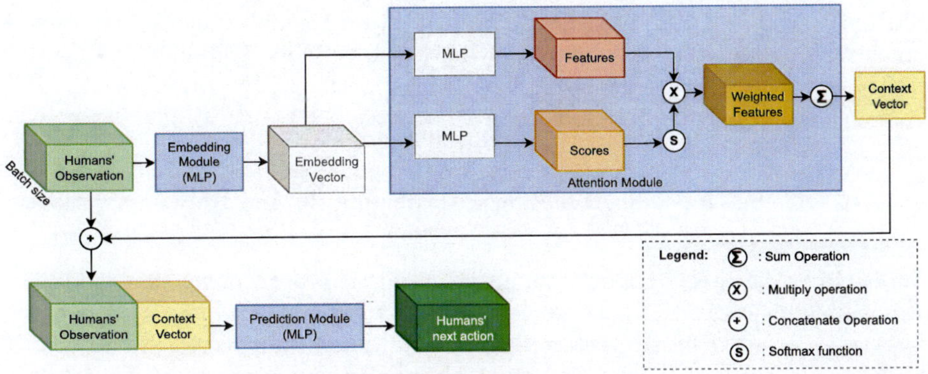

Fig. 1. Proposed World model

learning, we construct the SCG framework, which is described in Algorithm 1. The main purpose of SCG is to draw new samples from env' at every step of training π. So that, π is trained with a wide variety of valid data.

In the beginning, SCG takes in initial samples D, which are human trajectories from the target environment *env*, as shown on line 4. These data will be used to train the simulated one *env'* (line 7), which is described in detail in Sect. 2.3. Trained *env'* is then capable of generating infinite synthetic data samples D' (line 8). A key point to distinguish SCG from other synthetic data generation is its generating Algorithm 2. The *env'* randomly picks a part of a trajectory d_i from D to generate its future prediction g_i. Each pair $[d_i, g_i]$ form a synthetic training sample for π. As human motion is multimodal, given a partial history, there is no single accepted prediction. Therefore, using $d_i' = [d_i, g_i]$ as a training sample is reasonable. A main problem with SCG lies in the overfitting of π on $D' = \{d_i'\}$ over time. To mitigate it, we randomly reinitialize the weight θ and retrain *env'* at every training episode, as shown in line 6. Finally, a chosen existing navigation policy π is trained on D' (line 9). As SCG does not affect the original training process of π, within SCG we keep the training process of the selected π untouched.

2.3 Proposed World Model

The main purpose of the world model is to predict the next humans' position using current observations. In order to do that, the model takes humans' current position and velocity as the input, $O = \{O_i[p^t_{x,i}, p^t_{y,i}, v^t_{x,i}, v^t_{y,i}] : i = 0, \ldots, n\}$, and produces humans' next action, $A = \{A_i[v^{t+1}_{x,i}, v^{t+1}_{y,i}] : i = 0, \ldots, n\}$, which is then easily converted to the next humans' position. Besides, when humans move in dense areas, they might interact with each other, therefore, the model should pay more attention to those people. This motivates us to introduce the world model which consists of three components, as shown in Fig. 1:

Embedding Module. translates humans' state to high-dimensional vectors which will be easier for extracting moving features. This module is built with a multi-layer perception (MLP):

$$e_i = \phi_e(O_i; W_e) \tag{3}$$

where $\phi_e(.)$ is an embedding function with weights W_e and ReLU activations.

Attention Module. Taking human interaction into account, this module is responsible for creating a context vector, that will be attached to each human's observation. This context vector plays as additional information for the world model while predicting what it would observe in the next state. To do that, the Attention module takes features from the Embedding module and weighted them by score. At first, we implement an MLP layer for extracting features from the Embedding module:

$$h_i = \psi_h(e_i; W_h) \tag{4}$$

where $\psi_h(.)$ is a fully-connected layer with ReLU activations and weights W_h. Those features, h_i, are then, weighted by their attention scores, α, before summing up to create the context vector c:

$$\alpha_i = \psi_\alpha(e_i, W_\alpha) \tag{5}$$

$$c = \sum_{i=1}^{n} softmax(\alpha_i)h_i \tag{6}$$

where $\psi_\alpha(.)$ is a fully-connected layer with ReLU activations and weights W_α.

Prediction Module. predicts the next human action from the human's current observation, along with the context vector. This module is constructed by MLP layers:

$$A_i = f_A(I_i, c; W_A) \tag{7}$$

where $f_A(.)$ is an MLP with ReLU activations and W_A are the weights. To facilitate implementation, I_i is concatenated with the context vector c to create a single input. Finally, as the number of humans may vary, we encode those numbers as the batch size, which makes the model work normally with diversity in the number of inputs.

3 Experiments

Our framework undergoes evaluation using both simulation and real-world data training. We adopt SARL [4], a state-of-the-art policy network architecture that effectively navigates robots through crowds while considering crucial factors such as human-robot and human-human interactions, as well as the width of the robot and humans. Additionally, SARL takes into account dangerous situations

Table 1. Evaluation results on simulation training. "SARL": Original SARL. "SARL_SCG": SARL powered by SCG.

Setting	Success	Collision	Time	Reward
Number of humans: 5				
SARL	0.81	0.03	11.51	0.240
SARL_SCG	**0.97**	0.03	**9.64**	**0.345**
Number of humans: 10				
SARL	0.85	**0.00**	14.86	0.182
SARL_SCG	**0.96**	0.04	**12.05**	**0.261**
Number of humans: 15				
SARL	0.70	**0.10**	14.20	0.144
SARL_SCG	**0.87**	0.13	**12.68**	**0.207**

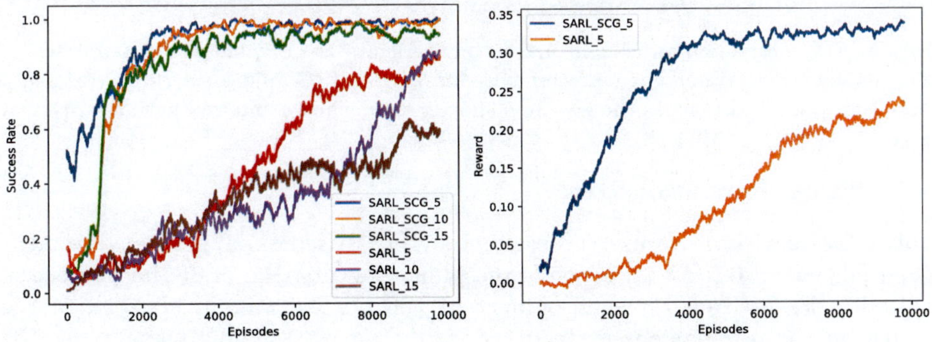

(a) Average success rate (b) Average cumulative reward

Fig. 2. Average success rate and cumulative reward on simulation training. "SARL_X": Original SARL in the environment of X humans. "SARL_SCG_X": SARL powered by SCG in the environment of X humans.

where the robot is in close proximity to humans. We employ SARL with the same settings as the original paper for both simulation and real-world data training, denoting the SARL model with SCG as SARL_SCG. Our world model implementation utilizes functions $\phi_e(.), \psi_h(.), \psi_\alpha(.), f_A(.)$, with hidden units of (150, 100), (100, 50), (100, 100), (150, 100, 100), respectively. The training phase consists of 10,000 episodes, consistent with the default setting in the SARL paper. Also in our experiments, we assume the robot is invisible to humans. This invisible setting is claimed as a clean test for validating SARL's ability in reasoning Human-Robot and Human-Human interactions [4]. Besides, in the visible setting, the robot can be easily avoided by humans, making it easier to navigate and harder to evaluate. To better evaluate the framework, we only use holonomic kinematics for the robot. A video with our simulation results can be found at this link[1].

[1] https://youtu.be/5PgvLuOzKAk.

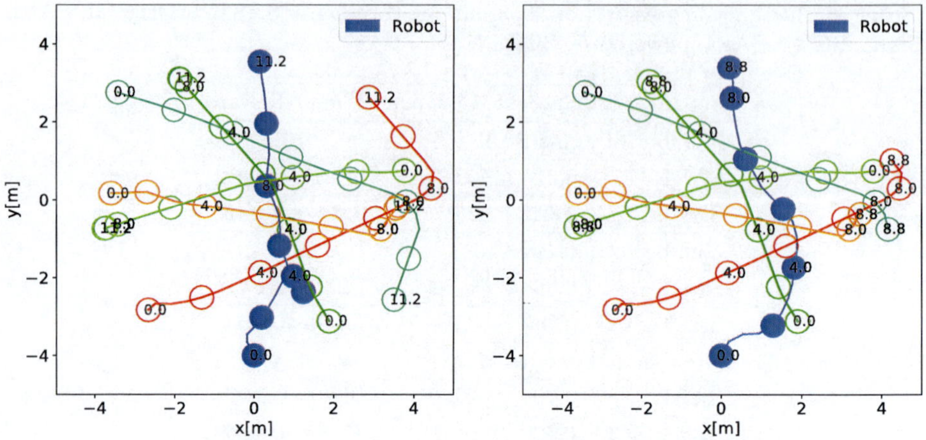

(a) SARL (b) SARL_SCG

Fig. 3. Trajectory comparison in a test case. Agents are demonstrated as circles at the labeled times. When encountering humans, SARL_SCG guides the robot through a smoother path and reaches the goal in a shorter time when compared with the original SARL

3.1 Using Simulation Data

Data Collection. We use the original Python-based simulation environment from [4], where ORCA [1] controls the human agents. We limit the number of episodes drawn from the environment to 150 during training, instead of the 10,000 in the original paper. In other words, no matter how many times the robot interacts with *env*, it can only collect experiences from these 150 episodes. We also generate another 100 episodes from *env* for testing. The experimental results of the test are shown in Table 1, Fig. 2, and Fig. 3.

Quantitative Evaluation. We analyze the performance of SCG by evaluating it with varying numbers of humans in the environment (5, 10, and 15). The original SARL experiences a significant reduction in success rate, attributed to insufficient training samples. Despite 10,000 training episodes, SARL only achieves an 81% success rate, whereas [4] reports it is capable of reaching 100%.

SARL_SCG outperforms the original SARL in success rate, navigation time, and average reward, across all evaluations. SARL_SCG achieves nearly perfect success rates of 97% and 96% in crowds of 5 and 10 humans, respectively, while the original SARL only reaches 81% and 85%. In a crowd of 15 humans, SARL_SCG maintains an 87% success rate, far superior to the normal SARL's 70% success rate. (See Table 1 for results.)

Qualitative Evaluation. We evaluated the effectiveness of SCG through qualitative assessments. Figure 2 shows that SARL_SCG achieves a much faster convergence speed in all scenarios. The average success rate and cumulative discounted reward almost reach their maximum values of approximately 100%

Table 2. Evaluation results on real data.

Setting	Success	Collision	Time	Reward
SARL	0.84	0.16	22.72	0.079
SARL_SCG	**0.92**	**0.08**	**20.49**	**0.107**

and 0.32, respectively, after only 5,000 episodes. In contrast, conventional SARL slowly improves and cannot even reach a 90% success rate after 10,000 episodes. In addition, SARL_SCG produces smoother trajectories that avoid collisions and reach the goal faster than SARL, as shown in Fig. 3.

3.2 Using Real Data

In real data training, we adopt Zara crowd datasets from Trajnet++ [12]. Initially, we joined consequence scenes into one and then divided them into overlapped sub-scenes. Each sub-scene contains 65 frames and starts later than the previous 5 frames. By doing that, we collected 114 sub-scenes, which we divides into two parts: 64 for training and 50 for testing. During the test phase, the robot's start position and goal are replaced by the human with the longest trajectory. The experimental results are shown in Table 2, Fig. 4, and Fig. 5

Evaluation with real data shows the same picture, where, the SCG framework has successfully provided SARL with sufficient training data to reach its potential success rate of approximately 92%. With the same training data, SCG has boosted SARL in terms of success rate and navigation time. This result also indicates the ability to apply SCG in real life data.

Qualitative evaluation of real data indicates that SARL_SCG has a better converging speed in comparison with SARL, as shown in Fig. 4. In terms of navigation trajectory, SARL_SCG successfully guide the robot to the goal without colliding with humans, while the robot with the original SARL hit humans at time 18.0 (Fig. 5).

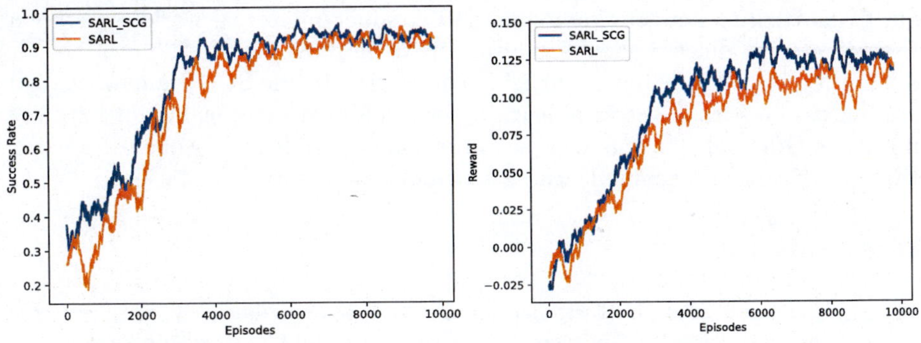

(a) Average success rate (b) Average cumulative reward

Fig. 4. Average success rate and cumulative reward on real data training.

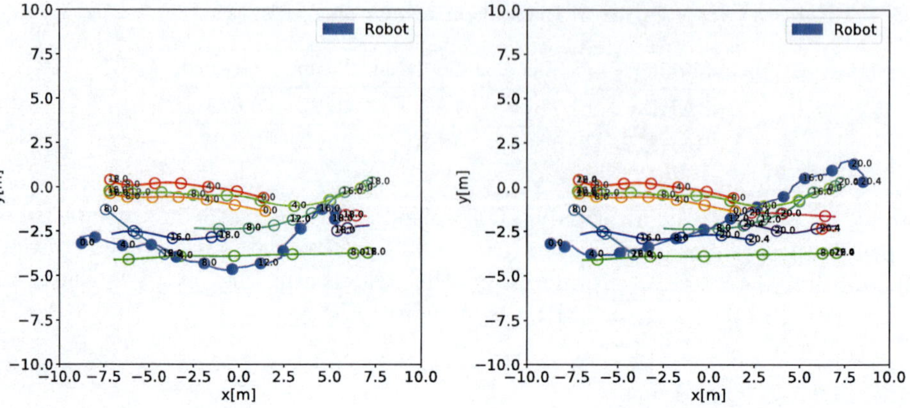

(a) SARL

(b) SARL_SCG

Fig. 5. Trajectory comparison in a test case. Agents are demonstrated as circles at the labeled times.

4 Conclusions and Future Work

We presented our proposed framework (SCG), a model-based reinforcement learning approach that operates directly on observations and efficiently learns navigation policy from little data. Our experiments in real data demonstrate that the framework can effectively learn to navigate through crowds with only approximately 2.5 min of observation, corresponding to 64 training episodes. While the framework can learn effectively from little data, it does not provide quick adaptation to different moving patterns. Our empirical experiments indicate that trained SARL in simulation performs relatively low in a real dataset, and vice versa. In this paper, our focus was to demonstrate the capability of SCG in a single environment.

In future work, we plan to use meta-learning techniques to quickly adapt our framework to new environments and conduct more real-world experiments to evaluate its effectiveness. We also aim to improve the training of the policy network by developing a learnable connection to the SCG framework, using approaches such as adversarial learning or reinforcement learning for joint optimization. Our goal is to enhance the performance and flexibility of our framework through advanced machine learning techniques and experimentation.

References

1. van den Berg, J., et al.: Reciprocal n-body collision avoidance. In: Springer Tracts in Advanced Robotics. Springer, Berlin, pp. 3–19 (2011). https://doi.org/10.1007/978-3-642-19457-3_1
2. Borenstein, J., Koren, Y.: Real-time obstacle avoidance for fast mobile robots.. IEEE Trans. Syst. Man Cybernet **19**(5), 1179–1187 (2011). https://doi.org/10.1109/21.44033

3. Borenstein, J., Koren, J.: The vector field histogram-fast obstacle avoidance for mobile robots. IEEE Trans. Robot. Autom. **7**(3), 278–288 (1991). https://doi.org/10.1109/70.88137
4. Chen, C., et al.: Crowd-robot interaction: crowd-aware robot navigation with attention-based deep reinforcement learning (2018). https://doi.org/10.48550/ARXIV.1809.08835. https://arxiv.org/abs/1809.08835
5. Chen, Y., et al.: Robot navigation in crowds by graph convolutional networks with attention learned from human gaze.. IEEE Rob. Autom. Lett. **5**(2) , 2754–2761 (2020). https://doi.org/10.1109/LRA.2020.2972868
6. Dosovitskiy, A., et al.: CARLA: an Open Urban Driving Simulator (2017). eprint: arXiv:1711.03938
7. Evans, M.J., Rosenthal, J.S.: Probability and Statistics, the Science of Uncertainty, p. 200. W. H. Freeman and Company, New York (2004)
8. Everett, M., Chen, Y.F., How, J.P.:Collision avoidance in pedestrian-rich environments with deep reinforcement learning. IEEE Access **9**, 10357–10377 (2021). https://doi.org/10.1109/access.2021.3050338
9. Holtz, J., Biswas, J.: SOCIALGYM: a Framework for Benchmarking Social Robot Navigation (2021). eprint: arXiv:2109.11011
10. Karnan, H., et al.: Socially compliant navigation dataset (SCAND): a large-scale dataset of demonstrations for social navigation. In: (2022). eprint: arXiv:2203.15041
11. Koenig, N., Howard, A.: Design and use paradigms for Gazebo, an open-source multi-robot simulator. In: 2004 IEEE/RSJ International Conference on Intelligent Robots and Systems (IROS). Vol. 3, Pp. 214924 (2004). https://doi.org/10.1109/IROS.2004.1389727
12. Kothari, P., Kreiss, S., Alahi, A.: Human trajectory forecasting in crowds: a deep learning perspective. In: IEEE Transactions on Intelligent Transportation Systems, pp. 1–15 (2021). https://doi.org/10.1109/TITS.2021.3069362
13. Martinez-Gonzalez, P., Oprea, S., Garcia-Garcia, A., Jover-Alvarez, A., Orts-Escolano, S., Garcia-Rodriguez, J.: UnrealROX: an extremely photorealistic virtual reality environment for robotics simulations and synthetic data generation. Virtual Reality **24**(2), 271–288 (2019). https://doi.org/10.1007/s10055-019-00399-5
14. Shah, S., et al.: AirSim: high-fidelity visual and physical simulation for autonomous vehicles (2017). eprint: arXiv:1705.05065

Danaflood: A Solution for Scalable Urban Street Flood Sensing

Tien Quang Dam[1], Duy Khanh Ninh[1(✉)], Anh Ngoc Le[2], Van Dai Pham[2], and Tran Duc Le[1]

[1] The University of Danang – University of Science and Technology, Danang, Vietnam
tiendam.vn@gmail.com, {nkduy,letranduc}@dut.udn.vn
[2] Department of Information Technology, Swinburne Vietnam, FPT University, Hanoi, Vietnam
{ngocla2,daipv11}@fe.edu.vn

Abstract. Urban flooding is difficult to predict, and most cities lack the tools to track its evolution automatically. Surveillance camera systems are available in nearly every city, but they lack a smart function that would send an alert in the event of an emergency. To detect street flooding alarmingly, we suggest a highly scalable intelligent system. This system can simultaneously produce high-resolution data for future use and send out high-abstract warning signals. The chosen deep convolutional neural network model, U-Net with backbone MobileNetV2, achieved a classification accuracy of 89.58% and flood image segmentation accuracy of 95.33%. The demo prototype model is deployed on a cloud instance, serving up to 100 camera points. This method would create a highly-scalable measurement of street flood conditions without requiring the installation of new on-site infrastructure.

Keywords: Multi-output deep learning model · Image segmentation · Image classification · Flood detection

1 Introduction

Apart from flooding in rural areas, urban flooding is caused by problems with drainage infrastructure, as pipe systems handle 45% of rain flow [16]. The more developed a city, the more vulnerable it is to urban flooding [12]. Vietnam, for example, is one of six countries that has been severely impacted by climate change and is vulnerable to factors related to rural and urban flooding [2]. Furthermore, increased data availability will result in more rational and data-driven planning decisions and more efficient and beneficial urban development. A low-cost tool for collecting such data will assist those areas in becoming more sustainable during climate change. Many cities worldwide, including Vietnam, have camera surveillance systems installed for security reasons. Despite the fact that this system is frequently connected to video recorders in the city's management center, it does not use any intelligent functionality. It is mostly operated

T. D. L. Nguyen et al. (Eds.): ICISN 2023, LNNS 752, pp. 546–555, 2023.
https://doi.org/10.1007/978-981-99-4725-6_65

by humans, which may result in missing information due to human inattention for an extended period of time [5]. Those camera systems were primarily used for an evidential recording of events but could not send warning signals. Additionally, providing reliable data in real-time may help in the effort required to deal with extreme weather. Therefore, developing an intelligent module for such camera systems would aid in providing alarm signals without requiring a large investment in infrastructure.

The main contributions of this work are:

1. Building a multi-task deep learning model that can simultaneously segment flooded areas in images and classify an intuitive measure of flooding level while remaining scalable and cost-effective.
2. Developing a source of information to assist citizens in choosing a suitable and safe commute route and minimizing damage caused by flooded streets. At the same time, high-resolution data should be collected for later use.

This study is divided into five sections: Sect. 2 discusses related research; Sect. 3 describes the methodology; Sect. 4 discusses the implementation, experiments, and findings; and Sect. 5 concludes the study.

2 Related Works

The target of a flood-related problem can be divided into four categories: (1) flood detection; (2) estimation of water level or depth; (3) inference of the impact area; and (4) risk assessment [8]. In this study, the authors concentrated on targets (1) and (2) to create data availability for later (3) and (4).

Some studies use image processing methods to detect the edge of a water body in images, such as Otsu thresholding and Canny edge detection, to detect changes in the water level in a river or lake. They tested in real-time with a low computational power edge device [18]. However, this method cannot adapt to changes in camera locations and lighting in the environment. Other approaches that rely on crowd-sourced data include social networks. The Mask-RCNN method has extremely high performance, predicting both water level and segmentation of flooded objects with high accuracy. Still, Mask-RCNN is a large model that requires a detailed dataset [3]. Furthermore, this method is dependent on third-party applications and raises legal concerns. The alarm signal may be outdated and easily influenced by incorrect user input. Some methods approach flood monitoring by using satellite or UAV data. This method may be an expensive option requiring a significant investment in equipment and infrastructure. Some countries and areas require special permission from a military agent to access UAVs or satellites [6,10]. The traditional sensor method would result in a higher resolution and more robust signal. This method may require the presence of an anchor object, such as a water level ruler, or the use of an ultrasound distance sensor. However, this involves the installation of new infrastructure and might not be suitable for street floods where it is not easy to install anything on the street [4,7]. Deep learning models can be used to solve the flood alarm problem as a classification or segmentation problem. A study proposed a massive flood warning system across Taiwan with over 2000 camera points [9]. But the project consumes 2000 GPU instances to serve as the computational core. Aside from that, using deep learning to segment and then compute a flood observing index is a good way to use a small model while still reporting an approximation of the scene's situation [11]. Those approaches have become a direct source of inspiration for this study.

3 Methodology

This paper delves into flood detection and reporting as a deep-learning segmentation and classification problem. However, the classification output will be derived from the segmentation model. Figure 1 depicts the entire pipeline and its expected output.

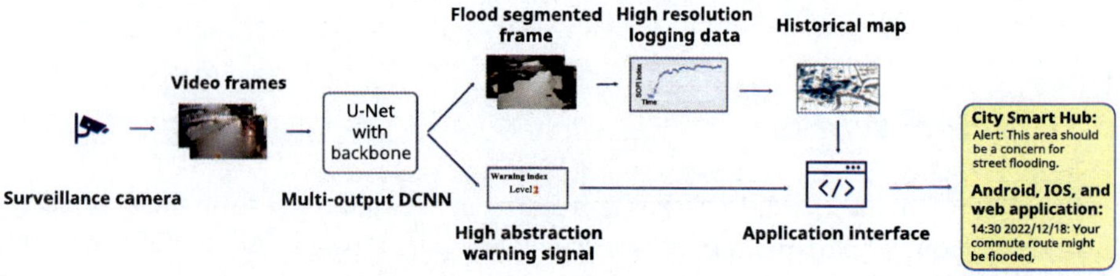

Fig. 1. Proposed pipeline with multi-output U-Net gives out two forms of useful data.

3.1 The Proposed Architecture for Multi-output Flooding Sensoring Model

Image segmentation is a branch of digital image processing and computer vision that aims to group pixels or regions in images; it can also be considered a classification problem that categorizes pixels. Three famous algorithms are usually used in flood area segmentation: U-Net, Mask-RCNN, and SegNet. In this study, the author chose U-Net because it is fast, easy to implement, and easy to expand and modify [11].

U-Net was created as a smaller and simpler model than Fully Convolutional Networks (FCN) to serve problems with limited data [13]. As the baseline method, a basic U-Net would be used in this study. It would be fed a Red Green Blue (RGB) image with a resolution of 512×512. The structure would have a depth of five, correlating with five down convolutions to extract features, so after each convolution, it would have dimensions of [16, 32, 64, 128, 256], and its spatial size would be respectively reduced by half. Each output would then be the input of the decoder's upsampling module as the residual connection, which aids the model's recovery in detail. Each down convolution includes a residual connection that boosts learning capacity. After each convolution, a batch normalization should be performed to improve training speed, followed by Rectified Linear Unit (ReLU) activation.

U-Net is also classified as a convolutional encoder-decoder structure, with the encoder on the left and the decoder on the right. Unlike Artificial Neural Network's (ANN's) AutoEncoder, U-Net provides more information in terms of detail recovery while requiring a smaller number of parameters in the model. This can be applied to problems requiring high-resolution input because there are fewer parameters to learn. This study will replace the encoder with various pre-trained computer vision models used as feature extractors. To determine which would fit in, a search experiment would be carried out.

Many popular computer vision models have been created and pre-trained using the ImageNet dataset[1], which contains 1000 classes and millions of samples. Those models can understand images' low- and high-level features, so it's beneficial to apply their knowledge to this smaller domain and capitalize on its ability to classify. As a result, the proposed model has two outputs: the classification head, which is the encoder's output, and the segmentation head, which is the decoder's output. Convert this approach into an end-to-end model that eliminates the need for post-processing using a multi-output model. The author understands the tradeoff between combining multioutput models and having two models work independently and can then tweak their output without retraining. The final proposed model is visualized in Fig. 2.

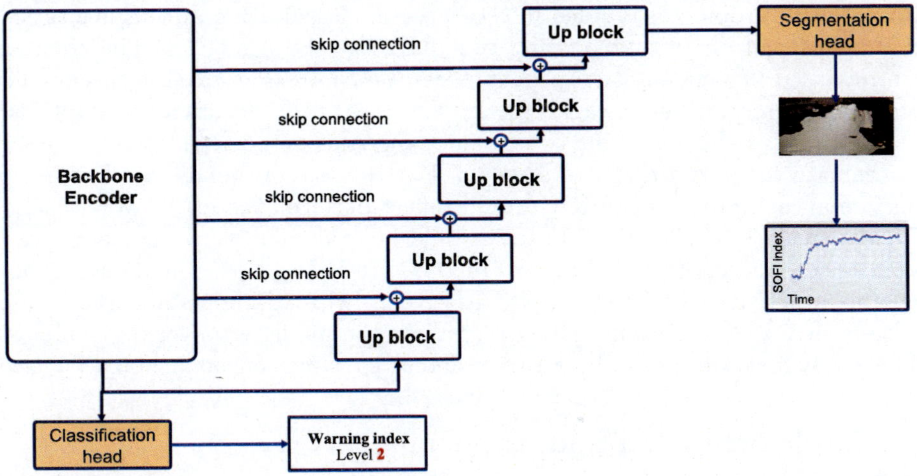

Fig. 2. Proposed pipeline with multi-output U-Net gives out two forms of useful data.

[1] https://www.image-net.org/.

3.2 Training Strategy

To train this multi-output model, the author chose dice loss[2] as the segmentation loss and cross-entropy loss as the classification loss. A dice score is a common way to compare the similarity of two images. This loss considers both local and generative similarity of the image and is computed as in Eq. 1. Using cross-entropy loss to optimize model performance in a classification task as in Eq. 2. Then, using a coefficient α to add weight to address the importance level of each task, combine two losses in Eq. 3. This study prioritizes classification tasks because the algorithm in this setting should achieve accuracy in alarm signals before logging data, then α was set to 1.5.

$$\mathcal{L}_{Dice} = 1 - \frac{2y\hat{y} + 1}{y + \hat{y} + 1} \tag{1}$$

$$\mathcal{L}_{Cross-entropy} = -\frac{1}{N}\sum_{1}^{N}[y_n \log \hat{y}_n + (1 - y_n)\log(1 - \hat{y}_n)] \tag{2}$$

$$\mathcal{L} = \mathcal{L}_{Dice} + \alpha\mathcal{L}_{Cross-entropy} \tag{3}$$

Using augmentation steps improves model performance, especially when there are few or many identical data points. It is recommended for U-Net architecture. Therefore, this training procedure would include augmentation phases such as random flip, scale, rotation, and brightness.

3.3 Metrics

Because this problem is divided into two parts, classification and segmentation, this model will be evaluated using two different sets of metrics. The common metrics used in segmentation problems can be counted as pixel accuracy, dice score (or f1 score), and mean IOU. In this case, pixel accuracy is simple but prone to class imbalance. The two retained metrics can be used interchangeably to evaluate the performance of a model. But the f1 score tends to reflect mean performance, and the mean IOU would reflect performance in the worst cases.

Meanwhile, the classification task would be evaluated using accuracy, precision, recall, and confusion matrices. In this, precision helps to understand how reliable a positive prediction is, whereas recall helps to understand the rate of false positives. A confusion matrix represents the model error for a given class. This study first considers all metrics to determine which encoder should be used.

4 Implementation and Experiment

4.1 Data

Collecting. Collecting 5089 image samples related to flooding from the Water Segmentation Dataset [17] (636 samples), European Flood 2013 [1] (3710 samples), Image Dataset for Roadway Flood [14] (441 samples), and 302 samples

[2] https://paperswithcode.com/method/dice-loss.

self-collected from 3 public street cameras in Danang, Vietnam, in October 2022. Then, select samples with camera angles and quality comparable to a standard street camera (high angle overview, mostly street in view). The author first collects a real camera scene for the test set, then combines it with similar images from the above dataset to build the training and validation sets.

Labelling. The data would then be labeled with two different types of labels: segmentation masks and classification labels. Although segmentation labeling is a laborious and time-consuming process, it can be sped up using the Interactive Segmentation Tool from Sofiiuk (2020) [15]. As defined in Table 1, samples would be classified into four classes. With that definition, the distribution of the collected dataset is indicated in Fig. 3. Classes 0 and 2 are abundant in the collected dataset, whereas classes 1 and 3 are scarce.

Table 1. Definition of classification label of camera flooding scene.

Label	Definition
0	No water
1	There is water in sight, but it is not affecting the vehicle route
2	Flooding streets, don't recommend pedestrians or vehicles entering the area
3	Dangerous flooding with a strong flow of water or very deep inundation

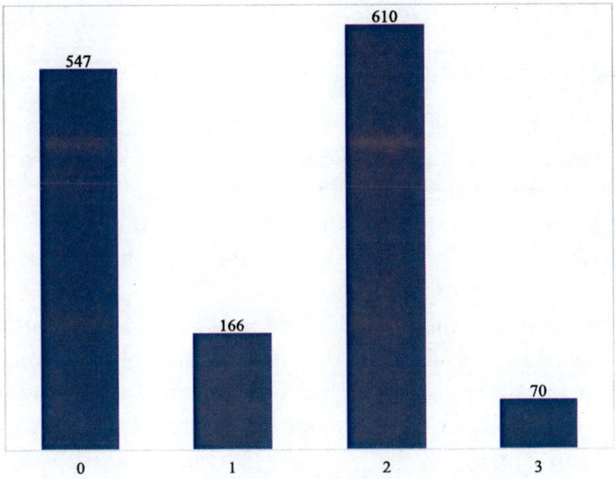

Fig. 3. Distribution of samples in the dataset according to inundation level.

4.2 Backbone Searching

On the prepared dataset, we run a backbone search; a backbone selection strategy favors well-known and small models. Table 2 shows that when it comes to classification metrics, the U-Net with Resnet18 performs well with an accuracy of 92%, and when it comes to segmentation metrics, the U-Net with VGG13_bn performs best. However, regarding inference speed and memory requirements, U-Net with the backbone MobileNet family is best, with MobileNetV2 performing well in both tasks. As a result, MobileNetV2 was chosen as the encoder backbone for U-Net to install Danaflood.

Table 2. U-Net backbone search result, trained with 20 epochs with the same setup. It can be concluded that changing the feature extraction encoder improved the model's performance on the test set. The inference performance was measured with an Nvidia GTX 1050, and the training performance was conducted on Google Colab's Tesla T4.

Backbone	#Params	FPS	Class Accuracy	Class Precision	Class Recall	Segmentation Accuracy	Segmentation F1	Segmentation MeanIOU
Original_unet	15M	14	0.3945	0.3516	0.3945	0.7683	0.5884	0.4169
Effientnet-b0	6.2M	16	0.7697	0.9478	0.7697	0.9486	0.885	0.7937
Effientnet-b2	10M	4	0.7278	0.7612	0.7278	0.9416	0.8695	0.7692
Resnet50	32.5M	10	0.8755	0.8867	0.8755	0.9376	0.8627	0.7585
Resnet34	24M	15	0.3347	0.4748	0.3347	0.8193	0.677	0.5117
Resnet18	14M	20	**0.9264**	0.8345	**0.9264**	0.9446	0.8743	0.7766
Mobilenet_v2	6.6M	20	0.8958	0.9093	0.8958	**0.9593**	0.8733	0.7751
Mobilenetv3_large_minimal_100	5.1M	21	0.7531	0.8184	0.7531	0.9404	0.8687	0.7678
Mobilenetv3_large_100	6.6M	19	0.8025	0.9445	0.8025	0.9538	0.8926	0.806
Mobilenetv3_small_minimal_100	5.1M	**24**	0.5526	0.8476	0.5526	0.9256	0.8376	0.7206
Mobilenetv3_small_100	3.5M	20	0.6923	0.8037	0.6923	0.9431	0.8717	0.7726
Mobilenetv3_small_075	2.8M	21	0.6566	0.661	0.6566	0.9371	0.8482	0.7364
Vgg13	18M	8	0.7651	**0.955**	0.7651	0.9559	0.8952	0.8103
Vgg13_bn	18M	7	0.7732	0.9459	0.7732	0.9568	**0.9009**	**0.8197**

4.3 Qualitative Evaluations

As shown in Fig. 4, the segmentation is not totally correct, sometimes mistaking road surfaces, and sidewalks as the water surface. As shown in Fig. 5, the SOFI in the camera scene when there is no flooding is normally less than 0.2, but it is greater than 0.4 when there is flooding. This index fluctuates due to objects or environmental factors, but it is still useful for recording flooding situations on the street.

On the other hand, the model performs well in classification, never misclassifying class 0 with other classes. Due to the imbalanced data and lack of training data for class 1, and also because it's hard for annotators to classify well between classes 1 and 2, the model works poorly with class 1. That can be seen in the confusion matrix in Fig. 6.

The final prototype of this study has been deployed on a cloud instance and can serve up to 100 camera points with an inference strategy that has a latency of 5 min.

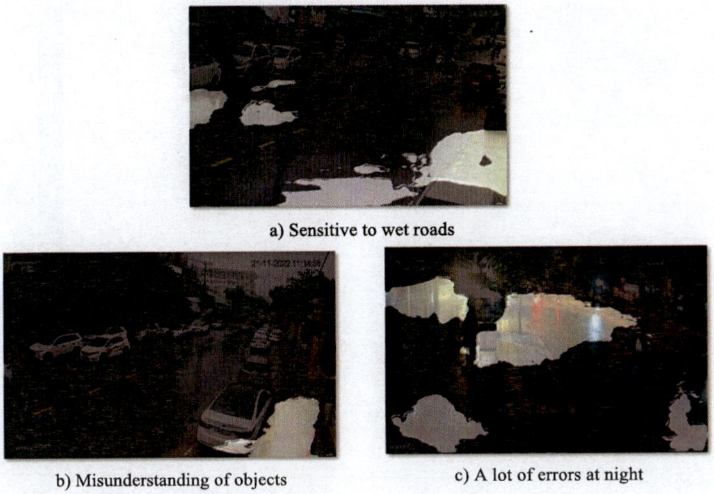

Fig. 4. Error cases of the current model.

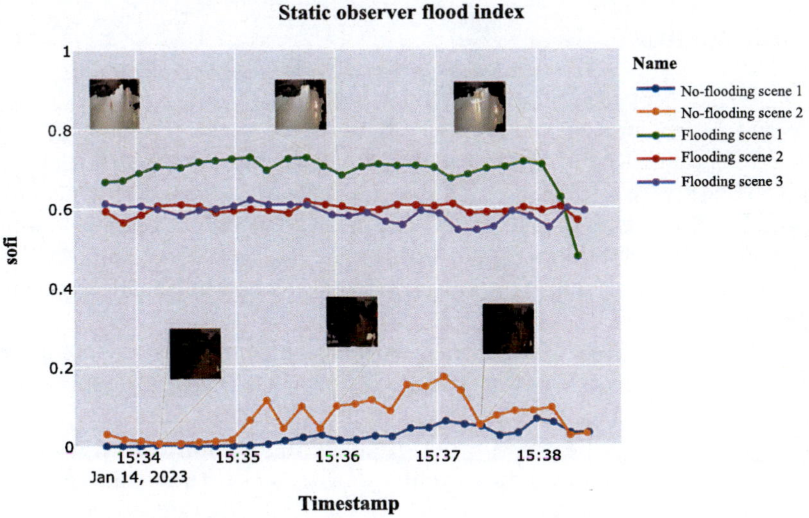

Fig. 5. Static observer flood index (SOFI) on different camera scenes indicates that SOFI can display a proxy of the flooded area in the camera scene.

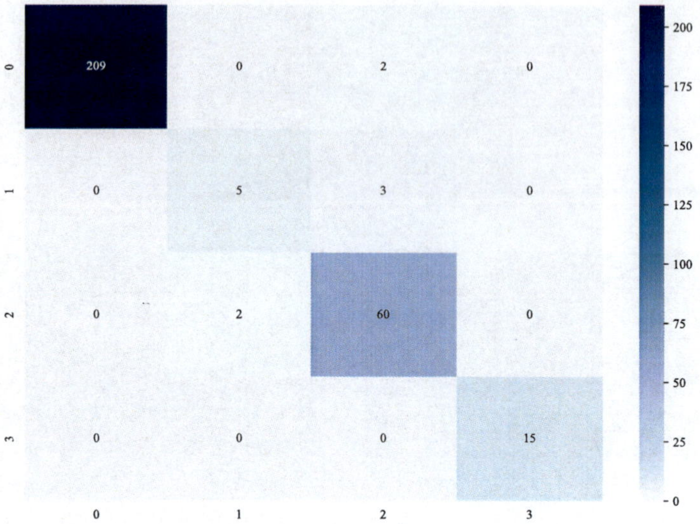

Fig. 6. Confusion matrix of the classification task.

5 Conclusion

This study has proposed a deep learning model with high scalability to detect and record flooding situations on the urban street without requiring to install any new on-site infrastructure but using street camera systems that are already available. The study result indicates that U-Net with a pretrained encoder backbone can improve its performance in both classification and segmentation tasks. Model U-Net with MobileNetV2 performed well on the test set and can be implemented on a low-cost budget, allowing citizens to commute safely during hazardous conditions.

Although this approach has a positive outcome, it does have some flaws, such as the fact that when the power grid goes down, the alarm system will fail; in this case, the camera system should have fallback energy to remain useful. This model can be optimized for better performance by upgrading to a more modern model and collecting more data to refine it with a real-time camera system for specific cities.

References

1. Barz, B., et al.: Enhancing Flood Impact Analysis using Interactive Retrieval of Social Media Images (August 2019). https://doi.org/10.5445/ksp/1000087327/06
2. Bronkhorst, B.V., Bhandari, P.: Climate Risk Country Profile: Viet Nam (November 2020). https://www.adb.org/publications/climate-risk-country-profile-viet-nam
3. Chaudhary, P., D'Aronco, S., Leitão, J.P., Schindler, K., Wegner, J.D.: Water level prediction from social media images with a multi-task ranking approach. ISPRS J. Photogram. Remote Sens **167**, 252–262 (2020). https://doi.org/10.48550/arxiv.2007.06749

4. Do, H.N., Vo, M.T., Tran, V.S., Tan, P.V., Trinh, C.V.: An early flood detection system using mobile networks. In: International Conference on Advanced Technologies for Communications 2016-January, pp. 599–603 (2016). https://doi.org/10.1109/ATC.2015.7388400
5. Hancke, G.P., de Silva, B.d.C., Hancke, G.P.: The role of advanced sensing in smart cities. Sensors (Basel, Switzerland) **13**(1), 393 (2013). https://doi.org/10.3390/S130100393
6. Ibrahim, N.S., Osman, M.K., Mohamed, S.B., Abdullah, S.H., Sharun, S.M.: The application of UAV images in flood detection using image segmentation techniques. Ind. J. Electr. Eng. Comput. Sci. **23**(2), 1219–1226 (2021). https://doi.org/10.11591/IJEECS.V23.I2.PP1219-1226
7. Karlo Tolentino, L.S., et al.: Real time flood detection, alarm and monitoring system using image processing and multiple linear regression. J. Comput. Innov. Eng. **7** (2022)
8. Khouakhi, A., Zawadzka, J., Truckell, I.: The need for training and benchmark datasets for convolutional neural networks in flood applications. Hydrol. Res. **53**(6), 795–806 2022). https://doi.org/10.2166/NH.2022.093/1048594/NH2022093.PDF
9. Lorini, V., Castillo, C., Nappo, D., Dottori, F., Salamon, P.: Social media alerts can improve, but not replace hydrological models for forecasting floods (December 2020). https://doi.org/10.48550/arxiv.2012.05852
10. Mateo-Garcia, G., et al.: Towards global flood mapping onboard low cost satellites with machine learning. Sci. Rep. **11**(1), 1–12 (2021). https://doi.org/10.1038/s41598-021-86650-z
11. Moy De Vitry, M., Kramer, S., Dirk Wegner, J., Leitao, J.P.: Scalable flood level trend monitoring with surveillance cameras using a deep convolutional neural network. Hydrol. Earth Syst. Sci. **23**(11), 4621–4634 (2019). https://doi.org/10.5194/HESS-23-4621-2019
12. Rentschler, J., et al.: Rapid Urban Growth in Flood Zones (2022). https://doi.org/10.1596/1813-9450-1014
13. Ronneberger, O., Fischer, P., Brox, T.: U-Net: convolutional networks for biomedical image segmentation. In: Navab, N., Hornegger, J., Wells, W.M., Frangi, A.F. (eds.) MICCAI 2015. LNCS, vol. 9351, pp. 234–241. Springer, Cham (2015). https://doi.org/10.1007/978-3-319-24574-4_28
14. Sazara, C., Cetin, M., Iftekharuddin, K.M.: Detecting floodwater on roadways from image data with handcrafted features and deep transfer learning. In: 2019 IEEE Intelligent Transportation Systems Conference, ITSC 2019, pp. 804–809 (October 2019). https://doi.org/10.1109/ITSC.2019.8917368
15. Sofiiuk, K., Petrov, I.A., Konushin, A.: Reviving iterative training with mask guidance for interactive segmentation (February 2021). https://doi.org/10.48550/arxiv.2102.06583
16. Tucci, C.: Cap-Net international network for capacity building in integrated water resources management. Tech. rep, World Meteorological Organization (2007)
17. Zaffaroni, M., Rossi, C.: Water segmentation dataset (December2019). https://doi.org/10.5281/ZENODO.3642406, https://zenodo.org/record/3642406
18. Zhang, Q., Jindapetch, N., Duangsoithong, R., Buranapanichkit, D.: A performance analysis for real-time flood monitoring using image-based processing. Ind. J. Electr. Eng. Comput. Sci. **17**(2), 793–803 (2019). https://doi.org/10.11591/IJEECS.V17.I2.PP793-803

Interactive Control Between Human and Omnidirectional Mobile Robot: A Vision-Based Deep Learning Approach

The Co Nguyen, Trung Nghia Bui, Van Nam Nguyen, Duy Phuong Nguyen, Cong Minh Nguyen, and Manh Linh Nguyen(✉)

Hanoi University of Scicence and Technology, Hanoi, Vietnam
linh.nguyenmanh@hust.edu.vn

Abstract. Nowadays, mobile robots have been popular not only in industrial applications such as materials transportation but also in non-industrial applications, e.g., human assistance. Among developed configurations, omnidirectional mobile robots have attracted great attention in recent times due to their superior maneuverability over their conventional counterparts. In this research, an application of a four mecanum-wheeled omnidirectional mobile robot (4-MWMR) in human assistance has been developed. By using image processing, the 4-MWMR is capable of following an authorized person, thereby assisting users in transporting large-size or heavy-weight materials. Good experimental results show the ability of the developed system to be used in practice.

Keywords: Omnidirectional mobile robot · Vision-based control · Deep learning

1 Introduction

Recently, the application of mobile robots has been exploded because of their maneuverability and high efficiency. Various prototypes and products have been developed to satisfy the strict requirements of users in terms of the high integration on perception, navigation, and manipulation [1]. Conventional mobile robots often use rubber wheels such as two-wheeled differential or car-like four-wheeled configuration. An obvious disadvantage of the above-mentioned configurations is that they are not omnidirectionality. To overcome this shortcoming, omnidirectional mobile robots (OMRs) with specialized wheels have been developed, with greater maneuverability, good adaptability in small spaces [2]. Because of the significant benefits, OMRs attract a lot of attention, not only in hardware but also in control design. Two OMR configurations that are widely used in research as well as in practical applications are 3-wheeled and 4-wheeled. The 3-wheeled configuration is often used to design small mobile robots with a light payload. For heavier transportation, the 4-wheeled configuration is preferred.

This research is funded by the Hanoi University of Science and Technology (HUST) under project number T2022-PC-005.

T. D. L. Nguyen et al. (Eds.): ICISN 2023, LNNS 752, pp. 556–565, 2023.
https://doi.org/10.1007/978-981-99-4725-6_66

In general, the control system of the OMRs can be categorized into low-level control, which handles the kinematic as well as the dynamic of the robot, and high-level control which mainly deal with navigation tasks based on vision and laser sensors. With the explosive development of artificial intelligence (AI) in this decade, in combination with vision sensors, the amount of information collected is sufficient for most of the problems related to mobile robotics, thus not only improving the ability to interact with users but also making the robot more and more intelligent. Therefore, high-level control systems based on AI have become more attractive in recent times [3].

To guarantee that mobile robots can move and perform their tasks in complex environments, most researchers focus on solving mapping and localization problems [4,5]. However, cooperating with human requires more than that. In this study, a mobile robot equipped with four mecanum omnidirectional wheels with the ability of person following, and hand pose-based control is developed. These features are important for the mobile robot to be used in many potential applications, such as elder person assistance, medical service support, robotic shopping cart, etc. Several results have been achieved with the human following mobile robot [6,7]. This research focuses on utilizing new achievements of deep neural network in image processing to solve problems such as face recognition, human and hand pose detection so that the mobile robot can recognize and tracks an admin, as well as manually controlled by hand pose. Experimental results getting from our prototype mobile robot show that all functions are performed well with high reliability.

2 System Description

2.1 Kinematic Model of 4-MWMR

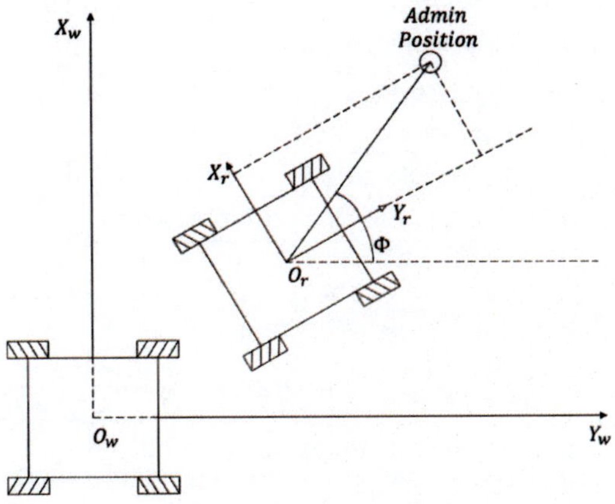

Fig. 1. Coordinate system assignment of a 4-MWMR

The typical configuration of a 4-MWMR is shown in Fig. 1. Define $X_w O_w Y_w$ and $X_r O_r Y_r$ as the inertial and body frame, respectively. The corresponding state-variables of the inertial and body frame are $[\dot{x}_w, \dot{y}_w, \dot{\Phi}]$ and $[\dot{x}_r, \dot{y}_r, \dot{\Phi}]$. Then, (1) and (2) describe the kinematic model and its inverson, respectively [8]:

$$\left[\dot{\theta}_1\ \dot{\theta}_2\ \dot{\theta}_3\ \dot{\theta}_4\right]^T = \frac{1}{R} J \left[\dot{x}_r\ \dot{y}_r\ \dot{\Phi}\right]^T \quad (1)$$

$$\left[\dot{x}_r\ \dot{y}_r\ \dot{\Phi}\right]^T = RJ^+ \left[\dot{\theta}_1\ \dot{\theta}_2\ \dot{\theta}_3\ \dot{\theta}_4\right]^T \quad (2)$$

with,

$$J = \begin{bmatrix} 1 & 1 & L+W \\ -1 & 1 & \text{-}L+W \\ 1 & 1 & \text{-}L+W \\ -1 & 1 & L+W \end{bmatrix} \quad (3)$$

and $J^+ = (J^T J)^{-1} J^T$ is the pseudo inverse matrix of J:

$$J^+ = \frac{1}{4} \begin{bmatrix} 1 & -1 & 1 & -1 \\ 1 & 1 & 1 & 1 \\ \frac{1}{L+W} & -\frac{1}{L+W} & -\frac{1}{L+W} & \frac{1}{L+W} \end{bmatrix} \quad (4)$$

The parameters of the prototype 4-MWMR used in our experimental system are provided in Table 1.

Table 1. Parameters of the 4-MWMR

Symbol	Desciption	Value	Unit
W	Half width of the platform	0.3	m
L	Half length of the platform	0.3	m
m	Total mass of the platform	40	kg
R	Wheel radius	0.076	m
D_θ	Wheel's viscous coefficient	0.1	-
f_i	Static friction of ith wheel	0.6	-
τ_i	Torque generated by wheel's motor	-	Nm
$\dot{\theta}_i$	angular velocity of $i^t h$ wheel	0	rad/s
Φ	Angle between body and inertial frame	-	rad

The movement of the mobile robot in the inertial reference frame can easily be obtained by the following transformation:

$$\left[\dot{x}_w\ \dot{y}_w\ \dot{\Phi}_w\right]^T = \Re(\Phi) \left[\dot{x}_r\ \dot{y}_r\ \dot{\Phi}_r\right]^T \quad (5)$$

where $\Re(\Phi)$ is the transformation matrix expressed by:

$$\Re(\Phi) = \begin{bmatrix} \cos(\Phi) & -\sin(\Phi) & 0 \\ \sin(\Phi) & \cos(\Phi) & 0 \\ 0 & 0 & 1 \end{bmatrix} \quad (6)$$

2.2 Experimental System Setup

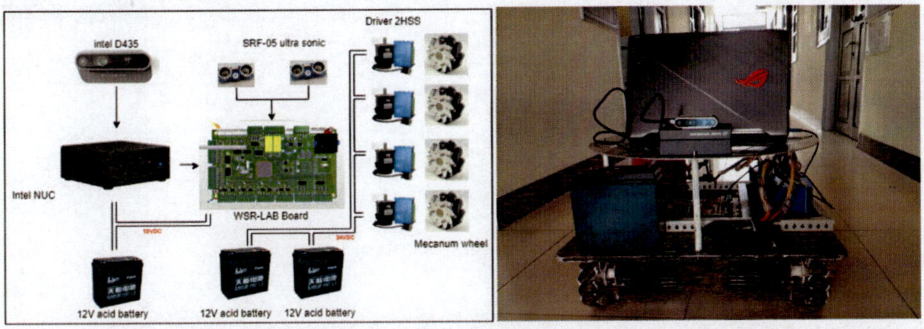

Fig. 2. Hardware of the mobile robot control system

The prototype 4-MWMR and its control system are shown in Fig. 2. Each wheel of the 4-MWMR is actuated by a hybrid step servo drive which can quickly perform commands of the high-level control system. Image and distance from target objects are collected and sent to the control system by an intel realsense D435i depth camera. A high-performance computer Intel NUC-11 NUC11TNHi70Z equipped with 11th generation Intel core processors is chosen to cope with complicated image processing algorithms in real-time. In addition, an interface board based on a 32bit real-time microcontroller STM32F103ZCT6 is specifically designed to receive the command of the main computer via a communication port and transform it into a high-frequency pulse to handle the servo drive. Besides, other tasks such as local control, and connecting to safety sensors can only be implemented in real-time based-on this board.

3 Control Design

3.1 Human Following Controller Design

Figure 3 shows the block diagram of the human following control system. The goal of the controller is to maintain a safety distance between the 4-MWMR and the admin. During human following, the heading control is employed, which means the reference distance $x^*_{h,r}$ and the rotation angle $\theta^*_{h,r}$ must be zero, while the distance along y axis, i.e., $y^*_{h,r}$, is kept constant. Since the closed-loop step-servo is used to move the 4-MWMR, and suppose that the slip phenomenon between the mecanum wheel and the moving surface is ignored, its is reasonable to assume that the relation between the real angular velocity of each wheel and

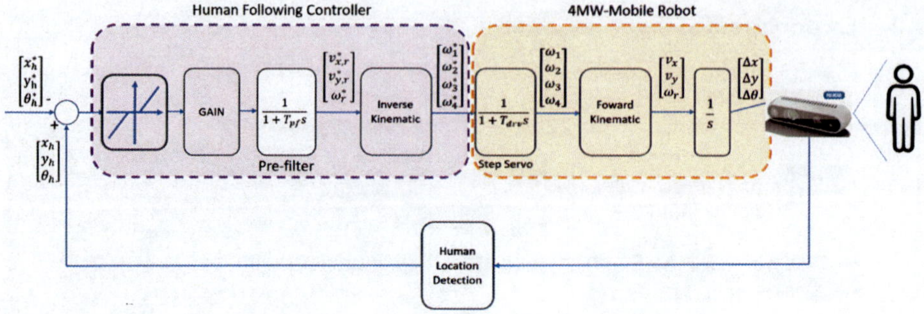

Fig. 3. "Follow me" control scheme

its reference can be expressed by a first order inertial with time constant T_{drv}:

$$\begin{bmatrix} \dot{\theta}_1 \\ \dot{\theta}_2 \\ \dot{\theta}_3 \\ \dot{\theta}_4 \end{bmatrix} = (\frac{1}{1 + T_{drv}s}) \begin{bmatrix} 1 & 0 & 0 & 0 \\ 0 & 1 & 0 & 0 \\ 0 & 0 & 1 & 0 \\ 0 & 0 & 0 & 1 \end{bmatrix} \begin{bmatrix} \dot{\theta}_1^* \\ \dot{\theta}_2^* \\ \dot{\theta}_3^* \\ \dot{\theta}_4^* \end{bmatrix} \tag{7}$$

Base on (2) and (7), the relation between the traveling distance of the mobile robot and its angular speed references is

$$\begin{bmatrix} \Delta x_r \\ \Delta y_r \\ \Delta \Phi_r \end{bmatrix} = \frac{1}{s} \frac{1}{(1 + T_{drv}s)} R J^+ \mathbf{I} \begin{bmatrix} \dot{\theta}_1^* \\ \dot{\theta}_2^* \\ \dot{\theta}_3^* \\ \dot{\theta}_4^* \end{bmatrix} \tag{8}$$

Due to the existance of an integral as seen in (8), a proportional controller is sufficient to guarantee the desired orientation $q_{h,r}^* = [x_{h,r}^*, y_{h,r}^*, \theta_{h,r}^*]^T$ of the admin in the 4-MWMR's body frame. The controller's gain K_P is chosen such that the 4-MWMR moves with the maximum velocity at maximum tracking error, which means:

$$\begin{aligned} 0 < x_{h,r}^* K_{Px} &\leq v_{xr,max} \\ 0 < y_{h,r}^* K_{Py} &\leq v_{yr,max} \\ 0 < \theta_{h,r}^* K_{P\theta} &\leq \omega_{r,max} \end{aligned} \tag{9}$$

where, $v_{yr,max}$, $v_{xr,max}$ and $\omega_{r,max}$ are the restrictions of the 4-MWMR's velocity. To limit the acceleration of the mobile robot as its target suddenly moves, a prefilter with a tunable time constant T_{pf} is inserted right behind the position controller as shown in Fig. 3.

4 Image Processing

As previously discussed, the behavior of the mobile robot is decided based on the actions of an admin. Hence, image processing in real-time is the most important task in this research. The three main tasks are: face recognition to recognize the admin, human presence and orientation detection which provides information for the heading controller designed above, and hand poses detection which allows the admin to move the 4-MWMR manually.

4.1 Face Recognition

Convolution neural network (CNN) is one of the most popular categories of neural networks that is especially designed to work with data which has some spatial properties, such as images and videos. The operating of a CNN is similar to a standard neural network. However, a key difference is that each unit in a CNN layer is a dimensional filter, i.e., convolution layer and pooling layer, which is able to find useful features within inputs, as well as reduce the size of parameters in comparison with a regular neural network.

In this research, the VGG-16 CNN created by the Visual Geometry Group at Oxford University is utilized in face recognition. The architecture used in our study is shown in Fig. 4. In comparison with the standard VGG-16 [9], a modification is made at the output layer. In detail, the three fully connected (FC) layers at the output of the standard VGG-16 are removed. Instead, one flatten and other FC layers are used to detect the probability p of human presence in the frame, as well as the coordinates $r = [x_1, x_2, y_1, y_2]$ of the box surrounding the detected face. The dataset used for face detection consists of 888 pictures which have been taken and labeled by ourselves. The training results of our model are quite good, i.e., the average loss over the entire training dataset is less than 0.1 whilst the achieved accuracy is greater than 0.95. After getting the face position from the previous step, the principal components analysis (PCA) is employed for face recognition in that region. The major advantage of the PCA is that it is a simple technique that is fast to compute which makes it suitable for our low cost and limited computational control system [10].

4.2 Human Position and Hand Poses Detection

Human position detection belong to the category of object detection, which is mainly based on machine learning and deep learning algorithms. In terms of detection accuracy and speed, deep learning-based detection algorithms such as Faster R-CNN [11], Single Shot Detectors (SSD) [12], and You Only Look Once (YOLO) [13] are better than their machine learning-based algorithms counterpart [14]. For our specific application, the detection algorithm must meet the requirements of real-time detection and ensure a certain detection accuracy. Hence, the MobileNet SDD pre-trained model with the ability to detect a large number of objects with different scales and different types in real time is chosen in this study [15–17]. The architecture of the human detection model is illustrated in Fig. 5. The final output is a vector $y^T = [x, y, w, h, p]$ which contains the location and the confidence measurement p of the final result.

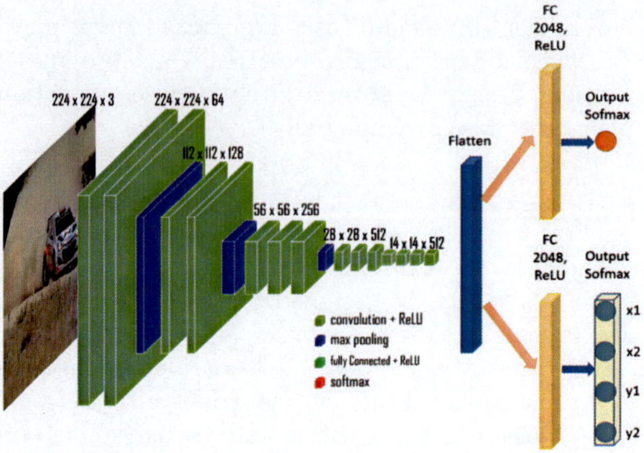

Fig. 4. VGG16-based face detection

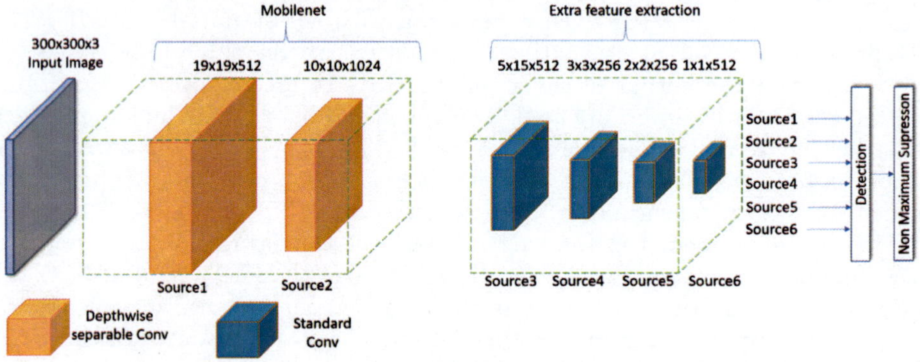

Fig. 5. Mobilenet-SSD's architecture for body position detection

Finally, the distance and orientation from the center of the detected box to the camera are directly measured by using depth camera D435i of Intel. The information of the distance is then sent to the heading-controller designed above.

To manually control the mobile robot by hand pose, the MediaPipe hand's module is utilized to detect 20 hand landmarks as shown in Fig. 6 [18]. Then from their relative positions to each other, basic commands such as start, stop, move left, move right, and rotate can be carried out.

5 Experimental Results

For face detection, the model shows quite good results where the average loss over the entire training dataset with 888 images is less than 0.1 and the achieved accuracy is greater than 0.95. An illustrative result is shown in Fig. 7(a), in which the confidence measurement is 86.7, and position of the bounding box is accurate. Similar result for face recognition using PCA algorithm is shown in Fig. 7(b).

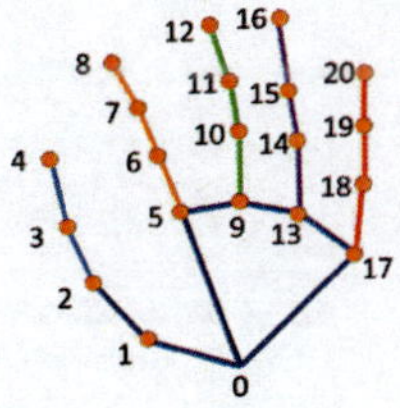

No.	Description	No.	Description
0	WRIST	11	MIDDLE_FINGER_DIP
1	THUMB_CMC	12	MIDDLE_FINGER_TIP
2	THUMB_MCP	13	RING_FINGER_MCP
3	THUMB_IP	14	RING_FINGER_PIP
4	THUMB_TIP	15	RING_FINGER_DIP
5	INDEX_FINGER_MCP	16	RING_FINGER_TIP
6	INDEX_FINGER_PIP	17	PINKY_MCP
7	INDEX_FINGER_DIP	18	PINKY_PIP
8	INDEX_FINGER_TIP	19	PINKY_DIP
9	MIDDLE_FINGER_MCP	20	PINKY_TIP
10	MIDDLE_FINGER_PIP		

Fig. 6. Hand landmarks

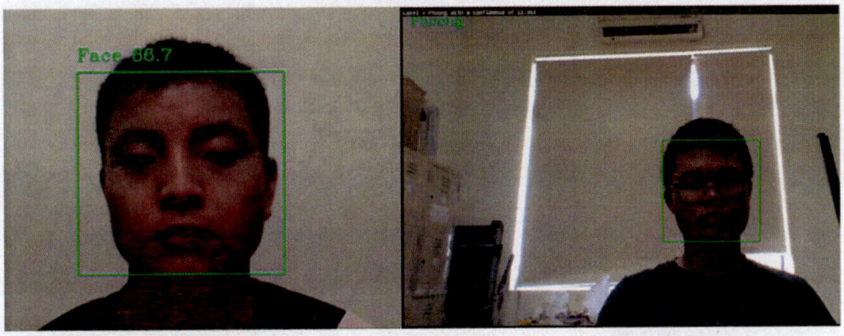

(a) Face detection (b) Face recognition

Fig. 7. Face detection and recognition result

Figures 8 shows the human as well as the hand pose detection results. Based on good prediction results, the mobile robot can perform well in both human following and hand pose control mode. In detail, the experiment results illustrated in Fig. 9 show that in steady-state, the 4-MWMR is able to track the admin and maintain a safety distance, i.e., $x = 0$, $y = 1000\,\text{mm}$ and $\theta \approx 0(rad)$. Thereby significantly improve the flexibility in human assisting operation. The basic operation of the developed system can be seen in the following link: https://www.youtube.com/watch?v=PPIrzwVJ3WI

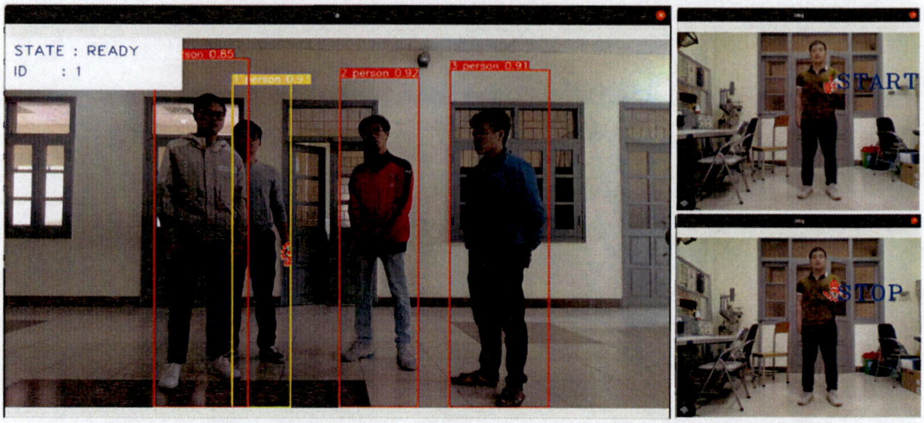

Fig. 8. Human detection based on Mobilenet SSD

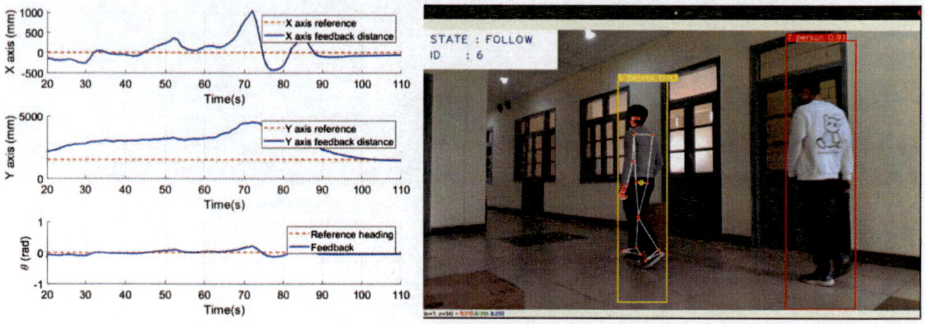

Fig. 9. Experimental results human following mode

6 Conclusions

In this research, an omnidirectional mobile robot aiming to assist humans in carrying materials is developed. By using various real-time image processing techniques such as VGG-16, mobilenet-SSD, main functions such as admin recognition, following, and hand pose-based control can be performed. In human following mode, a safety distance is guaranteed between the mobile robot and its admin. The admin can also switch the mobile robot from the following mode to hand pose mode to move it manually, which allows the mobile robot to operate in complex environments. In future works, collision avoidance and a mobile manipulator are going to be developed based on this platform.

Acknowledgements. This research is funded by the Hanoi University of Science and Technology (HUST) under project number T2022-PC-005.

References

1. Rubio, F., Valero, F., Llopis-Albert, C.: A review of mobile robots: concepts, methods, theoretical framework, and applications. Int. J. Adv. Robot. Syst. **16**(2) (2019). https://doi.org/10.1177/1729881419839596
2. Taheri, H., Zhao, C.X.: Omnidirectional mobile robots, mechanisms and navigation approaches. Mech. Mach. Theory **153**, 103958 (2020). https://doi.org/10.1016/j.mechmachtheory.2020.103958. ISSN 0094-114X
3. Cebollada, S., Payá, L., Flores, M., Peidró, A., Reinoso, O.: A state-of-the-art review on mobile robotics tasks using artificial intelligence and visual data. Expert Syst. Appl. **167**, 114195 (2021). https://doi.org/10.1016/j.eswa.2020.114195. ISSN 0957-4174
4. L. Payá, A. Gil, O. Reinoso: A state-of-the-art review on mapping and localization of mobile robots using omnidirectional vision sensors. J. Sens. 20 (2017). Article ID 3497650. https://doi.org/10.1155/2017/3497650
5. Panigrahi, P.K., Bisoy, S.K.: Localization strategies for autonomous mobile robots: a review. J. King Saud Univ. Comput. Inf. Sci. **34**(8), 6019–6039 (2022). https://doi.org/10.1016/j.jksuci.2021.02.015. ISSN 1319-1578
6. Gupta, M., Kumar, S., Behera, L., Subramanian, V.K.: A novel vision-based tracking algorithm for a human-following mobile robot. IEEE Trans. Syst. Man Cybern. Syst. **47**(7), 1415–1427 (2017). https://doi.org/10.1109/TSMC.2016.2616343
7. Jin, D., Fang, Z., Zeng, J.: A robust autonomous following method for mobile robots in dynamic environments. IEEE Access **8**, 150311–150325 (2020). https://doi.org/10.1109/ACCESS.2020.3016472
8. Yuan, Z., Tian, Y., Yin, Y., Wang, S., Liu, J., Wu, L.: Trajectory tracking control of a four mecanum wheeled mobile platform: an extended state observer-based sliding mode approach. IET Control Theory Appl. **14**, 415–426 (2020). https://doi.org/10.1049/iet-cta.2018.6127
9. https://www.geeksforgeeks.org/vgg-16-cnn-model/
10. Schenkel, T., Ringhage, O., Branding, N.: A Comparative Study of Facial Recognition Techniques: With focus on low computational power. Dissertation (2019)
11. Ren, S., He, K., Girshick, R., Sun, J.: Faster R-CNN: towards real-time object detection with region proposal networks. EEE Trans. Pattern Anal. Mach. Intell. **39**(6), 1137–1149 (2017)
12. Liu, W., et al.: SSD: single shot MultiBox detector. In: Leibe, B., Matas, J., Sebe, N., Welling, M. (eds.) ECCV 2016. LNCS, vol. 9905, pp. 21–37. Springer, Cham (2016). https://doi.org/10.1007/978-3-319-46448-0_2
13. Redmon, J., Divvala, S., Girshick, R., Farhadi, A.: You only look once: unified, real-time object detection. In: Proceedings of the IEEE Conference on Computer Vision and Pattern Recognition (CVPR), Las Vegas, USA, pp. 779–788 (2016)
14. Yinan, L.I.: A survey of research on deep learning target detection methods. China New Telecomm-unications **23**(9), 159–160 (2021)
15. Chiu, Y.-C., Tsai, C.-Y., Ruan, M.-D., Shen, G.-Y., Lee, T.-T.: Mobilenet-SSDv2: an improved object detection model for embedded systems. In: International Conference on System Science and Engineering (ICSSE) 2020, pp. 1–5 (2020). https://doi.org/10.1109/ICSSE50014.2020.9219319
16. Howard, A.G., et al.: MobileNets: efficient convolutional neural networks for mobile vision applications. http://arxiv.org/abs/1704.04861
17. Sandler, M., Howard, A., Zhu, M., Zhmoginov, A., Chen, L.-C.: MobileNetV2: inverted residuals and linear bottlenecks. http://arxiv.org/abs/1801.04381
18. https://google.github.io/mediapipe/solutions/hands.html

Intelligent Control for Mobile Robots Based on Fuzzy Logic Controller

Than Thi Thuong[1], Vo Thanh Ha[2](✉), and Le Ngoc Truc[3]

[1] Faculty of Electrical Engineering, University of Economics - Technology for Industries, Hai Bà Trưng, Vietnam
ttthuog.dien@uneti.edu.vn

[2] Faculty of Electrical and Electronic Engineering, University of Transport and Communications, Hanoi, Vietnam
vothanhha.ktd@utc.edu.vn

[3] Hung Yen University of Technology and Education, Hải Dương, Vietnam

Abstract. This paper recommends intelligent control for mobile robots based on fuzzy logic controllers (FLC). This controller is designed with only two input state variables, such as position error, position deviation derivative of the robot, and one output variable, velocity. The robot is moved according to the trajectories set by fuzzy selection rules with an 9×9 matrix. The proposed FLC controller is compared with classical PID controller. The robot with the FLC controller moves to follow the trajectory with lower error and faster setup time than the PID controller. The efficiency of this controller is demonstrated by MATLAB/Simulink.

Keywords: Mobile Robot · PID · Fuzzy Logic Control · FLC

1 Introduction

The mobile robot is an innovative solution for the future of digitization and industry 4.0. The self-propelled robot ensures the certainty and flexibility of the product. At the same time, it makes it easier to move goods inside factories and warehouses. Besides, robots also improve automation and solve production continuity problems [1, 2]. In the world, in recent decades, autonomous robot control has received extensive research and development attention, and many methods, from classical control to modern management, have been proposed to apply to self-propelled robots. Previously, most publications used the structure of two control loops as the outer kinematic loop uses the Lyapunov function to synthesize the position-tracking controller, and the dynamic inner circle controls the speed tracking. Many active loop control methods have been proposed, such as slip control [3–6] and backstepping control [7–9]. When the dynamic equation has uncertain parameters, adaptive management is included in the design [10–13]. The adaptive control combines with neurons to approximate the result. Unpredictable parts [14–16] and adaptive control combined with fuzzy logic [17–20] gave reasonable control quality, compensating for model error and system input noise.

T. D. L. Nguyen et al. (Eds.): ICISN 2023, LNNS 752, pp. 566–573, 2023.
https://doi.org/10.1007/978-981-99-4725-6_67

Although many advanced controllers have been researched and developed, traditional PID controllers are still chosen to be used in the problem of controlling orbital self-propelled robots because of their effectiveness of the controller. This ensures stability and traction. However, the accuracy achieved is not high. In the process of developing control techniques, Intelligent control, fuzzy logic is applied to work in many fields with the role of an observer. The fuzzy inference mechanism is considered a Simple and effective method for fine tuning classic controllers (Leonid Reznik, 1997; Jan Jantzen, 1998). Therefore, the sustainable controller FLC, when used to control the self-propelled robot, although it can.

This paper is organized into five main parts. Part 1 and part 2 present the introduction to the target study and kinematics and dynamics model. The fuzzy logic controller is designed in Sect. 3. Part 4 shows the simulation and simulation results. The last section is the conclusion.

2 Kinematic and Dynamic Model

2.1 Kinematic Model

The equation describing the kinematics of the mobile robot is expressed in Eq. (1) [1].

$$\dot{q} = \begin{bmatrix} \frac{r}{2}\cos(\theta) & \frac{r}{2}dcos(\theta) \\ \frac{r}{2}\sin(\theta) & \frac{r}{2}d\sin(\theta) \\ \frac{r}{2a} & \frac{r}{2a} \end{bmatrix} \begin{bmatrix} \dot{\varphi}_r \\ \dot{\varphi}_l \end{bmatrix} = \begin{bmatrix} \dot{x} \\ \dot{y} \\ \dot{\theta} \end{bmatrix} \tag{1}$$

where: r is right and left wheel radius; $2a$ is distance between the actuated wheels and the symmetry axis; $\dot{\varphi}_r$, $\dot{\varphi}_l$ are angular velocity of the right and left wheels; v, ω are Angular velocity of the right and left wheels; q is linear and angular velocities of robot; θ is orientation angle; $\dot{q}$ is robot speed.

2.2 Dynamic Model

The kinetic energy of the self-propelled robot is calculated by:

$$T_c = \frac{1}{2}m_c\vartheta_c^2 + \frac{1}{2}I_c\dot{\theta}^2 \tag{2}$$

$$T_{\omega R} = \frac{1}{2}m_\omega\vartheta_\omega^2 + \frac{1}{2}I_m\dot{\theta}^2 + \frac{1}{2}I_\omega\dot{\varphi}_r^2 \tag{3}$$

$$T_{\omega L} = \frac{1}{2}m_\omega\vartheta_\omega^2 + \frac{1}{2}I_m\dot{\theta}^2 + \frac{1}{2}I_\omega\dot{\varphi}_l^2 \tag{4}$$

where: where Tc is the kinetic energy of the DWMR without the wheels, $T_{\omega R}$ is the kinetic energy of the actuated wheels in the plane and $T_{\omega L}$ is the kinetic energy of all the wheels considering the orthogonal plane; m_c is mass of the robot without wheels and motors; m_ω is mass of each wheel and motor assembly; m_t is total mass of the DWMR; m_t is moment of inertia of the DWMR without wheels and motors about the vertical axis through P; I_c is moment of inertia of the DWMR without wheels and motors about the vertical axis through P; I_w is Moment of inertia of each wheel and motor about the wheel axis; I is Total inertia moment of the robot; $\dot{\varphi}_r$, $\dot{\varphi}_l$ are angular velocity of the right and left wheels; v, ω are Angular velocity of the right and left wheels; θ is orientation angle.

Mobile robot speed is calculated by:

$$\vartheta_i^2 = \dot{x}_i^2 + \dot{y}_i^2 \tag{5}$$

The coordinates of the wheels are therefore determined as follows:

$$\begin{cases} x_{\omega r} = x + a\sin\theta \\ y_{\omega r} = y + a\cos\theta \end{cases} \tag{6}$$

$$\begin{cases} x_{\omega l} = x - a\sin\theta \\ y_{\omega l} = y + a\cos\theta \end{cases} \tag{7}$$

From Eq. (2) to Eq. (7), the total kinetic energy:

$$T = \frac{1}{2}m_t\left(\left(\dot{x}^2 + \dot{y}^2\right) - \dot{y}d\dot{\theta}\cos(\theta) + m_t\dot{x}d\dot{\theta}\sin(\theta)\right) + \frac{1}{2}I\dot{\theta}^2 + \frac{1}{2}I_w\left(\dot{\varphi}_r^2 + \dot{\varphi}_l^2\right) \tag{8}$$

where: $m_t = m_c + 2m_\omega$; $I = m_c d^2 + I_c + 2m_\omega\left(d^2 + a^2\right) + 2I_m$ and $\dot{\theta} = \omega$

The robot's equation of motion is described by the system of equations:

$$\begin{cases} m\ddot{x} - m_c d\ddot{\theta}\sin\theta - md\dot{\theta}^2\cos\theta = F_1 - C_1 \\ m\ddot{y} - m_c d\ddot{\theta}\cos\theta - m_c d\dot{\theta}^2\sin\theta = F_2 - C_2 \\ -m_c d\sin\theta\ddot{x} + m_c d\cos\theta\ddot{y} + I\ddot{\theta} = F_3 - C_3 \\ I_\omega\ddot{\varphi}_r = \tau_r - C_4 \\ I_\omega\ddot{\varphi}_l = \tau_l - C_5 \end{cases} \tag{9}$$

The matrix linking the kinematic constraints:

$$\Lambda^T(q) = \begin{bmatrix} C_1 \\ C_2 \\ C_3 \\ C_4 \\ C_5 \end{bmatrix} \tag{10}$$

From Eqs. (9)- (10) The motion of the robot can be represented by the equation:

$$M(q)\ddot{q} + V(q,\dot{q}) + F(\dot{q}) + G(q) + \tau_d = B(q)\tau - \Lambda^T(q)^\lambda \tag{11}$$

where: $M(q)$ is positive inertia matrix; $V(q,\dot{q})$ is centripetal Matrix; $F(\dot{q})$ is surface friction; $G(q)$ is gravity acceleration matrix; τ_d is noise component; $B(q)$ is input matrix; $\Lambda^T(q)$ is binding matrix; λ is Lagrange multiplier vector.

2.3 Kinematic Model

The kinematic error model q_e of a self-propelled robot is a mathematical equation describing the deviation of the robot's position and posture, when the motion-controlled robot follows a desired trajectory ξd. The system of error function equations as follows:

$$\dot{q}_e = \begin{bmatrix} \dot{x}_e \\ \dot{y}_e \\ \dot{\theta}_e \end{bmatrix} = \begin{bmatrix} \cos(\theta_e) & 0 \\ \sin(\theta_e) & 0 \\ 0 & 1 \end{bmatrix} \begin{bmatrix} \vartheta_r \\ \omega_r \end{bmatrix} + \begin{bmatrix} -1 & y_e \\ 0 & -x_e \\ 0 & -1 \end{bmatrix} \begin{bmatrix} \vartheta \\ \omega \end{bmatrix} \tag{12}$$

3 Fuzzy Logic Controller Design

Use the inputs as bias and the difference derivative to configure the fuzzy logic controller, such as in Fig. 1 and Fig. 2. The transformed fuzzy block matches the data with the conditions of the given fuzzy rule. The output of the fuzzy set is converted to the clarity values through the centroid defuzzification method and converted into a control signal, as in Fig. 3. The FLC controller is controlled by rule table 1.

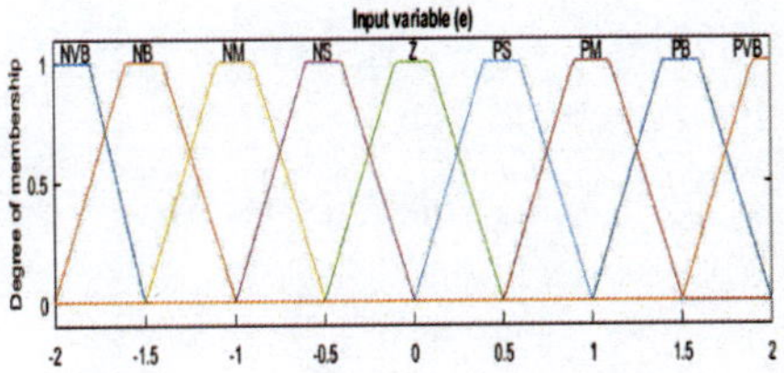

Fig. 1. Input of bias variable *e*

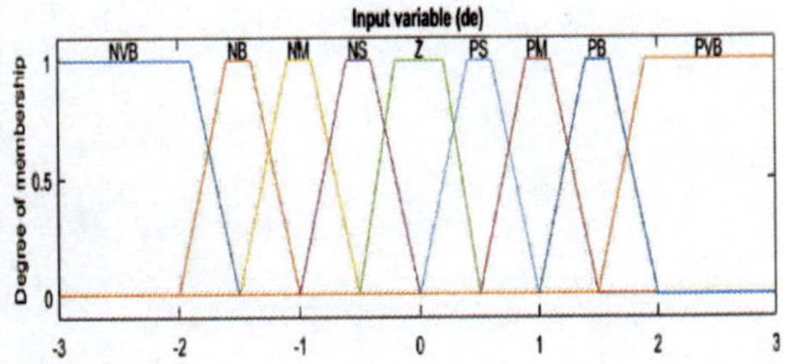

Fig. 2. Deviated variable derivative input *de*

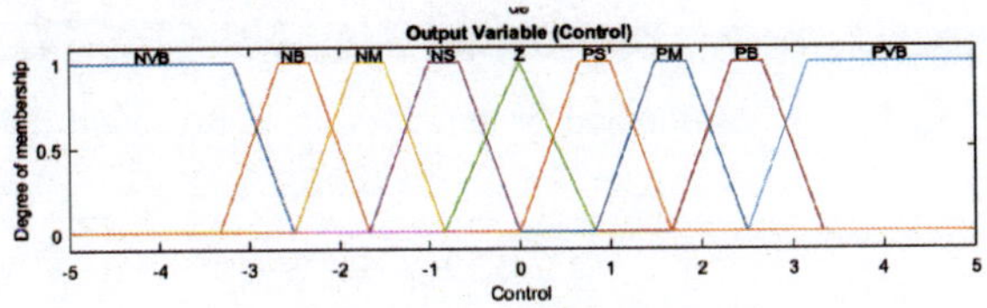

Fig. 3. Output variable

Table 1. The rule control for FLC controller

De/e	NVB	NB	NM	NS	Z	PS	PM	PB	PVB
NVB	PVB	PVB	PVB	PB	PM	PM	PS	Z	Z
NB	PVB	PVB	PB	PM	PS	PS	PS	Z	Z
NM	PVB	PB	PM	PS	PS	Z	Z	Z	NS
NS	PB	PM	PM	PS	PS	Z	Z	NS	NS
Z	PM	PM	PS	Z	Z	Z	NS	NS	NM
PS	PM	PS	PS	Z	NS	NS	NM	NM	NB
PM	PS	PS	Z	NS	NS	NM	NB	NB	NB
PB	PS	Z	Z	NS	NM	NM	NB	NVB	NVB
PVB	Z	Z	NS	NM	NM	NB	NB	NVB	NVB

4 Simulation Results on MATLAB/Simulink

The FLC controller is compared with PID controller. The parameters of the PID set are determined through the tuning simulation method on MATLAB/Simulink as $K_p = 0.7$; $K_I = 0.6$; $K_D = 0.01$.

Case 1: the trajectory is a circular orbit with radius 1, center is origin.

The results of the two controllers when the robot follows the same circular trajectory and the simulated response is shown in Figs. 4 and 5.

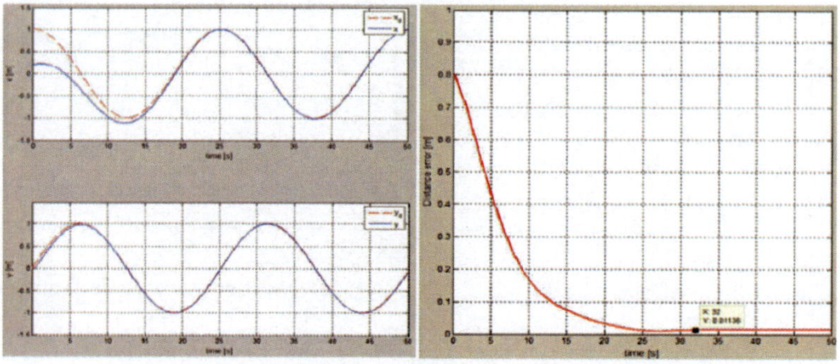

Fig. 4. x, y position and system error using PID controller

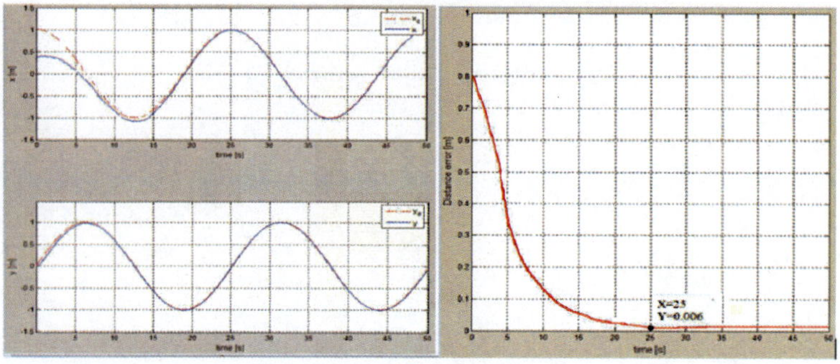

Fig. 5. X, Y position and system error using FLC controller

Based on the figure results in Figs. 4, 5, it is shown that both controllers respond to the stability of the moving robot following a preset trajectory. However, the robot position error is small for the PID controller (0.01), with a longer response time (32s). On the other hand, while the FLC controller system error is only 0.006, the response time is faster than the PID controller with 25s.

Case 2: The trajectory is simulated, which is the crackling trajectory.

The position response of the two controls is expressed in Figs. 6, 7. It is evident from the graph findings in Figs. 6, 7, that both controls react to the stability of a moving robot following a predetermined course. The PID controller, however, has a minor robot position error (0.01) and a slower reaction time (3s). However, even though the FLC controller's system error is just 0.006, it responds more quickly than the PID controller, which takes 2s.

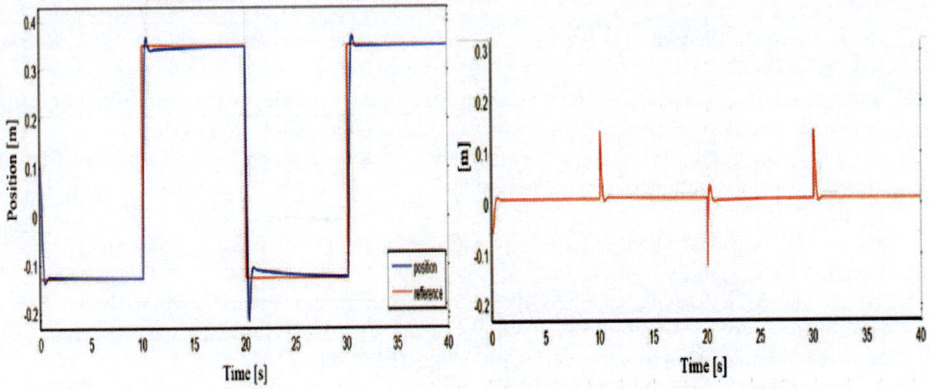

Fig. 6. Position and system error using traditional PID controller

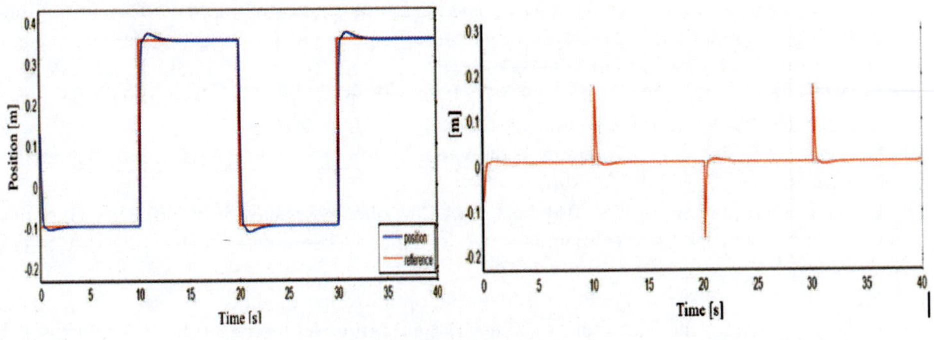

Fig. 7. Position and system error using FLC controller

5 Conclusion

The paper has proposed a kinematic and dynamic model for a mobile robot with a differential actuator based on the Lagrange dynamic approach. Mobile Robot is moved according to the trajectory set by PID and FCL controller. The FLC controller has the advantages of a simple design and better performance than the PID controller, with an orbital error of 0.006 and a setting time of 25s. However, to improve the moving quality of the mobile robot more accurately and faster, it is necessary to use intelligent control methods such as neural network, sliding mode controller with chattering controller, or hybrid controllers such as FLC controller combined with PID controller or sliding mode control connected with FLC controller.

References

1. Alexander, J.C., Maddocks, J.H.: On the kinematics of wheeled mobile robots. Int. J. Robot. Res. **8**(5), 15–27 (1989)
2. Barraquand, J., Latombe, J.: Nonholonomic multibody mobile robots: controllability and motion planning in the presence of obstacles. Algorithmica **10**(2), 121 (1993)
3. Campion, G., Bastin, G., d'Aandrea Novel, B.: Structural properties and classification of kinematic and dynamic models of wheeled mobile robots. IEEE Trans. Robot. Autom. **12**(1), 47–62 (1996)
4. Maaref, H., Barret, C.: Sensor-based navigation of a mobile robot in an indoor environment. Robot. Auton. Syst. **38**, 1–18 (2002)
5. Thongchai, S., Suksakulchai, S., Wilkes, D.M., Sarkar. N.: Sonar behavior -based fuzzy control for a mobile robot"**.** In: Proceedings of the 2000 IEEE International Conference on Systems, Man and Cybernetics, vol. 5, pp. 3532–3537 (2000)
6. Kolmanovsky, I., Harris McClamroch, N.: Developments in nonholonomic control problems. IEEE Control Syst. **15**(6), 20–36 (1995)
7. Lewis, F.L., Dawson, D.M., Abdallah, C.T.: Robot Manipulator Control: Theory and Practice,2nd edn. Marcel Dekker, Inc. (2003)
8. Li, Y.D., Zhu, L., Sun, M.: Adaptive neural-network control of mobile robot formations including actuator dynamics. In: Sensors, Measurement and Intelligent Materials, volume 303 ofApplied Mechanics and Materials, pp. 1768–1773. Trans Tech Publications (2013)
9. Li, Y.D., Zhu, L., Sun, M.: Adaptive RBFNN formation control of multi-mobile robots with actuator dynamics. Indo. J. Electr. Eng. **11**(4), 1797–1806 (2013)
10. DeCarlo, R.A., Zak, S.H., Drakunov, S.V.: Variable structure, sliding mode controller design. Control Handb. **57**, 941–951 (1996)
11. Derks, E.P.P.A., Pastor, M.S.S., Buydens, L.M.C.: Robustness analysis of radial base function and multilayered feedforward neural network models. Chemometr. Intell. Lab. Syst. **28**(1), 49–60 (1995)
12. Freire, F., Martins, N., Splendor, F.: A simple optimization method for tuning the gains of PID controllers for the autopilot of Cessna 182 aircraft using model-in-the-loop platform. J. Control Autom. Electr. Syst. **29**, 441–450 (2018)
13. Gao, W., Hung, J.C.: Variable structure control of nonlinear systems: a new approach. IEEE Trans. Ind. Electron. **40**(1), 45–55 (1993)
14. Lewis, F.L., Jagannathan, S., Yesildirek, A.: Neural Network Control of Robot Manipulator-sand Nonlinear Systems. Taylor & Francis, Ltd., 1 Gunpowder Square, London, EC4A 3DE (1999)

15. Lewis, F.L., Dawson, D.M., Abdallah. C.T., Robot Manipulator Control: Theory and Practice,2 edn. Marcel Dekker, Inc. (2003)
16. Li, Y., Qiang, S., Zhuang, X., Kaynak, O.: Robust and adaptive backstepping control for nonlinear systems using RBF neural networks. IEEE Trans. Ne ral Netw. **15**(3), 693–701 (2004)
17. Keighobadi, J., Mohamadi, Y.: Fuzzy sliding mode control of nonholonomic wheeled mobilerobot. In: Proceedings of the 9th IEEE International Symposium on Applied Machine Intelligence and Informatics—SAMI'2011, pp. 273–278. IEEE (2011)
18. Begnini, M., Bertol, D., Martins, N.: A robust adaptive fuzzy variable structure tracking control for the wheeled mobile robot: simulation and experimental results. Control Eng. Pract. **64**, 27–43 (2017)
19. Begnini, M., Bertol, D., Martins, N.: Design of an adaptive fuzzy variable structure compensatorfor the nonholonomic mobile robot in trajectory tracking task. Control Cybern. **47**, 239–275 (2018)
20. Begnini, M., Bertol, D., Martins, N.: Practical implementation of an effective robust adaptivefuzzy variable structure tracking control for a wheeled mobile robot. J. Intell. Fuzzy Syst. **35**, 1087–1101 (2018)

Check for updates

Optimal Navigation Based on Improved A* Algorithm for Mobile Robot

Thai-Viet Dang(✉) and Dinh-Son Nguyen

Hanoi University of Science and Technology, Hanoi, Vietnam
Viet.dangthai@hust.edu.vn

Abstract. Path planning is one of the core research direction of mobile robot navigation. First, a grid-based method transfers the complex environment to a simple grid-based map. Hence, the mobile robot's position is definitely determined in the grid map. To solve simultaneously the problems of the traditional A* path finding algorithm such as close distance from the obstacle, the shortest path, and path corners, the paper introduces the idea of the improved algorithm is to eliminate A* unnecessary nodes as for the purpose of reducing the computational scale. Then, the obtained path is smoothed by B-spline transition method. Hence, AMR's optimal obstacle avoidance strategy based on A* algorithm will be completely constructed. Finally, simulated results are shown to prove the feasibility of the proposed method.

Keywords: B-spline curve · Obstacle avoidance · Path planning · Mobile robot · RRT* algorithm

1 Introduction

With the advancement of mechatronics and informatics technologies during the Fourth Industrial Revolution, Autonomous Mobile Robots (AMRs) are widely used in outdoor/indoor environments [1, 2]. Path planning is the fundamental technology required to achieve optimal mobile robot navigation [3, 4]. The path planning environment of AMRs can be divided into global and local planning [5]. In fact, multi-objective optimization problems (MOOP) such as shortest length, smoothness, and obstacle avoidance will be simultaneously optimized.

Popular path planning algorithms for well-known environments include the Dijkstra algorithm [6, 7], the A* algorithm [8, 9], and the rapidly-exploring random tree (RRT) search algorithm [10, etc.]. Rachmawati et al. [6] introduced the Dijkstra algorithm for determining the shortest path. The performance of the Dijkstra algorithm was comparable to that of the A* algorithm in small scale environments but inferior in large scale environments. Wang et al. used a Heuristic optimization path-finding algorithm [7] to address this flaw. In addition, Yang et al. [8] increased the traditional A* algorithm's search direction from eight to sixteen in all surrounding neighborhoods that supported smoothing and optimizing the path of AMRs. However, increasing the number of neighbors would increase the search's computational steps. Wang et al. [9] proposed

T. D. L. Nguyen et al. (Eds.): ICISN 2023, LNNS 752, pp. 574–580, 2023.
https://doi.org/10.1007/978-981-99-4725-6_68

the EBHSA* algorithm to enhance the A* algorithm in terms of heuristic function optimization and smoothness. On the other hand, the issue of large memory usage in complex environments existed. Traish et al. presented Boundary Lookup Jump Point Search (BL-JPS) [11] searching for jump points during path expansion to remove redundant path points. Consequently, the search speed had been drastically increased. Yonggang et al. [5] devised a combination of the optimized A algorithm and the three-time Bezier curve to address the issue of numerous turning points and large turning angles.

The paper introduces the concept that the purpose of the enhanced algorithm is to eliminate A* unnecessary nodes in order to reduce the computational scale. Initially, a grid-based method is used to convert the dynamic environment into a simple grid-based model. The obtained path is then refined using the B-spline transition method. Consequently, the optimal obstacle avoidance strategy for AMR based on the A* algorithm will be constructed. Finally, simulated results are presented to demonstrate the method's applicability.

2 Proposed Method

Consider a mobile robot moving on a two-dimensional grid map containing models of obstacles (see Fig. 1). The white grid "0" is movable, while the black grid "1" is off-limits. S is start grid and G is goal grid. Path planning is innately viewed as a simultaneous multi-objective optimization problem (MOOP) with the shortest length, smoothness, and obstacle avoidance as the objectives. The environment is transferred to a simple grid model using the grid method. Utilizing the A* algorithm, the optimal path in a static grid-based map is determined.

Fig. 1. 30 × 30 grid based map

2.1 A* Algorithm

In search area, the conventional A* algorithm is a best-first algorithm that employs a heuristic search algorithm to evaluate the nodes surrounding the current position, as illustrated by Eq. 1. The search will continue until the target grid is located.

$$f(m) = h(m) + g(m) \tag{1}$$

where m is current node, f(m) is the cost evaluation function, h(m) is the estimated cost from m to G, g(m) is actual cost (path length) from m to next node. The common heuristic function distance is Manhattan distance as shown in Eq. 2.

$$h(m) = |x_G - x_b| + |y_G - y_b|, \tag{2}$$

(x_G, y_G) are the goal node's coordinates. (x_b, y_b) are any node's coordinates.

2.2 Obstacle Avoidance

To ensure the safe collision distance from the obstacle, a new environment model is introduced with additional risk grey zones around the black areas (see Fig. 2). So the cost evaluation function in Eq. 1 become to Eq. 3 as below:

$$f(m) = h(m) + g(m) + k(m), \tag{3}$$

where k(m) is risk cost function based on the additional distance from the obstacle.

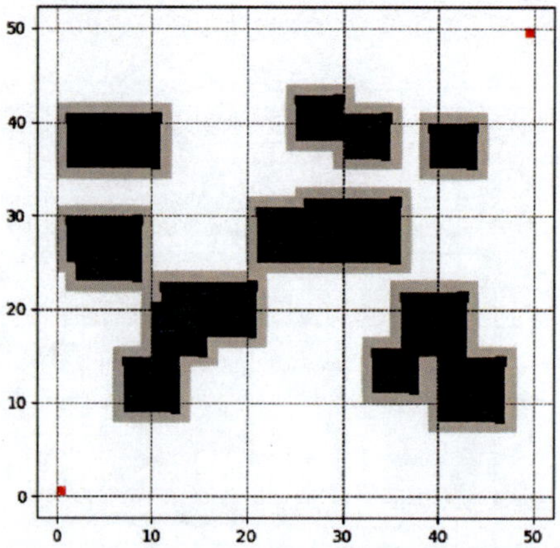

Fig. 2. New grid map with risk grey zones

In a complex environment, the search process will produce a large number of nodes, which will increase the computational scale, memory consumption, and inefficiency. As a result of redundant path points, the movement distance and computational scale will increase. Consequently, it is necessary to eliminate redundant points and smooth the path while simultaneously ensuring MOOPs.

2.3 Jump Point Search (JPS)

The output of the A^* algorithm is a group of path points P_i. With each set of three path points in Fig. 3, the line P_1P_3 can be connected. Then, check the condition for obstacle avoidance. If there is no obstacle grid on the line, the path P_1P_3 will be changed to $P_1P_2P_3$. The process of removing redundant points will be repeated until the target grid is located. Finally, the enhanced path planning would be obtained in its entirety. Obtaining optimal paths is considerably faster than using A* search.

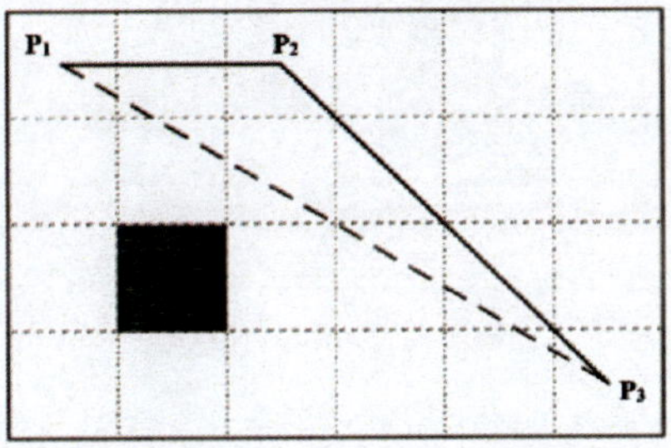

Fig. 3. Diagram of removing path redundant points

2.4 Smoothness

To assist the mobile robot navigation, a trajectory generation algorithm will be designed by the third-degree B-spline curve and straight line. The expression of the third-degree B-spline curve is shown in Eq. 4 (Fig. 4).

$$B(t) = \sum_{1=0}^{4} N_{i3}(u)pt_i. \quad (4)$$

where P_{ti} are the control path points and $N_{i3}(u)$ are the third degree B-spline basis functions. Therefore, the trajectory must be shorter than the planning path.

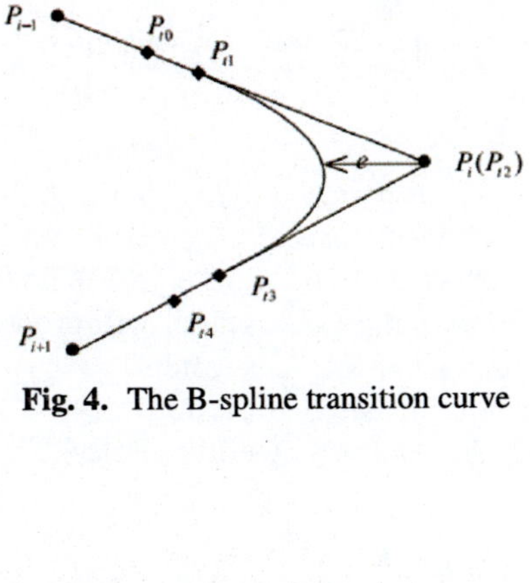

Fig. 4. The B-spline transition curve

3 Simulation Results

First, using mobile robot equipped with Lidar and embedded computer Jetson nano to build the grid map (see Fig. 5).

Fig. 5. Mobile robot

To verify the effectiveness of the improved algorithm, the following experiments were conducted, 50 × 50 grid environments (start point S (0,0) and goal point (50,50)) were used as test maps for simulation and compared with the conventional A* and improved A* algorithm. Intel(R) Core(TM) I7-8750h CPU @2.20ghz 2.21ghz,RaM 8.00 GB, 64 operating system, Windows 10 home English version computer and Visual studio 1.74 were used for experimental simulation (see Fig. 6).

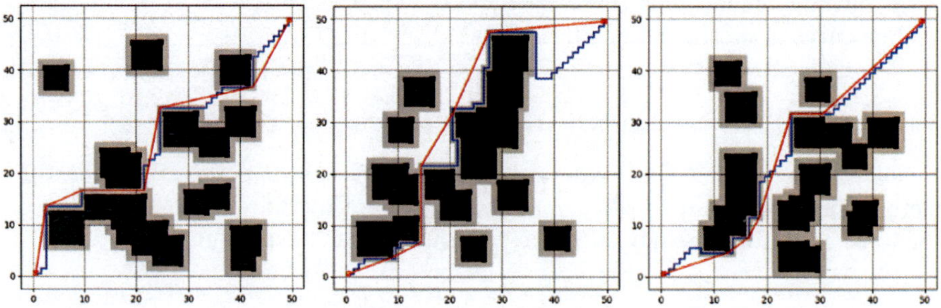

Fig. 6. Mobile robot path planning: blue line is A* algorithm and red line is improved A* algorithm.

In each of the three scenarios depicted in Fig. 6, the A* algorithm augmented with a safety cost function (gray zones) ensures a safe collision distance from the obstacle (see the blue path). The safety cost can be adjusted based on the severity of the collision with the obstacle. In addition, jump point search will produce the new red path after eliminating redundant path points from A*'s results.

In Fig. 7, a smoothed algorithm based on the third-degree B-spline curve will enhance the performance of the mobile robot's steering angles. The generation of trajectory

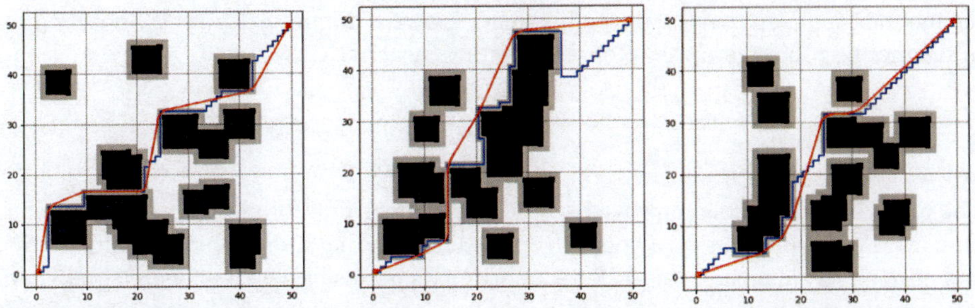

Fig. 7. Mobile robot path planning: blue line is A^* algorithm and red line is improved A^* algorithm combining JPS and smoothness.

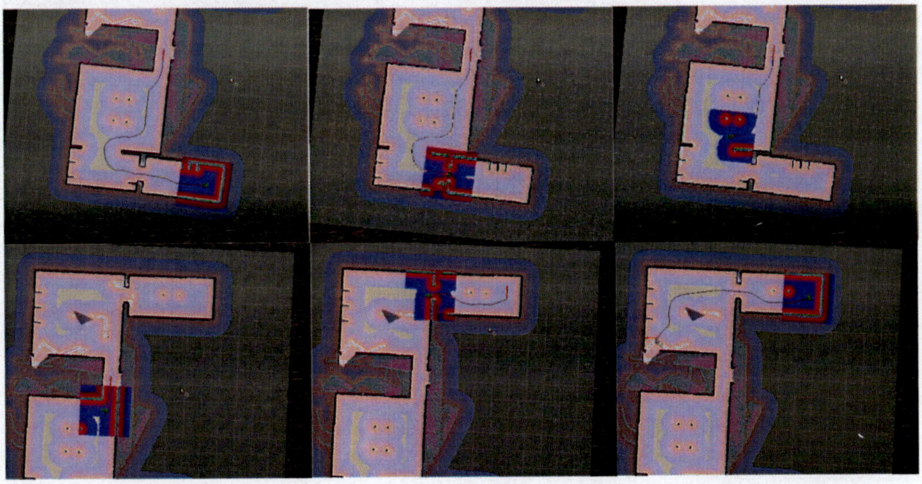

Fig. 8. The observation of the ROS mobile robot from the star grid to goal grid with the obtained path from the search area.

has met all MOOPs, including those listed below: shortest distance, smoothness, and avoidance of obstacles.

In Fig. 8, a mobile robot completely followed the trajectory generated by an improved smooth A* algorithm. In addition, the search area's safe cost ensured the mobile robot's robust movement while avoiding obstacles.

The mobile robot's optimal path navigation is created as follows: First, traditional A* path planning from start grid S to goal grid G is constructed using Manhattan distance. Then, utilizing an enhanced A* algorithm with safety cost (gray zones) to enhance the mobile robot's obstacle avoidance capability (see blue lines in Fig. 6). According to the results obtained, the number of points along Pi's path remains excessively high (see the grid on the blue lines). Therefore, a path-based JPS that eliminates redundant points will be developed (see redlines in Fig. 6). To enhance the ability to locate the path, the red line can traverse the gray areas without passing through any black areas. The authors then use the B-spline curve of third degree to generate Fig. 7's smoothed trajectory so that the mobile robot can move at the path's steering corner. Figure 8 depicts the observation of the mobile robot within the Robot Operating System (ROS).

4 Conclusions

The paper introduced the improved smoothed A* algorithm for mobile robot path planning in order to simultaneously optimize MOOPs including shortest length, smoothness, and obstacle avoidance. Using JPS, redundant path points in A*'s path were eliminated. In addition, the movement of mobile robots became safer with the addition of risk zones around obstacles. Simulations conducted in grid-based environments revealed the algorithm's advantages in terms of low memory consumption and optimization of computational calculation speed. Finally, path planning was conducted successfully in a practical ROS platform environment.

Acknowledgement. This research is funded by Hanoi University of Science and Technology (HUST) under project number T2022-PC-029.

References

1. Fragapane, G., Ivanov, D., Peron, M., Sgarbossa, F., Strandhagen, J.O.: Increasing flexibility and productivity in Industry 4.0 production networks with autonomous mobile robots and smart intralogistics. Ann. Oper. Res. **308**(1–2), 125–143 (2020). https://doi.org/10.1007/s10479-020-03526-7
2. Dang, T.V., Bui, N.T.: Obstacle avoidance strategy for mobile robot based on monocular camera. Electronics **12**(8), 1–20 (2023)
3. Patle, B.K., Pandey, A., Parhi, D.R., Jagadees, A.: A review: on path planning strategies for navigation of mobile robot. Def. Technol. **15**(4), 582–606 (2019)
4. Dang, T.V., Bui, N.T.: Multi-scale fully convolutional network-based semantic segmentation for mobile robot navigation. Electronics **12**(3), 1–18 (2023)
5. Yonggang, L., et al.: A mobile robot path planning algorithm based on improved A∗ algorithm and dynamic window approach. IEEE Access **10**, 57736–57747 (2022)
6. Rachmawati, D., Lysander, G.: Analysis of dijkstra's algorithm and A^* algorithm in shortest path problem. J. Phys. Conf. Ser. **1566**, 1–7, IOP Publishing, ICCAI 2019 (2020)
7. Wang, J., Zhang, X., Chen, B.: Heuristic optimization path-finding algorithm based on Dijkstra algorithm. J. Univ. Sci. Technol. Beijing **3**(3), 346–350 (2007)
8. Yang, J.M., Tseng, C.M., Tseng, P.S.: Path planning on satellite images for unmanned surface vehicles. Int. J. Naval Architect. Ocean Eng. **7**(1), 87–99 (2015)
9. Wang, H., Qi, X., Lou, S., Jing, J., He, H., Liu, W.: An efficient and robust improved A^* algorithm for path planning. Symmetry **13**(11), 1–19 (2021)
10. Zihan, Y., Linying X.: NPQ-RRT∗: An Improved RRT* Approach to Hybrid Path Planning. Complexity 2021, Article ID 6633878, 1–10 (2022)
11. Traish, J.M., Tulip, J.R., Moore, W.: Optimization using boundary lookup jump point search. IEEE Trans. Comput. Intell. AI iGames **8**(3), 268–277 (2016)

DTTP Model - A Deep Learning-Based Model for Detecting and Tracking Target Person

Nghia Thinh Nguyen[1], Duy Khanh Ninh[1](✉), Van Dai Pham[2], and Tran Duc Le[1]

[1] The University of Danang – University of Science and Technology, Danang, Vietnam
nghiathinh2000@gmail.com, {nkduy,letranduc}@dut.udn.vn
[2] Department of Information Technology, Swinburne Vietnam, FPT University, Hanoi, Vietnam
daipv11@fe.edu.vn

Abstract. Deep learning models have proven effective in various computer vision tasks, including object detection and tracking. In this paper, we propose a method for detecting and tracking a specific person in a video stream using deep learning models, called **DTTP** model. Our approach utilizes a combination of object detection and re-identification techniques to accurately track the target person across multiple frames after recognizing the face of that one. We evaluate our method on publicly available datasets and demonstrate its effectiveness in tracking a specific person with high accuracy and improving the model's speed by feature extraction of human pose and re-parameterization of deep learning models. Our DTTP model achieved 51.07 MOTA, 59.64 IDF1, and 47.31 HOTA on the test set of MOT17, which is higher accuracy and shows a favorable accuracy-speed trade-off compared to the state-of-the-art model like ByteTrack.

Keywords: human tracking · object detection · face recognition · human pose

1 Introduction

Tracking a particular person using deep learning models involves using computer vision and machine learning techniques to analyze video footage and identify individuals who can be wanted by law enforcement agencies or be required to pay attention. Some methods proposed for this task include using face recognition algorithms to match faces in video footage with a target person's database and object detection algorithms to identify and track individuals in a crowd. The traditional tracking method such as Kalman Filter [1], has limitations in handling complex situations, such as occlusions and changes in appearance. Model performance can be significantly degraded because people in a crowd can often obscure the view of the target person. Moreover, people may be at a different

T. D. L. Nguyen et al. (Eds.): ICISN 2023, LNNS 752, pp. 581–590, 2023.
https://doi.org/10.1007/978-981-99-4725-6_69

distance from the camera at particular times, making it difficult for the model to recognize the person consistently. Especially the change in facial features causes ambiguity in determining whether it is a target person.

This paper proposes a method for tracking a specific person in real-time video, combining deep learning models trained on large real-world datasets. The tracking target person model, called **DTTP** model, can detect and track the target person in a video stream by analyzing each frame and identifying the face's unique features. Our approach can handle occlusions and changes in appearance, such as clothing, distancing, or facial expressions, by updating the feature extraction of humans. Our main contributions including:

(1) Incorporating deep learning models to enhance the overall accuracy of detecting and tracking a specific person;
(2) Enhancing the tracking method by incorporating the use of feature extraction from the human pose model;
(3) Re-parameterizing the model to enhance its computational efficiency.

2 Related Works

Heads Tracking. Tracking a target person in complicated conditions, particularly in a crowd, is also considered a multiple object tracking (MOT) task. Our proposal is based on research related to this topic. First and foremost, a method involves detecting heads using a deep neural network and then tracking them using a skeleton graph [2], which is a graph-based representation of the body and head poses of pedestrians. Moreover, another solution proposes a new approach for detecting and tracking pedestrian heads in crowded environments using a Global Vision Transformer network [3]. The model can detect and track multiple pedestrian heads in real time, even in crowded and cluttered scenes. These can be difficult tasks in practice, as the heads of pedestrians may be obscured by other people or objects, and the movement of pedestrians in a crowded environment can be unpredictable. Additionally, the algorithm may struggle to accurately detect and track individual heads in real-world environments with varying lighting conditions.

Human Tracking. Some proposals are suggested to deal with the above problems. ByteTrack [4] is the state-of-the-art tracking model using a 3D human pose estimation model to predict the human pose of each frame and a motion segmentation algorithm to separate the person of interest from the background. Furthermore, a novel multi-pedestrian tracking method, BoT-SORT [5], is based on the SORT (Simple Online and Real-time Tracking) algorithm and aims to improve its robustness and association accuracy in crowded scenes. The method uses a deep learning-based detector to detect and track pedestrian bodies instead of heads - YOLOX [6] which increases the accuracy of the tracking model by free-anchor method and a Kalman Filter [1] to predict the location of each tracked person. The association of detections and tracks is performed using a modified version of the Hungarian algorithm, which considers appearance similarity and spatial proximity.

3 Methodology

In this approach, the primary objectives of our research endeavours Fig. 1 include addressing three key challenges:

Object Detection: The detection of the target individual is based on the location of their face and body within crowded environments.

Face Recognition: The extraction of facial features and comparison of similarities to locate the target individual.

Human Tracking: The identification of the target individual's face and body within the field of view of the camera and the improvement of the tracking model's precision through the utilization of feature extraction techniques on the human pose model.

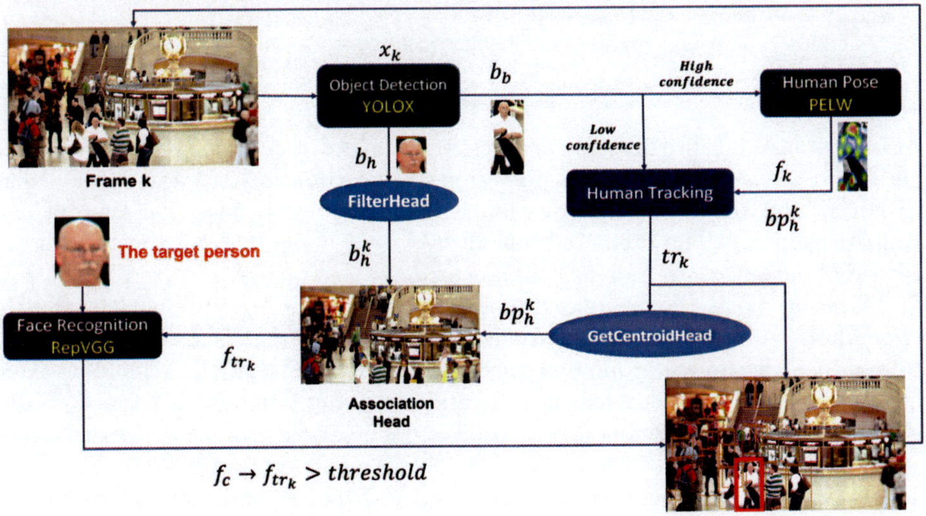

Fig. 1. Overview of ours DTTP model pipeline

3.1 Incorporating Deep Learning Models

Instead of utilizing the face and human detection models to tackle the problems of tracking the body and detecting the face of the target person, we combine two research to Object Detection tasks by using the YOLOX model [6]. Our outputs are separated into bounding boxes of the body b_b and the head b_h, which are defined as

$$b_b = [x_b, y_b, w_b, h_b], \tag{1}$$

$$b_h = [x_h, y_h, w_h, h_h]. \tag{2}$$

Because of the effects of crowd complexity on the accuracy of tracking heads, our tracking method is based on BoT-SORT [5] for tracking the target person.

The model applies the Kalman Filter [1] to predict the location of each tracked person x_k with input including the bounding box of the body and its velocity in the video frame k. Algorithm K based on the observation matrix is H, noise covariance R matrices and the covariance matrix P, defined as

$$x_k = [x_b, y_b, w_b, h_b, x'_b, y'_b, w', h'], \tag{3}$$

$$K = PH^T \left(HPH^T + R\right)^{-1}. \tag{4}$$

The Kalman Filter (KF) [1] goal is to try to estimate the state $x_k \in R$ given the measurements $z \in R$ by dividing into two-step predict and measurement, which creates new output in each video frame k as follows:

$$x_{k|k-1} = F_k x_{k-1|k-1}, \tag{5}$$

$$P_{k|k-1} = F_k P_{k-1|k-1} F_k^T + Q_k, \tag{6}$$

$$x_{k|k} = x_{k|k-1} + K(z_k + H_k x_{k|k-1}), \tag{7}$$

$$P_{k|k} = (I - K_k H_k) P_{k|k-1}. \tag{8}$$

At each step k, KF predicts the prior estimate of state $x_{k|k-1}$ and the covariance matrix $P_{k|k-1}$. KF updates the posterior state estimation $x_{k|k}$ given the observation z_k and the estimated covariance $P_{k|k}$, calculated based on the optimal Kalman gain K_k. The predicted output $x_{k|k-1}$ is computed with an IOU score with real output x_k to find cost distance cd_k.

However, with the complexity of crowds, such as occlusions, changes in appearance, and clothes color, which other tracking models like ByteTrack [4] cannot deal with, we propose a top-down human pose model called **PELW** model for extracting body features f_k, and bounding box head poses bp_h^k, which increase the model's confidence in tracking the target individual in the crowd and identify the face of them, respectively. We use euclidean distance ed_k to calculate the embedding distance of current features f_k and tracker features.

We chose the tracklets tr_k, which have the shortest distance score from forecast and reality to reduce error computed as

$$tr_k = \min(cd_k, ed_k). \tag{9}$$

Head detection based on bounding boxes extracted from the object detection model b_h^k causes confusion in black-haired identification because the target person can rotate the head against the camera direction. Therefore, our approach utilizes the histogram to filter out the exceptions based on a threshold, that is

$$b_h^k \rightarrow histogram(b_h^k) > threshold. \tag{10}$$

Regarding tracking and recognizing the target person's face in tr_k, we apply the Hungarian algorithm to associate b_h^k and bp_h^k, which increases credibility in identifying each person. The target person's facial feature extraction f_c is attached to f_{tr_k} during video streaming, defined as

$$f_{tr_k} \rightarrow associate(b_h^k, bp_h^k). \tag{11}$$

3.2 Data Argumentation for Face Recognition

There are many challenges for face recognition tasks in a crowd; thus, our proposal focuses on data augmentation, which can improve the model's performance. This is implemented by applying various transformation techniques to the images in the face datasets collected from the AFAD dataset[1], such as rotation, scaling, flipping, and adding noise and blur. These transformations create new, slightly different versions of the original images, allowing the model to learn to recognize the same face in different variations Fig. 2. Our approach also utilizes pre-processing techniques that enhance the performance of the face recognition model when implemented in a crowd by resizing the image to a fixed size, removing any noise, and using super-resolution from the OpenCV library[2] to improve the visibility of the face.

Fig. 2. the transform of faces using data augmentation

We use the **RepVGG** model [7] as a cutting-edge method, with a structural re-parameterization technique that allows us to decouple the training and inference time architectures. The facial feature extraction f_c of people must be higher than the threshold calculated during the training of the large datasets to identify the target person as

$$f_c \rightarrow f_{tr_k} > threshold. \tag{12}$$

Furthermore, by including a cosine term in the loss function, it aims to increase the angular margin between different classes using the ArcFace Loss function [8]. This term L encourages the feature vectors controlled by a scale parameter s for different classes to be more orthogonal to each other. It increases the inter-class correlation and the variance of the model's generalization by a margin parameter m, which is defined as

$$L_f = -\log(\cos(\theta + m)/s). \tag{13}$$

[1] https://afad-dataset.github.io/.
[2] https://opencv.org/.

3.3 Improving Re-id Tracking Method by Feature Extraction of Human Pose Model

In our efforts to enhance the performance of Re-identification (Re-id), we have employed the utilization of skeletal features f_k. This methodology in Fig. 3 is preferred over utilizing features from the entire image, as previously employed by other tracking models. The background portions of an image tend to exhibit variations over consecutive frames, thereby creating noise in the feature sequence, which is intended to identify individuals. In many instances, the backgrounds of two similar objects may lead to an indistinct differentiation between the two individuals.

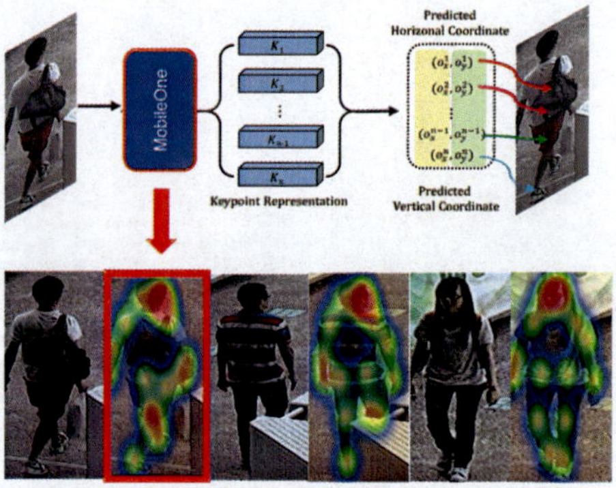

Fig. 3. Feature extraction of PEIW model

As seen in Figs. 3, the features f_k are primarily concentrated at positions such as the hands, feet, and body of the individual. These features f_k are subsequently updated in subsequent states k through the coefficient α to retain a portion of the information from the prior frame's feature, thereby avoiding loss of information when the individual is in motion as follows:

$$f_k = f_k\alpha + f_{k-1}(1-\alpha). \tag{14}$$

The Human Pose model, however, has a significant drawback. It incurs a relatively prolonged execution time because it requires the construction of backbone architecture with a substantial depth and the addition of semantically rich features after each block. As a trade-off, the processing speed is relatively slow. To mitigate this issue, we employ the MobileOne [9] as the backbone model, which possesses high accuracy. The model's output will be a set of coordinates (O_x, O_y) in Fig. 3 representing the positions of 17 points on the human body. From these points, we select a few points on the face bp_h^k to identify the target person's head. These points will be updated through the Mean Squared Error (MSE) loss function, which is defined as

$$L_p = \frac{1}{2*N} * \sum_{i=0}^{N} \left(x_p^i - x_{gt}^i\right)^2 + \left(y_p^i - y_{gt}^i\right)^2, \tag{15}$$

where N is the number of the parts of the human body, (x_p^i, y_p^i) is the predicted coordinates of the ith body part, (x_{gt}, y_{gt}) is the ground truth coordinates of the ith body part.

3.4 Re-parameterizing the DTTP Model to Enhance its Computational Efficiency

Due to the complexity and decreased processing speed that arises from the combination of multiple models, our solutions presented, such as YOLOX, PELW, and RepVGG, all utilize a technique of re-parameterizing the machine learning model for optimization. Re-parametrization is a technique used to modify the structure of deep learning models to make them more computationally efficient. The solution separates into two stages.

The training phase will consist of three branches, including two convolution layers of 3×3 ($W^{(3)}$) and 1×1 ($W^{(1)}$), as well as an identity layer ($W^{(0)}$). The parameters for each of these branches $\mu, \sigma, \gamma, \beta$ will be mean, standard deviation, scaling factor, and bias, respectively. These three layers will then pass through a batch normalization bn layer before being added together to produce the output, denoted as $M^{(2)}$. To maintain consistency in the dimensions of the layers, zero padding is applied to the $W'^{(1)}$ and $W'^{(0)}$ to match the 3×3 dimension of the conv3×3 layer $W'^{(3)}$ as follows:

$$M^{(2)} = bn\left(M^{(1)} * W^{(3)}, \mu^{(3)}, \sigma^{(3)}, \gamma^{(3)}, \beta^{(3)}\right) + bn\left(M^{(1)} * \right. \\ \left. W^{(1)}, \mu^{(1)}, \sigma^{(1)}, \gamma^{(1)}, \beta^{(1)}\right) + bn\left(M^{(1)}, \mu^{(0)}, \sigma^{(0)}, \gamma^{(0)}, \beta^{(0)}\right) \tag{16}$$

$$M^{(2)} = M^{(1)} * \left(W'^{(1)} + W'^{(3)} + W'^{(0)}\right) + \left(b'^{(0)} + b'^{(3)} + \right. \\ \left. b'^{(1)}\right) = M^{(1)} * W'^{(1+3+0)} + b'^{(1+3+0)}. \tag{17}$$

The implementation phase, however, only includes one 3×3 convolutional layer (W), and the other two layers are removed, and the output is wrapped with a Batch Normalization layer bn. Despite removing these two layers, the model's output remains unchanged compared to the architecture during the training phase $M^{(2)}$ as follows:

$$\begin{aligned} M^{(2)} &= bn(M^{(1)} * W, \mu, \sigma, \gamma, \beta)_{:,i,:,:} \\ &= \left(M^{(1)} * W'\right)_{:,i,:,:} + b'_i \\ &= M^{(1)} * W' + b. \end{aligned} \tag{18}$$

It can be observed that the number of parameters in both stages remains consistent.

4 Experiment

We conduct and evaluate **DTTP** model on the challenging benchmark datasets MOT17[3], which contains a total of 7 different sequences recorded in different scenarios such as urban, highway, and pedestrian areas. The datasets include video sequences with varying frame rates, object scales, and occlusions. Our test datasets are based on a public camera in the Philippines market, which has acquired is suitable for meeting our specific requirements, including high resolution, crowded environments, and relatively wide camera angles. Furthermore, we have utilized the COCO datasets[4], which is a comprehensive resource for large-scale object detection featuring over 330,000 images and 2.5 million object instances. Our object detection model focuses on two object categories: the head and body of humans. We trained a human pose model for extracting body features based on the HPII datasets[5], which contains over 40,000 images with annotations for the location of body joints. The dataset includes images of people in a variety of poses, including standing, sitting, and in motion, and includes a wide range of body types and ages. In Fig. 4, the ratio of the training set to test set in terms of the number of images and the number of instances including heads, persons per image is consistent. Therefore, the evaluation of the model will be more accurate and objective.

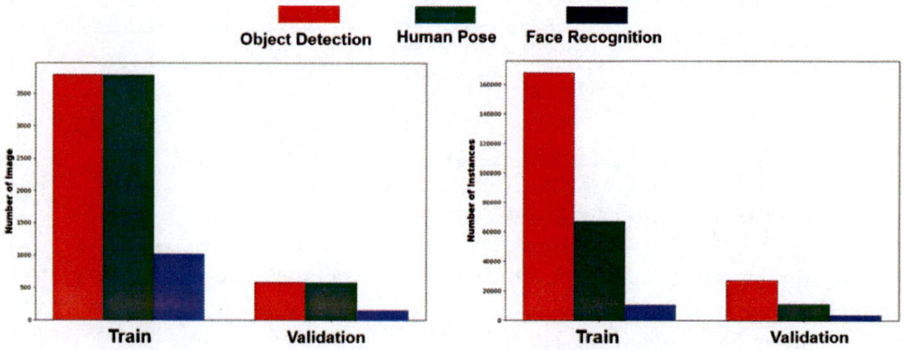

Fig. 4. The proportion of data distribution to solve three distinct tasks

For performance evaluation, the tracking performance is measured by standard metrics: Multiple Object Tracking Accuracy (MOTA) proposed by [10], the ratio of correctly identified detections IDF1 [10], and Higher Order Tracking Accuracy (HOTA) defined by [11].

---

[3] https://motchallenge.net/data/MOT17.
[4] https://cocodataset.org.
[5] http://human-pose.mpi-inf.mpg.de.

5 Result

The results shown in Table 1, our proposed DTTP surpasses ByteTrack model [4] by adding the feature extraction of the human body to improve the tracking target person. Additionally, it has been observed that when utilizing the YOLOX-s version for object detection, there is a substantial increase in the IDF1 index. This can be attributed to the significant difference in the number of parameters between YOLOX-nano and YOLOX, which greatly impacts the ability to identify the target person accurately. It is worth noting that the ByteTrack model consistently demonstrates high performance in terms of execution speed, with a larger number of frames per second (FPS) than the DTTP model. However, this disparity is minimal when utilizing the YOLOX-s version.

Table 1. The results of tracking models executed on NVIDIA GPU 1050 on MOT17 test set

Method	OD model	Image size	MOTA	IDF1	HOTA	FPS
ByteTrack	YOLOX-nano	416 × 416	26.29	37.29	31.54	14
DTTP	YOLOX-nano	416 × 416	26.21	37.45	31.52	5
ByteTrack	YOLOX-s	640 × 640	50.96	58.64	46.66	12
DTTP	YOLOX-s	640 × 640	**51.07**	**59.64**	**47.31**	**10**

Fig. 5. The DTTP model's full pipeline on the test set.

Figure 5 shows a series of steps in the full pipeline. DTTP model, which begins with the face of the target person and the video stream as inputs, is performed in sequential steps from left to right and top to bottom. Images 3 and 4 show that when pedestrians pass in front of the target person, their presence partially blocks the target person from view. Despite this occlusion, our proposed model can correctly identify and track the target person. Additionally, the model can maintain the tracking even when the target person turns away or moves far away from the camera in images 7 and 8.

6 Conclusion

This paper proposes a solution using deep learning techniques to detect and track a target person in a video stream. Despite the potential occlusion of other objects or crowds, the model can maintain accuracy in identifying and following the target. Our approach can be applied in various fields, such as surveillance, detecting robbery breaking into the house, or tracking wanted persons by law enforcement agencies. However, it's important to note that the model's performance may vary depending on factors such as image resolution, lighting conditions, and the specific architecture and parameters used in the model. In particular, our model cannot handle scenarios like individuals wearing masks in low-light conditions. Additionally, further research and development are needed to improve the robustness and accuracy of this technology.

References

1. Ali, N.H., Hassan, G.M.: Kalman filter tracking. Int. J. Comput. Appl. **89**(9) (2014)
2. Aziz, K.-E., Merad, D., Fertil, B., Thome, N.: Pedestrian head detection and tracking using skeleton graph for people counting in crowded environments. In: MVA, pp. 516–519 (2011)
3. Vo, X.-T., Hoang, V.-D., Nguyen, D.-L., Jo, K.-H.: Pedestrian head detection and tracking via global vision transformer. In: Sumi, K., Na, I.S., Kaneko, N. (eds.) Frontiers of Computer Vision, pp. 155–167. Springer, Cham (2022). https://doi.org/10.1007/978-3-031-06381-7_11
4. Zhang, Y., et al.: Bytetrack: multi-object tracking by associating every detection box. In: Avidan, S., Brostow, G., Cissé, M., Farinella, G.M., Hassner, T. (eds.) European Conference on Computer Vision, pp. 1–21. Springer, Cham (2022). https://doi.org/10.1007/978-3-031-20047-2_1
5. Aharon, N., Orfaig, R., Bobrovsky, B.Z.: Bot-sort: robust associations multi-pedestrian tracking, arXiv preprint arXiv:2206.14651 (2022)
6. Ge, Z., Liu, S., Wang, F., Li, Z., Sun, J.: Yolox: exceeding yolo series in 2021, arXiv preprint arXiv:2107.08430 (2021)
7. Ding, X., Zhang, X., Ma, N., Han, J., Ding, G., Sun, J.: RepVGG: making VGG-style convnets great again. In: Proceedings of the IEEE/CVF Conference on Computer Vision and Pattern Recognition, pp. 13733–13742 (2021)
8. Deng, J., Guo, J., Xue, N., Zafeiriou, S.: Arcface: additive angular margin loss for deep face recognition. In: 2019 IEEE/CVF Conference on Computer Vision and Pattern Recognition (CVPR), pp. 4685–4694 (2019)
9. Vasu, P.K.A., Gabriel, J., Zhu, J., Tuzel, O., Ranjan, A.: An improved one millisecond mobile backbone, arXiv preprint arXiv:2206.04040 (2022)
10. Bernardin, K., Stiefelhagen, R.: Evaluating multiple object tracking performance: the clear mot metrics. EURASIP J. Image Video Process. **2008**, 1–10 (2008)
11. Luiten, J., et al.: HOTA: a higher order metric for evaluating multi-object tracking. Int. J. Comput. Vision **129**(2), 548–578 (2021)

On the Principles of Microservice-NoSQL-Based Design for Very Large Scale Software: A Cassandra Case Study

Duc Minh Le[1(✉)], Van Dai Pham[1], Cédrick Lunven[2], and Alan Ho[2]

[1] Department of Information Technology, Swinburne Vietnam, FPT University, Hanoi, Vietnam
{duclm20,daipv11}@fe.edu.vn
[2] DataStax, Inc., Santa Clara, USA
{cedrick.lunven,alan.ho}@datastax.com

Abstract. Developing very large scale distributed software systems is challenging from both functional and data management perspectives. Methods based on Microservices Architecture (MSA) have gained popularity for addressing the functional challenges. On the other hand, cloud-aware, very large scale NoSQL data management systems have also proved their effectiveness in tackling data management's scalability challenges. Recent work have studied the combined approach for specific methods and systems. However, there has been no work that propose a complete method or study the underlying design principles. In this paper, we present the result of our initial research on this subject. We choose Cassandra as a case study as it is a popular system that supports cloud-aware, very-large-scale NoSQL data management. We propose the CaMSAndra software development method that combines the MSA and Cassandra methods. We define a UML metamodel for CaMSAndra and uses it as the basis for discussing the design principles. We analyse the relationship between bounded context and application workflow and, based on this, define a hierarchical service design that builds a service hierarchy by transforming an application workflow. We also discuss a data-driven cluster design in connection to the microservices. We demonstrate CaMSAndra with a well-known software domain called Hotel Reservation. We contend that our method is promising for developing very large microservice-based, NoSQL-based systems in general.

Keywords: Software Design · Microservices Architecture · NoSQL · Cassandra

1 Introduction

Developing very large scale distributed software systems is challenging from both functional and data management perspectives. Methods based on Microservices

T. D. L. Nguyen et al. (Eds.): ICISN 2023, LNNS 752, pp. 591–602, 2023.
https://doi.org/10.1007/978-981-99-4725-6_70

Architecture (MSA) have gained popularity for addressing the functional challenges. On the other hand, cloud-aware, very large scale NoSQL data management systems have also proved their effectiveness in tackling data management's scalability challenges. In fact NoSQL systems and their SQL counterparts have been studied in the context of the well-known CAP theorem [2]. This theorem basically states that distributed systems can at most achieve two out of the following three design properties: consistency (C), availability (A) and partition-tolerance (P). Most SQL-based systems are classified as CA with strict data consistency rules. In contrast, CP and AP systems both focus on partition-tolerance and provide non-strict forms of consistency. High consistency enforcement in very large scale distributed systems is extremely difficult to achieve without incurring some level of penalty in availability. In such systems, the continuity of operation in the face of failures and network changes is given a higher priority. Popular examples of AP systems include Cassandra, DynamoDB, CouchDB and Riak, while CP systems include MongoDB, Redis and HBASE. Among these, Cassandra, MongoDB and Redis are three popular very large scale NoSQL systems. In particular, Cassandra prioritises availability over consistency to provide what is known as eventual consistency [2].

Recent work have studied the combined approach for specific methods and systems. However, there has been no work that propose a complete method or study the underlying design principles. In this paper, we present the result of our initial research on this subject. We choose Cassandra as a case study for two main reasons. First, Cassandra is one of the most favourable NoSQL systems [1] that supports cloud-aware, very large data management scaling. Second, the authors have had extensive experiences in using and developing for this system. In particular, a leading multi-cloud, database-as-a-service version of Cassandra, named AstraDB, has been developed by Datastax.

We propose the CaMSAndra software development method that combines MSA and the Cassandra data modelling method. We define a UML metamodel for CaMSAndra and uses it as the basis to discuss the design principles. We analyse the relationships between the key concepts of the two component methods and discuss the principles for combining these concepts in the metamodel. In particular, we present a synthesis of bounded context and application workflow and, based on this, a hierarchical microservice design that transforms the application workflow to build the microservice hierarchy of a software. Further, we discuss a data-driven cluster design that explores the relationship between microservice and data distribution cluster in Cassandra. We demonstrate CaMSAndra with a well-known software domain called Hotel Reservation. We contend that our method is applicable to developing very large scale microservice- and NoSQL-based software systems in general.

The paper is structured as follows. Section 2 reviews the background knowledge and the related work. Section 3 presents an overview of our proposed our CaMSAndra method. Sections 4.1–4.3 discuss the core design principles in the context of CaMSAndra. Section 5 concludes the paper.

2 Background and Related Work

In this section, we introduce a motivating example and review a number of background concepts in the context of the related works. Contextualy, we define **very large-scale software** as a microservice-based software that stores a very large volume of data. The volume of data that is comparable to those managed by such global-scale software systems as Facebook and Netflix. In this context, we position our work as being related to Microservices Architecture and Cassandra.

2.1 Motivating Example: Hotel Reservation

To illustrate the concepts presented in this paper, we adopt the Hotel Reservation software example from Carpenter and Hewitt [5].

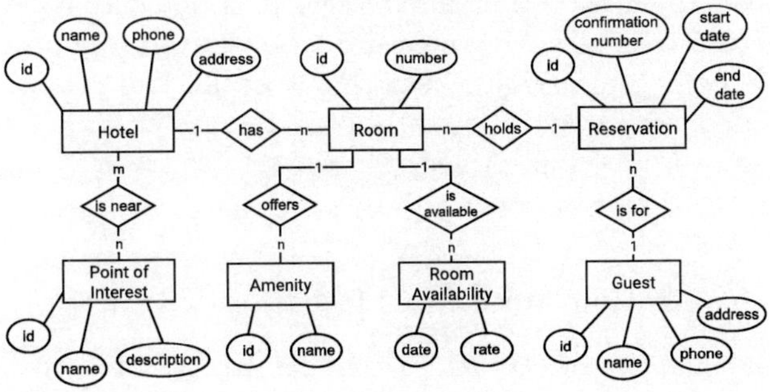

Fig. 1. Conceptual model of the Hotel Reservation domain (Adapted from [5]).

Figure 1 shows an entity relationship diagram (ERD) that represent the conceptual model of the hotel reservation domain. It consists of seven core concepts and the relationships between them. A hotel is located near some points of interests (POIs) and this relationship is used to answer user search queries about hotels and POIs. Once a hotel has been located, the user can proceed to make a reservation for the available rooms, each of which offers a number of amenities. Each reservation has a reservation period (start and end dates) and must include the guest details. A guest is the user who made the reservation.

2.2 Microservices Architecture

Microservices Architecture (**MSA**) is a modern scalable Internet-based software architecture. Each microservice represents a domain functionality that can be performed with a high degree of autonomy. In MSA, the software development process generally proceeds in a top down fashion, which starts with a high-level design to identify the bounded contexts. A **bounded context** defines the boundary of a microservice, which is represented by a domain model that captures the requirements of a business function or capability. Once the bounded contexts have been identified, the development process proceeds to tactical design to construct the domain model in each context. MSA has been discussed in the literature [3,4,9,10] to possess the following properties: (1) *service-based componentisation*, (2) *business-capability-driven*, (3) *distributed development*, (4) *modularity*, (5) *high autonomy*, (6) *infrastructure automation*, (7) *resilience*, (8) *observable*, and (9) *evolutionary design*.

Hierarchical Service Design. A number of recent works [4,8,13,15] have suggested to use a layered MSA style, in which microservices are organised into layers based on their domain dependencies. Two main benefits of this architecture style are that it helps (*i*) control the complexity of the system by reducing the service dependency and (*ii*) ease security enforcement. The former has recently been reported in [13] as a topic that requires further research. The latter is recently studied in the IoT context [11,14], where secured and resource-efficient access to the edge devices is a main concern. It is noted that both [11,14] use a form of layered, tree-based architecture to effectively organise services and manage the network complexity. A **service tree** [8] is a rooted tree in which the root node is a service and the non-root nodes are either another service or a non-service software module. An edge in the service tree represents functional dependency between its two nodes.

2.3 Cassandra Method for Data Modelling

Among the key concepts of Cassandra that are relevant to software design are query-driven data modelling with application workflow and peer-to-peer data distribution.

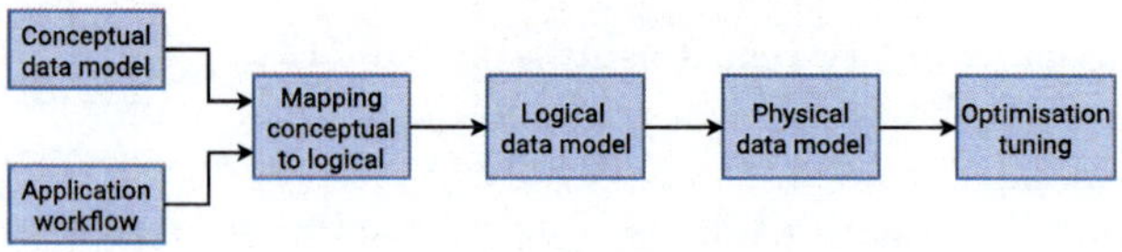

Fig. 2. Query-driven domain modelling method of Cassandra (Adapted from [5]).

As far as software development methodology is concerned, a unique feature of Cassandra-based system is its query-driven domain modelling. We prefer the more general term "domain modelling" to Cassandra's "data modelling" because, as will be explained below, in our view the method is actually a combination of data and behaviour modelling. In Cassandra, the idea is to combine the traditional conceptual data modelling (typically expressed in entity-relationship diagram (ERD) [6]) and the domain-specific behaviour requirements. These requirements constitute what is called in the Cassandra's literature the *application workflow*. We define **application workflow** as a layered, directed graph that represents a functional decomposition of a software, in which the bottom-level nodes are queries over the conceptual model of the software. In this paper, we will call this graph the **workflow model**.

To ease discussion in this paper we will refer to the Cassanda's domain modelling method simply as the **Cassandra method**. Further, we will refer to the aforementioned combined model of the Cassandra method as *Cassandra domain model*. When the context is clear, we will refer to this simply as **domain model**. Conceptually, we define the **Cassandra domain model** as a unified model consisting of a data model, describing the domain concepts and relationships among the concepts, and a relevant application behaviours that constitute a query-driven directed graph of functional decomposition over this data model.

For example, Figs. 1 and 3 show two component models that make up the domain model of hotel reservation. Figure 3, in particular, depicts the workflow model, expressed as a directed graph of function decomposition. Following the arrow paths lead us to the query functions that are defined in terms of the concepts in the conceptual model. For instance, the query function "*Q7. Find reservation by guest name*" defines a query on the two entities Reservation and Guest and the relationship between them. To ease discussion, we will refer to query function simply as **query**.

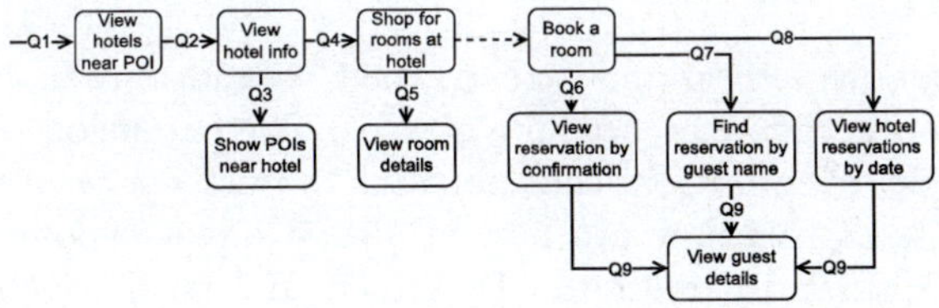

Fig. 3. Query-driven domain modelling for Hotel Reservation (Adapted from [5]).

Very Large Scale Data Management. As far as data management is concerned, Cassandra is well-known for its very large, horizontal scaling data management method. The method employs a peer-to-peer (P2P) data distribution scheme, which means that the overal storage capacity scale linearly with the number of nodes that participate in the system. Horizontal scaling is easier to manage than vertical scaling, which depends on a few high-performance server nodes. In Cassandra, data records (a.k.a *rows*) that share the same key prefix (called the *partition key*) form a data *partition*. Each partition is stored in a node that is responsible for the target token range containing the partition key. The partition key is hashed into token using a consistent hashing function.

DataStax Enterprise (DSE). DSE [7] and its cloud-based product line, named AstraDB[1], are highly scalable Cassandra-based data management systems. In particular, AstraDB is a multi-cloud database-as-a-service platform that consists

[1] https://www.datastax.com/products/datastax-astra.

in a suit of tools to ease system administration. These include DataStax's in-house tools as well as other open-source tools, notably those from the Apache Foundation.

3 Method Overview: CaMSAndra and the Metamodel

A key issue to address when designing microservice-based software with Cassandra is how to map the MSA concepts to the Cassandra method. Figure 4 shows a combined MSA-based and Cassandra-based software development method. We call this the Cassandra-MSA method or **CaMSAndra method** for short.

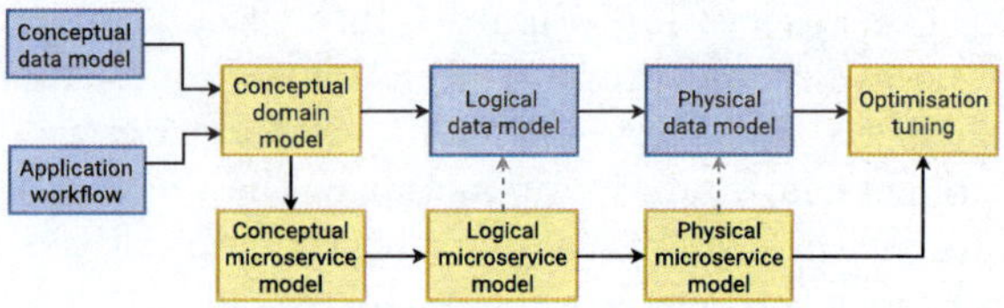

Fig. 4. The CaMSAndra method: MSA modelling with Cassandra.

CAMSAndra both revises and extends the Cassandra method shown in Fig. 2 to take into account the MSA design concerns. The extension involves adding a 3-component flow at the bottom that pertains to microservice construction. The three components of this flow are named after the corresponding three model versions of the Cassandra method. As shown in the figure, both logical and physical microservice models depend on the logical and physical data models for data storage design. The revision includes replacing the "Mapping conceptual to logical" component by the "Conceptual domain model" component, which explicitly reflects the existence of the Cassandra domain model. In addition, the "Optimisation and tuning" component is revised to consider both domain modelling and microservice modelling aspects.

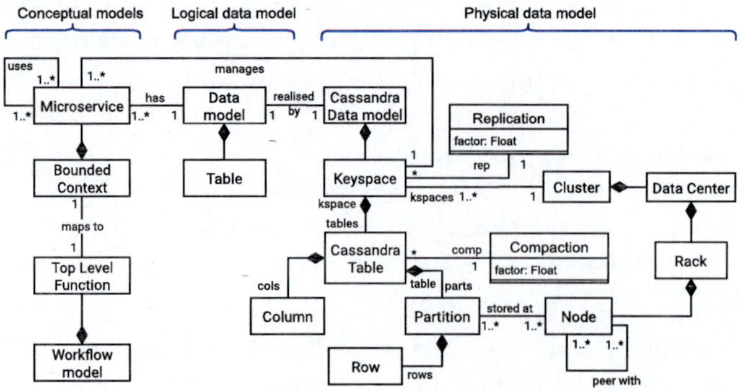

Fig. 5. The core CaMSAndra metamodel.

In the remainder of this paper, we discuss a number of core design principles in the context of the CaMSAndra method. Our discussion will focus on the relationships between key concepts of the Cassandra and MSA methods and how they lead to design insights for MSA-based software. To assist this investigation, we construct a metamodel that provides the foundational structure for the concepts under investigation. Figure 5 shows the UML diagram of the core metamodel of the CaMSAndra method. This model represents the concepts pertaining to microservices, workflow and data modelling components. The labelled curly brackets displayed at the top of the figure explains the connection between the metamodel and the models shown in Fig. 4. The two metamodel's substructures that pertain to logical and physical data models were constructed based on an analysis of Carpenter and Hewitt [5]. More specifically, within the scope of this paper we will investigate the principles that concern these essential relationships in Sect. 4.

To achieve preciseness, we use the Object Constraint Language (OCL) [12] to express the design rules associated with the metamodel. For conciseness, we use a short-hand notation to write OCL expressions on the model elements. For instance, the short-hand expression `Partition.table.cols->size()` (used in formula 1 of Sect. 4.3) means to apply the OCL's navigation rule to navigate from the context of a Partition to its associated Table and then to the set of Columns of this table and to perform the `size`() operation on this set.

4 Principles of CaMSAnda Design

4.1 Bounded Context Design with Workflow Model

Based on the definitions of bounded context and the Cassandra domain model, we map bounded context to a **top-level function (TLF)** of the workflow model that pertain to a well-defined domain behavior. Thus, as shown in Fig. 5, a workflow model consists in a set of TLFs. We argue that bounded context provides a necessary layer of abstraction on top of the workflow model that eases domain analysis and maintenance. Therefore, to extend domain requirements for new functions, we map them to relevant contexts or create new ones and add them to the corresponding workflow submodels. The general design rule is represented in Fig. 5, which states: one bound context per TLF.

For example, in the Hotel Reservation domain model shown in Fig. 3, we introduce two TLFs that represent two logical groupings of queries: hotel viewing and room booking. These TLFs represent the two bounded contexts of the domain. Figure 6 shows the bounded contexts of the example as dashed bounding boxes over two workflow submodels. The Hotel viewing context consists in the functions Q1-Q5, while the Room booking context consists in the functions Q6-Q9.

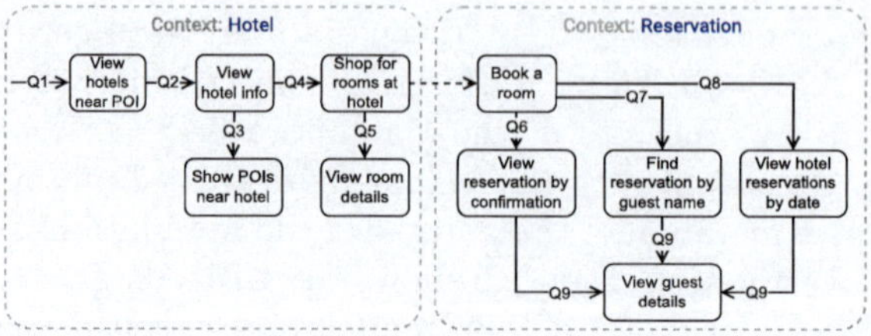

Fig. 6. Bounded contexts of Hotel reservation domain.

4.2 Hierarchical Microservice Design

After defining bounded contexts, the next step is to identify microservices. Typically, each bounded context contains one or more microservices. According to Carpent Hewitt [5], the logical data model (LDM) should be constructed from the domain model, and tables of the LDM are grouped to form the data boundaries of microservices. They recommend grouping denormalized tables representing the same data type to the same microservice.

We aim to generalize the identification of microservices in the conceptual modeling phase and establish specific rules for identifying them. We build on our recent work on hierarchical microservice design [8] and incorporate design considerations from the Cassandra method. The hierarchical workflow model of the Cassandra domain model serves as the basis for the service hierarchy. We convert the workflow model into a service tree using a 2-step procedure, which we call the **service construction procedure**:

1. **Determine a service** for each subset of functions that are associated to a main concept. This step generalises the service identification step described by Carpenter and Hewitt [5] to use the query functions in the workflow model. This helps move service identification step from the logical modelling phase to be performed earlier in conceptual modelling.
2. **Transform the service structure** (using the CRUD pattern) to expose the underlying concepts and, based on this, form a service tree. This step consolidates the functions to the underlying concept and makes this concept explicit in the service design. Focusing on the concept rather than the individual functions that it performs is necessary to avoid the anti-pattern of too-fine-grained service [10].

For example, Figs. 7(A) and 7(B) illustrate procedure. In Fig. 7(A), step 1 involves identifying three microservices based on the main concepts: Hotel, Point of Interest (POI), and Inventory. Step 2 involves transforming the Hotel service's structure to reveal the main concept and label it with the CRUD pattern. The Hotel service consists of the first 3 functions and two associated queries (Q1, Q2) that they serve. The POI service consists of one function that serves the query Q3, and the Inventory service consists of two functions that serve queries Q4 and Q5.

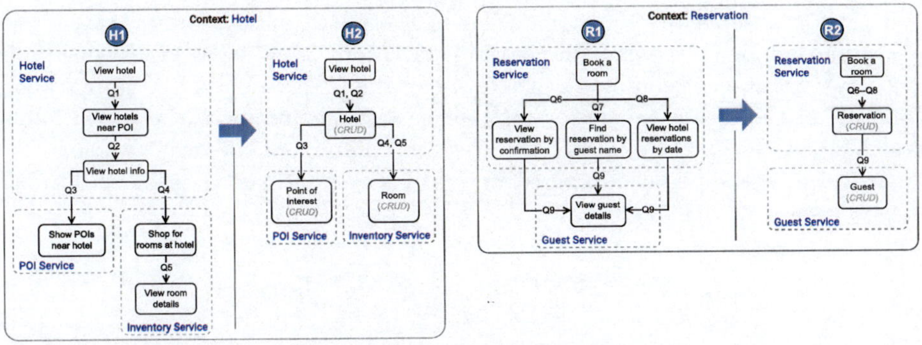

Fig. 7. (**A**) H1 → H2: Transforming Hotel services (LHS) to a service tree (RHS); (**B**) R1 → R2: Transforming Reservation services (LHS) to a service tree (RHS).

Similarly, Fig. 7(B) shows how the Reservation service tree is transformed using the service construction procedure. Step 1 involves identifying two services, Reservation and Guest. The Reservation service includes four functions and three queries (Q6-Q8), while the Guest service includes one function and query Q9. In step 2, the two services are transformed to reveal the underlying concepts and their CRUD operations.

4.3 Data-Driven Physical Design

In CaMSAndra, we observe that service autonomy for data distribution in Cassandra is limited to the cluster level, and beyond that lies the internal workings of the system. The physical data model in Fig. 5 shows that a microservice's data is stored in a keyspace within a single cluster, which can be either dedicated or shared [5]. However, microservices cannot control the placement of their data within specific data centers, racks, or nodes, which are internal to the Cassandra system and not the responsibility of the user application.

As far as microservice is concerned, therefore, an important cluster design concern is how to estimate a cluster's storage space based on the physical data model. To this end, we adapt the estimation technique presented in Chaps. 5 and 13 of Carpenter and Hewitt [5], which consists in 4 formulas for estimating the cluster size. Our contribution is to formulate the fourth formula (hinted at but not defined in [5]) and express all formulas more precisely using the CaMSAndra metamodel (see Fig. 5). In principle, the cluster size is estimated based on an indirect relationship that Cluster has with Partition, via Key Space and Cassandra Table, in the metamodel. Where suitable, we use OCL rules on the metamodel's structure to precisely express the formula terms.

Partition Size. Logically, the partition size (denoted by P_v) is determined by the number of cells (values) that it holds:

$$P_v = N_r \times N_g + N_s \tag{1}$$

Where: P_v: number of values (or cells) in the partition, N_r = `Partition.rows->size()`: number of rows of the current partition; N_c = `Partition.table.cols->size()`: number of columns of the owner table; N_{pk}: number of primary key columns of the owner table; N_s: number of static columns of the owner table; and $N_g = N_c - N_{pk} - N_s$: number of regular columns of the owner table.

Physically, the partition size (denoted by P_t) is measured based on formula 1 as follows:

$$\begin{aligned} P_t = & \sum_{1 \leq i \leq N_{pk}} \texttt{sizeOf}\left(c_{k_i}\right) + \sum_{1 \leq j \leq N_s} \texttt{sizeOf}\left(c_{s_j}\right) \\ & + N_r \times \left(\sum_{1 \leq k \leq N_g} \texttt{sizeOf}\left(c_{r_k}\right) + \sum_{1 \leq l \leq N_{pk}} \texttt{sizeOf}\left(c_{c_l}\right) \right) \\ & + P_v \times \texttt{sizeOf}\left(t_{avg}\right) \end{aligned} \tag{2}$$

Where: c_k, c_s, c_r, c_c: partition key columns, static columns, regular columns, and clustering columns (*resp.*); t_{avg}: the average number of bytes of metadata stored per cell; N_r, P_v: number of rows and logical partition size (*resp.*) as per formula 1; and `sizeOf`(): function that returns the size (in bytes) of the CQL data type of the involved columns.

Finally, taking into account the replication factor of each partition, the partition size's estimation becomes:

$$P = P_t \times R \times C \tag{3}$$

Where: P_t: the physical partition size as per formula 2; R = `Partition.table.kspace.rep.factor`: the replication factor of the keyspace containing the owner table of the partition; and C = `Partition.table.comp.factor` $\in \{2, 1.25\}$: compaction factor of the compaction strategy of the owner table.

Cluster Size. The total storage size of a cluster is determined by summing the sizes of all the partitions that are stored across all keyspaces and tables in that cluster. Assume that the usable storage space of the disk can be estimated with 90% of the disk size, the cluster size (denoted by S) is measured as follows:

$$S = \frac{\sum_{t \in \mathcal{P}} P_t}{90\%} \tag{4}$$

Where:
$\mathcal{P}$ = `Cluster.kspaces->collect(tables)->flatten()->collect(parts)->flatten()->asSet()`, i.e. the set of all the partitions across all tables and keyspaces of the cluster.

Discussion. We can extend the above technique to estimate the data storage size of a microservice and the storage size of a typical node. The relationship between Microservice and Keyspace in the metamodel would help define the former. On the other hand, the latter can be used, together with the cluster size, to estimate the number of nodes in a cluster. We plan to investigate these estimations in future work.

5 Conclusion

In this paper, we presented CaMSAndra method for developing very large scale microservice- and NoSQL-based software. Our method extended a state-of-the-art query-driven NoSQL data modelling method to take into account microservice design concerns. We constructed a UML metamodel for CaMSAndra and used it as the basis to discuss the core design principles. These principles arise out of the need to overcome the functional challenges associated with the metaconcepts and their relationships. Specifically, we presented a synthesis of bounded context and application workflow and, based on this, a hierarchical microservice design that transforms the application workflow to build the microservice hierarchy of a software. In addition, we discussed a data-driven cluster design that explores the relationship between microservice and data distribution cluster in Cassandra. We chose Cassandra as a case study as it is a popular system that supports cloud-aware, very-large-scale NoSQL data management. We demonstrated CaMSAndra with a well-known software domain called Hotel Reservation. We contend that our method is applicable to developing very large microservice- and NoSQL-based systems in general. Our plan for future work is to extend the method's scope to other comparable NoSQL systems that are in the CP and AP categories (e.g. MongoDB and Redis). We also plan to apply the method to develop real-world very large-scale software.

Acknowledgement. The authors wish to thank Jeffrey Carpenter [5] for granting us the permission to use the Hotel Reservation example in this paper. The authors would also like to thank the anonymous reviewers for their helpful feedbacks.

References

1. Ankomah, E., et al.: A comparative analysis of security features and concerns in NoSQL databases. In: Ahene, E., Li, F. (eds.) FCS 202. CCIS, vol. 1726, pp. 349–364. Springer, Singapore (2022). https://doi.org/10.1007/978-981-19-8445-7_22
2. Brewer, E.A.: Towards robust distributed systems. In: Proceedings of the 19th Annual ACM Symposium on Principles of Distributed Computing, PODC 2000, p. 7. ACM, New York (2000)
3. Bruce, M., Pereira, P.A.: Microservices in Action, 1st edn. Manning, Shelter Island (2018)
4. Carnell, J., Sánchez, I.H.: Spring Microservices in Action, 2nd edn. Manning, Shelter Island (2021)
5. Carpenter, J., Hewitt, E.: Cassandra: The Definitive Guide: Distributed Data at Web Scale, 3rd edn. O'Reilly Media, Sebastopol (2022)
6. Chen, P.P.S.: The entity-relationship model - toward a unified view of data. ACM Trans. Database Syst. **1**(1), 9–36 (1976)
7. Kashliev, A.: Storage and querying of large provenance graphs using NoSQL DSE. In: IEEE 6th International Conference on BigDataSecurity, HPSC and IDS, pp. 260–262 (2020)
8. Le, D.M.: Managing complexity in microservices architecture: a nested MultiTree domain-driven approach. In: Proceedings of Conference on APSEC 2022, Japan. IEEE Computer Society (2022)

9. Lewis, J., Fowler, M.: Microservices (2014). https://martinfowler.com/articles/microservices.html
10. Newman, S.: Building Microservices: Designing Fine-Grained Systems, 1st edn. O'Reilly Media, Beijing; Sebastopol (2015)
11. Oma, R., Nakamura, S., Duolikun, D., Enokido, T., Takizawa, M.: A fault-tolerant tree-based fog computing model. Int. J. Web Grid Serv. **15**(3), 219–239 (2019)
12. OMG: Object Constraint Language Version 2.4 (2014). http://www.omg.org/spec/OCL/2.4/
13. Waseem, M., Liang, P., Shahin, M.: A systematic mapping study on microservices architecture in DevOps. J. Syst. Softw. **170**, 110798 (2020)
14. Whaiduzzaman, M., Barros, A., Shovon, A.R., Hossain, M.R., Fidge, C.: A resilient fog-IoT framework for seamless microservice execution. In: 2021 IEEE International Conference on Services Computing (SCC), pp. 213–221 (2021). ISSN 2474-2473
15. Zhou, X., et al.: Benchmarking microservice systems for software engineering research. In: Proceedings of 40th International Conference on Software Engineering, ICSE 2018, pp. 323–324. ACM, New York (2018)

Check for updates

Policy Iteration-Output Feedback Adaptive Dynamic Programming Tracking Control for a Two-Wheeled Self Balancing Robot

Thanh Trung Cao, Van Quang Nguyen, Hoang Anh Nguyen Duc, Quang Phat Nguyen, and Phuong Nam Dao(✉)

School of Electrical and Electronic Engineering, Hanoi University of Science and Technology, Hanoi, Vietnam
nam.daophuong@hust.edu.vn

Abstract. This article discusses the trajectory tracking control requirement in a Two-Wheeled Self-Balancing Robot (TWSBR) with linearization and discretization after utilizing optimal control consideration. A proposed optimal control system that employs states observer to estimate state variables and Approximate/Adaptive Dynamic Programming (ADP) controller. The unification between tracking problem and output feedback ADP based optimal control design is guaranteed. Finally, simulation studies are used to validate the proposed control structure and demonstrate the performance of this control strategy.

Keywords: Adaptive/Approximate Dynamic Programming (ADP) · Two-Wheeled self balancing robot (TWSBR) · Output Feedback Controller · Policy Iteration (PI) algorithm

1 Introduction

Optimal control refers to solve control purpose after minimizing the cost function of dynamic systems. As we have all known that optimal control design is established to be derived from finding the solution of Hamilton-Jacobi-Bellman (HJB) partial derivative equation. However, direct solution of HJB equation is achieved impossibly and the indirect approach of employing adaptive dynamic programming (ADP) is feasible. The development of ADP technique for robotic systems has been considered sparingly in [1,2] with the disadvantage of utilizing Neural Networks (NNs). Based on linearization technique, it follows that the optimal control problem is developed for Robotics without using NNs [3]. Nevertheless, almost all ADP based optimal algorithm for Robotics has not been developed in the case of output feedback control. Motivated by the above consideration,

Supported by School of Electrical and Electronic Engineering, Hanoi University of Science and Technology.

T. D. L. Nguyen et al. (Eds.): ICISN 2023, LNNS 752, pp. 603–609, 2023.
https://doi.org/10.1007/978-981-99-4725-6_71

this article proposes an output feedback ADP strategy for a Two-Wheeled self balancing robot (TWSBR) to guarantee the unification between optimal control requirement and trajectory tracking performance. The following is how this article is organized. Section 2 discusses the preliminary and problem statement. In Sect. 3, the center algorithm is developed with state observer and Policy Iteration Output feedback. In Sect. 4, the simulation considerations are described to validate the proposed control algorithm.

2 Preliminary and Problem Statement

In the light of [3], the linearized dynamic model of a Two-Wheeled self balancing robot (TWSBR) system (Fig. 1) is described as follows:

$$\frac{d}{dt}x = \begin{bmatrix} 0&1&0&0&0&0 \\ 0&0&0&0&\zeta_1&0 \\ 0&0&0&1&0&0 \\ 0&0&0&0&0&0 \\ 0&0&0&0&0&1 \\ 0&0&0&0&\zeta_2&0 \end{bmatrix} x + \begin{bmatrix} 0&0 \\ \zeta_3&\zeta_3 \\ 0&0 \\ \zeta_4&-\zeta_4 \\ 0&0 \\ \zeta_5&\zeta_5 \end{bmatrix} u; \; \mathbf{y} = \begin{bmatrix} 1&0&0&0&0&0 \\ 0&0&1&0&0&0 \\ 0&0&0&0&1&0 \end{bmatrix}^T \eta \tag{1}$$

where

$$x = \begin{bmatrix} x' & \dot{x}' & \theta & \dot{\theta} & \phi & \dot{\phi} \end{bmatrix}^T; u = \begin{bmatrix} u_1 & u_2 \end{bmatrix}^T$$

and $\zeta_i, i = \overline{1,5}$ are obtained from [3]. Furthermore, in a considered (TWSBR) system, x', θ, ϕ are defined by the wheel-displacement, tilt angle, rotation angle, respectively and the parameters, physical variables of a considered TWSBR are also obtained from [3]. The control goal is to find the control inputs $u = \begin{bmatrix} u_1, u_2 \end{bmatrix}^T$ using PI Output feedback ADP method so that the wheel displacement, the tilt angle, and the rotation angle all converge to origin without knowing anything about the system dynamics. Furthermore this paper aims to solve the stabilization problem after linearizing mathematical model.

3 Policy Iteration-Output Feedback ADP Algorithm for a TWSBR

3.1 State Observer

We implement the observer design under the consideration of discrete time system (2) after discretizing the model (1):

$$\begin{aligned} x_{k+1} &= A_d x_k + B_d u_{k-d} \\ y_k &= C x_k \end{aligned} \tag{2}$$

According to [4,5], the state variables x_{k+d} and output $\bar{y}_{k-1,k-N}$ can be reconstructed in the time interval $[k-N-d, k+d]$ as

$$\begin{aligned} x_{k+d} &= P(N)\,\bar{x}_N + V(N)\,\bar{u}_{k-1,k-d-N} \\ \bar{y}_{k-1,k-N} &= U(N)\,\bar{x}_N + T(N)\,\bar{u}_{k-1,k-d-N} \end{aligned} \tag{3}$$

where

$$\bar{u}_{k-1,k-d-N} = \left[u_{k-1}^T, u_{k-2}^T, \ldots, u_{k-d-N}^T\right]^T \in \mathbb{R}^{m(N+d)}$$

$$\bar{y}_{k-1,k-N} = \left[y_{k-1}^T, y_{k-2}^T, \ldots, y_{k-N}^T\right]^T \in \mathbb{R}^{r(N+d)}$$

$$\bar{x}_N = \left[x_{k-d-1}^T, x_{k-d-2}^T, \ldots, x_{k-d-N}^T, x_{k-N}^T\right]^T \in \mathbb{R}^{n(N+1)}$$

$$P(N) = \left[0_{n \times nN} \; A_d^{N+d}\right] \in R^{n \times n(N+1)}$$

$$V(N) = \left[B_d \; A_d B_d \; \ldots \; A_d^{N+d-1} B_d\right] \in \mathbb{R}^{n \times m(N+d)}$$

$$U(N) = \begin{bmatrix} 0\,0 \ldots 0\,CA^{N-1} \\ 0\,0 \ldots 0\,CA^{N-2} \\ \ddots \; \vdots \quad \vdots \\ 0\,0 \; \ddots \; \vdots \quad \vdots \\ 0\,0 \ldots 0 \quad 0 \end{bmatrix} \in \mathbb{R}^{rN \times n(N+1)}$$

$$T(N) = \begin{bmatrix} \Omega & CB_d & CA_d B_d & \cdots & CA_d^{N-2} B_d \\ \Omega & 0 & CB_d & \cdots & CA_d^{N-3} B_d \\ \vdots & \vdots & \ddots & \ddots & \vdots \\ \Omega & \cdots & & 0 & CB_d \\ \Omega & 0 & 0 & 0 & 0 \end{bmatrix} \in \mathbb{R}^{rN \times m(N+d)}$$

with N being the observability index and $\Omega = 0_{r \times (m \times d + m)}$. It can be seen that $U(N)$ is full row rank and its pseudo left inverse is given as:

$$U^+(N) = U^T(N)\left(U(N)\,U^T(N)\right)^{-1} \tag{4}$$

Substituting (4) into the second equation in (3), $\bar{x}_N$ is computed as:

$$\bar{x}_N = U^+(N)\left(\bar{y}_{k-1,k-N} - T(N)\,\bar{u}_{k-1,k-d-N}\right) \tag{5}$$

Combining with the first equation in (3), it implies that:

$$\begin{aligned} x_{k+d} =& P(N)\,U^+(N)\left(\bar{y}_{k-1,k-N} - T(N)\,\bar{u}_{k-1,k-d-N}\right) \\ &+ V(N)\,\bar{u}_{k-1,k-d-N} \\ =& \left(V(N) - P(N)\,U^+(N)\,T(N)\right)\bar{u}_{k-1,k-d-N} \\ &+ P(N)\,U^+(N)\,\bar{y}_{k-1,k-N} \\ =& M_y \bar{y}_{k-1,k-N} + M_u \bar{u}_{k-1,k-d-N} \end{aligned} \tag{6}$$

3.2 PI Output Feedback Algorithm

In classic PI algorithm for the LQR problem, policy evaluation step is written as:

$$(A_d - B_d K_j)^T P_j (A_d - B_d K_j) - P_j + C^T Q_d C + K_j^T R_d K_j = 0$$

and then the policy improvement step can be obtained as

$$K_{j+1} = \left(R_d + B_d^T P_j B_d\right)^{-1} B_d^T P_j A_d \tag{7}$$

By defining $A_j = A_d - B_d K_j$, the discretized system (2) can be rewritten as

$$x_{k+1} = A_j x_k + B_d \left(K_j x_k + u_{k-d}\right) \tag{8}$$

With $\bar{K}_j = K_j \Theta$, $\bar{P}_j = \Theta^T P_j \Theta$, from (7) we have

$$\begin{aligned}
& x_{k+1}^T P_j x_{k+1} - x_k^T P_j x_k \\
=& \left(\bar{K}_j z_{k-d} + u_{k-d}\right)^T \left[\bar{H}_j^1 \; \bar{H}_j^2\right] \\
& \times \begin{bmatrix} -\bar{K}_j z_{k-d} + u_{k-d} \\ 2x_k \end{bmatrix} - \left(y_k^T Q y_k + z_{k-d}^T \bar{K}_j^T R \bar{K}_j z_{k-d}\right) \\
=& \left[u_{k-d}^T \otimes u_{k-d}^T - \left(z_{k-d}^T \otimes z_{k-d}^T\right)\left(\bar{K}_j \otimes \bar{K}_j\right)\right] vec\left(\bar{H}_j^1\right) \\
& + 2\left[\left(z_{k-d}^T \otimes z_{k-d}^T\right)\left(I_q \otimes \bar{K}_j\right) + \left(z_{k-d}^T \otimes u_{k-d}^T\right)\right] vec\left(\bar{H}_j^2\right) \\
& - \left(y_k^T Q y_k + z_{k-d}^T \bar{K}_j^T R \bar{K}_j z_{k-d}\right)
\end{aligned} \tag{9}$$

where $\bar{H}_j^1 = B_d^T P_j B_d$, $\bar{H}_j^2 = B_d^T P_j A_d \Theta$
For a sufficiently large positive integer s and $j = 0, 1, 2 \ldots$ we define

$$\phi_k^1 = \begin{bmatrix} u_{k_{j,0}-d} \otimes u_{k_{j,0}-d} - \left(\bar{K}_j \otimes \bar{K}_j\right)\left(z_{k_{j,0}-d} \otimes z_{k_{j,0}-d}\right) \\ \vdots \\ u_{k_{j,s}-d} \otimes u_{k_{j,s}-d} - \left(\bar{K}_j \otimes \bar{K}_j\right)\left(z_{k_{j,s}-d} \otimes z_{k_{j,s}-d}\right) \end{bmatrix}$$

$$\phi_k^2 = \begin{bmatrix} \left(I_q \otimes \bar{K}_j\right)\left(z_{k_{j,0}-d} \otimes z_{k_{j,0}-d}\right) + z_{k_{j,0}-d} \otimes u_{k_{j,0}-d} \\ \vdots \\ \left(I_q \otimes \bar{K}_j\right)\left(z_{k_{j,s}-d} \otimes z_{k_{j,s}-d}\right) + z_{k_{j,s}-d} \otimes u_{k_{j,s}-d} \end{bmatrix}$$

$$\phi_k^3 = \begin{bmatrix} z_{k_{j,0}-d}^T \otimes z_{k_{j,0}-d}^T - z_{k_{j,1}-d}^T \otimes z_{k_{j,1}-d}^T, \ldots, \\ z_{k_{j,s-1}-d}^T \otimes z_{k_{j,s-1}-d}^T - z_{k_{j,s}-d}^T \otimes z_{k_{j,s}-d}^T \end{bmatrix}^T$$

$$\Psi = \begin{bmatrix} y_{k_{j,0}}^T Q_d y_{k_{j,0}} + z_{k_{j,0}-d}^T \bar{K}_j^T R_d \bar{K}_j z_{k_{j,0}-d}, \ldots, \\ y_{k_{j,s}}^T Q_d y_{k_{j,s}} + z_{k_{j,s}-d}^T \bar{K}_j^T R_d \bar{K}_j z_{k_{j,s}-d} \end{bmatrix}^T$$

(9) can be rewritten as follows for any matrix $\bar{K}_j$ that stabilizes the system:

$$\Phi_j^P \begin{bmatrix} vec\left(\bar{H}_j^1\right) \\ vec\left(\bar{H}_j^2\right) \\ vec\left(\bar{P}_j\right) \end{bmatrix} = \psi \tag{10}$$

The triplet $\left(\bar{H}_j^1, \bar{H}_j^2, \bar{P}_j\right)$ in (9) can be uniquely solved (using the least square method) from Eq. (10). Then, $\bar{K}_{j+1}$ can be calculated from (7) as follows:

$$\bar{K}_{j+1} = \left(\bar{R} + \bar{H}_j^1\right)^{-1} \bar{H}_j^2 \tag{11}$$

PI Output Feedback ADP:

1. Step 1: Choose a sufficient small constant $\rho < 0$ and a stabilizing control matrix $\bar{K}_0$. On the time interval $[0, k_{0,0})$. $j \leftarrow 0$, use an initial stabilizing control law u_k
2. Step 2: While $\|\bar{P}_j - \bar{P}_{j-1}\| > \rho$, solve $u_k^j = -\bar{K}_j z_k + e_k$ on $[k_{j,0}, k_{j,s}]$ with e_k to be known as a suitable exploration noise.
3. Step 3: Calculate $\bar{P}$, $\bar{K}_{j+1}$ from (10) and (11), $j \leftarrow j+1$. Coming back Step 2.

Remark 1. This article discusses the extension of ADP technique in [3] with the addition of output feedback for a TWSBR. Moreover, this output feedback ADP strategy does not require the model parameters as above discussion.

4 Simulation Results

In this section, we discuss the outcomes of offline simulations. Applying the parameter values from [3], we obtain the explicit system matrices as:

$$A = \begin{bmatrix} 0 & 1 & 0 & 0 & 0 & 0 & 0 \\ 0 & 0 & 0 & 0 & 0 & -3.7706 & 0 \\ 0 & 0 & 0 & 0 & 1 & 0 & 0 \\ 0 & 0 & 0 & 0 & 0 & 0 & 0 \\ 0 & 0 & 0 & 0 & 0 & 0 & 1 \\ 0 & 0 & 0 & 0 & 0 & 68.9659 & 0 \end{bmatrix}$$

$$B = \begin{bmatrix} 0 & 0 \\ 0.599 & 0.599 \\ 0 & 0 \\ 1.0812 & -1.0812 \\ 0 & 0 \\ -5.776 & -5.776 \end{bmatrix}$$

Select the sampling period $h = 0.1$, the observability index $N = 2$ and the matrices

$$Q = \begin{bmatrix} 10 & 0 & 0 \\ 0 & 5 & 0 \\ 0 & 0 & 4 \end{bmatrix}; R = \begin{bmatrix} 1 & 0 \\ 0 & 1 \end{bmatrix}$$

It can be seen that Figs. 1 and 2 show the simulation results of not the convergence of learning algorithm but also tracking effectiveness.

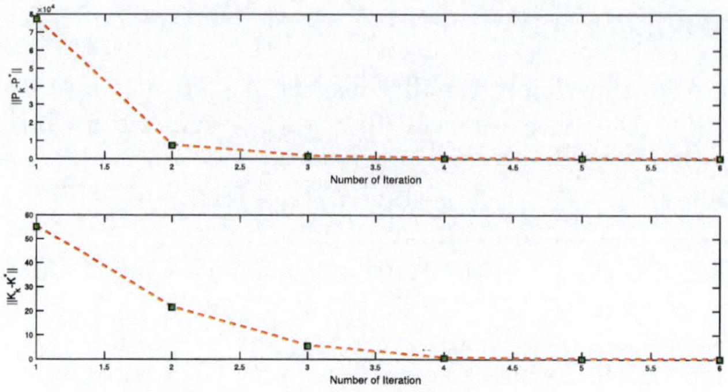

Fig. 1. Convergence matrix P in PI algorithm

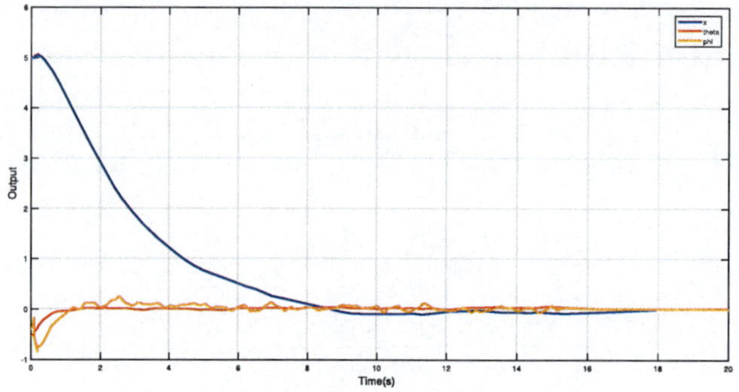

Fig. 2. The output of the system

5 Conclusions

In this article, a Policy Iteration (PI)-Output feedback Adaptive Dynamic Programming (ADP) is first proposed for a Two-Wheeled self balancing robot (TWSBR). For the control scheme, linearization and discretization are utilized to develop ADP technique without Neural Networks (NNs). Simulation studies further demonstrate the performance of the convergence of learning processes as well as the tracking problem.

References

1. Pham, T.L., Dao, P.N., et al.: Disturbance observer-based adaptive reinforcement learning for perturbed uncertain surface vessels. ISA Trans. **130**, 277–292 (2022)
2. Vu, V.T., Tran, Q.H., Pham, T.L., Dao, P.N.: Online actor-critic reinforcement learning control for uncertain surface vessel systems with external disturbances. Int. J. Control Autom. Syst. **20**(3), 1029–1040 (2022)
3. Dao, P.N., Liu, Y.-C.: Adaptive reinforcement learning strategy with sliding mode control for unknown and disturbed wheeled inverted pendulum. Int. J. Control Autom. Syst. **19**(2), 1139–1150 (2021)

4. Lewis, F.L., Vamvoudakis, K.G.: Reinforcement learning for partially observable dynamic processes: adaptive dynamic programming using measured output data. IEEE Trans. Syst. Man, Cybern. Part B **41**(1), 14–25 (2011). http://ieeexplore.ieee.org/document/5439950/
5. Huang, M., Jiang, Z.-P., Chai, T., Gao, W.: Sampled-data-based adaptive optimal output-feedback control of a 2-degree-of-freedom helicopter. IET Control Theory Appl. **10**(12), 1440–1447 (2016)

A Conceptual Model of Digital Twin for Potential Applications in Healthcare

Anh T. Tran[1], Duc V. Nguyen[2], Than Le[1,3], Ho Quang Nguyen[1](✉), Chi Hieu Le[4], Nikolay Zlatov[5], Georgi Hristov[6], Plamen Zahariev[6], and Vijender Kumar Solanki[7]

[1] Institute of Engineering and Technology, Thu Dau Mot University, Thu Dau Mot, Binh Duong, Vietnam
quangnh@tdmu.edu.vn
[2] HCMC University of Technology and Education, Ho Chi Minh City, Vietnam
[3] Artificial Intelligence Laboratory, Faculty of Information Technology, Ton Duc Thang University, Ho Chi Minh City, Vietnam
[4] University of Greenwich, Kent ME4 4TB, UK
[5] Bulgarian Academy of Sciences, Sofia, Bulgaria
[6] University of Ruse "Angel Kanchev", Ruse, Bulgaria
[7] CMR Institute of Technology, Hyderabad, TS, India

Abstract. Digital Twin (DT) is one of the important enabling technologies for Smart Manufacturing and Industry 4.0, with a huge potential for many impactful applications in healthcare and industries. This paper presents a conceptual model of a DT system, with a proof-of-concept (POC) prototype of a robot for demonstrations and further investigations of DT applications in telehealth and in-home healthcare. The successfully developed POC prototype were tested to evaluate time delay, and possible errors when operating and controlling the virtual and physical models of a robot. The proposed conceptual model of a DT system can be used for demonstrations about DT, with further developments for potential applications in healthcare and industries, especially when it is integrated with emerging technologies such as artificial intelligence, machine learning, big data analytics, smart sensors, augmented reality and virtual reality.

Keywords: Digital Twin · Industry 4.0 · Human-robot Interaction · ROS · Unity

1 Introduction

Digital Twin (DT) is one of the important enabling technologies for Smart Manufacturing and Industry 4.0, with a huge potential for many impactful applications in healthcare and industries [1–3]. A DT system basically has three main elements: the physical element, the digital or virtual element, and the connection element that connects the physical and the digital elements. A digital element is a digital replica of a physical element which is a potential and actual physical object such as devices, engineering systems and processes. With the use of real-world data as well as real-time simulation and optimization, DT systems can be used to control, monitor and predict how a product, a system or process performs and behaves, leading to enhancement of quality of decision-making and

T. D. L. Nguyen et al. (Eds.): ICISN 2023, LNNS 752, pp. 610–619, 2023.
https://doi.org/10.1007/978-981-99-4725-6_72

process optimization, as well as real-time control, monitoring and analysis of important performance indicators, and developing effective plans for predictive maintenance of products, equipment, production lines and facilities. Especially, with the availability of real-time data, interactions and responses, DT enables effective multidisciplinary collaborations and improved communications, to quickly and optimally make more informed decisions.

With the rapid advancements of digital transformation and smart technologies in recent decades, there have been more and more efforts and investments to develop impactful applications of DT and associated emerging technologies such as artificial intelligence (AI), machine learning (ML), big data analytics, smart sensors, augmented reality and virtual reality (AR/VR), especially applications in healthcare and medicine, including food and drug delivery, virus disinfection and telehealth services, as well as telesurgery or remote surgery and medical rehabilitation with robots [2–6].

This paper presents a conceptual model of a DT system, with development of a proof-of-concept (POC) prototype for demonstrations and further investigations of DT applications in healthcare and medicine, via the use of mobile robots with basic functions for applications in telehealth and in-home healthcare. A POC prototype of a DT system for a mobile robot was developed; it aims to effectively and conveniently manage and monitor operations of a robot for specific applications in telehealth and in-home healthcare, with the following key activities of design and development of DT, including (i) Building a physical element of a DT system; (ii) Building a virtual element of a DT system; (iii) Development of a connection element of a DT system, with data collection and transfer between the physical and the virtual elements; (iv) Development of a computation and simulation system; and (v) Local data collection, storage and cloud-based backups. The POC prototype of a DT system was successfully developed based on the latest development of simulations and interaction technologies, including AR/VR, a real-time 3D development platform (Unity) for building 2D and 3D applications such as games and simulations [7], and the open-source Robot Operating System (ROS) that helps build robotic systems [8].

2 Materials and Methods

2.1 Working Principle and a Structure of a Digital Twin System

A conceptual model with the fundamental elements of a DT system is presented in Fig. 1. There are three main elements: the physical element, the digital or virtual element, and the connection element that connects the physical and the digital elements. In this work, the physical and virtual elements are connected to each other through the TCP socket protocol suite. With the proposed DT system, the virtual model and the physical elements of a robot can interact with each other in which the data and/or information can be transferred and/or transmitted in two ways. When a robot operates in a certain environment, with the changes of a robot's state, such as a robot's location in its environment, the robot state data are collected by sensors, and they are then transferred and/or transmitted to a central processing unit (CPU). The collected data are optimized and processed at CPU, and they are transferred and/or to the virtual element of a robot, as the input data for the central processor of the virtual element which process and control the virtual robot in the virtual environment, and finally to display and store necessary information and data for specific applications, especially the important information and data for real-time control and monitoring of a robot.

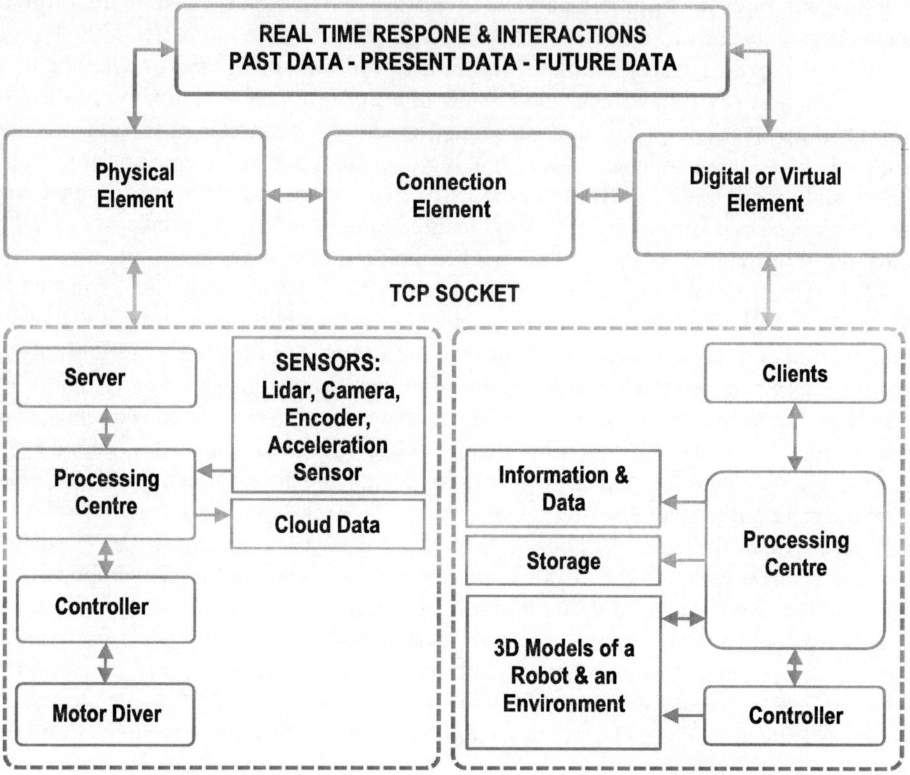

Fig. 1. A conceptual model with fundamental elements of a Digital Twin system.

Similarly, when controlling a virtual robot in the virtual element of a DT system, information and data of the state of a virtual robot are transferred to the processing centre, and they are then transmitted to the physical element of a DT system to control a physical robot. The data received from a virtual element of a DT system are processed, to generate signals to control the physical robot, such as to control movements of a robot.

The data and information are stored locally and on the cloud platform, and they are used to train algorithms for simulations and optimizations, especially to minimize errors between operations and controls of the physical and virtual models of a robot.

2.2 A Physical Element of a DT System – A Physical Robot Model

Figure 2 presents the physical and virtual models of a robot. A robot was built with a steel frame which has a rectangular shape with the size of 500x400x30 mm (Length x Width x Height). Four mecanum wheels are used for the driving function of a robot, and they are driven by four GP36 planetary geared motors. A lidar sensor is mounted on top of the robot for obtaining the best scanning view.

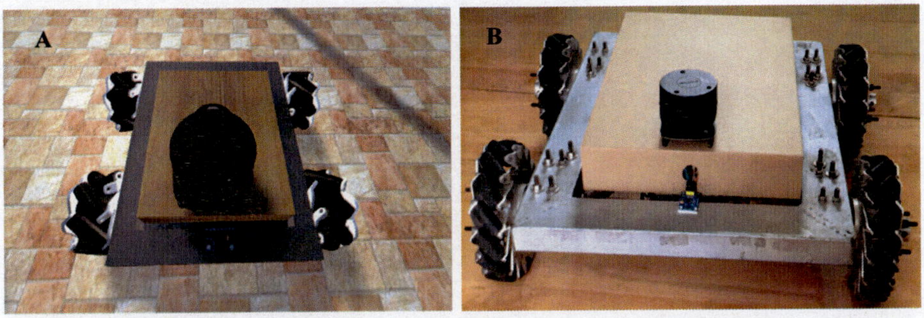

Fig. 2. A 3D model in the 3D virtual environment (A) and a physical model of a robot (B).

The instantaneous speed of a robot v is calculated as shown in Eq. (1) in which $\Delta(S)$ is the travel distance and $\Delta(t)$ is the travel time. The travel distance $\Delta(S)$ is calculated based on the number of the counted encoder pulses $\Delta(E)$ within the travel time $\Delta(t)$, the number of counted encoder pulses when the wheel turns with one revolution of the wheel, and the circumference of the wheel C. The number of counted encoder pulses when the motor turns with one revolution of the wheel is calculated based on the deceleration ratio (a) and the number of pulses when the encoder rotates one revolution (b).

$$v = \frac{\Delta(S)}{\Delta(t)} = \frac{\Delta(E) * C}{a * b * \Delta(t)} \tag{1}$$

In this study, the physical robot model is the mobile robot. The mobile robot uses the GP36 planetary geared motor with Encoder - DC Motor - 1:27-500CPR. The circumference of each wheel (C) is 40 cm. The following is the formula for calculating the speed of each wheel:

$$\mathrm{v_i} = \frac{\Delta(\mathrm{E}) * \mathrm{C}}{\mathrm{a} * \mathrm{b} * \Delta(\mathrm{t})} = \frac{(\mathrm{E_{cur}} - \mathrm{E_{pre}}) * 40}{500 * 27 * (\mathrm{t_{cur}} - \mathrm{t_{pre}})}, \mathrm{i} = [0, 1, 2, 3] \tag{2}$$

where $\mathbf{V_i}$ [cm/s] is the speed of the wheel, E_{pre} is the number of encoder pulses of the wheel at the previous measurement (t_{pre}), E_{cur} is the number of encoder pulses of the wheel at the current time (t_{cur}), a is the deceleration ratio, and b is the number of pulses when the encoder rotates one revolution. The acceleration of each wheel is calculated by the following formula:

$$\mathrm{a_i} = \frac{\Delta(\mathrm{v_i})}{\Delta(\mathrm{t})}, \mathrm{i} = [0, 1, 2, 3] \tag{3}$$

The robot uses the lidar sensor to scan for obstacles. The scanning radius is about 0.15–6 m with a 360-degree angle range. Data is transferred from the lidar to the central processor via the USB port. The data include the header, the scan start angle, the scan end angle, the angular distance between measurements, the time between measurements, the time between scans, the min range value, the range value maximum, and the data range. The distance is calculated using the following formula:

$$d = \frac{v * t}{2} = \frac{0.034 * t}{2} \tag{4}$$

where d [cm] is the distance from the lidar to the obstacle, t [μ s] is the time of the laser pulse emitted to the obstacle and back to the sensor, and v [cm/μ s] is the speed of sound.

2.3 A Virtual Element of a DT System – A Virtual Robot Model

The virtual model of a robot was built based on the physical model of the robot and then imported into the 3D virtual environment. The initial 3D virtual environment where an empty environment is unknown. Based on the data taken from the sensor of the physical model, the 3D virtual environment of the real environment is to be constructed.

2.4 Two-Way Data Transmission

The virtual model and the physical model of a robot are connected by the socket TCP protocol, with the built transmission pipeline. The place that sends or receives data is called a node. These nodes have to be named; however, it is importantly noted that, there should be no case in which the nodes have the same name. Nodes communicate with each other through pipes, called topics; and information travels in this pipeline, called messages. Sending and receiving messages between the nodes requires synchronisation between the data type and the topic.

2.5 Data Storage

Data is collected and stored locally and backed up in the cloud platform. The values of the sensors after being taken are sent to the central processor at the processing centre. The data will be processed and saved in the local storage. The cloud backup has a frequency of about 10 Hz, and it slows down the system processing speed. Therefore, the system backs up these data on the cloud platform after a period of time. A part of the data collected from the lidar sensor is shown in Fig. 3, and the data storage process is presented in Fig. 4. It is noted that, the graph of data was not sketched continuously. The data collected from the lidar are sketched out if the range is blocked by obstacles or they are left blank when nothing is in the scanning area of the lidar sensor.

2.6 A Mobile Application for Monitoring and Interacting with a Physical Robot Model of a DT System

In order to monitor and interact with a physical robot of a DT system, a mobile application was developed; It conveniently allows the user to control and monitor the activities of the physical robot through its virtual robot model. Users can control a physical robot to perform tasks such as delivery of food, water, or necessary items to the desired location for application cases of telehealth and in-home healthcare.

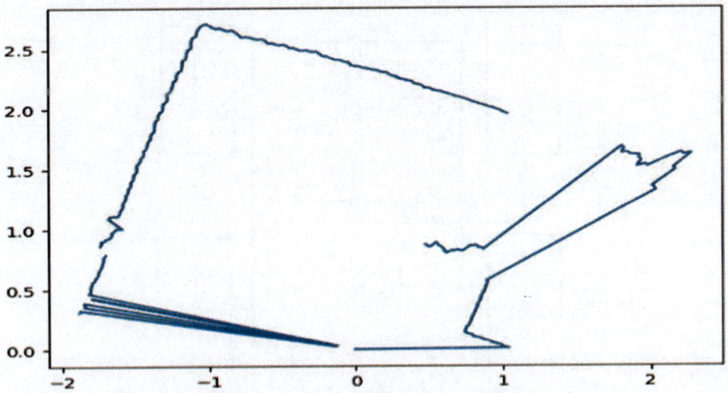

Fig. 3. A part of data collected from the lidar sensor.

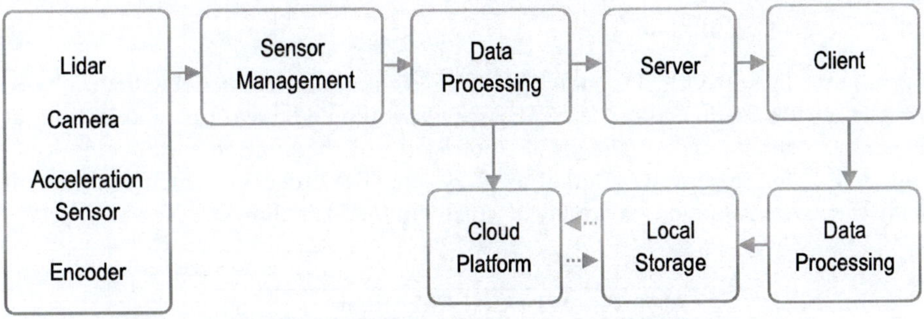

Fig. 4. A diagram of data storage and data processing for a DT system.

2.7 Analysis of Kinematics of a Robot

The robot uses four mecanum wheels, a type of omnidirectional wheel. That helps the robot move more flexibly than conventional wheeled robots. This is due to the roller mechanism with the shaft penetrating 45° to the main axis of rotation of the wheel. The calculation of robot positions is an important problem in autonomous robot systems; and development of the forward kinematics is necessary for calculating robot positions. The coordinate system is attached to the robot as shown in the Fig. 5. The speed of a robot is calculated based on Eq. (5). Where v_x is the speed of a robot in the x-axis, v_y is the speed of a robot in the y-axis, v_ω is the rotational speed of a robot in the z-axis, R is the vehicle radius, ω_i is the angular velocity of the wheel i (i = 1..4), and l_1 and l_2 are the distance between the wheel axle and the vehicle's center of gravity along the y and x axes.

$$\begin{bmatrix} v_x \\ v_y \\ v_\omega \end{bmatrix} = \frac{R}{4} \times \begin{bmatrix} 1 & -1 & 1 & -1 \\ 1 & 1 & 1 & 1 \\ \frac{1}{l_1+l_2} & -\frac{1}{l_1+l_2} & -\frac{1}{l_1+l_2} & \frac{1}{l_1+l_2} \end{bmatrix} \times \begin{bmatrix} \omega_1 \\ \omega_2 \\ \omega_3 \\ \omega_4 \end{bmatrix} \tag{5}$$

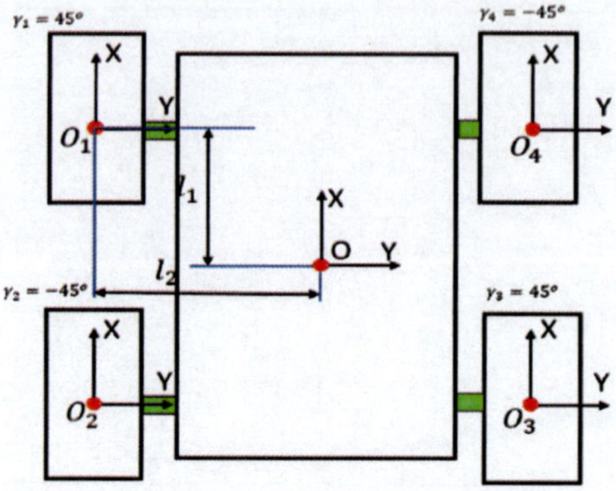

Fig. 5. A coordinate system of the robot OXY, and a coordinate system of the wheel O_iXY ($i = 1, 4$).

2.8 Latency Calculation

It is required that the virtual robot is operated and controlled in parallel with the physical robot in a DT system, and in the ideal cases, there should be real-time interactions and responses without a delay. The problem of latency is therefore important, and it is an important criterion for evaluation of a DT system. The process of calculating the time delay (latency calculation) is presented in Algorithm 1 as follows:

Algorithm 1: Latency Calculation

Input: t, t_{pre}, t_{sum}, Δ_{adv}, Δ_t, i;
$t_{pre} \leftarrow 0$;
$\Delta_{adv} \leftarrow 0$;
$i \leftarrow 0$;
$t_{sum} \leftarrow 0$;
$\Delta_t \leftarrow 0$;
while true **do**
  $t \leftarrow$ GET (time);
  $i \leftarrow i + 1$;
  $\Delta t \leftarrow t - t_{pre}$;
  $t_{sum} \leftarrow t_{sum} + t$;
  $\Delta_{adv} \leftarrow t_{sum}/i$;
  $t_{pre} \leftarrow t$;
end
Output: Δ_t, Δ_{adv};

Where t is the time at which the state data of the physical and the virtual robots are obtained. t_{pre} is the time at which the data was retrieved last time compared to the current one. t_{sum} is the total system uptime. i is the number of times the data is retrieved. Δ_t is the system latency and Δ_{adv} is the average system delay. The system runtime is obtained through the GET() function.

3 Results and Discussions

Experiments and demonstrations were done to test and evaluate the developed POC prototype of a DT system, with the focus on evaluation of the system latency and the possible errors between the virtual and physical robots when they are operated and controlled in both the virtual and the physical environments.

3.1 The Latency of a DT System

The process of transferring data between the physical and virtual robots in a DT system has a time delay. The system communicates through the WiFi network with a specific IP address. The frequency of the system is shown in Fig. 6, in which the average latency of the system is about 0.028 s and an average frequency is about 36 Hz. With this frequency, the system can satisfy the basic requirements for demonstrations of the POC prototype of a DT system.

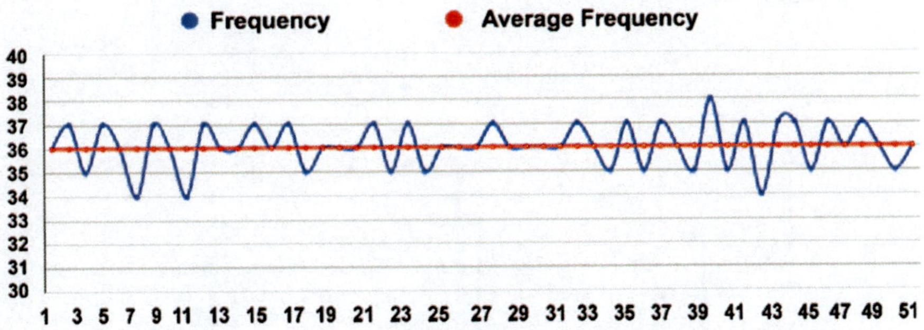

Fig. 6. A system frequency. The vertical axis is the frequency value (Hz) and the horizonal axis is the time value (s).

3.2 Analysis of Possible Errors When Operating and Controlling the Virtual and Physical Robots

The physical and the virtual robots are operated and controlled in parallel. However, there are problems related to acceleration, velocity, and errors of sensors, that lead to errors between the positions of the physical and virtual robots. The DT system was tested at speeds below 100 mm/s on the flat surface. Figure 7 presents errors between the positions of the physical and virtual robots.

Velocity calculations and simulations from sensor values are also performed to help the virtual robot behave as closely as possible to the physical robot. However, the simulation always has errors due to sensor errors, delay and moving terrain. The relative error of instantaneous velocity between the physical and virtual robots is calculated by the formula: $E_{relative} = 100 \times \frac{|v_{real} - v_{virtual}|}{v_{real}}$. The instantaneous velocity error is shown in Table 1 and Fig. 7. The error fluctuates about $\pm 5\%$ from the real instantaneous velocity. Under the above experimental conditions, the error is acceptable. However, when the condition occurs in the real environment, the error will change significantly. In order to minimize possible errors, simulations and AI/ML algorithms should be developed and implemented, especially to simulate and predict the next state of data to help reducing errors and gradually homogenising interactions and communications between the physical and virtual robots.

As shown in Table 1 about the error ratio between the instantaneous velocities of the physical and the virtual robots, the largest instantaneous velocity in the sample data is 33.61 mm/s and the smallest instantaneous velocity is 0.08 mm/s with the largest relative error of 3.23%. It is noted that, Column 2 in Table 1 is the time the instantaneous velocity data that are collected since the DT system starts operating.

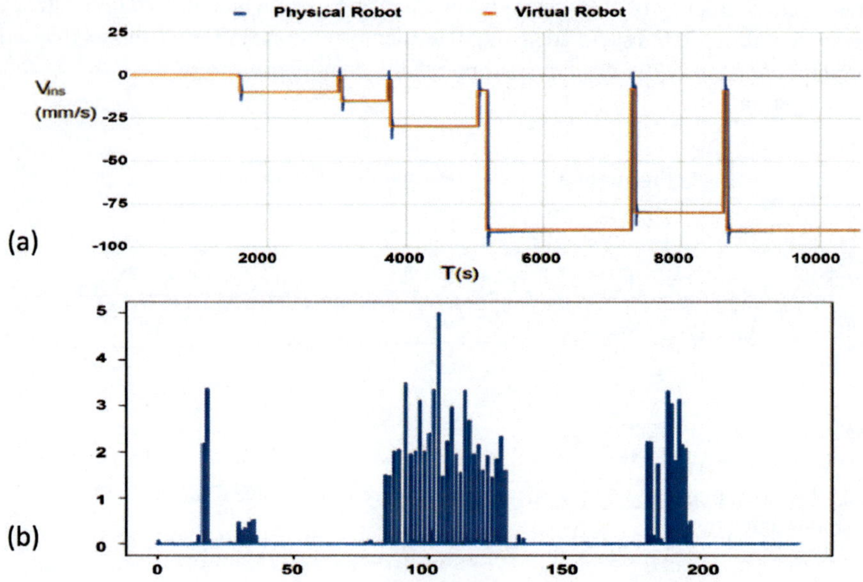

Fig. 7. The graphs of instantaneous velocity errors between of the physical and the virtual robots: (a) the instantaneous velocities, and (b) the error ratio between the instantaneous velocities.

4 Conclusion and Future Work

Under the impacts of Smart Manufacturing and Industry 4.0, there has been an emerging need to investigate and develop DT systems for different applications in healthcare and industries. In this paper, a conceptual model of a DM system, with a proof-of-concept (POC) prototype, were successfully developed and demonstrated, with the basic tests for further developments of potential applications, especially in telehealth and in-home healthcare, where the robots can be used for delivery of foods and drugs, virus disinfections, health monitoring and telehealth services. The future work which are well-aligned with this study include the following (i) integration of AR/VR to enhance interactions and user-experiences when using a DT system; (ii) development of a Simultaneous

Table 1. The errors between instantaneous velocities of the physical and the virtual robots.

STT	Time (s)	$V_{physical}$ [mm/s]	$V_{virtual}$ [mm/s]	Error [mm/s]	Relative error [%]
1	18.853	10.5523	10.2312	0.3211	3.0429
2	91.558	15.6166	16.1512	-0.5346	3.4232
3	98.763	34.8426	35.5456	-0.7031	2.0176
4	106.912	30.5213	31.1567	-0.6354	2.0818
5	143.295	33.6148	32.5345	-0.0587	0.1747
6	188.308	31.8922	30.8345	1.0576	3.2301
7	193.762	12.1184	11.7652	0.3532	2.9146
8	213.032	0.0797	0.0795	0.0002	0.2479

Localization and Mapping (SLAM) system for minimizing possible errors between the physical and virtual robots; (iii) integration of smart sensors and effective solutions for improving a system safety and security; (iv) full developments of a DT system with telepresence robots for specific applications in telehealth and in-home healthcare.

References

1. Wu, Y., Zhang, K., Zhang, Y.: Digital twin networks: a survey. IEEE Internet Things J. **8**(18), 13789–13804 (2021)
2. Liu, X., et al.: A systematic review of digital twin about physical entities, virtual models, twin data, and applications. Adv. Eng. Inform. **55**, 101876 (2023)
3. Mazumder, A., et al.: Towards next generation digital twin in robotics: trends, scopes, challenges, and future. Heliyon **9**(2), e13359 (2023)
4. Qadri, Y.A., et al.: The future of healthcare internet of things: a survey of emerging technologies. IEEE Commun. Surv. Tutor. **22**(2), 1121–1167 (2020)
5. Tamantini, C., et al.: A robotic health-care assistant for covid-19 emergency: a proposed solution for logistics and disinfection in a hospital environment. IEEE Robot. Autom. Mag. **28**(1), 71–81 (2021)
6. Raza, M., et al.: Telehealth technology: potentials, challenges and research directions for developing countries. In: Vo Van, T., Nguyen Le, T., Nguyen Duc, T. (eds.) Development of Biomedical Engineering in Vietnam. IFMBE Proceedings, vol. 63, pp. 523–528. Springer, Singapore (2018). https://doi.org/10.1007/978-981-10-4361-1_89
7. Unity, a real-time 3D development platform for building 2D and 3D application, like games and simulations.:www.unity.com. Accessed Feb 2023
8. ROS, the Robot Operating System for building robot applications. www.ros.org. Accessed Feb 2023

Different User Classification Algorithms of FFR Technique

Bach Hung Luu[1], Sinh Cong Lam[1](✉), Duc-Tan Tran[2], and Sanya Khruahong[3]

[1] Faculty of Electronics and Telecommunications, VNU University of Engineering and Technology, Hanoi, Vietnam
congls@vnu.edu.vn

[2] Faculty of Electrical and Electronic Engineering, Phenikaa University, Hanoi, Vietnam

[3] Faculty of Science, Naresuan University, Phitsanulok, Thailand

Abstract. Fractional Frequency Reuse (FFR) is a common technique of 4G, 5G and beyond cellular systems to improve the utilization of radio spectrum and Cell-Edge User (CEU) performance. Conventionally, there are various algorithms to identify the CEU in the FFR technique. This paper studies three well-known algorithms, called SINR-based, SNR-based and distance-based ones. Specially, to determine CEU, two first algorithms utilize the signal strength while the distance-based one follows the position of the user. The simulation results indicates that although the distance-based algorithm requires less signaling messages than others, it can derive the highest coverage probability.

Keywords: Fractional Frequency Reuse · User classification · SINR-based · SNR-based · distance-based

1 Introduction

In recent decades, an exponential increase in demand for high-speed data services has posed an urgent requirement to enhance the current spectrum efficiency and explore new frequency bands. The mid-band with frequencies from 3.5 GHz–6 GHz and millimeter wave with frequencies from 24–40 GHz are emerging as the most potential candidates for current 5G and future mobile communication systems [1]. These frequency bands can provide a much larger bandwidth in comparison with the current bandwidth with a frequency range from a few hundred MHz to a few GHz. However, a higher frequency usually suffers a higher path loss exponent which results in a higher penetration loss over the same distances. Furthermore, the far UE signals are usually blocked by a higher number of obstacles and consequently more scatterers and deeper fast-fading effects than the near UE. Consequently, there is a significant difference in Signal to Interference & Noise Ratio (SINR) between the near and far User Equipment (UE). Thus, the employment of advanced technique such as Fractional Frequency Reuse (FFR) is compulsory for 5G and beyond cellular systems to improve the performance of far UEs.

T. D. L. Nguyen et al. (Eds.): ICISN 2023, LNNS 752, pp. 620–628, 2023.
https://doi.org/10.1007/978-981-99-4725-6_73

In the early document in 2011 [2], 3GPP recommends that the 4G and later generations should implement the FFR technique to improve the average user performance, and then increase spectrum efficiency. Under the FFR technique, the users are usually classified into two Cell-Center User (CCUs) and Cell-Edge User (CEUs) by either downlink SNR, instantaneous SINR, or distance to the serving BS, which corresponds to SNR-based algorithm, SINR-based algorithm [3], and distance-based algorithm [4]. The SINR-based is considered the most popular FFR scheme since it can give a real-time adaptation to the instantaneous change of wireless conditions. Since this scheme also requires a lot of signaling messages and the classification procedure takes place at a very high density, the implementation of an SINR-based algorithm can cause signaling overload and network uncertainty. Thus, it seems to be impossible to deploy the SINR-based algorithm in practical networks. The SNR-based algorithm scheme requires fewer additional signaling messages since the downlink SNR value is regularly unchanged and should be estimated over a long time. Thus, this scheme is more practical than the SINR-based scheme. The distance-based scheme surpasses other schemes in 5G since this system supports UE's location determination through the downlink positioning reference signal (PRS) [5].

Research on the FFR schemes has been started in the last decade widely for 4G and recently developed for 5G and beyond. In [6], the uplink performance of different SINR-based sub-schemes in Full-Duplex and Half-Deplex small cell networks was analyzed. The paper also assumes that fractional power control is utilized to enhance the interference mitigation ability of FFR schemes. In [7], the general model of SINR-based schemes was introduced to improve the CEU performance in ultra-dense networks under stretched path loss model. The modified distance-based algorithm scheme was proposed for regular hexagonal heterogeneous cellular network layout [4] where each cell service area is divided into three layers. Through simulation, the authors proved that their proposed model can significantly improve the total network throughput compared to the regular distance-based algorithm. Since the downlink SINR strongly depends on path loss which is the function of the distance and path loss exponent, the farthest UE may not always have the lowest downlink SINR. Particularly, under the single slope model where the path loss exponent is a constant, the farthest UE is also the one with the lowest downlink SINR. However, under the 3GPP dual slope model where the path loss exponent depends on Light-of-Sight (LoS) and non-LoS probabilities, the farther UE can experience a higher SINR if it receives the signal from the LoS link. Thus, the SNR-based scheme is designed to overcome the limitation of the distance-based approach. However, the work on this scheme has not been deeply studied as two other schemes.

This paper aims to analyze and compare the performance of there FFR schemes, particularly SINR-based, SNR-based, and distance-based algorithm schemes in the 3GPP dual slope model.

2 System Model

The single-tier Poisson Point Process (PPP) mobile network is modeled in this section. The number of BSs is a Poisson Random variable with a mean of λ (BS/km^2), and their position is determined by the Spatial PPP. Without loss of generality, is also assumed that all users are randomly distributed and have connections to the nearest BSs. The Probability Density Function (PDF) of the distance between the user and the nearest BS is obtained from

$$f(r) = 2\pi\lambda r \exp\left(-\pi\lambda r^2\right) \tag{1}$$

The dual-sloped path loss model, which is characterized by the probability of LoS and nLoS occurrence, is utilized in analysis to estimate the downlink received signal power. As in the 3GPP document, the probability of LoS and nLoS can be simplified and determined as

$$p_l(r) = \exp\left(-\beta r\right) \tag{2}$$

$$p_n(r) = 1 - \exp\left(-\beta r\right) \tag{3}$$

For better expression, we assume that in each slope, the power loss over distance r is determined by the conventional path loss model. Particularly,

$$\text{For LoS} : r^{-\alpha_l} \text{ with probability } p_l(r) \tag{4}$$

$$\text{For nLoS} : r^{-\alpha_n} \text{ with probability } p_n(r) \tag{5}$$

where α_l and α_n are the path loss exponent of LoS and nLoS links, respectively.

Beside the path loss, the signal power over wireless channel is also affected by fast fading which is modeled as a random variable and officially called small-scale fading. Consequently, the power loss of the user signal over distance r is computed as the following equation

$$\mathcal{L}(r) = \begin{cases} g_l r^{-\alpha_l} \text{ with probability } p_l(r) \\ g_n r^{-\alpha_n} \text{ with probability } p_n(r) \end{cases} \tag{6}$$

where g_l and g_n are the fast fading of LoS and nLoS links. In this paper, g_l and g_n are supposed to have exponential distribution.

In the 3GPP cellular systems such as 4G and 5G, the BSs have permission to utilize the whole bandwidth. Thus, it is possible that more than one BSs transmit on the same sub-band on the downlink to serve its users. As the result, these users may experience inter-cell interference from all adjacent BSs. Thus, the downlink inter-cell interference of a user is given by

$$I = \sum_{k \in \theta_l} P_k g_{l,k} r_k^{-\alpha_l} + \sum_{k \in \theta_n} P_k g_{n,k} r_k^{-\alpha_n} \tag{7}$$

where P_k is the transmission power of BS k; θ_l and θ_n are the set of interfering BSs that affects the user on LoS and nLoS respectively.

The downlink SINR of the user under the effects of fast fading, dual-slope model, and Gaussian noise with a power of σ^2 is

$$SINR(P,r) = \begin{cases} \frac{Pg_l r^{-\alpha_l}}{I+\sigma^2} \text{ with probability } p_l(r) \\ \frac{Pg_n r^{-\alpha_n}}{I+\sigma^2} \text{ with probability } p_n(r) \end{cases} \quad (8)$$

where P is the serving power of the user.

2.1 FFR Design

In this section, the two-phase operation, which includes establishment and communication phase, is discussed in the case of SINR-based, SNR-based, and distance-based algorithm schemes. The establishment phase of there above FFR schemes is described as:

- *SINR-based algorithm*: In this scheme, the CCU and CEU are distinguished by the donwlink SINR in Eq. 8, where the user with a smaller downlink SINR in comparison with the SINR threhsold T_{SINR} is called CEU, and vice versa. Mathematically, the user is defined as the CCU with a probability of $\mathcal{P}_{SINR} = \mathbf{P}(SINR < T_{SINR})$.
  Using the definition of SINR in Eq. 8, the probability when the user at a distance r, $\mathcal{P}_{SINR}(r)$ is

$$\begin{aligned}\mathcal{P}_{SINR}(r) =& p_l(r)\mathbf{P}\left(\frac{Pg_l r^{-\alpha_l}}{I+\sigma^2} < T_{SINR}\right) + p_n(r)\mathbf{P}\left(\frac{Pg_n r^{-\alpha_n}}{I+\sigma^2} < T_{SINR}\right)\\ =& p_l(r)\mathbf{P}\left(g_l < \frac{T_{SINR}}{P} I r^{\alpha_l} + \frac{T_{SINR}}{\gamma} r^{\alpha_l}\right)\\ &+ p_n(r)\mathbf{P}\left(g_n < \frac{T_{SINR}}{P} I r^{\alpha_n} + \frac{T_{SINR}}{\gamma} r^{\alpha_n}\right)\end{aligned}$$

  where $\gamma = P/\sigma^2$. Since g_l and g_n are exponential random variables,

$$\begin{aligned}\mathcal{P}_{SINR}(r) =& p_l(r)\mathbf{E}\left[\exp\left(-\frac{T_{SINR}}{P} I r^{\alpha_l}\right)\exp\left(-\frac{T_{SINR}}{\gamma} r^{\alpha_l}\right)\right]\\ &+ p_n(r)\mathbf{E}\left[\exp\left(-\frac{T_{SINR}}{P} I r^{\alpha_n}\right)\exp\left(-\frac{T_{SINR}}{\gamma} r^{\alpha_n}\right)\right]\end{aligned}$$

- *SNR-based*: Instead of using SINR as the classification threshold, this approach utilizes the downlink SNR, which is obtain from Eq. 8 by ignoring interference element. The probability, that the user is called CCU if the SNR threshold is T_{SNR}, is $\mathcal{P}_{SNR} = \mathbf{P}(SNR < T_{SNR})$.
  When the user have a distance of r to its tagged BS, the probability $\mathcal{P}_{SNR}(r)$ can be obtained by the following mathematical transformation steps:

$$\begin{aligned}\mathcal{P}_{SINR}(r) =& p_l(r)\mathbf{P}\left(\frac{Pg_l r^{-\alpha_l}}{\sigma^2} < T_{SNR}\right) + p_n(r)\mathbf{P}\left(\frac{Pg_n r^{-\alpha_n}}{\sigma^2} < T_{SNR}\right)\\ =& p_l(r)\mathbf{P}\left(g_l < \frac{T_{SINR}}{\gamma} r^{\alpha_l}\right) + p_n(r)\mathbf{P}\left(g_n < \frac{T_{SINR}}{\gamma} r^{\alpha_n}\right)\end{aligned}$$

Since g_l and g_n are exponential random variables,

$$\begin{aligned}\mathcal{P}_{SINR}(r) =& p_l(r)\mathbf{P}\left(\frac{Pg_l r^{-\alpha_l}}{\sigma^2} < T_{SNR}\right) + p_n(r)\mathbf{P}\left(\frac{Pg_n r^{-\alpha_n}}{\sigma^2} < T_{SNR}\right) \\ =& p_l(r)\exp\left(-\frac{T_{SINR}}{\gamma}r^{\alpha_l}\right) + p_n(r)\exp\left(-\frac{T_{SINR}}{\gamma}r^{\alpha_n}\right)\end{aligned}$$

Thus, the probability $\mathcal{P}_{SNR}$ can be obtained by evaluating the expectation with respect to r

$$\mathcal{P}_{SNR} = \int_0^\infty r\exp(-\pi\lambda r^2)\mathcal{P}_{SNR}(r)dr \tag{9}$$

- *distance-based*: the 2-dimension distance between user' and BS's antenna is used for user classification. When the distance threshold R_0, the user at a distance r is called CCU with a probability of $\mathcal{P}_{dis} = \mathbf{P}(r < R_0)$. Since r follows the PDF in Eq. 1, the probability $\mathcal{P}_{dis}$ is

$$\mathcal{P}_{dis} = 2\pi\lambda R_0 \exp\left(-\pi\lambda R_0^2\right)$$

Regarding to the communication phase where the data transmission is performed between the user and target BS, all of these FFR schemes have the same procedure. Since the CCU and CEU have similar inter-cell interference statistics and be distinguished by the serving power, the downlink SINR of CCU and CEU are respectively $SINR(P, r)$ and $SINR(\phi P, r)$ where ϕ is the ratio between serving power of CEU and CCU, $\phi > 1$.

3 Coverage Probability

3.1 Definition

To evaluate the user performance, the coverage probability is usually used to illustrate how often the downlink SINR surpasses the coverage threshold $\hat{T}$. Under two-phase operation of FFR, the coverage probability is formulated as the conditional probability. Particularly, the coverage probability of CCU $\mathcal{P}_c$ and CEU $\mathcal{P}_e$ for different FFR schemes can be formulated as

- For CCU

$$\mathcal{P}_c(r) = \mathbf{P}\left(SINR'(P, r) > \hat{T} | \text{CCU condition}\right) \tag{10}$$

- For CEU

$$\mathcal{P}_e(r) = \mathbf{P}\left(SINR'(\phi P, r) > \hat{T} | \text{CEU condition}\right) \tag{11}$$

Thus, the average coverage of the user, that is either CCU or CEU, is

$$\begin{aligned}\mathcal{P}(r) =& \mathbf{P}\left(SINR'(P,r) > \hat{T}|\text{CCU condition}\right)\mathbf{P}\left(\text{CCU condition}\right)\\ &+ \mathbf{P}\left(SINR'(\phi P,r) > \hat{T}|\text{CEU condition}\right)\mathbf{P}\left(\text{CEU condition}\right)\end{aligned}$$

Employing the Bayes rules, the probability $\mathcal{P}(r)$ is expanded as

$$\begin{aligned}\mathcal{P}(r) =& \mathbf{P}\left(SINR'(P,r) > \hat{T}, \text{CCU condition}\right)\\ &+ \mathbf{P}\left(SINR'(\phi P,r) > \hat{T}, \text{CEU condition}\right)\end{aligned} \tag{12}$$

The coverage probability of the typical user in the network is obtained by evaluating the conditional probability with variable r

$$\mathcal{P} = 2\pi \int_0^\infty \lambda r \exp(-\pi\lambda r^2)\mathcal{P}(r)dr \tag{13}$$

3.2 Performance Comparison

In this section, Monte Carlo simulation is adopt to analyze and compare the coverage probability of different FFR schemes. The dual-slope with path loss exponent of LoS $\alpha_l = 3.5$ and nLoS $\alpha_n = 4.5$ are used.

Coverage Probability vs Coverage Threshold. In Fig. 1, we compare the user coverage probability of three different user classification cases: SINR-based, SNR-based and distance-based. The following parameters are selected for simulation: the density of BSs $\lambda = 2$ (BS/km^2), the signal to noise ratio SNR $= 10$ dB, the LoS parameter $\beta = 0.4$. As a shown in Fig. 1, the user coverage probability in case of SNR-based and SINR-based are very close together. This due to the assumption that all the channels and interfering sources have the same statistical properties. Thus, utilization of instantaneous SINR or SNR derive the similar results.

Coverage Probability vs User Classification Threshold. In this section, we compare the coverage probability of the CCU, CEU, and the typical user are analyzed with different user classification thresholds. For SINR-based and SNR-based algorithms, an increase in the SINR and SNR thresholds results in a higher probability that a user is defined as a CEU. In other words, when the threshold increases, only the users with very high SINR/SNR are called CCUs and more users with good SINR/SNRs are categorized as CEUs. Thus, the average coverage probability of CCUs and CEUs increases with thresholds in both cases of SINR-based and SNR-based algorithms as shown in Fig. 2.

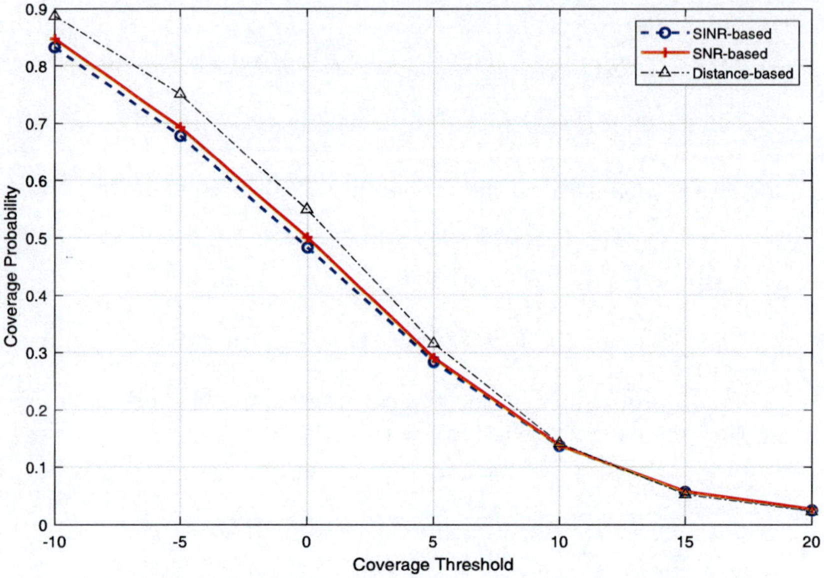

Fig. 1. Coverage probability vs coverage threshold

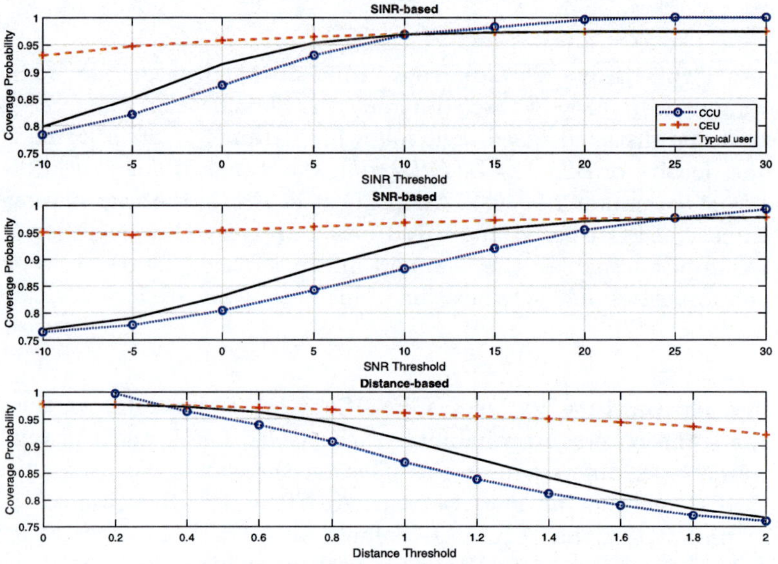

Fig. 2. Coverage probability vs thresholds

In contrast, when the distance threshold of the distance-based algorithm increases, the cell center area is expanded and more users at long distances from its serving BSs are called CCU. These users usually experience high penetration losses and low coverage probability. Thus, the coverage probability of the CCU decreases in this case. In addition, the expansion of the cell center area leads to the constriction of the cell edge area. Thus, only the users near cell borders, which usually achieve very low performance, are called CEUs. Consequently, the coverage probability of CEU reduces when the distance threshold increases.

Coverage Probability vs Density of BSs. In Fig. 3, we examine the effect of the density of BSs on the coverage probability between three algorithms with different values of density of BSs λ. For each value of λ, the SNR, SINR, and distance thresholds are selected so the ratio between CCUs and CEUs is equal to 1.

In general, there are increments in interfering power and the received signal of the user from serving BS when λ increases. However, when λ is large enough, interfering power rises faster than the desired signal. Thus, the user coverage probability of all three algorithms experiences declines.

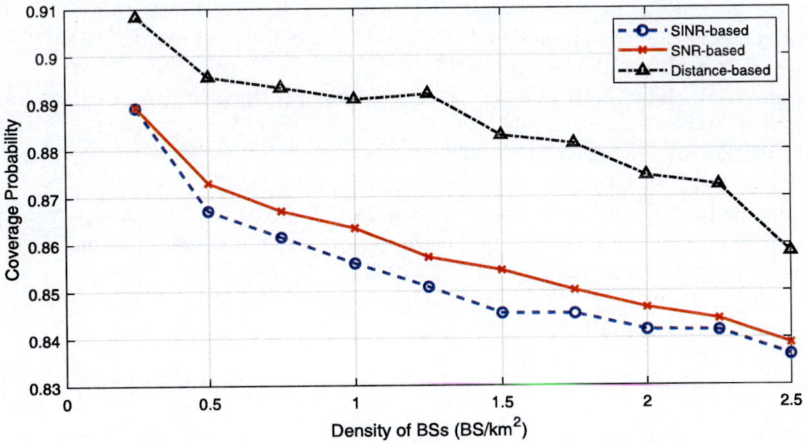

Fig. 3. User coverage probability vs density of BSs

An interesting result, that seen from Fig. 3, is that the distance-based algorithm can outperform both SINR-based and SNR-based algorithms. For example, when the density of BSs $\lambda = 1\ BS/km^2$, the user coverage probability in the distance-based system is 0.89 which is approximately 3% higher than others in SINR-based and SNR-based systems. However, the difference is small, so that there is required more works on the performance comparison for different network scenarios.

4 Conclusion

In this paper, the coverage probability of the typical user in the network utilizing the FFR technique with different user classification algorithms, particularly SINR-based, SNR-based, and distance-based, were studied. While the SINR-based and SNR-based algorithms utilize the received signal strength of the user to define it as CCU or CEU, CCU and CEU classification of the distance-based system is based on the distance from the user to its serving BS. The simulation results illustrate that SINR-based and SNR-based algorithms achieve similar performance which is a little bit lower than the distance-based one. In combination with fewer signaling requirements, it can be said that the distance-based algorithm is better than the SINR-distance and SNR-based ones.

References

1. 3GPP TS 38.101-1 version 15.3.0 Release 15: 5G; NR; User Equipment (UE) radio transmission and reception; Part 1: Range 1 Standalone (2018)
2. 3GPP TR 36.921 version 10.0.0 Release 10: Lte; evolved universal terrestrial radio access (E-UTRA); FDD home eNode B (HeNB) radio frequency (RF) requirements analysis (2019)
3. Hamza, A.S., Khalifa, S.S., Hamza, H.S., Elsayed, K.: A survey on inter-cell interference coordination techniques in OFDMA-based cellular networks. IEEE Commun. Surv. Tutor. **15**(4), 1642–1670 (2013)
4. Chang, S.-H., Kim, S.-H., Choi, J.P.: The optimal distance threshold for fractional frequency reuse in size-scalable networks. IEEE Trans. Aerosp. Electron. Syst. **56**(1), 527–546 (2020)
5. 3GPP TR 22.872: Study on positioning use cases (2019)
6. Firouzabadi, A.D., Rabiei, A.M., Vehkaperä, M.: Fractional frequency reuse in random hybrid FD/HD small cell networks with fractional power control. IEEE Trans. Wireless Commun. **20**(10), 6691–6705 (2021)
7. Lam, S.C., Tran, X.N.: Fractional frequency reuse in ultra dense networks. Phys. Commun. **48**, 101433 (2021)

Analyzing Information Security Among Nonmalicious Employees

Elerod D. Morris[1] and S. Raschid Muller[2](✉)

[1] Capitol Technology University, Laurel, MD 20708, USA
emorris@captechu.edu
[2] Arizona State University, Mesa, AZ 85212, USA
raschid.muller@asu.edu

Abstract. Insider threats pose a significant risk to organizational data security, and many organizations implement information security policies (ISPs) to reduce insider threats. This study used the unified theory of acceptance and use of technology 2 (UTAUT2) to examine factors that predict compliance among nonmalicious employees. A partial least squares structural equation modeling approach was used to examine survey data collected from N = 158 nonmalicious employees. The analysis indicated that social influence and facilitating conditions were the only UTAUT2 factors significantly predicting nonmalicious employees' compliance. The study's findings suggest that organizations should focus on building workplace cultures emphasizing ISP compliance's social importance.

Keywords: cybersecurity · UTAUT2 · insider threat · risk management · information security policies · information awareness

1 Introduction

This article examined the factors that predict nonmalicious employees' information security policy (ISP) compliance. This research topic was compliance as a strategy to combat unintentional insider threats. Policies are formal guidelines organizations use to govern employees' behaviors related to data access and use [7]. Organizations use policies to protect data from insider threats [10]. Insider threats are threats to the confidentiality, integrity, and availability of organizational data by individuals with authorized access to the organization's data or systems [17, 18], and [27]. Many scholars have argued that employees are the most significant threat to data security [14], and [21]. To combat insider threats associated with employees, organizations implement ISPs that identify and mandate data security procedures [7, 10], and [26]. The present study focused on a population of employees responsible for unintentional and nonmalicious insider threats based on Prabhu and Thompson's [25] taxonomy. The decision to focus on nonmalicious insider threats was based on the acknowledgment that "malicious insiders are more difficult to detect" [17].

Problem Statement - Insider threats are a significant general problem for organizations [3] and [12]. Insiders include trusted stakeholders such as corporate partners,

T. D. L. Nguyen et al. (Eds.): ICISN 2023, LNNS 752, pp. 629–636, 2023.
https://doi.org/10.1007/978-981-99-4725-6_74

employees, subcontractors, and vendors, and these insiders usually have authorized access to organizational information systems [21]. Researchers have identified insiders as the primary source of information security threats in organizations [3] and [12]. Organizations implement ISPs to protect against insider threats. Still, many employees fail to comply with these policies even when they have no malicious intent to harm their company, and the reasons are unknown [5] and [17].

2 Literature Review

This article examined factors associated with the unified theory of acceptance and use of technology 2 (UTAUT2) that predict nonmalicious employees' policy compliance. A quantitative, non-experimental research design and partial least squares structural equation modeling (PLS-SEM) analysis methods were used to examine the relationships between six predictor variables and the outcome variable behavioral intention to comply with policies. The six predictor variables drawn from the UTAUT2 (Fig. 1) were performance expectancy, effort expectancy, social influence, facilitating conditions, hedonic motivation, and habit [29].

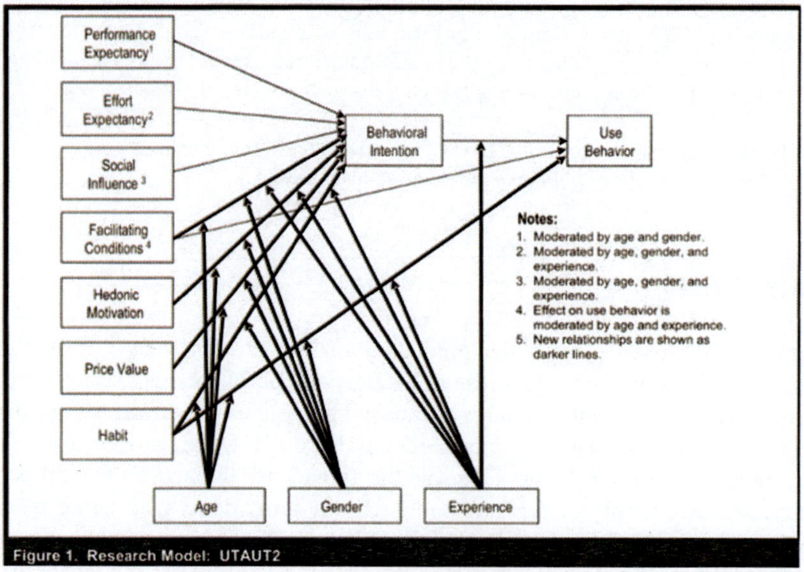

Fig. 1. Consumer Acceptance and Use of Information Technology: Extending the Unified Theory of Acceptance and Use of Technology, by V. Venkatesh, J. Y. L. Thong, and X. Xu, 2012, MIS Quarterly, 36(1), p. 160

2.1 Insider Threats

Scholars have argued that insiders represent the most significant threat to an organization's information security [14] and [21]. Insider threat has been defined within the literature as the intentional or unintentional actions of employees that result in unauthorized access to an organization's information assets [12, 31]. Wang et al. [31] noted that insiders, such as employees and organizational partners with legitimate access to an organization's systems, could accidentally expose the organization to cyber-attacks. Scholars do not agree on a formal definition of insider threats [5, 12, 19], and [30]. As a result, several taxonomies have been created to differentiate between types and categories of insider threats [2, 17], and [25].

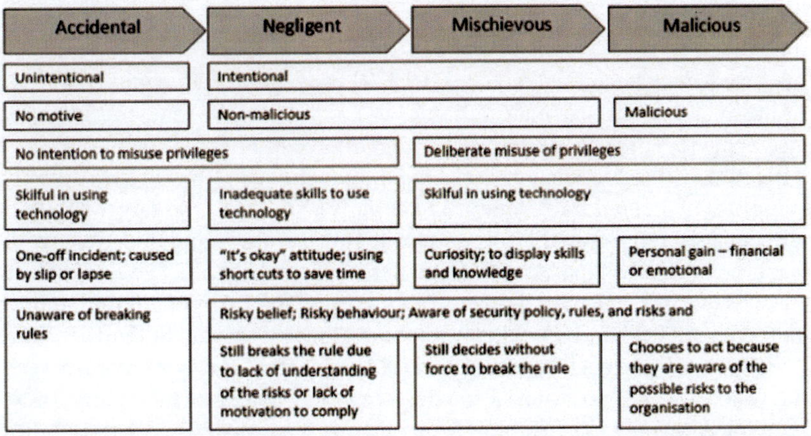

Fig. 2. A Unified Classification Model of Insider Threats to Information Security, by S. Prabhu and N. Thompson, 2020, ACIS 2020: 31st Australasian Conference on In- formation Systems, Wellington, New Zealand, p. 8

2.2 Types of Insider Threats

Prabhu and Thompson [25] conducted a thematic analysis of the scholarly literature to identify different types of insider threats. Despite the identified challenges, Prabhu and Thompson [25] developed a classification system based on the literature. They grouped insider threats into four categories: (a) accidental, (b) negligent, (c) mischievous, and (d) malicious. In addition to creating a classification system of insider threats, Prabhu and Thompson [25] also identified an insider typology that described the insider characteristics associated with each threat type. Figure 2 presents the characteristics of each type of insider. Hadlington [15] noted how human factors were closely linked with accidental insider threats. As indicated in Fig. 2, accidental insiders typically have no motive to harm the company, and are skilled in using technology but are unaware they have exposed the company to risk [25].

2.3 Information Security and ISP Compliance

The global economy has become dependent on information and information technology, so information security is a fundamental requirement of business functions in modern organizations [1]. Organizations generally develop formal policies and procedures governing employee actions to minimize risks associated with information security [9] and [24]. These policies and procedures are referred to as ISPs, and typically outline employees' roles and responsibilities regarding data and information handling [9]. It is critical to note that the development of an ISP does not guarantee employee compliance. Scholarly interest in the topic of policy compliance has grown significantly as information security becomes a more prominent concern for businesses [6, 7, 13, 17, 23], and [25].

3 Methodology

The study used Venkatesh et al.'s UTAUT2 [29] conceptual framework. The UTAUT2 incorporates intrinsic and extrinsic variables to predict users' behavioral intentions and actual use behaviors. The research used a quantitative, non-experimental design to test the significance of the relationships between the constructs of the UTAUT2 and nonmalicious employees' behavioral intentions to comply with ISPs. The target population for this study consisted of nonmalicious employees who work in organizations with ISPs. Participants comprised 84 males and 74 females, for a total of 158. Age categories ranged from 18-25 to over 65. Most participants (approximately 81 percent of the sample) were between the ages of 25 and 55. These numbers are very similar to the U.S. Bureau of Labor Statistics [30], where approximately 82 percent of the workforce ranged from 25 to 54 years old. A partial least squares structural equation modeling (PLS-SEM) analysis method was used to examine the predictive relationships with the study's independent and dependent variables. A PLS-SEM approach was selected because it allows researchers to identify complex predictive relationships between variables rather than just associations between variables [33].

4 Results

This study aimed to determine the extent that the unified theory of acceptance and use of technology 2 (UTAUT2) constructs could predict non-malicious employees' behavioral intentions to comply with information security policies. This non-experimental correlational research added to the body of literature on ISP compliance among non-malicious employees by addressing a gap in the body of knowledge identified by Homoliak et al. [17], Muller and Lind [22], and Prabhu and Thompson [25].

After establishing the reliability and validity of the model structure and the items supporting each construct, a path analysis model was created. Figure 3 presents the path model. The path analysis indicates that all the survey items significantly loaded in support of their associated construct. Black arrows represent the relationships between the individual items and the constructs. In Fig. 3, relationship paths between the constructs are illustrated using red and green lines. The red lines indicate a non-significant relationship, while the green lines represent a significant variable relationship. The results of the path analysis indicated that the only two significant variables that predict non-malicious employees' behavioral intentions to comply were social influence and facilitating conditions.

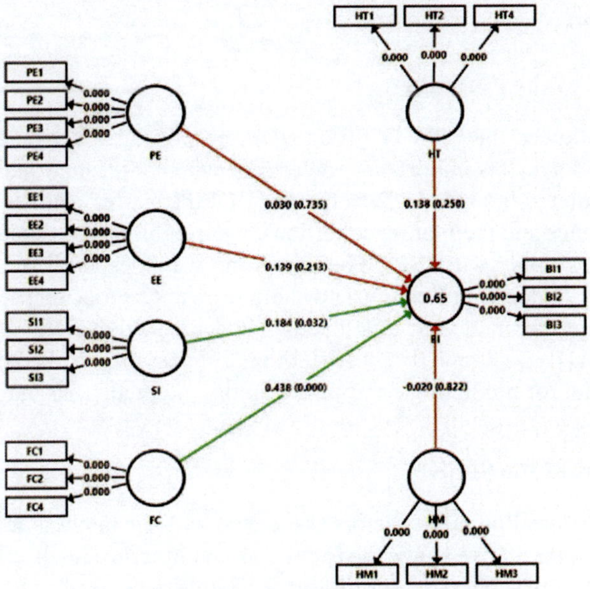

Fig. 3. Path Analysis Model

Table 1 presents the model's path coefficients, the t-statistics and p-values associated with each variable relationship, and the hypothesis testing result for each research question. The path coefficients indicate the strength of the relationship between the variables, and the p-values indicate whether a relationship is significant ($p < 0.05$). As indicated in both Fig. 3 and Table 2, the only significant relationships were between facilitating conditions and behavioral intention ($p = 0.000$) and between social influences and behavioral intention ($p = 0.032$). As a result, only the null hypotheses for Research Questions 3 and 4 could be rejected. Together these factors explained 65 percent ($R2 = 0.65$) of the variance in participants' behavioral intentions.

Table 1. BI = behavioral intention, EE = effort expectancy, FC = facilitating conditions, HM = hedonic motivation, HT = habit, PE = performance expectancy, and SI = social influence.

Path	Path Coefficient	T Statistics	P Values	Null Result
EE -> BI	0.139	1.247	0.213	Not Rejected
FC -> BI	0.438	4.733	0.000	Rejected
HM -> BI	-0.020	0.226	0.822	Not Rejected
HT -> BI	0.138	1.150	0.250	Not Rejected
PE ->BI	0.030	0.339	0.735	Not Rejected
SI ->BI	0.184	2.142	0.032	Rejected

5 Conclusion

5.1 Discussion of the Findings

This research followed the UTAUT2 framework proposed by Venkatesh et al. [29]. In analyzing the findings, it is important to determine whether the data analysis results supported or contradicted the UTAUT2 framework. Of the six independent variables, only two, social influence and facilitating conditions, were significant predictors of the behavioral intention to comply with ISPs. The remaining variables, performance expectancy, effort expectancy, hedonic motivation, and habit, were not significant predictors of policy compliance among nonmalicious employees. Thus, the study's findings only supported one-third of the UTAUT2 model. These findings suggest that the UTAUT2 may not be an effective model for predicting compliance among nonmalicious employees.

5.2 Limitations of the Study

The present study was limited by factors associated with the methodology and the target population. As a quantitative study, the focus was on numeric data that demonstrated the links between variables. No rich, descriptive, narrative data were collected, and participants could not share independent insights into their experiences with ISP compliance. Most ISP compliance research is quantitative, and several studies use the UTAUT2 [4, 20], and [23]. Another limitation of the present study associated with using the UTAUT2 was that participants' behavioral intentions were measured rather than actual compliance behaviors. Survey instruments measuring behavioral intentions are common in ISP literature [11, 23], and [20].

5.3 Recommendations for Further Research

First, more attention is needed to examine the behaviors of nonmalicious employees. The literature lacks a universal definition of insider threats, and several researchers have proposed frameworks for classifying insider threats [2, 3, 5, 12, 19, 25], and [31]. The present study relied on a classification framework proposed by Prabhu and Thompson [25]. However, further research should be conducted to determine the characteristics of nonmalicious versus malicious insider threats to help organizations understand the risk profiles of different insiders.

In conclusion, the findings indicated that among nonmalicious employees, only social influence and facilitating conditions were significant predictors of ISP compliance intentions. The remaining variables associated with the UTAUT2 were not significant. Information security is a priority for modern businesses, and many researchers have identified employees as one of the key insider threats because of their access to sensitive data [14, 18, 21], and [27]. Organizations implement end-user policies to safeguard data and reduce risk [10]. However, simply implementing policy does not guarantee compliance. Whether through in- intentional acts or negligence, employees can severely damage an organization by exposing it to cybersecurity threats [12, 28], and [30].

References

1. AlGhamdi, S., Win, K.T., Vlahu-Gjorgievska, E.: Information security governance challenges and critical success factors: systematic review. Comput. Secur. **99**(12), 102030 (2020). https://doi.org/10.1016/j.cose.2020.102030
2. AlMhiqani, M.N., et al.: A new taxonomy of insider threats: an initial step in understanding authorised attack. Int. J. Inf. Syst. Manage. **1**(4), 343–359 (2018). https://doi.org/10.1504/IJISAM.2018.094777
3. Alotaibi, M.J., Furnell, S., Clarke, N.: A framework for reporting and dealing with end-user security policy compliance. Inf. Comput. Secur. **27**(1), 2–25 (2019). https://doi.org/10.1108/ICS-12-2017-0097
4. Alqahtani, M., and Braun, R.: Reviewing influence of UTAUT2 factors on cyber security compliance: a literature review. J. Inf. Assur. Cyber Secur. **2021**, 666987 (2021b). https://doi.org/10.5171/2021.666987
5. Alsowail, R.A., Al-Shehari, T.: Empirical detection techniques of insider threat incidents. IEEE Access **8**, 7838–78402 (2020). https://doi.org/10.1109/ACCESS.2020.2989739
6. Amankwa, E., Loock, M., Kritzinger, E.: Establishing information security policy compliance culture in organizations. Inf. Comput. Secur. **26**(4), 420–436 (2018). https://doi.org/10.1108/ICS-09-2017-0063
7. Aurigemma, S., Mattson, T.: Exploring the effect of uncertainty avoidance on taking voluntary protective security actions. Comput. Secur. **73**(3), 219–234 (2018). https://doi.org/10.1016/j.cose.2017.11.001
8. Chen, L., Zhen, J., Dong, K., Xie, Z.: Effects of sanction on the mentality of information security policy compliance. Rivista Argentina de Clınica Psicologica. **29**(1), 39–49 (2020). https://doi.org/10.24205/03276716.2020.6
9. Chen, X., Wu, D., Chen, L., Teng, J.K.L.L.: Sanction severity and employees' information security policy compliance: investigating mediating, moderating, and control variables. Inf. Manage. **55**(8), 1049–1060 (2018). https://doi.org/10.1016/j.im.2018.05.011
10. Cram, W.A., Proudfoot, J.G., D'Arcy, J.: Organizational information security policies: a review and research framework. Eur. J. Inf. Syst. **26**(6), 605–641 (2017). https://doi.org/10.1057/s41303-017-0059-9
11. D'Arcy, J., Lowry, P.B.: Cognitive-affective drivers of employees' daily compliance with information security policies: a multilevel, longitudinal study. Inf. Syst. J. **29**(1), 43–69 (2019). https://doi.org/10.1111/isj.12173
12. Elifoglu, H., Abel, I., Tasseven, Q.: Minimizing insider. threat risk with behavioral monitoring. Rev. Bus. **38**(2), 61–74(2018). https://www.ignited.global/case/business/minimizing-insider-threat-risk-behavioural-monitoring
13. Glasofer, A., Townsend, A.B.: Determining the level of evidence: Nonexperimental research designs. Nursing Critical Care **15**(1), 24–27 (2020). https://doi.org/10.1097/01.CCN.0000612856.94212.9b
14. Gratian, M., Bandi, S., Cukier, M., Dykstra, J., and Ginther, A. (2018). Correlating human traits and cyber security behavior intentions. Computers and Security, 73(3), 345–358. https://doi.org/10.1016/j.cose.2017.11.015
15. Hadlington, Lee: The "human factor" in cybersecurity: Exploring the accidental insider. In: McAlaney, John, Frumkin, Lara A., Benson, Vladlena (eds.) Psychological and Behavioral Examinations in Cyber Security:, pp. 46–63. IGI Global (2018). https://doi.org/10.4018/978-1-5225-4053-3.ch003
16. Hina, S., Panneer Selvam, D.D.D., Lowry, P.B.: Institutional governance and protection motivation: theoretical insights into shaping employees' security compliance behavior in higher education institutions in the developing world. Comput. Secur. **87**(11), 101594 (2019). https://doi.org/10.1016/j.cose.2019.101594

17. Homoliak, I., Toffalini, F., Guarnizo, J., Elovici, Y., Ochoa, M.: Insight into insiders and it: a survey of insider threat taxonomies, analysis, modeling, and countermeasures. ACM Comput. Surv. **52**(2), 1–40 (2019). https://doi.org/10.1145/3303771
18. Ifinedo, P.: Effects of organization insiders' self-control and relevant knowledge on participation in information systems security deviant behavior. In: SIGMIS-CPR 2017: Proceedings of the 2017 ACM SIGMIS Conference on Computers and People Research, pp. 79–86. Association for Computing Machinery (2017). https://doi.org/10.1145/3084381.3084384
19. Kim, A., Oh, J., Ryu, J., Lee, K.: A review of insider threat detection approaches with IoT perspective. IEEE Access, **8**, 78847–78867 (2020). https://doi.org/10.1109/ACCESS.2020.2990195
20. Lee, H.-J., Kho, H.-S., Roh, E.-H., and Han, K.-S.: A study on the fac tors of experience and habit on information security behavior of new services – based on PMT and UTAUT2. J. Digital Contents Soc. **19**(1), 93–102 (2018). https://doi.org/10.9728/dcs.2018.19.1.93
21. Mamonov, S., Benbunan-Fich, R.: The impact of information security threat awareness on privacy-protective behaviors. Comput. Human Behav. **83**, 32–44 (2018). https://doi.org/10.1016/j.chb.2018.01.028
22. Muller, S.R., Burrell, D.N.: social cybersecurity and human behavior. Int, J. Hyperconnect. Internet of Things **6**(1), 1–13 (2022). https://doi.org/10.4018/IJHIoT.305228
23. Muller, S.R., Lind, M.L.: Factors in information assurance professionals' intentions to adhere to information security policies. Int. J. Syst. Softw. Secur. Protect. **11**(1), 17–32 (2020). https://doi.org/10.4018/IJSSSP.2020010102
24. Paananen, H., Lapke, M., and Siponen, M.: State of the art in information security policy development. Comput. Secur. **88**(1), 101608 (2020). https://doi.org/10.1016/j.cose.2019.101608
25. Prabhu, S., Thompson, N.: A unified classification model of in- sider threats to information security [paper presentation]. In: ACIS 2020: 31st Australasian Conference on Information Systems, Wellington, New Zealand (2020). http://hdl.handle.net/20.500.11937/81763
26. Rahimian, F., Bajaj, A., Bradley, W.: Estimation of deficiency risk and prioritization of information security controls: a data-centric approach. Int. J. Account. Syst. **20**, 38–64 (2016). https://doi.org/10.1016/j.accinf.2016.01.004
27. Safa, N.S., Maple, C., Watson, T., Von Solms, R.: Motivation and opportunity-based model to reduce information security insider threats in organisations. J. Inf. Secur. Appl. **40**(6), 247–257 (2018). https://doi.org/10.1016/j.jisa.2017.11.001
28. Theis, M.C., et al.: Common sense guide to mitigating insider threats (6th ed). Software Engineering Institute (2019). https://doi.org/10.1184/R1/12363665.v1
29. U.S. Bureau of Labor Statistics. (2022b, January 20). Labor force statistics from the current population survey: Employment status of the civilian noninstitutional population by age, sex, and race. https://www.bls.gov/cps/cpsaat11.htm
30. Venkatesh, V., Thong, J.Y.L., Xu, X.: Consumer acceptance and use of information technology: extending the unified theory of acceptance and use of technology. MIS Q. **36**, 157–178 (2012). https://doi.org/10.2307/41410412
31. Wang,X., Tan, Q., Shi, J., Su, S., Wang, M.: Insider threat detection us- ing characterizing user behavior. In: 2018 IEEE Third International Conference on Data Science in Cyberspace, 2018, pp. 476–482 (2018). https://doi.org/10.1109/DSC.2018.00077
32. Yang, J., Zhang, Y., Lanting, C.J.M.: Exploring the impact of QR codes in authentication protection: a study based on PMT and TPB. Wireless Pers. Commun. **96**(4), 5315–5334 (2017). https://doi.org/10.1007/s11277-016-3743-5
33. Zeng, N., Liu, Y., Gong, P., Hertogh, M., König, M.: Do right PLS and do PLS right: a critical review of the application of PLS-SEM in construction management research. Front. Eng. Manage. **8**(3), 356–369 (2021). https://doi.org/10.1007/s42524-021-0153-5

Evaluation System for Straight Punch Training: A Preliminary Study

Nguyen Phan Kien[1(✉)], Nguyen Viet Long[1], Doan Thi Anh Ngoc[1], Do Thi Minh Phuong[1], Nguyen Hong Hanh[1], Nguyen Minh Trang[1], Pham Thu Hien[1], Doan Thanh Binh[2], Nguyen Manh Cuong[3], and Tran Anh Vu[1]

[1] Hanoi University of Science and Technology, Hanoi, Vietnam
{ngoc.dta193273,Hanh.nh193251,Trang.nm193281, Hien.pt193254}@sis.hust.edu.vn, Vu.trananh@hust.edu.vn
[2] Electrics Power University, Hanoi, Vietnam
[3] Le Quy Don University, Hanoi, Vietnam
nmcuong@mta.edu.vn

Abstract. The punch is one of the major components of martial art that is related to kinematic indicators and impact forces. However, the impact forces of punching postures have not been fully investigated. Therefore, the aim of this research is to propose a new system to measure the acceleration and force created by the punch and monitor practitioners' stances during the punching process. A study was conducted in which six participants (3 females and 3 males) participated with at least 1 year of experience. Each participant performed 5 straight punches with the fist rotation and 5 straight punches without the fist rotation. Force was measured from the load cell, acceleration from an accelerometer, and stance was monitored via a motion analysis module using a computer. The experimental results from the proposed system showed that the hand acceleration and punch forces correlated strongly with an average acceleration of 28.3 m/s2 (without rotation) and 29.9 m/s2 (with rotation) producing an average force of 107.5 N and 139.9 N, respectively. These results show that the punching velocity had a great impact on the punching forces. The experiments also proved that the system can use to monitor the force, acceleration of the punch, and also the posture of practitioners when doing punches.

Keywords: Martial art · Straight punch · Motion tracking · Accelerometer · Load cell · MediaPipe

1 Introduction

There are many kinds of martial arts in the world nowadays that develop for many years with their own philosophy and style [3]. For example, Karate, which is an oriental art of self-defense, improves physical fitness and mental discipline in its practitioners and gained considerable popularity during the punch is a key to the development of practitioners in martial art. The punch in a martial art is used to create physical damage to an opponent. The right technique will lead to big damage. Besides that, having more

T. D. L. Nguyen et al. (Eds.): ICISN 2023, LNNS 752, pp. 637–647, 2023.
https://doi.org/10.1007/978-981-99-4725-6_75

knowledge of punching will improve tactical advantage, and score points against an opponent [7]. There are four types of basic punches which are the most often used attack form in martial art. They are straight punch, lunge punch, reverse punch, and jab punch [7]. Among these, the straight punch has the longest range and tends to be the most powerful. Straight punching techniques require the use of the entire body to apply optimum force for a very short period. So, to obtain more force at impact, the movement of the punch is needed to be considered [7]. From the normal fight stance, the attacker athlete must follow steps such as (1) lowering the center of mass of the whole body while extending the stance, (2) lunging towards the opponent, and (3) then extending the punching arm forward, executing the punch in the open unguarded part of the opponent's abdomen [5]. This punch is done based on the combination of a series of bodily movements, called the twisting of the hips and shoulder rotation. The kinematic movement of the straight punch technique is shown in the Fig. 1 below:

Fig. 1. Straight punch technique. Source. Choku-zuki (Straight Punch). 1 July 2018, https://bom.so/IXfGR2

In the research of Rathee et al. [16] the parameters of breaking a wooden board of two athletes have been studied. In this study, the deformation energy (DE) is the energy calculated in joules determined by the loss to deformation in the case of inelastic collision (show equation [1]):

$$D.E. = \frac{1}{2}((m_1 \times m_2)/(m_1 + m_2)) \times V_c^2 \quad (1)$$

- m_1 is the mass (kg) of the wooden board
- m_2 is the mass (kg) of the arm of the punching participant.
- V_c (ms^{-1}) is the velocity of the punch upon impact.

Based on the equation, DE will be increased by mainly the velocity of the punch. So the athletes have to train to increase the velocity of the punch if they want to create a bigger force. It means that the posture of the athletes also has to consider reducing the way they punch. Also, performing correct standing and movement is very important for the force to effectively impact and analyze the kinematic characteristics of punch type.

The assessment of the straight punch technique is very difficult because it is related to many factors such as motor characteristics, the technique of the athletes, tactics in combat, the psychological status of the athletes, and related methods [12, 13, 15, 17]. In fact, there is some research related to the relationship between punching force and postures [8]. Smith et al. [14] have proven that the force power of the punches is dependent on the athletes' skills. To measure the force, there is some measurement system has been developed to monitor them such as using a water-filled heavy bag [12], force plate [11], strain gauge-based measuring systems [5, 7], and an accelerometer-based measuring system. For example, in research [12], they use a punching bag with an embedded strain gauge to measure the force or use the BTS-3 boxing training simulator system in another research [6]. Suwarganda et al. [16] captured punching movement using 6 infrared Motion Analysis cameras and a Kistler force plate. Jacek Wasik et al. [16] applied reflective markers on fists to collect the punching kinematic data. However, these existing systems mainly focused on measuring and analyzing force. They do not focus on whole punching activities to find the reason to create the force in order to improve punching techniques. Hence, developing a new system that would meet the requirements of martial art disciplines was necessary.

The goal of our research is to propose a new system to detect punching stances, attacking acceleration, and punch force of karate practitioners for straight punch technique using instrumentation conducted by combining motion detection, an accelerometer, and a force measurement system using load-cell together. The system can evaluate the straight punch technique and give real-time feedback information to practitioners and instructors. Therefore, it will provide coaches and learners with a better understanding of the mechanics of the body in performing the straight punch technique as well as be used for training purposes and for technique evaluation.

2 Materials and Methods

2.1 Apparatus

The system architecture is presented in Fig. 2 In this diagram, the measuring system includes three parts as acceleration sensor to measure the acceleration of the hand movement, a force plate to measure the force, and the motion analysis to do pose detection. The motion analysis system uses Mediapipe Pose Detection algorithm [9] to detect the punch's angles.

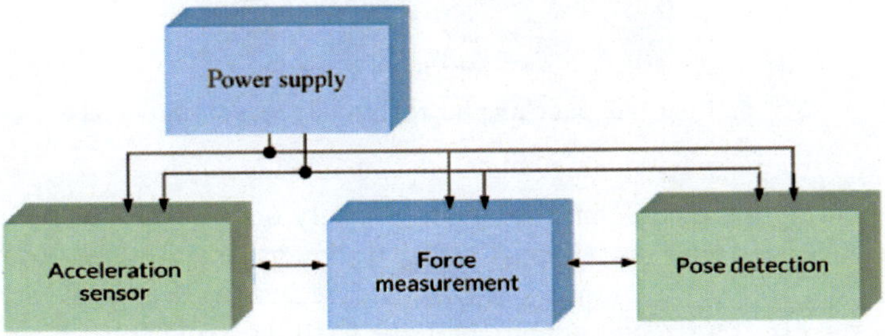

Fig. 2. Punch evaluation system diagram

Punching Force Measurement

For the punching force analysis of this system, two steel plates is cut to have a size of 20 × 15 × 0.4 cm and the 100-kg load cell was placed in between them to detect the force applied on the steel platform. The load cell after being modified had a sampling rate of 80 Hz. The load cell is placed in the middle of two platforms so that when the sensor receives the punching force, the load cell deforms and thereby obtains the punching force without deforming the contact surface. With the selected steel plate thickness of 0.4 cm, deformation of the steel surface for punching forces less than 5 kN will not occur. However, to ensure the safety of athletes, a pad with 3.0 cm thickness and a size of 20 × 15 cm is placed on the surface of the steel plate to reduce the direct impact of steel on the hand bone structure when punching. This will not reduce the measured punch force but only slow down the force acquisition by a negligible amount of time. The model of the load cell punching board is presented in Fig. 3. The photo of the load cell punching board respectively is shown in Fig. 4:

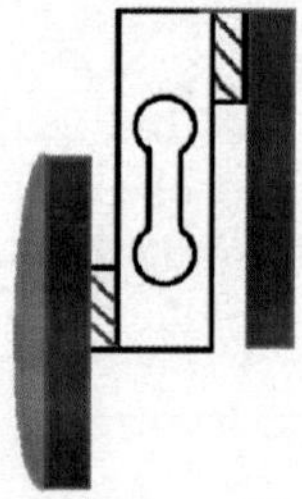

Fig. 3. Load cell punching board's model

Fig. 4. Load cell punching board. (a) Side-face (b) Front face

After securing the module and attaching it firmly on the wall, the entire load cell system was calibrated with and without the foam padding using overloads known masses. The setup of the loadcell calibration is shown in Fig. 5, we set up a simple pendulum test that acts as the fist of a straight punch. First, we fixed the load cell on a vertical plane, then hang the pendulum which is placed parallel to the plate of the loadcell board. The center of mass of the pendulum is lined in the center of the load cell. Second, we pulled the pendulum up to a known angle then release it and redo the experiment 5 times for each angle to take average value results. Repeat the calibration process with the other angle and other pendulum's mass. The main data obtained from the system were their punch's acceleration and the punch force displayed and recorded on the computer screen.

The impact force of the pendulum is determined by:

$$F_{Impact} = [(2kmgl(1 - cosX)]^{1/2} \quad (2)$$

- m is the mass of the pendulum (kg)
- g is gravity acceleration (m/s2)
- l is length odd wire(m)

X is angle between the wire and vertical axis.

The results of the load cell calibration is shown in Fig. 6.

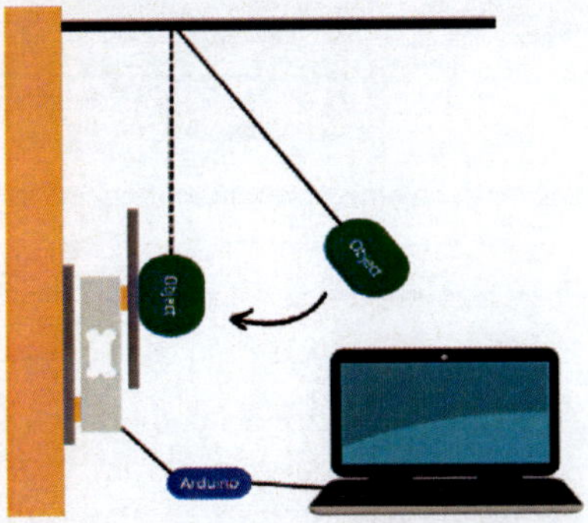

Fig. 5. Calibration setup of the load cell

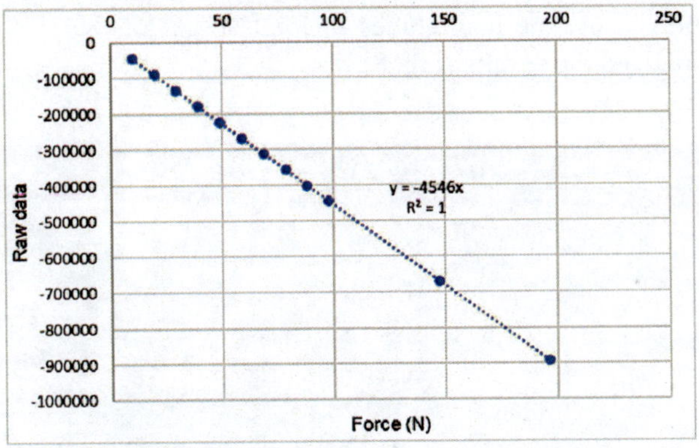

Fig. 6. Calibration curve of the load cell

Punching Acceleration Measurement

For the punching acceleration detection module, a tri-axial accelerometer was attached to the wrist of the subject to identify punch timing and measure the impact acceleration of the punch. The acceleration sensor used in this module is MPU6050 with a sampling rate of 1000 Hz. It's a 6-axis Motion Tracking chip that combines a 3-axis gyroscope, a 3-axis accelerometer, and a Digital Motion Processor (DMP). The accelerometer was then implanted in a small box with a size of 3 × 5 × 2 cm for convenience. Acceleration signals from the accelerometers were sent to the computer for analysis via a Bluetooth module. The acceleration measurement module is shown below (Fig. 7):

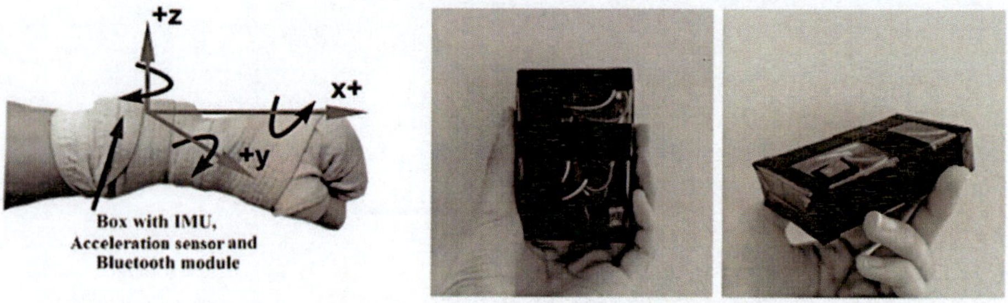

Fig. 7. The Accelerometer module attached to the wrist

Motion Analysis Module

For the motion analysis module, a smartphone's camera stabilized by a tripod was used to record videos of the punching process. A program was built with the support of MediaPipe [9] library for analyzing motion through these videos. MediaPipe is mainly focusing on data processing and building ML solutions across platforms [9]. Unfortunately, it is impossible to train MediaPipe-pose on a custom dataset. The training phase is usually done by TensorFlow. We just downloaded the trained model, then coded the get_angle function to get the magnitudes of joint angles of the arm in the punching motions and visualized the results (Fig. 8).

Fig. 8. Pose detection's result

2.2 Subjects

Six subjects (3 males, and 3 females) from the Bach Khoa Martial Art club were selected for this study. Table 1 shows the characteristic of the participants in the experiment. These subjects' experience in training ranged from 1 year to 3 years. All participants received and signed the informed consent form when joining these experiments.

Table 1. Participants' information

Subject	Height (m)	Weight (kg)	Experience (year)
1	1.76	63	1
2	1.78	67	3
3	1.62	55	3
4	1.63	51	1
5	1.59	50	1
6	1.75	62	1

The average height of male participants is 1.77 ± 0.10 m and the average weight is 65 ± 1 kg. The average height of female participants is 1.61 ± 0.20 m and the average weight is 64 ± 1 kg. Each participant was asked to punch the target in a testing sequence with straight punch techniques. The punches were performed from the normal standard stance with two parallel feet remaining in contact with the ground throughout the duration of the punches.

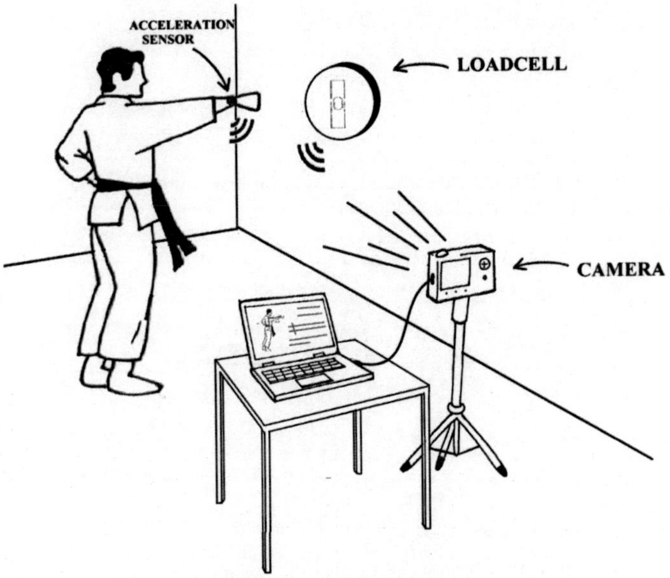

Fig. 9. Experiment setup

For the setup of the straight punch's measuring system, we connected the accelerometer and the load cell punching board to two computers via Bluetooth to display and record the results. The main data obtained from the system were their punch's acceleration and the punch force acted on the punching board. The accelerometer was fixed to the participant's gloves and the punching pad was attached vertically to the wall with the height adjusted to suitable for each subject. As the punch impacts the punching pad, it causes the steel plates to compress and produce an output voltage. This voltage unit is then converted to Newton when recording and analyzing data.

The motion analysis system used high-speed cameras to record the subject's movement. The camera was placed perpendicular to the plane of the subject's punch.

3 Results

3.1 Punch Acceleration

Measured data was accumulated to dataset X. Every sample has 2 columns (x-axis acceleration and angular acceleration) which have the graph as shown in the Fig. 10 and Fig. 11. At both graphs, the horizontal axis represents time (ms) and the vertical axis represents the magnitude of acceleration.

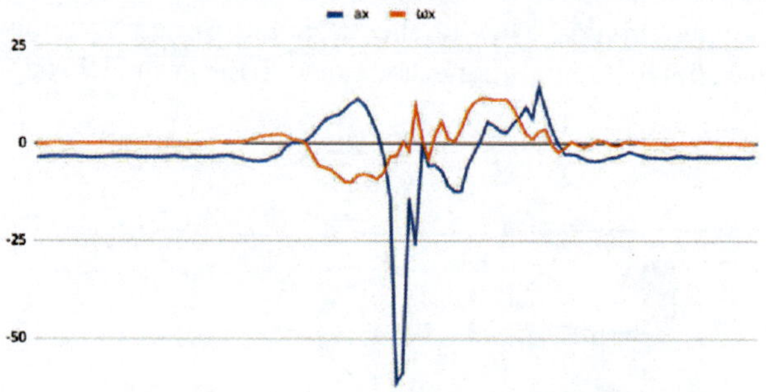

Fig. 10. Acceleration result's graph without rotation

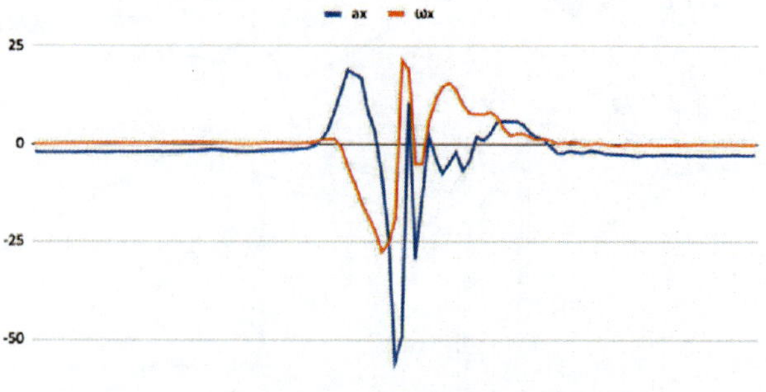

Fig. 11. Acceleration result's graph with rotation.

3.2 Punch Force

Tests were conducted to check the experimental values when punched with both types of punches which are straight punches with and without rotation. After checking the force trend generated by these punches, both punches were tested for achieving maximum average acceleration. Table 2 shows the values of force versus average acceleration performed by each participant. Table 3 shows the tests conducted to calculate the maximum acceleration that can be attained by both these punches.

Table 2. Average acceleration and corresponding forces

Force without rotation		Force with rotation	
Acceleration	**Force (N)**	**Accelearation**	**Force (N)**
20.6	76.4	25.1	111.9
26.3	136.6	28.4	168.3
21.4	80.2	27.8	115.6
20.0	65.5	22.3	80.8
20.6	73.7	19.6	90.1
25.6	104.6	26.3	132.6

Table 3. Maximum attainable acceleration (m/s2) of both punches

No	Without rotation	With rotation
1	22.4	26.2
2	27.2	31.7
3	23.3	30.7
4	20.6	23.3
5	21.1	21.1
6	26.8	28.2

3.3 Punch Angles

After conducting the punching test, we built a program using MediaPipe [9] to estimate the angle between the arm joints when punching. In this study, we performed three angles estimation: elbow angle in the preparation position, the angle between arm and torso, and elbow angle when punching out. Table 4 below shows the diverse angles measured by the program by 6 different people punching, combined with corresponding forces and acceleration.

By using a directional filter, we can effectively eliminate and mitigate the impact of the reflected wave, resulting in an improved image of the forward wave. This improved image closely resembles the forward wave prior to filtering, leading to an increase in diagnostic accuracy.

Table 4. Punching angles corresponding to maximum acceleration and force.

No	Angles	Acc. with rotation	Force with rotation	Acc. without rotation	Force without rotation
1	(84; 84; 178)	26.2	111.9	22.4	76.4
2	(86; 92; 178)	31.7	168.3	27.2	136.6
3	(92; 86; 176)	30.7	115.6	23.3	80.2
4	(82; 84; 168)	23.3	80.8	20.6	65.6
5	(82; 80; 170)	21.1	90.1	21.1	73.7
6	(84; 86; 176)	28.2	132.6	26.8	104.9

4 Discussion

4.1 Comparison of Punch Acceleration

As shown in Table 3, the maximum attainable accelerations of punches with rotation are higher than those without rotation. It shows that the straight punch without the fist rotation can attain a maximum acceleration of 28.3 m/s^2 while the one with rotation could attain an average high acceleration of 29.9 m/s^2. This small difference in acceleration shows that both punches attain almost the same average high acceleration.

This paper explores the effect of the reflected wave on the signal obtained in elastography. We propose the use of a directional filter to reduce its impact. Our results, obtained through simulation in Matlab, show that the filter is effective in minimizing the reflected wave. In future work, we aim to improve the filter to accommodate mutidirectional signals, thereby enhancing signal quality.

4.2 Comparison of Punch Force

The values obtained in Table 3 and Fig. 9 shows that the force generated by the punch with rotation is higher than the one without rotation when punched at the same acceleration. At all accelerations, the force of a straight punch with rotation is higher than the one without a fist rotation. This is completely consistent with the analysis in Daicu's study [2] when calculating the force generated by a straight punch with fist rotation and in the research of Bremer [1] when calculating the force of wrist rotation during a straight punch.

4.3 Comparison of Punch Acceleration

From the data in Table 4 of the angles measured by different punchers, it was shown that there will be an impact on the force and acceleration of the punch with different arm positions. Specifically, the experiment was done with Angle1, Angle2 angle of around 90°, and Angle3 angle of approximately 180, which will give the strongest punching force.

5 Conclusion

In this paper, we have presented a new measuring and monitoring system to evaluate the straight punch techniques. Our proposed system could give an evaluation of the punch techniques and real-time feedback information to athletes and coaches. The results getting from this system could be extremely valuable in characterizing each subject's technique, and in designing training programs as well as developing competitive strategies. In the future, we hope to improve the accuracy and feedback speed of our system to provide a better user experience. We also want to obtain more quantitative results by testing the system in a larger population and applying this approach not only in martial arts but also in other army training programs.

References

1. Bremer, A.K., Sennwald, G.R., Favre, P., Jacob, H.A.: Moment arms of forearm rotators. Clin Biomech. Bristol. Avon **21**(7), 683–691 (2006)
2. Diacu, F.: On the dynamics of karate. High School Math. Mag. PI Sky **6**, 23–32 (2003)
3. Chan, K., Pieter, W., Moloney, K.: Kinanthropometric profile of recreational taekwondo athletes. Biol. Sport **20**(3), 175–179 (2003)
4. Falco, C., et al.: Influence of the distance in a roundhouse kick execution time and impact force in Taekwondo. J. Biomech. **42**(3), 242–248 (2009)
5. Guidetilli, L., Musulin, A., Baldari, C.: Physiological factors in middleweight boxing performance. J. Sports Phys. Fitness **42**(3), 309–314 (2002)
6. Karpiłowski, B., Nosarzewski, Z., Staniak, Z.: A versatile boxing simulator. Biol. Sport **11**(2), 133–139 (1994)
7. Kumpf, C.: Wellness and Karate. Doctoral dissertation, Duquesne University, p. 1–9 (2018)
8. Li, Y., Yan, F., Zeng, Y., Wang, G.: Biomechanical analysis on roundhouse kick in taekwondo. In: Proceedings of the 23rd International Symposium on Biomechanics in Sports, Beijing, China, pp. 391–394 (2005)
9. Lugaresi, C., Tang, J.Q., Nash, H.: MediaPipe: A Framework for Building Perception Pipelines, pp. 7–11 (2019)
10. Nien, Y.H., Chuang, L.R., Chung, P.H.: The design of force and action time measuring device for martial arts. Int. Sport Eng. Assoc. **2**, 139–144 (2004)
11. Pedzich, W., Mastaler, A., Urbanick, C.: The comparison of the dynamics of selected leg strokes in taekwondo WTF. Acta Bioeng. Biomech. **8**(1), 83–90 (2006)
12. Pieter, F., Pieter, W.: Speed and force in selected taekwondo techniques. Biol. Sport **12**, 257–266 (1995)
13. Said, E., Ashker, S.: Technical performance effectiveness subsequent to complex motor skills training in young boxers. Eur. J. Sport Sci. **12**(6), 475–484 (2012)
14. Smith, M.S., Dyson, R.J., Hale, T., Janaway, L.: Development of a boxing dynamometer and its punch force discrimination efficacy. J. Sports Sci. **18**(6), 445–450 (2000)
15. Stephen, K.: The impact of martial arts training on adolescents. Faculty of Texas Tech University, pp. 1–3 (2003)
16. Suwarganda, E., Razali, R., Wilson, B., Ponniyah, A., Flyger, N.: Analysis of performance of the karate punch (Gyaku-zuki). In: 27 International Conference on Biomechanics in Sports, pp. 5–9 (2009)
17. Wasik, J.: Kinematic analysis of the side kick in Taekwondo. Acta Bioeng. Biomech. **13**(4), 71–78 (2011)

A Comparison of Deep Learning Models for Predicting Calcium Deficiency Stage in Tomato Fruits

Trung-Tin Tran[3](✉), Minh-Tung Tran[1], Van-Dat Tran[2], and Thu-Hong Phan Thi[3]

[1] FPT University-Swinburne Vietnam, 550000 Da Nang, Vietnam
[2] Institute for Research & Executive Education (VN-UK), Da Nang University, Da Nang 550000, Vietnam
[3] FPT University, Da Nang 550000, Vietnam
tintt19@fe.edu.vn

Abstract. Identifying and predicting nutritional deficiencies during the growing process of the tomato plant (*Solanum Lycopersicum L.)* is crucial since mineral nutrients are essential to plant growth. This paper aims to predict and recognize the nutrient deficiency occurring in tomato plants' flowering and fruiting stages by using two deep learning models, Yolov7 and ResNet50. The study focuses on predicting and classifying tomato plants' malnutrition stage with an essential mineral nutrient, calcium (Ca^{2+}). ResNet50 and Yolov7 are used to classify three stages of calcium deficiency in tomato fruits by analyzing the captured images of the development of tomato plants under greenhouse conditions. The dataset includes a total of 189 captured images that cover the different levels of calcium deficiency in tomato fruits. Of these, 80% (153 captured images) were used for the training dataset, and 20% (36 captured images) were applied to validate the test dataset. The purpose of this study is to recognize the stage of nutritional deficiencies in order to increase crop yields and prevent nutrient deficiency-related tomato diseases. By analyzing the tomato fruit images captured during tomato plant growth, the performance of ResNet50 and Yolov7 was validated, with accuracy rates of 97.2% and 85,8%, respectively.

Keywords: Tomato fruit · nutritional deficiency · stage of calcium deficiency · ResNet50 · Yolov7

1 Introduction

During the growth of tomato plants, calcium plays a crucial role in stimulating plant root growth and assisting in the formation of compounds that make up cell membranes, contributing to stronger plants. Additionally, calcium increases the activity of enzymes, neutralizes organic acids in plants, and enhances fruit sugar content, making them sweeter. However, plants with a calcium deficiency often display poor growth, stunted development, weak structure, and increased susceptibility to cracking. These symptoms include curled or scorched leaves with spots, inhibited buds, stunted root tips, wilting, and eventually, rotting or cracking of flowers and fruits, which ultimately impacts the yield and

T. D. L. Nguyen et al. (Eds.): ICISN 2023, LNNS 752, pp. 648–657, 2023.
https://doi.org/10.1007/978-981-99-4725-6_76

quality of tomato fruits, as shown in [1–3]. Generally, these symptoms are easily visible. By comparing these two models, it is possible to improve the performance of predicting and distinguishing nutrient deficiencies in the farming process.

Recent studies have utilized deep learning to indentify and analyze plant diseases. R. Thangaraj et al. surveyed deep learning methods for tracing and classifying crop disease and nutrient deficiency [4, 5]. Transfer learning has been identified as an effective method for predicting and classifying nutrient deficiencies in tomato plants and rice [6–8]. In one study, Ferentinos successfully identified leaf diseases in 29 crops using CNN-based models and 87,848 images, achieving a success rate of 99.53% [9]. Arantza Bereciartua-Pérez et al. also used a neural network model to classify plant diseases, including tomato plants, based on images captured using mobile devices [10]. Additionally, there are many approaches [11, 12] that use deep convolutional networks to detect plant disease.

As previously researched, the most studies focus on distinguishing and forecasting certain types of plant or crop illnesses, such as the impact of nutrient shortage symptoms on tomato plant/crop growth. The primary goal of this study is to compare and evaluate the performance of Yolov7 and ResNet50 models in identifying and predicting the essential mineral nutrient shortfall of Calcium. Several images of calcium nutrient-deficient stages were captured to collect the training, validation, and test datasets.

Based on the images captured in this study, we used ResNet50 and Yolov7 for training, recognition, and prediction of tomato nutrient status and evaluated the accuracy of these models in identifying and predicting nutrient deficiency using artificial intelligence (AI) systems. Unlike previous research, our study focused on assessing and forecasting nutrient deficiency status in tomato plants at different growth stages. The goal was to achieve high production yields and avoid the appearance of tomato diseases caused by nutrient deficiency by improving the predictive performance of deep learning models. We employed a modified structure of ResNet50 and Yolov7 to predict Calcium deficiency, and achieved an accuracy rate of 97.2% for the ResNet50 model and 85.8% for Yolov7.

This paper is structured as follows. Section 2 presents the dataset collection with Calcium deficiency for training model. ResNet50 and Yolov7 to predict and distinguish Calcium deficiency based on captured images of tomato fruits and the experimental result validation is demonstrated in Sect. 3. Section 4 shows discussion and validation and mentions the future work of author's research.

2 Data Collection

To train the DL model for this study, we set up experimental configurations in the greenhouse and captured images of tomato growth status to generate the training and validation datasets. We specifically collected images of tomato fruit at various stages of calcium deficiency during the growth process. We collected around 189 images, including three stages of calcium deficiency in tomato fruits. Figure 1 shows the captured images of tomato plants lacking calcium nutrients. As shown in Fig. 1, calcium deficiency symptoms generally appear on the surface of tomato fruit during growth. The signs of calcium deficiency include tomato fruits rotting or cracking on their surface, known as Blossom-end rot (BER), as described in [1]. Each stage of calcium deficiency symptoms is displayed in Fig. 1, with stage 3 being the most severe.

Fig. 1. Calcium deficiency symptom (BER)

For training and testing the performance of ResNet50 and Yolov7 models, we split the dataset into training, validation, and test with 80% (153 captured images), 20% (36 captured images), respectively.

3 Methodology and Result

Given the objective of recognizing and predicting nutrient deficiency symptoms using captured images of tomato fruits, the ResNet50 and Yolov7 models were utilized to classify and identify the stages of calcium deficiency.

3.1 Deep Convolutional Network

Accordingly, the study estimates forecasting and classifying the signs of Calcium nutrient shortage stages based on CNN with two models (ResNet50 and Yolov7). According to the study [13], Chien-Yao Wang and Alexey Bochkovskiy et al. developed the Yolov7 model with the new architecture for real-time object recognition. A study was performed by Delin Wu [14] using Yolov7 and data augmentation for the detection of Camellia oelifera fruit in complex scenes. The ResNet50 model was described by Brett Koonce [15]. Riaz Ullah Khan et al. [16] has been applied Yolov7 to evaluate the performance of the ResNet model based on image recognition. In the present study, we apply these two models to evaluate their performance and primary prediction rate based on the studies conducted by the authors. The accuracy of these two models in identifying, classifying, and determining calcium deficiency signs in tomato fruit.

Yolov7 Structure. The Yolov7 model is depicted in Fig. 2 for increasing training speed and good recognition results. Xiang Long et al. [17] developed an efficient and effective implementation of object detection using Yolov7.

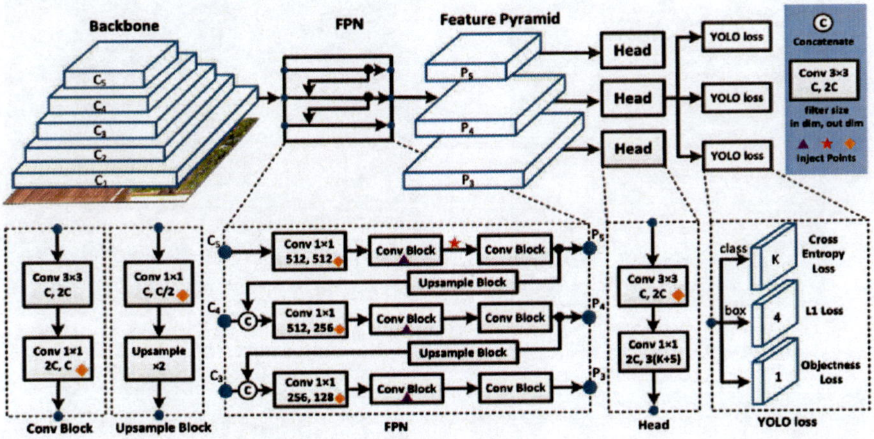

Fig. 2. Structure of Yolov7 [17].

ResNet50 Structure. We utilized the ResNet50 architecture, as shown in [15], for identifying, classifying, and determining the level of calcium deficiency in tomato fruit. Figure 3 illustrates ResNet50 architecture for this study.

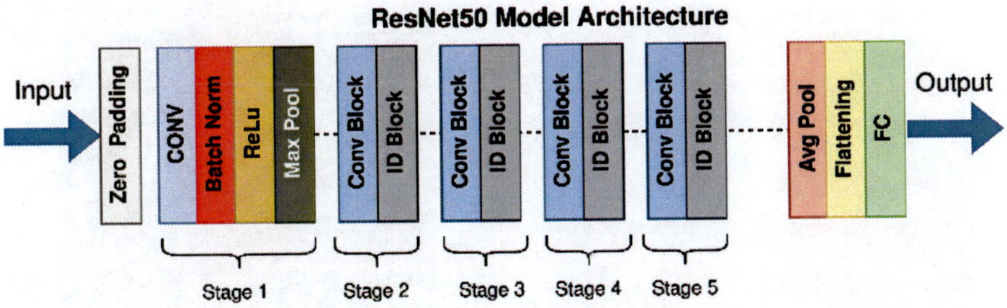

Fig. 3. Structure of ResNet50.

3.2 Validation Result

The ResNet50 and Yolov7 models were tested using the captured images from the test dataset and another source to determine their ability to identify and classify calcium deficiency stages. Figure 4 shows the results of ResNet50 in identifying and classifying calcium deficiency stages in tomato fruits.

Fig. 4. Recognition accuracy of alcium stages using ResNet50.

As shown in Fig. 4 and Table 1, the ResNet50 model achieved the highest accuracy of 100% for stage 1 in the test dataset. Similarly, the accuracy for stage 2 was 100% for the test dataset and 92% for the external dataset. However, this accuracy dropped to 97% for stage 3 and further to 85% for the external dataset. Overall, the ResNet50 model showed high and precise accuracy (97%) in predicting and classifying the three stages of calcium deficiency in tomato plants. In addition, it demonstrated promising results for different levels of calcium deficiency in tomato fruits.

Table 1. Comparison of the accuracy rate of two test dataset with ResNet50.

Stage of nutrient deficiency	Test dataset Accuracy	Figure	External dataset Accuracy	Figure
Calcium (Ca^{2+})	100%	Figure 4a-1	97%	Figure 4b-1
	100%	Figure 4a-2	92%	Figure 4b-2
	97%	Figure 4a-3	85%	Figure 4b-3

Figure 5 and Table 2 demonstrate that the performance of Yolov7 is acceptable but not highly accurate. The highest accuracy achieved is only 87% for the external dataset test. Specifically, Yolov7's accuracy for predicting stage 1 is the highest at 86% for the test dataset, while the accuracy for stage 2 is the lowest at 77%. The accuracy for stage 3 is 79%.

Table 2 demonstrates that the accuracy of Yolov7 in identifying and classifying tomato plants decreased with each stage. The highest is 86% in stage 1 and gradually decreases to 79% in stage 3 for the test dataset. Similarly, for the external dataset, the highest is 87% for stage 1 and gradually decreases to 55% for stage 3.

Through the use of two models, Yolov7 and Resnet 50, for predicting and classifying three stages of Calcium deficiency in tomato plants, in all respects, ResNet50 gives better and more accurate results in determining calcium deficiency classification of all three stages, compared with Yolov7. This suggests that ResNet50 can be a reliable choice for classification and prediction applications in detecting the symptoms of undernutrition in crops.

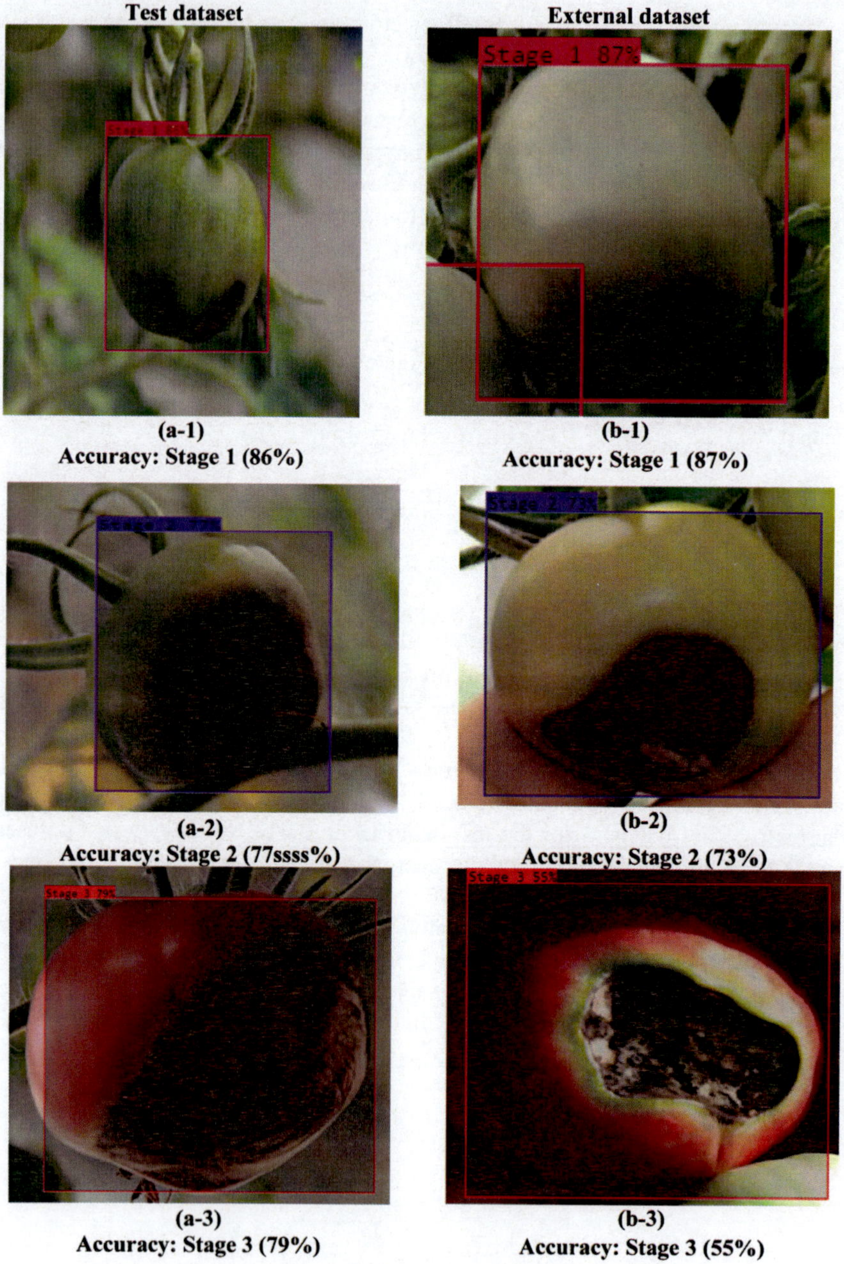

Fig. 5. Recognition accuracy of Calcium stages using Yolov7.

A confusion matrix is used to compare the ability to classify the state of calcium deficiencies when applying two modes, ResNet50 and Yolov7. Table 3 describes the

Table 2. Comparison of the accuracy rate of two test dataset with Yolov7.

Stage of nutrient deficiency	Test dataset Accuracy	Figure	External dataset Accuracy	Figure
Calcium (Ca^{2+})	86%	Figure 5a-1	87%	Figure 5b-1
	77%	Figure 5a-2	73%	Figure 5b-2
	79%	Figure 5a-3	55%	Figure 5b-3

results of the confusion matrix with ResNet50. Table 4 shows the results of the confusion matrix with Yolov7.

Table 3. Confusion matrix of ResNet50

	Stage 1	Stage 2	Stage 3
Stage 1	15	2	1
Stage 2	1	9	2
Stage 3	1	1	4
Total for Class	17	12	7
Accuracy	86.11%	83.33%	86.11%
Precision	0.83	0.75	0.67
Recall	0.88	0.75	0.57
F1 Score	0.86	0.75	0.62

Table 4. Confusion matrix of Yolov7

	Stage 1	Stage 2	Stage 3
Stage 1	11	3	2
Stage 2	4	8	2
Stage 3	2	1	3
Total for Class	17	12	7
Accuracy	69.44%	72.22%	80.56%
Precision	0.69	0.57	0.50
Recall	0.65	0.67	0.43
F1 Score	0.67	0.62	0.46

Comparing the indexes in Table 3 and Table 4, it was found that in this experiment, ResNet50 model achieved better results than Yolov7 model on the same validation dataset. In particular, ResNet50 received the same accuracy in all three levels of nutritional deficiencies of 86.11%, 83.33%, and 86.11%, respectively, for stage 1, stage 2, and stage 3. Meanwhile, Yolov7 only got 69.44%, 72.22%, and 80.56% accuracy rates for stage 1, stage 2, and stage 3. ResNet50's F1 score was 0.86 for stage 1, 0.75 for stage 2, and 0.62 for stage 3. With Yolov7, stage 1 has an F1 score of 0.67, stage 2 has 0.62, and stage 3 has a score of 0.46. Therefore, ResNet50 has better indicators than Yolov7 in this study.

4 Discussion and Conclusion

Previous studies mainly focused on using machine learning to identify and classify the deficiencies of some essential nutrients of plants, especially tomato plants. However, the accuracy of these methods is not always satisfactory, which is a critical issue in the agricultural field. To address this challenge, we aimed to compare and evaluate the effectiveness of various CNN-based models in predicting and classifying the early and late stages of nutritional deficiencies in tomato plants. In this study, we applied two different deep neural network models, ResNet50 and Yolov7, and improved their performance in predicting the expression of nutritional macronutrient deficiencies in greenhouse-grown tomato plants. Based on our findings, we draw the following conclusions:

- Both ResNet50 and Yolov7 models can predict, indicate, and classify the symptoms and levels of nutrition deficiency in tomato plant fruits.
- Our study showed that Yolov7 and ResNet50, produced the expected results in predicting calcium deficiency in tomato plants.

The experimental result indicates that the ResNet50 model outperformed the Yolov7 model in detecting calcium nutrient deficiency in tomato plants. Furthermore, ResNet50 model achieved a higher accuracy rate and F1 score in all three stages of nutrient deficiency compared to the Yolov7 model. These findings suggest that ResNet50 can be an effective tool for identifying and predicting plant nutrient deficiencies, which can help farmers improve crop quality and yield. However, further research is needed to validate the results on a larger dataset and crop types. In addition, other deep learning models and techniques could also be explored to improve the accuracy and efficiency of nutrient deficiency detection in plants.

The use of deep learning models in agriculture is expected to continue to improve, providing significant benefits to small tomato farms in Vietnam. Building on the performance of the two models mentioned in this study, the authors plan to develop a system for monitoring tomato growth processes in the Vietnam market. Moreover, the system can be extended and applied to various types of plants and diseases, ultimately leading to increased crop yields. This highlights the potential for further advancements in agriculture through the application of deep learning.

References

1. Ho, L.C., White, P.L.: A cellular hypothesis for the induction of blossom-end rot in tomato fruit. Ann. Bot. **95**, 571–681 (2005)

2. Morard, P., Pujos, A., Bernadac, A., Bertoni, G.: Effect on temporary calcium deficiency on tomato growth and mineral nutrition. J. Plant Nutr. **19**, 115–127 (2008)
3. Sethy, P.K., Barpanda, N.K., Rath, A.K.R., Behera, S.K.: Nitrogen deficiency prediction of rice crop based on convolutional neural network. J. Ambient. Intell. Humaniz. Comput. **11**(1), 5703–5711 (2020)
4. Thangaraj, R., Anandamurugan, S., Pandiyan, P., Kaliappan, V.K.: Artificial intelligence in tomato leaf disease detection: a comprehensive review and discussion. J. Plant Dis. Protect. **129**, 1–20 (2021). https://doi.org/10.1007/s41348-021-00500-8
5. Sowmiya, M., Krishnaveni, S.: Deep learning techniques to detect crop disease and nutrient deficiency-a survey. In: 2021 International Conference on System, Computation, Automation and Networking (ICSCAN) IEEE, Pudecherry, India (2021)
6. Barbedo, J.G.A.: Detection of nutrition deficiencies in plants using proximal images and machine learning: a review. Comput. Electron. Agric. **162**, 482–492 (2019)
7. Sharma, M., Nath, K., Sharma, R.K., Jyothi, C., Chaudhary, A.: Ensemble averaging of transfer learning models for identification of nutritional deficiency in rice plant. Electronics **11**(1), 148 (2022)
8. Waheed, H., Zafar, N., Akram, W., Manzoor, A., Gani, A., Islam, S.: Deep learning based disease, pest pattern and nutritional deficiency detection system for "Zingiberaceae" crop. Agriculture **12**(6), 742 (2022)
9. Ferentinos, K.P.: Deep learning models for plant disease detection and diagnosis. Comput. Electron. Agric. **145**, 311–318 (2018). https://doi.org/10.1016/j.compag.2018.01.009
10. BereciartuaPérez, A., Gómez, L., Picón, A., NavarraMestre, R., Klukas, C., Eggers, T.: Insect counting through deep learning-based density map estimation. Comput. Electron. Agric. **197**, 106933 (2022)
11. Kusanur, V., Chakravarthi, V.S.: Using transfer learning for nutrient deficiency prediction and classification in tomato plant. Int. J. Adv. Comput. Sci. Appl. **12**(10), 784–790 (2021)
12. Nayar, P., Chhibber, S., Dubey, A.K.: An efficient algorithm for plant disease detection using deep convolutional networks. In: 2022 14th International Conference on Computational Intelligence and Communication Networks (CICN). IEEE, Al-Khobar, Saudi Arabia (2022)
13. Wang, C.-Y., Bochkovskiy, A., Liao, H.-Y.M.: YOLOv7: Trainable bag-of-freebies sets new state-of-the-art for real-time object detectors (2022). https://doi.org/10.48550/arXiv.2207.02696
14. Delin, W., et al.: Detection of camellia oleifera fruit in complex scenes by using YOLOv7 and data augmentation. Appl. Sci. **12**(22), 11318 (2022)
15. Koonce, B.: ResNet50. Convolutional Neural Networks with Swift for TensorFlow, pp. 63–72. Apress, Berkeley (2021)
16. Khan, R.U., Zhang, X., Kumar, R., Aboagye, E.O.: Evaluating the performance of Resnet model based on image recognition. In: Proceedings of the 2018 International Conference on Computing and Artificial Intelligence. Chengdu, China, pp. 86–90 (2018)
17. Long, X., et al.: An effective and Efficient Implementation of Object Detector (2022).arXiv: 2007.12099

A Systematic Review on Crop Yield Prediction Using Machine Learning

Moon Halder[1], Ayon Datta[2], Md Kamrul Hossain Siam[3], Shakik Mahmud[4](✉), Md. Saem Sarkar[4], and Md. Masud Rana[5]

[1] American International University, Dhaka, Bangladesh
[2] Ahsanullah University of Science and Technology, Dhaka, Bangladesh
[3] Western Illinois University, Macomb, USA
[4] Japan-Bangladesh Robotics and Advanced Technology Research Center, Dhaka, Bangladesh
shakikmahmud@gmail.com
[5] Sher-e-Bangla Agricultural University, Dhaka, Bangladesh

Abstract. Machine learning is an essential tool for crop yield prediction. Crop yield prediction is a challenging task in the agriculture and agronomic field. In crop yield, many factors can impact crop yields such as soil quality, temperature, humidity, quality of the seeds, rainfall, and many more. To give an accurate yield prediction with the right machine learning algorithms we need to process a huge amount with the selections of impactful features. In this study, we performed a systematic literature review to select machine learning methods and features that can analyze large amounts of data and give more accurate results. We discuss the lacking's of existing research and generated a comparative analysis to give a clear aspect of the better solutions. From a critical evaluation and specific search criteria, we found – papers from AGORA that contain many more different databases such as MDPI, Tylor and Francis, IEEE, etc. From 660 we selected 50 papers from that number that were used more efficiently and gives accurate results with a thorough investigation with the help of our selection criteria and generic research questions that can filter and bring out the more relevant papers regarding these fields. From the selected papers we evaluate the methods, and geographical areas that have been selected for acquiring data analyzed the features, and have a thorough inspection of the selected factors that have the most impact on yield prediction. This study will help future researchers to give a clear understanding of existing research and guide them to generate a more effective model.

Keywords: Machine Learning · Deep Learning · Systematic Review · Artificial Neural Network · Agriculture

1 Introduction

Solid, feasible, and comprehensive food-producing methods are basic for accomplishing improvement objectives worldwide. Agronomic improvement is one of the effective ways to resolve extraordinary destitution, advance thriving, and maintain an anticipated 9.8 billion individuals by 2050 [1]. Development in the agronomy segment is about

T. D. L. Nguyen et al. (Eds.): ICISN 2023, LNNS 752, pp. 658–667, 2023.
https://doi.org/10.1007/978-981-99-4725-6_77

three times more successful in increasing salaries compared to other sectors. Agribusiness is additionally vital for financial development: accounting for about 4% of national income worldwide, and in some less developed countries, it may be more than 25% of total national income [2]. Machine learning has grown alongside big data advancements and ubiquitous computing to open up modern concerns, measurements, and data-driven forms of rural operative situations. Smart cultivating makes agribusiness more proficient and viable through exact and accurate algorithmic operation. A procedure that directs it is Machine Learning. Machine learning is all over all through the complete collecting cycle. With the help of computer vision technology, it begins to manage works like seed planting and breeding, soil quality checking, and water bolstering measurement and ends up choosing automated arms or robots for harvesting operations. The yield forecast is a well-known theme about accuracy farming because it characterizes mapping and supply management with the demand of crop harvesting. Modern methods have gone far beyond forward prediction based on historical information, and consolidate computational techniques to supply information and an exclusive multi-dimensional evaluation path of crop yield estimation for agriculturists. In this research, we are performing a systematic review on Crop yielding using machine learning which is able to compile, summarize, evaluate, and synthesize the accessible research information on this theme. One of the focal points of reviewing is that we will be able to pick up an opportunity to see early ranges of research headings and where the field may be heading before the work is published. We'll be able to get what reviewers seek in order to progress our claimed writing by conducting surveys ourselves and too by being able to see the perspectives of other analysts. This paper performed a systematic literature review with some selective criteria to connect more relevant works, nailed down the number of papers based on the recent review articles to prevent irrelevant studies, and sorted out 30 papers. We conducted reviews on those papers and compared and analyzed them to bring out the best work regarding crop yield prediction.

Moreover, this paper will provide a precise investigation of diverse applications, find out the holes and exhibit the distinctive forms of the characteristic of machine learning application that impacts crop yield prediction. This report accumulates the approach studies of different machine learning processes to provide a deeper understanding of crop-yielding research, the accuracy of various information based on the geological data, and its influence on the results along with the advantages and disadvantages. It'll also direct the would-be researchers for advanced inquiry in this field and will give a more profound understanding to scientists and policy creators about the influence of machine learning strategies on crop yield prediction.

2 Review of Literature

In both wealthy and developing nations, there is difficulty with the sustainable availability of nutritious food. Global food security has become more of a worry due to rapid urbanization, global warming, and the depletion of natural resources. The issue of food insecurity is further complicated by the fast population rise. In order to fulfill the food requirements of about 10 billion in 2050 [3], of which approximately 6.5 billion would live in metropolitan areas [4], food production must be raised by 70%. One of the criteria that we used to exclude publications from our examination of those that were retrieved was the fact that the publication was either a survey or a standard review piece. These articles that were omitted are, in reality, related works that are covered further on in this section. A review of research on nitrogen status estimate using machine learning was carried out by Chlingaryan and Sukkarieh [5]. The research comes to the conclusion that rapid advances in sensing technologies and machine learning techniques will lead to solutions in the agriculture industry that are more efficient and cost-effective. Elavarasan et al. conducted a literature review to investigate the machine learning models that have been connected with the prediction of agricultural yields based on meteorological factors. The research recommends doing a thorough search to identify other characteristics that might account for crop output [6]. A review article on the use of machine learning in the agriculture industry was written by the authors [7], and was published. Publications that focused on agricultural management, animal management, water management, as well as soil management were utilized for the analysis. Li, Lecourt, with Bishop conducted a review of research on how to determine the maturity of fruits in order to determine the best time to harvest the crop and provide a prediction of the production [8].

A Systematic Literature Review (SLR), as the name implies, must be systematic and encompass all previously published literature. This paper summarizes all of the published work on the uses of machine learning in crop yield prediction problems. We share our empirical findings and replies to the research issues posed in this review article in this study.

3 Research Methodology

It is possible to analyze the completeness or validity of a certain theory by doing a systematic literature review [9], which is an excellent method for evaluating theory or evidence in a particular field. Because they provide both impartiality and openness, the review procedures that Kitchenham and Charters [10] provided are suitable for the systematic literature review that we are doing. In the first step, the following hypotheses are specified. After the research topics have been finalized, suitable studies are chosen through the use of databases.

In this investigation, the AGORA research database was consulted. After identifying relevant research, these studies were filtered and evaluated based on predetermined criteria for inclusion and quality. After gleaning all of the pertinent information from the studies that were chosen, the data were finally retrieved and then synthesized in order to provide answers to the research questions. The strategy that we used may be broken down into three parts: the review of the plan, the review of the conduct, and the review of the report.

The planning of the review constitutes the first stage. During this phase of the process, research questions are formulated, a protocol is designed, as well as eventually, that protocol is verified to determine whether or not the technique can be implemented. Not only are the research topics established, but also the publication venues, beginning search strings, as well as publication selection criteria. After all of this information has been specified, the protocol will be changed yet again so that we can determine whether or not it constitutes an appropriate review process. Figure 1, is a representation of the internal

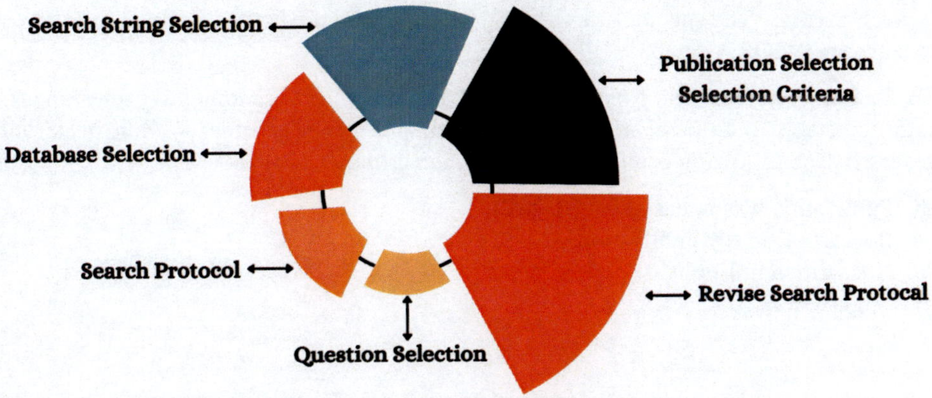

Fig. 1. Details working plan

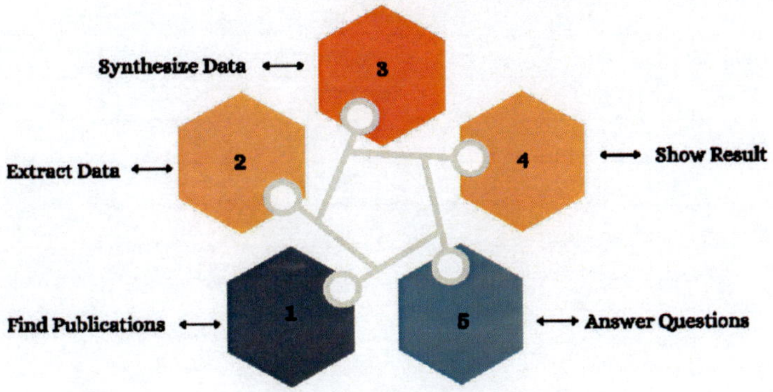

Fig. 2. Reporting steps

actions that are taken during the Plan Review stage. The next stage, which is shown in Fig. 2, is the process of performing the review. During the process of carrying out the evaluation, the publications were chosen by searching through each and every database. The data were extracted, which means that their information regarding authors, the year of publication, the kind of publication, and additional information pertaining to the study questions were kept. The data were then synthesized in order to present an overview of the relevant publications that have been published up to this point after all of the essential data had been appropriately extracted.

Research Questions: The following four research questions (RQs) have been established for this systematic literature review study.

i. Which branches of Machine Learning are being used for crop yield prediction?
ii. Which features have been used in the literature?
iii. Which evaluation approaches have been followed?
iv. Which algorithms/formulas/classifiers have been used?

Search String: The specific search string is 'crop yield prediction', and this string was used in the AGORA database.

Exclusion Criteria: The papers were assessed and ranked according to exclusion criteria to establish the parameters for the systematic review in order to omit irrelevant research. The following is a list of the exclusion criteria (EC) (Table 1):

i. Publication was released before 2022,
ii. Review of survey publication,
iii. Not written in English,
iv. Slightly related to ML.

Table 1. Papers are distributed through databases

Publisher Names	Num. of Papers	Publisher Names	Num. of Papers
MDPI	25	Hindawi	2
Taylor's & Francis	8	Springer	1
Elsevier	5	IEEE	1
Nature	3	Frontiers	2
CAAS	1	F1000	1
NBHAC-Napoca	1		

4 Result

Through the search string "Crop Yield Prediction Using ML" we retrieved 660 paper from AGORA, according to our exclusion criteria we select 50 research papers to conduct this study. Table 2 displays the publications that were chosen. The table displays the type of publication, references, and algorithms utilized in these publications (Fig. 3).

According to the analysis, most of the papers are based on deep learning, neural network, and supervised learning.

In order to address the second research question (RQ2), the characteristics of the machine learning algorithms used in the publications were analyzed and summarized. The results of this analysis are displayed in Table 2. Table 3 demonstrates that the most common features utilized are those related to temperature, rainfall, and soil type.

Evaluation criteria were chosen in order to answer research question three (RQ3). Table 4 lists all of the assessment parameters that were employed along with the frequency of use. The most often used parameter in the research is Mean Square Error (MSE), as seen in the Table 4.

General Discussion: The research being discussed is open to potential problems with validity, including external, construct, and reliability. The study used broad search criteria and returned a large number of publications, which addresses concerns about external

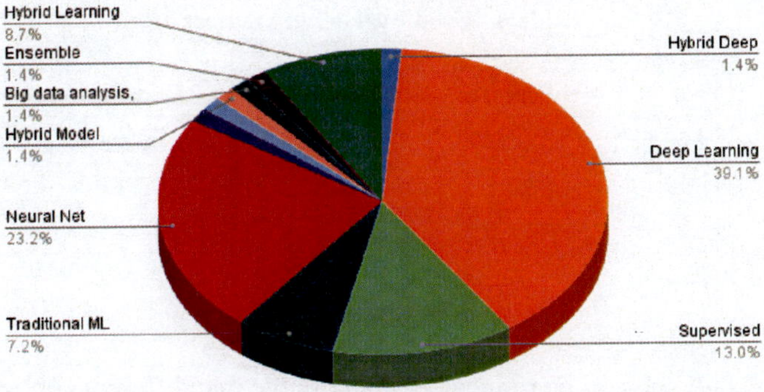

Fig. 3. Analysis result of RQ: 1

Table 2. Quantity of classification types (R1) and algorithms (R4).

Branches Type	No.	Algorithms	No.	Algorithms	No.
Deep learning	17	Random Forest	28	Decision Tree	7
Neural Network	13	ANN	16	Linear Regression	6
Supervised Learning	6	SVM	12	MLR	9
Hybrid Model	6	XGBoost	13	CNN	5
Traditional ML	5	SVR	14	DNN	2
Ensemble Learning	2	LASSO	8	ResNet	1
Reinforcement Learning	1	LSTM	8	SLR	1

Table 3. Used Features.

Features	No.	Features	No.	Features	No.
Weather	3	Seasosn	2	Yield	11
Soil	25	Climate	6	LAI	5
Area	13	Spatial Patterns	1	Satellite	7
Temperature	25	Vegetation Index	8	Hyperspectral Image Data	2
Humidity	8	VI	11	Soil Properties	2
Rainfall	17	NDVI	6	Soil Moisture	4
Region	1	Seasosn	2	Precipitation	8
Irrigation	7	Water	10	SPI	4
Meteorogical	3	Wind Speed	2		

and construct validity. The methodology used in the study has been clearly described, making it possible for the study to be replicated, and the results were found to be reliable.

Table 4. Used Evaluation Matrices.

Features	No.	Features	No.	Features	No.
MSE	44	Accuracy	4	R score	1
RMSE	38	SAE	1	MAPE	9
R2	39	Recall	3	PE	15
MAE	23				

Though if the study were to be replicated, it's possible that different publications would be selected, but it is unlikely that the overall findings would be affected.

Discussion for Article Search: There's a chance that essential publications have been overlooked. More synonyms may have been utilized, and a wide search could have yielded new research. However, a large number of papers were returned, suggesting that the search was sufficiently comprehensive.

5 Conclusion

The study found that the selected publications used various features depending on the scope of the research and data availability. Each paper researched yield prediction using machine learning but varied in the features used. The studies also varied in scale, location, and crop. The choice of features was dependent on the data available and the research goals. The study also found that models with more features did not always produce the best results for yield prediction. To determine the best-performing model, models with both more and fewer features should be tested. Many algorithms were used in the studies, and the results showed that no single model is the best, but some machine learning models are used more frequently than others. The most commonly used models were the random forest, neural networks, linear regression, and gradient boosting tree. Most of the studies tested multiple machine learning models to determine the best one for prediction. However, various types of algorithms are also used to solve this problem. We believe this study will pave the path for future research into the topic of agricultural yield prediction development. In our research plan, we intend to expand upon the findings of this study and concentrate on developing a DL-based crop production prediction model.

References

1. United Nations. World population projected to reach 9.8 billion in 2050, and 11.2 billion in 2100. United Nations (2017). https://www.un.org/en/desa/world-population-projected-reach-98-billion-2050-and-112-billion-2100. Accessed 18 Jan 2023
2. World Economic Situation and Prospects (WESP) - UN iLibrary. https://www.un-ilibrary.org/content/periodicals/24118370. Accessed 18 Jan 2023
3. World Health Organization. The State of Food Security and Nutrition in the World 2018: Building Climate Resilience for Food Security and Nutrition. Food and Agriculture Organization: Rome, Italy (2018)

4. Avtar, R., Tripathi, S., Aggarwal, A.K., Kumar, P.: Population–urbanization–energy nexus: a review. Resources **8**, 136 (2019)
5. Chlingaryan, A., Sukkarieh, S., Whelan, B.: Machine learning approaches for crop yield prediction and nitrogen status estimation in precision agriculture: a review. Comput. Electron. Agric. **151**, 61–69 (2018)
6. Elavarasan, D., Vincent, D.R., Sharma, V., Zomaya, A.Y., Srinivasan, K.: Forecasting yield by integrating agrarian factors and machine learning models: a survey. Comput. Electron. Agric. **155**, 257–282 (2018)
7. Liakos, K.G., Busato, P., Moshou, D., Pearson, S., Bochtis, D. Machine learning in agriculture: a review. Sensors **18**(8) (2018)
8. Li, B., Lecourt, J., Bishop, G.: Advances in non-destructive early assessment of fruit ripeness towards defining optimal time of harvest and yield prediction—a review. Plants **7**(1) (2018)
9. Tranfield, D., Denyer, D., Smart, P.: Towards a methodology for developing evidence-informed management knowledge by means of systematic review. Br. J. Manag. **14**, 207–222 (2003)
10. Kitchenham, B.A., Charters, S.: Guidelines for performing systematic literature reviews in software engineering. (EBSE 2007-001); Keele University: Keele, UK; Durham University: Durham, UK (2007)
11. Oikonomidis, A., Catal, C., Kassahun, A.: Hybrid deep learning-based models for crop yield prediction. Appl. Artif. Intell. **36**(1), 2031822 (2022)
12. Vashisht, S., Kumar, P., Trivedi, M.C.: Crop yield prediction using improved extreme learning machine. Commun. Soil Sci. Plant Anal. **54**(1), 1–21 (2023)
13. Bali, N., Singla, A.: Deep learning based wheat crop yield prediction model in Punjab region of North India. Appl. Artif. Intell. **35**(15), 1304–1328 (2021)
14. Gupta, S., et al.: Machine learning- and feature selection-enabled framework for accurate crop yield prediction. J. Food Qual. (2022)
15. Batool, D., et al.: A hybrid approach to tea crop yield prediction using simulation models and machine learning. Plants **11**(15), 2022 (1925)
16. Krithika, K.M., et al.: Models for feature selection and efficient crop yield prediction in the groundnut production. Res. Agric. Eng. **68**(3), 131–141 (2022)
17. Yli-Heikkilä, M., Wittke, S., et al.: Scalable crop yield prediction with sentinel-2 time series and temporal convolutional network. Remote Sens. **14**(17), 4193 (2022)
18. Pham, H.T., Awange, J., Kuhn, M.: Evaluation of three feature dimension reduction techniques for machine learning-based crop yield prediction models. Sensors **22**(17), 6609 (2022)
19. Cedric, L.S., et al.: Crops yield prediction based on machine learning models: case of West African countries. Smart Agric. Technol. **2**, 100049 (2022)
20. Septem Riza, L., et al.: Remote sensing and machine learning for yield prediction of lowland paddy crops. F1000 Res. (2022)
21. Cubillas, J.J., Ramos, M.I., Jurado, J.M., Feito, F.R.: A machine learning model for early prediction of crop yield, nested in a web application in the cloud: a case study in an olive grove in southern Spain. Agriculture **12**(9), 1345 (2022)
22. Liu, Y.: Rice yield prediction and model interpretation based on satellite and climatic indicators using a transformer method. Remote Sens. **14**(19), 5045 (2022)
23. Huang, H.: Developing a dual-stream deep-learning neural network model for improving county-level winter wheat yield estimates in China. Remote Sens. **14**(20), 5280 (2022)
24. Moot, D.J.: Simplified methods for on-farm prediction of yield potential of grazed lucerne crops in New Zealand. N. Z. J. Agric. Res. **65**(4–5), 252–270 (2021)
25. Parsaeian, M., Rahimi, M., Rohani, A., Lawson, S.S.: Towards the modeling and prediction of the yield of oilseed crops: a multi-machine learning approach. Agriculture **12**(10), 1739 (2022)

26. Ali, M., et al.: Coupled online sequential extreme learning machine model with ant colony optimization algorithm for wheat yield prediction. Sci. Rep. **12**(1), 5488 (2022)
27. Fei, S., Li, L., Han, Z., Chen, Z., Xiao, Y.: Combining novel feature selection strategy and hyperspectral vegetation indices to predict crop yield. Plant Methods **18**(1), 1–13 (2022)
28. Bian, C.: Prediction of field-scale wheat yield using machine learning method and multi-spectral UAV data. Remote Sens. **14**(6), 1474 (2022)
29. Cao, J.: Improving the forecasting of winter wheat yields in Northern China with machine learning-dynamical hybrid subseasonal-to-seasonal ensemble prediction. Remote Sens. **14**(7), 1707 (2022)
30. Kittichotsatsawat, Y., Tippayawong, N., Tippayawong, K.Y.: Prediction of arabica coffee production using artificial neural network and multiple linear regression techniques. Sci. Rep. **12**(1), 14488 (2022)
31. Khan, S.N.: A geographically weighted random forest approach to predict corn yield in the US corn belt. Remote Sens. **14**(12), 2843 (2022)
32. Tripathi, A., Tiwari, R.K., Tiwari, S.P.: A deep learning multi-layer perceptron and remote sensing approach for soil health based crop yield estimation. Int. J. Applied Earth Obs. Geoinf. **113**, 102959 (2022)
33. Khan, N., et al.: Prediction of oil palm yield using machine learning in the perspective of fluctuating weather and soil moisture conditions: evaluation of a generic workflow. Plants **11**(13), 1697 (2022)
34. Shen, Y., et al.: Improving wheat yield prediction accuracy using LSTM-RF framework based on UAV thermal infrared and multispectral imagery. Agriculture **12**(6), 892 (2022)
35. Srivastava, A.K., et al.: Winter wheat yield prediction using convolutional neural networks from environmental and phenological data. Sci. Rep. **12**(1), 3215 (2022)
36. Xu, W., et al.: Cotton yield estimation model based on machine learning using time series UAV remote sensing data. Int. J. Appl. Earth Obs. Geoinf. **104**, 102511 (2021)
37. Yildirim, T., et al.: Using artificial neural network (ANN) for short-range prediction of cotton yield in data-scarce regions. Agronomy **12**(4), 828 (2022)
38. Sun, Q., et al.: Coupling process-based crop model and extreme climate indicators with machine learning can improve the predictions and reduce uncertainties of global soybean yields. Agriculture **12**(11), 1791 (2022)
39. Saravia, D., et al.: Yield predictions of four hybrids of maize (Zea mays) using multispectral images obtained from UAV in the coast of Peru. Agronomy **12**(11), 2630 (2022)
40. Guan, H.: An improved approach to estimating crop lodging percentage with Sentinel-2 imagery using machine learning. Int. J. Appl. Earth Obs. Geoinf. **113**, 102992 (2022)
41. Song, C.: Development trends in precision agriculture and its management in china based on data visualization. Agronomy **12**(11), 2905 (2022)
42. Li, C., et al.: Improvement of wheat grain yield prediction model performance based on stacking technique. Appl. Sci. **11**(24), 12164 (2021)
43. Attia, A., et al.: Coupling process-based models and machine learning algorithms for predicting yield and evapotranspiration of maize in arid environments. Water **14**(22), 3647 (2022)
44. Kundu, S.G.: A ML-AI enabled ensemble model for predicting agricultural yield. Cogent Food Agric. **8**(1), 2085717 (2022)
45. Li, K.-Y., et al.: Toward automated machine learning-based hyperspectral image analysis in crop yield and biomass estimation. Remote Sens. **14**(5), 1114 (2022)
46. Raja, S.P., et al.: Crop prediction based on characteristics of the agricultural environment using various feature selection techniques and classifiers. IEEE Access **10**, 23625–23641 (2022)
47. Mokhtar, A., et al.: Using machine learning models to predict hydroponically grown lettuce yield. Sec. Technical Advances in Plant Science (2022)

48. Deng, Q., et al.: Winter wheat yield estimation based on optimal weighted vegetation index and BHT-ARIMA model. Remote Sens. **14**(9), 2022 (1994)
49. Gonzalez-Gonzalez, M.A., Guertin, D.P.: Seasonal bean yield forecast for non-irrigated croplands through climate and vegetation index data: geospatial effects. Int. J. Appl. Earth Obs. Geoinf. **105**, 102623 (2021)
50. Pang, A., Chang, M.W., Chen, Y.: Evaluation of random forests (RF) for regional and local-scale wheat yield prediction in southeast Australia. Sensors **22**(3), 717 (2022)
51. Barzin, R., Lotfi, H., Varco, J.J., Bora, G.C.: Machine learning in evaluating multispectral active canopy sensor for prediction of corn leaf nitrogen concentration and yield. Remote Sens. **14**(1), 120 (2021)
52. Ahmed, A.M., et al.: Kernel ridge regression hybrid method for wheat yield prediction with satellite-derived predictors. Remote Sens. **14**(5), 1136 (2022)
53. Zhao, Y.: Transfer learning based approach for yield prediction of winter wheat from planet data and SAFY model. Remote Sens. **14**(21), 5474 (2022)
54. Al-Adhaileh, M.H.: Artificial intelligence framework for modeling and predicting crop yield to enhance food security in Saudi Arabia. PeerJ Comput. Sci. **8**, e1104 (2022)
55. Ji, Z.: Prediction of corn yield in the USA corn belt using satellite data and machine learning: from an evapotranspiration perspective. Agriculture **12**(8), 1263 (2022)
56. Oliveira, M.F.D.: Training machine learning algorithms using remote sensing and topographic indices for corn yield prediction. Remote Sens. **14**(23), 6171 (2022)
57. Chidzalo, P., et al.: Trivariate stochastic weather model for predicting maize yield. J. Appl. Math. (2022)
58. Ding, Y., et al.: A study on cotton yield prediction based on the chlorophyll fluorescence parameters of upper leaves. Notulae Botanicae Horti Agrobotanici Cluj-Napoca **50**, 12775–12775 (2022)
59. Ganeva, D.: Phenotypic traits estimation and preliminary yield assessment in different phenophases of wheat breeding experiment based on UAV multispectral images. Remote Sens. **14**(4), 1019 (2022)
60. Gholizadeh, A.: Modeling the final fruit yield of coriander (Coriandrum sativum L.) using multiple linear regression and artificial neural network models. Arch. Agron. Soil Sci. **68**(10), 1398–1412 (2020)

A Next-Generation Device for Crop Yield Prediction Using IoT and Machine Learning

Md Kamrul Hossain Siam[2], Noshin Tasnia[1,3], Shakik Mahmud[1](✉), Moon Halder[4], and Md. Masud Rana[5]

[1] Japan-Bangladesh Robotics and Advanced Technology Research Center, Dhaka, Bangladesh
shakikmahmud@gmail.com
[2] Western Illinois University, Macomb, USA
[3] Jahangirnagar University, Savar Union, Bangladesh
[4] American International University, Dhaka, Bangladesh
[5] Sher-E-Bangla Agricultural University, Dhaka, Bangladesh

Abstract. This paper introduces a next-generation device for crop yield prediction that utilizes IoT and machine learning technologies. The device was implemented and tested, and it was found to have a high level of accuracy in predicting crop yields. It is a combination of three different machine learning models: Artificial Neural Network (ANN), Fuzzy Logic, and Support Vector Machine (SVM). The IoT sensors in the device gather data on various environmental and soil conditions such as temperature, humidity, and soil moisture, which is then fed into the machine learning models. The ANN is used to analyze the sensor data and extract features, the Fuzzy Logic model is used to handle uncertainty in the data and make predictions, and the SVM model is used for classification. The device was tested on various crops and it was observed that the accuracy of the predictions was good and the results were comparable to other state-of-the-art techniques. This technology has the potential to revolutionize the way farmers manage their crops and improve crop yields. It can also be used for crop forecasting, crop monitoring, and precision agriculture. By providing accurate and real-time information about crop yields, this device could help farmers make better decisions about their crops and increase their overall productivity and profitability.

Keywords: ANN · Fuzzy Logic · SVM · IoT · Crop Yield

1 Introduction

It is the innate habit of the soil to provide a certain proportion of nutrients for the growth and nutrition of the plant. It is called soil fertility. From sowing to growing, the seed depends on the preparation of the land. The soil prepared depends on the shape of the source or root and the moisture content. Soil moisture can be tested through Artificial Intelligence. Moisture object is the critical factor for the effects of physical and chemical aspects of food, determining how long we can preserve food nutrition.

Soil dampness (SM) is a massive segment of the hydrological cycle controlling run, vegetation creation, and evapotranspiration [1]. Soil dampness is a significant soil pointer

T. D. L. Nguyen et al. (Eds.): ICISN 2023, LNNS 752, pp. 668–678, 2023.
https://doi.org/10.1007/978-981-99-4725-6_78

to characterizing and recognizing horticultural dry spells. Assessment of soil dampness has applications for distinguishing beginning phase water deficiency conditions and advancing dry spell occasions for crop yield vulnerability and food security conditions, farming protection, policy making and dynamic, and harvest arranging, particularly for the dry and semi-parched pieces of the globe. The horticultural dry spell has a reactant elect that contributes to social and political struggles in agricultural nations. Accordingly, soil dampness demonstration and observation are of expanding interest. Checking the spatial and worldly varieties is essential for alleviating and adjusting to environmental changes to support editing frameworks and create exact farming and food security.

In recent years, remote sensing has applied to soul properties like soil moisture, roughness, and soil texture [2]. Our promising techniques are Soil digital photos that can predict soil moisture based on machine learning models and two nonlinear regression models. There are three ML techniques to predict the most accurate result placed on the soil surface. Those techniques are Artificial neural network (ANN), Fuzzy logic(FL), and support vector machine (SVM), and enable to make a decision is now the soil is the best for production.

2 Related Works

In recent years, researchers have been developing new models to predict crop yields and improve irrigation systems using IoT and machine learning techniques. Mohebbian et al. [4] introduced a model that predicts the conductivity of wastewater, which they applied for irrigation using a swarm optimization method and a decision tree. Kondaveti et al. [5] created a device that uses IoT to communicate with the device and uses time series analysis, ANN and multi-linear regression to predict rainfall, which helps to conduct irrigation systems more efficiently. Singh et al. [6] designed a device that gathers environmental data such as air humidity and soil moisture, and processes it with traditional ML techniques such as gradient boosted regression, random forest regression, elastic net regression and multi-linear regression to control the water pump of the irrigation system. Nishitha et al. [7] added a feature to their soil pH calculator model which calculates the necessary water and fertilizer quantity based on pH levels, which results in better crop growth and production, and reduces soil erosion. Akshay et al. [8] used the K-Nearest neighbor method, instead of K-mean and SVM, to process sensor data and achieve an accuracy of 93%. Laksiri et al. [9] found that linear regression was best for subsequent hour humidity and temperature prediction and sliding window model was best for soil moisture prediction. Klompenburg et al. [10] analyzed 80 articles and found that the most commonly used algorithm in the Machine Learning section was the artificial neural network (ANN) and the second most was linear regression. In the Deep Learning section, convolutional neural networks (CNN) were applied most, followed by long-short term memory (LSTM) and deep neural networks (DNN). The evaluation parameters used in these studies include root means square error, r-squared, mean absolute error, mean square error, mean fundamental percentage error, and reduced simple average ensemble.

3 Materials and Methods

3.1 Data Collection

Data collection is an essential step in any machine learning project, as it forms the foundation for training and testing the model. In this particular case, the training data was collected from various sources such as Kaggle and other internet sources, and then merged to create a comprehensive dataset. The testing data, on the other hand, was collected from sensors and field measurements, specifically for crop yield data. The process of data collection is crucial as it determines the quality and accuracy of the final model, thus care should be taken to ensure that the data is accurate, relevant, and unbiased.

3.2 Drip Irrigation Device Development Insights

The development of an automated intelligent drip irrigation system that aims to optimize water usage in agriculture by addressing issues such as poor irrigation techniques, water wastage, and crop degradation. Machine learning techniques are used to determine the most appropriate water level monitoring method, whether it be based on soil moisture or thermal imaging, and to predict crop wastage and growth based on human demand. The article focuses on the implementation of this device for farmers who rely on conventional irrigation methods and have limited knowledge of their soil's moisture level and weather patterns (Fig. 1).

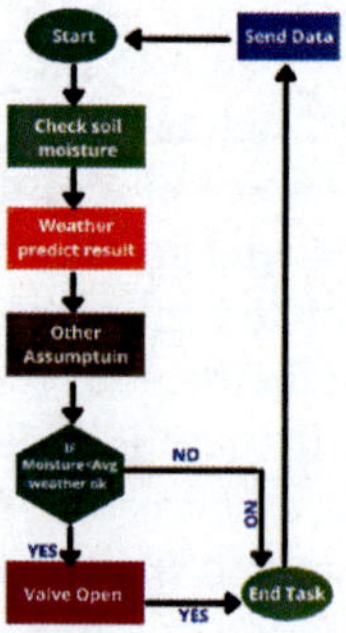

Fig. 1. Working Flowchart.

The proposed device aims to solve these problems by continuously monitoring soil moisture levels and adjusting the water valve accordingly. It also utilizes deep learning to predict weather patterns, allowing for watering during dry hours in anticipation of potential rain. The device also employs IoT technology to send all watering and soil history data to the cloud for analysis and record keeping.

A list of the main hardware components used in our experiment device is shown in Table 1.

Table 1. List of used components

SL	Names	SL	Names
1	Microcontroller	4	I2C and Display
2	GSM SIM 800L	5	Solar Panel
3	Soil Moisture Sensor	6	Solenoidal Valve

Hardware Implementation Setup. After a study, we choose those components (Arduino Mega, GSM, I2C, Soil Moisture Sensor, LED display, N channel Mosfet, Diode, Solenoid Valve, Switch, and LED) and connect them according to the circuit diagram Fig. 2.

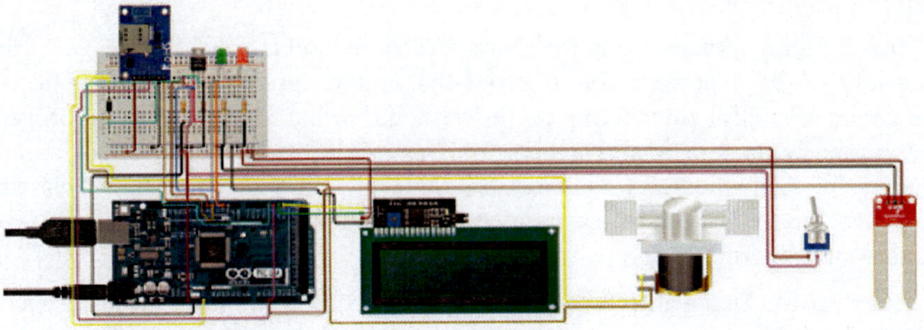

Fig. 2. Circuit Diagram

Network Architecture. It refers to the overall design and structure of a computer network. In the context of IoT, network architecture refers to the way devices, such as sensors and actuators, connect to a central server or hub, as well as the protocols and communication methods used to transmit data between these devices. In this case, the server was built on a PC using the PHP programming language, which is commonly used for web development and can be used to create web-based interfaces for controlling and interacting with IoT devices. The specific architecture of the network will depend on the needs of the particular IoT application, but it is important to ensure that it is secure, scalable, and able to handle the large amounts of data generated by IoT devices.

3.3 The Proposed ML Models

System Model. We chose fuzzy logic and ANN models because fuzzy logic can make decisions based on equivocal facts, while ANN can think to include human process without mathematical support. SVM will classify images. Image classification selects the crop from sensed data using various parameters. Hard classification—supervised and unsupervised classification—is utilized for remote sensing. Hard categorization does not assign land cover classes to pixels. Fuzziness causes jumbled pixels. Fuzzy logic uses categorization reasoning, whereas ANN (Artificial Neural Network) connects input and output data suited for human learning [3]. The SVM algorithm displays crop pictures and determines the optimal component values after gathering them.

Route Planning and Automated Driving. An Artificial Neural Network (ANN) model to make predictions about crop yields based on factors such as rainfall, temperature, and pH. The ANN model consists of three layers: the input layer, the hidden layer, and the output layer, with the prediction accuracy dependent on the number of neurons in the output layer. The distribution of rainfall, temperature, and watering can be uncertain and highly variable, making crop yields unpredictable, but ANN technology can help make early decisions about harvests and watering needs. Fuzzy logic can be used to adjust temperature and determine watering levels based on the exact moisture content of the soil, while the Support Vector Machine (SVM) classifier is utilized for infection identification in harvesting images. Overall, ANN technology is expected to produce less error than other techniques.

Soil Classification. Recognizing the characteristics of soil is crucial for reducing product quantity losses, especially for countries that export agricultural commodities. The classification of soil for engineering purposes should be based on mechanical properties such as permeability, stiffness, and strength. Understanding the type of soil is important for successful cultivation and construction. Support vector machine-based classification is used for soil types, which includes image acquisition, preprocessing, feature extraction, and classification.

Soil temperature Using our Sensor Device. Our device collects the data over the period, soil moisture, temperature, humidity, time duration & the status when the motor pump will be on or off (Fig. 3).

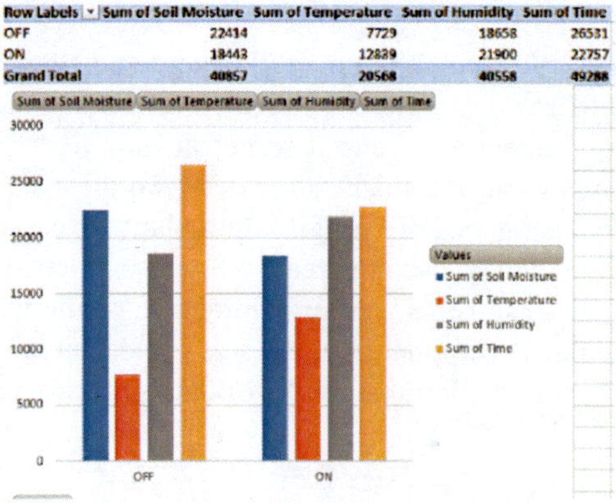

Fig. 3. Actual Soil Temperature using sensor device data

To find the accurate result of the applied machine learning technique, the mean squared error showing in the Table 2. The table showing the ANN better performance compares to Fuzzy logic (Table 3).

Table 2. Comparison of Mean squared error based on prediction algorithm using ANN & fuzzy logic

Parameter	Prediction soil temperature Using ANN	Prediction soil temperature using fuzzy logic
Mean squared Error	2.58	3.4

Table 3. Soil moisture & temperature are shown graphically

Day	1st	2nd	3rd	4th	5th	6th
Soil Moisture	54	12	34	7	50	4
Temperature	22.31	22.31	22.31	22.31	22.31	22.31
Humidity	70	40	35	18	23	52
Time	21	104	62	93	92	6
Status	ON	OFF	ON	OFF	OFF	ON
Prediction Soil Temperature	20.42	20.34	20.58	20.84	20.99	21.21
Prediction Soil Temp. Using Fuzzy logic	19.18	21.15	21.35	18.50	19.99	21.14

ML techniques predict Soil Moisture, Temperature, Production Rate, Damage to Crops, and Human Needs Based on Environmental and Weather Parameters & Data. This section of this paper predicted what factors should be needed for soil moisture, soil moisture based on air temperature, how many crops could be grown well, how many could be damaged, and whether our production rate will be able to complete human needs.

4 Result Analysis

We implement the device in a field (Fig. 4).

Fig. 4. Hardware deployment

To find the accurate result of ML techniques, R squared values and mean squared error are compared below Table 4, the result will show that the ANN model is better the comparison to the Fuzzy logic & SVM.

Table 4. Accuracy: prediction of soil moisture, temperature, production rate, damage crops.

Parameter	ANN	Fuzzy Logic	SVM
Mean Squared Error	4.04	19.26	40.51
Accuracy (R squared)	0.94	0.78	0.55

Now comparing to our device sensor, we compared all factors based on soil temperature, but if we don't count soil moisture then the performance comparison depends on R squared and MSE of predicted soil moisture using ANN and without examining the soil temperature as prediction parameter (Table 5).

Table 5. Predicted soil moisture considering and not considering soil temp.

Parameter	Considering Soil Temp	Not Considering Soil Temp
Mean Squared Error	4.04	5.78
Accuracy(R squared)	0.94	0.93

Support vector Regression Method & Result. SVM is one of the most popular ML tools for classification & regression [11]. The data we collect 85% of the data will be used as training data to build up the model and the remaining of the 15% data will be used as test data which we will classify & evaluate the performance of the model.

In the SVM there will be the following two models MLP & SVR. The Pearson correlation coefficient R is calculated the exact result is the predicted moisture and target moisture. When the values are enough to be close to a coefficient that will prove a good prediction model. The mean square error (MSE) between the predicted moisture, temperature, harvest production, and the measurement of these parameters is used to calculate network accomplishment and to compare different models. The more we get the low MSE Value, the more we get the accurate result (Fig. 5).

A binary classifier named SVM classifies data instances by constructing a linear separating hyperplane. The model identifies any departure from the observed data by a modest amount, using parameter values that minimize the sensitivity to errors in the case of SVM regression. The ensemble machine learning model is trained with collected real-time weather data to make an optimized decision, with an accuracy of 90%. The predicted soil moisture content is used to control the ON/OFF of the water pump. The classification accuracy for the models is, namely, SVM 87.5%, Naïve Bayes 76.4%, and KNN 70.8%. Real-time monitoring of temperature, humidity, and soil moisture content

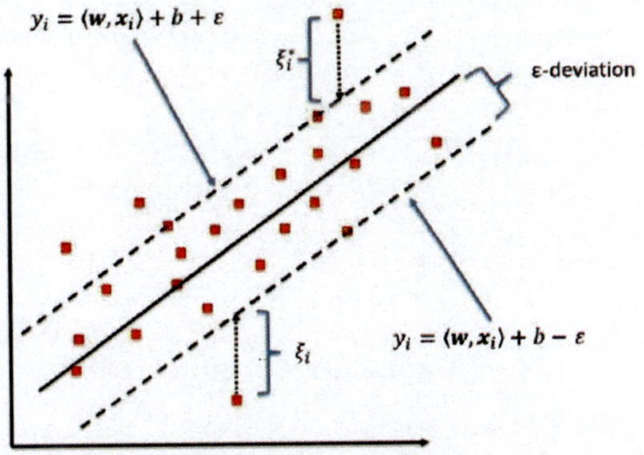

Fig. 5. e-intensive SVM

with infection detection on 2000 samples of plants, with a classification accuracy of 96%.

Rain prediction ANN Method & Analysis. We collected rainfall data from the Kaggle dataset 2008–2020 period. Additionally, time-series data on production harvesting for a 12 years epoch (2008–2020) was used to predict good harvest planning. This data was occupied to signify trends in interannual rainfall distribution and interseasonal one (i.e., major and minor seasons) for the above-mentioned 12 years period while assessing the pattern in annual rainfall distribution and crop yield (Fig. 6).

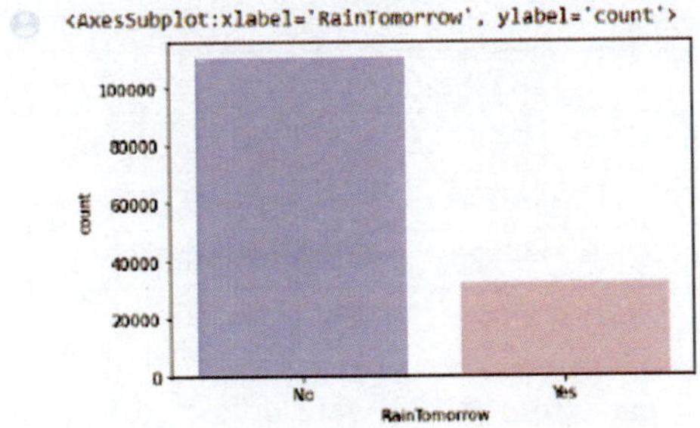

Fig. 6. Prediction of rain (testing day).

ANN model provides the aggregation of diverse components such as rainfall data and agronomic information to forecast how a specific plant could grow well or give the best production adapted over time & environment. Ricardian or cross-sectional approach [12, 13] is a similar model which is linked closely to a correlation between how potentially variable a particular landan is and the existing agro-climatic conditions (Fig. 7) (Tables 6 and 7).

Designing the Fuzzy Inference System. We selected two parts of Bangladesh, the northern & southern parts, to develop a fuzzy inference system (Fig. 8).

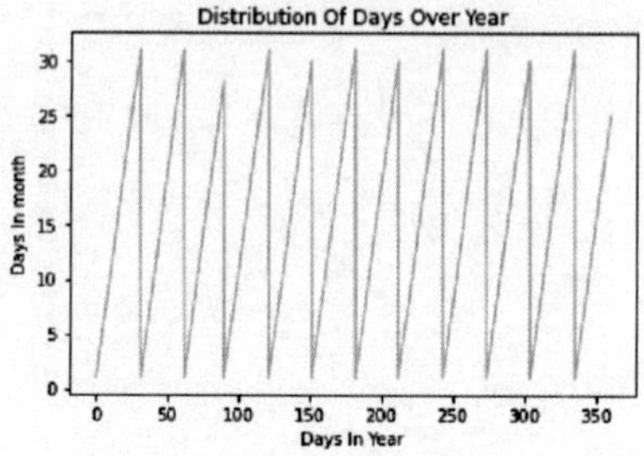

Fig. 7. Distribution of days over year.

Table 6. Comparing training & testing data we get the result of the classification report.

	Precision	Recall	F1- Score	Support
0	0.89	0.94	0.91	20110
1	0.71	0.51	0.58	5398
Accuracy			0.85	25508
Macro Avg	0.79	0.72	0.75	25508
Weighted Avg	0.85	0.85	0.84	25508

Table 7. Accuracy evaluation of model and prediction accuracy.

	Modeling	Accuracy	Prediction	Accuracy
Methods	mR^2	RMSE (g kg^{-1})	pR^2	RMSE (g kg^{-1})
ANN	0.446	7.388	0.724	4.713
SVM	0.257	8.556	0.732	4.638
Fuzzy Logic	0.156	9.120	0.349	7.224

Groundwater depth data from BWD were collected to develop a fuzzy inference system for the northern and southern parts of Bangladesh, and a comparison of the minimum, maximum, and average values was conducted.

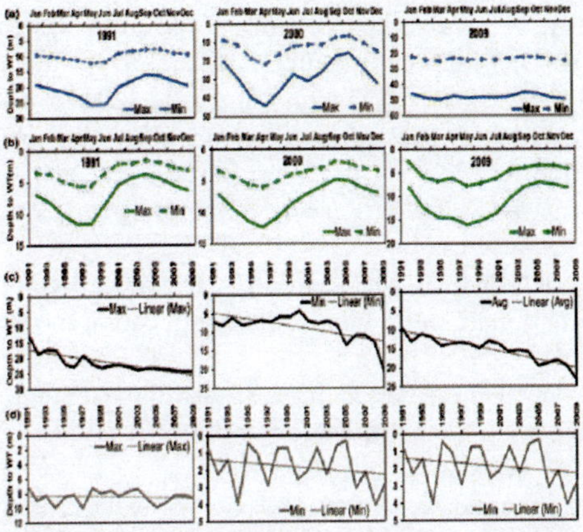

Fig. 8. Comparison of min, max, and average value depth to groundwater for the northern and southern parts. Data were collected from BWD.

5 Conclusion

The conclusion of this paper is that the next-generation device for crop yield prediction, which utilizes IoT and machine learning technologies, has been successfully implemented and tested. The device achieved a high level of accuracy in predicting crop yields, with an overall accuracy of 94%. The combination of three different machine learning models (ANN, Fuzzy Logic, and SVM) and the use of IoT sensors to gather data on environmental and soil conditions were key factors in the device's success. The device was tested on various crops and the results were comparable to state-of-the-art techniques, with the highest individual classification rate for rice at above 90% and the lowest for potatoes at close to 33%. The device has the potential to revolutionize the way farmers manage their crops and improve crop yields, and could also be used for crop forecasting, monitoring, and precision agriculture.

References

1. Hegazi, E.H., Samak, A.A., Yang, L., Huang, R., Huang, J.: Prediction of soil moisture content from sentinel-2 images using convolutional neural network (CNN). Agronomy **13**(3), 656 (2023)
2. Celik, M.F., Isik, M.S., Yuzugullu, O., Fajraoui, N., Erten, E.: Soil moisture prediction from remote sensing images coupled with climate, soil texture and topography via deep learning. Rem. Sens. **14**(21), 5584 (2022)
3. Beroho, M., et al.: Future scenarios of land use/land cover (LULC) based on a CA-Markov simulation model: case of a Mediterranean Watershed in Morocco. Rem. Sens. **15**(4), 1162 (2023)
4. Mohebbian, M., Vedaei, S.S., Bahar, A.N., Wahid, K.A., Dinh, A.: Times series prediction used in treating municipal wastewater for plant irrigation. In: 2019 IEEE Canadian Conference of Electrical and Computer Engineering (CCECE), pp. 1–4 (2019)

5. Kondaveti, R., Reddy, A., Palabtla, S.: Smart irrigation system using machine learning and IOT. In: 2019 International Conference on Vision Towards Emerging Trends in Communication and Networking (ViTECoN), pp. 1–11 (2019)
6. Singh, G., Sharma, D., Goap, A., Sehgal, S., Shukla, A.K., Kumar, S.: Machine learning based soil moisture prediction for Internet of Things based smart irrigation system. In: 2019 5th International Conference on Signal Processing, Computing and Control (ISPCC), pp. 175–180 (2019)
7. Nishitha, N., Vasuda, R., Poojith, M., Ramesh, T.K.: Irrigation monitoring and controlling system. In: 2020 International Conference on Communication and Signal Processing (ICCSP), pp. 853–857 (2020)
8. Akshay S., Ramesh, T.K.: Efficient machine learning algorithm for smart irrigation. In: 2020 International Conference on Communication and Signal Processing (ICCSP), pp. 867–870 (2020)
9. Laksiri, H.G.C.R., Dharmagunawardhana, H.A.C., Wijayakulasooriya, J.V.: Design and optimization of IoT based smart irrigation system in Sri Lanka. In: 2019 14th Conference on Industrial and Information Systems (ICIIS), pp. 198–202 (2019)
10. van Klompenburg, T., Kassahun, A., Catal, C.: Crop yield prediction using machine learning: a systematic literature review. Comput. Electron. Agric. **177**, 105709 (2020). https://doi.org/10.1016/j.compag.2020.105709
11. Das, P., Jha, G.K., Lama, A., Parsad, R.: Crop yield prediction using hybrid machine learning approach: a case study of lentil (Lens culinaris Medik.). Agriculture. **13**(3), 596 (2023)
12. Oo, A.T., Van Huylenbroeck, G., Speelman, S.: Measuring the economic impact of climate change on crop production in the dry zone of Myanmar: a Ricardian approach. Climate **8**(1), 9 (2020). https://doi.org/10.3390/cli8010009
13. Baylie, M.M., Fogarassy, C.: Examining the economic impacts of climate change on net crop income in the Ethiopian Nile basin: a Ricardian fixed effect approach. Sustainability **13**(13), 7243 (2021)

Improved EfficientNet Network for Efficient Manifold Ranking-Based Image Retrieval

Hoang Van Quy[1(✉)], Pham Thi Kim Dzung[2], Ngo Hoang Huy[3], and Tran Van Huy[1]

[1] Hong Duc University, Thanh Hoa, Vietnam
{hoangvanquy,tranlehuy}@hdu.edu.vn
[2] Swinburne Vietnam, FPT University, Hanoi, Vietnam
dungptk8@fe.edu.vn
[3] CMC University, Hanoi, Vietnam
nhhuy@cmc-u.edu.vn

Abstract. The Efficient Manifold Ranking (EMR) is a scalable graph-based ranking algorithm that is applied widely in Content-based Image Retrieval (CBIR). However, the effectiveness of an EMR algorithm depends on (1) the feature extraction technique applied to images to extract feature vectors and (2) the relational graph architecture of anchor points built inside the EMR. To address the first problem, EfficientNet-B7 + is proposed in this article which is fine-tuned from a pre-trained model of EfficientNet and is used to extract deep feature vectors of images. Regarding the second problem, we adopt the relational graph architecture of lvdc-EMR, in which the anchor points of the graph are generated by a variant of the Fuzzy C-Mean (FCM) clustering algorithm that was developed by our research team. The experiments conducted on three benchmark datasets Logo2K +, VGGFACE2-S, and Corel30K bring the mean image retrieval accuracy to 88%, demonstrating the effectiveness of our proposed method. Comparing the average values while retrieving using lvdc-EMR, the proposed EfficientNet-B7 + obtains from 4% to 6% better than the original EfficientNet-B7.

Keywords: Content-based Image Retrieval · Deep Feature · Efficient Manifold Ranking · EfficientNet · Fuzzy C-Mean Clustering

1 Introduction

In CBIR, the similarity between a given query image and each image in the database is calculated via similarity measurements based on the feature representation of the images for retrieval [1]. EMR is a promising approach applied in CBIR that allows ranking effectively the similarity of images [2]. The effectiveness of an EMR algorithm depends on the feature extraction technique applied to the images for extracting their features and the relational graph architecture of anchor points built inside the EMR. Thanks to calculating only based on the selected anchor points of the images instead of looking across the entire database, EMR increases the computing speed of image retrieval, reduces the storage capacity of databases, and enhances the accuracy of manifold ranking. Another

T. D. L. Nguyen et al. (Eds.): ICISN 2023, LNNS 752, pp. 679–684, 2023.
https://doi.org/10.1007/978-981-99-4725-6_79

variant of EMR is lvdc-EMR [3] that was developed by our research team. The algorithm of lvdc-EMR contains a relational graph architecture where the anchor points of the graph are generated by a variant of the FCM clustering algorithm. As the quality of anchor points raises higher, the retrieval accuracy will also be higher.

In terms of image feature representation, there are two main types of features [1]: low-level features that represent texture, color, and shape characteristics of images; and high-level features that demonstrate high semantic characteristics of images. High-level features can be generated from convolutional neural networks (CNNs), deep neural networks (DNNs), recurrent neural networks (RNNs), auto-encoder, deep hashing, etc. in which CNN features extracted entirely from neural networks are dominant [4]. CNN features represent a powerful representative capability for images that are used to solve image retrieval problems for high performance [5].

For each type of image feature representation, the corresponding feature extraction techniques are also developed. Most of the current works directly use the available CNN models, which have been pre-trained on classification tasks, as feature extraction tools for CBIR systems [5]. However, there always exists the different purposes between the image classification and image retrieval problems, namely classifying aims at class discrimination in datasets, while image retrieval focuses on the similarity of all images in a dataset with a given image. Directly using the pre-trained CNN models for image retrieval will lead to limited performance. Consequently, it is a natural selection of fine-tuning pre-trained CNN models to receive more robust image representations or use transfer learning for trainings that better meets the actual tasks related to image retrieval.

In recent studies, there are many deep neural networks that carry information about the complexity of images that is meaningful to people [6], such as VGG16, ResNet-v2–152, EfficientNet, etc. The authors in [7] show that EfficientNets is a family of models that achieve much better accuracy and efficiency than other networks. A variant the EfficientNet-B7 achieves state-of-the-art 84.3% top-1 accuracy on the ImageNet dataset while being 8.4x smaller and 6.1x faster on inference than the best existing ConvNet. This is the reason why in this article we choose the pre-trained model of EfficientNet-B7 network as an initial neural network for fine-tuning.

To solve the mentioned limitation of pre-trained models, in this article, we apply the fine-tuning technique on EfficientNet-B7 by subtracting the last classification layer, then replacing some layers, and finally using the transfer learning technique to train. The deep features extracted from this fine-tuned network were used for the experiments with lvdc-EMR, demonstrating the feasibility and effectiveness of our proposed method.

The attributions of this paper are as follows:

- Propose a fine-tuning method for a pre-trained CNN network, thereby obtaining high-level features which represent better image discrimination than the features extracted at the last layer of the original network.
- Use the high dimensional features extracted from the proposed EfficientNet-B7 + network as input for manifold ranking with lvdc-EMR to obtain higher image retrieval accuracy.
- Conduct the experiments on three benchmark datasets and compare the image retrieval accuracy.

The rest of the article is organized as follows. Section 2 reviews the work relating to fine-tuning pre-trained models and lvdc-EMR. A fine-tuned model EfficientNetB7 + is proposed in Sect. 3. Section 4 provides the experimental results, and the conclusion follows in Sect. 5.

2 Related Work

2.1 EfficientNet Family

The authors of the EfficientNet network argue that to achieve better accuracy, the network can be extended by increasing the number of layers or making each layer wider, or the input image has a higher resolution or a combination of all these factors. Neural Architecture Search is used to extend the underlying network architecture EfficientNet-B0. Thereby EfficientNet-B0 has been expanded from EfficientNet-B1 to EfficientNet-B7. The results show that EfficientNets achieve higher efficiency and accuracy than other DNNs [10]. Furthermore, EfficientNet has significantly reduced the number of network training parameters. At the present time, the EfficientNet-B7 is the latest fine-tuning model that was trained on ImageNet with 14 million images labeled on 21K groups and achieves the best accuracy on many datasets [7]. In this paper, we also propose a new member of the EfficientNet family, called EfficientNet-B7 +, and use it as a feature extractor for the CBIR system.

2.2 Lvdc-EMR

The original EMR is considered a manifold ranking method that successfully uses the traditional clustering algorithm K-means for determining anchor points [2]. Finding appropriate anchor points to build an EMR Graph is the prerequisite for improving the performance of CBIR systems by reducing processing time and the applicability of EMR to large databases.

Image retrieval meets challenges when performed on large datasets. One of the challenges is the manifold properties of the data as it does not fully conform to the Gaussian distribution [8]. In this situation, K-means is a hard clustering algorithm that may not be efficient. In addition, K-means based on Euclidean distance directly calculates the straight-line distances among data points in high-dimensional space. This makes K-means may not completely suitable for clustering data with manifold structure.

To replace K-means in EMR, FCM - a powerful soft clustering algorithm was proposed. However, to enable image retrieval to perform on large data of manifold properties, lvdc-EMR used an improved lvdc-FCM [3] which admits the lvdc condition (short for a large number of image vectors, a large number of vector dimensions, and a large number of clusters). The output of lvdc-FCM is the selected higher-quality anchor points. The calculation of the cluster centers A_c is done by Eq. 1, where only n_{best} feature vectors are selected to participate in updating the new center values.

$$\forall c = \overline{1, C}, A_c = \frac{\sum_{i=1}^{n} \mu_{c,i}^{p} E_i}{\sum_{i=1}^{n} \mu_{c,i}^{p}} \tag{1}$$

, where the membership matrix $\mu_{c,i}$ is equal to

$$\mu_{c,i} = \begin{cases} max\left(\dfrac{1}{\sum_{\substack{1 \leq c\prime \leq C \\ i \in n_{best}(A_{c\prime})}} \left(\frac{||E_i - A_c||}{||E_i - A_{c\prime}||}\right)^{\frac{2}{m-1}}}, \mu_{\varepsilon} \right) & for\, i \in n_{best}(A_c; n_b) \\ 0 & for\, i \notin n_{best}(A_c; n_b) \end{cases}$$

As a result, lvdc-EMR derives high-quality anchor points, thus building a more optimal relationship graph than the original EMR. The results of this article also show that lvdc-EMR increases the effectiveness of CBIR in terms of accuracy.

3 Proposed EfficientNet

The evolution of deep learning models aims to achieve higher accuracy with faster processing time for the machine learning problem. As introduced in Sect. 2.1, eight different models from B0 to B7 form a family of EfficientNets that have evolved using the fine-tuning technique. The proposed fined-tuned EfficientNet-B7 + network belongs to this family and has been retrained using transfer learning [9]. Transfer learning allows us to save time and inherit a high degree of image discrimination of fully extracted features from the pre-trained EfficientNet-B7 model. Adding layers helps EfficientNet-B7 + increase the discriminant level of the image even further and is then retrained to adapt to a particular dataset. The model obtained after training is used as a feature extractor to return higher accuracy when solving machine learning problems.

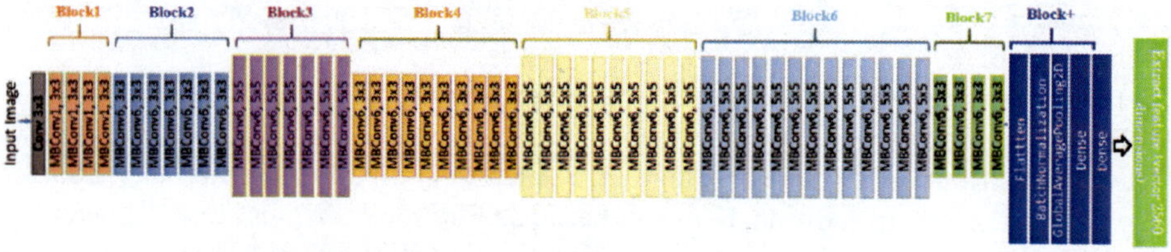

Fig. 1. EfficientNet-B7 + is fine-tuned from EfficientNet-B7

Figure 1 depicts the structure of the proposed EfficientNet-B7 + network where the last classification layer of the original one is removed and replaced by a new block, called Block +. This block of EfficientNet-B7 + includes five new layers, i.e. Flatten, BatchNormalization, Glob-alAveragePooling2D, and Denses. Block + receives the feature vectors inferred from EfficientNet-B7's layers and runs them through a layer using the batch operator to normalize so that these vectors maintain the mean output close to 0 and the output standard deviation close to 1. Next, applying the Global Average Pooling operator is to reduce the trainable parameters of the model. Finally, use the Dense operator to obtain the output feature vectors with the desired number of dimensions.

4 Experiment

To evaluate the effectiveness of the proposed Efficient B7 +, we select three benchmark image datasets Logo2K +, VGGFACE2-S, and Corel30K for conducting the experiments in this article. The first dataset Logo2K + contains 22725 logo images of 441 different brands that are placed in 441 groups. The second dataset VGGFACE2-S is more complex and consists of 61750 images that are divided into 494 groups. The last dataset Corel30K has 30887 images with the most complex visualization concepts divided into 306 items by themes and semantic sense. The dimensionality of the feature vectors is quite large at 2560. The parameters $\mu_\varepsilon = 10^{-6}$ and the different value of n_{best} when using the lvdc-FCM for each dataset to build the EMR graph, i.e., $n_{best} = 300$ for Logo2K + and Corel30K, $n_{best} = 600$ for VGGFACE2-S. To objectively evaluate the effectiveness, we use a mean average precision (*mAP*) (Table 1).

Table 1. Retrieval accuracy of EfficientNet-B7+.

lvdc-EMR	L-Normalization	Logo-2K + C = 10000 $n_{best} = 300$	VGGFACE2-S C = 10000 $n_{best} = 600$	Corel30K C = 10000 $n_{best} = 300$
EfficientNet-B7	L2-Axis = 1	79,81%	85,71%	78,35%
(2560 dimensions)	L2-Axis = 0	80,28%	86,19%	79,48%
	L1-Axis = 1	80,48%	86,37%	79,89%
	L1-Axis = 0	**81,27%**	**87,15%**	**80,39%**
EfficientNet-B7 +	L2-Axis = 1	83,81%	91,80%	83,81%
(2560 dimensions)	L2-Axis = 0	84,22%	92,78%	84,19%
	L1-Axis = 1	84,78%	93,12%	84,38%
	L1-Axis = 0	**85,89%**	**93,68%**	**84,85%**

The parameter C - number of anchor points, is chosen appropriately so that the accuracy of the retrieval is the maximum value for each case. An lvdc-EMR is used to rank the image database according to each given query image. The comparison results between the original network and the fine-tuned network EfficientNet-B7 + show a clear difference in retrieval accuracy, with a large gap of over 4%. In particular, with the VGGFACE2-S dataset, the accuracy is the highest, up to 93.68%. While with the Corel30K dataset, which has high visual complexity, the proposed fine-tuned model also obtains over 4.5%. Even with a large C parameter and a very large size of feature vectors (2560-dimensional vectors), the lvdc-EMR algorithm is still efficient in terms of clustering speed and retrieval accuracy.

5 Conclusion

This article proposes a method of using deep features to increase the efficiency of the manifold ranking algorithm for CBIR. The first contribution of the article is to pose fine-tuning an original pre-trained deep neural network, thereby holding feature vectors with a high number of dimensions but with good image discrimination. The second contribution is to build a ranking model with lvdc-EMR on these feature vectors to obtain high retrieval accuracy. In the future, we will use several deep feature image descriptors to create an effective similarity measure between the query image and the images in the database.

References

1. Afshan, L., et al.: Content-based image retrieval and feature extraction: a comprehensive review. Math. Probl. Eng. **2019**, 9658350, 21 (2019). https://doi.org/10.1155/2019/9658350
2. Xu, B., Bu, J., Chen, C., Wang, C., Cai, D., He, X.: EMR: a scalable graph-based ranking model for content-based image retrieval. IEEE Trans. Knowl. Data Eng. **27**, 102–114 (2015)
3. Quy, H.V., Huy, T.V., Huy, N.H., Tuyet, D.V., Ablameyko, S.: A modified efficient manifold ranking algorithm for large database image retrieval. Nonlinear Phenom. Complex Syst. **23**(1), 79–89 (2020)
4. Sahu, M., Dash, R.: A survey on deep learning: convolution neural network (CNN). In: Mishra, D., Buyya, R., Mohapatra, P., Patnaik, S. (eds.) Intelligent and Cloud Computing. SIST, vol. 153, pp. 317–325. Springer, Singapore (2021). https://doi.org/10.1007/978-981-15-6202-0_32
5. Chen, X., Li, Y.: Deep feature learning with manifold embedding for robust image retrieval. Algorithms **13**(12), 318 (2020)
6. Saraee, E., Jalal, M., Betke, M.: Visual complexity analysis using deep intermediate-layer features. Comput. Vis. Image Underst. **195**, 102949 (2020)
7. Tan, M., Le, Q.: Efficientnet: Rethinking model scaling for convolutional neural networks. In: International Conference on Machine Learning, pp. 6105–6114. PMLR (2019)
8. Trung, H.X., Tuyet, D.V., Huy, N.H., Ablameyko, S., Cuong, N.Q., Quy, H.V.: A novel non-gaussian feature normalization method and its application in content-based image retrieval. Nonlinear Phenom. Complex Syst. **22**(1), 1–17 (2019)
9. Pan, W.: A survey of transfer learning for collaborative recommendation with auxiliary data. Neurocomputing **177**, 447–453 (2016). https://doi.org/10.1016/j.neucom.2015.11.059
10. Kamble, R., Samanta, P., Singhal, N.: Optic disc, cup and fovea detection from retinal images using U-Net++ with EfficientNet encoder. In: Huazhu, F., Garvin, M.K., MacGillivray, T., Yanwu, X., Zheng, Y. (eds.) Ophthalmic Medical Image Analysis, pp. 93–103. Springer International Publishing, Cham (2020). https://doi.org/10.1007/978-3-030-63419-3_10

Author Index

T. D. L. Nguyen et al. (Eds.): ICISN 2023, LNNS 752, pp. 685–688, 2023.
https://doi.org/10.1007/978-981-99-4725-6